Wine Microbiology and Biotechnology

WINE
MICROBIOLOGY AND BIOTECHNOLOGY

Edited by

Graham H. Fleet
Department of Food Science and Technology
The University of New South Wales
Sydney, Australia

harwood academic publishers
Switzerland • Australia • Belgium • France • Germany • Great Britain
India • Japan • Malaysia • Netherlands • Russia • Singapore • USA

Harwood Academic Publishers

Private Bag 8
Camberwell, Victoria 3124
Australia

3-14-9, Okubo
Shinjuku-ku, Tokyo 169
Japan

58, rue Lhomond
75005 Paris
France

Emmaplein 5
1075 AW Amsterdam
Netherlands

Glinkastrasse 13–15
O-1086 Berlin
Germany

5301 Tacony Street, Drawer 330
Philadelphia, Pennsylvania 19137
United States of America

Post Office Box 90
Reading, Berkshire RG1 8JL
Great Britain

Front cover photograph: Scanning electron micrograph of *Saccharomyces cerevisiae* after autolysis for five days. Cell magnification 13,000 times. Yeast autolysis contributes to the chemical and sensory properties of wines and has a key role in the production of sparkling wines—Chapter 7.

Library of Congress Cataloging-in-Publication Data

Wine microbiology and biotechnology / edited by Graham H. Fleet.
p. cm.
Includes bibliographical references and index.
ISBN 3-7186-5132-7
1. Wine and wine making—Microbiology. I. Fleet, G. H. (Graham H.), 1946– .
QR151.W62 1992
663'.203—dc20 92-14987
CIP

 Printed in Singapore by Kyodo Printing Co (S'pore) Pte Ltd.

CONTENTS

PREFACE

The bioconversion of grape juice into wine is no longer the simple yeast fermentation elucidated by Louis Pasteur a little over 125 years ago. Developments during the last 20–25 years have clearly shown that fermentation and production of good quality wines is a complex ecological and biochemical process involving the interaction of many microbial species, represented by fungi, yeasts, lactic acid bacteria, acetic acid bacteria and, occasionally, other bacteria such as species of *Bacillus, Clostridium* and *Streptomyces*. Adding to this complexity are the contributions of killer or zymocidal yeasts, and bacteriophages (viruses) of lactic acid bacteria and acetic acid bacteria. During this same 20–25 year period, there have been major advances in fermentation technology, enzyme technology and technologies for the genetic engineering of microorganisms, all of which are now impacting on winemaking practices. These recent developments in wine microbiology and biotechnology have now reached a stage of scientific maturity and practical significance that they warrant collective documentation and presentation in one volume. Apart from documenting recent advances, however, this book has two other important objectives. These are:

- to emphasize the diversity of microorganisms associated with the total process of winemaking and to expand thinking beyond the view that *Saccharomyces cerevisiae* is the only significant wine microorganism, and
- to move discussions on the microbial ecology, biochemistry and technology of winemaking to include more modern concepts of biotechnology and molecular biology.

This book contains 17 chapters written by an international collection of contributors who are recognized authorities in their field. Authors were requested to write their chapter in the form of a comprehensive review, emphasizing latest findings and directions for future research, but also providing relevant background information. In dealing with their particular topic, authors were also requested to link the scientific fundamentals of microbiology, biochemistry and biotechnology to the more practical concerns of wine production and wine quality. I believe that the authors have done an excellent job in meeting these demands as well as meeting the book objectives already mentioned. I warmly thank the authors for supporting these wishes, and for their cooperative spirit in meeting other editorial requests.

The first chapter outlines the diversity of microbial species (yeasts, fungi, bacteria and viruses) associated with winemaking, indicating current trends in their taxonomy and in procedures for their isolation, enumeration and identification. The next five chapters focus on the growth and biochemical reactions of yeasts during alcoholic fermentation with special emphasis on their metabolism of grape juice sugars, nitrogen compounds, organic acids and sulfur compounds. The autolysis of yeast cells is a biochemical reaction that has been much overlooked in the context of winemaking. Chapter 7 considers the occurrence, mechanism and significance of autolysis of wine yeasts with special reference to champagne fermentations. Killer yeasts are a new concept in wine fermentation and Chapter 8 examines the occurrence, properties, and practical importance of these yeasts. The final chapter specifically devoted to wine yeasts considers basic strategies and prospects for their improvement by modern genetic technologies. The malolactic fermentation highlights the importance of lactic acid bacteria in wine production and is discussed in Chapter 10 along with new thoughts about the molecular basis of this reaction. Fungi have fascinating ecological and biochemical roles in the production of highly prized botrytized wines, and this topic is given detailed consideration in Chapter 11. An outstanding wine can easily be ruined through the use of contaminated cork closures and this possibility is discussed in a novel chapter on the microbiology and chemistry of cork taints. The use of sulfur dioxide in winemaking continues to come under increasing scrutiny because of its adverse effect on some consumers, and justifies Chapter 13 which takes a renewed look at the action of this agent on wine yeasts and bacteria. The microbial spoilage of wines by undesirable species of yeasts and bacteria has been known since the pioneering studies of Pasteur, but rarely discussed with the biochemical depth and practical advice given in Chapter 14. The use of commercial preparations of yeasts and lactic acid bacteria in the wine industry is now widespread and has focused increasing interest on how these commercialized species and strains are selected and cultivated on a large scale. This information is covered in Chapter 15. The next chapter looks at developments in fermentation technology with particular consideration of the prospects for using high cell density reactors and immobilized cell reactors in wine fermentations. The final chapter reviews the activities of enzymes in winemaking, covering the use of exogenous enzymes for juice extraction, clarification and flavor development as well as the significance of enzymes produced by the wine microorganisms.

In summary, this book draws upon the expertise of leading researchers from key wine producing countries of the world to provide a comprehensive account of the microbiology and biotechnology of modern winemaking. It is hoped that it will serve as a most useful text and reference source for students of winemaking courses, the practising winemaker and the academic involved in teaching and research.

Finally, I would like to acknowledge the willing and professional efforts of Christine Locke and Sharon Debreczeni as word processor operators, and to give special thanks to my family, Jo, Anika and Vanessa for their understanding and support.

Graham H. Fleet

LIST OF CONTRIBUTORS

Pierre Barre, Institut des Produits de la Vigne, Institut National de la Recherche Agronomique, 34060 Montpellier Cedex 1, France

Linda F. Bisson, Department of Viticulture and Enology, University of California, Davis, CA, 95616, USA

Bruno Blondin, Ecole Nationale Supérieure Agronomique, 34060 Montpellier Cedex 1, France

Rose-Marie Canal-Llaubères, Application and Development Department, Novo Nordisk Ferment Ltd, Neumatt, CH-4243 Dittingen, Switzerland

Claudine Charpentier, Laboratoire d'Oenologie, Université de Bourgogne, Dijon, France

Richard Degré, Lallemand Inc., 1620 Prefontaine, Montreal, Quebec, Canada, H1W 2N8

Sylvie Dequin, Institut des Produits de la Vigne, Institut National de la Recherche Agronomique, 34060 Montpellier Cedex 1, France

Charles Divies, Microbiology Laboratory, ENSBANA, University of Bourgogne, Dijon, France

Bernard J. Donèche, Institut d'Oenologie, Université de Bordeaux II, 351, Cours de la Libération, 33405 Talence, France

Michel Feuillat, Laboratoire d'Oenologie, Université de Bourgogne, Dijon, France

Graham H. Fleet, Department of Food Science and Technology, The University of New South Wales, PO Box 1, Kensington, New South Wales, 2033, Australia

Gillian M. Heard, Department of Food Science and Technology, The University of New South Wales, PO Box 1, Kensington, New South Wales, 2033, Australia

Thomas Henick-Kling, Cornell University, Department of Food Science and Technology, New York State Agricultural Experiment Station, Geneva, NY, USA

Paul A. Henschke, The Australian Wine Research Institute, PO Box 197, Glen Osmond, South Australia 5064, Australia

Vladimir Jiranek, The Australian Wine Research Institute, PO Box 197, Glen Osmond, South Australia 5064, Australia

Terry H. Lee, The Australian Wine Research Institute, PO Box 197, Glen Osmond, SA 5064, Australia

Ferdinand Radler, Institut für Mikrobiologie und Weinforschung der Johannes Gutenberg-Universität, Postfach 3980, D-6500 Mainz, Germany

Doris Rauhut, Forschungsanstalt Geisenheim, Institut für Oenologie und Getränkeforschung, Fachgebiet Mikrobiologie und Biochemie, von-Lade-Straße 1, D-6222 Geisenheim, Germany

Patrizia Romano, Department for the Protection and Improvement of Food and Agricultural Products, University of Bologna, Via F.11i Rosselli 107, 42100 Reggio Emilia, Italy

Kenichi Shimizu, Research Laboratory, Sainte Neige Wine Co. Ltd, 107–1 Kamikanogawa, Yamanashi-shi, Yamanashi 405, Japan

Robert F. Simpson, The Australian Wine Research Institute, PO Box 197, Glen Osmond, SA 5064, Australia

Wolf-Rüdiger Sponholz, Forschungsanstalt Geisenheim, Fachgebiet Mikrobiologie und Biochemie, D-6222 Geisenheim, Germany

Giovanna Suzzi, Department for the Protection and Improvement of Food and Agricultural Products, University of Bologna, Via F.11i Rosselli 107, 42100 Reggio Emilia, Italy

Françoise Vezinhet, Institut des Produits de la Vigne, Institut National de la Recherche Agronomique, 34060 Montpellier Cedex 1, France

Chapter 1

THE MICROORGANISMS OF WINEMAKING—ISOLATION, ENUMERATION AND IDENTIFICATION

GRAHAM H. FLEET

Department of Food Science and Technology, The University of New South Wales, PO Box 1, Kensington, New South Wales 2033, Australia

Chapter Outline

INTRODUCTION

Present understanding of the microbiology of winemaking has its origins in the pioneering studies of Louis Pasteur, just over 125 years ago (Amerine, 1985). Pasteur showed that yeasts were responsible for the alcoholic fermentation of grape juice into wine and that certain species of bacteria could grow in wine causing its spoilage. Since Pasteur, the microbiology of winemaking has been studied extensively and, increasingly, these studies are revealing the complexity of the ecology involved. Table 1.1 gives some idea of the diversity of microorganisms that are significant in wine production. Essentially, the overall process is a complex ecological and biochemical interaction between species of yeasts, bacteria, fungi and their viruses. The ability to measure the presence and population of these species on grapes, in the wine, on the surfaces of winery equipment, in commercial yeast preparations, in fining agents, in enzymes, on corks etc is fundamental to understanding the science of winemaking and to the implementation of quality assurance and total quality management programs. This chapter gives a general account of the methods used for the isolation, enumeration and identification of microorganisms significant in winemaking.

YEASTS

Yeasts are significant in winemaking because; (i) they conduct the alcoholic fermentation, (ii) they can spoil wines during conservation in the cellar and after packaging, and (iii) they affect wine quality through autolysis. Current taxonomies recognise the existence of about 500 species of yeasts, of which only 15–20 are relevant to winemaking. The essential features of these wine yeasts are

TABLE 1.1 The diversity and significance of microorganisms associated with winemaking.

Microbial group	*Significance*
Yeasts	Alcoholic fermentation; spoilage; autolysis; deacidification
Killer yeasts	Alcoholic fermentation; spoilage control
Lactic acid bacteria	Malolactic fermentation; spoilage; autolysis
Acetic acid bacteria	Spoilage; stuck fermentations
Fungi	Botrytized wines; spoilage; corky taints
Bacillus, *Clostridium* spp.	Spoilage
Actinomyces, *Streptomyces* spp.	Earthy, corky taints
Bacteriophages	Disrupt malolactic fermentation

outlined by Ribéreau-Gayon *et al.* (1975), Amerine *et al.* (1980), Benda (1982), Lafon-Lafourcade (1983) and Zambonelli (1988).

The main reference texts on the classification and taxonomic descriptions of yeasts are those of Kreger-van Rij (1984) and Barnett, Payne and Yarrow (1990). Shorter, more practical descriptions of the classification and general properties of yeasts have been written by Kreger-van Rij (1987), Lachance (1987) and Kurtzman (1990). Barnett (1992) has reviewed the taxonomy of *Saccharomyces* in an attempt to rationalize the confusion that has developed over the classification and nomenclature of species in this genus.

Isolation and Enumeration

Plating of samples onto an agar medium is the standard procedure for isolating and enumerating yeasts from grapes, grape juice, wine and winery environments. Both pour and spread plating techniques are used, but there is increasing evidence that spread plating gives more accurate data (King *et al.*, 1986). Other useful methods for enumerating yeasts are membrane filtration, most probable number and microscopic examination. Aspects of these methods have been described by Mossel, Dijkmann and Koopmans (1980), Jarvis and Williams (1987) and Fleet (1992).

A variety of plating media has been used for the isolation of yeasts from wines, and these include grape juice agar, malt extract agar, yeast-malt (YM) agar and other nutrient agars (Lafon-Lafourcade and Joyeux, 1979; Fleet *et al.*, 1984; Mora, Barbas and Mulet, 1990). These are non-selective media that allow the growth of all yeast species associated with the fermentation. A limitation in their use, however, occurs with samples containing mixtures of yeast species at significantly different populations. In these circumstances, colonies of the most prevalent species dominate on the plate, preventing observation of those species that are present in the sample at lesser populations. Thus, in samples of grape juice or early-ferment wine when species of *Kloeckera*, *Hanseniaspora* and *Candida* dominate the flora, it can be difficult, if not impossible, to find colonies of *Saccharomyces cerevisiae* on plates of these media, even though this species may be present. Similarly, these media may not provide reliable data on the counts of *Kloeckera*, *Hanseniaspora* or other non-*Saccharomyces* species in wine samples taken after the second day of fermentation because they now occur in much lower numbers than dominating *S. cerevisiae*. To overcome such limitations, it is necessary to use plating media that will selectively suppress the growth of the dominant species, without inhibiting those present in lower numbers.

Heard and Fleet (1986) showed that lysine agar can be used most effectively to selectively isolate and enumerate the populations of non-*Saccharomyces* species (e.g. *Kl. apiculata*, *C. stellata*) during wine fermentations. Predominating *S. cerevisiae* does not interfere with the growth of these yeasts on this medium because it is unable to use lysine as a nitrogen source and, consequently, does not grow. The successful use of lysine agar to specifically monitor the growth of non-*Saccharomyces* yeasts during wine fermentation has now been demonstrated by other researchers (Moore, Johnson and Morris, 1988; Martinez, Millan and Ortega, 1989; Mora *et al.*, 1990; Mora and Mulet, 1991). Kish, Sharf and Margalith (1983) reported the use of a nutrient medium containing ethanol (12%) and sodium metabisulfite (0.015%) for the selective enumeration of *Saccharomyces* species during wine fermentation. Because of their lesser tolerance of ethanol and sulfur dioxide, species of *Kloeckera* and *Hanseniaspora* in particular, and other non-*Saccharomyces* yeasts have difficulty growing in this medium, therefore making it useful to monitor the growth of *S. cerevisiae* during the early stages of fermentation. Pitt and Hocking (1985) have described the use of a medium containing elevated concentrations of sorbate or benzoate for the selective isolation of *Zygosaccharomyces bailii*, a species often associated with wine spoilage. Biggy agar can be used to monitor the presence of hydrogen sulfide (H_2S) producing yeasts. This medium contains bismuth sulfite which reacts with H_2S-producing colonies causing them to appear brown-black (Rupela and Tauro, 1984; Martinez *et al.*, 1989).

Bacterial growth on yeast isolation plates may be a problem when examining samples of grapes or musts. Such growth can be controlled, without detriment to yeast development, by incorporating oxytetracycline (100 mg/l) or other bacterial antibiotic into the medium composition (Fleet, 1992). Fungal contamination of yeast isolation plates can also be a problem with some grape samples. Some authors have added sodium propionate (0.1–0.2%) or biphenyl (0.02%) to the medium to control this interference (Goto and Yokotsuka, 1977; Lafon-Lafourcade and Joyeux, 1979; Moore *et al.*, 1988; Fleet, 1992).

The detection of killer toxin producing yeasts is done by the seeded agar plate technique, using malt extract agar buffered at pH 4.5 with 0.5 M phosphate citrate buffer. Approximately 10^5 cells/ml of the sensitive strain are suspended in 15 ml of pre-sterilized, molten (45°C) medium and then the mixture is poured into a petri dish and allowed to solidify. Strains of the killer yeast are streak inoculated onto the surface of the seeded agar which is then incubated at 20°C for 48 h. Killer activity is recognized by inhibition of growth (clear zones) of the seeded strain in the region surrounding growth of the killer strain (Figure 1.1) (Rosini, 1983; Young, 1987; Heard and Fleet, 1987). Heard and Fleet (1987) noted that some killer strains of *S. cerevisiae* could be distinguished from sensitive strains by their differences in colony morphology on malt extract agar. Jacobs and van Vuuren (1991) observed that colony color on WL medium could be used to differentiate killer and sensitive strains of *S. cerevisiae*.

Although enumeration of yeasts by agar plating is suitable for ecological studies, it is too slow to meet the requirements of modern quality assurance programs. Generally, it is necessary to incubate agar plates for up to four days before reliable data on counts are obtained. More rapid, automated methods are being developed for the enumeration of yeast populations, and these include the application of impedance, ATP-bioluminescence, and fluorescence microscopy technologies (see review of Fleet, 1992). In some instances, these technologies have been applied to wine yeasts (Lonvaud-Funel and Joyeux, 1982; Thomas and

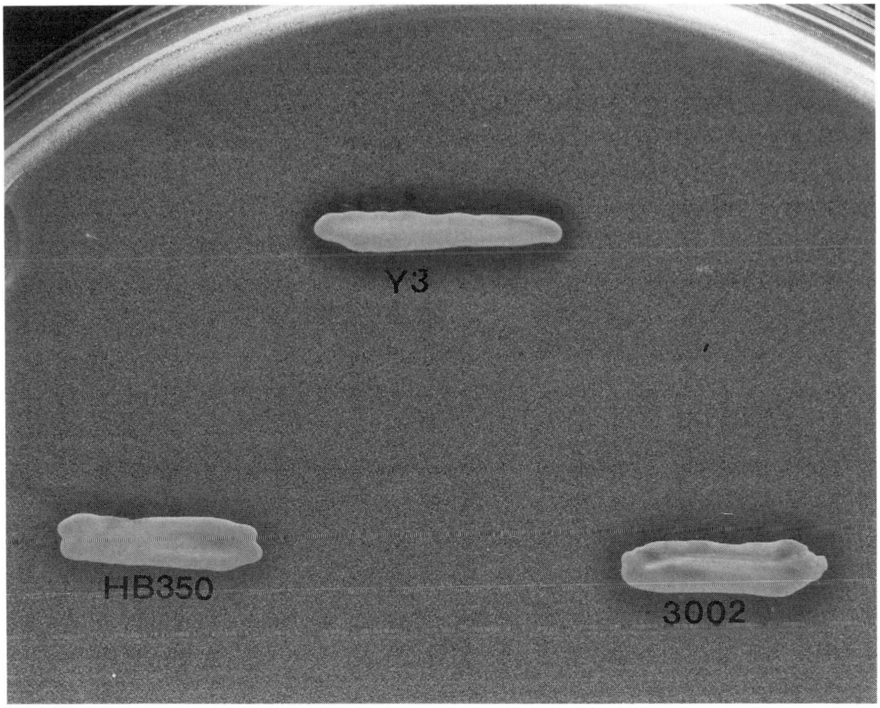

FIGURE 1.1 Malt extract agar plate with seeded growth of a sensitive strain of *Saccharomyces cerevisiae* and streak growth of several killer strains of *Saccharomyces cerevisiae*. Killer activity is indicated by the clear zones.

Ackerman, 1988; Henschke and Thomas, 1988; Thomas, 1988) but further research and development of these applications are required.

Identification

Species

Yeasts are classified into genera and species on the basis of their morphological, physiological, biochemical and, more recently, genetic properties (Kreger-van Rij, 1987; Kurtzman, 1990). They are identified by conducting the tests and following the classification schemes described in Kreger-van Rij (1984) and Barnett *et al.* (1990). Lin and Fung (1987) have given a critical assessment of some of these tests. Generally, it is necessary to conduct about 50 to 100 tests to reliably identify most yeasts to species level. Consequently, the work-load is demanding and one to two weeks are often needed to obtain a final result. Moreover, accurate interpretation of the data requires considerable expertise, and is further complicated by many changes to the generic and specific names of yeasts that have occurred over the years. These changes have arisen as a result of the newer molecular approaches (nucleic acid hybridization) to taxonomy (Kurtzman and Phaff, 1987), and from the desire of taxonomists to describe the most realistic phylogenetic classifications (Barnett, 1992).

Wine yeasts have not escaped the influences of changing classifications and nomenclature, which have caused a certain degree of confusion to the practising winemaker. Well-known wine species, such as *Saccharomyces bayanus*, *S. capensis* and *S. chevalieri*, considered to be taxonomically similar, have been described within the one species, *S. cerevisiae* (Kreger-van Rij, 1984). Controversy surrounds this classification and further revisions may be expected in the future. Martini and Martini (1990) and Barnett (1992) have presented enlightening discussions of the taxonomy of *Saccharomyces* species. Other relevant changes include the reclassification of *Saccharomyces bailii* as *Zygosaccharomyces bailii*, *Saccharomyces rosei* as *Torulaspora delbrueckii* and *Torulopsis stellata* as *Candida stellata* (Kreger-van Rij, 1984). The confusion over yeast nomenclature is also fuelled by the differing philosophical approaches to taxonomy expressed in the two major reference works Kreger-van Rij (1984) and Barnett *et al.* (1990).

The laborious, time-consuming and costly nature of identifying yeasts by conventional procedures has led to the development of shortened, more convenient identification schemes that are based on yeast response to a few carefully selected tests, and on the use of commercially available miniaturized diagnostic kits (see reviews of Deak and Beuchat, 1987; Lin and Fung, 1987).

Deak and Beuchat (1987) and Fung and Liang (1990) have described identification systems for food spoilage yeasts based on their responses to some 15 to 25 key tests. Although the reliability of these systems needs further verification (Rohm and Lechner, 1990), they could be of use for rapidly identifying wine yeasts. Rodriguez (1987) has described a simplified system for identifying wine spoilage yeasts, based on their growth in lysine agar, pantothenate-free agar, and two other media containing either cycloheximide or 5% sodium chloride.

Several ready-to-use diagnostic kits have been developed for the rapid identification of clinically significant yeast species (Lin and Fung, 1987; Deak and Beuchat, 1988). Generally, they exploit yeast growth response to a very limited range of carbon and nitrogen compounds, and give results within 48 h. Computer assisted interpretation of data is available for some kits. Of these systems, the API 20C and API 20C Auxanogram kits have been used with considerable advantage and success for the identification of wine yeasts (Lafon-Lafourcade and Joyeux, 1979; Cuinier and Leveau, 1979; Subden, Cornell and Noble, 1980; Fleet *et al.*, 1984). The API 50 CH (Schmidt, 1980; Ison, 1987) and the API-ATB 32 C (Rohm, Lechner and Lehner, 1990) systems extend the range of tests available, but the application of these expanded systems to wine yeasts has not been evaluated. Heard and Fleet (1990) concluded that additional tests must be conducted in conjunction with the API systems in order to obtain reliable identification of wine yeasts. They reported the development and evaluation of a novel system that enables all of the fermentation and assimilation tests listed in Kreger-van Rij (1984) and Barnett *et al.* (1990) to be done in the wells of a microtitre tray. In addition, vitamin growth tests, other growth tests and sporulation tests could be conducted in the wells. Using this system, reliable identification of wine yeasts was obtained within 48 h (Heard and Fleet, 1990). When data from the trays are processed using the computer program of Barnett *et al.* (Barnett, Payne and Yarrow, Yeast Identification: PC Program—Version 2, August 1990) a most convenient and rapid system for identifying wine yeasts is obtained (Fleet, 1992).

Strains

Years of experience have shown that wine quality is significantly affected by the particular strain of *S. cerevisiae* conducting the fermentation. Thus, important quality parameters such as volatile acidity, sulfide production, aromatic character and dryness of fermentation can be determined by the strain of *S. cerevisiae* (Kunkee, 1984; Subden, 1984; Thornton and Bunker, 1989). In addition, there may be important strain influences within species such as *Kloeckera apiculata*, *Candida stellata* and other non-*Saccharomyces* yeasts that also contribute to some fermentations (Fleet, 1990, 1992). Consequently, the ability to differentiate between strains is becoming an important requirement of modern winemaking. Because conventional identification schemes do not differentiate between strains, several molecular technologies are being developed and evaluated for this purpose (Degré *et al.*, 1989a).

Electrophoretic profiles of either extracellular, or intracellular proteins or isoenzymes of wine isolates of *S. cerevisiae* are sufficiently discriminating to allow strain differentiation (Leveau, Bouix and Cuinier, 1980; Cuinier, Bouix and Leveau, 1980; Subden *et al.*, 1982; Van Vuuren and Van Der Meer, 1987; Degré *et al.*, 1989a). Chromatographic profiles of the fatty acids comprising cell lipids is also sufficiently discriminating to allow differentiation of wine strains of *S. cerevisiae* (Tredoux *et al.*, 1987; Oosthuizen *et al.*, 1987; Malfeito-Ferreira, Aubyn and Loureiro, 1989). However, the profiling of proteins and fatty acids requires a strong degree of control and standardization because the protein and lipid composition of yeasts vary with culture conditions and age.

The possibility of using DNA-restriction analysis for differentiating wine strains of *S. cerevisiae* looks extremely promising (Dubourdieu *et al.*, 1987; Degré *et al.*, 1989a, b; Querol *et al.*, 1992). In this method, restriction endo-nucleases are used to fragment either chromosomal or mitochondrial DNA into short segments that are then separated by electrophoresis. The pattern of separated segments of the DNA gives a "fingerprint" of the strain (Meaden, 1990). The electrophoretic profiles of chromosomes of wine strains of *S. cerevisiae* have been reported by Blondin and Vezinhet (1988), Petering, Langridge and Henschke (1988) and Degré *et al.* (1989a) and these appear sufficiently discriminating to allow strain differentiation.

Although nucleic acid probe and ELISA technologies are well developed for diagnosing certain species and strains of bacteria, their specific application to the diagnosis of food and beverage yeasts is still lacking. However, Kuniyuki,

Rous and Sanderson (1984) have reported an ELISA assay for *Brettanomyces* species that cause wine spoilage.

LACTIC ACID BACTERIA

Lactic acid bacteria are significant in winemaking because they cause wine spoilage and are responsible for the malolactic fermentation. At the end of malolactic fermentation, wines often remain in contact with greater than 10^7 cells/ml of lactic acid bacteria. Autolysis of those bacteria is another mechanism, yet unexplored, by which they could affect wine quality. Aspects of the occurrence and growth of lactic acid bacteria in wines as well as their contribution to the malolactic fermentation have been reviewed by Wibowo *et al.* (1985), Davis *et al.* (1985b) and Kunkee (1991); see also Chapter 10. Lactic acid bacteria found in wines are listed in Table 1.2 and are classified in the genera *Leuconostoc*, *Pediococcus* and *Lactobacillus*. Taxonomic descriptions of these species are given by Garvie (1986) and Kandler and Weiss (1986). Other very good discussions of the taxonomy, ecology and biochemistry of these bacteria are given by Holzapfel and Schillinger (1992) for *Leuconostoc*, by Hammes, Weiss and Holzapfel (1992) for *Lactobacillus* and by Weiss (1992) for *Pediococcus*.

Isolation and Enumeration

Procedures for the isolation and enumeration of lactic acid bacteria from wines have been discussed by Wibowo *et al.* (1985). Sharpe (1979) and Garvie (1984) give more general accounts of procedures for handling these bacteria. Lactic acid bacteria have been isolated from grapes and wines by either of three procedures: (i) enrichment of the sample in broth culture followed by plating of this culture onto an agar medium; (ii) direct inoculation of the sample (appropriately diluted) onto an agar medium, and (iii) membrane filtration of the

TABLE 1.2 Species of lactic acid bacteria and acetic acid bacteria found in wines.

Lactic acid bacteria
Leuconostoc oenos
Pediococcus pentosaceus, P. parvulus, P. damnosus
Lactobacillus plantarum, L. casei, L. brevis, L. fermentum
L. buchneri, L. trichodes, L. hilgardii, L. fructivorans
Acetic acid bacteria
Gluconobacter oxydans
Acetobacter aceti, A. pasteurianus

sample followed by incubation of the membrane on an agar medium (Wibowo *et al.*, 1985). The most useful and preferred procedure is direct inoculation (spread plating) onto an agar medium because the colonies which develop can be enumerated, differentiated by their appearance, and isolated for identification.

Numerous media have been used for the isolation and enumeration of lactic acid bacteria from grape and wine samples and these include de Man, Rogosa Sharpe (MRS) agar, tomato juice (TJ) agar, Irrmann agar, tomato juice wine agar, yeast extract agar and various modifications of MRS agar to incorporate either grape juice, tomato juice, cysteine, malic acid and various sugars (Wibowo *et al.*, 1985). Nakagawa agar and sucrose agar, developed for the isolation of brewery lactic acid bacteria could be useful for wine lactic acid bacteria (Casey and Ingledew, 1981). Pan, Lee and Fleet (1982) compared several media for the isolation and enumeration of lactic acid bacteria from wines and concluded that no one medium would reliably recover all of the species present in any one sample. They recommended that simultaneous platings of samples onto MRS, TJ and Irrmann agars were necessary to obtain reliable ecological data. However, for routine analyses this approach is not practical and, as a compromise that gives very good results, we plate samples onto MRS agar which is fortified with 20% tomato juice and pH adjusted to 5.5. Some recent studies suggest that it may be better to supplement the MRS agar with apple juice (Kelly, Asmundson and Hopcroft, 1989; Champagne, Gardner and Lafond, 1989; Champagne and Gardner, 1990).

The addition of cycloheximide (100 mg/l) or pimaricin (50 mg/l) to the plating medium is necessary to inhibit the growth of yeasts and moulds. Incubation of plates under an atmosphere enriched in carbon dioxide or nitrogen can encourage faster growth of the colonies of lactic acid bacteria (Lafon-Lafourcade and Joyeux 1979; van der Westhuizen *et al.*, 1981; Pan *et al.*, 1982; Kelly *et al.*, 1989).

Identification

Wine isolates of lactic acid bacteria that are Gram positive and catalase negative are simply classified to genus by conducting two tests: microscopic observation of cell morphology; and examination for gas production from glucose fermentation. Species of *Leuconostoc* are cocco-bacilli or cocci in pairs or chains that produce gas from glucose (heterofermentative); *Pediococcus* species are cocci in pairs or tetrads that do not produce gas from glucose (homofermentative) and *Lactobacillus* species are rods that may or may not produce gas from glucose. Pilone, Clayton and van Duivenboden (1991) suggest that mannitol production from fructose (*Leuconostoc* and heterofermentative *Lactobacillus*) and ammonia production from arginine (*Leuconostoc* and heterofermentative *Lactobacillus*) are additional simple tests that could be used in the initial stages of identification. Identification of species requires further testing for a wide range of properties according to the standard taxonomic descriptions for *Leuconostoc* (Garvie, 1986), *Pediococcus* (Garvie, 1986) and

Lactobacillus (Kandler and Weiss, 1986). The conventional approach to such testing is lengthy and tedious but has become more convenient in recent years through the availability of miniaturized API 50 CH test galleries. Using these galleries, identification of isolates is made on the basis of their carbohydrate fermentation profile. Computer programs are available to facilitate the interpretation of these data (Cox and Thomsen, 1990). These galleries are widely used to identify wine lactic acid bacteria (Lafon-Lafourcade and Joyeux 1979; Wibowo *et al.*, 1985; Davis *et al.*, 1988; Tracey and Britz, 1987; Milliere *et al.*, 1989) although some authors have questioned their reliability (Pardo *et al.*, 1988; Jensen and Edwards, 1991). The fermentation reactions given in these galleries depend upon the pH of the basal medium and the time of incubation and, consequently, these variables must be carefully controlled (Davis *et al.*, 1988; Jensen and Edwards, 1991). However, it is also important to realise that carbohydrate fermentation reactions are not the same for all strains within a species. Using API 50 CH galleries, Davis *et al.* (1988) demonstrated significant variability in the fermentation reactions given by different strains of *Leuconostoc oenos* and by different strains of *Pediococcus parvulus*.

Because of its importance in winemaking, *Leuc. oenos* has been the subject of particularly detailed taxonomic study (Garvie and Farrow, 1980; Tracey and Britz, 1987; Millerie *et al.*, 1989; Dicks, van Vuuren and Dellaglio, 1990). These studies demonstrate the uniqueness of this wine organism but also confirm considerable strain heterogeneity within the species. In addition to fermentation reactions, strains of *Leuc. oenos* show interesting variation in colony and cell morphology, susceptibility to bacteriophages (Davis *et al.*, 1988) and in fatty acid composition of their cellular lipids (Tracey and Britz, 1989).

Modern molecular methods based on nucleic acid properties are impacting upon the taxonomy and identification of lactic acid bacteria (Stackebrandt and Teuber, 1988; Collins-Thompson, Slade and Goethals 1991) and are being used to examine wine organisms. Dicks *et al.* (1990) have confirmed the taxonomic status of *Leuc. oenos* by measuring DNA-hybridization and DNA base composition properties. Plasmid DNA has been found in some isolates of *Pediococcus pentosaceus* and *P. parvulus* (Graham and McKay 1985; Daeschel and Klaenhammer, 1985; Raccach, 1987) and *Leuc. oenos* (Janse *et al.*, 1987; Sgorbati, Palonzona and Ercole, 1987; Cavin *et al.*, 1988) and may be useful in profiling strains of these species. Lonvaud-Funel, Biteau and Fremaux (1989) described the first DNA-probe for the specific identification of *Leuc. oenos*. Subsequently, Lonvaud-Funel, Joyeux and Ledaux (1991) described procedures for producing DNA probes for *Leuc. oenos*, *Leuc. mesenteroides*, *P. pentosaceus*, *P. damnosus*, *Lactobacillus hilgardii*, *L. brevis*, *L. plantarum* and *L. casei* and showed how these probes could be used to specifically monitor the growth of these individual species during wine fermentations.

Bacteriophages

The destruction of lactic acid bacteria by bacteriophages and the disruption of milk fermentations have been known in the dairy industry for a long time. Wine lactic acid bacteria are susceptible to phage attack, also, with the consequence that the malolactic fermentation may be prevented, delayed or disrupted. In some cases, the ecology of the fermentation can be transformed—the desired strain of *Leuc. oenos* being destroyed by the phage and the fermentation being completed by a less desired strain or less desired species of *Pediococcus* or *Lactobacillus*. Aspects of bacteriophages in enology have been reviewed by Gnaegi *et al.* (1984) and Sozzi and Mignot (1988).

The occurrence of bacteriophages in wines or isolates of wine lactic acid bacteria can be demonstrated by several procedures. Sozzi, Maret and Poulin (1976) demonstrated the presence of phage in wine by electron microscopy. Samples of wine suspected of having a faulty malolactic fermentation were filtered to remove microbial cells, and then subjected to ultracentrifugation to sediment and concentrate the phage particles. The sedimented material was then examined under the transmission electron microscope for the presence of particles with characteristic phage morphology (Figure 1.2). In another approach, samples of the suspect wine are filter-sterilized to remove yeast and bacterial cells, and the filtered wine is then incubated with broth cultures of the host bacterium (e.g. *Leuc. oenos*). A decrease in turbidity of the host culture, indicating phage attack and lysis of the bacterial cells, is evidence of phage presence in the wine. Because the phage-bacterial interaction is somewhat specific, it is necessary to use a range of host strains for this procedure (Gnaegi *et al.*, 1984; Henick-Kling, Lee and Nicholas, 1986a, b). Davis, Silveira and Fleet (1985) described a direct plating procedure for detecting phage in wine samples. Undiluted wine samples (0.1 ml) are spread inoculated over the surface of plates of either MRS or TJ agars and incubated at 25–30°C for several days. Because the sample is undiluted, lactic acid bacteria grow as a dense array of colonies over the surface of the plate. Phage presence is indicated by circular zones of clearing where bacterial growth has been inhibited (Figure 1.2). Any plating medium which supports the growth of wine lactic acid bacteria can be used with this procedure, but it is necessary to incorporate cycloheximide (100 mg/l) into the medium formulation to prevent the growth of yeasts. Davis *et al.* (1985a) also described a variation of this method whereby filter-sterilized samples of wine are spot inoculated onto plates of media that have been seeded with the host bacterium. After incubation, these plates are examined for phage plaques (zones of clearing in the lawn of bacterial growth) at the sites of wine inoculum. This procedure was also used by Nel *et al.* (1987) to isolate bacteriophages from wine.

Lee (1978) described a procedure for detecting lysogenic phages in isolates of wine lactic acid bacteria. Cultures of the organism were induced to lysis by treatment with mitomycin C, and samples of the lysates were examined under the electron microscope for the presence of phage particles. Phages were

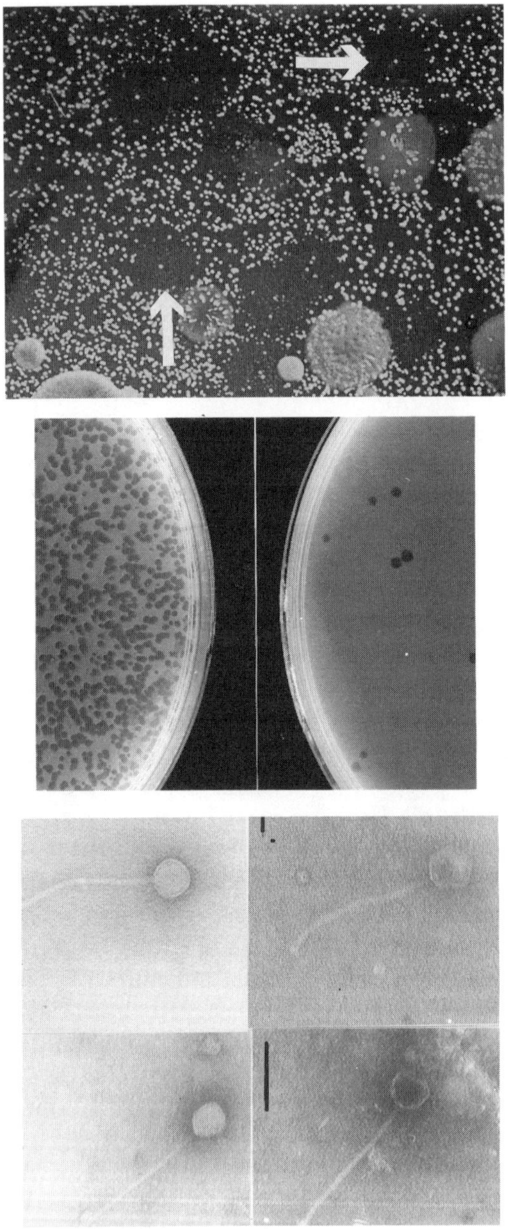

FIGURE 1.2 Bacteriophages of *Leuconostoc oenos*. (a) MRS-tomato juice agar plate spread inoculated with 0.1 ml of wine from interrupted malolactic fermentation. The small colonies are *Leuconostoc oenos*. The circular zones where colony growth is inhibited represent bacteriophage activity, (b) Bacteriophage plaques in a lawn of growth of *Leuconostoc oenos* in MRS-tomato juice agar, (c) Electron micrographs of bacteriophages of *Leuconostoc oenos* (Davis *et al*., 1985a).

detected in species of *Lactobacillus* and *Pediococcus*, but not in the strains of *Leuc. oenos* that were tested.

After detection in wine samples, phage presence can be confirmed by the standard plaque assay procedure using agar plates seeded with an appropriate strain or species of the host bacterium (Figure 1.2) (Davis *et al.*, 1985a, Nel *et al.*, 1987). Plaque formation and clarity are enhanced by the inclusion of calcium chloride in the assay medium and incubation of the plates at 30°C. This procedure can be used in experiments to quantitatively measure the influences of environmental factors (e.g. pH, sulfur dioxide concentration, ethanol, temperature) on bacteriophage development in wines (Davis *et al.*, 1985a; Henick-Kling *et al.*, 1986a, b; Nel *et al.*, 1987).

ACETIC ACID BACTERIA

Acetic acid bacteria are Gram negative, rod-shaped microorganisms that have the unique metabolic capacity to oxidize ethanol to acetic acid. They are classified into two genera *Acetobacter* and *Gluconobacter*, the taxonomic properties of which are described by De Ley, Gillis and Swings (1984). More general descriptions of these bacteria are given by Swings (1992). The main species associated with winemaking are listed in Table 1.2.

Acetic acid bacteria cause the vinegary spoilage of wine through the oxidation of ethanol to acetaldehyde and acetic acid. Until recently, this was thought to be their only relevance in winemaking. Now, it is considered that these bacteria can affect wine quality by mechanisms other than the oxidation of ethanol. Acetic acid bacteria are common inhabitants of grapes, especially if they are damaged or infected with *Botrytis cinerea* (Joyeux, Lafon-Lafourcade and Ribéreau-Gayon, 1984a, b). Excessive growth of these bacteria on grapes gives juices with significantly altered chemical composition, especially increases in the concentrations of acetic acid, gluconic acid and dihydroxyacetone (Joyeux *et al.*, 1984b; Sponholz and Dittrich, 1985; Eschenbruch and Dittrich, 1986; Drysdale and Fleet, 1989). Such changes, as well as subsequent growth of these bacteria in the grape juice, can affect the growth of yeasts during alcoholic fermentation and, possible, the conduct of the malolactic fermentation (Joyeux *et al.*, 1984b; Grossman and Becker, 1984; Drysdale and Fleet, 1989). In addition, significant populations of *Acetobacter aceti* and *A. pasteurianus* are found in wines during conservation in barrels, even after long periods of anaerobic storage. The enological importance of this occurrence is unclear as it is generally believed that these bacteria do not exhibit good survival under anaerobic conditions (Joyeux *et al.*, 1984a; Drysdale and Fleet, 1985; Drysdale, 1988). Further details about the occurrence, growth and significance of acetic acid bacteria in wines are reviewed by Lafon-Lafourcade and Joyeux (1981) and Drysdale and Fleet (1988).

Isolation, Enumeration, Identification

Acetic acid bacteria are isolated from wines by direct (spread) inoculation of samples over the surface of an appropriate agar medium. The plates are incubated aerobically at 25–30°C for 48–72 h and the colonies counted. Drysdale and Fleet (1988) have discussed the range of media that have been used to isolate acetic acid bacteria from wines. Glucose-yeast extract—calcium carbonate—agar, as described by Carr and Passmore (1979), has been used most effectively to survey the presence of acetic acid bacteria in Australian wines (Drysdale and Fleet, 1985; Drysdale, 1988). The calcium carbonate gives the medium a milky-white appearance and serves to neutralize acid produced by colonies of acetic acid bacteria. These colonies are easily recognized because they are surrounded by a clear zone due to solubilization of the calcium carbonate. Subsequently, these colonies must be confirmed as comprising Gram negative, catalase positive, rods. It is necessary to incorporate either cycloheximide or pimaricin into the medium to inhibit the growth of yeasts. Wine lactic acid bacteria appear not to grow very well on this medium and, consequently, do not interfere with the isolation of acetic acid bacteria. If their growth becomes a problem, it may be controlled by the addition of streptomycin (25 mg/l) which, according to Lafon-Lafourcade and Joyeux (1979), does not affect the growth of acetic acid bacteria. Alternatively, it may be necessary to use a medium such as yeast extract—ethanol agar which will selectively support the growth of acetic acid bacteria, but not lactic acid bacteria (De Ley *et al.*, 1984). Etani *et al.* (1985) and Mariette *et al.* (1991) describe the use of an agar medium comprising nutrients, glucose, acetic acid and ethanol for selectively isolating acetic acid bacteria.

Isolates of acetic acid bacteria are identified to species level by conducting a few specific cultural tests, especially tests for ability to oxidize D,L-lactate, oxidize ethanol, form water soluble pigments, oxidize glycerol, produce various ketogluconic acids from glucose, and use ethanol and sodium acetate as carbon sources. In particular, species of *Acetobacter* can completely oxidize ethanol through acetic acid to carbon dioxide and water, whereas species of *Gluconobacter* are unable to oxidize ethanol past acetic acid. More advanced techniques for differentiating species of these bacteria include numerical analysis of phenotypic properties (Gosselé *et al.*, 1982), electrophoretic profiles of proteins (Gosselé *et al.*, 1983), distribution of respiratory quinones (Yamada and Kondo, 1984), nucleic acid hybridization (Yamada *et al.*, 1981, 1984) and plasmid profiles (Fukaya *et al.*, 1985; Teuber, Sievers and Andersen, 1987; Mariette *et al.*, 1991).

Bacteriophages which attack species of *Gluconobacter* and *Acetobacter* have been described (Robakis *et al.*, 1985; Stamm *et al.*, 1988) but their enological significance has not received consideration.

BACILLUS AND OTHER BACTERIAL SPECIES

There are scattered reports in the literature that *Bacillus* and other bacterial species can survive and grow in wines and possibly contribute to spoilage (Chapter 14). Gini and Vaughn (1962) isolated *Bacillus* species, including *B. coagulans*, *B. circulans* and *B. subtilis*, from spoiled California sweet wines and demonstrated their ability to grow to 10^6–10^7 cells/ml when reinoculated into wine. Murrell and Rankine (1979) isolated *B. megaterium* from a bottle of spoiled brandy. We have isolated *B. coagulans* and *B. badius* from wine corks and demonstrated their ability to grow when inoculated into red and white wines (Lee *et al.*, 1984). Rare cases of wine spoilage by species of *Clostridium* have been noted (Chapter 14).

Bacillus species can be isolated by plating wine samples onto a nutrient agar medium (e.g. plate count agar) to which cycloheximide has been added to prevent yeast growth. The plates are incubated aerobically at 25–30°C for 24–48 h. Colonies that are comprised of Gram positive rods, are catalase positive and produce endospores are considered as species of *Bacillus*. These isolates can be classified to species by further testing and reference to detailed descriptions of the *Bacillus* genus as given by Claus and Berkeley (1986), Ash *et al.* (1991) and Slepecky and Hemphill (1992). Species of *Clostridium* are isolated by the same procedure except that the plates must be incubated under anaerobic conditions. Isolates are classified to species according to the descriptions given by Cato, George and Finegold (1986) and Hippe, Andreesen and Gottschalk, 1992).

Fleet *et al.* (1984) briefly noted the isolation of *Actinomyces* and *Streptomyces* species from wines during fermentation and conservation. The characteristic powdery colonies of these bacteria along with their earthy or musty odor developed on the plates of MRS-tomato juice agar used for the isolation of lactic acid bacteria. These colonies did not develop quickly but became evident after storage of the plates at room temperature for two to three weeks. The isolation of these bacteria from wines is most unusual and difficult to explain, but their finding is recorded here in the hope that they may be verified by other researchers. Lefebvre *et al.* (1983) isolated *Streptomyces* species from wine cork and implicated these bacteria as a potential source of corky taints in wines. Amerine *et al.* (1980) have noted the association of species *Actinomyces* and *Streptomyces* with wooden barrels. General information about these bacteria is given by Schaal (1992) and Korn-Wendisch and Kutzner (1992).

FUNGI

There are several mechanisms by which fungi (other than yeasts) can impact on wine quality. The contamination and growth of fungi on grapes is well acknowledged. The phenomena of "pourriture noble" and "pourriture grisse", as caused by *Botrytis cinerea*, are extensively discussed in the literature (Ribéreau-Gayon, Ribéreau-Gayon and Seguin, 1980; Bayonove, 1989; Chapter 11), but there are many other fungal species that also attack grapes, thereby affecting the chemical and biochemical properties of the juice. These species cover a range of genera, including *Penicillium*, *Aspergillus*, *Mucor*, *Rhizopus*, *Cladosporium*, *Alternaria*, *Uncinula*, *Plasmopora* and others (Amerine *et al.*, 1980; McGrew, 1982; Kovac, Lavollay and Vzbaski, 1983; Nair, 1985; Ravji, Rodriguez and Thornton, 1988; Pearson, 1990). Although fungal mycelium and fungal spores will be significant contaminants of the juice, the conditions of anaerobosis, increasing ethanol concentration and presence of sulfur dioxide are strong deterrents to their growth during wine fermentation and conservation (Pardo *et al.*, 1989; Querol, Jiménez and Huerta, 1990). However, it is now realized that several fungal species can grow under anaerobic conditions and these can cause spoilage of fruit juices, including grape juice (Pitt and Hocking, 1985; Hocking, 1988).

Fungal contamination of wine corks and of wooden barrels can be a significant cause of flavor taints in finished wine. The fungi use the cork or wood as growth substrates, generating metabolic end-products that are subsequently leached into the wine. Fungi isolated from these sources include species of *Penicillium*, *Aspergillus*, *Trichoderma*, *Cladosporium*, *Paecilomyces* and *Monilia* (Davis, Fleet and Lee, 1981; Lefebvre *et al.*, 1983; Castera-Rossignol, 1983; Chapter 12).

Procedures for the isolation, enumeration and identification of fungi from foods and beverages have undergone extensive revision in recent years and are discussed in detail by Pitt and Hocking (1985), King *et al.* (1986), Jarvis and Williams (1987), Samson and van Reenen-Hoeckstra (1988) and Samson *et al.* (1992). Direct plating of sample suspensions onto dichloran-rose bengal-chloramphenicol (DRBC) agar is now widely accepted as the preferred method for isolation of fungi from foods and beverages. The plates are incubated aerobically at 20–25°C for 72–96 h and the numbers and types of colonies recorded. The advantage of this medium over others that have been used in the past (e.g. acidified potato dextrose agar) lies in the combined use of dichloran and rose bengal to limit the spreading-lateral growth of fungal colonies without inhibiting their initial development (King, Hocking and Pitt, 1979). Without this control, the plates may be overgrown by a few dominant colonies and the presence of other fungal species in the sample would be missed. Bacterial growth on DRBC agar is prevented by the antibiotic, chloramphenicol. However, yeasts are able to grow and, in some circumstances, this can interfere with the growth of other fungi. A medium that selectively controls the growth of yeasts without inhibiting fungi would be most useful, especially for isolating fungi from grapes.

Fungal isolates from DRBC agar are purified by streaking onto a medium such as malt extract agar, and then identified. Identification to genus and species is largely based on morphological criteria as determined by observations under the microscope. Pitt and Hocking (1985) and Samson and van Reenen-Hockstra (1988) are key references for the identification of fungi isolated from foods and beverages.

CONCLUSION

The taxonomy of wine microorganisms is not static and is subject to on-going revision as new methods, especially those involving molecular or genetic technologies, become available and provide the basis for more accurate classifications. The ability to distinguish between strains of key wine microorganisms, such as strains of *S. cerevisiae* or strains of *Leuc. oenos*, has emerged as an important need in modern winemaking. Techniques based on electrophoretic profiles of chromosomes, electrophoretic profiles of restriction enzyme digests of nucleic acids, and on the use of specific nucleic acid probes offer promising prospects for strain differentiation but further research and substantiation of these approaches are required. In the future, it may be possible to have nucleic acid probe and ELISA assays for the detection of specific wine or grape spoilage yeasts, bacteria and fungi, but such assays need to be developed. At present, computer assisted schemes for the rapid identification of wine yeasts and lactic acid bacteria are a reality. Impedance, bioluminescence, flow cytometric and filtration-microscopy technologies provide opportunities for the rapid, automated enumeration of wine microorganisms but, once again, further research and validation of these methodologies are required.

REFERENCES

Amerine, M.A. (1985) Winemaking. In *Worlds Debt to Pasteur*, edited by H. Koprowski and S.A. Plotin, pp. 67–81, New York: Alan R. Liss Incorporated.

Amerine, M.A., Berg, H.W., Kunkee, R.E., Ough, C.S., Singleton, V.L. and Webb, A.D. (1980) *The Technology of Winemaking*, 4th edition. Westport, Connecticut: AVI Publishing Company.

Ash, C., Farow, J., Wallbanks, S., and Collins, M.D. (1991) Phylogenetic heterogeneity of the genus *Bacillus* revealed by comparative analysis of small-subunit-ribosomal RNA sequences. *Letters in Applied Microbiology*, **13**, 207–211.

Barnett, J.A. (1992) The taxonomy of the genus *Saccharmoyces* Meyen *ex* Reess: a short review for non-taxonomists. *Yeast*, **8**, 1–23.

Barnett, J.A., Payne, R.W. and Yarrow, D. (1990) *Yeasts, characteristics and identification*, second edition. Cambridge: Cambridge University Press.

Bayonove, C. (1989) Incidences des attaques parasitaires fongiques sur la composante qualitative du raisin et des vins. *Revue Francaise d'Oenologie*, **116**, 29–39.

Benda, I. (1982) Wine and brandy. In *Prescott and Dunn's Industrial Microbiology*, fourth edition. edited by G. Reed. pp. 293–402. Westport, Connecticut: AVI Publishing Company.

Blondin, B. and Vezinhet, F. (1988) Identification de souches de levure oenologiques par leurs caryotypes obtenus en electrophorèse en champ pulsé. *Revue Francaise d'Oenologie*, **28**, 7–11.

Carr, J.G. and Passmore, S.M. (1979) Methods for identifying acetic acid bacteria. In *Identification Methods for Microbiologists*, edited by F.A. Skinner and D.W. Lovelock, pp. 33–47. London: Academic Press.

Casey, G.P. and Ingledew, W.M. (1981) The use and understanding of media used in brewing bacteriology. II Selective media for the isolation of lactic acid bacteria. *Brewers Digest* March, 38–45.

Castera-Rossignol, A. (1983) Contrôle microbiologigue des bouchons. Bouchons steriles. Conditions de conservation de bouchons. *Connaissance de la Vigne et du Vin*, **17**, 189–193.

Cato, E., George, W.L. and Finegold, S.M. (1986) Genus *Clostridium*. In *Bergey's Manual of Systematic Bacteriology*, volume 2, edited by P.H.A. Sneath, N.S. Mair and M.E. Sharpe, pp. 1141–1200. Baltimore: Williams and Wilkins.

Cavin, J.F., Schmitt, P., Arias, A., Lin, J. and Divies, C. (1988) Plasmid profiles in *Leuconostoc* species. *Microbiologie-Aliments Nutrition*, **6**, 55–62.

Champagne, C.P. and Gardner, N. (1990) Effet du milieu et de la méthode de dénombrement sur la détermination de la population viable de cultures de *Leuconostoc oenos*. *Sciences des Aliments*, **10**, 899–905.

Champagne, C.P., Gardner, N. and Lafond, A. (1989) Production of *Leuconostoc oenos* in apple juice media. *Lebensmittel Wissenschaft und Technologie*, **22**, 376–381.

Claus, D. and Berkeley, R.C.W. (1986) The genus *Bacillus*. In *Bergey's Manual of Systematic Bacteriology*, volume 2, edited by P.H.A. Sneath, N.S. Mair and M.E. Sharpe, pp. 1105–1139. Baltimore: Williams and Wilkins.

Collins-Thompson, D., Slade, P.J. and Goethals, M. (1991) Use of low molecular mass RNA profiles to identify lactic acid bacteria and related organisms associated with foods. *International Journal of Food Microbiology*, **14**, 135–144.

Cox, R.P. and Thomsen, J.K. (1990) Computer-aided identification of lactic acid bacteria using the API 50 CHL system. *Letters in Applied Microbiology*, **10**, 257–259.

Cuinier, C., Bouix, M., Leveau, J.Y. (1980) Méthode d'étude de l'origine des levures en oenologie. In *Current Developments in Yeast Research, Advances in Biotechnology*, edited by G.G. Stewart and I. Russell, pp. 505–508. Toronto: Pergamon Press.

Cuinier, C. and Leveau, J.Y. (1979) L'identification des levures des vignoble et des vins, méthode rapide à l'aide de la gallerie API 20C. *Vignes et Vins*, **283**, 44–49.

Daeschel, M. and Klaenhammer, T. (1985) Association of 13.6-megadalton plasmid in *Pediococcus pentosaceus* with bacteriocin activity. *Applied and Environmental Microbiology*, **50**, 1538–1541.

Davis, C.R., Fleet, G.H. and Lee, T.H. (1981) The microflora of wine corks. *The Australian Grapegrower and Winemaker*, **208**, 42, 44.

Davis, C., Silveira, N. and Fleet, G.H. (1985a) Occurrence and properties of bacteriophages of *Leuconostoc oenos* in Australian wines. *Applied and Environmental Microbiology*, **50**, 872–876.

Davis, C.R., Wibowo, D., Eschenbruch, R., Leet, T.H. and Fleet, G.H. (1985b) Practical implications of malolactic fermentation: a review, *American Journal of Enology and Viticulture*, **36**, 290–301.

Davis, C.R., Wibowo, D., Fleet, G.H. and Lee, T.H. (1988) Properties of wine lactic acid bacteria: their potential enological significance. *American Journal of Enology and Viticulture*, **39**, 137–142.

Deák, T. and Beuchat, L.R. (1987) Identification of foodborne yeasts. *Journal of Food Protection*, **50**, 243–264.

Deák, T. and Beuchat, L.R. (1988) Evaluation of simplified and commercial systems for identification of foodborne yeasts. *International Journal of Food Microbiology*, **7**, 135–145.

Degré, R., Thomas, D.Y., Ash, J., Mailhiot, K., Morin, A. and Dubord, C. (1989a) Wine yeasts strain identification. *American Journal of Enology and Viticulture*, **40**, 309–315.

Degré, R., Thomas, D.Y., Frenette, J. and Mailhiot, K. (1989b) Identifying wine yeast strains by their DNA sequence using DNA probes. *Revue Francaise d'Oenologie*, **119**, 23–26.

DeLey, J., Gillis, M. and Swings, J. (1984) Family Acetobacterinceae. In *Bergeys Manual of Systematic Bacteriology* (ninth edition), volume 1, edited by N.R. Krieg and J.G. Holt, pp. 267–299. Baltimore: Williams and Wilkins.

Dicks, L.M.T., Van Vuuren, H.J.J. and Dellaglio, F. (1990) Taxonomy of *Leuconostoc* species, particularly *Leuconostoc oenos*, as revealed by numerical analysis of total soluble cell protein patterns, DNA-base compositions, and DNA-DNA hybridizations. *International Journal of Systematic Bacteriology*, **40**, 83–91.

Drysdale, G. (1988) Occurrence and growth of acetic acid bacteria in Australian wines. Master of Science Thesis. Kensington, Australia: University of New South Wales.

Drysdale, G. and Fleet, G.H. (1985) Acetic acid bacteria in some Australian wines. *Food Technology Australia*, **37**, 17–20.

Drysdale, G. and Fleet, G.H. (1988) Acetic acid bacteria in winemaking: a review. *American Journal of Enology and Viticulture*, **39**, 143–154.

Drysdale, G. and Fleet, G.H. (1989) The effect of acetic acid bacteria upon the growth and metabolism of yeasts during the fermentation of grape juice. *Journal of Applied Bacteriology*, **67**, 471–481.

Dubourdieu, D., Sokol, A., Zucca, J., Thalouain, P., Dattee, A. and Aigle, M. (1987) Identification des souches de levures isolées de vins par l'analyse de leur ADN mitochondrial *Connaissance de la Vigne et du Vin*, **21**, 267–278.

Eschenbruch, B. and Dittrich, H.H. (1986) Metabolism of acetic acid bacteria in relation to their importance to wine quality *Zentralblatt Fur Mikrobiologie*, **141**, 279–289.

Etani, E., Ohmori, S., Masai, H. and Suzuki, K. (1985) *Acetobacter polyoxygenes* sp nov., a new species of an acetic acid bacterium useful for producing vinegar with high acidity. *Journal of General and Applied Microbiology*, **31**, 475–490.

Fleet, G.H. (1990) Growth of yeasts during wine fermentations *Journal of Wine Research*, **1**, 211–223.

Fleet, G.H. (1992) Spoilage yeasts. *Critical Reviews in Biotechnology*, **12**, 1–44.

Fleet, G.H., Lafon-Lafourcade, S. and Ribereau-Gayon, P. (1984) Evolution of yeasts and lactic acid bacteria during fermentation and storage of Bordeaux wines. *Applied and Environmental Microbiology*, **48**, 1034–1038.

Fukaya, M., Iwata, T., Etani, E., Masai, H., Ouozumi, T. and Beppu, T. (1985) Distribution and characterization of plasmids in acetic acid bacteria. *Agricultural and Biological Chemistry*, **49**, 1349–1355.

Fung, D.Y. and Liang, C. (1990) Critical review of isolation, detection and identification of yeasts from meat products. *Critical Reviews in Food Science and Nutrition*, **29**, 341–360.

Garvie, E.I. (1984) Separation of species of the genus *Leuconostoc* and differentiation of the leuconostocs from other lactic acid bacteria. In *Methods in Microbiology*, volume 16, edited by T. Bergan, pp. 147–178. London: Academic Press.

Garvie, E.I. (1986) *Leuconostoc, Pedicococcus*. In *Bergey's Manual of Systematic Bacteriology*, Volume two, edited by P.H. Sneath, N.S. Mair and M.E. Sharpe, pp. 1071–1079. Baltimore: Williams and Wilkins.

Garvie, E.I. and Farrow, J.P.E. (1980) The differentiation of *Leuconostoc oenos* from non-acidophilic species of *Leuconostoc* and the identification of five strains from the American Type Culture Collection. *American Journal of Enology and Viticulture*, **31**, 154–157.

Gini, B. and Vaughn, R.H. (1962) Characteristics of some bacteria associated with the spoilage of California dessert wines. *American Journal of Enology and Viticulture*, **13**, 20–31.

Gnaegi, F., Cazelles, O., Sozzi, T. and d'Amiro (1984) Connaissance sure les bacteriophages de *Leuconostoc oenos* et progrès dan la maitrîse de la fermentation malolactique des vins. *Revue Suisse Viticulture Arborculture Horticulture*, **16**, 59–65.

Gosselé, F., Swings, J., Kersters, K. and DeLey, J. (1982) Numerical analysis of Frateur's phenotypic data on acetic acid bacteria. *Antonie van Leeuwenhoek*, **48**, 289–295.

Gosselé, F., Swings, J., Kersters, K. and DeLey, J. (1983) Numerical analysis of phenotypic features and protein gel electropherograms of *Gluconobacter* Asai 1935 emend. *International Journal of Systematic Bacteriology*, **33**, 65–81.

Goto, S. and Yokotsuka, I. (1977) Wild yeast populations in fresh grape musts of different harvest times. *Journal of Fermentation Technology*, **55**, 417–422.

Graham, D. and McKay, L. (1985) Plasmid DNA in strains of *Pediococcus cerevisiae* and *Pediococcus pentosaceus*. *Applied and Environmental Microbiology*, **50**, 532–534.
Grossman, M.K. and Becker, R. (1984) Investigations on bacterial inhibition of wine fermentation. *Kellerwritschaft*, **10**, 272–275.
Hammes, W.P., Weiss, N. and Holzapfel, W. (1992) The genera *Lactobacillus* and *Carnobacterium*. In, *The Prokaryotes*, second edition, edited by A. Balows *et al*., pp. 1535–1594. New York: Springer-Verlag.
Heard, G.M. and Fleet, G.H. (1986) Evaluation of selective media for enumeration of yeasts during wine fermentation. *Journal of Applied Bacteriology*, **60**, 477–481.
Heard, G.M. and Fleet, G.H. (1987) Occurrence and growth of killer yeasts during wine fermentation. *Applied and Environmental Microbiology*, **53**, 2171–2174.
Heard, G.M. and Fleet, G.H. (1990) A convenient microtitre tray procedure for yeast identification. *Journal of Applied Bacterilogy*, **68**, 447–45.
Henick-Kling, T., Lee, T.H. and Nicholas, D. (1986a) Inhibition of bacterial growth and malolactic fermentation in wine by bacteriophage. *Journal of Applied Bacteriology*, **61**, 287–294.
Henick-Kling, T. and Lee, T.H. and Nicholas, D. (1986b) Characterization of the lytic activity of bacteriophages of *Leuconostoc oenos* isolated from wine. *Journal of Applied Bacteriology*, **61**, 525–534.
Henschke, P.A. and Thomas, D.S. (1988) Detection of wine-spoiling yeasts by electronic methods. *Journal of Applied Bacteriology*, **64**, 123–133.
Hippe, H., Andreesen, J.R. and Gottschalk, G. (1992) The genus *Clostridium*-nonmedial. In *The Prokaryotes*, second edition, edited by A. Balows *et al*., pp. 1800–1866. New York: Springer-Verlag.
Hocking, A.D. (1988) Responses of fungi to modified atmospheres. *CSIRO Food Research Quarterly*, **48**, 56–65.
Holzapfel, W.H. and Schillinger, U. (1992) The genus *Leuconostoc*. In, *The Prokaryotes*, second edition, edited by A. Balows *et al*., pp. 1508–1534. New York: Springer Verlag.
Ison, R.W. (1987) A revised method for the application of API 50 CH carbohydrate kits to yeasts. *Letters in Applied Microbiology*, **4**, 9–12.
Jacobs, C.J. and van Vuuren, H.J.J. (1991) Effects of different killer yeasts on wine fermentations. *American Journal of Enology and Viticulture*, **42**, 295–300.
Janse, J.B.H., Wingfield, B.D., Pretorius, I.S. and van Vuuren, H.J. (1987) Plasmids in *Leuconostoc oenos*. *Plasmid*, **17**, 173–175.
Jarvis, B. and Williams, A.P. (1987) Methods for detecting fungi in foods and beverages. In *Food and Beverage Mycology* edited by L.R. Beuchat, pp. 599–617. New York: Van Nostrand Reinhold.
Jenson, K.A. and Edwards, C.G. (1991) Modification of the API rapid CH system for characterization of *Leuconostoc oenos*. *American Journal of Enology and Viticulture*, **42**, 274–277.
Joyeux, A., Lafon-Lafourcade, S. and Ribéreau-Gayon, P. (1984a) Evolution of acetic acid bacteria during fermentation and storage of wine. *Applied and Environmental Microbiology*, **48**, 153–156.
Joyeux, A., Lafon-Lafourcade, S. and Ribéreau-Gayon, P. (1984b) Metabolism of acetic acid bacteria in grape must: consequences in alcoholic and malolactic fermentation. *Sciences des Aliments*, **4**, 247–255.
Kandler, O. and Weiss, N. (1986) *Lactobacillus*. In *Bergey's Manual of Systematic Bacteriology*, Volume two, edited by P.H. Sneath, N.S. Mair, and M.E. Sharpe, pp. 1208–1234. Baltimore: Williams and Wilkins.
Kelly, W.J., Asmudson, R.V. and Hopcroft, D.H. (1989) Growth of *Leuconostoc oenos* under anaerobic conditions. *American Journal of Enology and Viticulture*, **40**, 277–282.
King, A.D., Hocking, A.D. and Pitt, J.I. (1979) Dichlaran-rose bengal medium for enumeration and isolation of moulds from foods. *Applied and Environmental Microbiology*, **37**, 959–964.

King, A.P., Pitt, J.I., Beuchat, L.R. and Corry, J. (1986) *Methods for the Mycological Examination of Food.* New York: Plenum Press.

Kish, S., Sharf, R. and Margalith, P. (1983) A note on a selective medium for wine yeasts. *Journal of Applied Bacteriology,* **55**, 177–179.

Korn-Wendisch, F. and Kutzner, H.J. (1992) The family *Streptomycetaceae*. In, *The Prokaryotes*, second edition, edited by A. Balows *et al.*, pp. 921–995. New York: Springer Verlag.

Kovac, V., Lavollay, J. and Vzbaski, L.J. (1983) Quelques moisissures présentant un intérêt oenologique, *Penicillium* et *Cladosporium*. *Bulletin de l'Office International de la Vin*, **56**, 420–432.

Kreger-van Rij, N. (1984) *The Yeasts, a Taxonomic Study,* Amsterdam: Elsevier.

Krejer-van Rij, N. (1987) Classification of yeasts. In *The Yeasts*, Volume 1, second edition, *Biology of Yeasts*, edited by A.H. Rose and J.S. Harrison, pp. 5–62. London: Academic Press.

Kuniyuki, A.H., Rous, C. and Sanderson, J.L. (1984) Enzyme linked immunosorbent assay (ELISA) detection of *Brettanomyces* contaminants in wine production. *American Journal of Enology and Viticulture*, **35**, 143–145.

Kunkee, R.E. (1984) Selection and modification of yeasts and lactic acid bacteria for wine fermentation. *Food Microbiology,* **1**, 315–332.

Kunkee, R.E. (1991) Some roles of malic acid in the malolactic fermentation in winemaking. *FEMS Microbiology Reviews*, **88**, 55–72.

Kurtzman, C.P. (1990) Classification and general properties of yeasts. In *Yeast Biotechnology and Biocatalysis* edited by H. Verachtert and R. De Mot, pp. 1–34. New York: Marcel Dekker.

Kurtzman, C. and Phaff, H.J. (1987) Molecular taxonomy. In *The Yeasts*, Volume 1, second edition, *Biology of Yeasts*, edited by A.H. Rose and J.S. Harrison, pp. 63–94. London: Academic Press.

Lachance, M. (1987) Approaches to yeast identification. In *Yeast Biotechnology,* edited by D.R. Berry, I. Russell and G.G. Stewart, pp. 33–52. London: Allen and Unwin.

Lafon-Lafourcade, S. (1983) Wine and brandy. In *Biotechnology,* Volume 5, *Food and Feed Production with Microorganisms*, edited by G. Reed, pp. 81–164. Weinheim: Verlag Chemie.

Lafon-Lafourcade, S. and Joyeux, A. (1979) Techniques simplifées pour le denombrement et l'identification des microorganismes vivants dans les mouts et les vins. *Connaissance de la Vigne et du Vin*, **4**, 295–310.

Lafon-Lafourcade, S. and Joyeux, A. (1981) Les bactéries acetiques du vin. *Bulletin de l'Office International du Vin*, **608**, 803–829.

Lee, A. (1978) Bacteriophages associated with lactobacilli isolated from wine. In *Proceedings of the Fifth International Oenological Symposium* (Auckland New Zealand) edited by E. Lemperle and J. Frank, pp. 287–295. Breisach: International Association of Modern Winery Technology and Management.

Lee, T.H., Simpson, R.F., van De Peer, J., Fleet, G.H., Davis, C., Daly, N. and Yap, A. (1984) Microbiology of wine corks. In *Advances in Viticulture and Oenology for Economic Gain*, edited by T.H. Lee and T.C. Somers, pp. 435–450. Adelaide: The Australian Wine Research Institute.

Lefebvre, A., Riboulet, J.M., Boidron, J.N. and Ribéreau-Gayon, P. (1983) Incidence des microorganismes du liége sur les alterations el olfactives du vin. *Sciences des Aliments*, **3**, 265–278.

Leveau, J.Y., Bouix, M. and Cuinier, C. (1980) Differenciation fine des levures par electrophorese et immunoélectrophorese des proteines cytoplasmiques. In *Current Developments in Yeast Research. Advances in Biotechnology* edited by G.G. Stewart and I. Russell, pp. 387–392. Toronto: Pergamon Press.

Lin, C.C.S. and Fung, D.Y. (1987) Conventional and rapid methods for yeast identification. *Critical Reviews in Microbiology,* **14**, 273–289.

Lonvaud-Funel, A., Biteau, N. and Fremoux, C. (1989) Identification de *Leuconostoc oenos* par utlisation de sones d'ADN. *Sciences des Aliments*, **9**, 533–541.
Lonvaud-Funel, A. and Joyeux, A. (1982) Application de bioluminescence pour le dénombrement des microorganismes vivants dans les vins. *Connaissance de la Vigne et du Vin*, **16**, 241–256.
Lonvaud-Funel, A., Joyeux, A. and Ledoux, O. (1991) Specific enumeration of lactic acid bacteria in fermenting grape must and wine by colony hybridization with non-isotopic DNA probes. *Journal of Applied Bacteriology*, **71**, 501–508.
Malfeito-Ferreira, M., Aubyn, A. and Loureiro, V. (1989) Long chain fatty acid composition as a tool for differntiating spoilage wine yeasts. *Mycotaxon*, **36**, 35–42.
Mariette, I., Schwarz, E., Vogel, R.F. and Hammes, W.P. (1991) Characterization by plasmid profile analysis of acetic acid bacteria from wine, spirit and cider acetators for industrial vinegar production. *Journal of Applied Bacteriology*, **71**, 134–138.
Martinez, J., Millán, C. and Ortega, J.M. (1989) Growth of natural flora during the fermentation of inoculated musts from "Pedro Ximenez" grapes. *South African Journal of Enology and Viticulture*, **10**, 31–35.
Martini, A. and Martini, A.V. (1990) Grape must fermentation: past and present. In *Yeast Technology*, edited by J.F.T. Spencer and B.M. Spencer, pp. 105–123. Berlin: Springer-Verlag.
McGrew, J.R. (1982) Fungal diseases: a factor in vine culture. *Developments in Industrial Microbiology*, **23**, 87–90.
Meaden, P.J. (1990) DNA fingerprinting of brewers' yeast: current perspectives. *Journal of the Institute of Brewing*, **96**, 195–200.
Millan, C. and Ortega, J.M. (1988) Production of ethanol, acetaldehyde and acetic acid in wine by various yeast races: role of alcohol and aldehyde dehydrogenases. *American Journal of Enology and Viticulture*, **39**, 107–112.
Milliere, J.B., Mathot, A.G., Schmitt, P. and Divies, C. (1989) Phenotypic characterization of *Leuconostoc* species. *Journal of Applied Bacteriology*, **67**, 529–542.
Moore, K.J., Johnson, M.G. and Morris, J.R. (1988) Indigenous yeast microflora on Arkansas white Riesling (*Vitis vinifera*) grapes and in model must systems. *Journal of Food Science*, **53**, 1725–1728.
Mora, J., Barbas, J.I. and Mulet, A. (1990) Growth of yeast species during the fermentation of musts inoculated with *Kluyveromyces thermotolerans* and *Saccharomyces cerevisiae*. *American Journal of Enology and Viticulture*, **141**, 156–159.
Mora, J. and Mulet, A. (1991) Effects of some treatments of grape juice on the population and growth of yeast species during fermentation. *American Journal of Enology and Viticulture*, **42**, 133–136.
Mossel, D.A.A., Dijkmann, K.E. and Koopmans, M. (1980) Experience with methods for the enumeration and identification of yeasts occurring in foods. In *Biology and Activities of Yeasts* edited by F.A. Skinner, S.M. Passmore and R.R. Davenport, pp. 279–288. London: Academic Press.
Murrell, W.G. and Rankine, B.C. (1979) Isolation and identification of a sporing *Bacillus* from bottled brandy. *American Journal of Enology and Viticulture*, **30**, 247–249.
Nair, N.G. (1985) Fungi associated with bunch rot of grapes in the Hunter Valley. *Australian Journal of Agricultural Research*, **36**, 435–442.
Nel, L., Wingfield, B.D., van der Meer, L.J., van Vuuren, H.J.J. (1987) Isolation and characterization of *Leuconostoc oenos* bacteriophages from wine and sugar cane. *FEMS Microbiology Letters*, **44**, 63–67.
Oosthuizen, A., Koch, J.F., Viljoen, B.C., Muller, H.B. and Lategan, P.M. (1987) The value of long chain fatty acid composition in the identification of some brewery yeasts. *Journal of the Institute of Brewing*, **93**, 174–176.
Pan, C.S., Lee, T.H. and Fleet, G.H. (1982) A comparison of five media for the isolation of lactic acid bacteria from wines. *The Australian Grapegrower and Winemaker* **220**, 42, 44, 46.

Pardo, I., Garcia, M.J., Zuniga, M. and Urubura, F. (1988) Evaluation of the API 50 CHL system for identification of *Leuconostoc oenos*. *American Journal of Enology and Viticulture*, **39**, 347–350.

Pardo, I., Garcia, M.J., Zuniga, M. and Uruburu, F. (1989) Dynamics of microbial populations during fermentation of wines from the Utiel-Requena region of Spain, *Applied and Environmental Microbiology*, **55**, 539–541.

Pearson, R.C. (1990) Current research on grape fungal diseases and their control in New York. *The Australian and New Zealand Wine Industry Journal*, **5**, 206–209.

Petering, J., Langridge, P. and Henschke, P. (1988) Fingerprinting of wine yeasts. The application of chromosome electrophoresis. *The Australian and New Zealand Wine Industry Journal*, **3**, 48–52.

Pilone, G.J., Clayton, M.G. and van Duivenboden, R.J. (1991) Characterization of wine lactic acid bacteria: single broth culture for tests of heterofermentation, mannitol from fructose and ammonia from arginine. *American Journal of Enology and Viticulture*, **42**, 153–157.

Pitt, J.I. and Hocking, A.D. (1985) *Fungi and Food Spoilage*. Sydney: Academic Press.

Querol, A., Huerta, T., Barrio, E. and Ramon, D. (1992) Dry yeast strain for use in fermentation of Alicante wines: selection and DNA patterns. *Journal of Food Science*, **57**, 183–185.

Querol, A., Jiménez, M. and Huerta, T. (1990) Microbiological and enological parameters during fermentation of musts from poor and normal grape harvests in the region of Alicante (Spain). *Journal of Food Science*, **55**, 1603–1606.

Raccach, M. (1987) Pediococci and biotechnology. *CRC Critical Reviews in Microbiology*, **14**, 291–309.

Ravji, R.G., Rodriguez, S.B. and Thornton, R.J. (1988) Glycerol production by four common grape molds. *American Journal of Enology and Viticulture*, **39**, 77–82.

Ribéreau-Gayon, J., Peynaud, E., Ribéreau-Gayon, P. and Sudraud, P. (1975) *Traité d'Oenologie. Sciences et Techniques du Vin*, Tome 2, Paris: Dunod.

Ribéreau-Gayon, J., Ribéreau-Gayon, P. and Seguin, G. (1980) *Botrytis cinerea* in enology. In, *The Biology of Botrytis*, edited by J.R. Coley-Smith, K. Verhoeff and W.R. Jarvis, pp. 251–274. London: Academic Press.

Robakis, N.K. *et al.* (1985) Isolation and characterization of two phages for *Gluconobacter oxydans*. *Journal of General Microbiology*, **131**, 2467–2473.

Rodriguez, S.B. (1987) A system for identifying sopilage yeast in packaged wine. *American Journal of Enology and Viticulture*, **38**, 273–276.

Rohm, H. and Lechner, F. (1990) Evaluation and erliability of a simplified method for identification of foodborne yeasts. *Applied and Environmental Microbiology*, **56**, 1290–1295.

Rohm, H., Lechner, F. and Lehner, M. (1990) Evaluation of API ATB 32C system for the rapid identification of foodborne yeasts. *International Journal of Food Microbiology*, **11**, 215–224.

Rosini, G. (1983) The occurrence of killer character in yeasts. *Canadian Journal of Microbiology*, **29**, 1462–1464.

Rupela, O.P. and Tauro, P. (1984) Isolation and characterization of low hydrogen sulfide producing wine yeast. *Enzyme and Microbial Technology*, **6**, 419–421.

Samson, R.A., Hocking, A.D., Pitt, J.I. and King, A.D. (1992) *Modern Methods in Food Mycology*. Amsterdam: Elsevier.

Samson, R.A. and van Reenen-Hoekstra, E.S. (1988) *Introduction to Food-Borne Fungi*, third edition, Delft: Centraalbureau Voor Schimmecultures.

Sattler, K., Babel, W. and Wunsche, I. (1990) Acetic acid bacteria—a group of microorganisms possessing both an important technological tradition and perspective—a survey of some novel aspects. *Zentralblatt fur Mikrobiologie*, **145**, 555–562.

Schaal, K.P. (1992) The genera *Actinomyces*, *Arcanobacterium* and *Rothia*. In, *The Prokaryotes*, second edition, edited by A. Balows *et al.*, pp. 850–905. New York: Springer-Verlag.

Schmidt, J.L. (1980) Evaluation d'une méthode rapide d'assimilation des sucres (API 50 CH) en vue de l'identification des levures. *Annale Technologie de Agricole*, **29**, 47–52.

Sgorbati, B., Palenzona, D. and Ercoli, L. (1987) Characterization of the pesticide-resistance plasmid BL34 from *Leuconostoc oenos*. *Microbiologie-Ailments—Nutrition*, **5**, 295–301.

Sharpe, M.E. (1979) Identification of lactic acid bacteria. In, *Identification Methods for Microbiologists*, second edition, edited by F.A. Skinner and D.W. Lovelock, pp. 233–259. London: Academic Press.

Slepecky, R.A. and Hemphill, H.E. (1992) The genus *Bacillus*-nonmedical. In, *The Prokaryotes*, second edition, edited by A. Balows, pp. 1663–1696. New York: Springer-Verlag.

Sozzi, T., Maret, R. and Poulin, J.M. (1976) Mise en évidence de bacteriophages dans le vin. *Experentia*, **32**, 568–569.

Sozzi, T. and Mignot, O. (1988) Bacteriophage in oenology. *Bulletin de l'Office International du Vin*, **61**, 705–716.

Sponholz, W.R. and Dittrich, H.H. (1985) Uber die Herkunft von Gluconsaure. 2-und 5-oxo-gluconsaure sowrie Glucuron-und Galacturonsaure in Mosten and Weinen, *Vitis*, **24**, 51–58.

Stackebrandt, E. and Teuber, M. (1988) Molecular taxonomy and phylogenetic position of lactic acid bacteria. *Biochimie*, **70**, 317–324.

Stamm, W.W., Kittelmann, M., Follman, H. and Trüper, H.G. (1988) The occurrence of bacteriophages in spirit vinegar fermentation. *Applied Microbiology and Biotechnology*, **30**, 41–46.

Subden, R.E. (1984) Current developments in wine yeasts. *Critical Reviews in Biotechnology*, **5**, 49–65.

Subden, R.E., Cornell, R. and Noble, A.C. (1980) Evaluation of API 0C clinical yeast identification system for must and wine yeast identification. *American Journal of Enology and Viticulture*, **31**, 364–366.

Subden, R.E., Irwin, D., Cunningham, J.D. and Meiering, A.G. (1982) Wine yeast isozymes. Genetic differences in 18 stock cultures. *Canadian Journal of Microbiology*, **28**, 1047–1050.

Swings, J. (1992) The genera *Acetobadter* and *Gluconobacter*. In *The Prokaryotes*, second edition, edited by A. Balows, pp. 2268-2286. New York: Springer-Verlag.

Teuber, M., Sievers, and Andersen, A. (1987) Characterization of the microflora of high acid submerged fermenters by distinct plasmid profiles. *Biotechnology Letters*, **9**, 265–268.

Thomas, D.S. (1988) Electropositively charged filters for the recovery of yeasts and bacteria from beverages. *Journal of Applied Bacteriology*, **65**, 35–41.

Thomas, D.S. and Ackerman, J.C. (1988) A selective medium for detecting yeasts capable of spoiling wine. *Journal of Applied Bacteriology*, **65**, 299–308.

Thornton, R.J. and Bunker, A. (1989) Characterization of wine yeasts for genetically modifiable properties. *Journal of the Institute of Brewing*, **95**, 181–184.

Tracey, R.P. and Britz, T.J. (1987) A numerical study of *Leuconostoc oenos* strains from wine. *Journal of Applied Bacteriology*, **63**, 523–532.

Tracey, R.P. and Britz, T.J. (1989) Cellular fatty acid composition of *Leuconostoc oenos*. *Journal of Applied Bacteriology*, **66**, 445–456.

Tredoux, H.G., Kock, J.L.F., Lategan, P.M. and Muller, H.B. (1987) A rapid identification technique to differentiate between *Saccharomyces cerevisiae* strains and other yeast species in the wine industry. *American Journal of Enology and Viticulture*, **38**, 161–164.

van Vuuren, H.J.J. and van der Meer, L. (1987) Fingerprinting of yeasts by protein electrophoresis. *American Journal of Enology and Viticulture*, **38**, 49–53.

van der Westhuizen, L.M., Agenbach, W.A., Loos, M.A. and Schoombee, N.F. (1981) Comparison of procedures for isolation of malolactic bacteria from wine. *American Journal of Enology and Viticulture*, **32**, 168–170.

Weiss, N. (1992) The genera *Pediococcus* and *Aerococcus*. In *The Prokaryotes*, second edition, edited by A. Balows, pp. 1502-1507. New York: Springer-Verlag.

Wibowo, D., Eschenbruch, R., Davis, C., Fleet, G.H. and Lee, T.H. (1985) Occurrence and growth of lactic acid bacteria in wine—a review. *American Journal of Enology and Viticulture*, **36**, 302–313.

Yamada, Y., Ishikawa, T., Yamashita, M., Tahara, Y., Yamasato, K. and Kaneko, T. (1981) Deoxyribonucleic acid base composition and deoxyribonucleic acid homology in acetic acid bacteria, especially in the polarly flagellated intermediate strains. *Journal of General and Applied Microbiology*, **27**, 465–475.

Yamada, Y., Itakura, N., Yamashita, M. and Tahara, Y. (1984) Deoxyribonucleic acid homologies in strains of *Gluconobacter* species. *Journal of Fermentation Technology*, **62**, 595–600.

Yamada, Y. and Kondo, K. (1984) *Gluconoacetobacter*, a new sub-genus comprising the acetate-oxidizing acetic acid bacteria with ubiquinone-10 in the genus *Acetobacter*. *Journal of General and Applied Microbiology*, **30**, 297–303.

Young, T.W. (1987) Killer yeasts. In *The Yeasts*, Volume two, *Yeasts and the Environment*, second edition edited by A.H. Rose and J.S. Harrison. pp. 131–164. London: Academic Press.

Zambonelli, C. (1988) *Microbiologia Biotecnologie dei Vini*. Bologna: Edagricole.

Chapter 2

YEASTS—GROWTH DURING FERMENTATION

GRAHAM H. FLEET and GILLIAN M. HEARD

Department of Food Science and Technology, The University of New South Wales, PO Box 1, Kensington, New South Wales 2033, Australia

Chapter Outline

INTRODUCTION

The fermentation of grape juice into wine is a complex biochemical process involving interactions between yeasts, bacteria, fungi and their viruses. The alcoholic fermentation is the most important reaction, and the key role of yeasts in conducting this fermentation has been recognized and studied since the pioneering investigations of Louis Pasteur some 140 years ago (Amerine, 1985). During fermentation, yeasts utilize sugars and other constituents of grape juice as substrates for their growth, converting these to ethanol, carbon dioxide and other metabolic end-products that contribute to the chemical composition and sensory quality of the wine. The concentrations of substrates utilized and products generated depend upon the species of yeast present, and the quantitative extent of their growth. To fully understand the role of yeasts in fermentation, it is essential to know: (i) the taxonomic identity of each species contributing to the fermentation; (ii) the kinetics of growth of each species; (iii) the biochemical properties of these yeasts and the chemical changes they produce; and (iv) the influences of vinification factors upon the kinetics of growth and chemical change. Although wine yeasts have been extensively investigated over many years, these studies have lacked a quantitative focus. This chapter reviews the growth of yeasts in wines from a quantitative perspective, dealing mostly with studies reported during the last 10 years. Excellent descriptive accounts of earlier studies have been reported by Ribéreau-Gayon *et al.* (1975), Kunkee and Amerine (1970), Kunkee and Goswell (1977), Benda (1982), Lafon-Lafourcade (1983) and Zambonelli (1988).

ORIGIN OF WINE YEASTS

Yeasts responsible for the alcoholic fermentation originate from either of three sources: (i) the surface of the grape; (ii) surfaces of winery equipment; and (iii) inoculum cultures. Grapes have long been considered the primary source of indigenous wine yeasts. Many studies in different countries have described the isolation and identification of yeasts from grape surfaces and these have been reviewed by Amerine and Kunkee (1968), Kunkee and Amerine (1970) and Kunkee and Goswell (1977). Unfortunately, very few of these reports provide sound quantitative data about the populations of individual species on the grapes. In a provocative discussion of the origin of indigenous wine yeasts, Martini and Martini (1990) describe most of these ecological studies as "redundant and monotonous" and misleading in their conclusions because selective enrichment methods, rather than direct plating methods, were often used to make the yeast isolations. In addition, they stressed that isolation methods should incorporate a thorough rinsing procedure (e.g. with sonication) to ensure complete dislodgement of yeasts from the grape surface. Significant studies that have provided some reliable quantitative data on the yeast ecology of grapes are those of Barnett *et al.* (1972), Davenport (1974), Sapis-Domercq *et al.* (1977), Goto and Yokotsuka (1977), Bureau *et al.* (1982), Rosini, Federici and Martini (1982) and Parish and Carroll (1985). The conclusions of these studies and earlier qualitative studies can be summarized as follows: (i) mature, sound grapes, aseptically crushed, will give a must with a total yeast population between 10^3–10^5 colony forming units (cfu) per ml; (ii) apiculate yeasts of the genera *Kloeckera* (especially *Kl. apiculata*) and *Hanseniaspora* are the predominant species on the surface of the grape, accounting for about 50–75% of the total yeast population; (iii) significant by their presence, but at lesser populations than *Kloeckera/Hanseniaspora* species, are species of *Candida* (especially *C. stellata* and *C. pulcherrima*), *Cryptococcus, Rhodotorula, Pichia, Kluyveromyces* and *Hansenula*; and (iv) fermentative species of *Saccharomyces* (e.g. *S. cerevisiae*) occur at very low populations on sound, undamaged grapes (less than 50 cfu/ml) and are rarely isolated by direct plating methods. Earlier conclusions that *S. cerevisiae* was prevalent on grapes are now considered erroneous because enrichment methods, rather than direct plating methods, were used for its isolation (Martini and Martini, 1990).

Numerous factors affect the total yeast population and the relative proportions of individual species on grapes. These factors include temperature, rainfall and other climatic influences (Kunkee and Goswell, 1977; Querol, Jimenez and Huerta, 1990; Longo *et al.*, 1991), degree of maturity at harvest (Rankine and Lloyd, 1963; Goto and Yokotsuka, 1977; Rosini *et al.*, 1982; Martinez, Millan and Ortega, 1989), use of fungicides (Sapis-Domercq *et al.*, 1977; Bureau *et al.*, 1982), physical damage due to mould, insect or bird attack (Guerzoni and Marchelli, 1987; Longo *et al.*, 1991) and grape variety. However, insufficient quantitative data are available to be able to draw general conclusions about these influences. The need for better study of the effects of these factors on the yeast ecology of grapes will become more apparent as winemakers recognize the special contributions that particular indigenous species can make to wine fermentations.

The surfaces of winery equipment (e.g. crushers, presses, tanks, fermentors, barrels, pipes, hoses, pumps, filtration and bottling units) that come into contact with grape juice and wine become locations for the development of a so called residential or winery yeast flora. The extent of this development (quantitative studies are lacking) will depend upon the nature of the surface and the degree to which the particular surface has been cleaned and sanitized. Key studies which demonstrate the association of yeasts with these surfaces are those of Peynaud and Domercq (1959), Ribéreau-Gayon and Peynaud (1960), Minárik (1969), Belin (1979) and Rosini (1984). *Saccharomyces cerevisiae* is prevalent on these surfaces, demonstrating the selective effects of grape juice and wine as growth substrates. Other species of *Saccharomyces*, as well as species of *Candida, Pichia, Hansenula* and *Brettanomyces* are also found at these locations. Following from the studies of Rosini (1984) who used a marked strain of *S. cerevisiae* to trace its origin, Martini and Martini (1990) concluded that winery surfaces were the main source of indigenous *S. cerevisiae* in fermentations. Such *S. cerevisiae* could comprise those strains originating from grapes, or starter cultures used in previous vintages. Once again, properly designed experiments that demonstrate the quantitative contribution of winery surfaces to the yeast populations of grape juices and wines are absent.

Obviously, starter cultures used to inoculate the juice will be a principal source of yeasts. Presently, various strains of *S. cerevisiae* are used for this purpose (Kraus, Reed and Villettaz, 1983; Kunkee, 1984; Reed and Nagodawithana, 1988), but the future could see the development and use of other species for this purpose. Even with the addition of starter inoculum, it should not be forgotten that indigenous species originating from the grape and winery surfaces will be present in the juice/must. Active dry yeast preparations are not necessarily pure and can contain a proportion of non-*Saccharomyces* species (Radler and Lotz, 1990).

GROWTH KINETICS

During fermentation, the viable population of yeasts in the grape juice increases from initial values of 10^4–10^6 cfu/ml to 10^8–10^9 cfu/ml. This growth follows a pattern typical of microorganisms grown in batch culture and is characterized by a lag phase, exponential phase, stationary phase and decline or death phase. The kinetic characteristics of this growth can be described in terms of: (i) duration of the lag phase; (ii) growth rate; (iii) maximum cell population formed; and (iv) duration of the stationary and decline phases. These characteristics vary with the fermentation, being influenced by a range of factors as listed in Table 2.1. They can be monitored by measurement of viable yeast counts, cell dry weight, microscopic (total) yeast count, carbon dioxide (CO_2) production, ethanol production and fermentable sugar utilization (Bely, Sablayrolles and Barre, 1990). However, cessation of yeast growth as determined by viable counts does not necessarily correspond with a cessation in metabolic activity of the cells. Several studies have clearly shown that even when the fermentation has reached the stationary-decline phases, and the yeast cells appear to be non-proliferating, there is still substantial metabolic activity. Most notably, residual sugars continue to be fermented with the production of CO_2 and ethanol (Ribéreau-Gayon, 1985; Ough, Davenport and Joseph, 1989; Larue and Lafon-Lafourcade, 1989; Bely *et al.*, 1990, see also Figure 2.1). This could be a stage

TABLE 2.1 Factors affecting yeast growth during alcoholic fermentation.

Clarification of grape juice
Addition of sulfur dioxide
Temperature of fermentation
Composition of grape juice
Inoculation with selected yeasts
Interactions with other microorganisms

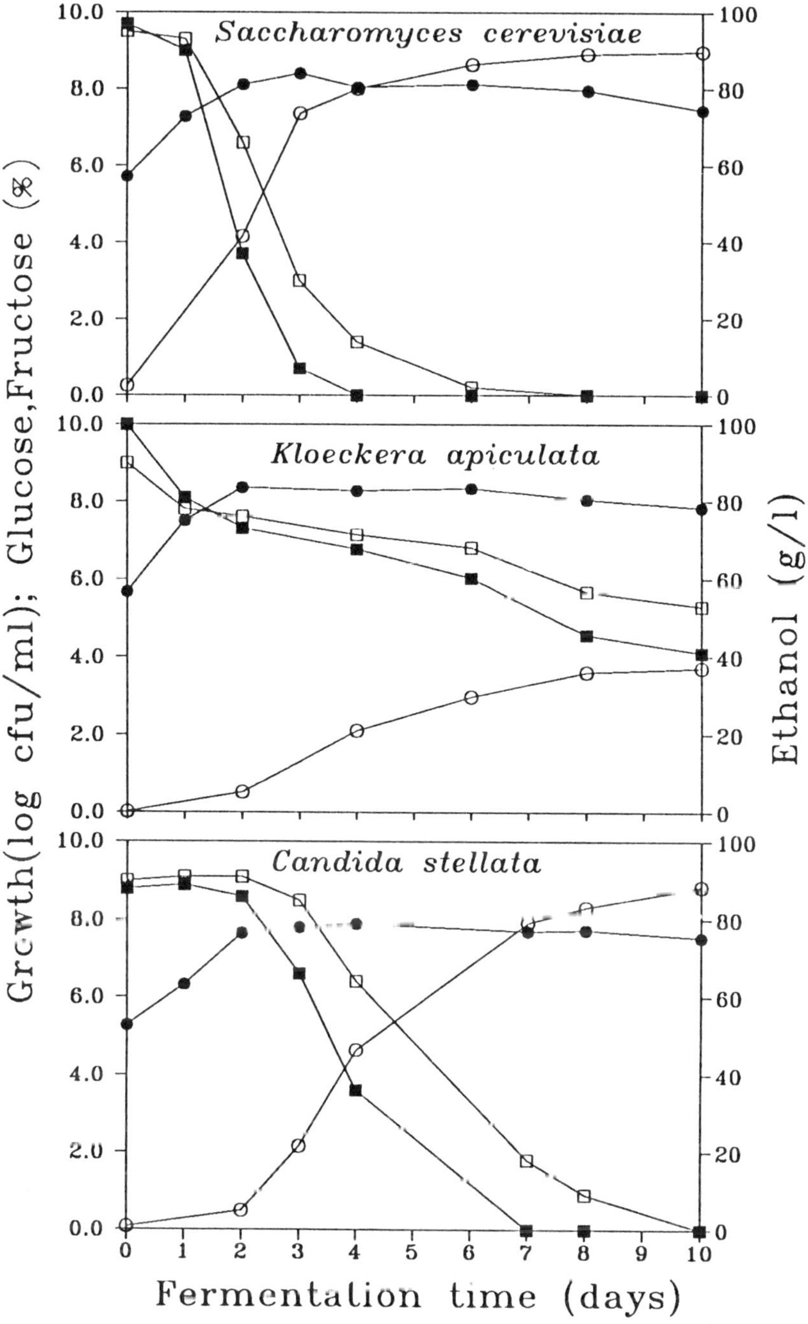

FIGURE 2.1. Kinetics of growth, sugar utilization and ethanol production by three species of wine yeast during fermentation of a synthetic grape juice medium. Colony forming units ●; ethanol O; glucose ■; fructose ❑. (Charoenchai and Fleet, unpublished data).

when significant quantities of secondary end-products such as acetaldehyde, esters and higher alcohols are produced, as suggested by the data of Millan and Ortega (1988) and Edwards *et al.* (1990). More information is required about the biochemical activities of non-proliferating yeast cells, keeping in mind that the exponential phase of fermentation is relatively short (a few days) and that the stationary-decline phases are significantly longer.

In concluding this section, it needs to be remembered that the growth characteristics of the total yeast population represent the sum of contributions from individual species and strains, all of which will have their own growth profile, kinetic characteristics and particular metabolic influences. Figure 2.1 compares the growth kinetics of *S. cerevisiae*, *Kl. apiculata* and *C. stellata* in a model grape juice medium.

Growth of Individual Species

Qualitative, semi-quantitative studies

It is well established from many years of research that wine fermentations, as conducted by traditional methods (without inoculation), are not the action of a single species or single strain of yeast. Rather, the final product results from the combined action of several yeast species which grow more or less in succession throughout the fermentation. The fermentation is initiated by the growth of various species of *Kloeckera, Hanseniaspora* and *Candida* and, to a lesser extent, species of *Hansenula* and *Pichia*. Their growth is generally limited to the first two or three days of fermentation, after which they die off. At this time, more strongly fermenting and more ethanol tolerant species of *Saccharomyces* take over the fermentation, subsequently predominating until its completion. The disappearance from the fermentation of *Kloeckera, Hanseniaspora, Candida* and other non-*Saccharomyces* species is generally considered to occur because of their weaker ethanol tolerance and inability to survive the increasing concentrations of ethanol produced by *Saccharomyces* species. Even within the species of *Saccharomyces*, there has been distinct recognition of a range of fermentative abilities and degrees of ethanol tolerance, leading to the taxonomic descriptions of species such as *Saccharomyces cerevisiae* var. *ellipsoides, S. bayanus* and *S. oviformis*. Current yeast taxonomy has all of these species lumped under the one description as *S. cerevisiae* (Kreger-van Rij, 1984; Zambonelli, Romano and Suzzi, 1989; Martini and Martini, 1990). Because of its almost invariable dominance of wine fermentations, *S. cerevisiae* has become universally accepted as the principal wine yeast, with the recognition that some strains within this species produce better wines than others. The concept of inoculating grape juice with selected starter cultures of *S. cerevisiae* to encourage rapid, consistent fermentations has become widely accepted within the wine industry. Along with this technology, came the general assumption that the inoculated strain of *S. cerevisiae* would dominate the fermentation and overwhelm any contribution from indigenous yeasts.

The general conclusions described in the previous paragraph have evolved from an enormous body of descriptive ecological research that has been well reviewed (Kunkee and Amerine, 1970; Ribéreau-Gayon *et al.*, 1975; Benda, 1982; Lafon-Lafourcade, 1983; Zambonelli, 1988). In addition, they have been supported by more recent and more precise semi-quantitative studies (Table 2.2) where various authors have reported the frequency of isolation of different yeast species at different stages during fermentation.

Quantitative studies

Quantitative studies of yeast growth have provided more precise knowledge about the ecology of fermentation. Fleet, Lafon-Lafourcade and Ribéreau-Gayon (1984) conducted quantitative ecological studies at two *premier crus* and one unclassified château in Bordeaux, France. Each of these wineries produced wines by traditional fermentation, that is without inoculation of *S. cerevisiae*. Growth of each species of yeast associated with the fermentations was monitored each day or second day for several weeks. In the two red and two white wines examined, there was a successional development of yeast species, progressing from *Kloeckera/Hanseniaspora* and *Candida* species through to *S. cerevisiae*, thereby confirming the conclusions of earlier qualititative and semi-quantitative investigations. However, the study of Fleet *et al.* (1984) is significant because it revealed, probably for the first time, the extent that the non-*Saccharomyces* species contributed to the fermentation. These species (especially *Kl. apiculata* and *C. stellata*) survived during the fermentation for longer periods than previously thought and grew to maximum populations of 10^6–10^7 cfu/ml. Such growth was considered to be quantitatively significant and likely to influence the chemical composition of the wine. Moreover, alterations to grape juice composition caused by these yeasts were likely to influence the growth kinetics and biochemical behavior of *S. cerevisiae*. Heard and Fleet (1985, 1986) conducted quantitative studies, similar to the Bordeaux experiments, at several commercial wineries in Australia. Both red wine and white wine fermentations were examined as well as fermentations which were allowed to

TABLE 2.2 Summary of semi-quantitative and quantitative studies on the growth of yeast species during wine fermentation.

Semi-quantitative Studies

Sapis-Domercq (1970); Minarik (1971); Cuinier and Guerineau (1976); Goto and Yokotsuka (1977); Sapis-Domercq *et al.* (1977); Belin (1979); Goto (1980); Bureau *et al.* (1982); Feuillat (1983); Valade and Moulin (1984); Parrish and Carroll (1985); Guijo, Millan and Ortega (1986); Moore *et al.* (1988); Mora *et al.* (1988); Querol *et al.* (1990); Longo *et al.* (1991).

Quantitative Studies

Fleet *et al.* (1984); Heard and Fleet (1985, 1986, 1988a, b); Pardo *et al.* (1989); Martinez *et al.* (1989); Mora *et al.* (1990); Mora and Mulet (1991).

develop by traditional means (uninoculated) and those which were inoculated with *S. cerevisiae*. Growth profiles for *Kl. apiculata, Candida* species (*C. stellata, C. krusei* and *C. colliculosa*) and *S. cerevisiae* were similar to those found in the Bordeaux studies. Interestingly, growth of the non-*Saccharomyces* yeasts also occurred in fermentations inoculated with *S. cerevisiae*, leading Heard and Fleet (1985) to conclude that indigenous yeasts could still be significant participants in inoculated fermentations. Subsequent quantitative studies of yeast growth during fermentations conducted in Spain (Martinez *et al.*, 1989; Mora, Barbas and Mulet, 1990; Mora and Mulet, 1991) and in Arkansas, USA (Moore, Johnson and Morris, 1988) have confirmed the general conclusions of Fleet and coworkers about the relative contributions of non-*Saccharomyces* yeasts and *S. cerevisiae*. The data of Mora *et al.* (1990) confirmed the significant contribution that *C. stellata* can make to fermentations and also indicated the potential for growth of *Kluyveromyces thermotolerans*. The role of indigenous *Pichia membranaefaciens* as a participant in wine fermentations warrants special mention and further study because it is considered a film-yeast and its growth under somewhat anaerobic fermentations is unexpected (Rankine, 1966). Fleet *et al.* (1984) noted its growth during the later stages of fermentation of some red wines. Mora and Mulet (1991) have recorded a similar observation and the data of Moore *et al.* (1988) and Martinez *et al.* (1989) also suggest a contribution from this yeast.

On the basis of the qualitative and quantitative studies to date (Table 2.2), Fleet (1990) has proposed that the growth of yeasts during the alcoholic fermentation of wines could be generalized as shown in Figure 2.2.

FACTORS AFFECTING YEAST GROWTH DURING FERMENTATION

Many winemaking variables affect fermentation and wine quality (Table 2.1) (Ribéreau-Gayon *et al.*, 1975; Amerine *et al.*, 1980; Farkas, 1988; Zambonelli, 1988; Cantarelli, 1989). A good understanding of the effect of these factors upon the overall conduct of the fermentation will emerge only after the kinetics of yeast growth are quantitatively measured as described in Figures 2.1 and 2.2. Such data, when correlated with biochemical analyses of the fermenting juice, will then indicate how these variables determine wine quality.

Clarification of Grape Juice

The procedures used to clarify juices, especially for white wine fermentations, are certain to influence the populations of indigenous yeasts in the juice and their subsequent contribution to the fermentation. There are two possibilities; clarification could remove indigenous yeasts and, therefore, decrease their contribution to fermentation, or it could encourage the selective growth of some species, thereby enhancing their contribution to the fermentation. The final outcome will depend on the procedure used for clarification. Such procedures

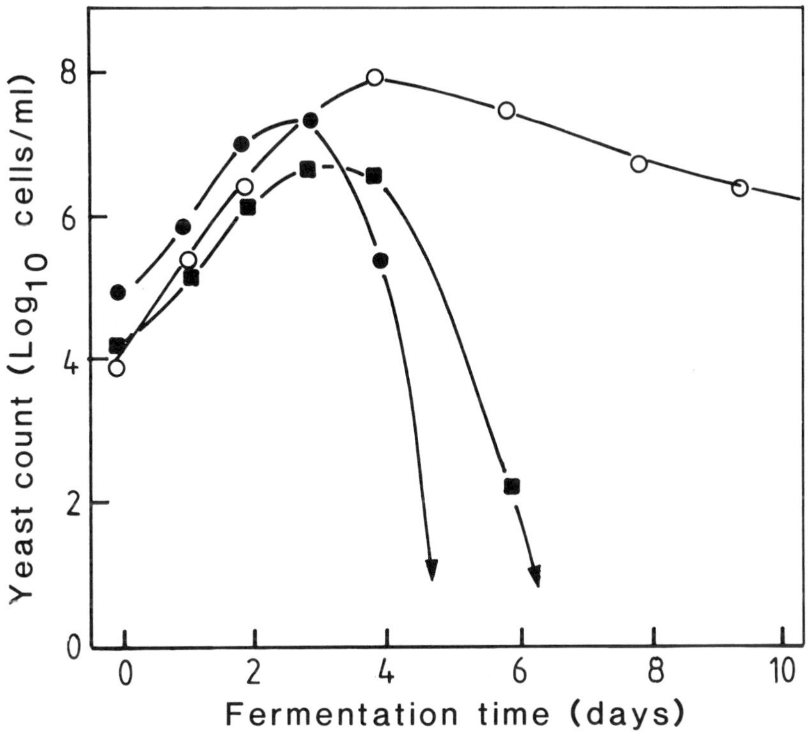

FIGURE 2.2. Generalized growth of yeast species during alcoholic fermentation of wine. *Saccharomyces cerevisiae* O; *Kloeckera/Hanseniaspora* species ●; *Candida* species ■. Variations will occur in the initial and maximum populations for each species. (Fleet, 1990).

include cold-settling, enzyme treatment, centrifugation and filtration. Centrifugation and filtration are likely to remove yeast cells, therefore, decreasing or even eliminating any contribution from indigenous yeasts. In contrast, cold-settling presents opportunities for the growth of indigenous yeasts, especially those species that grow well at low temperatures. Obviously, the time and temperature of this operation will affect the ecological outcome. To date, the influences of these juice treatments on the ecology of fermentation have not been sufficiently studied. Mora and Mulet (1991) noted an initial decrease in the populations of several yeast species during cold settling and bentonite treatments of Chenin blanc and Prensal musts. However, growth of some species (e.g. *Kl. apiculata, C. stellata, P. membranaefaciens*) was observed during the later stages of settling, enabling them to make significant contributions to subsequent juice fermentations which were conducted after inoculation with *S. cerevisiae*. Compared with other yeasts, *S. cerevisiae* seemed to be more susceptible to population decrease during cold settling.

Sulfur Dioxide (SO_2)

Addition of SO_2 to grapes or juice for controlling oxidation reactions and restricting the growth of indigenous microflora is a well established practice in winemaking (Beech and Thomas 1985; Chapter 13). As an antimicrobial, SO_2 can be expected to have the following effects on the kinetics of yeast growth during fermentation: (i) increase the lag phase and delay the onset of fermentation; (ii) decrease the growth rate and increase the time to complete fermentation; (iii) accelerate the decline/death phase; and (iv) selective effects upon the species or strains which grow and contribute to the fermentation. The extent of these influences will range from negligible to significant, depending upon the concentration of added SO_2, the composition of the juice which will affect the proportion of SO_2 in the active, undissociated state, and the SO_2-tolerance of the species or strains of yeasts present in the juice. These considerations are derived from an understanding of the action of SO_2 on microorganisms, in general (Rose and Pilkington, 1989). Despite the significant use of SO_2 in winemaking, there are surprisingly few well-controlled studies that have examined its action on the survival and growth of yeasts during fermentation.

The growth studies of Goto (1980) and Heard and Fleet (1988a) clearly show that addition of 50 mg/l or more SO_2 (total) to grape juice decreases the initial population of yeasts by about 10 fold, giving a lag phase of about 1–2 days before the onset of fermentation. A similar trend is evident from the data of Moore *et al.* (1988). The general view that addition of SO_2 to grape juice will selectively restrict the growth of indigenous non-*Saccharomyces* species and encourage the selective growth of *S. cerevisiae* is not necessarily supported by the data of quantitative growth studies. We have consistently found growth of indigenous yeasts (*Kloeckera* and *Candida* species) in commercial wine fermentations where the usual concentrations of SO_2 (50–100 mg/l total) have been added to the juice (Fleet *et al.*, 1984; Heard and Fleet, 1985, 1986). Similar conclusions are apparent form the data of Goto (1980), Moore *et al.* (1988), Martinez *et al.* (1989), Mora *et al.* (1990) and Mora and Mulet (1991). In more specific studies, growth of *Kl. apiculata* was not inhibited by total SO_2 concentrations of 100–150 mg/l (Heard and Fleet, 1988a). These findings raise questions about the efficacy of SO_2 in controlling indigenous yeasts and challenge one of the basic reasons for using SO_2 in winemaking. More detailed, systematic examination of the action of SO_2 on the growth of wine yeasts is required; see also Chapter 13.

Apart from affecting the ecology of fermentation, SO_2 could also affect the metabolic activity of the participating yeasts and, therefore, the chemical composition of the wine. Wines fermented in the presence of SO_2 had higher concentrations of acetaldehyde, for example, than those fermented in its absence (Herraiz *et al.*, 1989).

Temperature

The temperature at which the alcoholic fermentation is conducted affects: (i) the rate of yeast growth and, consequently, the duration of fermentation; (ii) the extent to which different yeast species contribute to the fermentation; and (iii) the biochemical reactions of the yeasts which, ultimately, determine the chemical composition and sensory quality of the wine.

The rate of yeast growth and alcoholic fermentation increases as the temperature increases, with maximum rates generally occurring at temperatures between 20 and 25°C (Amerine *et al.*, 1980; Benda, 1982). Such conclusions have been demonstrated experimentally in a series of studies by Ough and coworkers who examined the fermentation rates of grape juice over the range 10–33°C, after inoculation with *S. cerevisiae*. Their studies included examination of juice prepared from several red and white grape varieties, juice adjusted to pH values of 3.0, 3.5 and 4.0, juice adjusted to ethanol concentrations of 0–8% and juice treated with bentonite (Ough, 1964, 1966a, b; Ough and Groat, 1978). Red wines are generally fermented at 20–30°C, while white wines are often fermented in the range 10–20°C. In recent years, there has been a preference by some winemakers to ferment white wines at the lower temperatures to enhance the production and retention of flavor volatiles (Killian and Ough, 1979; Kunkee, 1984). Such trends have required the selection and use of strains of *S. cerevisiae* which exhibit good growth rates at low temperatures. It must be stressed that the fermentation, itself, generates heat. Unless cooling is applied, the fermentation will not be isothermic and a temperature increase of 5–10°C may occur from the beginning to end of fermentation (Sablayrolles, Barre and Grenier, 1987; Sablayrolles and Barre, 1989). Cooling control is particularly important if there is a possibility of the temperature increasing above 30°C, where sensitivity of *S. cerevisiae* to ethanol is enhanced and the fermentation may stop, prematurely (van Uden, 1989).

The concept that temperature may affect the ecology of wine fermentation is relatively new. Sharf and Margalith (1983) examined the mixed growth of *S. cerevisiae* and *Kl. apiculata* in a synthetic medium at 10, 20 and 30°C. *Kloeckera apiculata* dominated the fermentations at 10 and 20°C, whereas *S. cerevisiae* was the dominant species at 30°C. The authors suggested that *Kl. apiculata* had better ability than *S. cerevisiae* to grow at lower temperatures. In laboratory studies with grape juice, Heard and Fleet (1988b) examined the growth kinetics of several wine strains of *S. cerevisiae*, *Kl. apiculata* and *C. stellata* in mixed culture at 10, 20 and 25°C. At 25°C, the profile of yeast growth followed that illustrated in Figure 2.2, with *S. cerevisiae* being the dominant species and *Kl. apiculata* and *C. stellata* dying off at earlier stages. However, with fermentations conducted at 10°C, *Kl. apiculata* and *C. stellata* did not die off, but remained at high populations (10^7–10^8 cfu/ml) throughout the fermentation. Moreover, in some trials at 10°C, *Kl. apiculata* replaced *S. cerevisiae* as the dominant species. Subsequent studies by Gao and Fleet (1988) showed that *Kl. apiculata* and *C. stellata* exhibited increased tolerance of ethanol at 10–15°C, supporting more general observations that the ethanol tolerance of yeasts is very much influenced by temperature (Casey and Ingledew, 1986; D'Amore and Stewart, 1987). Thus, increased ethanol tolerance of the non-*Saccharomyces* yeasts at low temperatures appears to be a major factor accounting for their stronger if not dominant contribution to low-temperature fermentations. In addition, it seems that some strains of these yeasts may exhibit faster growth rates than *S. cerevisiae* at lower temperatures. A systematic comparison of the growth rates of the various species of wine yeasts at different temperatures has not been reported and would be a worthy research topic.

Although more ecological studies are required, especially of commercial fermentations, it is not unreasonable to propose from the available data (Table 2.2), that fermentations conducted below 20°C are more likely to have a stronger contribution from the growth of indigenous, non-*Saccharomyces* species, whereas those conducted above 20°C will be less affected by these yeasts. Such ecological influences should be reflected in the chemical composition and sensory properties of the wine. Several studies have shown that strains of *Kl. apiculata, C. stellata, C. pulcherrima* and *C. krusei*, for example, have the potential to produce greater or lesser concentrations of volatiles such as higher alcohols, acetic acid, esters, and acetaldehyde than *S. cerevisiae* (Tables 2.3 and 2.4).

In addition to its influences on growth rate and fermentation ecology, temperature also affects the biochemical activities of yeasts and their production of secondary metabolites. Such influences have been studied for *S. cerevisiae*, (e.g. Bisson, 1984; Cottrell and McLellan, 1986; Parrish and Carroll, 1987), but not for the non-*Saccharomyces* species. Low-temperature fermentations, characterized by a stronger growth of non-*Saccharomyces* species, are likely to have decreased concentrations of ethanol. Sugars metabolized by these yeasts are not converted to the high concentrations of ethanol, as occurs with *S. cerevisiae* (Heard and Fleet, 1988b; Mateo *et al.*, 1991). Also, removal of sugars from the juice by the growth of these yeasts leaves decreased amounts for conversion to ethanol by *S. cerevisiae*.

Composition of the Juice

The chemical and physical composition of the juice affects the rate and completeness of fermentation as well as the concentration of many aroma and flavor constituents in the wine. Important variables that are likely to affect yeast growth are; sugar concentration, supply of nitrogenous substrates, presence of adequate vitamins, concentration of dissolved oxygen, and concentration of insoluble solids. Other factors that require consideration include fungicide residues from the grapes and any yeast inhibitory or stimulatory substances produced by the growth of fungi or bacteria on the grapes prior to harvest.

In most instances, grape juices possess all the nutrients necessary for yeast growth and complete fermentation, although it is important to remember that juice composition will vary according to grape variety, soil condition, use of fertilizers and maturity at harvest. In addition, processing operations such as

TABLE 2.3 Concentrations of ethanol and higher alcohols produced by different species of wine yeasts.

Species	*Ethanol %*	*Propanol mg/l*	*Isobutanol mg/l*	*Isoamyl alcohol mg/l*	*2-Phenylethanol mg/l*
Saccharomyces cerevisiae	6–23	9–170	5–78	17–330	5–83
Kloeckera apiculata	5–6.5	16–32	4–38	4–17	21–27
Candida krusei	1–6.5	1–23	38–106	22–100	–
Candida stellata	4.5–6.5	4–8	13–21	–	6–11
Hansenula anomala	0.2–4.5	3–15	18–29	11–25	27
Metschnikowia pulcherrima	2	<1–43	37–123	21–243	22

Data abstracted from Ferreira (1959); Guymon, Ingraham and Crowell (1961); Van Zyl, De Vries and Zeeman (1963); Rankine (1967); Bertrand (1968); Sponholz and Dittrich (1974); Soufleros and Bertrand (1979); Shimazu and Watanabe (1981); Shinohara (1984); Cottrel and McLellan (1986); Cabrera *et al.* (1988), Herraiz *et al.* (1990); Mateo *et al.* (1991).

TABLE 2.4 Concentrations of acetic acid, esters and acetaldehyde produced by different species of wine yeasts.

Species	*Acetic acid g/l*	*Ethyl acetate mg/l*	*Isoamylacetate mg/l*	*Acetaldehyde mg/l*
Saccharomyces cerevisiae	0.3–0.8	10–100	0.1–16	50–120
Kloeckera apiculata	1–2.5	25–375	0.4–5	6–40
Candida krusei	1.0	220–730	< 0.1	–
Candida stellata	1–1.3	7–25	0.1–0.4	–
Hansenula anomala	1–2	137–2150	1–11	–
Metschnikowia pulcherrima	0.1–0.15	150–382	0.1–0.8	23–40

Data abstracted from Cantarelli (1966); Usseglio-Tomasset (1967); Bertrand (1968); Sponholz and Dittrich (1974); Soufleros and Bertrand (1979); Shimazu and Watanabe (1981); Milan and Ortega (1988); Cabrera *et al.* (1988); Herraiz *et al.* (1990); Mateo *et al.* (1991).

pressing, clarification, ion-exchange, desulfiting and blanketing with nitrogen or carbon dioxide can affect composition. In various contexts, the articles of Kunkee and Amerine (1970), Benda (1982), Lafon-Lafourcade (1983), Barre (1988) and Zambonelli (1988) have considered the influences of these variables on juice fermentability and the growth rate of *S. cerevisiae*. Notably absent, however, are discussions of the effects of juice composition upon the contributions of non-*Saccharomyces* to the fermentation. It must be stressed that any significant growth of *Kloeckera*, *Hanseniaspora*, *Candida* or other non-*Saccharomyces* species during juice settling or during the initial stages of fermentation will alter juice composition and affect its suitability as a substrate for subsequent growth of *S. cerevisiae*. Thus, the growth of these species could change the glucose to fructose ratio in the juice, or remove nitrogenous and vitamin nutrients, but such effects have not been investigated.

The concentration of fermentable hexoses in grape juices varies between 125–250 g/l, depending upon grape variety and environmental influences (Benda, 1982). In extreme cases where grapes have been infected with *Botrytis cinerea* ("pourriture noble"), hexose concentrations up to 400 g/l have been encountered (Lafon-Lafourcade, 1983). The growth reaction of yeasts varies with sugar concentration (Fleet, 1992). Thus, it is likely that the initial concentration of sugar in grape juice will selectively influence the species and strains of yeast responsible for the fermentation. Such selective influences of sugar concentration could be amplified at the lower fermentation temperatures (e.g. 15°C) (Monk and Cowley, 1984). However, the impact of sugar concentration on the ecology of wine fermentations has not been examined in any scientific depth. There is some suggestion that the presence and growth of *C. stellata*, for example, may be encouraged in juices with high sugar concentration (Benda, 1982; Lafon-Lafourcade, 1983).

Pure culture studies with *S. cerevisiae* convincingly show that the rate and completeness of fermentation decrease as the initial concentration of sugar in the juice increases above 200 g/l. Such high concentrations affect yeast growth by increasing the lag phase, decreasing growth rate and decreasing the maximum population of cells produced (Lafon-Lafourcade, 1983; Monk and Cowley, 1984; Nishino, Miyazaki and Tohjo, 1985).

Pentose sugars occur in grape juices at concentrations of 0.1–5 g/l (Amerine *et al.*, 1980). It has been accepted for a long time that yeasts, in general, do not ferment pentoses (Rose, 1987). However, this conclusion may need to be reviewed in light of increasing evidence that some strains of yeasts can ferment these sugars, especially xylose, into ethanol. Although *S. cerevisiae* is unable to ferment pentoses such as xylose, rhamnose and arabinose, other yeasts including species of *Candida*, *Hansenula* and *Metschnikowia*, have been found to ferment xylose to ethanol under anaerobic conditions (Rose, 1987). The concentrations of individual pentose sugars in grape juice and wines have not been the subject of sufficiently detailed analysis to make firm judgements about their fate during alcoholic fermentation. In a related context, the fate of grape pectins during alcoholic fermentation also demands more careful scrutiny.

Strains of indigenous non-*Saccharomyces* could produce pectinolytic enzymes that degrade these polysaccharides and, also, there is new information that some strains of *S. cerevisiae* produce extracellular polygalacturonases (McKay, 1990; Fleet, 1992).

Free amino acids and ammonium ions are the principal nitrogen sources used by yeasts for growth during alcoholic fermentation. For a long time, it was considered that most grape juices contained these nitrogen substrates at concentrations far in excess of that required for maximum yeast growth and maximum fermentation rate. Recent evidence, however, suggests that this is not always the case, and that some juices (especially those that have been heavily processed or clarified) may not have sufficient nitrogen nutrients to allow maximum yeast growth (Ingledew and Kunkee, 1985; Salmon, 1989; Bely, *et al.*, 1990; Chapter 4). Moreover, the nitrogen demand by yeasts increases with increasing sugar concentration in the juice, and can vary significantly (e.g. from 460–780 mg/l of free amino nitrogen) with the strain of *S. cerevisiae* (Jiranek, Langridge and Henschke, 1989). Thus, supplementation of juices with various yeast foods or diammonium phosphate is now common practice to ensure that nitrogen availability is not a factor that limits yeast growth (Monk, 1982; Monk and Costello, 1984; Ingledew and Kunkee, 1985). To date, the studies on nitrogen requirements by wine yeasts have been conducted with strains of *S. cerevisiae*. Nothing has been published about the nitrogen demands of the non-*Saccharomyces* species or about removal of nitrogen from the juice by these yeasts prior to growth of *S. cerevisiae*.

Further consideration needs to be given to the breakdown and utilization of grape juice proteins as a source of nitrogen for yeast growth. Contrary to general opinion, some strains of *S. cerevisiae* appear to produce extracellular proteolytic enzymes (Feuillat, 1984; Fleet, 1992). There is a good possibility that some strains of the non-*Saccharomyces* wine yeasts are proteolytic (Fleet, 1992) and, already, there is one firm report that a strain of *Kl. apiculata* gives significant production of an extracellular protease (Lagace and Bisson, 1990).

Under most circumstances, grape juices contain sufficient concentrations of vitamins (inositol, thiamine, biotin, pantothenic acid, nicotinic acid) to permit maximum growth of *S. cerevisiae* (Monk, 1982; Ough *et al.*, 1989). Vitamin losses may occur in heavily processed juices (e.g. desulfited or fined juices), in which case supplementation has been observed to increase the growth rate of *S. cerevisiae* (Monk 1984; Monk and Costello, 1984). Species of non-*Saccharomyces* wine yeasts (e.g. *Kl. apiculata*, *C. stellata*, *M. pulcherrima*) are far more demanding of vitamins than *S. cerevisiae* (Kreger-van Rij, 1984) and, consequently, this may be a factor which limits their contribution to fermentation.

Controlled aeration of the grape juice prior to fermentation or during the early stages of fermentation is stimulatory to yeast growth and fermentation (Ingledew and Kunkee, 1985; Sablayrolles and Barre, 1986) but more precise study of this effect is needed. Further studies are also required of the effects of grape solids and other particulate matter on the rate of fermentation. Yeast growth and fermentation rates are significantly increased by the presence of these materials (Ough and Groat, 1978; Barre, 1988; Edwards *et al.*, 1990). Consideration needs to be given to the possibility that different yeast species may be selectively adsorbed to such particulate material and that a biofilm of immobilized yeast cells probably develops around each particle. The metabolic characteristics of these biofilms are possibly different to yeast cells in suspension.

The pH of grape juice varies between 3.0 and 4.0, depending upon the concentrations of tartaric and malic acids. Although the rates of growth and juice fermentation by *S. cerevisiae* are decreased as the pH is decreased from 3.5 to 3.0 (Ough 1966a, b), it is not known how juice pH affects the growth of non-*Saccharomyces* species and their contribution to the overall fermentation. There is some indication from the studies of Heard and Fleet (1988b) that their growth and contribution is stronger at pH 3.5 than at pH 3.0.

Juices containing residual fungicides may be prone to altered alcoholic fermentation, depending upon the concentration and type of fungicide present. Such residues could delay the onset of yeast growth, decrease the rate of growth, decrease the yield of growth and accelerate death of the yeast cells after growth has ceased. As a consequence of these influences, slower, incomplete fermentations may result (Beuchat, 1973; Sapis-Domercq *et al.*, 1977; Bureau *et al.*, 1982; Gnaegi and Aerny, 1984; Lemperle, 1988; Girand, Blazy-Maugen and Michel, 1989). The data of Sapis-Domercq *et al.* (1977) and Bureau *et al.* (1982) also suggest that fungicide residues could impact on the ecology of the fermentation. Quantitative growth studies are required, but it would appear that different yeasts may be influenced to different extents by these residues, and this may affect their contribution to the fermentation.

Inoculation with Selected Yeasts

One of the most significant discoveries in winemaking is that alcoholic fermentation can be encouraged and better controlled by inoculation of the grape juice with selected cultures of *S. cerevisiae*. Comparisons between inoculated and uninoculated fermentations were made by Rankine and Lloyd (1963), Meyer (1981) and Valade and Moulin (1984) and they consistently showed that the lag phase was decreased and that the fermentation was completed faster after inoculation. Moreover, the wines produced by inoculation were of consistent acceptable quality. In many countries, now, inoculation is the rule rather than the exception and use of this concept has become attractive and convenient by the commercial availability of active dried cultures of *S. cerevisiae*, specifically selected and prepared for wine fermentations (Kraus, Reed and Villettaz, 1983; Barre and Vezinhet, 1984; Reed and Nagodawithana, 1988).

From an ecological viewpoint, it is expected that the inoculated strain of *S. cerevisiae* will suppress and outgrow indigenous non-*Saccharomyces* species and indigenous strains of *S. cerevisiae*, and dominate the fermentation. Although these expectations and assumptions have been widely accepted by winemakers, they have until recently escaped rigorous scientific verification. In studies conducted at several Australian wineries, Heard and Fleet (1985) showed that growth of *Kl. apiculata* and *Candida* species was not suppressed in fermentations inoculated with commercial strains of *S. cerevisiae*. Both red and white wines fermented after inoculation with selected strains of *S. cerevisiae* gave profiles of yeast growth similar to those shown in Figure 2.2. Similar findings have now been reported for several inoculated fermentations conducted in Spain (Martinez *et al.*, 1989; Mora *et al.*, 1990). It may be concluded, therefore, that inoculation of the grape juice with high initial populations of *S. cerevisiae* will not necessarily prevent the growth of indigenous non-*Saccharomyces* yeasts, and that these species may also contribute to the overall fermentation. Heard and Fleet (1985) also questioned the assumed dominance of inoculated strains of *S. cerevisiae* over indigenous strains of this species, highlighting the fact that conventional cultural techniques for isolating and identifying wine yeasts do not distinguish between strain types. Using electrophoretic methods to distinguish between strains of *S. cerevisiae*, Bouix, Leveau and Cuinier (1981) in France had earlier demonstrated that, in some wineries, the inoculated strain did not dominate the fermentation. Loiseau *et al.* (1987) and Delteil and Aizac (1988) addressed this question by inoculation with genetically marked strains of *S. cerevisiae* whose growth could be followed independently of indigenous strains. They concluded that dominance of the inoculated strain was not always assured, and depended on the specific conditions of vinification. Using hydrogen sulfide production as a marker to differentiate between strains of *S. cerevisiae*, Martinez *et al.* (1989) demonstrated that the inoculated strain tended to dominate, but that indigenous strains remained in significant numbers throughout the fermentation. Further studies on the questions of species and strain dominance are needed but, for the present, it must be accepted that inoculation of any particular strain of *S. cerevisiae* does not necessarily guarantee its dominance or exclusive contribution to the fermentation. Significant factors that will influence the outcome will be the population of indigenous yeasts already in the juice and the extent to which they have adapted to grow in that juice.

Stuck Fermentations

A sporadic but serious problem in winemaking is the premature cessation of yeast growth and alcoholic fermentation, giving a wine with residual, unfermented sugar and less than expected concentration of ethanol. Such fermentations are commonly referred to as being "stuck" or "sluggish" and aspects relating to their occurrence and prevention have been reviewed by Larue and Lafon-Lafourcade (1989) and Munoz and Ingledew (1990). Factors considered to cause this problem include: excessive clarification and processing of the juice; fermentation temperature too high; juice deficiency in nutrients or factors required for maximum yeast growth and survival (especially in juices with high initial sugar concentration); presence of fungicide residues; and influences of acetic-acid bacteria. Proposals to overcome stuck fermentations include: controlled aeration of the juice or wine; addition of nitrogen containing yeast foods (Ingledew and Kunkee, 1985); and addition of yeast cell-wall hulls (Lafon-Lafourcade, Geneix and Ribéreau-Gayon, 1984).

Researchers at Bordeaux have proposed that medium chain-length fatty acids such as decanoic and octanoic acids have a key role in causing stuck fermentations. These acids are produced by *S. cerevisiae* during fermentation and, above certain concentrations, they become toxic to its continued growth and survival (Lafon-Lafourcade *et al.*, 1984; Larue and Lafon-Lafourcade, 1989; Edwards *et al.*, 1990). The possibility that indigenous non-*Saccharomyces* wine yeasts might also produce these acids has not been considered. Removal of these acids from wines can be done by adsorption with yeast cell-wall hulls (yeast ghosts) (Larue and Lafon-Lafourcade, 1989). Moreover, juices supplemented with yeast hulls exhibit more complete fermentation and less tendency to become stuck (Munoz and Ingledew, 1989a; Edwards *et al.*, 1990). However, adsorption of toxic acids may not be the only mechanism by which these hulls overcome stuck fermentations. According to Munoz and Ingledew (1989b, 1990), these hulls may also serve as a source of sterols which are essential for the sustained growth of yeasts under conditions of limited oxygen supply. Prospects for the commercial use of yeast hulls in winemaking seem encouraging but further studies on their action are needed, especially to resolve unclear conclusions about their potential "side-effects" on the conduct of the malolactic fermentation and on wine sensory qualities (Reed and Nagodawithana, 1988; Edwards *et al.*, 1990). Commercially available yeast hulls appear to be crude preparations of yeast cell walls (mostly polysaccharide) that would be contaminated with significant amounts of cellular and membranous lipid (Fleet, 1991). Such preparations quickly become rancid on storage, and could easily introduce undesirable taints to wine flavor.

Interaction of Wine Yeasts with Other Microorganisms

Species of fungi, acetic acid bacteria and lactic acid bacteria occur naturally on grapes where their populations are determined by the amount of rainfall prior to grape harvest, degree of physical damage to the berry, use of fungicides, and time between harvest, crushing and fermentation. Growth of these species on the grapes or in the grape juice prior to fermentation will alter the chemical composition of the juice and generate products that could affect subsequent growth of wine yeasts. One could also consider, here, certain yeast–yeast interactions.

The mycology of grapes has been poorly investigated, particularly in a quantitative sense. Although there is significant knowledge about the ecology and biochemical activities of *Botrytis cinerea* on grapes (Ribéreau-Gayon, Ribéreau-Gayon and Seguin, 1980; Chapter 11) other fungal species such as those of *Aspergillus*, *Penicillium*, *Mucor* and *Alternaria* may occur and grow

(Davenport, 1974; Nair, 1985). Individually or collectively, the growth of such species can significantly affect the composition and properties of the grape juice (Bayonove, 1989). Moreover, there are several reports that extracts of grapes infected with *B. cinerea* will either inhibit (Ribéreau-Gayon, 1985) or activate (Reed and Nagodawithana, 1988) alcoholic fermentation, especially in juices of high sugar concentration.

As reviewed by Lafon-Lafourcade and Joyeux (1981) and Drysdale and Fleet (1988), the significance of acetic acid bacteria in winemaking extends beyond their traditional role of causing "vinegary spoilage" of the final product. The species of concern are *Acetobacter pasteurianus*, *A. aceti* and *Gluconobacter oxydans*. These bacteria can develop on grapes, especially if they are damaged, and contaminate the juice at populations up to 10^6 cfu/ml. Early studies by Vaughn (1938) and Gilliland and Lacey (1964) indicated that some strains of *Acetobacter* were antagonistic to the growth of yeasts, including wine yeasts. Decreased fermentation of grape juice by *S. cerevisiae* was noted by Joyeux and Lafon-Lafourcade (1984) when *A. aceti* and *G. oxydans* had been grown in the juice prior to inoculation with the yeast. Similar observations were reported by Grossman and Becker (1984) who suggested that anti-yeast factors other than acetic acid could be produced by acetic acid bacteria. Drysdale and Fleet (1989) reported a detailed study of the combined growth of acetic acid bacteria (*A. aceti*, *A. pasteurianus* and *G. oxydans*) and wine yeasts (*S. cerevisiae*, *Kl. apiculata* and *C. stellata*) in grape juice. Their data on the growth kinetics of the different species and chemical changes caused to juice composition clearly showed that these bacteria readily grew in conjunction with the yeasts, limiting the ability of the latter to conduct a complete alcoholic fermentation.

Lactic acid bacteria are an integral component of the ecology of winemaking (Wibowo *et al.*, 1985). The principal species associated with wines are *Leuconostoc oenos*, *Pediococcus parvulus*, *P. pentosaceus* and various species of *Lactobacillus*. These bacteria occur in grape juice at low populations (approximately 10^3 cfu/ml) but, generally, die off during alcoholic fermentation (Fleet *et al.*, 1984). Consequently, they exert little or no effect on yeast growth. However, factors that delay the onset of alcoholic fermentation could encourage the growth of these species and their utilization of grape juice components that, otherwise, would be available to yeasts. By this mechanism, therefore, they could restrict yeast growth and disturb the alcoholic fermentation (Beelman *et al.*, 1982; King and Beelman, 1986). In addition, their production of acetic acid or other secondary metabolites at this stage could also affect yeast growth (Beelman *et al.*, 1982). For these reasons, some authors caution against the inoculation of malolactic bacteria prior to alcoholic fermentation (Lafon-Lafourcade, 1983; Ribéreau-Gayon, 1985).

Species of *Bacillus* (Lee *et al.*, 1984), *Streptomyces* (Fleet *et al.*, 1984) as well as some fungi such as *Trichosporon* and *Aureobasidium* species may be significant in wine fermentations but their influences remain totally unexplored.

Because a variety of different yeast species and strains are associated with wine fermentations, there is significant scope for yeast–yeast interactions. The best known of such interactions is the activity of killer yeasts (Young, 1987). It is now well documented that killer strains of *S. cerevisiae* may be operative during alcoholic fermentation and can destroy sensitive wine strains of this species (Heard and Fleet, 1987; Fleet, 1990). Moreover, wine species of *Candida*, *Pichia*, *Hanseniaspora* and *Hansenula* that show killer activity against wine strains of *S. cerevisiae* have been reported (Fleet, 1990). The interactions of killer yeasts during wine fermentations are outside the scope of this chapter and are discussed in detail in Chapter 8. Apart from the killer effect, there are early reports (cited by Benda, 1982) and a more recent finding (Herraiz *et al.*, 1990) that, under some circumstances, the growth of *Kl. apiculata* may have inhibitory effects on the growth of *S. cerevisiae*. These interesting observations need verification and further study.

AUTOLYSIS

Autolysis of yeast cells occurs at the end of fermentation, when their growth has ceased. It is not a rapid event, but it occurs gradually as the yeast cells die. During autolysis, intracellular organization is lost and cell proteins, lipids, nucleic acids and polysaccharides undergo endogenous enzymatic degradation into products that diffuse into the external environment (Hough and Maddox, 1970; Arnold, 1981; Chapter 7). Surprisingly, cell wall integrity and basic cell shape are not lost but wall porosity is altered to permit diffusion of autolytic products into the extracellular environment (Charpentier *et al.*, 1986). These products include peptides, amino acids, fatty acids, nucleotides and nucleosides that can affect wine sensory quality and/or serve as nutrients that promote the growth of either spoilage or malolactic bacteria (Wibowo *et al.*, 1985). Yeast autolysis will also release enzymes such as esterases, glycosidases, proteinases and lipases that, over time, could modify wine composition.

Yeast autolysis is greatest during storage of wine on its lees and accounts for increases in the concentration of wine amino acids and peptides observed at this stage of production (Ferrari and Feuillat, 1988). The character of high quality champagne produced by the "methode champenoise" has been correlated with the occurrence of yeast autolysis during ageing in bottles and concomitant breakdown of yeast proteins, lipids and nucleic acids (Troton *et al.*, 1989; Leroy *et al.*, 1990). Despite the inevitability of yeast autolysis during winemaking, and its potential impact on wine quality, there is surprisingly little information about factors that initiate this process, its kinetics and biochemistry (Arnold, 1981). Virtually all of the information on yeast autolysis relates to *S. cerevisiae*. Nothing is known about the autolytic behavior of other wine yeasts such as species of *Kloeckera*, *Hanseniaspora*, *Hansenula* and *Candida*. Detailed aspects about yeast autolysis during winemaking are discussed in Chapter 7.

CONCLUSION

Saccharomyces cerevisiae, although heralded as the principal wine yeast, is not the only species which contributes to wine fermentations. Other species also make a contribution, although often in a transient way. The fermentation can be manipulated to encourage the dominance of *S. cerevisiae*, but there can be no assurance that any inoculated strain will necessarily dominate over indigenous strains—unless the latter are first removed or inactivated. Many variables affect the growth and dominance of yeasts during wine fermentation, but precise information about the influence of these factors will become evident only after quantitative ecological study. The use of "marked" strains of *S. cerevisiae* (Delteil and Aizac, 1988; Petering, Henschke and Langridge, 1991) that are easily distinguished from indigenous strains will certainly facilitate the progress of these studies.

More specific information is required about the growth kinetics and metabolic reactions of non-*Saccharomyces* yeasts (e.g. species of *Kloeckera*, *Hanseniaspora*, *Candida*, *Kluyveromyces*, etc.) during grape juice processing and fermentation. It is certain that these species will have particular influences on the chemical composition and sensory quality of wines. Such influences need to be better identified and appropriate control strategies developed. It would be unfortunate, however, if these influences were considered only in a negative context. To the contrary, these non-*Saccharomyces* yeasts introduce a certain dimension of ecological and biochemical diversity into winemaking and, presumably, this diversity is transferred to the sensory quality of wines. For the creative, forward-looking winemaker, such diversity will offer opportunities for innovation and gainful exploitation.

REFERENCES

Amerine, M.A. (1985) Winemaking. In *Worlds Debt to Pasteur* edited by H. Koprowski and S.A. Plotin, pp. 67–81, New York: Alan R. Liss Incorporated.

Amerine, M.A., Berg, P., Kunkee, R.E., Ough, C.S., Singleton, V.L. and Webb, A.D. (1980) *The Technology of Winemaking* 4th edition. Westport Connecticut: AVI Publishing Company.

Amerine, M.A. and Kunkee, R.E. (1968) Microbiology of winemaking. *Annual Review of Microbiology*, **22**, 323–358.

Arnold, W.N. (1981) Autolysis. In *Yeast Cell Envelopes. Biochemistry, Biophysics and Ultrastructure*. Volume 2, edited by W.N. Arnold, pp. 129–137. Boca Raton: CRC Press.

Barnett, J.A., Delaney, M.A., Jones, E., Magson, A.B. and Winch, B. (1972) The numbers of yeasts associated with wine grapes of Bordeaux. *Archives of Microbiology*, **83**, 52–55.

Barre, P. (1988) Control of alcoholic fermentation in winemaking. *Proceedings of the Eighth International Biotechnology Symposium*, pp. 899–907, Paris.

Barre, P. and Vezinhet, F. (1984) Evaluation towards fermentation with pure culture yeasts in winemaking. *Microbiological Sciences*, **1**, 159–163.

Bayonove, C. (1989) Incidences des attaques parasitaires fongiques sur la composante qualitative du raisin et des vins. *Revue Francaise d'Oenologie*, **116**, 29–39.

Beech, F.W. and Thomas, S. (1985) Action microbienne de l'anhydride sulfureux. *Bulletin de l'Office International du Vin*, **58**, 564–581.
Beelman, R.B., Keen, R.M., Banner, M.J. and King, S.W. (1982) Interactions between wine yeast and malolactic bacteria under wine conditions. *Developments in Industrial Microbiology*, **23**, 107–121.
Belin, J.M. (1979) Contribution a l'étude des levures des chais-taxonomie, repartition des levures. *Mycopathologia*, **67**, 67–81.
Bely, M., Sablayrolles, J.M. and Barre, P. (1990) Description of alcoholic fermentation kinetics: its variability and significance. *American Journal of Enology and Viticulture*, **41**, 319–324.
Benda, I. (1982) Wine and brandy. In *Prescott and Dunns Industrial Microbiology* fourth edition, edited by G. Reed, pp. 293–402. Westport: AVI Publishing Company.
Bertrand, A. (1968) Utilisation de la chromatographic en phase gazeuse pour le dosage des constituents volatils du vin. *Connaissance de la Vigne et du Vin*, **2**, 175–207.
Beuchat, L.R. (1973) Inhibitory effects of benomyl on *Saccharomyces cerevisiae* during peach fermentation. *American Journal of Enology and Viticulture*, **24**, 110–115.
Bisson, J. (1984) Effet de la temperature de fermentation sur le type et la qualitie des vins blanc secs. *Revue Francaise d'Oenologie*, **93**, 21–24.
Bouix, M., Leveau, J.Y. and Cuinier, C. (1981) Determination de l'origine des levures de vinification, par une methode de differentiation fine des sourches. *Connaissance de la Vigne et du Vin*, **15**, 41–52.
Bureau, G., Brun, D., Vigues, A., Maujean, A., Vesselle, G. and Feuillat, M. (1982) Etude d'une microflore levurienne champenoise. *Connaissance de la Vigne et du Vin*, **16**, 15–32.
Cabrera, M.J., Moreno, J., Ortega, J.M. and Medina, M. (1988) Formation of ethanol, higher alcohols, esters and terpenes by five yeast strains in musts from Pedro Ximenez grapes in various degrees of ripeness. *American Journal of Enology and Viticulture*, **39**, 283–289.
Cantarelli, C. (1966) Influence des températures de fermentation et de conservation du vin et des vins spéciaux. *Bulletin de l'Office International du Vin*, **39**, 1191–1205.
Cantarelli, C. (1989) Factors affecting the behaviour of yeast in wine fermentation. In *Biotechnology Applications in Beverage Production*, edited by C. Cantarelli and G. Lanzarini, pp. 127–151. London: Elsevier.
Casey, G.P. and Ingledew, W.M. (1986) Ethanol tolerance in yeasts. *CRC Critical Reviews in Microbiology*, **13**, 219–280.
Charpentier, C., Nguyen van Long, T., Bonaly, R. and Feuillat, M. (1986) Alteration of cell wall structure in *Saccharomyces cerevisiae* and *Saccharomyces bayanus* during autolysis. *Applied Microbiology and Biotechnology*, **24**, 405–413.
Cottrell, T.H.E. and McLellan (1986) The effect of fermentation temperature on chemical and sensory characteristics of wines from seven white grape cultivars grown in New York State. *American Journal of Enology and Viticulture*, **37**, 190–194.
Cuinier, C. and Guerineau, L. (1976) Evaluation de la microflora au cours de la vinification des vins de Chinon. *Vignes et Vins*, **269**, 29–41.
D'Amore, T. and Stewart, G. (1987) Ethanol tolerance of yeast. *Enzyme and Microbial Technology*, **9**, 322–330.
Davenport, R. (1974) Microecology of yeasts and yeast-like organisms associated with an English vineyard. *Vitis*, **13**, 290–301.
Delteil, D. and Aizac, T. (1988) Comparison of yeast inoculation techniques by use of a "marked" yeast strain. *The Australian and New Zealand Wine Industry Journal*, **3**, 53–56.
Drysdale, G.S. and Fleet, G.H. (1988) Acetic acid bacteria in winemaking—a review. *American Journal of Enology and Viticulture*, **39**, 143–154.
Drysdale, G.S. and Fleet, G.H. (1989) The effect of acetic acid upon the growth and metabolism of yeasts during the fermentation of grape juice. *Journal of Applied Bacteriology*, **67**, 471–481.

Edwards, C.G., Beelman, R.B., Bartley, C.E. and McConnell, A.L. (1990) Production of decanoic acid and other volatile compounds and the growth of yeast and malolactic bacteria during vinification. *American Journal of Enology and Viticulture*, **41**, 48–56.

Farkas, J. (1988) *The Technology and Biochemistry of Wine*. Montreux: Gordon and Breach Science Publishers.

Ferrari, G. and Feuillat, M. (1988) L'elevage sur lie des vins blancs de Bourgogne I. Etude des composés azotés, des acides gras et analyse sensurielle des vins. *Vitis*, **27**, 183–197.

Ferreira, J.D. (1959) The growth and fermentation characteristics of six yeast strains of several temperatures. *American Journal of Enology and Viticulture*, **10**, 1–7.

Feuillat, M. (1983) Essais de levurage en vinification en blanc. *Revue Francaise d'Oenologie*, **91**, 69–87.

Feuillat, M. (1984) Mise en evidence de souches de levures a caractère proteolytique au cours la fermentation alcoolique de moût de raisin. *Bulletin de l'Office International du Vin*, **645**, 878–890.

Fleet, G.H. (1990) Growth of yeasts during wine fermentations. *Journal of Wine Research*, **1**, 211–223.

Fleet, G.H. (1991) Cell walls. In *The Yeasts*, Volume 4, second edition, edited by A.H. Rose and J.S. Harrison, pp. 199–277. London: Academic Press.

Fleet, G.H. (1992) Spoilage yeasts. *CRC Critical Reviews in Biotechnology*, **12**, 1–44.

Fleet, G.H., Lafon-Lafourcade, S. and Ribéreau-Gayon, P. (1984) Evolution of yeasts and lactic acid bacteria during fermentation and storage of Bordeaux wines. *Applied and Environmental Microbiology*, **48**, 1034–1038.

Gao, C. and Fleet, G.H. (1988) The effects of temperature and pH on the ethanol tolerance of the wine yeasts. *Saccharomyces cerevisiae*, *Candida stellata*, and *Kloeckera apiculata*. *Journal of Applied Bacteriology*, **65**, 405–410.

Gilliland, R.B. and Lacey, J.P. (1964) Lethal action by an *Acetobacter* on yeasts. *Nature*, **202**, 727–728.

Girond, S., Blazy-Maugen, F. and Michel, G. (1989) Influence de quelques pesticides viticoles sur les levures et la fermentation. *Revue Francaise d'Oenologie*, **119**, 14–22.

Gnaegi, F. and Aerny, J. (1984) Influence des fungicides inhibiteurs de la biosynthese des sterols sur la fermentation alcoolique et la qualité du vin. *Bulletin de l'Office International du Vin*, **646**, 995–1005.

Goto, S. (1980) Changes in the wild yeast flora of sulphited grape musts. *Journal of Institute of Enology and Viticulture Yamanashi*, **15**, 29–32.

Goto, S. and Yokotsuka, I. (1977) Wild yeast populations in fresh grape musts of different harvest times. *Journal of Fermentation Technology*, **55**, 417–422.

Grossman, M.K. and Becker, R. (1984) Investigations on bacterial inhibition of wine fermentation. *Keller-Wirtschaft*, **10**, 272–275.

Guerzoni, E. and Marchetti, R. (1987) Analysis of yeast flora associated with grape sour rot and of chemical disease markers. *Applied and Environmental Microbiology*, **53**, 571–576.

Guijo, S., Millán, C. and Ortega, J.M. (1986) Fermentative features of vinification and maturation yeasts isolated in the Montilla-Moriles region of Southern Spain. *Food Microbiology*, **3**, 133–142.

Guymon, J.F., Ingraham, J.L. and Crowell, E.A. (1961) Influence of aeration upon the formation of higher alcohols by yeasts. *American Journal of Enology and Viticulture*, **12**, 60–66.

Heard, G.M. and Fleet, G.H. (1985) Growth of natural yeast flora during the fermentation of inoculated wines. *Applied and Environmental Microbiology*, **50**, 727–728.

Heard, G.M. and Fleet, G.H. (1986) Occurrence and growth of yeast species during fermentation of some Australian wines. *Food Technology in Australia*, **38**, 22–25.

Heard, G.M. and Fleet, G.H. (1987) Occurrence and growth of killer yeasts during wine fermentation. *Applied and Environmental Microbiology*, **53**, 2171–2174.

Heard, G.M. and Fleet, G.H. (1988a) The effect of sulfur dioxide on yeast growth during natural and inoculated wine fermentation. *The Australian and New Zealand Wine Industry Journal*, **3**, 57–60.
Heard, G.M. and Fleet, G.H. (1988b) The effects of temperature and pH on the growth of yeast species during the fermentation of grape juice. *Journal of Applied Bacteriology*, **65**, 23–28.
Herraiz, T., Martin-Alvarez, P.J., Reglero, G., Herraiz, M. and Cabezudo, M.D. (1989) Differences between wines fermented with and without sulphur dioxide using selected yeasts. *Journal of the Science of Food and Agriculture*, **49**, 249–258.
Herraiz, T., Reglero, G., Herraiz, M., Martin-Alvarez, P.J. and Cabezudo, M. (1990) The influence of yeast and type of culture on the volatile composition of wines fermented without sulfur dioxide. *American Journal of Enology and Viticulture*, **41**, 313–318.
Hough, J.S. and Maddox, I.S. (1970) Yeast autolysis. *Process Biochemistry*, **210**, 50–53.
Ingledew, W.M. and Kunkee, R.E. (1985) Factors influencing sluggish fermentation of grape juice. *American Journal of Enology and Viticulture*, **36**, 65–76.
Joyeux, A. and Lafon-Lafourcade, S. (1984) Metabolisme des bactéries acetique dans le moût de raisin. *Sciences des Aliments*, **4**, 247–255.
Jiranek, V., Langridge, P. and Henschke, P. (1990) Nitrogen requirement of yeast during fermentation. In *Proceedings of the Seventh Australian Wine Industry Technical Conference*, edited by P.J. Williams, D.M. Davidson, and T.H. Lee, pp. 166–171. Adelaide: Australian Wine Research Institute.
Killian, R.E. and Ough, C.S. (1979) Fermentation esters—formation and retention as affected by fermentation temperature. *American Journal of Enology and Viticulture*, **30**, 301–305.
King, S.W. and Beelman, R.B. (1986) Metabolic interactions between *Saccharomyces cerevisiae* and *Leuconostoc oenos* in a model grape juice/wine system. *American Journal of Enology and Viticulture*, **37**, 53–60.
Kraus, J.K., Reed, G. and Villettaz, J.C. (1983) *Levures* seches actives de vinification. 1. Fabrication et characteristiques. *Connaissance de la Vigne et du Vin*, **17**, 93–103.
Kreger-van Rij, N.J.W. (1984) *The Yeasts—a Taxonomic Study*, third edition, Amsterdam: Elsevier.
Kunkee, R.E. (1984) Selection and modification of yeasts and lactic acid bacteria for wine fermentation. *Food Microbiology*, **1**, 315–332.
Kunkee, R.E. and Amerine, M. (1970) Yeasts in winemaking. In *The Yeasts*, Volume 3, *Yeast Technology* edited by A.H. Rose and J.S. Harrison, pp. 5–72. London: Academic Press.
Kunkee, R.E. and Goswell, R.W. (1977) Table wines. In *Alcoholic Beverages*, edited by A.H. Rose, pp. 315–385. London: Academic Press.
Lafon-Lafourcade, S. (1983) Wine and brandy. In *Biotechnology* Volume 5. *Food and Feed Production with Microorganisms* edited by H.J. Rehm and G. Reed, pp. 81–163. Weinheim: Verlag Chemie.
Lafon-Lafourcade, S., Geneix, C. and Ribéreau-Gayon, P. (1984) Inhibition of alcoholic fermentation of grape must by fatty acids produced by yeasts and their elimination by yeast ghosts. *Applied and Environmental Microbiology*, **47**, 1246–1249.
Lafon-Lafourcade, S. and Joyeux, A. (1981) Les bactéries acetiques du vin. *Bulletin de l'Office International du Vin*, **54**, 803–829.
Lagace, L.S. and Bisson, L.F. (1990) Survey of yeast acid proteases for effectiveness of wine haze reduction. *American Journal of Enology and Viticulture*, **41**, 147–151.
Larue, F. and Lafon-Lafourcade, S. (1989) Survival factors in wine fermentation. In *Alcohol Toxicity in Yeasts and Bacteria*, edited by N. Van Uden, pp. 193–215. Boca Raton: CRC Press.
Lee, T.H., Simpson, R.F., Van de Peer, J.M., Fleet, G.H., Davis, C., Daly, N. and Yap, A. (1984) Microbiology of wine corks. In *Advances in Viticulture and Oenology for Economic Gain* edited by T.H. Lee and T.C. Somers, pp. 435–450. Adelaide: Australian Wine Research Institute.

Lemperle, E. (1988) Fungicide residues in musks and wines. In *Proceedings of the Second International Symposium on Cool Climate Viticulture and Oenology*, edited by R.E. Smart, R.J. Thornton, S.B. Rodriguez and J.E. Young, pp. 211–218. Auckland: New Zealand Society for Viticulture and Oenology.

Leroy, M.J., Charpentier, M., Duteurtre, B., Feuillat, M. and Charpentier, C. (1990) Yeast autolysis during champagne ageing. *American Journal of Enology and Viticulture*, **41**, 21–28.

Loiseau, G., Vezinhet, F., Valade, M., Vertes, A., Cuinier, C. and Delteil, D. (1987) Contrôle d'efficacite de levurage por la mise en ovure de souches de levures oenologique marquées. *Revue Francaise d'Oenologie*, **106**, 29–36.

Longo, E., Cansado, J., Agrelo, D. and Villa, T.G. (1991) Effect of climatic conditions on yeast diversity in grape justs from Northwest Spain. *American Journal of Enology and Viticulture*, **42**, 141–144.

Martinez, J., Millan, C. and Ortega, J.M. (1989) Growth of natural flora during the fermentation of inoculated musts from "Pedro Ximenez" grapes. *South African Journal of Enology and Viticulture*, **10**, 31–35.

Martini, A. and Martini, A.V. (1990) Grape must fermentation past and present. In *Yeast Technology*, edited by J.F.T. Spencer and D.M. Spencer, pp. 105–123. Berlin: Springer-Verlag.

Mateo, J.J., Jimenez, M., Huerta, T. and Pastor, A. (1991) Contribution of different yeasts isolated from musts of monastrell grapes to aroma of wines. *International Journal of Food Microbiology*, **14**, 153–160.

McKay, A.M. (1990) Degradation of polygalacturonic acid by *Saccharomyces cerevisiae*. *Letters in Applied Microbiology*, **11**, 41–45.

Meyer, J.P. (1981) Le levurage. *Revue Francaise d'Oenologie*, **86**, 65–69.

Millan, C. and Ortega, J.M. (1988) Production of ethanol, acetaldehyde and acetic acid in wine by various yeast races: role of alchol and aldehyde dehydrogenase. *American Journal of Enology and Viticulture*, **39**, 107–112.

Minarik, E. (1969) Eoclogy of yeasts and yeasts like microorganisms in secondary habitats in Czechoslovakia. *Antonie van Leeuwenhoek. Supplementary Yeast Symposium*, **35**, D7–8.

Minarik, E. (1971) Etude de la flore levurienne des regions viticoles peripheriques en Tchecosloviquie. *Connaissance de la Vigne et du Vin*, **2**, 185–198.

Monk, P.R. (1982) Effect of nitrogen and vitamin supplementation on yeast growth and rate of fermentation of Rhine Riesling grape juice. *Food Technology in Australia*, **34**, 328–332.

Monk, P.R. and Costello, P.J. (1984) Effect of ammonium phosphate and vitamin mixtures on yeast growth in preserved grape juice. *Food Technology in Australia*, **36**, 25–28.

Monk, P. and Cowley, P.J. (1984) Effect of nicotinic acid and sugar concentration of grape juice and temperature on accumulation of acetic acid during yeast fermentation. *Journal of Fermentation Technology*, **62**, 515–521.

Moore, K.J., Johnson, M.G. and Morris, J.R. (1988) Indigenous yeast microflora on Arkansas White Riesling (*Vitis vinifera*) grapes and in model must systems. *Journal of Food Science*, **53**, 1725–1728.

Mora, J., Barbas, J.I. and Mulet, A. (1990) Growth of yeast species during the fermentation of musts inoculated with *Kluyveromyces thermotolerans* and *Saccharomyces cerevisiae*. *American Journal of Enology and Viticulture*, **41**, 156–159.

Mora, J., Barbas, J.I., Ramis, B. and Mulet, A. (1988) Yeast microflora associated with some Majorcan musts and wines. *American Journal of Enology and Viticulture*, **39**, 344–346.

Mora, J. and Mulet, A (1991) Effects of some treatments of grape juice on the population and growth of yeast species during fermentation. *American Journal of Enology and Viticulture*, **42**, 133–136.

Munoz, E. and Ingledew, W.M. (1989a) An additional explanation for the promotion of a more rapid complete fermentation by yeast hulls. *American Journal of Enology and Viticulture*, **40**, 61–64.

Munoz, E. and Ingledew, W.M. (1989b) Effect of yeast hulls on stuck and sluggish fermentations: importance of the lipid component. *Applied and Environmental Microbiology*, **55**, 1560–1564.

Munoz, E. and Ingledew, W.M. (1990) Yeast hulls in wine fermentations—a review. *Journal of Wine Research*, **1**, 197–210.

Nair, N.G. (1985) Fungi associated with bunch rot of grapes in the Hunter Valley. *Australian Journal of Agricultural Research*, **36**, 435–442.

Nishino, H., Miyazaki, S. and Tohjo, K. (1985) Effect of osmotic pressure on the growth rate and fermentation activity of wine yeasts. *American Journal of Enology and Viticulture*, **36**, 170–174.

Ollivier, C., Stonestreet, T., Larue, F. and Dubourdieu, D. (1987) Incidence de la composition colloidale des mouts blancs sur leur fermentescibilité. *Connaissance de la Vigne et du Vin*, **21**, 59–70.

Ough, C.S. (1964) Fermentation rates of grape juice. I. Effects of temperature and composition on white juice fermentation rates. *American Journal of Enology and Viticulture*, **15**, 167–177.

Ough, C.S. (1966a) Fermenrtation rates of grape juice. II. Effects of initial Brix, pH and fermentation temperature. *American Journal of Enology and Viticulture*, **17**, 20–26.

Ough, C.S. (1966b) Fermenrtation rates of grape juice. III. Effects of initial ethyl alcohol, pH and fermentation temperature. *American Journal of Enology and Viticulture*, **17**, 74–81.

Ough, C.S., Davenport, M. and Joseph, K. (1989) Effects of certain vitamins on growth and fermentation rate of several commercial active dry wine yeasts. *American Journal of Enology and Viticulture*, **40**, 208–213.

Ough, C.S. and Groat, M.L. (1978) Particle nature, yeast strain and temperature interactions on the fermentation rates of grape juice. *Applied and Environmental Microbiology*, **35**, 881-885.

Pardo, I., Garcia, M.I., Zuniga, M. and Uruburu, F. (1989) Dynamics of microbial populations during fermentation of wines from the Utiel-Requena region of Spain. *Applied and Environmental Microbiology*, **55**, 539–541.

Parrish, M.E. and Carroll, D.E. (1985) Indigenous yeasts associated with muscadine (*Vitis rotundifolia*) grapes and musts. *American Journal of Enology and Viticulture*, **36**, 165–170.

Parrish, M.E. and Carroll, D.E. (1987) Fermentation characteristics of *Saccharomyces cerevisiae* isolates from *Vitis rotundifolia* grapes and musts. *American Journal of Enology and Viticulture*, **38**, 45–48.

Petering, J.E., Henschke, P.A. and Langridge, P. (1991) The *Escherichia coli* β-glucuronidase gene as a marker for *Saccharomyces* yeast strain identification. *American Journal of Enology and Viticulture*, **42**, 6–12.

Peynaud, E. and Domercq, S. (1959) A review on microbiological problems in wine making in France. *American Journal of Enology and Viticulture*, **10**, 69–77.

Querol, A., Jiménez, M. and Huerta, T. (1990) Microbiological and enological parameters during fermentation of musts from poor and normal grape-harvests in the region of Alicente (Spain). *Journal of Food Science*, **55**, 1603–1606.

Radler, F. and Lotz, B. (1990) The microflora of active dry yeast and the quantitative hcanges during fermentation. *Wein Wissenschaft*, **45**, 114–122.

Rankine, B.C. (1966) *Pichia membranaefaciens*, a yeast causing film-formation and off-flavour in table wine. *American Journal of Enology and Viticulture*, **17**, 82–86.

Rankine, B.C. (1967) Formation of higher alcohols by wine yeasts and relationship to taste thresholds. *Journal of the Science of Food and Agriculture*, **18**, 583–589.

Rankine, B.C. and Lloyd, B. (1963) Quantitative assessment of dominance of added yeast in wine fermentations. *Journal of the Science of Food and Agriculture*, **14**, 793–798.

Reed, G. and Nagodawithana, T.W. (1988) Technology of yeast usage in wine making. *American Journal of Enology and Viticulture*, **39**, 83–90.

Ribéreau-Gayon, P. (1985) New developments in wine microbiology. *American Journal of Enology and Viticulture*, **36**, 1–10.

Ribéreau-Gayon, J., Peynaud, E., Ribéreau-Gayon, P. and Sudraud, P. (1975) *Sciences et Techniques du Vin*, Tome 2, Dunod: Paris.

Ribéreau-Gayon, J., Ribéreau-Gayon, P. and Seguin, G. (1980) *Botrytis cinerea* in enology. In *The Biology of Botrytis*, edited by J.R. Coley-Smith, K. Verhoeff and W.R. Jarvis, pp. 251–274. New York: Academic Press.

Rose, A.H. (1987) Responses to the chemical environment. In *The Yeasts*. Volume 2, second edition, edited by A.H. Rose and J.S. Harrison, pp. 5–40. London: Academic Press.

Rose, A.H. and Pilkington, B.J. (1989) Sulphite. In *Mechanisms of Action of Food Preservation Procedures*, edited by G.W. Gould, pp. 201–216. London: Elsevier.

Rosini, G. (1984) Assessment of dominance of added yeast with fermentation and origin of *Saccharomyces cerevisiae* in wine making. *Journal of General and Applied Microbiology*, **30**, 249–256.

Rosini, G., Federici, F. and Martini, A. (1982) Yeast flora of grapes during ripening. *Microbial Ecology*, **8**, 83–89.

Sablayrolles, J.M. and Barre, P. (1986) Evaluation des besoins en oxygéne de fermentations alcooliques en conditions oenologiene simulées. *Sciences des Aliments*, **6**, 373–383.

Sablayrolles, J.M. and Barre, P. (1989) Automated control of the temperature during alcoholic fermentation in oenological conditions. *Sciences des Aliments*, **9**, 239–251.

Sablayrolles, J.M., Barre, P. and Grenier, P. (1987) Design of a laboratory automatic system for studying alcoholic fermentations in anisothermal enological conditions. *Biotechnology Technologies*, **1**, 181–184.

Salmon, J.M. (1989) Effect of sugar transport inactivation in *Saccharomyces cerevisiae* on sluggish and stuck enological fermentations. *Applied and Environmental Microbiology*, **55**, 953–958.

Sapis-Domercq, S. (1970) Nouvelles études de la microflora en vinification. *Connaissance de la Vigne et du Vin*, **4**, 45–61.

Sapis-Domercq, S., Bertrand, A., Mur, F. and Sarre, C. (1977) Influence des produits de traitment de la vigne sur la microflore levurienne. Experimentation 1976. *Connaissance de la Vigne et du Vin*, **11**, 227–242.

Sharf, R. and Margalith, P. (1983) The effect of temperature on spontaneous wine fermentation. *European Journal of Applied Microbiology and Biotechnology*, **17**, 311–313.

Shimazu, Y. and Watanabe (1981) Effects of yeast strains and environmental conditions on formation of organic acids in must during fermentation. *Journal of Fermentation Technology*, **59**, 27–32.

Shinohara, T. (1984) L'importance des substances volatiles du vin. Formation et effets sur la qualité. *Bulletin de l'Office International du Vin*, **57**, 606–618.

Soufleros, E. and Bertrand, A. (1979) Rôle de la souche de levure dans la production des substances volatiles au cours de la fermentation su jus de raisin. *Connaissance de la Vigne et du Vin*, **13**, 181–198.

Sponholz, W. and Dittrich, H.H. (1974) Die Bildung von SO_2-bindenden Garungs-Nebenproduckten höheren Alkoholen und Estern bei ernigen Reinzuchthefestammen und bei einigen für die Weinbereitung wichtigen wilden Hefen. *Dic Wein Wissenschaft*, **29**, 301–314.

Troton, D., Charpentier, M., Robillard, B., Calvayrac, R. and Duteurtre, B. (1989) Evolution of lipid contents of champagne wine during the second fermentation of *Saccharomyces cerevisiae*. *American Journal of Enology and Viticulture*, **40**, 175–182.

Usseglio-Tommasset, L. (1967) Gli acidi volatili omologhi dell'acetico nei fermentati con specie diverse di lieviti. *Atti Accad. Ital. Vite Vino*, **19**, 165–168.

Valade, M. and Moulin, J.P. (1984) Le levurage en vinification en blanc. *Revue Francaise d'Oenologie*, **93**, 43–56.

Van Uden, N. (1989) Effect of alcohols on the temperature relations of growth and death in yeasts. In *Alcohol toxicity in Yeasts and Bacteria*, edited by V. Van Uden, pp. 77–88. Boca Raton: CRC Press Inc.

Van Zyl, J.A., De Vries, M.J. and Zeeman, A.S. (1963) The microbiology of South African winemaking. III. The effect of different yeasts on the composition of fermented musts. *South African Journal of Agricultural Science*, **6**, 165–180.

Vaughn, R. (1938) Some effects of association and competition on *Acetobacter. Journal of Bacteriology*, **36**, 357–367.

Wibowo, D., Eschenbruch, R., Davis, C.R., Fleet, G.H. and Lee, T.H. (1985) Occurrence and growth of lactic acid bacteria in wine—a review. *American Journal of Enology and Viticulture*, **36**, 302–313.

Young, T.W. (1987) Killer yeasts. In *The Yeasts*, Volume 2, second edition, edited by A.H. Rose and J.S. Harrison, pp. 131–164. London: Academic Press.

Zambonelli, C. (1988) *Microbiologia e Biotecnologia dei Vini*. Bologna: Edagricole.

Zambonelli, C., Romano, P. and Suzzi, G. (1989) Microorganisms of wine. In *Biotechnology Applications in Beverage Production*, edited by C. Cantarelli and G. Lanzarini, pp. 17–30. London: Elsevier Applied Science.

Chapter 3

YEASTS—METABOLISM OF SUGARS

LINDA F. BISSON

Department of Viticulture and Enology, University of California, Davis, CA 95616, USA

Chapter Outline

INTRODUCTION

The metabolism of sugars in the yeast *Saccharomyces cerevisiae* has received much scientific attention, largely due to the genetic and biochemical tractability of this organism. Recent excellent reviews have appeared detailing the analysis of mutants affecting glycolysis (Fraenkel, 1982;1986), and of energy metabolism in yeast (Gancedo and Serrano, 1989), to which the reader is referred. It is perhaps paradoxical that the original impetus to study sugar metabolism in *Saccharomyces* arose from the importance of this organism in several industrial processes such as baking, brewing and wine production, yet very little characterization of the pathways under these fermentative conditions has been undertaken. This review will focus primarily upon sugar metabolism during fermentation of grape juice.

The conversion of grape juice sugars, glucose and fructose, to ethanol is of central importance in wine production. The final concentration of ethanol is, of course, dependent upon the initial concentration of sugars in the must or juice, as well as upon fermentation temperature since more ethanol will be lost from warmer, faster fermentations. Therefore, sugar content of grapes is a significant factor in choosing the time of harvest. The amount of sugar remaining in a wine after fermentation is an important stylistic consideration. Residual sugar is desirable in a dessert or sweet wine and undesirable in most table or dry wines but from a sensory perspective, final sugar, acid and alcohol levels must be in balance in the finished product. The nutrient composition of the grape juice or must will affect both the rate of fermentation and, therefore, the temperature of the fermentation, and the length of time to completion (dryness) of the fermentation. A better understanding of the factors controlling sugar utilization and fermentation rate will allow greater control over fermentation parameters and better diagnosis of problem fermentations.

Before discussing the routes of sugar metabolism and their regulation, a brief description of the growth medium, grape juice, and of the events of a typical must fermentation will be given. Since grape juice does not contain significant quantities of mannose, galactose, maltose or raffinose, all readily metabolized by *Saccharomyces*, these sugars will not be discussed. Emphasis will be placed upon the step of sugar uptake as this is the major site of control of utilization of substrates under anaerobic conditions.

Grape Juice Composition

An appreciation of the nutrient composition of the medium is important in understanding sugar metabolism during grape juice fermentation because nutrient availability impacts metabolic activities. Grape juice composition varies with the grape variety, time of harvesting, soil composition and climate, and as a function of vineyard treatments such as fertilization and irrigation. In general, grape juice destined for table wine contains 140 to 260 g/l of total sugar, with sugar content increasing as a function of length of time to harvest following veraison (Amerine *et al.*, 1980). Sucrose is cleaved to glucose and fructose in the grape berry; thus, the bulk of the juice sugar is comprised of an equimolar concentration of these two sugars with only trace amounts of sucrose. Pentoses also occur in low concentration, but these do not serve as growth or energy substrates for *Saccharomyces*. Nitrogen content varies greatly, depending upon the viticultural region and vineyard practices. Grape juice may contain an excess of nitrogen or it may be nitrogen deficient, resulting in a stuck or sluggish fermentation (Ingledew and Kunkee, 1985). In a survey of California juices, the average juice was found to contain 0.44 g/l of ammonia equivalents (Monteiro and Bisson, 1991), with 0.5 g/l reported by Agenbach (1977) as being necessary to achieve maximal cell production during fermentation. Nitrogen content impacts on the rate of sugar utilization in two ways, namely, by affecting the total number of cells present and by regulating carbon flux through glycolysis. The major nitrogen containing compounds present are the amino acids alanine, arginine, aspartate, asparagine, glutamate, glutamine, proline, serine and threonine, and ammonia (Kliewer, 1970; Kluba *et al.*, 1978). Proline cannot be metabolized as a nitrogen source under anaerobic conditions because molecular oxygen is required by proline oxidase, the first enzymatic step in the catabolic pathway for proline degradation. Grape must also contain phosphate and other yeast micronutrients such as vitamins, minerals and trace salts, generally in sufficient quantities to allow maximal cell yield (Ough *et al.*, 1989). However, some juices may be deficient in one or more of these components depending upon soil composition or microbial attack of the fruit, as in the case of berries severely infected with *Botrytis* or other molds (Amerine *et al.*, 1980). While the typical juice might not be deficient in micronutrient growth factors, it may lack components required for ethanol tolerance and maintenance of sugar utilization rates in stationary phase. This class of components constitutes the so-called survival factors, sterols and unsaturated fatty acids, required for maintenance of plasma membrane integrity (Lafon-Lafourcade *et al.*, 1979; Traverso-Rueda and Kunkee, 1982; reviewed in Larue and Lafon-Lafourcade, 1989). Manipulation of grape juice composition, such as the addition of sulfur dioxide (SO_2), can influence the spectrum of end products produced during sugar catabolism, as described later. The pH of grape juice varies, between 3.0 and 4.0, and is determined by the concentrations of the two principal organic acids, tartaric and malic acids.

Overview of Grape Juice Fermentation

The typical grape juice fermentation is anaerobic. Molecular oxygen available in the processed must or juice is rapidly consumed by the microorganisms present. Although virtually undetectable in the grape microbial flora, *Saccharomyces* rapidly dominates the fermentation, with the finished wine displaying an essentially pure culture of this yeast (Amerine *et al.*, 1980). Inoculation of the fermentation with *Saccharomyces* allows more rapid domination of the fermentation by this organism. It is clear that *Saccharomyces* is well-adapted to growth on high sugar-containing media under anaerobic conditions.

Yeast cells proliferate during grape juice fermentation, their kinetics of growth following a typical microbial growth curve. For cells not pre-adapted to juice, there is a lag phase, followed by exponential growth to a final population of approximately 2×10^8 cells/ml at stationary phase. Juices are either uninoculated (natural fermentation) or inoculated, generally at 10^5–10^6 cells/ml. We have found that nitrogen consumption occurs early in the fermentation, typically before 5% ethanol has accumulated in the medium. Following nitrogen depletion, the cells usually undergo two more rounds of cell division. Approximately 50% of the total available sugar is usually consumed during the growth phase, the remainder being fermented by stationary phase cells. A typical fermentation profile is presented in Figure 3.1 (data from Monteiro, 1991), displaying sugar concentration as change in specific gravity (°Brix), culture absorbance, and ethanol concentration (%v/v). These data are taken from a fermentation of Chenin blanc juice by a commercial yeast, Prise de Mousse (Red Star), maintaining fermentation at 30°C. A significant amount of ethanol is produced by stationary phase cells under these conditions. Thus, significant sugar metabolism occurs in the absence of growth. Under these conditions, the sugar is more correctly thought of as being consumed exclusively as an energy source, not as a carbon source for the production of biomass.

GLYCOLYSIS

The major route of glucose and fructose utilization in *Saccharomyces* is glycolysis. The glycolytic pathway can be considered as the steps leading from glucose (or fructose) to pyruvate. The enzymes of the glycolytic pathway have

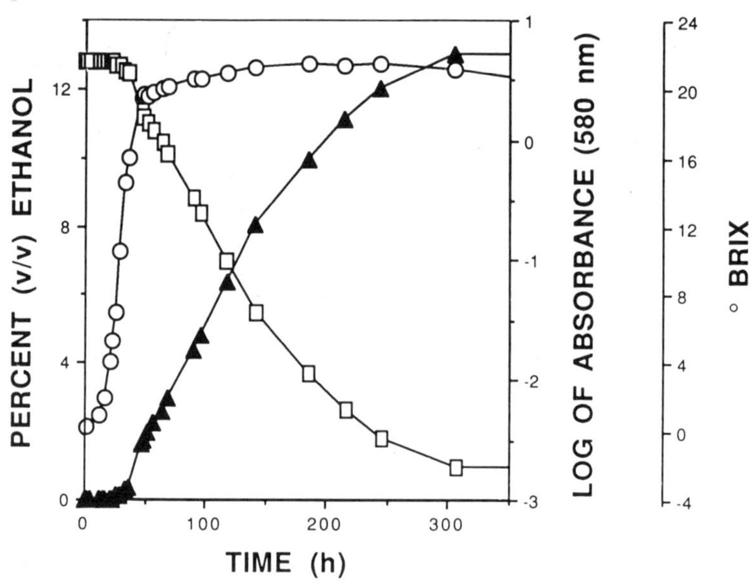

FIGURE 3.1 Fermentation profile of Chenin blanc juice fermented by an industrial strain of *Saccharomyces cerevisiae*, Prise de Mousse. Ethanol, % v/v ▲; yeast growth (absorbance) ○; sugar concentration (BRIX) □

been studied in detail. Mutant strains defective in the structural genes for each of these enzymes are available (Lam and Marmur, 1977; Clifton *et al.*, 1978; Ciriacy and Breitenbach, 1979; Lobo, 1984; Aguilera and Zimmermann, 1986; Rodicio and Heinisch, 1987), and most of these genes have been cloned and sequenced (Kawasaki and Fraenkel, 1982; reviewed in Fraenkel, 1986; Gancedo and Serrano, 1989; Schaaff *et al.*, 1989). Four enzymes appear to be in single copy in the haploid genome: phosphoglucose isomerase, triosephosphate isomerase, phosphoglycerate mutase and pyruvate kinase. Multiple forms exist for three enzymes: glyceraldehyde-3-phosphate dehydrogenase (three isozymes) (Holland *et al.*, 1983), enolase (two isozymes) (Holland *et al.*, 1981; Entian *et al.*, 1987) and hexokinase/glucokinase (three isozymes) (Maitra, 1970; Colowick, 1973; Gancedo *et al.*, 1977). Single gene mutants deficient in almost 99% of aldolase activity have been isolated (Lobo, 1984); however, it has not clearly been demonstrated that this mutation, *fba1*, defines the structural gene for this enzyme. Phosphofructokinase is composed of 8 subunits, $\alpha_4\beta_4$, with an approximate molecular weight of 800 000 (Kopperschalger *et al.*, 1977; Tijane *et al.*, 1982). Genetic data suggested that either subunit alone was sufficient for at least partial phosphofructokinase activity, as *pfk1* (α subunit) and *pfk2* (β subunit) single mutants retain phosphofructokinase activity while the double mutant, *pfk1 pfk2* does not (Clifton and Fraenkel, 1982; Navon *et al.*, 1979; Heinisch, 1986). The *PFK1* and *PFK2* genes are thought to be related by a gene duplication event (Heinisch *et al.*, 1989). Nadkarni *et al.* (1982) isolated an additional mutation, *pfk3*, thought to encode for a subunit, necessary for activity in the *pfk1* or *pfk2* mutants. The activity of this subunit may be induced by glucose. One form of phosphofructokinase may be particulate within the cell (Nadkarni *et al.*, 1982).

Transport of Sugars into the Cell

The first step of glycolysis is transport of the sugar into the cell. Sugars may enter cells in one of three ways: simple diffusion, facilitated (or carrier-mediated) diffusion and active transport. Glucose, fructose and mannose are transported via facilitated diffusion, a non-concentrative process not requiring expenditure of energy, and are thought to share the same transporter molecules (reviewed in Cooper, 1982). Sucrose is cleaved outside of the cells by a secreted invertase and the monosaccharides produced are transported into the cell. The disaccharide maltose is transported into the cell via an active mechanism, one that requires energy and is concentrative. Transport of sugar is thought to be a key site for the control of glycolytic flux under anaerobic conditions.

Sugar transport in *Saccharomyces* is quite complex, now known to involve multiple carriers (Kruckeberg and Bisson, 1990). There appears to be at least two kinds of uptake, one having a low K_m (1–2 mM for glucose; 5–7 mM for fructose and 7–10 mM for mannose) or high affinity for the sugars and the other having low affinity with K_m values approximately 10-fold higher for each sugar (10–20 mM for glucose; 50–70 for fructose; 70–100 for mannose) (Serrano and

DelaFuente, 1974; Bisson and Fraenkel, 1983a). Mutants weak or defective in hexokinase activity and unable to phosphorylate a given sugar display only the high K_m system for that sugar (Bisson and Fraenkel, 1983a; Lang and Cirillo, 1987). *Saccharomyces* possess three enzymes catalyzing the phosphorylation of glucose at the six position: hexokinase PI (A), hexokinase PII (B) and glucokinase. All three enzymes will utilize mannose as substrate, but only the hexokinases can efficiently phosphorylate fructose (Maitra, 1970). Mutants lacking hexokinase PI and PII lose low K_m uptake of fructose, but retain it for glucose and mannose. Triple kinase mutants that lack glucokinase activity in addition to hexokinase PI and PII do not display a low K_m system for either sugar. The differences in the kinetic parameters of sugar uptake and phosphorylation may explain why, in most fermentations, glucose appears to disappear prior to fructose.

Fuhrmann *et al.* (1989) investigated glucose transport in vesicles prepared from wild-type and triple kinase mutant strains, and found that the K_m for glucose uptake in vesicles prepared from derepressed wild-type cells was identical to that of the mutants, having a value of 1 mM. From this observation, they concluded that the effect of kinase activity *in vivo* was an artifactual one. This conclusion was based upon kinetic modeling of the uptake process assuming simple carrier kinetics. However, the key experiment in proposing a role of sugar kinases in transport other than phosphorylation and trapping of the sugar came from studies using 6-deoxyglucose, a non-phosphorylatable glucose analog (Bisson and Fraenkel, 1983b). In wild-type cells, 6-deoxyglucose uptake is biphasic, displaying systems of low and high K_m while in triple kinase mutants only a single system of high K_m is observed. This finding implies some fundamental kinetic difference in the two strains as a consequence of the loss of kinase activity, not explainable on the basis of sugar entrapment as neither strain can "trap" this sugar. Carruthers and Melchior (1983), in investigating the kinetics of glucose uptake in ghosts of red blood cells, also found that in ghosts the glucose transporter behaves as a simple carrier (with K_m for entry equal to Km for exit) as was subsequently observed by Fuhrmann (Fuhrmann *et al.*, 1989) with yeast. However, when the ghosts were made in the presence of cytoplasmic components, the kinetics of uptake changed, and efflux K_m was no longer equal to the influx value. This agrees with the asymmetry of the red blood cell transporter kinetics previously observed *in vivo* (reviewed in Widdas, 1988). The work of Carruthers and Helgerson (1989) suggests a role of ATP in regulating the kinetics of glucose transport. Thus, in eukaryotic cells such as *S. cerevisiae*, the control of sugar uptake appears to be very complex and as yet not reproducible in an *in vitro* system.

Work on the kinetics of glucose uptake in *Saccharomyces* under different growth conditions revealed that the high K_m system (low affinity) was constitutive during growth, decreasing in activity during stationary phase, while the low K_m (high affinity) system was repressible by high glucose concentration in the medium (Bisson and Fraenkel, 1984; Bisson, 1988). Fuhrmann *et al.*, (1989) also found that the K_m of transport in vesicles derived from cells grown under glucose-repressing conditions was high (approximately 10 mM), while that of vesicles from cells grown under derepressing conditions was 1 mM, values in excellent agreement with what was observed *in vivo*.

Glucose transport has been examined in an industrial strain of *S. cerevisiae* (UCD522) during grape juice fermentation (McClellan *et al.*, 1989). In laboratory media, this strain displayed transport kinetics identical to laboratory strains. Under grape juice conditions, repression of the low K_m transport system was apparent until glucose became exhausted in the fermentation. In contrast to the kinetics in laboratory media, a striking substrate inhibition was observable. Mechanistically, substrate inhibition is thought to indicate multiple points of attachment of the substrate, such that at high substrate concentration more than one molecule attempts to bind simultaneously, inhibiting transporter activity. However, this phenomenon was only observed in cells grown anaerobically in grape juice; it was not observed in cells grown anaerobically in a rich medium containing high sugar concentration (McClellan *et al.*, 1989).

A gene encoding a putative glucose transporter has been identified, *SNF3*, from a screen of mutants deficient in sucrose utilization (Bisson *et al.*, 1987; Neigeborn and Carlson, 1984). The SNF3 protein has extensive structural homology to the family of mammalian glucose carriers (Celenza *et al.*, 1988). Genes capable of complementing a null mutation of the *SNF3* locus were identified (Bisson *et al.*, 1987). Two of these have been sequenced, *HXT2* (Kruckeberg and Bisson 1990), and *HXT1* (Lewis and Bisson, 1991). Sequence analysis of these two genes also revealed extensive homology to the mammalian glucose transporters, and both seem to encode a system of high affinity. Thus, there are at least three genes specifying low K_m glucose transporters in *Saccharomyces*, and possibly more (Kruckeberg and Bisson, 1990). The physiological roles of each of these carriers are largely unknown; all are expressed under some growth conditions. A high K_m, low affinity transporter has not yet been identified. However, there is some evidence for an interconversion of the high and low affinity states (Serrano and DelaFuente, 1974; Ramos *et al.*, 1988), depending upon the metabolic condition of the cells. This finding of a family of transporter genes in *Saccharomyces* that may be expressed simultaneously clearly indicates the complexity of the process for transporting sugar into the cell.

Phosphorylation of Sugars

The next step in glycolysis is phosphorylation of the sugar which, in *Saccharomyces*, is catalyzed by one of three kinase enzymes as mentioned already. Hexokinase PII appears to be constitutive, and is the major isozyme expressed during fermentation at high sugar concentrations. The ratio of phosphorylation of glucose to fructose is about unity (1:1) for hexokinase PII (Colowick, 1973). Hexokinase PI has been described as being glucose repressible, not expressed until glucose is depleted in the medium (Kopperschlager and Hoffmann, 1969; Gancedo *et al.*, 1977; Muratsubaki and

Katsume, 1979). However, we recently generated a fusion of the bovine ß casein gene to the *HXK1* promoter (Jimenez-Flores *et al.*, 1990), transformed a haploid strain capable of conducting a grape juice fermentation with this plasmid, and investigated the time of expression of casein during grape juice fermentations. The heterologous protein was expressed at the mid-log stage of growth, when the cells had depleted the medium of amino acids, but when sugar concentration was still quite high, 10–15% w/v (Fong and Bisson, unpublished data). The fructose to glucose phosphorylation ratio for hexokinase is approximately 3:1, and, since glucose is consumed at a faster rate than fructose from mixtures of the two sugars (D'Amore *et al.*, 1989), the concentration of fructose relative to glucose is elevated towards the end of fermentation. Under these conditions, expression of hexokinase PI might be advantageous for maintenance of overall fermentation rate. However, this observation raises questions about the nature of the glucose repression of expression of this enzyme observed in laboratory media. Expression of other glucose repressible functions in the natural environment needs to be characterized.

Mutant analysis implied a role of hexokinase PII in glucose repression (Entian and Mecke, 1982; Entian and Frohlich, 1984). Entian *et al.*, (1987) showed that kinase and repressor activities of hexokinase PII could be separated, based upon mutational analysis of the gene, meaning that mutations affecting repression but not catalytic activity could be isolated. On the other hand, Ma *et al.* (1989a,1989b) could not reproduce this result in a similar analysis of hexokinase PII mutants, reporting that there was a strong correlation between residual kinase activity in their mutants and glucose repression. Strain differences may explain these seemingly contradictory findings. The role of hexokinase PII in glucose repression remains unresolved at this point. Recently, Herrero *et al.* (1989) demonstrated a casein-kinase II like activity of Hexokinase PII, which may be related to its possible role in glucose repression.

Glucose-6-Phosphate to Ethanol

The next step in the pathway for glucose utilization is the conversion of glucose-6-phosphate to fructose-6-phosphate, catalyzed by phosphoglucose isomerase. Mutants deficient in phosphoglucose isomerase have been identified, and the gene cloned and sequenced (Aguilera and Zimmermann, 1986; Green *et al.*, 1988). Mutants lacking phosphoglucose isomerase activity are unable to grow on glucose, suggesting that very little carbon catabolism occurs via the pentose phosphate pathway in *Saccharomyces*, in contrast to what has been observed for other yeasts (Caubet *et al.*, 1988). In a radiorespirometric analysis of glucose metabolism in *Saccharomyces*, Bruinenberg *et al.* (1986) concluded that the activity of the pentose phosphate pathway was sufficient to produce enough NADPH and pentose phosphates as needed for biosynthesis, but that this pathway appeared to play no role in sugar catabolism in this organism.

Phosphofructokinase, the second irreversible step of the pathway, is thought to be a key regulatory enzyme based upon the number of allosteric effectors that have been described. Activity is modulated by AMP, ADP, ATP, P_i fructose-6-phosphate, fructose-1,6-diphosphate and fructose-2,6-diphosphate, and many other metabolites (Sols, 1981). Because of the abundance of effectors of phosphofructokinase activity, this enzyme is thought to be an important site of control of glycolytic flux, especially under certain growth conditions (see later).

Triosephosphate isomerase, glyceraldehyde-3-phosphate dehydrogenase, phosphoglycerate kinase, phosphoglycerate mutase, and enolase catalyze freely reversible steps, common to both glycolysis and gluconeogenesis. The structural genes for these enzymes share at least one common transcriptional control element, encoded by the *GCR1* gene (Clifton and Fraenkel, 1981). Mutations in these genes have been isolated which affect growth on glucose, as expected (reviewed in Fraenkel, 1986). For plasmids carrying an essential glycolytic gene, expression levels may affect plasmid copy number (Piper and Curran, 1990).

The last irreversible step of the glycolytic pathway is pyruvate kinase, also under control by allosteric effectors (Rhodes *et al.*, 1986). This enzyme might also be involved in overall control of the pathway (Den Hollander *et al.*, 1986a; 1986b).

Pyruvate to Ethanol

Fermentation generates reducing power in the form of NADH which must be reoxidized to NAD^+. In yeast under anaerobic conditions, this is accomplished in two steps, namely, decarboxylation of pyruvate by pyruvate decarboxylase to give acetaldehyde, and subsequent reduction of acetaldehyde to ethanol by alcohol dehydrogenase. Isozymes exist for both enzymatic steps. Pyruvate decarboxylase exists in two forms, a major (*PDC1*) and a minor (*PDC5*) form (Hohmann and Cederberg, 1990; Seeboth *et al.*, 1990). Deletion of the *PDC1* gene resulted in an 80% decrease in total pyruvate decarboxylase activity, while deletion of *PDC5* did not appreciably affect activity levels (Hohmann and Cederberg, 1990). Furthermore, *PDC5* seems to be expressed only in strains carrying *pdc1* null mutations (Hohmann and Cederberg, 1990; Seeboth *et al.*, 1990). The function of the *PDC5* enzyme in wild-type cells is not clear. Perhaps the appropriate environmental or physiological conditions necessary for expression of *PDC5* have not yet been discovered.

Alcohol dehydrogenase has four isozymes, ADH1, ADHII, ADHIII and ADHIV (Lutsdorf and Megnet, 1968; Ciriacy, 1975; Paquin and Williamson, 1986; Walton *et al.*, 1986). ADHI is involved in anaerobic fermentation, and is cytoplasmic. ADHII is repressed by high glucose concentration, is also cytoplasmic, and is believed to be involved in consumption of ethanol; however, ADHII can substitute for ADHI in fermentative growth if a mutation allowing its expression in the presence of glucose occurs (reviewed in Blumberg *et al.*, 1987). ADHIII is a mitochondrial enzyme, also thought to function during growth on ethanol. Overexpression of ADHIII does not compensate for loss of ADHI and ADHII for fermentative growth (Blumberg *et al.*, 1987). Recently, a new alcohol dehydrogenase activity, ADHIV, was discovered as conferring resistance to antimycin A when overexpressed, complementing loss of ADHI under fermentative conditions (Paquin and Williamson, 1986; Walton *et al.*, 1986). This enzyme has a somewhat different structure and substrate specificity than the other yeast ADH isozymes (Drewke and Ciriacy, 1988). The physiological role of this isozyme in wild-type cells is unknown. Under aerobic growth conditions, pyruvate may be further metabolized via the tricarboxylic acid cycle. This cycle is not operative under conditions of anaerobic fermentation, due to glucose repression of mitochondrial activity (Polakis and Bartley, 1965; Polakis *et al.*, 1965; Beck and von Meyenburg, 1968; Chapman and Bartley, 1968; Wales *et al.*, 1980).

Regulation of Pathway

Control of carbon flux through the glycolytic pathway is complex. Den Hollander and co-workers (Den Hollander *et al.*, 1986a; 1986b; Reibstein *et al.*, 1986; Campbell-Burk *et al.*, 1987) in a series of elegant studies investigated glycolysis in *S. cerevisiae* under aerobic and anaerobic conditions using ^{13}C and ^{31}P NMR spectroscopy combined with determination of the ratio of different end-products produced using ^{14}C-labelled glucose. These authors were able to simultaneously assess flux of sugar carbon through glycolysis, the pentose phosphate pathway, to storage carbohydrate and macromolecules, and to other end products such as glycerol. These studies revealed that the major point of control of sugar utilization is at the step of transport or phosphorylation, or both. They were unable to differentiate between uptake and subsequent phosphorylation of glucose in their analysis. Under some conditions of growth, control of flux may be mediated by modulation of phosphofructokinase activity, or by a step later in glycolysis, presumably pyruvate kinase. Ingram and co-workers (Dombek and Ingram, 1988; Alterthum *et al.*, 1989) determined that the decline in fermentation rate of glucose, observed in batch cultures of *Saccharomyces*, could be explained by the accumulation of AMP and subsequent inhibition of hexokinase activity. The freely reversible enzymatic steps, those used for gluconeogenesis as well as glycolysis, are not thought to be major points for control of flux of either pathway. Mutations of phosphoglycerate mutase that affect the Keq for the reaction have been described (Gautan, 1988). These mutants were either unaffected in glycolytic rate but impaired for gluconeogenesis, or impaired for glycolysis, but unaffected for gluconeogenesis. Mutants defective in phosphofructokinase or pyruvate kinase activity displayed defects in glucose uptake, while mutants blocked at phosphoglucose isomerase or phosphoglycerate mutase did not affect glucose transport unless cells had been incubated with this sugar prior to assay of uptake (L. Bisson, unpublished observations). Alonso *et al.* (1989) reported a 60% decrease in the total glucose transport activity of a *pgi1* mutant following incubation of the strain in the presence of glucose. Obviously, metabolites accumulate under conditions of arrest of glycolysis that feedback and control the activity of the hexose carriers.

Loss of activity of any of the three irreversible steps of glycolysis, sugar phosphorylation, phosphofructokinase and pyruvate kinase, immediately impacts the activity of the glucose transporters. Indeed, there may be no one true rate limiting step in this pathway, but a coordinated balancing of uptake and flux through both higher and lower glycolysis. There is mounting evidence that the glycolytic enzymes exist as a multienzyme complex in eukaryotic cells (Green *et al.*, 1965; Walsh *et al.*, 1989), and that certain kinds of regulatory interactions may be possible which cannot be duplicated with purified enzymes in an *in vitro* system. One possible explanation for the existence of isozymes is that they have a key role in establishing the proper complex—i.e. glycolytic or gluconeogenic. Overproduction of one of eight enzymes of glycolysis, including hexokinase, pyruvate kinase and phosphofructokinase, did not affect the rate of glucose utilization (Schaaff *et al.*, 1989), again suggesting that no one step is rate limiting. The model that emerges from these investigations is one in which no single enzyme or step mediates flux exclusively, but an alteration in flux is translated into a compensating alteration in glucose uptake. Thus, sugar transport may be viewed as the primary site of rate control, with exclusion of the substrate perhaps being the best mechanism to prevent the occurrence of any imbalance in upper and lower glycolysis, that is, in ATP consumption and production.

Control of glycolytic flux is also modulated by the availability of other nutrients, in particular by nitrogen. Nitrogen limitation causes a loss of transporter activity (Busturia and Lagunas, 1986). Salmon (1989) has shown that the inactivation of glucose uptake resulting from nitrogen limitation causes a decrease in overall glycolytic rate. It is not known how nitrogen physiologically modulates carrier activity. However, ammonium nitrogen (NH_4^+) is an allosteric effector of both phosphofructokinase and pyruvate kinase (Sols, 1981; Rhodes *et al.*, 1986), so that its effect on transporter activity may be indirect.

MINOR END PRODUCTS OF SUGAR METABOLISM

Organic Acids and Glycerol

When grape must or juice is inoculated with aerobically-grown *Saccharomyces*, ethanol is not immediately produced (Pena *et al.*, 1972; reviewed in Whiting, 1976). In respiring cells, pyruvate decarboxylase and alcohol dehydrogenase (ADHI) activities are not highly expressed, both of these enzymes being inducible by glucose (Denis *et al.*, 1983; Rieger *et al.*, 1983; Schmitt *et al.*, 1983; Sharma and Tauro, 1986). As a consequence, compounds other than ethanol are produced at the beginning of fermentation of grape juice. Glycerol, pyruvate and succinate are formed at this time, as are other organic acids (Ribéreau-Gayon *et al.*, 1956a,b). Dihydroxyacetone phosphate can be reduced to glycerol phosphate, converting one molecule of NADH to NAD^+. Glycerol phosphate is then dephosphorylated yielding glycerol. While this pathway allows regeneration of NAD, no ATP is produced. A ratio between pyruvate and glycerol seems to be maintained as these substances are found in equimolar concentration in the medium. Both may be reconsumed later in fermentation.

Factors Affecting Production

The major factor directing the formation of other end products of sugar metabolism is the enzymatic machinery in the yeast cells at the time of their inoculation into grape must. Thus, the end products appearing in fermentation with an aerobically grown starting inoculum will be quite different from those coming from a pre-adapted inoculum such as cells actively growing in grape juice. Pyruvate carboxylase, catalyzing the conversion of pyruvate to oxalacetic acid, is thought to compete with pyruvate decarboxylase for substrate (Van Urk *et al.*, 1988; reviewed in Whiting, 1976). Under conditions of limited pyruvate decarboxylase activity, pyruvate carboxylase would have the kinetic advantage, directing carbon into the TCA cycle (Van Urk *et al.*, 1988). The organic acids produced during adaptation to glucose are those found as intermediates in the TCA cycle (Suomalainen and Keranen, 1967; Van Urk *et al.*, 1988). Production of oxalacetic acid from pyruvate is energy consuming; pyruvate carboxylase requires ATP. It is known that, at this time of adaptation to glucose, protein synthesis is occurring, and presumably transcription as well, both of which are strongly energy consuming. For each molecule of pyruvate and accompanying glycerol produced, no net ATP would be formed. For each pyruvate entering the TCA cycle via pyruvate carboxylase, one ATP would be consumed. Net ATP would have to come from the limited respiration still occurring under these conditions. Indeed, glycerol and succinate formation do seem to be stimulated by aeration (Ribéreau-Gayon *et al.*, 1956a,b). However, grape juice fermentations rapidly become anaerobic, and nitrogen-gas flushed grape must can be fermented starting with aerobically grown yeast. Aerobically grown yeast contain PEP carboxykinase activity, an enzyme involved in gluconeogenesis. In contrast to the other exclusively gluconeogenic enzymatic steps, PEP carboxykinase catalyzes a reversible reaction. It is interesting that, in *Saccharomyces*, this enzyme requires ATP/ADP as cofactor, instead of ITP/IDP or GTP/GDP as in mammalian cells. If this reaction ran in the direction of oxalacetic acid, a molecule of ATP would be generated. In this case, entrance to the TCA cycle acids would be energy generating, not energy consuming. NAD^+ could then be regenerated from NADH by the activity of malate dehydrogenase or succinate dehydrogenase or both. It is known that the succinate appearing early in fermentation arises from glucose in the medium (Thoukis *et al.*, 1965). PEP carboxykinase is inactivated by glucose (Muller *et al.*, 1981), but may function initially in energy generation until inactivated, perhaps when pyruvate decarboxylase activity is elevated sufficiently. The possible role of PEP carboxykinase in the energetically favorable production of oxalacetic acid merits study.

The mitochondrial TCA cycle enzymes are glucose repressed as well as glucose inactivated, and are therefore not expressed during fermentative growth (Polakis and Bartley, 1965; Polakis *et al.*, 1965; Beck and von Meyenburg, 1968; Chapman and Bartley, 1968; Duntze *et al.*, 1969; Wales *et al.*, 1980). However, there exists cytoplasmic forms of several of these activities that are expressed under these conditions. Two enzymatic steps of the TCA cycle, those interconverting α-ketoglutarate and succinate, do not appear to have cytoplasmic forms. Thus, carbon flow from oxalacetate can be viewed as following either an oxidative path to α-ketoglutarate or a reductive one to succinate. The yeast strain, size of the inoculum, pH and composition of the must have strong effects on the levels of organic acids produced (reviewed in Whiting, 1976).

The main volatile acid in fermented beverages is acetic acid. Because of the negative sensory attributes of this acid, its production during grape juice fermentation is highly undesirable. Acetic acid appears to be formed early in the fermentation (Whiting, 1976). The major factor in acetate production seems to be strain background (Shimazu and Watanabe, 1981). In comparing anaerobic growth of two strains of *Saccharomyces*, Verduyn *et al.* (1990) found a striking strain difference in acetate production. The strain not excreting acetate was found to possess higher levels of acetyl-CoA synthetase activity, in agreement with earlier observations (Postma *et al.*, 1989). Other, less well-characterized factors also seem to influence acetate production (Shimazu and Watanabe, 1981). Investigation of the production of acetic acid is complicated by the fact that other organisms, such as acetic acid bacteria and the yeast *Brettanomyces*, are present during typical must fermentation and these are capable of producing this compound in high concentration under appropriate conditions. It is important to note that the presence of other microorganisms and the resulting competition with *Saccharomyces* may impact the metabolic activities of this organism and thus affect the end products produced (Drysdale and Fleet, 1989).

The metabolic inhibitor SO_2 is often added to must as an antimicrobial agent or antioxidant. Its effectiveness as an antimicrobial agent is highly questionable (King *et al.*, 1981) because strains of *Saccharomyces* are resistant to this compound due to detoxification. Within the cell, SO_2 binds to acetaldehyde (Stratford and Rose, 1986; Suzzi *et al.*, 1985), and the acetaldehyde-SO_2 complex is probably excreted from the cells. This prevents regeneration of NAD^+ from acetaldehyde and influences the spectrum of end products early in fermentation, resulting in increased pyruvate and glycerol production (reviewed in Whiting, 1976). Sulfur dioxide also forms a complex with thiamine, depleting the cells of this vitamin. Thiamine is required for decarboxylation reactions, such as that catalyzed by pyruvate decarboxylase.

GLOBAL REGULATION OF CARBON METABOLISM

Mechanisms for the global regulation of carbon metabolism in yeast have been described. Glucose repression (also called carbon catabolite repression) functions at the level of transcription to prevent the synthesis of mitochondrial enzymes, proteins involved in respiratory metabolism, and enzymes needed for catabolism of other non-glucose substrates. Numerous mutants affecting glucose repression have been isolated and reviewed (Gancedo and Gancedo, 1986). A related phenomenon, glucose inactivation, also occurs. Enzymes not needed for glucose metabolism are inactivated, disappearing more rapidly from the cells than can be explained by normal turnover and inhibition of *de novo* synthesis. The mechanism of glucose inactivation is unknown, but seems to involve phosphorylation of the target proteins followed by their proteolysis (Neeff *et al.*, 1978; Muller *et al.*, 1981; Mazon *et al.*, 1982; Tortora *et al.*, 1983).

The availability of molecular oxygen also plays a role in control of sugar metabolism. The Pasteur effect has been described as the inhibition of fermentation in favor of respiration in the presence of oxygen. Although this phenomenon was initially described in *Saccharomyces*, it operates in this yeast only under very special circumstances of nitrogen limitation (reviewed in Lagunas, 1986). Den Hollander *et al.* (1986a) found that the appearance of the Pasteur effect in *Saccharomyces* was strongly dependent upon the respiratory capacity of the cells at the time, thus no effect was observed in glucose repressed cells. Other non-*Saccharomyces* yeasts show a strong Pasteur effect.

OTHER YEASTS ASSOCIATED WITH WINE

Investigation of glycolysis in other yeasts associated with grape juice fermentation has not been extensive. However, several recent reviews have considered carbon metabolism in different genera of yeast (Wohrer, 1980; Fiechter *et al.*, 1981; Kapelli, 1986). One yeast commonly associated with wine spoilage, *Brettanomyces*, will ferment glucose faster in the presence of molecular oxygen than in its absence. The physiological basis of this phenomenon, termed the "Custers Effect" in this yeast, is being investigated (Gaunt *et al.*, 1988). These yeasts produce large amounts of acetic acid during fermentation, and the requirement for oxygen may relate to the end-products formed.

CONCLUSION

The catabolic pathways of sugar metabolism in *Saccharomyces* have been well characterized both biochemically and genetically. Many details of regulation of carbon flux through these pathways are being elucidated. Most of these studies have been conducted on yeast grown under laboratory conditions, which differ significantly from the natural environment. No doubt some interesting insights will be gained concerning the mechanisms of control of sugar utilization once this process is thoroughly investigated during growth and fermentation of natural substrates.

REFERENCES

Agenbach, W. A. (1977) A study of must nitrogen content in relation to incomplete fermentations, yeast production and fermentation. *Proceedings of the South African Society for Enology and Viticulture*, **1977**, 66–87.

Aguilera, A. and Zimmermann, F. K. (1986) Isolation and molecular analysis of the phosphoglucose isomerase structural gene of *Saccharomyces cerevisiae*. *Molecular and General Genetics*, **202**, 83–98.

Alonso, A., Pascual, C., Romay, C., Herrera, L. and Kotyk, A. (1989) Inhibition of hexose transport by glucose in a glucose-6-phosphate isomerase mutant of *Saccharomyces cerevisiae*. *Folia Microbiology*, **34**, 273–278.

Alterthum, F., Dombek, K. M. and Ingram, L. O. (1989) Regulation of glycolytic flux and ethanol production in *Saccharomyces cerevisiae*: Effects of intracellular adenine nucleotide concentrations on the *in vitro* activities of hexokinase, phosphofructokinase, phosphoglycerate kinase and pyruvate kinase. *Applied and Environmental Microbiology*, **55**, 1312–1314.

Amerine, M. A., Berg, H. W., Kunkee, R. E., Ough, C. S., Singleton, V. L. and Webb, A. D. (1980) *The Technology of Wine Making*, 4th edn, pp. 1–794. Westport, Connecticut: AVI Publishing.

Beck, C. and von Meyenburg, H. K. (1968) Enzyme pattern and aerobic growth of *Saccharomyces cerevisiae* under various degrees of glucose limitation. *Journal of Bacteriology*, **96**,479–486.

Bisson, L. F. (1988) High-affinity glucose transport in *Saccharomyces cerevisiae* is under general glucose repression control. *Journal of Bacteriology*, **170**, 4838–4845.

Bisson, L. F. and Fraenkel, D. G. (1983a) Involvement of kinases in glucose and fructose uptake by *Saccharomyces cerevisiae*. *Proceedings of the National Academy of Science, USA*, **80**, 1730–1734.

Bisson, L. F. and Fraenkel, D. G. (1983b) Transport of 6-deoxy-D-glucose in *Saccharomyces cerevisiae*. *Journal of Bacteriology*, **155**, 995–1000.

Bisson, L. F. and Fraenkel, D. G. (1984) Expression of kinase-dependent uptake in *Saccharomyces cerevisiae*. *Journal of Bacteriology*, **159**, 1013–1017.

Bisson, L. F., Neigeborn, L., Carlson, M. and Fraenkel, D. G. (1987) The *SNF3* gene is required for high-affinity glucose transport in *Saccharomyces cerevisiae*. *Journal of Bacteriology*, **169**, 1656–1662.

Blumberg, H., Hartshorne, T., Irani, M. H., Taguchi, A., Taylor, W. E., Yu, J. and Young, E. T. (1987) Regulation of alcohol dehydrogenase gene expression in *Saccharomyces cerevisiae*. In *Biological Research on Industrial yeasts, Volume III* edited by G. G. Stewart, I. Russell, R. D. Klein, and R. R. Heibsch, pp. 1–10, Boca Raton, Florida: CRC Press.

Bruinenberg, P. M., Waslander, G. W., van Dijken, J. P. and Scheffers, W. A. (1986) A comparative radiorespirometric study of glucose metabolism in yeasts. *Yeast*, **2**, 117–121.

Busturia, A. and Lagunas, R. (1986) Catabolite inactivation of the glucose transport system in *Saccharomyces cerevisiae*. *Journal of General Microbiology*, **132**, 379–385.

Campbell-Burk, S. L., den Hollander, J. A., Alger, J. R. and Shulman, R. G. (1987) ^{31}P NMR saturation-transfer and ^{13}C NMR kinetic studies of glycolytic regulation during anaerobic and aerobic glycolysis. *Biochemistry*, **26**, 7493–7500.

Carruthers, A. and Helgerson, A. L. (1989) The human erythrocyte sugar transporter is also a nucleotide binding protein. *Biochemistry*, **28**, 8337–8346.

Carruthers, A. and Melchior, D. L. (1983) Asymmetric or Symmetric? Cytosolic modulation of human erythrocyte hexose transfer. *Biochimica et Biophysica Acta*, **728**, 254–266.
Caubet, R., Guerin, B. and Guerin, M. (1988) Comparative studies on the glycolytic and hexose monophosphate pathways in *Candida parapsilosis* and *Saccharomyces cerevisiae*. *Archives of Microbiology*, **149**, 324–329.
Celenza, J. L., Marshall-Carlson, L. and Carlson, M. (1988) The yeast *SNF3* gene encodes a glucose transporter homologous to the mammalian protein. *Proceedings of the National Academy of Science, USA*, **85**, 2130–2134.
Chapman, C. and Bartley, W. (1968) The kinetics of enzyme changes in yeast under conditions that cause the loss of mitochondria. *Biochemical Journal*, **107**, 455–465.
Ciriacy, M. (1975) Genetics of alcohol dehydrogenase in *Saccharomyces cerevisiae*. Isolation and genetic analysis of *adh* mutants. *Mutation Research*, **29**, 315–326.
Ciriacy, M. and Breitenbach, I. (1979) Physiological effects of seven different blocks in glycolysis in *Saccharomyces cerevisiae*. *Journal of Bacteriology*, **133**, 152–160.
Clifton, D. and Fraenkel, D. G. (1981) The *gcr* (glycolysis regulation) mutation of *Saccharomyces cerevisiae*. *The Journal of Biological Chemistry*, **256**, 13074–13078.
Clifton, D. and Fraenkel, D. G. (1982) Mutants of yeast phosphofructokinase. *Biochemistry*, **21**, 1935–1942.
Clifton, D., Weinstock, S. G. and Fraenkel, D. G. (1978) Glycolysis mutants in *Saccharomyces cerevisiae*. *Genetics*, **88**, 1–11.
Colowick, S. P. (1973) The Hexokinases. In *The Enzymes*, 3rd edition, edited by P. D. Boyer, **9**, 1–48, New York: Academic Press.
Cooper, T.G. (1982) Transport in *Saccharomyces cerevisiae*. In *The Molecular Biology of the Yeast Saccharomyces cerevisiae: Metabolism and Gene Expression*, Edited by J.N. Strathern, E.W. Jones and J.R. Broach, pp. 399–461. Cold Spring Harbor, New York: Cold Spring Harbor Laboratory.
D'Amore, T., Russell, I. and Stewart, G. G. (1989) Sugar utilization by yeast during fermentation. *Journal of Industrial Microbiology*, **4**, 315–324.
Den Hollander, J. A., Ugurbil, K., Brown, T. R., Bednar, M., Redfield, C. and Shulman, R. G. (1986a) Studies of anaerobic and aerobic glycolysis in *Saccharomyces cerevisiae*. *Biochemistry*, **25**, 203–211.
Den Hollander, J. A., Ugurbil, K. and Shulman, R. G. (1986b) ^{31}P and ^{13}C NMR studies of intermediates of aerobic and anaerobic glycolysis in *Saccharomyces cerevisiae*. *Biochemistry*, **25**, 212–219.
Denis. C. L., Ciriacy, M. and Young, E. T. (1983) mRNA levels for the fermentative alcohol dehydrogenase of *Saccharomyces cerevisiae* decrease upon growth on a non-fermentable carbon source. *The Journal of Biological Chemistry*, **258**, 1165–1171.
Dombek, K. M. and Ingram, L. O. (1988) Intracellular accumulation of AMP as a cause for the decline in rate of ethanol production by *Saccharomyces cerevisiae* during batch fermentation. *Applied and Environmental Microbiology*, **54**, 98–104.
Drysdale, G.S. and Fleet, G.H. (1989) The effect of acetic acid bacteria upon the growth and metabolism of yeasts during the fermentation of grape juice. *Journal of Applied Bacteriology*, **67**, 471–481.
Drewke, C. and Ciriacy, M. (1988) Overexpression, purification and properties of alcohol dehydrogenase IV from *Saccharomyces cerevisiae*. *Biochimica et Biophysica Acta*, **950**, 54–60.
Duntze, W., Neumann, D., Gancedo, J. M., Atzpodien, W. and Holzer, H. (1969) Studies on the regulation and localization of the glyoxalate cycle enzymes in *Saccharomyces cerevisiae*. *European Journal of Biochemistry*, **10**, 83–89.
Entian, K.-D. and Frohlich, K.-U. (1984) *Saccharomyces cerevisiae* mutants provide evidence of hexokinase PII as a bifunctional enzyme with catalytic and regulatory domains for triggering carbon catabolite repression. *Journal of Bacteriology*, **158**, 29–35.

Entian, K.-D. and Mecke, D. (1982) Genetic evidence for a role of hexokinase isozyme PII in carbon catabolite repression in *Saccharomyces cerevisiae*. *The Journal of Biological Chemistry*, **257**, 870–874.

Entian, K.-D., Meurer, B., Kohler, H., Mann, K.-H. and Mecke, D. (1987) Studies on the regulation of enolases and compartmentation of cytosolic enzymes in *Saccharomyces cerevisiae*. *Biochimica et Biophysica Acta*, **923**, 214–221.

Fiechter, A., Fuhrmann, G. F. and Kappeli, O. (1981) Regulation of glucose metabolism in growing yeast cells. *Advances in Microbial Physiology*, **22**, 123–183.

Fraenkel, D. G. (1982) Carbohydrate metabolism. In *The Molecular Biology of the Yeast Saccharomyces: Metabolism and Gene Expression*, edited by J. N. Strathern, E. W. Jones and J. R. Broach, pp. 1–37, Cold Spring Harbor, New York: Cold Spring Harbor Laboratory.

Fraenkel, D. G. (1986) Mutants in glucose metabolism. *Annual Reviews of Biochemistry*, **55**, 317–337.

Fuhrmann, G. F., Volker, B., Sander, S. and Potthast, M. (1989) Kinetic analysis and simulation of glucose transport in plasma membrane vesicles of glucose-repressed and derepressed *Saccharomyces cerevisiae* cells. *Experientia*, **45**, 1018–1023.

Gancedo, J. M., Clifton, D. and Fraenkel, D. G. (1977) Yeast hexokinase mutants. *The Journal of Biological Chemistry*, **252**, 4443–4444.

Gancedo, J. M. and Gancedo, C. (1986) Catabolite repression mutants of yeast. *FEMS Microbiology Reviews*, **32**, 179–187.

Gancedo, C. and Serrano, R. (1989) Energy-yielding metabolism. In *The Yeasts*, Volume 3, 2nd Edition, edited by A. H. Rose and J. S. Harrison, pp. 205–259. New York: Academic Press.

Gaunt, D. M., Degan, H. and Lloyd, D. (1988) The influence of oxygen and organic hydrogen acceptors on glycolytic carbon dioxide production in *Brettanomyces anomalus*. *Yeast*, **4**, 249–256.

Gautan, N. (1988) Mutated forms of phosphoglycerate mutase in yeast affect reversal of metabolic flux. *The Journal of Biological Chemistry*, **263**, 15400–15406.

Green, D. E., Murer, E., Hultin, H. O., Richardson, S. H., Salmon, B., Brierly, G. P. and Baum, H. (1965) Association of integrated metabolic pathways with membranes I. Glycolytic enzymes of the red corpuscle and yeast. *Archives of Biochemistry and Biophysics*, **112**, 635–647.

Green, J. B. A., Wright, A. P. H., Cheung, W. Y., Lancashire, W. E. and Hartley, B. S. (1988) The structure and regulation of phosphoglucose isomerase in *Saccharomyces cerevisiae*. *Molecular and General Genetics*, **215**, 100–106.

Heinisch, J. (1986) Construction and physiological characterization of mutants disrupted in the phosphofructokinase genes of *Saccharomyces cerevisiae*. *Current Genetics*, **11**, 227–234.

Heinisch, J., Ritzel, R. G., Von Borstel, R. C., Aguilera, A., Rodicio, R. and Zimmermann, F. K. (1989) The phosphofructokinase genes of yeast evolved from two duplication events. *Gene*, **78**, 309–321.

Herrero, P., Fernandez, R. and Moreno, F. (1989) The hexokinase PII of *Saccharomyces cerevisiae* is a protein kinase. *Journal of General Microbiology*, **135**, 1209–1216.

Hohmann, S. and Cederberg, H. (1990) Autoregulation may control the expression of yeast pyruvate decarboxylase structural genes *PDC1* and *PDC5*. *European Journal of Biochemistry*, **188**, 615–621.

Holland, M. J., Holland, J. P., Thill, G. P. and Jackson, K. A. (1981) The primary structures of two yeast enolase genes. *The Journal of Biological Chemistry*, **256**, 1385–1395.

Holland, J. P., Labieniec, L., Swimmer, C. and Holland, M. J. (1983) Homologous nucleotide sequences at the 5' termini of messenger RNAs synthesized from the yeast enolase and glyceraldehyde-3-phosphate dehydrogenase gene families. *The Journal of Biological Chemistry*, **258**, 5291–5299.

Ingledew, W. M. and Kunkee, R. E. (1985) Factors influencing sluggish fermentations of grape juice. *American Journal of Enology and Viticulture*, **36**, 65–76.
Jimenez-Flores, R., Richardson, T. and Bisson, L. F. (1990) Expression of bovine ß-casein in *Saccharomyces cerevisiae* and characterization of the protein product produced *in vivo*. *Journal of Agricultural and Food Chemistry*, **38**, 1134–1141.
Kappeli, O. (1986) Regulation of carbon metabolism in *Saccharomyces cerevisiae* and related yeasts. *Advances in Microbial Physiology*, **28**, 181–209.
Kawasaki, G. and Fraenkel, D. G. (1982) Cloning of yeast glycolysis genes by complementation. *Biochemical and Biophysical Research Communications*, **108**, 1107–1112.
King, A. D., Ponting, J. D., Sanshuck, D. W., Jackson, R. and Mihara, K. (1981) Factors affecting death of yeast by sulfur dioxide. *Journal of Food Protection*, **44**, 92–97.
Kliewer, W. M. (1970) Free amino acids and other nitrogenous fractions in wine grapes. *Journal of Food Science*, **35**, 17–21.
Kluba, R. M., Mattick, L. R. and Hackler, L. R. (1978) Changes in the free and total amino acid composition of several *Vitis labruscana* grape varieties during maturation. *American Journal of Enology and Viticulture*, **29**, 102–111.
Kopperschlager, G., Bar, J., Nissler, K. and Hoffmann, E. (1977) Physiochemical parameters and subunit composition of yeast phosphofructokinase. *European Journal of Biochemistry*, **81**, 317–325.
Kopperschlager, G. and Hoffmann, E. (1969) Uber multiple formen der hexokinase in hefe. *European Journal of Biochemistry*, **9**, 419–423.
Kruckeberg, A. L. and Bisson, L. F. (1990) The *HXT2* gene of *Saccharomyces cerevisiae* is required for high affinity glucose transport. *Molecular and Cellular Biology*, **10**, 5903-5913.
Lafon-Lafourcade, S., Larue, F. and Ribereau-Gayon, P. (1979) Evidence for the existence of "survival factors" as an explanation for some peculiarities of yeast growth, especially in grape must of high sugar concentration. *Applied and Environmental Microbiology*, **38**, 1069–1073.
Lagunas, R. (1986) Misconceptions about the energy metabolism of *Saccharomyces cerevisiae*. *Yeast*, **2**, 221–228.
Lam, K. B. and Marmur, J. (1977) Isolation and characterization of *Saccharomyces cerevisiae* glycolytic pathway mutants. *Journal of Bacteriology*, **130**, 746–749.
Lang, J. M. and Cirillo, V. P. (1987) Glucose transport in a kinaseless *Saccharomyces cerevisiae*. *Journal of Bacteriology*, **169**, 2932–2937.
Larue, F. and Lafon-Lafourcade, S. (1989) Survival factors in wine fermentation. In *Alcohol toxicity in yeasts and bacteria*, edited by N. van Uden, pp. 193–215, Boca Raton, Florida: CRC Press, Inc.
Lewis, D.A. and Bisson, L.F. (1991) The *HXT1* gene product of *Saccharomyces cerevisiae* is a new member of the family of hexose transporters. *Molecular and Cellular Biology*, **11**, 3804-3813.
Lobo, Z. (1984) *Saccharomyces cerevisiae* aldolase mutants. *Journal of Bacteriology*, **160**, 222–226.
Lutsdorf, V. and Megnet, R. (1968) Multiple forms of alcohol dehydrogenase in *Saccharomyces cerevisiae*. *Archives for Biochemistry and Biophysics*, **126**, 933–944.
Ma, H., Bloom, L. M., Walsh, C. T. and Botstein, D. (1989a) The residual enzymatic phosphorylation activity of hexokinase II mutants is correlated with glucose repression in *Saccharomyces cerevisiae*. *Molecular and Cellular Biology*, **9**, 5643–5649.
Ma, H., Bloom, L. M., Zhu, Z., Walsh, C. T. and Botstein, D. (1989b) Isolation and characterization of mutations in the *HXK2* gene of *Saccharomyces cerevisiae*. *Molecular and Cellular Biology*, **9**, 5630–5642.
Maitra, P. K. (1970) A glucokinase from *Saccharomyces cerevisiae*. *The Journal of Biological Chemistry*, **245**, 2423–2431.

Mazon, M., Gancedo, J. M. and Gancedo, C. (1982) Inactivation of yeast fructose 1,6 bisphosphatase. *The Journal of Biological Chemistry*, **257**, 1128–1130.

McClellan, C. J., Does, A. L. and Bisson, L. F. (1989) Characterization of hexose uptake in wine strains of *Saccharomyces cerevisiae* and *Saccharomyces bayanus*. *American Journal of Enology and Viticulture*, **40**, 9–15.

Monteiro, F.F. (1991) Amino acid utilization and urea formation by *Saccharomyces* during fermentations, PhD Thesis, University of California, Davis.

Monteiro, F. F. and Bisson, L. F. (1991) Biological assay of nitrogen content of grape juice and prediction of sluggish fermentations. *American Journal of Enology and Viticulture*, **42**, 47–57.

Muller, M., Muller, H. and Holzer, H. (1981) Immunochemical studies on catabolite inactivation of phosphoenolpyruvate carboxykinase in *Saccharomyces cerevisiae*. *The Journal of Biological Chemistry*, **256**, 723–727.

Muratsubaki, H. and Katsume, T. (1979) Distribution of hexokinase isozymes depending upon carbon source in *Saccharomyces cerevisiae*. *Biochemical and Biophysical Research Communications*, **86**, 1030–1036.

Nadkarni, M., Lobo, Z. and Maitra, P. K. (1982) Particulate phosphofructokinase of yeast: Physiological studies. *FEBS Letters*, **147**, 251–255.

Navon, G., Shulman, R. G., Yamame, T., Eccleshal, T. R., Lam, K. B., Batonofky, J. J. and Marmur, J. (1979) Phosphorous-31 nuclear magnetic resonance studies of wild-type and glycolytic pathway mutants of *Saccharomyces cerevisiae*. *Biochemistry*, **18**, 4487–4499.

Neeff, J., Hagele, E., Bauhaus, J., Heer, U. and Mecke, D. (1978) Evidence for catabolite degradation in the glucose-dependent inactivation of yeast cytoplasmic malate dehydrogenase. *European Journal of Biochemistry*, **87**, 489–495.

Neigeborn, L. and Carlson, M. (1984) Genes affecting the regulation of *SUC2* gene expression by glucose repression in *Saccharomyces cerevisiae*. *Genetics*, **115**, 247–253.

Ough, C. S., Davenport, M. and Joseph, K. (1989) Effects of certain vitamins on growth and fermentation rate of several commercial active dry wine yeasts. *American Journal of Enology and Viticulture*, **40**, 208–213.

Paquin, C. E. and Williamson, V. M. (1986) Ty insertion at two loci account for most of the spontaneous antimycin A resistance mutations during growth at 15°C of *Saccharomyces cerevisiae* strains lacking *ADHI*. *Molecular and Cellular Biology*, **6**, 70–79.

Pena, A., Cinco, G., Gomez-Puyon, A. and Tuena, M. (1972) Effect of the pH of the incubation medium on glycolysis and respiration in *Saccharomyces cerevisiae*. *Archives of Biochemistry and Biophysics*, **153**, 413–425.

Piper, P. W. and Curran, B. P. G. (1990) When a glycolytic gene on a yeast 2micronORI-*STB* plasmid is made essential for growth its expression level is a major determinant of plasmid copy number. *Current Genetics*, **17**, 119–123.

Polakis, E. S. and Bartley, W. (1965) Changes in the enzyme activities of *Saccharomyces cerevisiae* during aerobic growth on different carbon sources. *Biochemical Journal*, **97**, 284–297.

Polakis, E. S., Bartley, W. and Meek, G. A. (1965) Changes in the respiratory enzymes during the aerobic growth of yeast on different carbon sources. *Biochemical Journal*, **97**, 298–302.

Postma, E., Verduyn, C., Scheffers, W. A. and Van Dijken, J. P. (1989) Enzymatic analysis of the Crabtree effect in glucose-limited chemostat cultures of *Saccharomyces cerevisiae*. *Applied and Environmental Microbiology*, **55**, 468–477.

Ramos, J., Szkutnicka, K. and Cirillo, V. P. (1988) Relationship between the low- and high-affinity glucose transport systems in *Saccharomyces cerevisiae*. *Journal of Bacteriology*, **170**, 5375–5377.

Reibstein, D., den Hollander, J. A., Pilkis, S. J. and Shulman, R. G. (1986) Studies on the regulation of yeast phosphofructo-1-kinase: Its role in aerobic and anaerobic glycolysis. *Biochemistry*, **25**, 219–227.

Rhodes, N., Morris, C. N., Ainsworth, S. and Kinderlerer, J. (1986) The regulatory properties of yeast pyruvate kinase. *Biochemical Journal*, **234**, 705–715.

Ribéreau-Gayon, J., Peynaud, E. and Lafon, M. (1956a) Investigations on the origin of secondary products of alcoholic fermentation, Part I. *American Journal of Enology and Viticulture*, **7**, 53–61.

Ribéreau-Gayon, J., Peynaud, E. and Lafon, M. (1956b) Investigations on the origin of secondary products of alcoholic fermentation, Part II. *American Journal of Enology and Viticulture*, **7**, 112–118.

Rieger, M., Kapelli, O. and Fiechter, A. (1983) The role of limited respiration in the incomplete oxidation of glucose by *Saccharomyces cerevisiae*. *Journal of General Microbiology*, **129**, 653–661.

Rodicio, R. and Heinisch, J. (1987) Isolation of the yeast phosphoglyceromutase gene and construction of deletion mutants. *Molecular and General Genetics*, **206**, 133–140.

Salmon, J. M. (1989) Effect of sugar transport inactivation in *Saccharomyces cerevisiae* on sluggish and stuck enological fermentations. *Applied and Environmental Microbiology*, **55**, 953–958.

Schaaff, I., Heinisch, J. and Zimmermann, F. K. (1989) Overproduction of glycolytic enzymes in yeast. *Yeast*, **5**, 285–290.

Schmitt, H. D., Ciriacy, M. and Zimmermann, F. K. (1983) The synthesis of yeast pyruvate decarboxylase is regulated by large variations in the messenger RNA level. *Molecular and General Genetics*, **192**, 247–252.

Seeboth, P. G., Bohnsack, K. and Hollenberg, C. P. (1990) *pdc1*^{0} Mutants of *Saccharomyces cerevisiae* give evidence for an additional structural *PDC* gene: Cloning of *PDC5*, a gene homologous to *PDC1*. *Journal of Bacteriology*, **172**, 678–685.

Serrano, R. and DelaFuente, G. (1974) Regulatory properties of the constitutive hexose transport in *Saccharomyces cerevisiae*. *Molecular and Cellular Biochemistry*, **5**, 161–171.

Sharma, S. and Tauro, P. (1986) Control of ethanol production by yeast: Role of pyruvate decarboxylase and alcohol dehydrogenase. *Biotechnology Letters*, **8**, 735–738.

Shimazu, Y. and Watanabe, M. (1981) Effects of yeast strains and environmental conditions on formation of organic acids in must fermentation. *Journal of Fermentation Technology*, **59**, 27–32.

Sols, A. (1981) Multimodulation of enzyme activity. *Current Topics in Cellular Regulation*, **19**, 77–101.

Stratford, M. and Rose, A.H. (1986) Transport of sulphur dioxide by *Saccharomyces cerevisiae*. *Journal of General Microbiology*, **132**, 1–6.

Suomalainen, H. and Keranen, A. J. A. (1967) Keto acids formed by baker's yeast. *Journal of the Institute of Brewing*, **73**, 477–484.

Suzzi, G., Romano, P. and Zambonelli, C. (1985) *Saccharomyces* strain selection in minimizing SO_2 requirement during vinification. *American Journal of Enology and Viticulture*, **36**, 199–202.

Thoukis, G., Ueda, M. and Wright, D. (1965) The formation of succinic acid during alcoholic fermentation. *American Journal of Enology and Viticulture*, **16**, 1–8.

Tijane, M. N., Chaffotte, A. F., Yon, J. M. and Laurent, M. (1982) Separation and chemical differentiation of α and β subunits in yeast phosphofructokinase. *FEBS Letters*, **148**, 267–270.

Tortora, P., Burlini, N., Leoni, F. and Guerritore, A. (1983) Dependence on cyclic AMP of glucose induced inactivation of yeast gluconeogenic enzymes. *FEBS Letters*, **155**, 39–42.

Traverso-Rueda, S. and Kunkee, R. E. (1982) The role of sterols on growth and fermentation of wine yeasts under vinification conditions. *Developments in Industrial Microbiology*, **23**, 131–143.

Van Urk, H., Mak, P. R., Scheffers, W. A. and Van Dijken, J. P. (1988) Metabolic responses of *Saccharomyces cerevisiae* CBS 8066 and *Candida utilis* CBS 621 upon transition from glucose limitation to glucose excess. *Yeast*, **4**, 283–292.

Verduyn, C., Postma, E., Scheffers, W. A. and Van Dijken, J. P. (1990) Physiology of *Saccharomyces cerevisiae* in anaerobic glucose-limited chemostat cultures. *Journal of General Microbiology*, **136**, 359–403.

Wales, D. S., Cartledge, T. G. and Lloyd, D. (1980) Effects of glucose repression and anaerobiosis on the activities and subcellular distribution of tricarboxylic acid cycle and associated enzymes in *Saccharomyces carlsbergensis*. *Journal of General Microbiology*, **116**, 93–98.

Walsh, J. L., Keith, T. J. and Knull, H. R. (1989) Glycolytic enzyme interactions with tubulin and microtubules. *Biochimica et Biophysica Acta*, **999**, 64–70.

Walton, J. D., Paquin, C. E., Kaneko, K. and Williamson, V. M. (1986) Resistance to Antimycin A in yeast by amplification of *ADH4* on a linear, 42 kb palindromic plasmid. *Cell*, **46**, 857–863.

Whiting, G. C. (1976) Organic acid metabolism of yeasts during fermentation of alcoholic beverages: A review. *Journal of the Institute of Brewing*, **82**, 84–92.

Widdas, W. F. (1988) Old and new concepts of the membrane transport for glucose in cells. *Biochimica et Biophysica Acta*, **947**, 385–404.

Wohrer, W., Forstenlehner, L. and Rohr, M. (1980) Evaluation of the Crabtree effect in different yeasts grown in batch and continuous culture. In *Current Developments in Yeast Research*, edited by G. G. Stewart and I. Russell, pp. 405–441. New York: Pergamon Press.

Chapter 4

YEASTS—METABOLISM OF NITROGEN COMPOUNDS

PAUL A. HENSCHKE and VLADIMIR JIRANEK

The Australian Wine Research Institute, PO Box 197, Glen Osmond, South Australia 5064, Australia

Chapter Outline

Chapter Outline (Cont.)

Chapter Outline (Cont.)

INTRODUCTION

Nitrogen compounds are essential to the growth and metabolism of yeasts. Of the nutrients utilized by yeasts during the fermentation of grape must, nitrogen is, quantitatively, second only to carbon. The nitrogen fraction is complex and variable, being composed mainly of the amino acids, ammonium ions, peptides and protein. The kinetics of fermentation and the elaboration of aroma and health-related metabolic by-products will be determined, in part, by the qualitative and quantitative composition of the must nitrogen fraction.

The wide variation in nitrogen composition of grape must and its association with fermentation problems have become appreciated only during the last decade or so. Historically, nitrogen metabolism by yeasts was of no special interest to winemakers, despite the work of early researchers, such as Castor (1953), Cantarelli (1957) and Ough (1964) (reviewed by Kunkee and Amerine, 1970), and an early recognition of its significance by the brewing industry (Rainbow, 1969; Hardwick, 1983). Because nitrogen was not a component of must or wine that was routinely analyzed, it was not linked to any particular enological problem. Recognition of the importance of nitrogen in the fermentation of grape must has largely occurred through empirical observation. Problems linked to the nitrogen composition of must now include "sluggish" and "stuck" fermentation (reviewed by Kunkee, 1991), formation of reduced-sulfur compounds, especially hydrogen sulfide (reviewed by Henschke and Jiranek, 1991), and the potential formation of ethyl carbamate from metabolically produced urea (Ough, 1991). In a positive context, there is potential for manipulating the fermentation bouquet of wine by controlling the amount and type of nitrogen source, but this is not yet well understood (Rapp and Versini, 1991).

Wine is the result of a complex interaction between yeast, grape must and physical conditions. In order to ameliorate or eliminate fermentation problems and to enhance wine aroma, it is necessary to not only improve our understanding of yeast nitrogen metabolism and physiology but, also, to increase our knowledge of the influence of viticultural and enological practices on the nitrogen composition of grapes and must. Specifically, it is essential to establish the quantitative importance of nitrogen on fermentation kinetics and the metabolic reactions involved in the generation of aroma and other important compounds. Such need is becoming increasingly relevant in a commercial context because winemakers are looking for new strategies for manipulating the composition and flavor of wine, and for better ways of controlling and predicting the behavior of fermentation.

This chapter will review the importance of nitrogen from source, to transport, metabolic fate and influence on the fermentation carried out by the yeast *Saccharomyces cerevisiae*. Unfortunately, there is very little information about the nitrogen metabolism of non-*Saccharomyces* species of wine yeasts. Basic aspects of yeast metabolism will be briefly mentioned and the reader referred to appropriate texts and reviews. Emphasis will be placed on the influence of amount and type of nitrogen on the kinetics of yeast growth and fermentation, and on the generation of important metabolic by-products.

NITROGEN SOURCES

Grape Must Composition

Grape must contains the full complement of nutrients necessary to support yeast growth during the wine fermentation. The main source of carbon and energy for growth is provided by the grape sugars usually present in large excess (often exceeding 20% w/v) to that needed for maximal growth. By contrast, the total nitrogen content (N) of grape juice ranges 40-fold from 60 to 2400 mg N/l (Amerine *et al.*, 1980; Ough and Amerine, 1988) and, therefore, can be growth limiting. Also, nitrogen is present as a complex range of compounds (Table 4.1). Trace amounts of nitrogen compounds such as vitamins, nucleotides (Amerine *et al.*, 1980) and nitrates (Ough and Crowell, 1980) are present. Various amines may also be present, some resulting from microbial activity (Daudt and Ough, 1980; Radler and Fäth, 1991). Amino acids, ammonium, peptides and proteins are, quantitatively, the most important nitrogen compounds.

Ammonium ion concentration ranges between 19 and 240 mg N/l (mean 96 mg N/l) in Californian grapes (Ough, 1969), between 43 and 115 mg N/l in Stellenbosch varieties (Ough and Kriel, 1985), 17 to 156 mg N/l (mean 55 mg N/l) in French grape juices (Bely, Sablayrolles and Barre, 1991) and 21 and 241 mg N/l (mean 69 mg N/l) in Australian grape juice (Gockowiak and Henschke, unpublished data).

Surveys of amino acid content of grape juices from different viticultural regions have revealed immense variation in both the amino acids present and the concentrations in which they occur (Table 4.2a). Free α-amino nitrogen concentration of ninety musts comprising 25 varieties from four French viticultural regions ranged from 28 to 336 mg N/l with a mean value of 120 mg N/l (Bely, Sablayrolles and Barre, 1991). Despite these differences, proline, arginine, alanine and glutamate and to a lesser extent glutamine, serine, threonine and

TABLE 4.1 Nitrogenous fractions of grape juice.

Nitrogen fraction	*Concentration in grape juice (mg/l)*									
	1[a]		*2*		*3*		*4*		*5*	
Ammonia	10–120	(2)	45–99	(6)	45–89	(4)	7–127	(15)	0–146	(25)
Amino Acid	170–1120[b]	(2)	–		704–1070[c]	(4)	19–144	(10)	–	
Amino	–		101–168	(6)	46–81	(4)	14–176	(15)	15–182	(22)
Amide[d]	10–40	(2)	–		–		–		–	
Humin[e]	5–20	(2)	–		–		–		–	
Polypeptide	–		–		–		10–70	(10)	38–132	(5)
Hexosamine	–		–		–		–		18–29	(2)
Protein	10–100	(2)	–		–		–		28–97	(5)
Residual	100–200	(2)	–		–		–		–	
TOTAL	305–1600	(2)	358–570	(6)	322–490	(4)	98–618	(15)	98–1130	(28)

Numbers in parentheses indicate the number of samples analysed.

[a]References: (1) Hennig, 1945 cited by Koch and Sajak, 1959; (2) Lafon–Lafourcade and Peynaud, 1959; (3) Lafon–Lafourcade and Guimberteau, 1962; (4) Bizeau, 1963; (5) Cordonnier, 1966.

[b]Dipeptides included

[c]Arginine, proline, serine and threonine.

[d]Asparagine and glutamine.

[e]Tryptophan and tyrosine.

TABLE 4.2A Average amino acid composition of grape juices.

Amino acid	*Concentration in grape juice (mg/l)*[a]								
	1[b]	*2*	*3*	*4*	*5*	*6*	*7*	*8*	
Origin	*CA, USA*	*France*	*OH, USA*	*Germany*	*NY, USA*	*France*	*CA, USA*	*Australia*	*Mean*
Alanine	–	–	304	181	596	164	192	124	153
γ–Amino butyric acid	–	–	–	–	81	44	–	87	75
Arginine	403	327	176	820	402	481	339	320	394
Asparagine	–	–	–	–	194[c]	–	170[d]	6	21
Aspartate	52	2	61	62	37	119	33	37	56
Citrulline	–	–	–	–	2	–	7	3	4
Cysteine	–	–	–	–	–	7	–	–	7
Cystine	4	0	8	–	–	–	–	–	6
Glutamate	687	173	144	132	71	352	170[d]	71	151
Glutamine	–	–	–	–	194[c]	–	198	101	118
Glycine	7	22	8	11	8	62	–	3	18
Histidine	92	11	17	62	20	42	35	15	28
Isoleucine	66	7	7	79	33	47	26	17	30
Leucine	62	20	14	108	43	78	33	18	40
Lysine	16	16	14	37	1	64	7	4	20
Methionine	14	1	7	24	12	7	12	4	7
Ornithine	–	–	–	–	4	7	10	4	6
Phenylalanine	51	5	–	103	58	55	37	12	32

TABLE 4.2A (Cont.) Average amino acid composition of grape juices.

Amino acid		*Concentration in grape juice (mg/l)*[a]								
		1[b]	*2*	*3*	*4*	*5*	*6*	*7*	*8*	
Origin		*CA, USA*	*France*	*OH, USA*	*Germany*	*NY, USA*	*France*	*CA, USA*	*Australia*	*Mean*
Proline		–	266	198	484	85	383	1427	–	651
Serine		–	69	302	385[e]	33	103	133[f]	65	104
Threonine		–	258	66	152	76	122	133[f]	70	91
Tryptophan		47	1	–	1	17	–	–	6	7
Tyrosine		20	0	–	25	16	37	19	6	16
Valine		60	6	21	79	49	66	43	25	40
TOTAL	(mg/l)	1581	1184	1347	2745	1838	2240	2721	998	2075
	(mg N/l)	265	208	212	522	340	374	427	199	290
Nº samples analysed		7	8	8	30	3	80	45	203	384

[a]Values quoted were calculated from published data except for those sourced from references 1,2, and 3.
[b]References: (1) Castor, 1953; (2) Récolte, 1957 cited by Lafon–Lafourcade and Peynaud, 1959; (3) Gallander, Cahoon and Beelman, 1969; (4) Dittrich and Sponholz, 1975; (5) Kluba, Mattick and Hackler, 1978; (6) Ooghe and Kastelijn, 1988; (7) Huang and Ough, 1989; (8) Gockowiak and Henschke, unpublished data.
[c]Asparagine and glutamine.
[d]Asparagine and glutamate.
[e]Serine and amides.
[f]Serine and threonine.

γ-amino butyric acid frequently predominate. By accounting for between 60 and 90% of total juice nitrogen (Kliewer, 1969), the amino acids alone represent the main source of nitrogen for yeast growth.

For methodological reasons few reliable reports are available on the peptide and protein content of grape musts (Waters, 1991). Cordonnier (1966) estimated peptides at 20.8% of total nitrogen. Protein content has been reported to vary between 1–13.1% (Koch and Sajak, 1959; Cordonnier, 1966; Tarantola, 1970; Anelli, 1977; Amerine *et al.*, 1980; Correa *et al.*, 1988).

The nitrogen composition and metabolism of *Vitis* species, especially *V. vinifera*, has been the subject of a Symposium (Rantz, 1991).

Factors affecting composition

Only recently has interest turned towards the nitrogen composition of grapes and its relationship to fermentation and wine composition. Because the maturity of grapes for harvest is judged by criteria other than nitrogen content (Sinton *et al.*, 1978; du Plessis, 1984), must nitrogen may be far from optimal, resulting in fermentation faults and wine of lesser quality. Understanding the nitrogen requirements of fermentation will enable nitrogen fertilizer addition in the vineyard to be managed to benefit wine production.

The composition and concentration of nitrogen compounds in grape berries prior to crushing are subject to various viticultural influences. The amino acids, in particular, can vary according to grape variety (Castor, 1953; Récolte, 1957, cited by Lafon-Lafourcade and Peynaud, 1959; Gallander, Cahoon and Beelman, 1969; Dittrich and Sponholz, 1975; Anelli, 1977; Kluba, Mattick and Hackler, 1978; Etiévant *et al.*, 1988; Ooghe and Kastelijn, 1988; Huang and Ough, 1989; Henick-Kling and Larsson, 1991; Huang and Ough, 1991; Table 4.2b), berry maturation (Lafon-Lafourcade and Guimberteau, 1962; Kliewer, 1968; Agenbach, 1977; Kluba, Mattick and Hackler, 1978), nitrogen fertilization (Agenbach, 1977; Bell, Ough and Kliewer, 1979; Ough and Bell, 1980; Peacock, Christensen and Broadbent, 1989; Bell, 1991; Goldspink and Gordon, 1991), rootstock (Ough and Tabacman, 1979; Monk, Hook and Freeman,1987), fungal infection (Dittrich and Sponholz, 1975; Dittrich, Sponholz and Göbel, 1975), soil type (Rankine *et al.*, 1971; Etiévant *et al.*, 1988; Huang and Ough, 1989), method of harvest (Cantagrel, Symonds and Carles, 1982), and 'Carbonic Maceration' (Flanzy, 1982).

Following crushing, the must is immediately inoculated with yeast, or alternatively it may be processed further. Examples of these processes include pressing (Bergner and Haller, 1969; Amerine and Joslyn, 1970), enzymic treatment, clarification by 'cold-settling' (Vos and Gray, 1979; Houtman and du Plessis, 1981), floatation, centrifugation or filtration (Houtman and du Plessis, 1981), fining with various adsorbants (Colagrande, Griselli and Del Re, 1973; Houtman and du Plessis, 1981), ion-exchange (du Plessis, 1963; Dittrich *et al.*, 1975), heating (Mayer, Pause and Vetsch, 1973) and preservation and storage (Grechko and Emel'yanova, 1984; Monk, 1984; Monk and Costello, 1984). The composition of must nitrogen may be altered directly by these treatments or indirectly because the time and temperature regimes employed may permit

TABLE 4.2B Average amino acid composition (mg/l) of Australian grape varieties[a].

Amino acid	*Cabernet Sauvignon*	*Chardonnay*	*Pinot Noir*	*Riesling*	*Sauvignon Blanc*	*Shiraz*	*Traminer*	*Mean*
Alanine	106	181	100	78	257	64	106	124
γ-Amino butyrate	113	87	69	98	105	86	76	87
Arginine	279	314	351	211	591	298	536	320
Asparagine	5	11	2	1	6	3	7	6
Aspartate	23	34	32	52	36	27	34	37
Citrulline	1	3	2	1	32	3	2	3
Glutamate	43	83	55	67	91	71	71	71
Glutamine	45	149	54	48	181	92	169	100
Glycine	1	5	1	1	1	3	2	3
Histidine	7	15	7	13	37	20	23	15
Isoleucine	14	18	17	18	15	21	12	17
Leucine	23	18	16	16	19	27	18	18
Lysine	2	2	3	7	3	5	5	4
Methionine	1	4	2	5	2	5	4	4
Ornithine	1	3	2	7	2	3	3	4
Phenylalanine	10	18	5	8	17	16	10	12
Serine	38	100	42	61	45	39	48	65
Threonine	42	73	68	69	74	61	95	70
Tryptophan	1	6	5	6	5	9	11	6
Tyrosine	2	6	2	3	24	14	5	6
Valine	26	28	20	21	38	30	22	25
Total (mg/l)	782	1157	857	789	1581	897	1259	997
(mg N/l)	159	222	182	146	339	179	279	199
Nº of samples	9	66	36	48	8	21	15	203

[a]Gockowiak and Henschke, unpublished data.

considerable growth of various microorganisms with concomitant utilization of nitrogen. Enzymic treatment, cold-settling, and juice storage are especially important in this regard. A recent trend towards low or nil addition of sulfites to musts could encourage the growth of microorganisms. Little or no quantitative data exists on the effects of many of these treatments on microbial growth or nitrogen depletion. The nitrogen content of grapes at harvest will, therefore, only serve as a guide for that available at inoculation.

In view of the data of Heard and Fleet (Fleet, 1990) it is evident that wild yeasts can play a much greater role in juice processing and fermentation than was previously appreciated. In this respect, the demonstrated proteolytic activity of these yeasts (Lagace and Bisson, 1990) may be important in providing an additional source of assimilable nitrogen. This observation warrants further investigation because of the potential importance of such activities in the later, nitrogen depleted, stages of fermentation.

Nitrogen supplements

Must is frequently supplemented with assimilable nitrogen, largely as a precaution against the development of nitrogen-related fermentation problems. Salts of ammonium, generally diammonium phosphate (DAP), are widely used. Under European law, additions of DAP up to 300 mg/l are permissible, in the U.S.A. up to 950 mg/l may be added (Higgins, 1990), whereas, in Australia the maximum addition is indirectly limited by a maximum wine phosphate concentration of 400 mg of Pi (inorganic phosphate)/l. Generally, DAP (200 mg/l) is added to must prior to inoculation with yeast, with further additions (100 mg/l) being done in response to hydrogen sulfide formation (Henschke and Jiranek, 1991).

Urea and several commercial "yeast foods" are alternative nitrogen supplements (Ingledew, Sosulski and Magnus, 1986). The apparent involvement of urea in ethyl carbamate formation in wine (Ingledew, Magnus and Patterson, 1987; Ough, Crowell and Mooney, 1988; Monteiro, Trousdale and Bisson, 1989) has led to its use being prohibited in most wine-producing countries. The value of yeast foods is also questionable in light of analysis which revealed them to be highly variable in assimilable nitrogen content and some to possess relatively high concentrations of heavy metals (Ingledew *et al.*, 1986) which are restricted by legislation in some countries.

Yeast hull (ghost) preparations, widely recognized to stimulate fermentation (Munoz and Ingledew, 1990), were thought to function by providing trace amounts of assimilable nitrogen and by adsorbing toxic medium-chain fatty acids from the medium (Lafon-Lafourcade, Geneix and Ribéreau-Gayon, 1984; Munoz and Ingledew, 1989). Evidence now indicates that their stimulatory mechanism involves the provision of the 'oxygen-substitutes', sterols and unsaturated fatty acids (Munoz and Ingledew, 1990).

Assay Methods

The diversity of nitrogen compounds present in grape must presents a practical analytical problem. Accurate quantification can be achieved by specific methods but these are not suited to rapid, routine use. Commonly used methods involve the chemical determination of the free α-amino nitrogen (FAN) group using either the ninhydrin (Lie, 1973) or 2,4,6-trinitrobenzene sulfonic acid (TNBS) (Crowell, Ough and Bakalinsky, 1985) procedures. While these methods give a 'chemical' value for nitrogen this does not necessarily equate with the 'biological value' which is a measure of the assimilable nitrogen. Discrepancies arise from non-α-amino nitrogen present in the nitrogen rich amino acids, arginine, lysine, glutamine, asparagine and histidine, which give an underestimate of nitrogen content. The secondary amino acid, proline, although not usually utilized under enological conditions (Pekur, Bur'yan and Pavlenko, 1981; Ingledew *et al.*, 1987), may interfere. The presence of large peptides and protein, which are not assimilated by most species of yeast (Becker, Naider and Katchalski, 1973; Lagace and Bisson, 1990), may also interfere in analysis by these methods. The term 'assimilable nitrogen', which is the sum of total amino acids minus proline, as proposed by Ough *et al.* (1990), will be used in this review. Assimilable nitrogen will be expressed as mg N/l (calculated from amino acid content), or as mg FAN/l when only free α-amino nitrogen values are available.

Several attempts have been made to devise a method to measure assimilable nitrogen. *Saccharomyces cerevisiae* is capable of assimilating most nitrogen compounds present in must with the exception of proline, peptides exceeding about five residues, and proteins. Monteiro and Bisson (1991a) investigated a bioassay approach in which the extent of yeast growth was used to indicate the nitrogen content of a medium. A linear relationship was found between biomass and the addition of known amounts of ammonium ions to a nitrogen deficient test medium. By introducing a sample of grape juice into the test medium in place of ammonium and quantifying the biomass yield, it was possible to determine the assimilable nitrogen content, expressed as 'ammonium ion equivalents'. Despite some limitations, this approach shows promising results and warrants further development.

Another method is designed around an apparatus for automatically monitoring fermentation by weight loss from CO_2 evolution (Bely, Sablayrolles and Barre, 1990a; 1990b). From an investigation of 40 grape musts, assimilable nitrogen concentration (53–411 mg FAN/l) was found to highly correlate with maximum CO_2 production rate (r=0.965) and total fermentation time (r=0.931), the latter by an inverse exponential relationship. Because the maximum rate of CO_2 production occurs during the first third of fermentation, the initial nitrogen concentration can be estimated during the early stages of fermentation. This system is discussed in greater detail in a later section.

METABOLISM

The general aspects of nitrogen metabolism in yeasts have been reviewed by Suamalainen and Oura (1971), Cooper (1982), Jones and Fink (1982), Davis (1986), Large (1986), Nykänen (1986), Pierce (1987), Hinnebusch (1988), Cartwright *et al.* (1989), O'Connor-Cox and Ingledew (1989), Bisson (1991) and Schwencke (1991).

The uptake and metabolism of nitrogen compounds by *S. cerevisiae* depends not only on the strain and its physiological condition, but also on the chemical and physical properties of its environment.

Saccharomyces cerevisiae can grow on a diverse range of nitrogen compounds, including ammonium, urea, amino acids, small peptides, purine- and pyrimidine-based compounds (Cooper, 1982; Large, 1986). Grape must contains varying proportions of all these compounds, which are potentially utilizable by the cell. Before utilization can proceed, the nitrogen compound must be transported into the cell. Depending on the type of nitrogen compound to be accumulated, it may be utilized as follows:

1. Without modification, e.g. an amino acid is incorporated directly into protein;
2. As a source of nitrogen, e.g. an amino acid is degraded to liberate nitrogen for the biosynthesis of other nitrogen cell constituents; the carbon skeleton may be excreted from the cell;
3. As a source of carbon, e.g. the carbon component of an amino acid is released and used for the biosynthesis of other carbon cell constituents (Large, 1986).

The metabolic efficiency of a nitrogen source depends on the expression, regulation and efficiency of transport systems as well as the regulation and energetics of subsequent catabolic and anabolic processes. Consequently, growth, fermentation rate and biomass yield will depend on both the quantity and nature of the nitrogen source(s) available.

Assimilable Nitrogen Sources

The majority of the nitrogen compounds found in grape must are utilizable by yeast as sole sources of nitrogen. Of the amino acids, lysine, cysteine, proline and to a variable extent, histidine and glycine are notable exceptions (Watson, 1976; Pekur *et al.*, 1981; Cooper, 1982; Large 1986; Ingledew *et al.*, 1987). Through direct incorporation, they nonetheless fulfill a role in protein synthesis. Large peptides and proteins do not support growth since *S. cerevisiae* appears to be neither capable of hydrolyzing or accumulating these macromolecules. Their exploitation may, however, be facilitated by other yeast species present in fermenting must which possess extracellular proteolytic activity (Nelson and Young, 1986; Lagace and Bisson, 1990).

The inability of certain amino acids to meet biosynthetic nitrogen requirements is a consequence of conditional or permanent metabolic blocks preventing the partial or complete degradation of these compounds. Proline is utilized only to a limited extent under enological conditions due to nitrogen catabolite inhibition of the proline permease (Lasko and Brandriss, 1981; Horák and Ríhová, 1982) and a limited supply of molecular oxygen essential for the activity of proline oxidase (Duteurtre, Bourgeosis and Chollot, 1971; Mantachian, 1984). The inability of *S. cerevisiae* to grow solely on lysine is thought to result from toxicity of α-amino adipate semialdehyde, a catabolite of lysine which accumulates in the absence of better nitrogen sources (Zaret and Sherman, cited by Cooper, 1982). Cysteine can be degraded to either hydrogen sulfide, pyruvate and ammonium by cysteine desulfhydrase, or to hydrogen sulfide and serine by the reverse reaction of cysteine synthetase (Aida, Tokuyama and Uemura, 1969; Tokuyama *et al.*, 1973). However, at levels needed for appreciable growth cysteine shows toxic effects similar to other sulfhydryl compounds (Maw, 1965). Histidine is apparently not degraded by many yeasts (Cooper, 1982; Large, 1986).

Grape juice contains concentrations of nucleic acid derivatives apparently too low to support significant growth, although these compounds, thymine or thymidine excepted (Wickner, 1975), can be utilized for their nitrogen by yeast (Lee, 1987).

Order of amino acid and ammonium uptake

Selective accumulation of various amino acids and ammonium occurs in *S. cerevisiae* (Table 4.3). Castor (1953) observed differences in the timing, rate and extent of uptake of individual amino acids during fermentation of grape must. The important amino acids, as evident by their uptake coinciding with cell growth were arginine, glutamate, valine, isoleucine, leucine, histidine, aspartate, tryptophan, phenylalanine and methionine, with 70–95% removal of these acids occurring by the cessation of growth. Arginine and histidine formed the greatest residual concentrations, but the initial concentration of arginine in the must was 5–10 times greater than that for the other amino acids. Initiation of the uptake of valine and tyrosine was delayed and small amounts of these amino acids together with isoleucine, leucine and tryptophan were released back into the medium during the final stage of fermentation. Glycine, lysine and cysteine were not significantly utilized. Overall, the accumulation of amino acids and ammonium (Castor and Archer, 1959), albeit at different rates, was restricted to the growth phase, after which no important changes were detected except for a small efflux of some amino acids near the end of fermentation.

The general observations of Castor have been confirmed by other investigations (Pekur *et al.*, 1981; Ingledew and Kunkee, 1985; Monk *et al.*, 1987; Ough *et al.*, 1990; Usseglio-Tomasset and Bosia, 1990; Bisson, 1991), but the overall conclusions are limited by the fact that grape juice is difficult to define or control. A large number of juices must be studied before generalizations can be drawn. We have attempted to overcome such limitations by employing model

TABLE 4.3 Kinetics of absorption of nitrogen compounds during fermentation.

	1[a]		2			3		4		
Test medium	*Grape juice*		*Wort*			*Defined maltose*		*GJM*[b]		
Nitrogen compound	*Hours to 50% depletion*	*Relative position*	*Hours to 50% depletion*	*Relative position*	*Group*	*Hours to 50% depletion*	*Relative position*	*Hours to 50% depletion*	*Relative position*	*Group*
Alanine	–	–	32	9	C	11	7	29 (>50)	11 (13)	B
Arginine	33	8	17	2	A	3	1	15 (21)	2 (7)	A
Asparagine	–	–	–	–	A	–	–	18 (18)	5 (4)	A
Aspartate	24	5	20	4	A	8	4	17 (17)	4 (3)	A
Glutamate	27	7	17	2	A	7	3	21 (22)	7 (8)	B
Glutamine	–	–	–	–	A	–	–	17 (19)	4 (5)	A
Glycine	>130	12	37	11	C	13	9	>42 (>50)	13 (13)	C
Histidine	34	9	27	7	B	–	–	27 (24)	10 (10)	B
Isoleucine	15	1	27	7	B	10	6	19 (20)	6 (6)	A
Leucine	20	2	26	6	B	9	5	19 (19)	6 (5)	A
Lysine	120	11	17	2	A	3	1	14 (15)	1 (1)	A
Methionine	21	3	21	5	B	7	3	17 (18)	6 (4)	A
Ammonium	–	–	30	8	C	16	11	– (16)	– (2)	A
Phenylalanine	25	6	32	9	C	12	8	22 (23)	8 (9)	B
Serine	–	–	19	3	A	6	2	17 (17)	4 (3)	A
Threonine	–	–	16	1	A	6	2	16 (16)	3 (2)	A
Tryptophan	22	4	43	13	C	18	12	24 (25)	9 (11)	C
Tyrosine	35	10	38	12	C	>21	13	32 (>50)	12 (13)	C
Valine	25	6	33	10	B	14	10	24 (27)	9 (12)	B

Values quoted were calculated or derived from published data except those sourced from 4.
[a]References: (1) Castor, 1953; (2) Jones and Pierce, 1964; (3) Egbosimba, Okafor and Slaughter, 1988; (4) Jiranek, 1992.
[b]Chemically-Defined Grape Juice Medium; refer to Table 4.4; Values in parentheses determined in GJM supplemented with ammonium (155 mg N/l).

media which permit precise definition of chemical composition and experimental conditions (Henschke, 1990; Jiranek, Langridge and Henschke, 1990); it is then necessary to verify the findings in grape juices. Model media have been in use for some time. Yeast Carbon Base (Difco, USA), for example, supplemented with amino acids as required has proven convenient but is a poor model for grape juice. The full composition of a chemically-defined grape juice medium based on typical grape juice analysis is given in Table 4.4. The amino acid composition given is in excess of that needed by wine yeasts when fermentation is conducted under anaerobic conditions (Jiranek *et al.*, 1990).

Using the Chemically-Defined Grape Juice medium supplemented with a near excess of amino acids (4.87 g amino acids/l), Jiranek *et al.* (1990) investigated the kinetics of amino acid utilization in the presence and absence of ammonium ions by wine yeast under semi-anaerobic conditions. In the absence of ammonium ions three patterns of utilization were distinguished on the basis of timing and extent of amino acid uptake (Table 4.3; Jiranek, 1992). Initiation of uptake and removal of Group A constituents was rapid, while that of Group B was delayed. Near or complete uptake was observed for amino acids of both groups. Commencement of uptake of Group C compounds coincided with the depletion of those of Group A, and rarely attained completion. Surprisingly, these observations were similar to those made for wort fermentation (Table 4.3) (Jones and Pierce, 1964; Maule *et al.*, 1966, Palmqvist and Äyräpää, 1969). A major difference relates to the timing of ammonium accumulation, which was rapid under enological conditions but fell into Group C under brewing conditions. The delayed uptake of ammonium has been attributed to the influence of maltose on the ammonium permease (Egbosimba, Okafor and Slaughter, 1988).

The sequence of amino acid accumulation is influenced by the presence and relative abundance of various nitrogen compounds, especially the ammonium ion, and possibly, the total nitrogen concentration (Rose, 1987). In model media fermentations (4.87 g amino acids/l; 785 mg N/l) supplemented with ammonium ions (155 mg N/l), the apparent sequence of amino acid utilization was no longer evident (Jiranek *et al.* 1990). The initial accumulation of ammonium ions from the medium commenced most rapidly at the expense of both Group A and B components, the accumulation of which was slightly delayed and largely incomplete. The accumulation of arginine and alanine by *S. cerevisiae* strains 72 and 77 was most delayed by ammonium, while that of lysine was least affected (Jiranek, 1992). In general, these observations are supported by those made in grape juices either supplemented with or containing ammonium ions (Monk *et al.*, 1987; Ough *et al.*, 1990; Bisson, 1991).

No systematic studies on the influence of total nitrogen content on amino acid accumulation appear to have been done. An insight is provided by an extensive investigation by Ough and colleagues (1991) involving 11 Californian grape juices having a wide range of total assimilable nitrogen. For example, the sequence of accumulation of nine key amino acids is contrasted for two juices containing moderate (Juice A, 20.56 mM, 378 mg N/l) and high (Juice K; 30.34

TABLE 4.4 Composition of a chemically-defined grape juice medium.

Component	*Amount per litre*	*Component*		*Amount per litre*
Carbon sources[a]		**Nitrogen sources**[b]	**% Nitrogen**[c]	
Glucose	200 g	Alanine	15.7	100 mg
Salts		Arginine	32.2	750 mg
K Tartrate	5 g	Asparagine	21.2	150 mg
L-Malic acid	3 g	Aspartic acid	10.5	350 mg
Citric acid	0.2 g	Glutamic acid	9.5	500 mg
K_2HPO_4	1.14 g	Glutamine	19.2	200 mg
$MgSO_4.7H_20$	1.23 g	Glycine	18.6	50 mg
$CaCl_2.2H_2O$	0.44 g	Histidine	27.1	150 mg
Trace minerals		Isoleucine	10.7	200 mg
$MnCl_2.4H_2O$	198.2 μg	Leucine	10.7	300 mg
$ZnCl_2$	135.5 μg	Lysine	19.2	250 mg
$FeCl_2$	32.0 μg	Methionine	9.4	150 mg
$CuCl_2$	13.6 μg	Phenylalanine	8.5	150 mg
H_3BO_3	5.7 μg	Proline	12.2	500 mg
$Co(NO_3)_2.6H_2O$	29.1 μg	Serine	13.3	400 mg
$NaMoO_4.2H_2O$	24.2 μg	Threonine	11.8	350 mg
KIO_3	10.8 μg	Tryptophan	13.7	100 mg
Vitamins		Tyrosine	7.7	20 mg
Myo-Inositol	100 mg	Valine	12.0	200 mg
Pyridoxine.HCl	2 mg	Diammonium		
Nicotinic acid	2 mg	phosphate[d]	21.2	
Ca Pantothenate	1 mg	**Lipids**[e]		
Thiamin.HCl	0.5 mg	Ergosterol		10 mg
p-amino benzoic acid	0.2 mg	Tween 80®		0.5 ml
Riboflavin	0.2 mg			
Biotin	0.125 mg			
Folic acid	0.2 mg			

[a]Equimolar concentrations of glucose and fructose can be used; adjust pH to 3.2 or 3.5 as required.
[b]Medium given contains 4.87 g amino acids/l (785 mg N/l).
[c]Nitrogen content of source.
[d]Ammonium chloride may be used; typically 100 mg/l (26.2 mg N/l) for grape juice.
[e]Optional. Disperse 100 mg ergosterol in 5 ml Tween 80®, dissolve in hot ethanol and make up to 20 ml for addition (0.5 ml) to starter culture medium containing reduced sugar concentration (100 g sugars/l) (Jiranek, Langridge and Henschke, 1990).

mM, 738 mg N/l) assimilable nitrogen. For a given strain of yeast in the moderate-N juice, essentially all (about 90%) of the assimilable nitrogen was accumulated by mid-term fermentation (at about 12–16°Brix). By contrast, in the high-N juice, amino acid accumulation ceased earlier in fermentation (at about 20–16°Brix), leaving greater residual amounts. The most striking differences between the moderate- and high-N juices were observed in the kinetics of ammonium and arginine accumulation by the yeasts. In both juices, ammonium accumulation initially paralleled that of the key amino acids. In the moderate-N juice, arginine then accumulated at a lower rate to exhaustion, whereas, in the high-N juice accumulation of this amino acid was reduced such that there was either no further change or an efflux, giving high residual arginine after fermentation. In the latter, ammonium continued to be utilized at a low rate until it became depleted. Depending on the strain of yeast, residual arginine ranged from 225–350 mg N/l in the high-N juice fermentations. As the two juices initially differed widely in the relative concentrations of ammonium and arginine, the differences in amino acid accumulation patterns may reflect the effects of these compounds on amino acid transport (see section on **Transport**). Ammonium is well known to exert an inhibitory effect on the transport of many amino acids (refer to Table 4.5; Grenson and Hou, 1972; Wiame, Grenson and Arst, 1985). These observations have important implications for the use of ammonium based supplements for adjusting total nitrogen content of musts in relation to the evolution of urea and aroma compounds (see later sections).

As for the assimilable amino acids, the concentration of proline initially decreased concomitantly with the bulk of amino acids, but then, depending on the yeast strain, either remained static or increased in concentration, in some cases greater than that initially present in the parent must. The excretion of proline was greater in the moderate-N juice fermentation. Under anaerobic conditions, some proline is incorporated directly into protein, none is metabolized (Cooper, 1982; Ingledew *et al.*, 1987), and some is excreted from yeast as the result of arginine catabolism (Ough *et al.*, 1991; see Figure 4.9).

Amino acid utilization profile

Nitrogen utilization and efflux are dynamic and continuous processes which proceed throughout fermentation. However, once growth has ceased and the fermentation has become 'dry', changes in nitrogen composition of the wine are small. The nitrogen balance at the end of fermentation when related to initial must composition can provide both qualitative and quantitative information on yeast nitrogen demand during fermentation.

The pattern of amino acids utilized during fermentation reflects both the initial distribution of nitrogen compounds as well as strain dependent preferences. When yeasts were presented with a mixture of all amino acids in excess in a model medium, the most important source of nitrogen was arginine, which provided 30–50% of the total nitrogen utilized (Figure 4.1). Lysine, serine, threonine, leucine, aspartate and glutamate were the next most accumulated compounds, while glycine and tyrosine, tryptophan and alanine were the least

TABLE 4.5 Amino acid transport mechanisms of *Saccharomyces* yeasts.

Amino acid	*Transport system*	*Ammonium sensitive*[a]	K_m (μM)	V_{max}[b]	*Reference*
L-Alanine	GAP	Yes			1
γ-Amino butyric acid	Proline permease	Yes	12		2,3
	GABA permease	Yes	100		2,3
	GAP	Yes			2
L-Arginine	Arginine permease	No	10	12	4,5
	GAP	Yes	7.6	23	1,5
L-Asparagine	Glutamine permease	No	350	33[c]	6,7
	GAP	Yes			6,7
L-Citrulline	GAP	Yes	80		1
L-Glutamic acid	Dicarboxylic amino-acid permease 1	No	17		7,8
	Dicarboxylic amino-acid permease 2	Yes			7,8
	GAP	Yes	1000	20[c]	7
L-Glutamine	Glutamine permease 1	No			9
	Glutamine permease 2	No			9
	GAP	Yes	40		1,9
L-Glycine	Glycine permease	No	400	2	10
	GAP	Yes	14 000	2	1,10
L-Histidine	Histidine permease 1	No	17	11	11
	Histidine permease 2	No	4000	28	11
	GAP	Yes	25	16	1
L-Leucine	Leucine permease	No	1000		12,13,14
	GAP	Yes			12,14
L-Lysine	Lysine permease	No	25	8.1	15
	Arginine permease	No	200	17	15
	GAP	Yes	3.1		1

TABLE 4.5 (Cont.) Amino acid transport mechanisms of *Saccharomyces* yeasts.

Amino acid	*Transport system*	*Ammonium sensitive*[a]	K_m (µM)	V_{max}[b]	*Reference*
L-Methionine	Methionine permease 1	No	12	13	16
	Methionine permease 2	No	770	13	16
	GAP	Yes			1
L-Proline	Proline-imino acid permease	Yes	31	40[d]	17,18,19
	GAP	Yes	>2500	160[d]	17,20
L-Serine	specific permease	Partial	250		21,22
	GAP	Yes	500		1,21,22
L-Threonine	specific permease	No			13
	GAP	Yes			23
L-Tryptophan	Arom permease	No	410	1.2[c]	23,24,25
	GAP	Yes	10		1,23
L-Valine	GAP	Yes			1

References: (1) Grenson, Hou and Crabeel, 1970; (2) Grenson *et al.*, 1987; (3) McKelvey, Rai and Cooper, 1990; (4), Grenson *et al.*, 1966; (5), Chan and Cossins, 1976; (6) Gregory, Keenan and Rose, 1982; (7) Darte and Grenson, 1975; (8) Joiris and Grenson, 1969; (9) Grenson and Dubois, 1982; (10) Ballarin-Denti *et al.*, 1984; (11) Crabeel and Grenson, 1970; (12) Bussey and Umbarger, 1970; (13) Gits and Grenson, 1969; (14) Ramos, de Bongioanni and Stoppani, 1980; (15) Grenson *et al.*, 1966; (16) Gits and Grenson, 1967; (17) Horák and Ríhová, 1982; (18) Magaña-Schwencke and Schwencke, 1969; (19) Magaña-Schwencke, Kuznar and Schwencke, 1973; (20) Lasko and Brandriss, 1981; (21) Shukla, Viswanathan and Singh, 1982; (22) Verma, Rao and Prasad, 1984; (23) Wiame *et al.*, 1985; (24) Greasham and Moat, 1973; (25) Kotyk and Dvoráková, 1990.
[a]Activity in ammonium grown cells; [b]nmol/mg protein.min; [c]nmol/g dry weight.min; [d]nmol/g dry weight.s; GAP – General amino acid permease.

utilized. This pattern of nitrogen utilization was qualitatively similar irrespective of yeast strain (three strains), degree of aeration (anaerobic and semi-aerobic), and the sugar (200 and 250 g/l) and ammonium (0 and 155 mg N/l) concentration of the medium (Jiranek, Langridge and Henschke, 1990), as was the pattern of utilization of the remaining amino acids.

Somewhat comparable data were obtained by Dittrich and Sponholz (1975) in a study with 12 grape musts (Figure 4.2). The main differences were in the extent of arginine and lysine utilization. Incomplete arginine utilization was presumably due to the inhibition of its uptake by ammonium ions (Grenson, Hou and Crabeel, 1970). Lysine was only present at low concentration in the grape juices

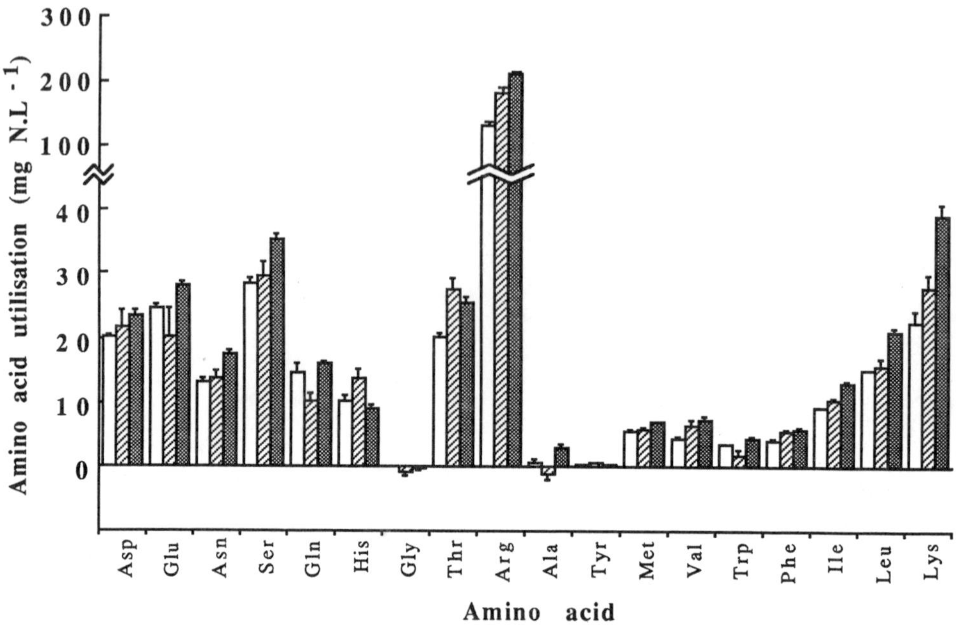

FIGURE 4.1 Profile of amino aid utilization by *Saccharomyces cerevisiae* AWRI strains 65 (□), 72 (▨) and 77 (▩) during the fermentation of a chemically-defined grape juice medium (Table 4.4) at 25°C. Fermentations were initiated with a headspace of nitrogen and conducted as described previously (Jiranek *et al.*, 1990).

investigated and did not meet yeast demand. The amides, glutamine and asparagine, unfortunately were not quantified independently of serine. The most preferred nitrogen sources after arginine were ammonium, serine, threonine leucine and histidine. Comparison with the data of Ooghe and Kastelijn (1988) emphasizes the variability between strains used for fermenting grape musts. In a study with Californian musts, Ough *et al.* (1991) noted the inability of yeasts to fully utilize the high concentrations of arginine present in many high nitrogen musts, while depleting ammonium and most other amino acids. Although further research is needed, these studies demonstrate that amino acids vary in their relative importance as nitrogen sources for yeasts, a fact that should be considered when applying processes which modify the amino acid profile of musts.

Transport

Ammonia and urea

The accumulation of the ammonium ion by *S. cerevisiae* has been investigated through use of the non-metabolizable ammonium analogue, methylamine (Roon, Larimore and Levy, 1975; Dubois and Grenson, 1979). Two specific permeases have been identified which have substrate affinities (K_m) of 0.2 and 2 mM and

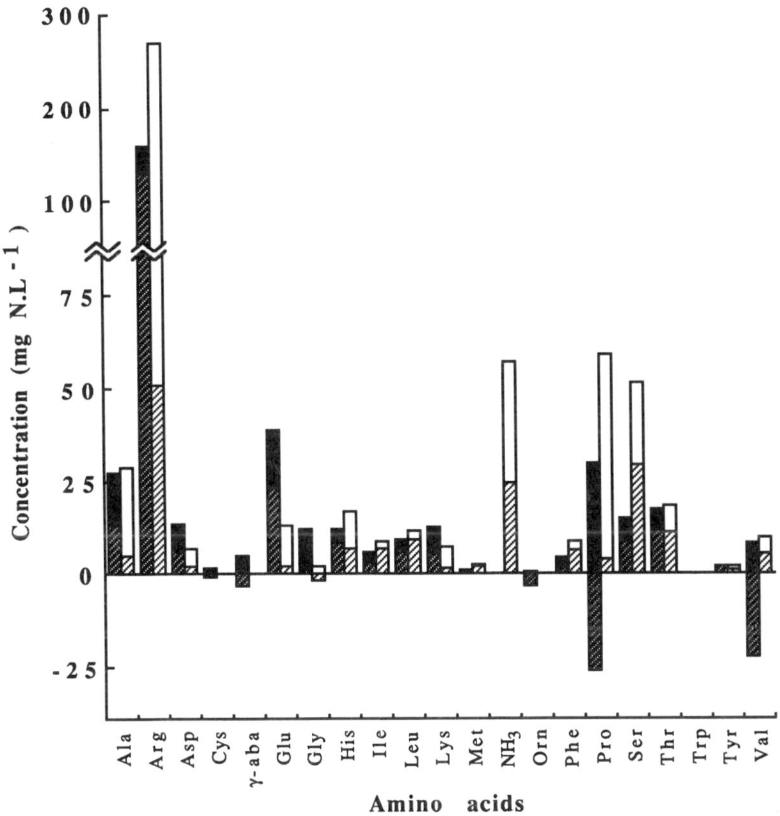

FIGURE 4.2 Profile of amino acid utilization by strains of *Saccharomyces cerevisiae* during fermentation of grape musts. Evolution of amino acids (cross hatching, ▨ and ▨) is compared with initial concentrations in the parent grape juices (solid bars). Values derived from Ooghe and Kastelijn (1988) (black bars, ■) are the mean of 24 samples and Dittrich and Sponholz (1975) (white bars, □) of 30 samples. Serine and amides were not determined separately in the latter analysis.

maximal accumulation rates (V_{max}) of 50 and 20 nmol.mg protein^{-1} min^{-1}, respectively (Dubois and Grenson, 1979). Transport activity is non-competitively inhibited by several amino acids and requires the presence of a fermentable or oxidizable energy source (Roon *et al.*, 1975, Egbosimba and Slaughter, 1987). Urea is transported *via* two mechanisms in this yeast. The first, is an active transport system which has an apparent K_m of 14 μM, is sensitive to nitrogen repression, and depends on both energy metabolism and exposure to oxaluric acid, a gratuitous inducer of allantoin degradative enzymes (Cooper and Sumrada, 1975). The second system is one of passive or facilitated diffusion which operates at external urea concentrations of greater than 0.5 mM, and is probably bidirectional in function.

Amino acids

Comprehensive reviews on the amino acid transport mechanisms of *S. cerevisiae* have been published by Eddy (1982), Wiame *et al.* (1985), Horák (1986), Cartwright *et al.* (1989) and Henschke and Rose (1991). Briefly, at least 11 transport systems specific for a single or a small related group of L-amino acids have been identified. Their substrate affinities and accumulation rates are given in Table 4.5. In addition to these systems both D- and L-isomers of the neutral and basic amino acids, and proline to a limited extent, are transported by the group-specific transport system, the 'general amino acid permease' (GAP) (Grenson *et al.*, 1970). The GAP appears to operate as a 'nitrogen scavenger' system, being expressed only when the readily assimilated nitrogen sources have been utilized (Woodward and Cirillo, 1977).

It has been suggested that the kinetics of utilization and the described grouping of amino acids may be partly explained by the appearance and activity of amino acid transport systems (Rose and Keenan, 1981; Rose, 1987). During the initial stage of fermentation, permeases specific for a small number of amino acids mediate the accumulation of most of the Group A and B acids. Differences in transport kinetics of elements of these groups are possibly due to competition for permease sites by the transported amino acids, and permease characteristics, namely, timing of expression, substrate affinities (K_m), maximal transport rates (V_{max}) and transinhibition (Roon *et al.*, 1975; Woodward and Cirillo, 1977; Rose and Keenan, 1981; Cooper, 1982; Eddy, 1982; Knatchbull and Slaughter, 1987).

The GAP is the principle transport mechanism for amino acids during the later stages of the fermentation. Inhibited and repressed by ammonium (Grenson and Hou, 1972; Wiame *et al.*, 1985), GAP activation appears to coincide with the depletion of ammonium (Rose, 1987). Accumulation of Group C amino acids is believed to be achieved entirely *via* the GAP.

Peptides and proteins

No information is available on the removal of peptides from grape musts, although studies in wort (Calderbank, Rose and Tubb, 1985) suggest that they may be utilized by yeasts. The transport of di- and tripeptides only is thought to be accomplished by a general peptide-transport system (Marder, Becker and Naider, 1977; Moneton, Sarthou and Le Goffie, 1986a; 1986b), although the utilization of peptides up to five residues has been reported (Becker *et al.*, 1973; Naider, Becker and Katchalski, 1974). The importance of peptides to yeast metabolism remains to be clarified. The utilization of large peptides or proteins depends on the ability of yeast either to transport these compounds or degrade them extracellularly. Neither of these mechanisms have been reported in *S. cerevisiae* (Rosi, Costamagna and Bertuccioli, 1987; Sturley and Young, 1988; Lagace and Bisson, 1990). However, some non-*Saccharomyces* species isolated from fermenting must possess significant acidic protease activity (Lagace and Bisson, 1990).

Other nitrogen compounds

As discussed by Large (1986), a broad range of nitrogen compounds can be metabolized by yeasts. Transport of these compounds will be the single most important criterion determining their utilization. D-Amino acids (Rytka, 1975), *S*-adenosylmethionine (Murphy and Spence, 1972), nucleotides and nucleic acid derivatives (Chen, van Gheluwe and Buday, 1973; Lee, 1987) and vitamins including thiamin, riboflavin and biotin (Iwashima, Nichino and Nose, 1973; Perl, Kearney and Singer, 1976; Rogers and Lichstein, 1969) can be accumulated from the medium by *S. cerevisiae*. The uptake of other vitamins is implied by their essential requirement for growth.

Regulation

The activity of transport systems can be regulated by several mechanisms which vary in their specificity, response time and level of influence (i.e., transcription, translation or precursor/active enzyme). The four that will be discussed are feedback inhibition, transinhibition, inactivation/reactivation and repression.

Feedback inhibition is the process by which an accumulated substrate inhibits further accumulation by the same transport carrier. It is evident as a progressive decrease of substrate accumulation rate. The substrate carrier is prevented from returning to the external plasma membrane surface by continuing to be occupied by substrate which is present in high intracellular concentrations. Where internal concentrations of substrate are kept low by catabolism or compartmentation into subcellular organelles, feedback inhibition is considered to be of little consequence.

Transinhibition refers to the competition for uptake exhibited between substrates which do not share a common transport mechanism. Cooper and Sumrada (Cooper, 1982) provide the example of asparagine inhibiting the accumulation of allantoin, a structurally and metabolically unrelated nitrogen compound. Although repression may play a role, the rapidity of the inhibition seems to relegate repression to being of importance only in the long term. Instead, it is suggested that inhibition is the manifestation of competition for a component common to both transport systems (Roon, Meyer and Larimore, 1977; Cooper, 1982). This common component could be the plasma membrane ATPase which regulates the flow of protons from the cell and may possess a finite capacity. The transport of amino acids and ammonium is coupled to proton movements, as is maltose but not glucose. Transinhibition has been invoked to explain the delayed transport of ammonium ions in maltose worts when compared with that in glucose worts (Egbosimba and coworkers, 1987, 1988). Jiranek *et al.* (1990) also reported the preferential transport of ammonium over amino acids in a chemically defined grape juice medium containing glucose as the sole carbon source (Table 4.3). Transinhibition may also explain the preferential uptake of amino acids that cotransport single protons per substrate molecule, often by specific permeases, over those which cotransport two protons, as taken up by the GAP (Eddy, 1978). However, a mechanism involving the finite capacity of the ATPase fails to explain the relative indifference of amino acid transport order to wide differences in amino acid composition (Pierce, 1987).

Inactivation-reactivation processes have been described for several ammonium-sensitive permeases of *S. cerevisiae*, including the general amino-acid permease (GAP) (Grenson and Acheroy, 1982; Grenson, 1983a, b). During growth on a poor nitrogen source such as proline, the permease is synthesized through expression of the structural gene, *GAP1*. Permease activity is subsequently determined according to the net dominance of reactivator and inactivator proteins. Thus, in proline grown cells, the product of the constitutively expressed *npr1* [nitrogen permease reactivator] gene will dominate over the *npi1* and *npi2* [nitrogen permease inactivator; formerly *mut2* and *mut4*] gene products to yield GAP activation. In the presence of ammonium ions, the balance is shifted towards the inactivator proteins resulting in permease inactivation through stoichometric binding to the PGR site of the molecule (Vandenbrol *et al.*, 1987). Efficient nitrogen sources such as ammonium, or derivatives thereof, are thought to favor the dominance of the negative effect by either inactivating the NPR1 protein, repressing its synthesis or escalating the NPI1-NPI2 system.

In addition to the above mechanism, both ammonium-sensitive permeases and degradative enzymes could be regulated by nitrogen catabolite repression (NCR). Repression may be acting at the level of transcription, translation or processing of mRNA or precursor protein. Wiame *et al.* (1985) concluded that the product of the *gdhCR* gene, GDHCR, could be a candidate for 'general repressor', exerting negative control on synthesis of all but one of the enzymes, permeases and regulatory proteins which are subject to NCR. Activation of the GDHCR molecule to the 'activated GDHCR repressor' required glutamine as an effector. Failure of NCR to operate in *gdhA*$^-$ (NADP$^+$-linked glutamate dehydrogenase lacking) mutants even upon glutamate supplementation (Grenson and Hou, 1972) suggested that a complex formed from this protein together with the ammonium ion and α-ketoglutarate might also act as an effector of GDHCR in wild-type cells.

Factors affecting nitrogen accumulation

Many factors affect the of assimilation of nitrogen compounds from a mixture. These include culture conditions, medium composition and yeast strain.

pH

Active transport systems are affected at low pH, a fact attested to by reports that the pH optima of several permeases generally are well above the pH of grape juice (Dubois and Grenson, 1979; Gregory, Keenan and Rose, 1982; Ballarin-Denti *et al.*, 1984; McKelvey, Rai and Cooper, 1990; Kotyk and Dvoráková, 1990). The high pH optimum for ammonium transport (pH 6–6.5; Dubois and Grenson, 1979) is of particular interest because of the low pH of grape musts and the use of diammonium phosphate as a nitrogen supplement. Monteiro and Bisson (1991) noted biomass yield was a function of pH for yeast grown on ammonium, a 50% reduction at pH 3.25 being observed when compared with growth obtained at pH 4.0. Biomass yield on the amino acids alanine, asparagine, arginine, glutamate and glutamine as sole nitrogen sources was, however, little affected over the same range of pH, but was reduced at pH 2.75. The influence of pH on the uptake of amino acids in grape musts needs to be verified.

Ethanol toxicity

Yeast tolerance to ethanol is a complex phenomenon, the fundamentals of which are discussed by Jones (1989) and van Uden (1989), and enological significance reviewed by Larue and Lafon-Lafourcade (1989) and Kunkee (1991).

By increasing membrane permeability to protons (Leão and van Uden, 1984a; Cartwright *et al.*, 1986; Petrov and Okorokov, 1986) and/or inhibiting the proton-pumping membrane ATPase's (van Uden, 1989), ethanol dissipates the proton-motive force across the plasma membrane. Active transport mechanisms are therefore diminished or rendered inoperative (Thomas and Rose, 1979; Leão and van Uden, 1983, 1984b; Ferreras, Iglesias and Girbes, 1989). Inefficient uptake of amino acids is exacerbated by leakage of accumulated nitrogen compounds due to ethanol-modified permeability of cell membranes (Vas, 1953). Ethanol may also affect active transport by a mechanism involving reduction in the number of carriers located in the plasma membrane (Ferreras *et al.*, 1989).

Studies on the rates of accumulation of amino acids and sugar have shown a differential non-competitive inhibition by ethanol (Cartwright *et al.*, 1989; van Uden, 1989). The transport of ammonium and of glycine, which occurs by the general amino acid permease, shows a 50% inhibition (I_{50}) of the maximum rate (V_{max}) at 1.33 and 1.21 M ethanol, respectively, while glucose transport is less sensitive (I_{50}= 1.44 M) to ethanol inhibition (van Uden, 1989). Thus, active transport processes, such as amino acid accumulation (Borst-Pauwels, 1981), are inhibited to a greater degree than glucose transport, which is a facilitated diffusion process (Eddy, 1982). A progressive reduction in the transport activity of the general amino acid permease as ethanol accumulates has been confirmed by Iglesias *et al.*, (1990). They suggested that ethanol deactivated the general amino acid permease through the selective inhibition of a component(s) of the transport protein. These observations help to explain the relatively greater sensitivity of yeast growth to ethanol compared with fermentation activity.

Temperature

Because carrier-mediated active transport is temperature dependent, the rate of accumulation of amino acids is decreased at lower temperatures. Furthermore, different activation energies of maximum transport rates for different amino acids indicated interaction between the transport complex and the physical state of the surrounding membrane (Horák and Kotyk, 1977). The quantitative demand for amino acids will, in part, be affected by transport function. Data on the influence of temperature on the sequence of amino acid accumulation are conflicting (Castor and Archer, 1959; Ough *et al.*, 1991), but this may reflect the different experimental conditions. Ough's group shows a lower rate and extent of uptake, relative to the progress of fermentation at 15°C compared with 20°C. This could largely be accounted for by a lesser accumulation of arginine, alanine and glutamine. Ammonium accumulation, however, seemed to be largely independent of temperature.

Carbon dioxide pressure

Carbon dioxide pressure reduces the rate and extent of amino acid absorption from wort (Kumada *et al.*, 1975) or grape juice (Pekur *et al.*, 1981); in the latter case by about 50%. Specifically, rates of valine, leucine and isoleucine absorption from wort were progressively reduced with increasing CO_2 pressure (Knatchbull and Slaughter, 1987). It is noteworthy that the α-keto derivatives of these amino acids are commonly excreted by yeasts during fermentation indicating that they are in excess (see below). The ability of yeasts to fix CO_2 provides an avenue through which exogenous carbon dioxide can enter carbon metabolism (Slaughter, 1989), possibly impinging on α-keto acid formation. Thus, at higher CO_2 pressures, cellular pools of these α-keto acids and hence their corresponding amino acids may become elevated resulting in the inhibition of the respective transport mechanisms and, inturn, reduced uptake of exogenous amino acid.

Degree of aeration

The indirect affect of oxygen on amino acid transport function by modification of membrane composition is considered below. As previously mentioned, the utilization of proline is affected by the degree of aeration of the medium (Pekur*et al.*, 1981; Ingledew *et al.*, 1987). In addition, under anaerobic conditions, proline and urea are sometimes observed to be released into the medium (Henschke and Ough, 1991; Ough *et al.*, 1991), possibly as a result of arginine catabolism (Brandriss and Magasanik, 1980; Monteiro and Bisson, 1991). Aeration of grape must provides a way of utilizing this often concentrated nitrogen source in grape musts, but the apparent adverse effects of aeration on wine flavor deter adoption of this approach (Houtman and du Plessis, 1986b).

Plasma membrane composition

Growth of yeast under strictly anaerobic conditions induces an absolute requirement for the membrane structural lipids, sterols and unsaturated fatty acids (Andreasen and Stier, 1953, 1954; Radler, 1978; Ratledge and Evans, 1989). Aerobically propagated yeast may accumulate over 2% of its dry mass as sterols. During anaerobic growth, these aerobically-synthesized membrane lipids become greatly diluted through cell division (Haukeli and Lie, 1976; Aries and Kirsop, 1977), around 0.02% sterol per yeast dry mass being growth limiting for wine yeasts (Strydom, Kirschbaum and Tromp, 1982). Growth is limited to several generations, unless small amounts of a sterol, e.g. ergosterol, and an unsaturated fatty acid, e.g. Tween 80®, a source of oleic acid, are supplied exogenously (David and Kirsop, 1973). Evidence suggests that changes in membrane composition under fermentative growth reduces efficiency of transport systems (Henschke and Rose, 1991), nitrogen carriers included, especially in the presence of ethanol, and cell survival is compromised (Lafon, 1978; Lafon-Lafourcade, 1986).

Sterols largely fulfill a non-specific structural role in yeast membranes (Nes, 1973; Rattray, Schibeci and Kidby, 1975). Function of the main sterol, ergosterol, may be comparable to cholesterol in mammalian cell membranes whose presence influences phospholipid/protein interactions and, in turn, the activity of membrane-bound enzymes such as permeases (Papahadjopoulos, Cowden and Kimelberg, 1973). One of the few investigations into the relationship between plasma membrane sterol composition and the uptake of nitrogenous compounds found a positive correlation (Haukeli and Lie, 1976). Proudlock, Haslam and Linnane (1969) proposed that sterols merely stabilized membrane structure, thereby maintaining high permeability integrity.

The role of phospholipid fatty-acyl residues in membrane structure and function have been well reviewed (Rattray *et al.*, 1975; Prasad and Rose, 1986; Henschke and Rose, 1991). As for sterols, the growth requirement for fatty acids by yeasts appears fairly non-specific (Light, Lennarz and Bloch, 1962). However, at the level of transport efficiencies, the degree of unsaturation of the fatty-acyl residues, an oxygen dependent process, is important. By using the rate of L-alanine accumulation as a measure of GAP activity, Calderbank *et al.* (1984) observed that anaerobically grown yeasts more rapidly acquired GAP activity when enriched in the doubly unsaturated linoleyl rather than the singly unsaturated oleyl fatty-acyl residues. Similarly, enrichment with linoleyl residues led to elevated rates of L-arginine (Keenan and Rose, 1979) and peptide (Calderbank *et al.*, 1985) accumulation, and also offered some protection against the toxic effects of ethanol (Thomas and Rose, 1979). The influences on transport were attributed to changes in membrane fluidity in the vicinity of permease proteins.

It is not known if all transport systems are equally affected by changes in membrane lipid composition. Nonetheless, the importance of aerobic starter-culture propagation for ensuring optimal nitrogen accumulation and retention becomes evident.

Strain of yeast

Jiranek, Langridge and Henschke (1991) highlighted differences between strains of *S. cerevisiae* in the pattern of amino acids utilized and total nitrogen demand and, in turn, fermentation efficiency under nitrogen-limited conditions in a chemically-defined grape juice medium. Differences in the kinetics of individual amino acid accumulation were also evident for a few amino acids (Jiranek, 1992). Compared with strain AWRI 72, strain AWRI 77 displayed a greater sensitivity of tyrosine accumulation to ammonium supplementation, while lower rates of glycine and alanine accumulation were observed for unsupplemented fermentations. A low rate and extent of alanine accumulation during grape juice fermentation formed the basis for differentiating strain Epernay 2 from three others, especially at low temperature (15.5°C cf. 20°C) (Ough *et al.*, 1991). Epernay 2 was further distinguished from other strains by releasing large amounts of arginine back into the medium after initially absorbing the amino acid from both a moderate- and high-nitrogen grape juice. Strain Montrachet, a high nitrogen demanding yeast, exhibited the greatest rate of accumulation of most amino acids and ammonium relative to the amount of sugar fermented. In general, quantitative differences were much greater than qualitative ones.

Wild yeast growth

Little specific information is available for non-*Saccharomyces* species associated with grape must fermentation. These yeasts compete with the inoculated strain for nitrogen; conditions which favor their growth will reduce the available nitrogen content, and may, therefore, limit growth of the inoculated strain (see earlier section). As already mentioned, the growth of some non-*Saccharomyces* species during fermentation can result in the elaboration of acidic protease activity (Lagace and Bisson, 1990) which may degrade large peptides and proteins to smaller sized products. The inoculated yeast may, therefore, benefit by direct utilization of the released assimilable nitrogen. Another possibility concerns the fate of wild yeast populations present in the early stages of fermentation (Fleet, 1990). Proteolysis associated with the early autolysis of these yeasts could release amino acids and other nitrogenous compounds back into the wine, in a manner similar to that observed in wine conserved on yeast lees (Stuckey *et al.*, 1991).

Catabolism

The specific order of amino acid uptake suggests that the full complement of amino acids required mainly for protein synthesis will not always come from exogenous sources. The biosynthesis of deficient amino acids will be necessary despite the availability of exogenous supplies. Jones, Pragnell and Pierce (1969) confirmed this by tracing the fate of ^{15}N-labelled amino acids following their absorption from a wort containing all amino acids. The label was randomly distributed throughout all cellular amino acids. Furthermore, amino acids biosynthesis continued despite their accumulation from the medium in amounts exceeding metabolic requirements. Thus, of the three possible fates of an accumulated nitrogen compound, that is, direct incorporation, use as a nitrogen source or use as a carbon source, the second option, degradation of the compound and use of the nitrogen component as a biosynthetic precursor, is the more important outcome.

Labelling of the carbon skeletons of amino acids with ^{14}C revealed little interconversion between amino acids, the majority of carbon skeletons found in yeast proteins being derived either from the exogenous amino acid or from α-keto acids arising from sugar metabolism (Jones *et al.*, 1969). As a consequence of the contribution of carbon skeletons from sugar metabolism, a considerable proportion of the deaminated derivatives of accumulated amino acids is excreted into the medium. In contrast, during nitrogen starvation, the carbon skeletons of only hydrophobic amino acids are excreted, whereas those of basic amino acids are retained (Woodward and Cirillo, 1977). The decarboxylation and an NADH-linked reduction of the deaminated derivatives of amino acids prior to their excretion is an important route for the formation of flavor active higher alcohols (Sentheshanmuganathan, 1960; Nykänen, 1986; Rapp and Versini, 1991; see later section).

Large (1986) summarizes the nitrogen catabolic pathways of yeasts (Figure 4.3). Essentially, all nitrogen compounds accumulated are degraded to one of two end-products, ammonium or glutamate. Through the action of NADP-dependent glutamate dehydrogenase (NADP-GDH), catalyzing the formation of glutamate from ammonium and α-ketoglutarate, and NAD-dependent glutamate dehydrogenase (NAD-GDH) catalyzing the reverse reaction, these end-products of nitrogen catabolism are interconverted (Holzer and Schneider, 1957; Heirholzer and Holzer, 1963). An alternative pathway has been reported in *S. cerevisiae* (Roon, Even and Larimore, 1974), *Saccharomycodes ludwigii* and

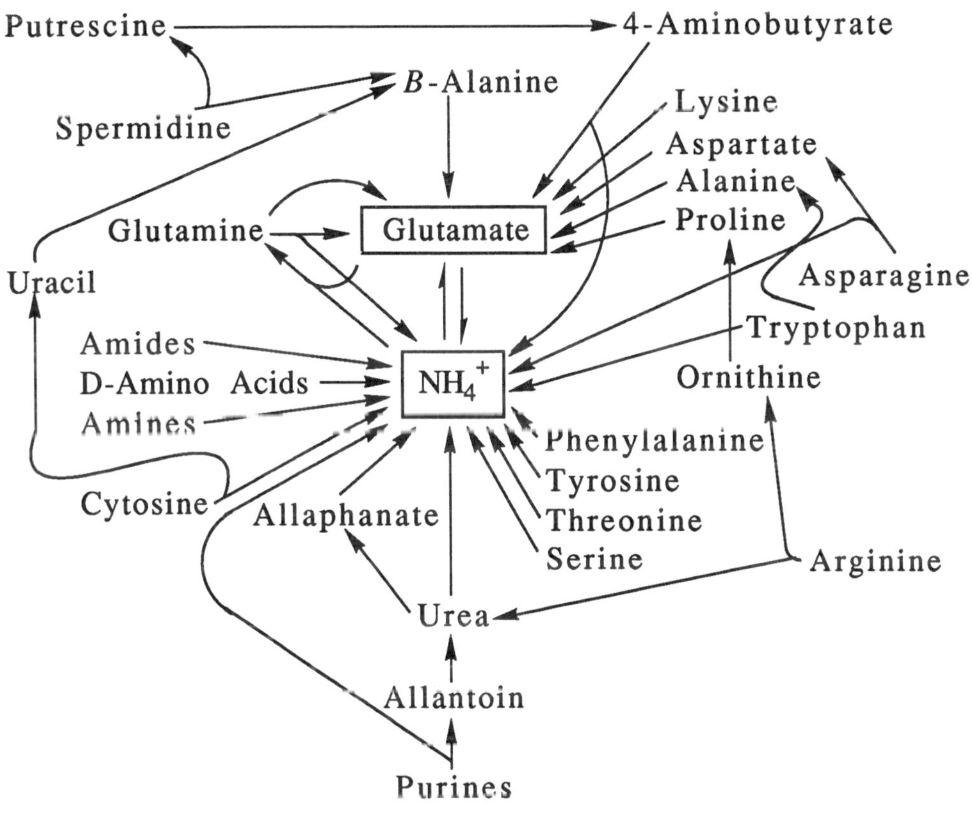

FIGURE 4.3 Degradation of nitrogenous compounds by yeasts. [Redrawn from Large (1986), 'Degradation of organic nitrogen compounds by yeasts.' *Yeast*, **2**, 1–34. Copyright John Wiley & Sons, Ltd, and reprinted with permission.]

Schizosaccharomyces spp., (Johnson and Brown, 1974). Based on the low activities of the enzymes glutamate synthase and glutamine synthetase, this system is considered to fulfill an auxiliary role. Through their interconversion, ammonium and glutamate, together with glutamine, act as the ultimate precursors of all nitrogen biosynthetic reactions.

Relative efficiencies of nitrogen sources

Saccharomyces cerevisiae has no absolute requirement for amino acids, and can grow on ammonium, urea and most amino acids as sole sources of nitrogen. A good nitrogen source has been described as one which is: rapidly and economically transported; readily degraded to ammonium and/or glutamate; and without toxic effects on the cell (Cooper, 1982). The wide use of ammonium in industrial fermentations is a reflection of its cost effectiveness and its metabolic efficiency. Determinations of growth rate (μ) in a chemically defined medium have identified ammonium, glutamine, arginine, glutamate and asparagine ($\mu = 0.37 - 0.47\ h^{-1}$) as good sources of nitrogen, whereas, glycine, threonine, isoleucine, valine and tryptophan ($\mu = 0.08 - 0.12\ h^{-1}$) are poor sources (Watson, 1976). Cooper (1982) reports similar findings for five laboratory strains of *S. cerevisiae* as summarized in Table 4.6.

Ammonium and glutamate are good nitrogen sources because they directly enter the intracellular pool of biosynthetic precursors. The ability of arginine to support high growth rates presumably stems from the fact that it is rich in nitrogen and requires the symport of only one proton, making it economical to transport. The enological importance of arginine is evident by observations which show this amino acid, often the most predominant assimilable nitrogen source in Australian juices, to alone satisfy 30–50% of yeast total nitrogen requirements (Figure 4.1). Nevertheless, the degradation of arginine, which proceeds via proline formation, must be energetically taxing, a fact likely to explain its delayed and reduced utilization from ammonium supplemented fermentations. Some strains excrete urea while metabolizing high concentrations of arginine; the urea is reabsorbed and degraded further when the concentration of assimilable nitrogen becomes low (Henschke and Ough, 1991; Monteiro and Bisson, 1991b) due to nitrogen derepression (Cooper and Sumrada, 1975).

Typically, the ‘preferred’ amino acids, namely those largely taken up early in fermentation, support high growth rates when supplied as sole nitrogen sources. Thus, the sequence of amino acid accumulation and the extent to which individual amino acids contribute to total nitrogen requirements are functions of the closeness with which the amino acid matches the criteria of a good nitrogen source. However, mixtures of amino acids, as occur in musts, generally support faster rates of growth and fermentation than do single compounds. The functioning of a number of transport systems increases the rate and diversity of nitrogen accumulation and reduces the need for amino acid catabolism for biosynthesis of deficient compounds.

TABLE 4.6 Mean doubling times for the growth of five laboratory strains of yeast on various compounds as sole source of nitrogen.

Nitrogen compound	*Doubling time (min)*
Glutamine	137
Ammonia	142
Asparagine	146
Urea	158
Glutamate	159
Serine	163
Arginine	169
Aspartate	170
Alanine	187
Phenylalanine	190
Ornithine	203
Citrulline	208
Allantoin	208
Valine	225
Leucine	257
Proline	260
Tyrosine	265
Isoleucine	280
Threonine[1]	298
Tryptophan	298
Allantoate	322
Methionine[a]	349
Histidine[2]	475
Glycine[3b]	608
Lysine	no growth
Cysteine	no growth

Values were calculated from the data of Cooper (1982). Yeasts were cultured in Wickerham's medium. Superscripted numbers indicate the number of yeasts which failed to grow on the respective nitrogen compound.
[a]Mean of four strains only; [b]Limited growth.

Integration and regulation

In addition to regulation through enzyme synthesis and activity (see earlier section), nitrogen metabolism is also controlled and integrated through processes of compartmentation and the accumulation and mobilization of intracellular reserves.

With compartmentation, access of the substrate to its enzyme is restricted by mechanisms which prevent entry of the substrate into the cell or by subcellular partitioning of the substrate or the enzyme in vacuoles. With regard to entry to the cell, the availability of good nitrogen sources rapidly inhibits the transport systems for poorer nitrogen sources (Grenson and Hou, 1972; Roon *et al.*, 1975; Courchesne and Magasanik, 1983). The failure of substrate to enter the cell not only prevents its degradation but prevents the needless induction of further enzymic activity. Proline oxidase and ornithine transaminase are almost completely regulated through inducer/substrate exclusion (Brandriss and Magasanik, 1979b; Deschamps, Dubois and Wiame, 1979).

The amino acid/H^+ antiport of 10 amino acids into the vacuole is mediated by at least seven transport systems (Sato, Ohsumi and Anraku, 1984). Sequestration of amino acids into the vacuole enables these compounds to be accumulated in excess of immediate metabolic requirements and also affords these reserves protection from the degradative machinery which is induced as a result of their initial accumulation into the cytosol. Subcellular partitioning has been reported for arginine metabolism. Catabolism of this amino acid ordinarily proceeds *via* the formation of proline which in turn is catabolized to yield glutamate (Brandriss and Magasanik, 1980). Isolation of the synthetic pathway in the cytosol and the degradative pathway in the mitochondrion circumvents the potential for futile cycling and results in the net catabolism of arginine (Cooper, 1982).

The immediate metabolic demands for nitrogen are catered for by the cytoplasmic nitrogen pool, central to which is the glutamate and ammonium/α-ketoglutarate interconversion. During nitrogen excess, accumulation of assimilable nitrogen should not only satisfy metabolic demands, but also enable the establishment and maintenance of cellular amino acid reserves. Under nitrogen limited conditions, mobilization of these reserves will occur. Therefore, as the supply of exogenous nitrogen diminishes, a corresponding decline in cytoplasmic nitrogen levels will relieve catabolite repression. The induction and reactivation of the high affinity general amino acid permease along with other ammonia-sensitive permeases (see Table 4.5) and intracellular proteolytic (Betz, 1976; Achstetter and Wolf 1985; Schwencke, 1991) and allantoin- or arginine-degrading enzymes will result (Whitney and Magasanik, 1973; Bossinger, Lawther and Cooper, 1974; Sumrada and Cooper, 1978). While some of the permeases or degradative enzymes may already be present, others will need to be synthesized. Vacuolar nitrogen may be mobilized to fuel this *de novo* synthesis, which may be assisted by nucleic acid recycling (Mounolou, 1971; Dombek and Ingram, 1988). Nitrogen mobilization is typically pre-empted by an elevation in arginase activity (Whitney and Magasanik, 1973; Sumrada and Cooper, 1978).

Differential extraction of cellular amino acid pools has revealed that the proportion of these compounds located within the vacuole can be substantial (Wiemken and Durr, 1974; Messenguy, Colin and Ten Have, 1980; Kitamoto *et al.*, 1988). With the exception of glutamate, aspartate and leucine, the proportion of free cellular amino acids sequestered to the vacuolar pool is typically greater than 50%, or of the order of 90% for arginine, histidine, lysine and ornithine. The vacuole also contains the products of adenine and guanine catabolism, allantoin and allantoate, S-adenosyl methionine, and several enzymes associated with the degradation of cellular constituents (Schlenk, Dainko and Svihla, 1970; Zacharski and Cooper, 1978; Achstetter and Wolf, 1985).

A failure of induced transport and catabolic processes to yield sufficient nitrogen from extracellular sources or the intracellular turn-over of proteins will lead to a critical nitrogen limitation within the cell. Under these conditions, the role of vacuolar nitrogen will shift from one of a source of nitrogen for the induced nitrogen-acquiring processes to one of a nitrogen source for the sustenance of metabolism *per se* (Cooper, 1982; Large, 1986; Schwencke, 1991).

The ability of a growing culture of yeast to continue cell division for one or two generations after transfer from a rich medium to distilled water is well known, yet the time period for which intracellular nitrogen pools are able to sustain yeast metabolism upon depletion of exogenous supplies has not been fully characterized. Yeast strain, growth phase, growth conditions and medium composition prior to depletion are likely to be key factors in this regard. Watson (1976) observed the total cellular nitrogen pool to be reduced to approximately 45% of its initial size after only 15 min incubation in a nitrogen-free medium. A similar reduction was seen specifically for the vacuolar arginine pool after a one hour incubation of *S. cerevisiae* X2180-1A cells in nitrogen-free medium (Kitamoto *et al.*, 1988).

Cellular reserves appear to have a variable ability to obviate metabolic imbalances of enological importance. While glucose catabolism is sustained for several days following nitrogen starvation (Busturia and Lagunas, 1986; Salmon, 1989), excessive formation of hydrogen sulfide is observed within 30–60 min after nitrogen depletion (Henschke and Jiranek, 1991; Jiranek, 1992).

The nitrogen anabolic pathways are well reviewed by Jones and Fink (1982) and Hinnebusch (1988), and so will not be considered herein.

ENOLOGICAL ASPECTS OF NITROGEN METABOLISM

The more serious fermentation problems encountered in contemporary winemaking, especially slow or incomplete fermentations and hydrogen sulfide formation, had only been sporadically researched until the last decade or so (Agenbach, 1977; Eschenbruch, Bonish and Fisher, 1978; Vos and Gray, 1979; Ingledew and Kunkee, 1985; Kunkee, 1991). Indeed, it would appear that stuck fermentation, except that caused by extreme temperature and excessive sugar concentration, is a relatively recent phenomenon due to changes in vineyard fertilizer usage and table wine technology towards low temperature, low solids and low oxygen tension fermentation (Agenbach, 1977; Amerine *et al.*, 1980; Houtman and du Plessis, 1986; Allen and Auld, 1988; Henschke, 1990; Kunkee, 1991). Lately, attention has focused on the formation of ethyl carbamate from urea, a product of yeast nitrogen metabolism (Ough, 1991). In a positive context, the possibility of enhancing the fermentation bouquet of wine through manipulating the nitrogen content of the must has attracted interest for some time (Vos, 1981; Rapp and Versini, 1991; Spayd *et al.*, 1991).

An imbalance in the supply and demand of nitrogen plays an important role in the development of fermentation problems. A deficit of assimilable nitrogen can, in part, be the cause of inadequate fermentation rate and excessive hydrogen sulfide formation. In fact, deliberate reduction of nitrogen content to prevent the incomplete fermentation of sugar, by successive cycles of yeasting and subsequent removal, is exploited in the production of the sweet wines of Asti Spumante (Tarantola, 1967). On the other hand, enhanced wine flavor and urea formation appear to be linked to higher concentrations of assimilable nitrogen and particular amino acids.

This section will consider the role of assimilable nitrogen in fermentation kinetics, generation of positive and negative aroma compounds, accumulation of urea/ethyl carbamate and future fermentation management strategies.

Quantitative Aspects of Fermentation

The nitrogen content of yeast is about 6.0% by wt (Rosén, 1989) of which the proteinaceous structural and metabolic components represent 10–30% (Salmon, 1989). During the yeast growth phase, nitrogen is rapidly assimilated along with the bulk of other nutrients necessary for anabolic activity. The minor by-products of sugar and nitrogen metabolism that affect the composition and flavor of wine are excreted during this phase. Throughout the non-proliferating cell phase, during which much of the sugar is fermented, nitrogen requirements are relatively small; protein synthesis decreases to less than 10% that of growing cells and protein turnover increases.

Nitrogen metabolism promotes growth essentially by supplying the precursors for protein and nucleic acid synthesis. A rich supply of nitrogen will allow high rates of growth and biomass yield, and stimulates fermentation activity and the formation of end-products of metabolism. Ultimately, growth will become restricted by a limitation of other nutrients or accumulation of toxic products. However, a limited supply of nitrogen or an inability to accumulate it will directly restrict growth and metabolic activity. While, from the point of view of fermentation kinetics, nitrogen excess would be desirable, it is less clear to what extent nitrogen composition and availability affects the composition and sensory aspects of wine. Thus, conditions permitting 'maximum' rates may not necessarily equate with 'optimal' conditions for best wine composition.

The quantitative role of assimilable nitrogen in the fermentation of grape must will be considered in terms of yeast requirements, and the kinetics of growth and fermentation.

Yeast nitrogen requirements

Yeast growth depends on the availability of nutrients including nitrogen. In an early study into the total nitrogen balance of grape must fermentation, Cantarelli (1957) found that a minimum of 200 mg N/l was necessary for adequate yeast growth and fermentation. A more detailed investigation was undertaken by Agenbach (1977), who reported a minimum requirement of free amino nitrogen (FAN) of 70 to 140 mg/l for complete fermentation of musts with initial sugar concentrations ranging between 183 and 258 g sugar/l. Bely *et al.* (1990a) found that the minimum FAN concentration of 140 mg/l was needed for satisfactory fermentation in a study of 40 musts having sugar concentrations ranging from 160–240 g/l. This concentration has been substantiated by model fermentation studies (Rapp and Franck, 1971; Reed and Peppler, 1973; Bely *et al.*, 1990a; Table 4.7). By comparison, normal gravity brewing worts where the initial sugar concentration is around 120 g/l, require a minimum of 100–150 mg FAN/l for satisfactory fermentation (Pierce, 1987; O'Connor-Cox and Ingledew, 1989). Unfortunately, there is no clear or agreed definition of minimum or satisfactory fermentation performance.

Agenbach (1977) further found that increasing available nitrogen stimulated growth and fermentation, but the proportion of nitrogen assimilated followed an approximate exponential relationship. Thus, the proportion of nitrogen assimilated only increased by a small extent from 200 to 260 mg FAN/l when the available nitrogen concentration increased nearly three-fold from 300 to 800 mg FAN/l. Increasing the initial sugar concentration from 208 to 258 g/l increased the proportion of nitrogen assimilated by about 10% at the high end of the available nitrogen range.

Jiranek, Langridge and Henschke (1990) reported a systematic study of nitrogen demand of Australian *S. cerevisiae* wine yeasts as influenced by some fermentation conditions. Eight strains were cultivated in an air-saturated synthetic grape juice medium containing 200 g glucose/l, in which all amino acids were present to excess. Demand for assimilable nitrogen ranged from 330 to 470 mg N/l (mean of 400 mg N/l) at a fermentation temperature of 25°C. Increasing the initial glucose concentration from 200 to 250 g/l in a synthetic grape juice gave an increase in demand of up to 60 mg N/l, depending on the strain examined. This compares with an increase of about 30 mg FAN/l for similar variations in sugar concentration in grape musts having similar initial nitrogen concentrations as reported by Agenbach (1977). The extra consumption of nitrogen may reflect a greater need for maintenance of carbon catabolic machinery as the result of extended fermentation. Of the factors studied, fermentation in the presence of air had the greatest effect on nitrogen demand. When the headspace of the fermentation flask initially contained air instead of gaseous nitrogen, and the medium was saturated with air (about 8 mg O_2/l) the mean nitrogen demand of the strains increased from 400 to 658 mg N/l and from 430 to 833 mg N/l for initial sugar concentrations of 200 and 250 g/l, respectively. The magnitude of this oxygen effect was also observed in grape musts (Ingledew and Kunkee, 1985).

TABLE 4.7 Nitrogen utilization by yeasts during fermentation.

Nitrogen utilized for complete fermentation				
Total		*Minimum*	*Initial sugar*	
mg/l	*mg N/l*	*mg FAN/l*	*concentration (g/l)*	*Reference*[a]
–	–	150[b]	112–124	1
916	174	–	190	2
615[f]	144[f]	–	221	2
–	–	70[k]	183	3
–	–	130/130/140[k]	208/233/258	3
283	–	–	232	4
878[g]	–	–	216	5
5297	795	–	220	6
–	26–495[e]	–	220	6
–	–	113–190[b]	112	7
654	159	–	–	8
–	330–470[c,h]	–	200/250	9
–	460–920[c,i]	–	200/250	9
–	300[e]	140	160–240	10
–	400/429[j]	–	210–258	11

[a]References: (1) Enari, 1974; (2) Dittrich and Sponholz, 1975; (3) Agenbach, 1977; (4) Bell, Ough and Kliewer, 1979; (5) Vos, Zeeman and Heyman, 1978; (6) Ingledew and Kunkee, 1985; (7) Pickerbell, 1986; (8) Ooghe and Kastelijn, 1988; (9) Jiranek *et al.*, 1990; (10) Bely *et al.*, 1990a, 1991; (11) Ough *et al.*, 1991.
Studies were conducted in grape juices except: [b]wort; [c]chemically-defined grape juice media, refer to Table 4.4; [d]defined media; [e]mg Free α-amino nitrogen/l; [f]Juice from *Botrytis* infected fruit; [g]Concentration for maximal fermentation rate; [h]Initially air-saturated medium, 8 yeast strains; [i]Initially headspace and medium (ratio 1:1) contained air, 6 yeast strains; [j]15/20°C fermentation temperature, respectively, 4 yeast strains; [k]Nitrogen-depleted juice supplemented with diammonium phosphate (mg N/l = mg FAN/l).

The stimulatory effects of air on nitrogen assimilation, yeast growth and fermentation rate in grape must of low and moderate nitrogen contents are shown in Table 4.8. The oxygen substitutes, ergosterol and the source of oleic acid, Tween 80®, restored nitrogen assimilation to that stimulated by aeration. Oxygen stimulates cell growth by allowing growth limiting sterol and unsaturated fatty acid synthesis to operate (Radler, 1978; Strydom, Kirschbaum and Tromp, 1982; Ingledew and Kunkee, 1985; Larue and Lafon-Lafourcade, 1989; Mauricio, Guijo and Ortega, 1991). The main effect of sterol addition is on the retention of cell viability in the latter stages of fermentation enabling higher rates of fermentation (Ribéreau-Gayon, 1985; Larue and Lafon-Lafourcade, 1989).

TABLE 4.8 The effect of air and 'survival factors' on assimilation of nitrogen from a grape must. [Adapted from Ingledew and Kunkee (1985) by Cantarelli (1989)].

Atmosphere	*Additions*	*Nitrogen utilization (mg FAN/l)*	*Cell count ($\times 10^7$/l)*	*Fermentation rate (g sugar/100 ml.24h)*
Air	None	87	35.0	1.80
Nitrogen	None	65	1.3	0.50
Air	YE	314	36.0	4.00
Nitrogen	YE	107	3.0	1.22
Nitrogen	YE + SF	293	32.0	3.60

YE: yeast extract; SF: survival factors (Tween 80®, ergosterol).
Musts without and with YE addition contained 99 and 386 mg FAN/l, respectively.

The addition of grape solids also reduces fermentation time (Tromp, 1984; Houtman and du Plessis, 1986b; Groat and Ough, 1978), possibly by providing the assimilable lipids, sterols and unsaturated fatty acids, that stimulate growth (Radler, 1978; Houtman and du Plessis, 1986b). Both of these effects increase the demand for assimilable nitrogen (Tromp, 1984; Ingledew and Kunkee, 1985; Houtman and du Plessis, 1986b) but it is not clear whether this is a cause or effect. The enhancement of fermentation rate observed in the latter stages of fermentation may result from improved nitrogen assimilation. This is essential to allow protein synthesis necessary for fermentation and energy metabolism to continue. Despite the importance and variability of oxygen in must (0–8 mg/l at air saturation), it is not commonly quantified in commercial practice.

The variability in nitrogen demand by different strains of *S. cerevisiae* wine yeasts has been confirmed in grape musts by Ough *et al.* (1991). Calculation of mean values for four Californian wine yeasts from 11 grape musts having moderate to high amino acid concentration (range of 141–738 mg assimilable N/l), are presented in Table 4.9. The musts had a mean sugar concentration of 21.3°Brix. The maximum values for Montrachet are likely to be higher because four of the musts used in this study appeared to contain insufficient assimilable nitrogen for this particular strain. Montrachet was the highest nitrogen requiring strain studied, consuming from 81 to 122 mg N/l more than the lowest nitrogen requiring yeast, Epernay 2, depending on the temperature. These values are similar to those determined in a synthetic grape juice medium by Jiranek *et al.* (1990) (Table 4.7).

The data of Ough *et al.* (1991) confirm the view that fermentations conducted at higher temperature consume greater amounts of assimilable nitrogen (Table 4.9). Epernay 2 showed the greatest response to temperature, and Flor the least. No explanation, apart from higher metabolic rate, has been provided for the greater assimilation of nitrogen at higher temperature. Since white musts are fermented down to 8°C and red musts up to 32°C, it is clearly important to investigate nitrogen demand over this range.

TABLE 4.9 Calculated mean nitrogen concentration (mg N/l) assimilated by four wine yeasts during fermentation of 11 grape musts at two temperatures[a].

Strain	*15.5°C*	*20°C*
Epernay 2	329	392
Flor	396	402
Prise de Mousse	422	449
Montrachet	451	473
Mean	400	429

[a]Nitrogen consumption was calculated from the data of Ough *et al.* (1991). Mean total and assimilable (total amino acids—proline) amino acid concentrations were 552 and 473 mg N/l, respectively.

As a consequence of strain variation in nitrogen utilization, wines produced from grape musts with assimilable nitrogen exceeding 300–500 mg N/l may contain residual nitrogen (Table 4.10). The residual amount will depend on the strain of yeast, must composition and fermentation conditions. The highest nitrogen requiring strains, Montrachet and Prise de Mousse, produced the lowest residual nitrogen. Consequently, these strains gave lower correlations between initial and residual nitrogen contents because a greater proportion of wines contained little or no residual nitrogen. The main residual amino acids were proline, arginine, glutamine and alanine, and ammonium, largely reflecting their high initial concentration (84% of total N) in the juices before fermentation.

Residual nitrogen in wine is of considerable importance to subsequent microbial stability, especially for wines containing residual fermentable sugar. Alternatively, wines in which secondary fermentation, either *tirage* or malolactic fermentation, is to be induced require residual assimilable nitrogen; lactic acid bacteria have specific requirements (Garvie, 1967).

Yeast growth

Yeast growth is usually described in terms of the kinetic parameters associated with the different stages of growth: lag, exponential, linear, stationary and decline phases. Those affected by the nitrogen supply are (i) specific growth rate (linear phase), (ii) the time when specific growth rate begins to decline, and (iii) maximum yield of yeast (biomass) produced and maximum cell density (Monk, 1982; Monk and Storer, 1986; Bezenger and Navarro, 1988; Bely *et al.*, 1990a, b).

Agenbach (1977) investigated the relationship between assimilable nitrogen, yeast growth and fermentation activity in a nitrogen depleted grape must

TABLE 4.10 Correlation (r^2) between initial assimilable nitrogen concentration (mg N/l) of 11 grape juices and residual nitrogen[a] of corresponding wines made with four yeasts at two temperatures[b].

Strain	*15.5°C*		*20°C*	
	% Residual	r^2	*% Residual*	r^2
Epernay 2	23.2	0.571	15.5	0.485
Flor	16.8	0.619	16.0	0.414
Prise de Mousse	12.2	0.506	10.6	0.427
Montrachet	10.8	0.504	9.0	0.248
Mean	15.8	0.593	12.8	0.456

[a]Total amino acids minus proline nitrogen.
[b]Nitrogen consumption was determined from the data of Ough *et al.*, (1991). Mean and range of assimilable nitrogen (- proline) concentration of the 11 juices was 473 and 141–738 mg N/l, respectively.
($r^2 > 0.332$ significant at $P = 0.05$).

supplemented with diammonium phosphate for the purpose of defining minimum requirements for fermentation (Table 4.11). The biomass formed (2.06–2.85 g dry mass/l) under these conditions depended to some extent on the initial sugar concentration of the must. Mean fermentation rate was related to biomass formation. Partial inhibition of yeast growth occurred when initial sugar concentration approached 258 g/l. Increased osmotic stress is known to reduce growth rate and cell numbers (Nishino, Miyazaki and Tohjo, 1985). Proportionally more nitrogen was assimilated per biomass formed in response to increased sugar concentration such that the estimated cellular nitrogen to carbon ratio increased. Temperature stress, although reducing viability (Ribéreau-Gayon, 1985), also increased nitrogen demand (Table 4.9) suggesting

TABLE 4.11 Effect of sugar concentration on minimum nitrogen assimilation, biomass yield and mean fermentation rate in a nitrogen-depleted Colombard grape must[a] supplemented with diammonium phosphate (Agenbach, 1977).

Sugar conc. (g/l)	*Nitrogen assimilated (mg FAN/l)*	*Biomass yield (g dry mass/l)*	*Ratio of N assimilated: biomass (%w/w)*	*Mean fermentation rate (g CO_2/l.Day)*
183	71.8	2.06	3.3	8.4
208	127.5	2.63	4.9	12.3
233	128.6	2.85	4.5	11.5
258	143.2	2.38	6.0	9.1

[a]Must supplemented with micro-nutrients (0.56% Yeast Carbon Base); Fermentation temperature was 25°C.

that higher amounts of nitrogen are needed by the cell to sustain metabolic function under stress conditions.

Yeast biomass was exponentially related to the assimilable nitrogen concentration of a must (Cantarelli, 1957; Agenbach, 1977) (Figure 4.4). Thus, biomass formation was only stimulated by 14–40%, depending on initial sugar concentration, when a must containing minimum nitrogen (140 mg FAN/l) was supplemented with a large excess (807.5 mg N/l). Nitrogen (as urea) toxicity to growth occurred at concentrations exceeding 1000 mg N/l. Fermentation rate, however, was comparatively more highly stimulated (82–134%). Bely *et al.* (1990a) obtained an exponential relationship (r=0.811) between maximum cell density (X_{max}) and assimilable nitrogen content (N) of 40 musts fermented with one strain of yeast (K1 ICV-INRA) at 24°C:

$$\text{Log}\, X_{max} = 0.553\, \text{Log}\, N + 0.817$$

The musts, which varied in sugar concentration between 160–240 g/l had initial assimilable nitrogen (α-amino acid and ammonium) concentrations of 53–411 mg FAN/l. It was suggested that other must nutrients (or inhibitors) in grape musts also influenced yeast growth.

Agenbach (1977) noted that the nitrogen consumed was not directly proportional to that available. Investigation of the relationship between assimilated nitrogen and fermentation kinetic parameters showed that biomass yield was exponentially related while fermentation rate was directly related (r=0.94) to assimilated nitrogen. Thus, maximum biomass yield (about 3 g dry mass/l) occurred when the concentration of assimilated nitrogen was maximal at 300 mg N/l, although available nitrogen exceeded 500 mg N/l (Figure 4.4). Maximum fermentation rate was about 21 g CO_2/l per day which occurred when assimilated nitrogen was maximal (300 mg N/l). This finding indicated that nitrogen affects yeast growth and fermentation by different mechanisms, as earlier shown by Cantarelli (1957). Of practical importance is the fact that fermentation is stimulated by nitrogen at concentrations which fail to further stimulate growth.

As shown by Agenbach (1977), biomass yield only responds to nitrogen supplementation when the initial concentration is under a threshold value. Bely *et al.* (1990a) investigated the response of yeast growth to the timing of nitrogen addition (Figure 4.5). Maximum yeast growth was obtained when additions were made to a nitrogen deficient must at any time after inoculation until late log phase. After this phase, potential biomass yield was rapidly reduced to control levels. Interestingly, high fermentation rates were maintained for nitrogen additions made well into stationary phase at which time stimulation of yeast growth had markedly decreased. This may be explained by the greater sensitivity of amino acid accumulation compared with sugar transport to inhibition by ethanol (Cartwright *et al.*, 1989; van Uden, 1989; Iglesias *et al.*, 1990).

The complexity of nitrogen transport function in yeast may explain the influence of the nitrogen source on yeast growth in grape musts. Cantarelli

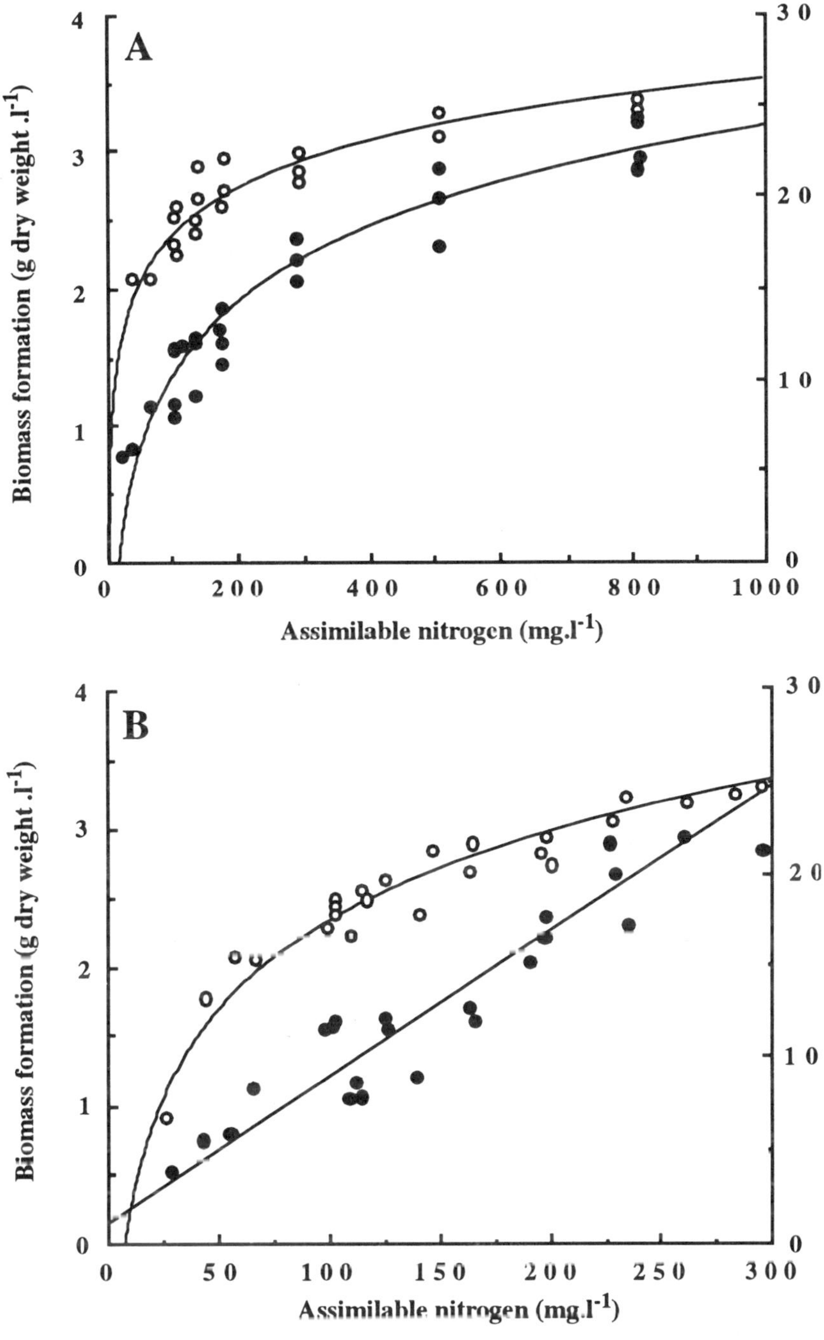

FIGURE 4.4 Relationship between assimilable nitrogen (A) or nitrogen assimilated (B) and yeast biomass yield (o) and fermentation rate (•) in grape musts [Redrawn from Agenbach (1977) with permission from South African Journal of Enology and Viticulture].

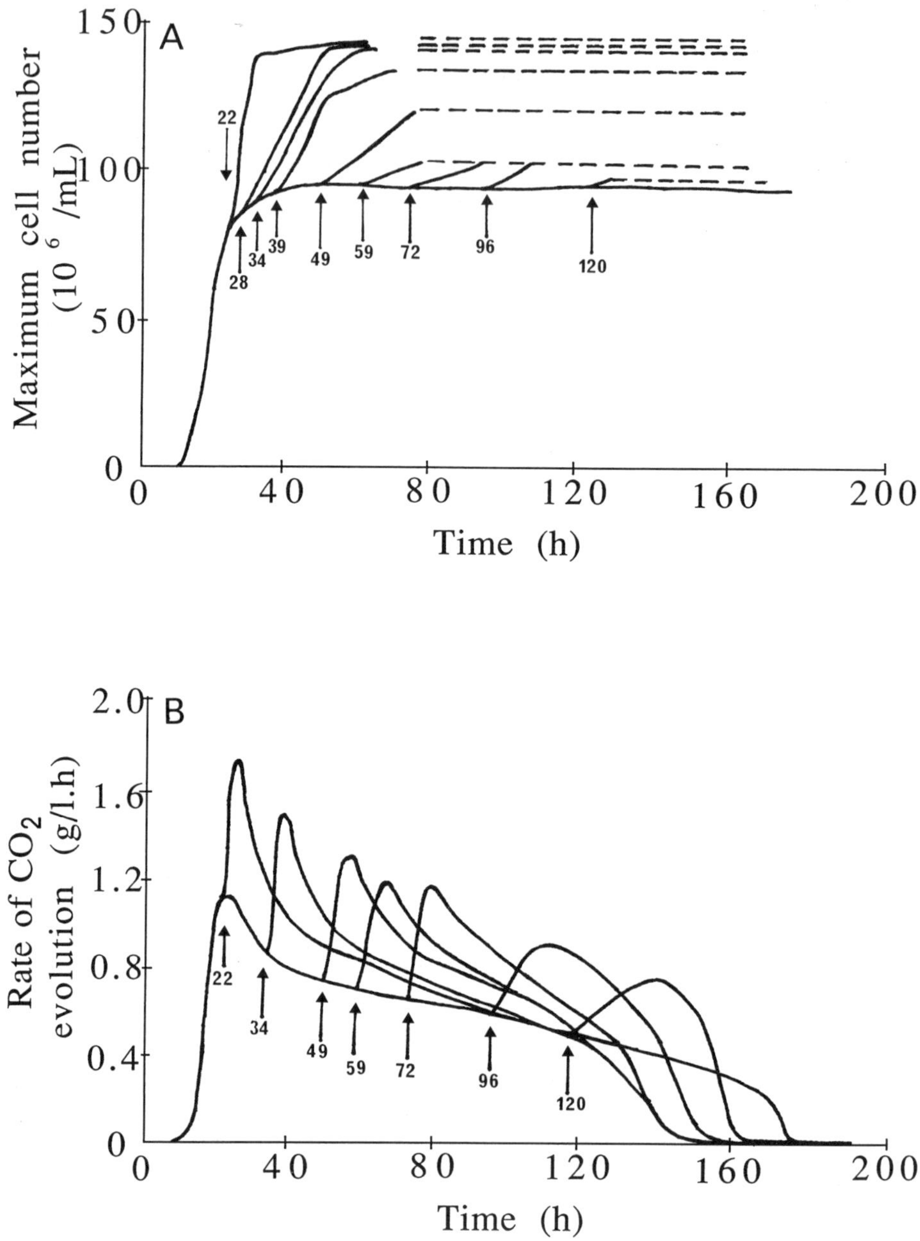

FIGURE 4.5 Effects of diammonium phosphate supplementation (63 mg N/l) on fermentation of a chemically-defined grape juice medium (initial concentration of 120 mg FAN/l) at 24°C. Arrows indicate timing of supplement in hours on yeast growth (A) and fermentation rate (B). [Redrawn from Bely *et al.* (1990a) with permission from the Society of Fermentation Technology, Japan.]

(1957) reported the effects of various nitrogen supplements, ammonium salts (45 mg N/l) and organic nitrogen compounds (100 mg N/l), on yeast biomass formation and fermentation rate in a heat sterilized grape juice medium (296 mg total N/l). Relative to the unsupplemented control which produced 1.84 g dry mass/l, ammonium salts stimulated biomass formation by 13–98%, while fermentation rate was stimulated by 7–16%. Aspartic acid, arginine, glutamic acid, asparagine and urea, stimulated biomass formation by 22–77%. Fermentation rate was more highly stimulated relative to yeast growth with organic nitrogen sources compared with the inorganic nitrogen compounds, ranging from 13–79% over the control. In general, the nitrogen sources varied extensively in the degree of stimulation of biomass formation and fermentation activity.

Complex nitrogen supplements may stimulate yeast growth to a greater extent than single sources. Monk (1982) studied the rate of yeast growth in Australian Riesling grape juices of low nitrogen concentration (85–133 mg FAN/l). The exponential rate of growth was not significantly influenced by nitrogen supplementation. The linear phase of growth, however, varied between 40–300% of the unsupplemented control (85 mg FAN/l) depending on the type and concentration of nitrogen source added. Maximum growth rates were achieved at supplement concentrations of about 2 g/l: glutamic acid (190 mg N/l) stimulated growth rate by 50%; diammonium phosphate (420 mg N/l) by 80%; and an amino acid mixture (320 mg N/l) by 130% of the control juice. Monteiro and Bisson (1991) also noted a greater growth response to a complex (peptone) source compared with ammonium and urea, the difference being more obvious at higher concentrations of ethanol. Complex nitrogen sources may be stimulating growth at later stages of fermentation by satisfying transport systems which are less inhibited by ethanol.

Fermentation rate

Numerous investigations have reported a link between fermentation rate and the nitrogen concentration of grape must and other media, but fewer have studied kinetic aspects. The minimum and optimum concentrations of assimilable nitrogen for growth and fermentation are summarized in Table 4.7.

In general, biomass formation is an exponential function of the concentration of assimilable nitrogen. Fermentation rate, however, appears to follow a different relationship with nitrogen and is not, therefore, a direct function of biomass yield. Cantarelli (1957) observed a log function relating fermentation rate and total nitrogen concentration, giving a complex relationship between yeast biomass yield and fermentation rate. Agenbach (1977) described exponential functions relating the concentration of assimilable nitrogen with biomass yield or fermentation rate, but the functions were quantitatively different (Figure 4.4). The function relating maximum rate of CO_2 production (g/l.h) and nitrogen (N; mg FAN/l) as determined by Bely *et al.* (1990a) from a study of 40 musts fermented at 24°C (r=0.965) is given by:

$$\text{Log}\,(dCO_2/dt)_{max} = 0.843\,\text{Log}\,N - 1.72$$

Determination of the actual concentration of nitrogen assimilated by yeast provides a clearer picture of the different roles of nitrogen in yeast growth and fermentation activity. While an exponential function relates assimilated nitrogen with biomass yield (r=0.94), a direct function relates assimilated nitrogen with fermentation rate (about 8 g CO_2/l per day per 100 mg FAN/l assimilated; r=0.94) (Figure 4.4).

This important observation that assimilable nitrogen imposes a differential effect on yeast growth and fermentation activity has been confirmed by others. That is, below about 140 mg N/l, growth and fermentation rates become inadequate for practical purposes while above 400 mg N/l assimilable nitrogen strongly stimulates fermentation rate (and reduces total time) over and above that of simply stimulating biomass yield or growth rate (Amerine and Ough, 1963; Ough and Amerine, 1966; Agenbach, 1977; Vos, Crous and Swart, 1980; Monk, 1982; Egbosimba and Slaughter, 1987; Monk *et al.*, 1987; Bely *et al.*, 1990a, b). Ough and coworkers showed a linear relationship between fermentation rate and total nitrogen concentration of must (Ough and Amerine, 1966), and also yeast viability (Amerine and Ough, 1963)

Biomass yield, when expressed as g protein/l, also closely paralleled the nitrogen content (expressed as mg N/l) of the supplemented must (Monk, 1982). Resultant increases in growth rate and biomass yield were accompanied by greater rates of fermentation. Fermentation rate was highly correlated with nitrogen content of the supplemented must, but was less well related to biomass yield expressed as protein.

Modelling nitrogen effects

Bezenger *et al.* (1986) and Bezenger and Navarro (1987, 1988) have extended the models of the enological fermentation, proposed by Holzberg *et al.*, (1967) and Boulton (1980), to account for the influence of nitrogen on fermentation kinetics. Experiments were performed in a synthetic grape juice medium with initial total nitrogen concentrations ranging from 70–266 mg N/l. Initial nitrogen concentration did not affect the maximum specific yeast growth rate, but did affect the growth yield and also affected the subsequent rate of growth decline. Low initial nitrogen concentration, caused nitrogen exhaustion and growth rate declined rapidly. Higher initial concentrations of nitrogen delayed or prevented nitrogen exhaustion and maximum growth rate was sustained for a longer period before growth rate declined at a lower rate. The concentration of alcohol that had accumulated at the time the maximum specific growth rate declined, was related to initial nitrogen concentration; it was suggested that ethanol tolerance was, therefore, a function of nitrogen concentration. This claim obviously needs confirmation by direct measurement of ethanol tolerance (see for example Juroszek, Feuillat and Charpentier, 1987); it is likely that yeast growth and not fermentation activity is primarily inhibited by ethanol since growth rate declined more sharply than fermentation rate (van Uden, 1989).

Maximal carbon dioxide evolution rate was directly proportional to initial nitrogen concentration of the medium. In addition, cellular nitrogen content and the specific rate of CO_2 production peaked early, during the first tenth of total fermentation time, and then decreased to a lower value. The specific rate of CO_2 production was linearly related to the specific growth rate. Overall, yeast growth and, CO_2 and ethanol production could be described as a function of initial nitrogen concentration (Bezenger and Navarro, 1987, 1988). In summary, the model accounted for temporal aspects of the changing nitrogen concentration on fermentation activity as a function of initial nitrogen concentration; a high initial nitrogen concentration stimulates and sustains maximum CO_2 production rate, intermediate concentrations have a neutral effect while a low initial concentration reduces CO_2 production rate and shortens the phase at maximum rate. It was concluded, in confirmation of that by Saito and Slaughter (1984), that the stimulation of fermentation rate by nitrogen was through stimulation of protein synthesis, and not activation of glycolytic enzymes *per se*.

Weight loss as a result of CO_2 evolution was used by Bely *et al.* (1990a, b) to model the kinetics of grape must fermentation. Based on the initial sugar concentration of the must, the sugar consumed and ethanol produced could be calculated from the amount of CO_2 released according to a relationship described by El Haloui, Picque and Corrieu (1988). Based on the behavior of selected kinetic parameters of fermentation, three phases of complete fermentation were recognized; lag phase; growth phase, during which the maximum specific rate of CO_2 production, the maximum rate of CO_2 production and maximum cell population occurred and; stationary phase, during which fermentation activity decreased although viability usually remained high (95%).

Total fermentation time depended on the initial concentration of assimilable nitrogen according to an inverse exponential relationship ($r=0.931$), a correction being made for differences in initial sugar concentration. Maximum cell population was less well related ($r=0.811$), but maximum CO_2 production rate was highly correlated ($r=0.965$) with nitrogen by an exponential relationship. Thus, an important relationship between maximum CO_2 production rate and fermentation duration was established, and is discussed later in relation to predicting sluggish fermentations.

'Sluggish' and 'stuck' fermentations

A primary objective of making dry table wine is to achieve a better than 98% transformation of grape sugar into alcohol and carbon dioxide at a controlled rate through the fermentative activity of wine yeasts. Generally, fermentation proceeds at a rate greater than required and must be controlled, most commonly by refrigeration. Occasionally, despite the imposition of measures known to stimulate fermentation, such as warming, addition of ammonium and vitamin supplements, and pumping over (limited aeration), the rate of fermentation becomes unacceptably slow, or may even cease. These fermentations are often referred to as 'sluggish' or 'stuck', respectively. Apart from Houtman and du Plessis (1986), few investigators have objectively defined sluggish fermentation. Fermentations which exceeded 25–55 days for initial sugar concentrations of 150–240 g/l, respectively, were considered to be retarded (slow), while those exceeding 55–85 days for these initial sugar concentrations, were considered lagging (sluggish). Strictly speaking, stuck fermentation implies total cessation of fermentation, measured as sugar utilization, CO_2 evolution or ethanol accumulation.

The commercial implications of sluggish and stuck fermentations arise from the fact that fermentor space is not efficiently utilized, and the wine is at risk of spoilage due to the low rate of protective CO_2 evolution and residual sugar content. Progress in the elimination of this economically important problem appears to be somewhat sluggish itself. Probable reasons are the sporadic occurrence, a large number of different causes and few convenient predictive tests. Furthermore, fermentations, once having stuck are difficult to restart. Indeed, a final but generally successful option involves blending a stuck wine back into a vigorously fermenting must. Therefore, in practice the cause and effect relationship for sluggish or stuck fermentations is not usually verified.

Many conditions have been described that may result in an incomplete or severely reduced rates of fermentation, including inadequate assimilable nitrogen (Table 4.12). The relative incidence of these conditions is not known.

The dependence of yeast growth and fermentation activity on the concentration of assimilable nitrogen has led several investigators to define the nitrogen requirements of yeast under enological and brewing conditions (Table 4.7). The minimum required for a satisfactory rate of fermentation in clarified juice is considered to be about 140 mg FAN/l. However, optimum or maximum fermentation rate requires a high concentration of 800–900 mg N/l, of which only 400–500 mg N/l is assimilated.

Grape musts with initial concentrations of nitrogen below the threshold value of 140 mg FAN/l can be protected from a retarded rate of fermentation by supplementation with ammonium salts. However, once a fermentation has become sluggish or stuck as the result of nitrogen deficiency, supplementation may not necessarily restore the sugar catabolism to an acceptable rate, despite the fact that an addition of nitrogen early during fermentation is effective (Salmon, 1989; Bely *et al.*, 1990a). In severe cases, re-inoculation is often necessary to restart fermentation (Amerine *et al.*, 1980).

A combination of kinetic and physiological investigations has provided a metabolic basis for this apparent paradox. Two effects are observed, one relates to the population-dependent rate of sugar catabolism and the other to the specific sugar catabolic rate of the cell. A low initial concentration of nitrogen acts by limiting growth rate and biomass formation resulting in a low rate of sugar catabolism (Agenbach, 1977; Monk, 1982; Bely *et al.*, 1990a). The second effect concerns the reduction in the specific sugar catabolic activity of the cell, which results from exhaustion of the nitrogen supply.

A reduced rate of fermentation resulting from exhaustion of the nitrogen supply, led Lagunas *et al.* (1982) to propose this phenomenon as an explanation for stuck fermentation. Subsequently, Busturia and Lagunas (1986) and Salmon (1989) observed that nitrogen starved cells undergo arrest of protein synthesis. Unable to replace the rapidly turned-over sugar transport systems, a rapid

TABLE 4.12 Suggested causes of sluggish and stuck fermentation.

Condition	*Reference*
Nutritional deficiency	
Assimilable nitrogen deficiency	Bizeau (1963); Agenbach (1977); Vos *et al.* (1978); Wenzel and Dittrich (1978); Vos and Gray (1979); Monk (1982); van Rooyen and Tromp (1982); Tromp (1984); Ingledew and Kunkee (1985); Busturia and Lagunas (1986); Thornton and Rodriguez (1987); Salmon (1989); Bely *et al.* (1990a, b); Jiranek *et al.* (1991); Kunkee (1991); Sponholz (1991)
Deficiency of oxygen or ergosterol and unsaturated fatty acids	Aries and Kirsop (1977); Larue *et al.* (1978); Houtman and du Plessis (1986); Mauricio *et al.* (1991)
Vitamin deficiency	Monk and Costello (1984)
Mineral deficiency	Monteiro and Bisson (1991)
Inhibitory substances	
Sugar	Ingledew and Kunkee (1985)
Ethanol	Thomas and Rose (1979); Leão and van Uden (1983, 1984a); van Uden (1989)
Fatty acids	Lafon-Lafourcade *et al.* (1984); Larue and Lafon-Lafourcade (1989)
Acetic acid accumulation	Pampulha and Loureiro (1989); Pinto *et al.* (1989)
Killer toxins	van Vuuren and Wingfield (1986)
High SO_2/petite yeast induction	Lehmann (1987)
Pesticide residues	Lemperle (1988)
Physiological condition	
Poorly propagated yeast starter cultures	Monk and Storer (1986); Larue and Lafon-Lafourcade (1989); Mauricio *et al.* (1991)
Technological practices	
High clarification/ low solids	Wucherpfennig and Bretthauer (1970); Ribéreau-Gayon, Lafon-Lafourcade and Bertrand (1975); Groat and Ough (1978); Vos and Gray (1979); Houtman *et al.* (1980a); Houtman and du Plessis (1981); van Rooyen and Tromp (1982)
Excessive temperature	Amerine *et al.* (1980); Barillère *et al.* (1983); Rozes *et al.* (1988); van Uden (1989)
Temperature-shocking	Rozes *et al.* (1988)
Inadequate agitation/ premature flocculation	Groat and Ough (1978); Ingledew and Kunkee (1985)

decline in sugar catabolism results. While the majority of metabolically important proteins are relatively stable to nitrogen starvation, glucose transport systems are an exception with half-lives of about 6 hours in fermenting cells (Busturia and Lagunas, 1986). The ATPase, which energizes the active transport of amino acids and ammonium, has a longer half-life of about 11 hours in nitrogen starved growing cells. Moreover, its activity (V_{max}) is stimulated, possibly to prevent cell death from intracellular acidification which results from glycolytic activity (Benito, Portillo and Lagunas, 1992).

The timing of ammonium additions on the restoration of metabolic events associated with fermentation activity has been investigated by Salmon (1989). Complete inactivation of glucose transport occurred about 50 hours after nitrogen starvation. Supplementation with ammonium only restored yeast growth, protein synthesis and CO_2 evolution when the initial ammonium starvation had occurred during the active phase of yeast growth. Beyond this stage, only a transitory increase in CO_2 evolution resulted. This observation corresponds with the inability to sometimes restart a stuck fermentation with ammonium additions when less than 50% of unfermented sugar remains.

The importance of ammonium as a stimulant of fermentation as distinct from stimulating biomass formation has been well documented (Cantarelli, 1957; Agenbach, 1977; Yoshino and Murakami, 1982; Saito and Slaughter, 1984; Egbosimba and Slaughter, 1987; Salmon, 1989; Monteiro and Bisson, 1991). The investigations of Salmon (1989) have provided an insight into the mechanism by comparing the dynamics of the intracellular ammonium pool, protein synthesis and biomass in relation to extracellular nitrogen availability. In the presence of a minimal supply of nitrogen (as ammonium; 150 mg N/l), the cell maintained a near constant pool of ammonium (1–2 mM) in the face of continuous protein synthesis as the supply became exhausted. It was suggested that nitrogen for protein synthesis was supplied by drawing on the vacuolar reserve and then ultimately through biomass cycling observed during the final half stage of fermentation. Fermentation completed in about 100 hours, compared with about 60 hours under near-optimum conditions provided by 390 mg N/l. Under severe nitrogen limitation (23.5 mg N/l), the ammonium pool increased sharply to about 5.5 mM after the extracellular concentration became exhausted, and then declined sharply. Protein synthesis continued until the intracellular pool became exhausted at which time the rate of fermentation significantly diminished. Less than 80 g sugar/l had been consumed in 190 hours at 24°C.

Thus, it is clear that an adequate rate of protein synthesis is necessary to maintain an adequate rate of sugar catabolism. Since the intracellular pool of ammonium appears to stimulate protein synthesis, any factors affecting its regeneration will ultimately compromise fermentation rate. It is, therefore, important that cell transport function is optimized through adequate use of oxygen or its substitutes as well as ensuring an adequate supply of assimilable nitrogen.

Ammonium stimulation of fermentation activity may be related to its preferential accumulation by the cell and its central role in anabolic metabolism.

Apart from stimulating protein and nucleic acid synthesis, and hence, maintaining high activity of rapidly turned-over metabolic proteins, several other functions have been suggested. Evidence for the stimulation/regulation of glycolysis by ammonium was suggested by a transitory increase in the evolution rate of CO_2 when it was added to fermenting cells (Lloyd, Kristenson and Degn, 1983). This notion is supported by the observation of Yoshino and Murakami (1982) that ammonium ions activated phosphofructokinase and pyruvate kinase in permeabilized cells. In addition, at high concentrations of ammonium, Slaughter and Housten (1978) and Devine and Slaughter (1980) found a greater efficiency of ethanol production at the expense of higher alcohol production in glucose fermenting cells. In a systematic study of biomass production and fermentation stimulants, Monteiro and Bisson (1991) found that fermentation activity responded to ammonium to a much greater extent than to salts and trace elements, whereas the reverse was the case for biomass yield.

Predictive approaches

Several biological approaches have been recently proposed for predicting low-nitrogen related fermentation faults.

As mentioned already, Monteiro and Bisson (1991) have described a bioassay for determining the assimilable nitrogen fraction of grape must. Comparison of the ammonium ion equivalent of different musts determined by bioassay with that of the amino acid content was generally good. For example, Chenin Blanc gave a correlation coefficient (r^2) of 0.81, French Colombard 0.85, but Grenache gave only 0.63. While it might be anticipated that the generally regarded better nitrogen sources would be highly correlated with yeast growth, that was not necessarily the case. Furthermore, these apparent discrepancies were associated with the variety of must. Evidence suggested that each juice exhibited specific deficiencies in other nutrients which influenced the efficiency of utilization of certain amino acids, and hence affected yeast growth. In addition, ethanol concentration and pH were shown to influence the correlations between individual amino acids and yeast growth, indicating that these factors can significantly affect amino acid utilization. Clearly, such a test has the potential to provide valuable information on the nutrient status of a must, thereby predicting fermentation problems.

A second bioassay for detecting must nitrogen deficiency is based on a laboratory fermentation system that automatically measures weight loss due to CO_2 evolution (Bely *et al.*, 1990a). A most important practical relationship was noted between maximum rate of CO_2 production and fermentation duration. The normalized CO_2 production rate, while peaking early in fermentation as for normal fermentations, decreased more rapidly in the case of sluggish fermentations. Because maximum CO_2 production rate can be determined during the first third of the fermentation duration, measurement of this parameter provides an early prediction of sluggish or incomplete fermentation.

Bely *et al.* (1990a) observed a departure from linearity of the slope of maximum CO_2 production rate when plotted against must assimilable nitrogen.

This point of inflexion, which occurred at about 140 mg FAN/l, corresponded to a maximum CO_2 production rate of 1.3 g CO_2/l per hour at 24°C. Fermentations giving a maximum production rate less than this threshold value became protracted or incomplete. These threshold values were confirmed by determining the reduction in fermentation duration as a function of nitrogen additions made to nitrogen deficient musts. Below about 140 mg FAN/l large reductions in fermentation duration were obtained. Above the threshold value only relatively small reductions in fermentation duration were observed, while above 300 mg FAN/l no significant reductions were obtained (Bely *et al.*, 1991). An inflexion was also noted at a maximum CO_2 production rate of 1.25 g CO_2/l per hour when reduction in fermentation time, in response to nitrogen addition, was plotted as a function of maximum CO_2 production rate. The minimum and optimum assimilable nitrogen values agree well with other determinations (Table 4.7).

This method for detecting sluggish fermentation has several important advantages. The test conditions are similar to actual commercial conditions, and can predict a nitrogen deficiency during the first third of fermentation duration. The test monitors fermentation activity, not just nitrogen content or yeast growth, this being especially important in the light of the fact that growth and fermentation are stimulated by nitrogen and other nutrients in different ways. In addition, the quantity and quality of must nitrogen varies considerably, further eroding the relationship between assimilable nitrogen and yeast growth.

Prevention

As mentioned already, the addition of diammonium phosphate is a widely used procedure for ensuring complete fermentation, being first recommended for enological use by Audounaud in 1898 (cited by de Castella, 1909). In this context, there are two important variables; amount added and timing of addition.

From an investigation of the effect of ammonium supplementation on rate and duration of fermentation of 50 grape musts ranging in FAN and ammonium concentrations from 406–1116 mg N/l and soluble solids of 15.5–25.7°Brix, Vos *et al.* (1980) established an optimal FAN to °Brix ratio of 43.9 for maximal fermentation rate. This corresponds to an initial must nitrogen concentration of 800–900 mg FAN/l required for maximum fermentation rate as shown by Agenbach (1977). Calculation of the concentration of ammonium needed for maximum fermentation rate is given by the formula:

$$NH_3\ (\text{mg N/l}) = \frac{43.9 - \text{FAN/°Brix}}{0.108}$$

The data of Bely *et al.* (1991) showed little reduction in fermentation duration to nitrogen supplementation when the initial nitrogen concentration of musts exceeded 300 mg FAN/l. Although the addition of higher nitrogen concentrations could sustain higher fermentation rates, the time for complete fermentation was not greatly reduced and there is a risk that the wine will contain higher concentrations of residual nitrogen leading to decreased microbiological stability.

Bely *et al.* (1990a) have established some important principles with respect to timing of nitrogen addition. Neither the timing of addition nor the number of additions influenced the effectiveness of nitrogen supplementation, providing that it was made before the mid point of fermentation. Effectiveness progressively decreased as nitrogen was added later during the fermentation, leading Bely *et al.* (1991) to recommend addition near the mid-point of fermentation. The response to nitrogen addition could be monitored by the maximum rate of CO_2 production, allowing adequate time for a suitable addition to be made for achieving a useful reduction in fermentation time. The observations that complex nitrogen sources stimulate growth and fermentation to a greater extent than ammonium when added later in fermentation (Monk, 1982; Monteiro and Bisson, 1991) suggests that other nitrogen compounds should be evaluated for management of sluggish or stuck fermentations. This difference in response may be related to the expression of nitrogen transport systems later in fermentation that are less inhibited by ethanol as discussed in a previous section. The consequences of these treatments on wine aroma would need to be examined.

Fermentation rate was noted to increase shortly after the addition of nitrogen, irrespective of the timing of addition (Salmon, 1989). An important practical consideration for isothermal fermentation is that the increase of the instantaneous fermentation rate was greatest at the start of fermentation, but decreased as fermentation proceeded. Split additions or a single addition at the mid point of fermentation would place least stress on energy demand of the cooling system.

Oxygen and the oxygen substitutes, ergosterol and unsaturated fatty acids, are strong stimulants of fermentation (David and Kirsop, 1972, 1973; Houtman and du Plessis, 1981; Strydom *et al.*, 1982; Lafon-Lafourcade and Ribéreau-Gayon, 1979; Larue *et al.*, 1982; Kunkee, 1991). Ingledew and Kunkee (1985) showed that oxygen not only increased the efficiency of sugar catabolism but also of nitrogen utilization (Table 4.8). In a nitrogen-limited must, under anaerobic conditions nitrogen utilization (65.7%) and yeast growth (about 1.3×10^7 cells/ml) were low, whereas under aerobic conditions FAN utilization (87.9%) and growth (about 3×10^8 cells/ml) were highly stimulated. Although the cells from the aerobic fermentation could be expected to have a lower nitrogen content, fermentation rate was much greater and nearer completion. Supplementation of the low nitrogen must with the oxygen substitutes, ergosterol and Tween 80® (oleic acid source) and yeast extract (287 FAN/l) under anaerobic conditions produced similar nitrogen utilization, cell density and fermentation rate as for the aerobic fermentation supplemented with yeast extract alone. The role of sterols in transport function is not clear but unsaturated fatty acids activate certain amino acid transport systems (Cartwright *et al.*, 1989; Henschke and Rose, 1991). Ingledew and Kunkee (1985) concluded that, in order to reduce problems with sluggish and stuck fermentations, it is necessary to use a high rate of inoculation (10^7 cells/ml) with a highly active starter culture (aerobically propagated), add the maximum allowable concentration of assimilable nitrogen, and aerate the fermentation during the initial stage.

Strain selection

The genetic diversity in the regulation of nitrogen uptake and its metabolism predicts that yeast strains will vary in their demand for both total and individual nitrogen compounds. Different combinations of nitrogen sources will yield differences in growth and fermentation rates, and biomass formation. It should, therefore, be possible to select or breed strains for optimal fermentation activity under nitrogen limited conditions.

As noted already, Jiranek *et al.* (1990) identified wide variability in the nitrogen requirements (330–470 mg N/l) of nine wine yeasts when fermenting a synthetic grape juice medium in which all amino acids were present in excess. Furthermore, the palate of amino acids utilized depended on the strain and fermentation conditions but were broadly similar (Jiranek, 1992). In a further study, Jiranek *et al.* (1991) compared the fermentation rates of four different nitrogen-requiring strains (330–470 mg N/l) in moderate (390 mg N/l) and low nitrogen (78 mg N/l)-containing media. Maximum fermentation rates were 4–5 g glucose/l per hour and 0.7–1.25 g/l per hour in the two media at 25°C, respectively, the rates being approximately proportional to initial nitrogen concentration. However, strain differences in fermentation rate were greater under conditions of limited nitrogen availability than with moderate availability. One strain, AWRI 72, fermented at a rate nearly twice that of the slowest strain, despite no significant differences in yeast growth rate or biomass yield. This observation encourages the search for more-nitrogen-efficient yeasts and the development of better strains by genetic techniques as one means of reducing the incidence of sluggish fermentations of nitrogen deficient musts. Such an approach would also reduce the need for nitrogenous yeast foods.

Thus, strains vary in the efficiency of nitrogen utilization in terms of growth and fermentation rate at optimum and limiting concentrations of nitrogen. The mechanism of enhanced fermentation efficiency of some strains under nitrogen limitation is not clear, but the key rate limiting step(s), such as sugar transport (Busturia and Lagunas, 1986), may be more stable to inactivation by nitrogen deficiency. Strain differences in ATPase activation by nitrogen starvation (Benito *et al.*, 1992) may exist. Differences may also exist in the ability to recycle or scavenge residual nitrogen. While in commercial terms several of the low nitrogen fermentations exhibited extremely sluggish rates, none of the fermentations failed to complete, that is, became stuck. Bezenger and Navarro (1988) and Bely *et al.* (1990a) also failed to observe incomplete sugar catabolism resulting from a nitrogen deficit, which contrasts with the findings of Busturia and Lagunas (1986) and Salmon (1989). The former investigations employed air-saturated musts suggesting that nitrogen deficiency may not in itself lead to a complete inhibition of sugar catabolism, but may do so in combination with a membrane lipid deficiency. It should be noted that only the fermentation conducted at less than 37 mg N/l by Salmon (1989) appears to have stuck. This could be regarded as an extreme case of nitrogen deficit (Monk, 1982; Ingledew and Kunkee, 1985; Gockowiak and Henschke, 1992).

Analysis of the data of Ough *et al.* (1991) highlights the variation in fermentation activity of yeast strains in response to nitrogen (Table 4.13). The strong fermentation yeasts, Prise de Mousse and Montrachet, show a significant dependence on must nitrogen concentration, whereas the poorer fermentation strains appear not to be affected by nitrogen, at least over the range 373–706 mg N/l. Of the major nitrogen sources present in the musts examined, arginine but not ammonium concentration was highly correlated with fermentation rate; this may reflect the higher mean contribution to nitrogen by arginine, being 279 compared with 97 mg N/l. This contrasts with an earlier report by Ough and Kunkee (1968) in which ammonium concentration correlated with fermentation rate.

Comparison of fermentation rates (Table 4.13) with the concentration of nitrogen assimilated (Table 4.9) shows a direct relationship. Montrachet is exceptional in having the greatest nitrogen demand but a comparatively lesser fermentation rate; this is to be expected in view of the complex regulation of sugar and nitrogen accumulation and metabolism. On the basis of these limited data, it can be concluded that the lower nitrogen requiring yeasts also possess lower fermentation rates, suggesting that they would not be suitable for the fermentation of nitrogen limited musts. This was observed in the lowest nitrogen must (141 mg N/l) studied. Therefore, strain evaluation must be done under nitrogen limiting conditions, but other factors, such as, oxygen and the quality of nitrogen source is likely to be important.

TABLE 4.13 Correlation (r^2) between average fermentation rate (°Brix/day) and assimilable nitrogen, arginine or ammonium concentration (mg N/l) of grape juice[a].

Nitrogen source[b]	*Total assimilable nitrogen*				*ARG*	NH_4^+
Strain	*15.5°C*		*20°C*		*15.5°C*	*15.5°C*
	Ferm. rate	r^2	*Ferm. rate*	r^2	r^2	r^2
Epernay 2	1.35	0.039	2.07	0.359	0.227	0.039
Flor	1.94	0.179	3.06	0.030	0.148	0.170
Prise de Mousse	2.41	0.403	3.58	0.490	0.204	0.235
Montrachet	2.17	0.829	3.37	0.772	0.571	0.720
Mean	1.97	0.789	3.02	0.773	0.850	0.209

[a]Fermentation rates were calculated from the data of Ough *et al.*, (1991); the nine grape juices had a mean of 473 and ranged 373–706 mg assimilable N/l.
[b]Nitrogen component(s) considered in determining correlation coefficients.
($r^2 > 0.399$ significant at $P = 0.05$).

Yeast Flavor Compounds

The secondary metabolites of grape are responsible for the principal flavor compounds in grape must, and provide the basis of 'varietal character' (Schreier, 1979; Baumes *et al.*, 1986; Williams, Sefton and Wilson, 1988; Rapp and Versini, 1991). Fermentation increases the chemical and flavor complexity of wine by assisting the extraction of compounds from solids present in grape must, modifying some grape-derived compounds, and producing yeast metabolites (Figure 4.6). For convenience, the yeast derived flavor compounds can be placed into three groups: the major volatile products of fermentation; the trace 'fermentation bouquet' compounds; and the undesirable or 'negative' aroma compounds.

The major volatile products of yeast metabolism, ethanol, glycerol and carbon dioxide are only indirectly affected by nitrogen metabolism. These compounds make a relatively small, but, nonetheless, fundamental contribution to wine flavor. The main groups of compounds that form the 'fermentation bouquet', are the organic acids, higher alcohols and esters and, to a lesser extent, aldehydes, and are influenced to various degrees by the nitrogen source (Rapp and Versini, 1991). The catabolism of amino acids to yield the keto-acids and their corresponding alcohols provides an example of the direct involvement of nitrogen metabolism in this group. The most deleterious 'negative' aroma compounds are the reduced sulfur compounds, hydrogen sulfide, organic

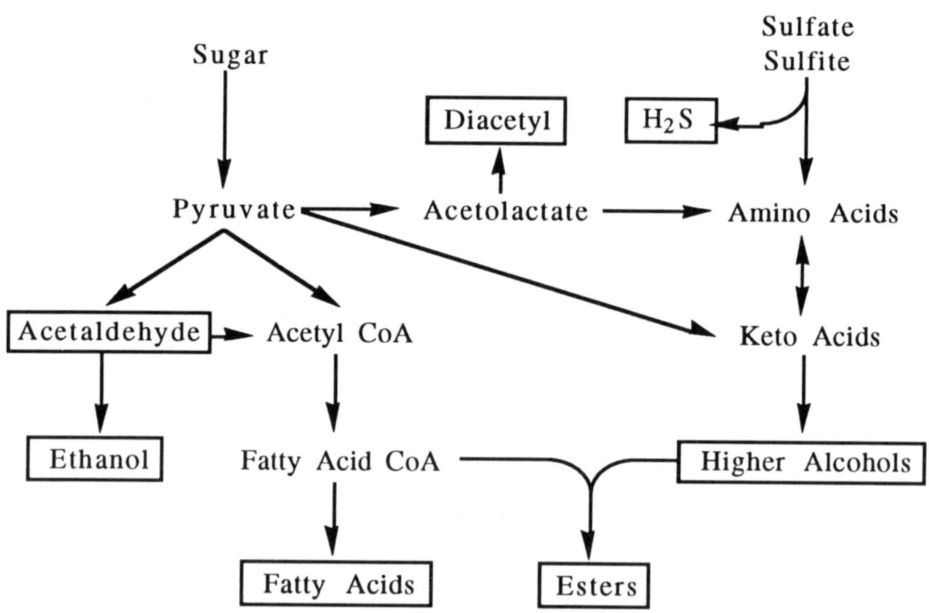

FIGURE 4.6 Derivation of flavor compounds from sugar, amino acids and sulfur metabolism by yeasts.

sulfides and thiols (mercaptans). Hydrogen sulfide formation is regulated by nitrogen availability (Vos and Gray, 1979; Monk, 1982; Henschke and Jiranek, 1991). When present in excess concentrations, the fermentation bouquet compounds, acetaldehyde, acetic acid, ethyl acetate, higher alcohols, and diacetyl included, may also be regarded as undesirable.

The significance of yeast metabolites in the aroma of wine has been a driving force in the selection of strains incapable of producing negative aroma compounds (Rankine, 1968, 1977; Zeeman *et al.*, 1982; Romano *et al.*, 1985; Henschke, 1990; Subden, 1991), and determining the factors which affect their formation. The aroma profile produced by yeast strains appears to be less important to wine flavor than is the case for brewing. Beer flavor is profoundly affected by the balance of esters, higher alcohols, organic acids and diacetyl. These compounds, however, do seem to be more important in young fruity wines, such as 'nouveau' style, and neutral flavored table wines. Isoamyl acetate, while not present in the must, is the main compound responsible for the characteristic fruity aroma of wine made from *Vitis Vinifera* cv. Pinotage (van Wyk *et al.*, 1977).

General aspects of yeast metabolite formation

Nitrogen compounds affect the sensory properties of wine by either stimulating yeast growth and fermentation activity, hence concomitantly increasing metabolite formation, or more directly through the provision of metabolite precursors or the regulation of their formation. Alternatively, nitrogen compounds themselves, such as the products of cell autolysis, may contribute to or modify wine flavor (Feuillat and Charpentier, 1982; Miller *et al.*, 1985; Voilley *et al.*, 1988; Stuckey *et al.*, 1991). Assigning a relative importance to these mechanisms is difficult since few systematic or model investigations appear to have been made (Ribéreau-Gayon and Peynaud, 1966, 1975; Nykänen, 1986; Rapp and Versini, 1991).

The affect of must nitrogen on the sensory properties of wine has been studied indirectly by monitoring the effects of nitrogen fertilizer application to vines (Samvelyan, Khachatryon and Martirosyan, 1972; Morozov, 1973; Sinton *et al.*, 1978; Bell, Ough and Kliewer, 1979; Bravdo and Hepner, 1987; Dukes *et al.*, 1991; Spayd *et al.*, 1991) and directly, by supplementing grape must with assimilable nitrogen (Agenbach, 1977; Vos, Zeeman and Heyman, 1978; Vos, 1981; Tromp, 1984). Nitrogen fertilizer applications generally increase the assimilable nitrogen content of grape juice, though this is dependent on soil attributes, fertilization regime and physiological aspects of the vine. According to published accounts, wines arising from nitrogen fertilized vines are preferred to those from the non-fertilized reference controls, but the effects of long term fertilization or of trials conducted on rich soils is not well known.

Sinton *et al.* (1978) obtained a high correlation (r^2=0.750) between must nitrogen content and wine taste intensity, irrespective of aroma type. In a study of some 50 juice/wine pairs over two vintages, Vos *et al.* (1978) and Vos (1981) recorded a strong positive relationship between must nitrogen (FAN), fermentation rate, and wine sensory quality, established by preference ranking or the scoring of wine attributes. Once musts attained a nitrogen content exceeding 800 mg N/l, either naturally or by supplementation with diammonium phosphate, no further increase in fermentation rate or wine quality was observed. A wide variation in sensory quality was nonetheless seen for wines produced from musts having similar nitrogen contents. Tromp (1984) observed a reduction in total fermentation time of 13.5 days from 65.5 days and a 10.3% increase in wine sensory rating in response to the addition of 250 mg N/l (as diammonium phosphate) to a must containing 270 mg N/l α-amino nitrogen and grape solids. When the solids were omitted from the must, nitrogen supplementation only decreased fermentation time by 6.2 days, and no improvement in wine quality (1.9%) was observed. Overall, little can be concluded with certainty from these investigations, except that must nitrogen may be involved in enhancing wine flavor under certain circumstances, and that systematic studies are needed to explore this possibility.

The affect of α-amino nitrogen compounds, especially ammonia and amino acids, on the formation of the major fermentation by-products has been studied in model media containing single or groups of compounds (summarized by Ribéreau-Gayon and Peynaud, 1966; Nykänen, 1986; Rapp and Versini, 1991). Both the type and concentration of nitrogen source(s) influenced the formation of the principal by-products, glycerol, acetic acid, acetaldehyde and 2,3-butanediol. Differences in the growth efficiency of sole nitrogen sources together with their influence on the composition of the cellular nitrogen pool (Watson, 1976), may in part, explain the observed changes in the fermentation products. These data serve to highlight the importance of individual amino acids, but, extrapolation to the more complex enological fermentation is difficult.

The specific influences of nitrogen on accumulation of the various groups of compounds constituting the fermentation bouquet will be dealt with briefly. The principal correlations reported between must nitrogen content and aroma compounds are positive for acetates, ethyl esters, and 1-propanol, and negative for the remaining higher alcohols (Rapp and Versini, 1991).

Keto-acids and higher alcohols

The principal higher alcohols produced by yeast are the aliphatic alcohols n-propanol, isobutanol (2-methyl-1-propanol), active amyl alcohol (2-methyl-1-butanol), isoamyl alcohol (3-methyl-1-butanol), and the aromatic alcohols hexanol and 2-phenylethanol (Table 4.14). Although, they exhibit a harsh, unpleasant aroma, at the concentrations generally found in wine, below 300 mg/l, they are usually considered desirable (Rapp and Versini, 1991). Yeast strain (Peynaud and Guimberteau, 1962; Rankine, 1967; Simpson, 1979; Zeeman *et al.*, 1982; Cabrera *et al.*, 1988), yeast growth (Quain, 1988), ethanol production (see Cabrera *et al.*, 1988), fermentation temperature (Ough *et al.*, 1966; Rankine, 1967; Ciolfi, Castina and Di Stefano, 1985), must pH (Rankine, 1967), aeration (Guymon, Ingraham and Crowell, 1961; Crowell and Guymon, 1963) and level of solids

TABLE 4.14 Metabolites of amino acids and related compounds.

Amino acids	*Keto acids*	*Aldehydes*	*Alcohols*
Alanine	2-Ketopropionic acid	Acetaldehyde	Ethanol
Serine	3-Hydroxy-2-keto-propionic acid	Glyoxal	Glycol
2-Amino-butyric acid	2-Ketobutyric acid	Propionaldehyde	1-Propanol
–	–	Butyraldehyde	1-Butanol
Valine	2-Ketoisovaleric acid	Isobutyraldehyde	2-Methyl-1-propanol
Isoleucine	2-Keto-3-Methyl-valeric acid	2-Methylbutyr-aldehyde	2-Methyl-1-butanol
Leucine	2-Ketoisocaproic acid	Isovaleraldehyde	3-Methyl-1-butanol
–	–	Hexanal	1-Hexanol
Phenylalanine	3-Phenyl-2-keto-propionic acid	–	Phenethyl alcohol
Tyrosine	3-(4-hydroxyphenyl)-2-ketopropionic acid	–	Tyrosol
Tryptophan	–	–	Tryptophol

(Crowell and Guymon, 1963; Groat and Ough, 1978; Klingshirn, Liu and Gallander, 1987) affect the production of each alcohol in different ways. Furthermore, grape variety and maturity influence their concentrations (Rankine, 1967; Rankine and Pocock, 1969; Cabrera *et al.*, 1988), most likely through qualitative as well as quantitative differences in amino acid composition (Sinton *et al.*, 1978; Ough and Bell, 1980; Vos, 1981; Herraiz *et al.*, 1989; Rapp and Versini, 1991).

When the amino acids shown in Table 4.14 are catabolized by the Ehrlich mechanism for their nitrogen component, the carbon skeletons, formed by trans- or de-amination, are excreted. The resultant α-keto acids may be excreted directly or decarboxylated to the corresponding aldehyde and then reduced to the alcohol. This latter process appears to fulfil an important metabolic role in fermentation, that is, 'fine-tuning' the reoxidation of NADH (van Dijken and Scheffers, 1986; Quain, 1988). In the absence of the appropriate amino acids, as may occur at the commencement of fermentation, the α-keto acids and corresponding higher alcohols are formed from sugar metabolism via pyruvate (Chen, 1978; MacDonald *et al.*, 1984; Nykanen, 1986).

In a simple medium containing ammonium as the sole nitrogen source, higher alcohols are formed at relatively low concentrations by the anabolic pathway (Herraiz *et al.*, 1989). Supplementation with precursor amino acids, or growth in complex media, produces a substantial increase in the concentration of the higher alcohols, excluding n-propanol, due to their additional formation by the Ehrlich mechanism (Herraiz *et al.*, 1989). Progressive increases in the assimilable nitrogen content of the medium reduce the concentration of total higher alcohols. For example, Ough and Bell (1980) showed that increasing must nitrogen content over the range 287–766 mg N/l decreased the concentration of total higher alcohols, isobutyl, isoamyl and active amyl alcohols. This was confirmed by Vos (1981), who also showed a decrease of hexanol and phenylethanol. Conversely, n-propanol was positively correlated to the total nitrogen content of must in both studies.

The inverse relationship between must assimilable nitrogen content and the concentration of total higher alcohols may result from nitrogen repression of transamination, and hence reduced activity of the Ehrlich pathway (Large, 1986). n-Propanol is primarily formed from sugar metabolism, by the condensation of pyruvate and acetyl CoA, since its Ehrlich mechanism precursor, α-aminobutyric acid, does not occur in must. Fermentation studies using dual-labelled amino acids and sugars have demonstrated that no simple relationships exist between concentration of the parent amino acid and the corresponding alcohol (Chen, 1978). Nevertheless, availability and timing of amino acid uptake will influence higher alcohol formation and affect wine flavor (Ribéreau-Gayon and Peynaud, 1960).

Fatty acids and esters

The short chain fatty acids, acetic, propanoic and butanoic, and lactic acid, are by-products of fermentation. Quantitatively, acetic acid is the most important, ranging from <0.2 to >2 g volatile acidity/l (acetic acid is the major component of volatile acidity) with a mean of 0.54 g/l (n=1650) in Australian table wine (J. M. Eglinton, unpublished data). Depending on the style of wine, acetic acid becomes objectionable near its flavour threshold of 0.7 to 1.1 g/l (Corison *et al.*, 1979). Strain of yeast and extracellular lipids are the main factors affecting its concentration in wine (Shimazu and Watanabe, 1981; Henschke and Dixon, 1990; Delfini *et al.*, 1989). The longer straight-chain fatty acids (C8 to C12) are intermediates in the formation of long chain acids (C16 and C18) which are structural components of cell membranes. Concentration of the shorter chain acids increases while the longer chain acids decreases during fermentation. Few studies report an influence of assimilable nitrogen on fatty acid accumulation in wine. Three reports show an inverse relationship on acetic acid accumulation (Agenbach, 1977; Tromp 1984; Dukes *et al.*, 1991). Similarly, the concentration of succinic acid decreases while that of lactic and tartaric acid increased (Dukes *et al.*, 1991). Malic and citric acid concentrations were not significantly affected. No explanations were given.

Two major groups of esters are formed during fermentation, the ethyl esters of straight-chain fatty acids and acetates of the higher alcohols. Some of the fatty-acid CoA derivatives which react with malonyl-CoA during chain elongation, take part in ester formation. Reaction with an alcohol forms the fatty acid ester. Ester formation is related to the availability of the alcohol and the yeast demand for fatty acid synthesis (Nordström, 1964; Nykänen, 1986). Its formation is positively correlated with must total nitrogen (Bell *et al.*, 1979), as are the ethyl esters of the longer chain fatty acids, ethyl hexanoate, ethyl octanoate and ethyl decanoate (Ough and Lee, 1981). The two acetate esters, isoamyl and hexyl acetate were also positively correlated while phenylethyl acetate was negatively correlated with must nitrogen. It is generally believed that esters contribute most to the desirable aspects of the fermentation bouquet of wine and account much for the greater aroma intensity of wines derived from musts of higher nitrogen content. Must derived from nitrogen fertile vineyards show a lesser variation in wine ester content (Houtman, Morais and du Plessis, 1980). The shorter chain fatty acid ethyl esters generally contribute more to wine aroma than the less volatile longer chain esters (Simpson, 1979). The concentrations of acetate esters are relatively more variable in wine, being more dependent on yeast strain and sugar concentration than the fatty acid esters.

Carbonyl compounds

The carbonyl compounds, acetaldehyde, other short chained aliphatic aldehydes and diacetyl constitute the most volatile aroma fraction produced by yeast during fermentation. Acetaldehyde and diacetyl are the most important sensory carbonyl compounds produced, typically ranging from 10–200 mg/l and 0.05–5 mg/l wine, respectively. Acetaldehyde is characterized by a sour, green apple aroma and diacetyl by a buttery aroma. The aliphatic carbonyls are intermediates in the formation of higher alcohols from amino acids and sugar (Table 4.14). Conversely, the aliphatic carbonyls are intermediates in the biosynthesis of branched chain amino acids from sugar metabolism. Diacetyl is formed by the decarboxylation of 2-acetolactate, and hexanal is derived from C6 lipid metabolism. As for the aliphatic higher alcohols, the availability of the respective amino acid affects the formation and excretion of the related aliphatic aldehyde; a nitrogen deficiency increasing aldehyde content (Dukes *et al.*, 1991) by way of increased 2-keto acid formation. Appreciable concentrations of carbonyl compounds are only found during the most active phase of fermentation, and are reduced to alcohols as the source of fermentable sugar becomes depleted. Yeast strain, fermentation temperature, aeration and sulfur dioxide are also important factors affecting formation of carbonyl compounds (Nykänen, 1986; Rapp and Versini, 1991).

Diacetyl synthesis is correlated with the availability of the amino acids, specifically valine and threonine (Dufour, 1989; O'Connor-Cox and Ingledew, 1989). In low nitrogen media, valine synthesis is activated leading to formation of the intermediate, 2-acetolactate, which spontaneously forms diacetyl by an oxidative decarboxylation. Providing yeast is still active while diacetyl is formed, it is normally reduced to 2,3-butanediol via acetoin which has a low aroma threshold. Under conditions of nitrogen sufficiency, threonine, a potent repressor of valine uptake, reduces diacetyl precursor formation.

A nitrogenous carbonyl compound, 2-acetyltetrahydropyridine, which has a very low aroma threshold (1.6 ng/l), is responsible for the mousy taint in wine

(Heresztyn, 1986). It is produced by strains of at least five species of *Dekkera* (*Brettanomyces*) (P. R. Grbin, unpublished data) from the metabolism of lysine and ethanol by an unknown mechanism. The dependence on lysine suggests that fermentation yeasts having a high lysine demand may be useful in limiting formation of the taint.

Hydrogen Sulfide

Hydrogen sulfide (H_2S), an unpleasant aroma compound with a low sensory threshold (10–100 μg/l; Acree, Sonoff and Spittstoesser, 1972; Rankine, 1963), can be formed in excessive amounts by yeast during the fermentation of grape must. Of the several mechanisms accounting for its formation (Monk, 1986; Henschke and Jiranek, 1991; Rauhut, Chapter 6), only that involving yeast nitrogen metabolism will be discussed.

Wineries commonly observe two broad phases of H_2S production (Henschke and Jiranek, 1991). The first phase occurs during active fermentation and is responsive to supplementation with assimilable nitrogen. By contrast, the second phase is most frequently observed near to depletion of sugar from the must and proceeds in the presence of assimilable nitrogen. Formation of H_2S in the latter phase is potentially the more serious since little is lost by entrainment with fermentation gases.

Inorganic sulfur sources

As a product of the Sulfate Reduction Sequence (SRS), H_2S is an intermediate in the biosynthesis of all sulfur-containing compounds, including the amino acids, methionine and cysteine, and the methyl-group donor, *S*-adenosylmethionine. In the first of a series of regulated steps of this pathway, sulfate is accumulated from the medium, and then reduced to sulfite following two steps of activation with ATP (Figure 4.7). In the final step, sulfite is reduced by sulfite reductase to sulfide. At this point, sulfide is combined enzymatically with the nitrogen-containing carbon precursors, *O*-acetyl serine and *O*-acetyl homoserine, to form cysteine and ultimately methionine, respectively. OAS-OAH sulfhydrylase catalyzes both condensation reactions (reviewed by Yamagata, 1989).

The principal regulatory molecules acting on the SRS and sulfur-amino acid biosynthetic pathways are the end-products, cysteine and methionine (de Vito and Dreyfuss, 1964; Dott and Truper, 1978, 1979), and particularly the derivatives of the latter, *S*-adenosyl methionine (Cherest *et al*., 1973) and methionyl-tRNA (Cherest, Surdin-Kerjan and de Robichon-Szulmajster, 1971; Surdin-Kerjan, Cherest and de Robichon-Szulmajster, 1973; Yamagata, 1987). Principally by repression, cysteine and methionine effectively control their own metabolic pools. Since only limited amounts of cysteine (0–5 mg/l) and methionine (0–25 mg/l) are present in grape juice (Table 4.2A), the SRS will be activated to produce sulfide whenever there is a metabolic demand for these amino acids.

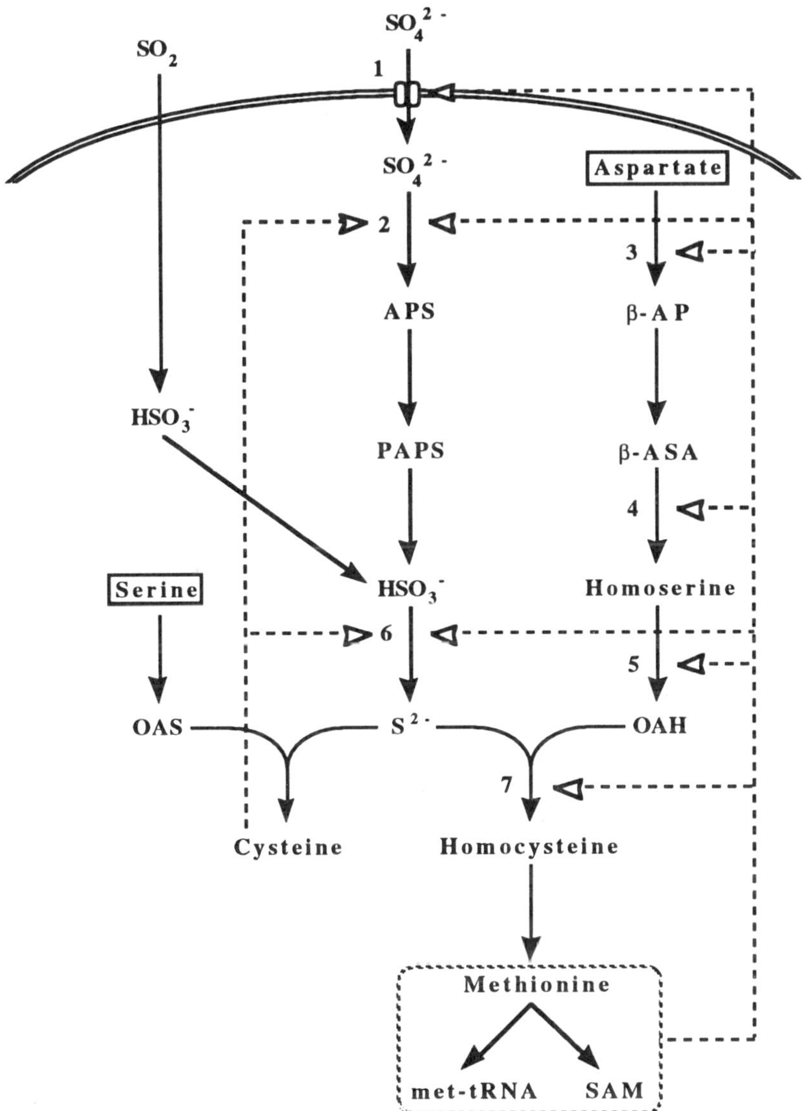

FIGURE 4.7 Repressive regulation of the Sulfate Reduction Sequence and methionine/cysteine biosynthetic pathways. Repression of enzyme synthesis ——> ; Nitrogen-containing compounds derived from the cytosolic nitrogen pool or exogenous sources are boxed (□); Enzymes: (1) sulfate permease; (2) ATP-sulfurylase; (3) PAPS kinase; (4) PAPS reductase; (5) serine acetyltransferase; (6) sulfite reductase; (7) OAS-OAH sulfhydrylase; (8) homocysteine methyltransferase; (9) aspartate kinase; (10) aspartate-semialdehyde dehydrogenase; (11) homoserine dehydrogenase; (12) homoserine acetyltransferase; Abbreviations: APS, adenosyl 5'-phosphosulfate; β-AP, β-aspartyl phosphate; PAPS, 3'-phosphoadenosyl 5'-phosphosulfate; β-ASA, β-aspartate semialdehyde; OAS, *O*-acetylserine; OAH, *O*-acetylhomoserine; SAM, *S*-adenosylmethionine; met-tRNA, methionyl-tRNA.

In the event of a general deficiency of nitrogen coinciding with a demand for protein synthesis, an intracellular depletion of nitrogen will ultimately result. A concomitant deficiency of the nitrogenous precursors of sulfur-amino acid biosynthesis, and in turn methionine, cysteine and their regulatory derivatives, will culminate in derepression of the SRS. Hence, H_2S production will be activated (Hartnell and Spedding, 1979; Vos and Gray, 1979).

Investigations by Stratford and Rose (1985) and Jiranek (1992) in grape juice media, show that when cells grow on sulfate under nitrogen starvation conditions, H_2S production is small and short lived. As an explanation of this observation we have proposed that although the SRS is fully derepressed, in the wine strains studied the flux of sulfate to sulfide is low (possibly to conserve ATP energy) (Henschke and Jiranek, 1991). This may occur by feedback mechanisms limiting sulfate accumulation and/or activation (de Vito and Dreyfuss, 1964; Hawes and Nicholas, 1973; Heinzel and Trüper, 1976), or the failure of sulfate uptake during nitrogen deficiency, analogous to that described for glucose transport (Busturia and Lagunas, 1986; Salmon, 1989).

A high and sustained rate of H_2S production is observed in response to a nitrogen deficiency when cells are grown in the presence of sulfite. Because sulfite freely diffuses into the cell (Stratford and Rose, 1986), it essentially by-passes all regulatory steps normally controlling its production and, therefore, supply to the reductase enzyme. Consequently, sulfite of exogenous origin will be directly reduced to sulfide whenever sulfite reductase is activated (Stratford and Rose, 1985; Jiranek, 1992). Under conditions of nitrogen starvation, sulfite reductase enzyme has a half-life of several hours (Jiranek, 1992). It is noteworthy that sulfite concentrations in musts are invariably high through its widespread use by winemakers or its microbial formation (Eschenbruch, 1973; Suzzi, Romano and Zambonelli, 1985), making this inorganic sulfur compound the principal precursor for sulfide formation. Furthermore, cellular sulfite, by taking the form of bisulfite at cytosolic pH, can exceed extracellular concentration by as much as 60-fold, thereby enhancing H_2S formation (Stratford and Rose, 1986).

Nitrogen supplementation rapidly suppresses the evolution of H_2S by nitrogen starved cells (Vos and Gray, 1979; Monk, 1982; Stratford and Rose, 1985; Henschke and Jiranek, 1991; Jiranek, 1992; Figure 4.8). The immediacy of response suggests that the most plausible mechanism involves sequestration of sulfide by replenishment of the nitrogen precursors, *O*-acetyl serine and *O*-acetyl homoserine (Jiranek, 1992). A long term control over H_2S formation is provided by end product repression of the SRS enzymes as described previously.

This mechanism is further supported by the H_2S production response of nitrogen starved cells to supplementation by individual amino acids as nitrogen sources. The readily assimilated amino acids were as effective as ammonium in suppressing H_2S production, while those which are poor nitrogen sources or are not degraded for nitrogen by *S. cerevisiae*, proline, lysine, histidine, glycine, valine, methionine, tryptophan and threonine, only partially suppressed production (Henschke and Jiranek, 1991). Threonine is known to regulate

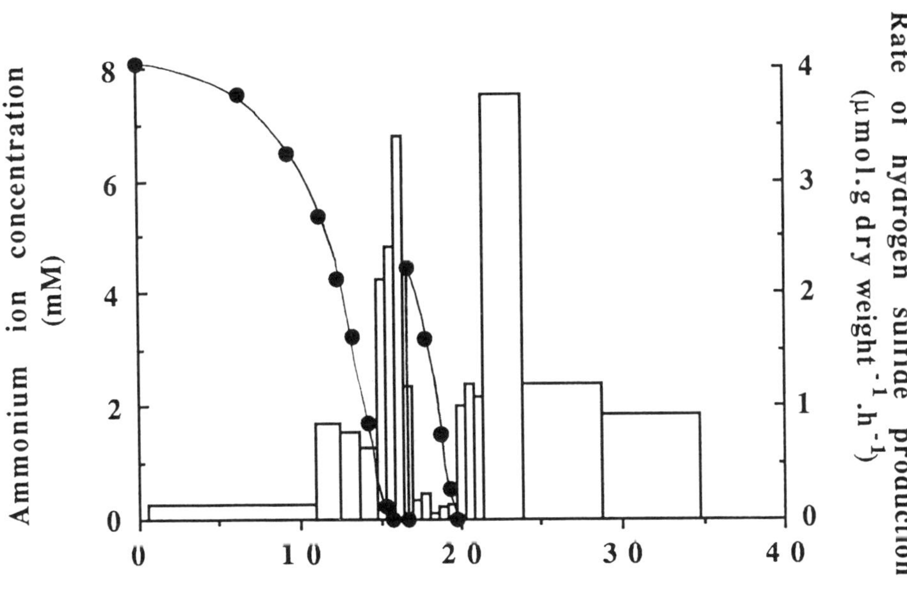

FIGURE 4.8 Relationship between ammonium ion concentration (o) and hydrogen sulfide production (histogram) by *Saccharomyces cerevisiae* AWRI 72 in an air-saturated chemically-defined grape juice medium (Table 4.4) at 25°C. The medium contained ammonium ions (8 mM) as sole nitrogen source and sodium metabisulfite (0.26 mM). Ammonium ions were added at 16.5 hours to a concentration of 4.15 mM. Fermentations were sparged continuously with a stream (300 ml/min) of oxygen-free nitrogen (Jiranek, 1992, with permission from the American Society for Enology and Viticulture).

homoserine synthesis, and cysteine is degraded to H_2S, possibly by cysteine desulfhydrase (Aida *et al.*, 1969; Tokuyama *et al.*, 1973). The ability of amino acids to contribute to the nitrogen pool and to supply the immediate precursors which combine with sulfide will prevent the formation of free H_2S.

Hydrogen sulfide formed at the end of fermentation shows little response to the addition of assimilable nitrogen. We have observed that, in some instances, aeration or supplementation with lipids during the latter stage of fermentation substantially reduced H_2S production. Together with the impaired ability of cells to accumulate nitrogen in the presence of a high concentration of ethanol (van Uden, 1989; Iglesias *et al.*, 1991), we have suggested that intracellular depletion of nitrogen may account for this observation (Henschke and Jiranek, 1991; Jiranek, 1992).

Strain variability in H_2S production is well known (Rankine, 1968; Eschenbruch *et al.*, 1978). Mutant strains of *S. cerevisiae* have been isolated that produce only trace amounts of hydrogen sulfide irrespective of the source of sulfur or the nitrogen status of the must. A unigenic trait was shown by genetic analysis, suggesting a leaky mutation for sulfite reductase activity (Zambonelli, Mutinelli and Pacchetti, 1975).

Organic sulfur sources

In the first report linking the depletion of grape must nitrogen with production of H_2S, Vos and Gray (1979) attributed its formation to the catabolism of sulfur-amino acids. The liberation of sulfide through the degradation of cysteine (Maw, 1965; Wainwright, 1971; Eschenbruch *et al.*, 1978) by yeast cysteine desulfhydrase (Aida *et al.*, 1969; Tokuyama *et al.*, 1973) had been frequently demonstrated and was assumed to be the mechanism responsible. Since the cysteine content of grape must is generally very low, Vos and Gray suggested that significant amounts of cysteine arose from the breakdown of must proteins by proteolytic activity elaborated by yeast in response to nitrogen starvation. Recent evidence does not support this hypothesis. Strains of *S. cerevisiae* do not appear to produce significant proteolytic activity (Lagace and Bisson, 1990), although the involvement of non-*Saccharomyces* strains has not been fully defined. Even complete proteolysis of must proteins would not release sufficient cysteine (Waters, 1991) to account for the levels of H_2S produced, and concomitantly released methionine could be expected to inhibit sulfide formation by way of repression of the SRS. Finally, H_2S production occurs in chemically defined media in the absence of protein.

Urea and Ethyl Carbamate

With recent advances in analytical techniques, it has become possible to detect and identify a myriad of compounds in wine which previously escaped detection. One such compound is ethyl carbamate (urethane). This compound occurs naturally in many fermented foods and beverages as the result of metabolic activity (Ough, 1976). Because ethyl carbamate is a suspected human carcinogen, regulatory authorities in several countries have recently imposed a voluntary restriction on the maximum concentration permitted in wine (Ough, 1991). This action has prompted investigations into the origin of ethyl carbamate during wine fermentation.

Precursor compounds

Carbamyl compounds are known to react with ethanol to form ethyl carbamate (Ough, 1976; Ough, Crowell and Gutlove, 1988); these compounds exist as metabolic intermediates in grape and yeast metabolism. The release of these compounds from autolyzing or ethanol killed cells as may occur during fortified wine production is one mechanism suggested for higher concentrations found in these wines (Monteiro, Trousdale and Bisson, 1989). Studies in model solutions established that urea, with a lesser contribution by citrulline and N-carbamyl α-amino acids, reacts with ethanol in acidic conditions to form ethyl carbamate (Ough, 1976; Ough *et al.*, 1988):

$$NH_2\text{-}CO\text{-}NH_2 + C_2H_5OH \rightarrow NH_2\text{-}CO\text{-}O\text{-}CH_2\text{-}CH_3$$

Approximately 0.8 mg ethyl carbamate/l is formed in a model solution of urea and ethanol (12% vol.) at pH 3.2 and 71°C after 48 hours. Freshly made wines do not contain significant concentrations of ethyl carbamate, but it forms at the rate of about 4.8 μg/year at 13°C from an initial concentration of 10 mg urea/l (Ough, Stevens and Almy, 1989). The remaining reactants in wine have not been verified (Ough *et al.*, 1990; Tegmo-Larsson and Henick-Kling, 1990).

In the past, an obvious source of urea in wine arose through the use of urea-containing yeast foods as a fermentation stimulant (Ingledew *et al.*, 1987). The most significant source of urea formed during the fermentation of grape must is the enzymic cleavage of arginine by arginase, which releases urea and ornithine (Ough, Crowell and Mooney, 1988; Figure 4.9). Monteiro *et al.* (1989) confirmed this mechanism using radiolabelled arginine. Urea may be reabsorbed by the cell *via* active and facilitated diffusion transport systems, subject to inhibition by good nitrogen sources, and further degraded to ammonium and CO_2 by urea carboxylase and allophanate hydrolase (Cooper and Sumrada, 1975; Davis, 1986).

Yeast strain

The importance of yeast strain in determining the formation of urea in wine has been established from surveys and trial fermentations. Based on a survey of over 400 wines produced in California from grape musts having assimilable nitrogen concentrations ranging from <1000 to >2000 mg/l, one yeast strain, Prise de Mousse, consistently produced low concentrations of urea (mean=1.37 mg/l; N=141), while Montrachet (mean=4.29 mg/l; N=55), French White (mean=8.03; N=15) and Pasteur (mean=6.39; N=57) produced high concentrations (Ough *et al.*, 1990). Red wines contained less urea than white wines for a given concentration of total assimilable nitrogen. This was explained on the basis that the major red wine variety in California, Cabernet Sauvignon, contains a lower ratio of arginine to proline compared with the major white variety, Chardonnay, (Huang and Ough, 1991) therefore reducing the propensity to form urea. The importance of strain has been confirmed by trial fermentations conducted in various juices containing a wide range of assimilable nitrogen concentrations (Table 4.15; Ough *et al.*, 1991). Except for the low urea producing strain, urea production increased with fermentation temperature. This may be the result of increased nitrogen metabolism at higher temperature (see Table 4.9).

Total nitrogen and arginine

Under usual fermentation conditions in musts with low to moderate concentrations of assimilable nitrogen (<1000 mg/l), the accumulation of arginine by yeast proceeds early during the growth phase with little or no urea excretion into the fermentation medium (Ough *et al.*, 1990; Ough, 1991; Table 4.16). In musts containing higher concentrations of assimilable nitrogen, yeasts tend to show an incomplete or delayed uptake of arginine and the urea-producing strains excrete urea into the fermentation medium (Ough *et al.*, 1990;

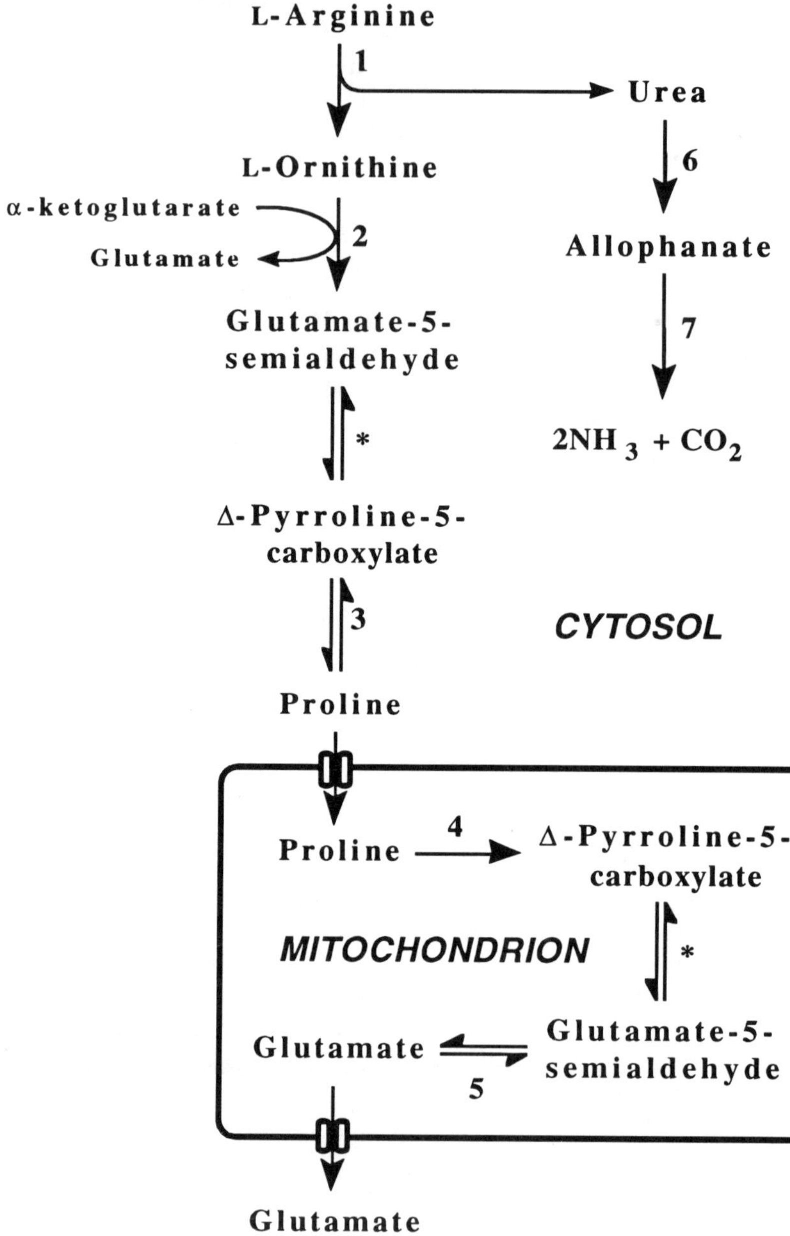

FIGURE 4.9 Arginine catabolism and urea formation in the yeast *Saccharomyces cerevisiae*.
Enzymes: (1) arginase; (2) ornithine aminotransferase; (3) Δ-pyrroline-5-carboxylate reductase; (4) proline oxidase; (5) Δ-pyrroline-5-carboxylate dehydrogenase; (6) urea carboxylase; (7) allophanate hydrolase; (*) spontaneous.

TABLE 4.15 Effect of yeast strain and temperature on accumulation of urea in wine (μM) and, correlation (r^2) between urea concentration and initial concentration (mM) of either total assimilable nitrogen, arginine (ARG), ammonium (NH_4^+), glutamine (GLN), alanine (ALA) or citrulline (CIT) of grape juice[a].

Nitrogen component[b] *(mM)*	*Total assimilable nitrogen*				*ARG*	*NH_4^+*	*GLN*	*ALA*	*CIT*
	22.0				*4.98*	*6.9*	*2.05*	*2.49*	*0.10*
Strain	*15.5°C*		*20°C*		*15.5°C and 20°C data combined*				
	Urea	r^2	Urea	r^2	r^2	r^2	r^2	r^2	r^2
Epernay 2	55.6	0.023	90.6	0.006	0.018	0.044	0.111	0.410	0.033
Flor	11.8	0.593	15.5	0.395	0.175	0.307	0.266	0.295	0.138
Prise de Mousse	11.1	0.539	10.1	0.590	0.359	0.658	0.434	0.229	0.257
Montrachet	29.8	0.429	37.6	0.417	0.198	0.285	0.356	0.359	0.154
Mean	27.1	0.058	38.5	0.034	0.040	0.269	0.385	0.693	0.017

[a]Calculated from the data of Ough *et al.*, (1991); Total assimilable nitrogen concentration of 11 grape juices ranged from 8.72 (141) to 30.34 mM (738 mg N/l). Mean urea concentrations are 1.7 and 2.4 mg/l at 15.5 and 20°C, respectively.
[b]Nitrogen component(s) considered in determining correlation with urea concentrations. ($r^2 > 0.332$ significant at $P = 0.05$).

Ough, 1991; Table 4.15; Table 4.16). Urea formation by Epernay 2, the highest producer, is not correlated with nitrogen concentration, while the other strains show different degrees of correlation. Epernay 2 and Montrachet tend to form large concentrations of urea early in fermentation which, to a greater extent for Epernay 2, are not remetabolized (Ough *et al.*, 1991). These yeasts may have impaired or readily inhibited urea transport or metabolic capabilities. On the other hand, urea production by Prise de Mousse and Flor correlate with must nitrogen (Table 4.15) and wine residual nitrogen concentrations (see Table 4.10; r^2=0.651 and 0.607 for the two strains, respectively). The known inhibition of urea transport and catabolism by good nitrogen sources may explain this occurrence (Cooper and Sumrada, 1975). It has therefore been suggested that the metabolism of arginine in the presence of high nitrogen content leads to urea excretion by susceptible strains, the remetabolism of which may only occur once the total assimilable nitrogen content becomes low later during fermentation (Henschke and Ough, 1991).

The interaction of other predominant grape amino acids with arginine and urea formation and metabolism is extremely complex as shown in Table 4.15. Glutamine is known to inhibit ammonium transport (Dubois *et al.*, 1974), for example, and ammonium inhibits arginine metabolism (Middelhoven, 1964). The inhibition of urea formation in a susceptible juice by high concentrations of added ammonium was suggested to operate by the inhibition of arginine catabolism (Henschke and Ough, 1991). Thus, the balance of these nitrogen sources in relation to arginine can produce complex results. At very high amino acid concentrations, yeast growth and arginine catabolism become inhibited resulting in urea excretion during the final stage of fermentation (Ough, 1991; Table 4.16).

Aerobic conditions

Initial aeration of arginine-rich musts also stimulates urea production by urea excreting strains (Henschke and Ough, 1991). The mechanism of action by oxygen has not been determined, but may relate to a more rapid accumulation of nitrogen needed for anabolic reactions that are stimulated by oxygen. The surplus urea is excreted and because of the prevailing rich nitrogen conditions, it is only partially remetabolized (Cooper and Sumrada, 1975). This suggestion was supported by the observation that intensive aeration lead to the early disappearance of the accumulated urea, possibly through increased demand for nitrogen. Under the conditions studied, addition of a high concentration of ammonium to the must suppressed urea excretion, through its inhibition of arginine metabolism (Middelhoven, 1964).

Future prospects

Although further work is needed to define more precisely urea and ethyl carbamate formation during fermentation and maturation, it is clear that the concentration of these compounds may be reduced by the choice of low-urea excreting strains, and the avoidance of stimulating yeast biomass production in

TABLE 4.16 The influence of must assimilable nitrogen concentration on the metabolism of arginine and excretion of urea by yeast. (Adapted from Ough, 1991).

Must nitrogen content	*Yeast strain*	*Arginine conc. (mg/l)*		*Urea conc. (mg/l)*	
		Juice	*Wine*	Maximum[a]	*Wine*
Low	UPS	200	Nil	Nil	Nil
Medium-high	71B	1500	100	Nil	Nil
	Montrachet	1500	Nil	6	Nil
	Epernay	1500	100	4	3
Very high	71B	1500	750	Nil	Nil
	Montrachet	1500	600	6	4
	Epernay	1500	750	3.5	3
Very high	UPS	2000	1600	3	3

[a]Maximum concentration of urea formed at any stage of fermentation.
UPS, Urea producing strain not specified.

musts containing a high concentration of arginine. The judicious use of ammonium salts, and possibly asparagine and glutamine, may also afford some protection (see Sumrada and Cooper, 1978). Furthermore, modification of the gene (*CAR1*) regulating arginase expression may be used to limit urea excretion (Sumrada and Cooper, 1982; Suizu *et al.*, 1990). Hydrolysis of wine urea by an acid urease treatment appears promising (Ough and Trioli, 1988).

REFERENCES

Achstetter, T. and Wolf, D.H. (1985) Proteinases, proteolysis and biological control in the yeast *Saccharomyces cerevisiae*. *Yeast*, **1**, 139–157.

Acree, T.E., Sonoff, E.P. and Spittstoesser, D.F. (1972) Effect of yeast strain and type of sulfur compound on hydrogen sulfide production. *American Journal of Enology and Viticulture*, **23**, 6–9.

Agenbach, W.A. (1977) A study of must nitrogen content in relation to incomplete fermentations, yeast production and fermentation activity. In *Proceedings of the South African Society for Enology and Viticulture* (Cape Town, Nov., 1977), pp. 66–88. Stellenbosch: South African Society of Enology and Viticulture.

Aida, K., Tokuyama, T. and Uemura, T. (1969) The role of cysteine desulfhydrase and cystein sythase in the evolution of hydrogen sulfide in pantothenic acid deficient yeast, In: *Proceedings of the Third International Symposium on Yeasts* (The Hague, 1969), *Antonie van Leeuwenhoek*, **35** Supplement.

Allen, M.S. and Auld, P.W. (1988) Stuck chardonnay ferments: experience in the Hunter and Mudgee regions. *The Australian Grapegrower and Winemaker*, **April**, 9–11.

Amerine, M.A. and Joslyn, M.A. (1970) Composition of grapes. In *Table Wines: The Technology of their Production*, 2nd edn., pp. 233–290. Berkeley, California: University of California Press.

Amerine, M.A. and Ough, C.S. (1963) Studies with controlled fermentation. VIII factors affecting aldehyde accumulation. *American Journal of Enology and Viticulture*. **15**, 23–33.

Amerine, M.A., Berg, H.W., Kunkee, R.E., Ough, C.S., Singleton, V.L. and Webb, A.D. (1980) The composition of grapes and wines. In *The Technology of Wine Making*, 4th edn. Westport, Connecticut: AVI Publishing Company.

Andreasen, A.A. and Stier, T.J.B. (1953) Anaerobic nutrition of *Saccharomyces cerevisiae*. I. Ergosterol requirement for growth in a defined medium. *Journal of Cellular and Comparative Physiology*, **41**, 23–36.

Andreasen, A.A. and Stier, T.J.B. (1954) Anaerobic nutrition of *Saccharomyces cerevisiae*. II. Unsaturated fatty acid requirement for growth in a defined medium. *Journal of Cellular and Comparative Physiology*, **43**, 271–281.

Anelli, G. (1977) The protein of musts. *American Journal of Enology and Viticulture*, **28**, 200–203.

Aries, V. and Kirsop, B.H. (1977) Sterol synthesis in relation to growth and fermentation by brewing yeasts inoculated at different concentrations. *Journal of the Institute of Brewing*, **83**, 220–223.

Ballarin-Denti, A., den Hollander, J.A., Sanders, D., Slayman, C.W. and Slayman, C.L. (1984) Kinetics and pH-dependence of glycine-proton symport in *Saccharomyces cerevisiae*. *Biochimica et Biophysica Acta*, **778**, 1–16.

Barillere, J.M., Bidan, P. and Dubois, C. (1983) Thermal resistance of yeasts and lactic acid bacteria isolated from wine (french). *Bulletin de l'Office International de la Vigne et du Vin*, **56**, 327–351.

Baumes, R., Cordonnier, R., Nitz, S. and Drawert, F. (1986) Identification and determination of volatile constituents in wines from different vine cultivars. *Journal of Food Science and Agriculture*, **37**, 927–943.

Becker, J.M., Naider, F. and Katchalski, E. (1973) Peptide utilization in yeast: Studies on methionine and lysine auxotrophs of *Saccharomyces cerevisiae*. *Biochimica et Biophysica Acta*, **291**, 388–397.

Bell, A.A., Ough, C.S. and Kliewer, W.M. (1979) Effects on must and wine composition, rates of fermentation, and wine quality of nitrogen fertilization of *Vitis vinifera* var. Thompson seedless grapevines. *American Journal of Enology and Viticulture*, **30**, 124–129.

Bell, S.-J. (1991) The effect of nitrogen fertilization on growth, yield, and juice composition of *Vitis vinifera* cv. Cabernet Sauvignon grapevines. In *Proceedings of the International Symposium on Nitrogen in Grapes and Wine* (Seattle, 1991) Edited by J. Rantz. pp. 206–210. Davis, CA.: American Society for Enology and Viticulture.

Bely, M., Sablayrolles, J.-M. and Barre, P. (1990a) Automatic detection of assimilable nitogen deficiencies during alcoholic fermentation in enological conditions. *Journal of Fermentation and Bioengineering*, **70**, 246–252.

Bely, M., Sablayrolles, J.-M. and Barre, P. (1990b) Description of alcoholic fermentation kinetics: Its variability and significance. *American Journal of Enology and Viticulture*, **41**, 319–324.

Bely, M., Sablayrolles, J.-M. and Barre, P. (1991) Automatic detection and correction of assimilable nitogen deficiency during alcoholic fermentation under enological conditions. In *Proceedings of the International Symposium on Nitrogen in Grapes and Wine* (Seattle, 1991) Edited by J. Rantz. pp. 211–214. Davis, CA: American Society for Enology and Viticulture.

Benito, B., Portillo, F. and Lagunas, R. (1992) In vivo activation of the yeast plasma membrane ATPase during nitrogen starvation. *Federation of European Biochemical Societies*, **300**, 271–274.

Bergner, K.G. and Haller, H.E. (1969) Das verhalten der freien aminosauren von weissenweinen in verlauf der garung, bei ausbau, lagerung und umgarung. *Mitteil Klosterneuburg*, **19**, 264–288.

Betz, H. (1976) Inhibition of protein synthesis stimulates intracellular protein degradation in growing yeast cells. *Biochemical and Biophysical Research Communications*, **72**, 121–130.

Bezenger, M.-C. and Navarro, J.-M. (1987) Influence de l'azote sur la fermentation alcoolique en milieu modele simulant les conditions de l'oenologie. *Sciences des Aliments*, **7**, 41–60.

Bezenger, M.-C. and Navarro, J.-M. (1988) Alcoholic fermentation: Model accounting for initial nitrogen influence. *Biotechnology and Bioengineering*, **31**, 747–749.

Bezenger, M.-C., Navarro, J.-M., Sablayrolles, J.M. and Barre, P. (1986) Prevision du deroulement en isotherme de la fermentation alcoolique d'un millieyu modele simulant les conditions de l'oenologie: influence des parametres azote et temperature. *Science des Aliments*, **6**, 185–200.

Bisson, L.F. (1991) Influence of nitrogen on yeast and fermentation of grapes. In *Proceedings of the International Symposium on Nitrogen in Grapes and Wines* (Seattle, WA,1991), edited by J. Rantz, pp. 172–184. Davis, CA: American Society for Enology and Viticulture.

Bizeau, C. (1963) Étude des facteurs limitant la croissance des levures dans les moûts de clairette, et de muscat blanc a petits grains. *Annales de Technologie Agricole*, **12**, 247–276.

Borst-Pauwels, G.W.F.H. (1981) Ion transport in yeast. *Biochimica et Biophysica Acta*, **650**, 88–127.

Bossinger, J., Lawther, R.P. and Cooper, T.G. (1974) Nitrogen repression of the allantoin degradative enzymes in *Saccharomyces cerevisiae. Journal of Bacteriology*, **118**, 821–829.

Boulton, R. (1980) The prediction of fermentation behaviour by a kinetic model. *American Journal of Enology and Viticulture*, **31**, 40–44.

Brandriss, M.C. and Magasanik, B. (1979a) Genetics and physiology of proline utilisation in *Saccharomyces cerevisiae*: enzyme induction by proline. *Journal of Bacteriology*, **140**, 498–503.

Brandriss, M.C. and Magasanik, B. (1979b) Subcellular compartmentation in control of converging pathways for proline and arginine metabolism in *Saccharomyces cerevisiae. Journal of Bacteriology*, **145**, 1359–1364.

Brandriss, M.C. and Magasanik, B. (1980) Proline: an essential intermediate in arginine degradation in *Saccharomyces cerevisiae. Journal of Bacteriology*, **143**, 1403.

Bravdo, B.-A. and Hepner, Y. (1987) Irrigation management and fertigation to optimise grape composition and vine performance. In *Grapevine Canopy and Vigor Management*. Edited by W.M. Kliewer. pp. 49–67. Acta Hortic. 206.

Bussey, H. and Umbarger, H.E. (1970) Biosynthesis of the branched-chain amino acids in yeast: A trifluoroleucine-resistant mutant with altered regulation of leucine uptake. *Journal of Bacrteriology*, **103**, 286–294.

Busturia, A. and Lagunas, R. (1986) Catabolite inactivation of the glucose transport system in *Saccharomyces cerevisiae. Journal of General Microbiology*, **132**, 379–385.

Cabrera, M.J., Moreno, J., Ortega, J.M. and Medina, M. (1988) Formation of ethanol, higher alcohols, esters and terpenes by five yeast strains in musts from Pedro Ximenez grapes in various degrees of ripeness. *American Journal of Enology and Viticulture*, **39**, 283–287.

Calderbank, J., Rose, A.H. and Tubb, R.S. (1985) Peptide removal from all-malt adjuncts worts by *Saccharomyces cerevisiae* NCYC 240. *Journal of the Institute of Brewing*, **91**, 321–324.

Calderbank, J., Keenan, M.H.J., Rose, A.H. and Holman, G.D. (1984) Accumulation of amino acids by *Saccharomyces cerevisiae* Y185 with phospholipids enriched in different fatty-acyl residues: A statistical analysis of data. *Journal of General Microbiology*, **130**, 2817–2824.

Cantarelli, C. (1957) On the activation of alcoholic fermentation in wine making. *American Journal of Enology and Viticulture*. **8**, 113–120, 167–175.

Cantarelli, C. (1989) Factors affecting the behaviour of yeast in wine fermentation. In *Biotechnology applications in beverage production*. Edited by C. Cantarelli and G. Lanzarini, pp. 127–151. London and New York: Elsevier Applied Science.

Cantagrel, R., Symonds, P. and Carles, J. (1982) Composition en acides amines du mout en fonction du cepage et de la technologie et son influence sur la qualite de vin. *Science des Aliments*, **2**, 109–142.

Cartwright, C.P., Juroszek, J.-R., Beaven, M.J., Ruby, F.M.S., de Morais, S.M.F. and Rose, A.H. (1986) Ethanol dissipates the proton-motive force across the plasma membrane of *Saccharomyces cerevisiae. Journal of General Microbiology*, **132**, 369–377.

Cartwright, C.P., Rose, A.H., Calderbank, J. and Keenan, M.H.J. (1989) Solute transport. In *The Yeasts* Vol. 3, 2nd edition, Edited by A.H. Rose and J.S. Harrison, pp. 5–56. London: Academic Press.

Castor, J.G.B. (1953) The free amino acids of musts and wines. II. The fate of amino acids of must during alcoholic fermentation. *Journal of Food Research*, **18**, 146–151.

Castor, J.G.B. and Archer, T.E. (1959) The free amino acids of musts and wines III: Effect of added ammonia and of fermentation temperature on the fate of amino acids during fermentation. *Journal of Food Research*, **24**, 167–175.

Chan, P.Y. and Cossins, E.A. (1976) General properties and regulation of arginine transporting systems in *Saccharomyces cerevisiae. Plant and Cell Physiology*, **17**, 341–353.

Chen, E.C.-H. (1978) The relative contribution of Ehrlich and biosynthetic pathways to the formation of fusel alcohols. *Journal of the American Society of Brewing Chemists*, **36**, 39–43.

Chen, E.C.-H., van Gheluwe, G. and Buday, A. (1973) Effects of mashing temperature on the nitrogenous constituents of wort and beer. In *Proceedings of the American Society of Brewing Chemists*, Annual meeting, New Orleans, Louisiana, (April 28–May 3, 1973). pp. 6–10.

Cherest, H., Surdin-Kerjan, Y. and de Robichon-Szulmajster, H. (1971) Methionine-mediated repression in *Saccharomyces cerevisiae*: a pleiotropic regulatory system involving methionyl transfer ribonucleic acid and the product of the gene eth2. *Journal of Bacteriology*, **106**, 758–772.

Cherest, H., Surdin-Kerjan, Y., Antoniewski, J. and de Robichon-Szulmajster, H. (1973) *S*-adenosyl methionine-mediated repression of methionine biosynthetic enzymes in *Saccharomyces cerevisiae. Journal of Bacteriology*, **114**, 928–933.

Ciolfi, C., Castino, M. and Di Stefano, R. (1985) Studio sulla riposta metabolica di lieviti di specie diverse fermenrani un unico mosto a temperature comprese fra 10 e 40°C. Nota II. *Rivista di Viticoltura e di Enologia*, **10**, 489–507.

Colagrande, O.F., Griselli, F. and Del Re, A.A. (1973) Etude des phenomenes d'echange lors de l'emploi des bentonites œnologiques I. Les echanges de sodium et de proline. *Connaissance de la Vigne et du Vin*, **7**, 93–106.

Cooper, T.G. (1982) Nitrogen metabolism in *Saccharomyces cerevisiae*. In *The Molecular Biology of the Yeast Saccharomyces.* Metabolism and Gene Expression,edited by J. N. Strathern, E. W. Jones and J. B. Broach. pp. 39–99. Cold Spring Harbor Laboratory, New York .

Cooper, T.C. and Sumrada, R. (1975) Urea transport in *Saccharomyces cerevisiae. Journal of Bacteriology*, **121**, 571–576.

Cordonnier, R. (1966) Etude des proteins et des substances azotees. Leur evolution au cours des traitements oenologique. Conditions de le stabilite proteique des vius. *Bulletin de l'Office International de la Vigne et du Vin*, **39**, 1475–1489.

Corison, C.A., Ough, C.S., Berg, H.W. and Nelson, K.E. (1979) Must acetic acid and ethyl acetate as mold and rot indicators in grapes. *American Journal of Enology and Viticulture*, **30**, 130–134.

Correa, I., Polo, M.C., Amigo, L. and Ramos, M. (1988) Séparation des protéines des moûts de raison au moyen de techniques électrophorétiques. *Connaissance de la Vigne et du Vin*, **22**, 1–9.

Courchesne, W.E. and Magasanik, B. (1983) Ammonia regulation of amino acid permeases in *Saccharomyces cerevisiae. Molecular and Cellular Biology*, Apr.1983, 672–683.

Crabeel, M. and Grenson, M. (1970) Regulation of histidine uptake by specific feedback inhibition of two histidine permeases in *Saccharomyces cerevisiae. European Journal of Biochemistry*, **14**, 197–204.

Crowell, E. A. and Guymon, J.F. (1963). Influence of aeration and suspended material on higher alcohols, acetoin and diacetyl during fermentation. *American Journal of Enology and Viticulture*, **14**, 214–222.

Crowell, E.A., Ough, C.S. and Bakalinski, A. (1985) Determination of alpha amino nitrogen in musts and wines by TNBS method. *American Journal of Enology and Viticulture*, **36**, 175–177.

Darte, C. and M. Grenson. (1975). Evidence for three glutamic acid transporting systems with specialized physiological functions in *Saccharomyces cerevisiae. Biochemical and Biophysical Research Communications*, **67**, 1028–1033.

Daudt, C.E. and Ough, C.S. (1980) A method for detecting volatile amines in grapes and wines. *American Journal of Enology and Viticulture*, **31**, 356–359.

David, M.H. and Kirsop, B.H. (1972) The varied response of brewing yeasts to oxygen and sterol treatment. In *Proceedings of the American Society of Brewing Chemists*, **30**, 14–17.

David, M.H. and Kirsop, B.H. (1973) Yeast growth in relation to the dissolved oxygen and sterol content of wort. *Journal of the Institute of Brewing*, **79**, 20–25.

Davis, R.H. (1986) Compartmental and regulatory mechanism in the arginine pathways of *Neurospora crassa* and *Saccharomyces cerevisiae*. *Microbiological Review*, **50**, 280–313.

de Castella, F. (1909) The nitrogen food of yeast and its bearing on the making of full-bodied dry wines. *Journal Department Agriculture Victoria*, **7**, 1–4.

Delfini, C., Garcia-Moruno, E., Pagliara, A. and Contiero, M. (1989) Technological and biological factors causing the production of large amounts of acetic acid by wine yeasts during alcoholic fermentation. Proceedings of the VII International Symposium on Yeasts, Leuven, 18–22 September, 1989, pp. 30–31.

Deschamps, J., Dubois, E. and Wiame, J.-M. (1979) L ornithine transaminase EC-2.6.1.13 synthesis in *Saccharomyces cerevisiae* regulation by inducer exclusion. *Molecular and General Genetics*, **174**, 225–232.

Devine, S.J. and Slaughter, J.C. (1980) The effect of medium composition on the production of ethanol by *Saccharomyces cerevisiae*. *FEMS Microbiology Letters*, **9**, 19–21.

de Vito, P.C. and Dreyfuss, J. (1964) Metabolic regulation of adenosine triphosphate sulfurylase in yeast. *Journal of Bacteriology*, **88**, 1341–1348.

Dittrich, H.H. and Sponholz, W.-R. (1975) Die aminosäureabnahme in Botrytis-infizierten Traubenbeeren und die bildung höherer alhohole in deisen Mosten bei ihrer vergärung. *Wein-Wissenschaft*, **30**, 188–210.

Dittrich, H.H., Sponholz, W.R. and Göbel, H.G. (1975) Vergleichende untersuchugen von mosten und weinen aus gesunden und aus Botrystis-infizierten Traubenbeeren. II. Modellversuche zur veränderung des mostes durch Botrystis-infektion und ihre konsequenzen für die neben-produkbildung bei der gärung, *Vitis*, **13**, 336–347.

Dombek, K.M. and Ingram, L.O. (1988) Intracellular accumulation of AMP as a cause for the decline in rate of ethanol production by *Saccharomyces cerevisiae* during batch fermentation. *Applied and Environmental Microbiology*, **54**, 98–104.

Dott, W. and Trüper, H.G. (1978) Sulfite formation by wine yeast. VI Regulation of biosynthesis of NADPH and BV-dependent sulfite reductases. *Archives of Microbiology*, **118**, 249–251.

Dott, W. and Trüper, H.G. (1979) Sulfite formation by wine yeast. VII Regulation of the enzymatic steps between sulfide and sulfur-containing amino acids. *Archives of Microbiology*, **121**, 251–253.

Dubois, E. and Grenson, M. (1979) Methylamine/ammonia uptake systems in *Saccharomyces cerevisiae*: Multiplicity and regulation. *Molecular and General Genetics*, **175**, 67–76.

Dubois, E., Grenson, M. and Wiame, J.M. (1974) The participation of the anabolic glutamate dehydrogenase in the nitrogen catabolite repression of arginase in *Saccharomyces cerevisiae*. *European Journal of Biochemistry* **48**, 603–616.

Dufour, (1989) Effect of wort amino acids on beer quality. *Louvain Brewing Letters* **2**, 11–19.

Dukes, B., Goldspink, B., Elliott, J. and Frayne, R. (1991) Time of nitrogenous fertilisation can reduce fermentation time and improve wine quality. In *Proceedings of the International Symposium on Nitrogen in Grapes and Wines* (Seattle, WA,1991), edited by J. Rantz, pp. 249–254. Davis, CA: American Society for Enology and Viticulture.

du Plessis, C.S. (1963) The ion exchange treatment (H cycle) of white grape juice prior to fermentation 1. The effect upon fermentation. *South African Journal for Enology and Viticulture*, **6**, 561–580.

du Plessis, C.S. (1984) Optimum maturity and quality parameters in grapes: A review. *South African Journal for Enology and Viticulture*, **5**, 35–42.

Duteurtre, B., Bourgeosis, C. and Chollot, B. (1971) Study of the assimilation of proline by brewing yeast. *Journal of the Institute of Brewing*, **77**, 28.

Eddy, A.A. (1978) Proton-dependent solute transport in microorganisms. *Current Topics in Membrane Transport*, **10**, 279–360.

Eddy, A.A. (1982) Mechanisms of solute transport in selected eukaryotic microorganisms. *Advances in Microbial Physiology*, **24**, 1–78.

Egbosimba, E.E. and Slaughter, J.C. (1987) The influence of ammonium permease activity and carbon source on the uptake of ammonium from simple defined media by *Saccharomyces cerevisiae. Journal of General Microbiology*, **133**, 375–379.

Egbosimba, E.E., Okafor, E.L.C. and Slaughter, J.C. (1988) Control of ammonium uptake from malt extract medium by *Saccharomyces cerevisiae. Journal of the Institute of Brewing*, **94**, 249–252.

El Haloui, N., Picque, D. and Corrieu, G. (1988) Alcoholic fermentation in wine-making:on line measurement of density and carbon dioxide evolution. *Journal of Food Engineering*, **8**, 17–30.

Enari, T.-M. (1974) Amino acids, peptides and proteins. *European Brewery Convention. Wort symposium zeit.* Monograph **1**, 1974, 73–88.

Eschenbruch, R. (1973) On the sulfite formation and sulfite consumption of wine yeasts. *Die Wynboer*, **506**, 42–44.

Eschenbruch, R., Bonish, P. and Fisher, B.M. (1978) The production of H_2S by pure culture wine yeast. *Vitis*, **17**, 67–74.

Etiévant, P., Schlich, P., Bouvier, J-C., Symonds, P. and Bertrand, A. (1988) Varietal and geographic classification of french red wines in terms of elements, amino acids and aromatic alcohols. *Journal of the Science of Food and Agriculture*, **45**, 25–41.

Ferreras, J.M., Iglesias, R. and Girbes, T. (1989) Effect of the chronic ethanol action on the activity of the general amino-acid permease from *Saccharomyces cerevisiae var. ellipsoideus. Biochimica et Biophysica Acta*, **979**, 375–377.

Feuillat, M. and Charpentier, C. (1982) Autolysis of yeasts in champagne. *American Journal of Enology and Viticulture*, **33**, 6–13.

Flanzy, C. (1982) Métabolisme des substances azotées au cours de la vinification par macération cabonique. *Science des Aliments*, **2**, 165–188.

Fleet, G.H. (1990) Which yeast species really conduct the fermentation? In *Proceedings of the seventh Australian wine industry technical conference*; 14–16 August 1989, Adelaide, SA, edited by P.J. Williams, D. Davidson and T.H. Lee. pp. 153–156. Adelaide, SA: Australian Industrial Publishers.

Gallander, J.F., Cahoon, G.A. and Beelman, R.B. (1969) Free amino acids in musts of eight eastern grape varieties. *American Journal of Enology and Viticulture*, **20**, 140–145.

Garvie, E.I. (1967) The growth factor and amino acid requirements of species of the genus *Leuconostoc*, including *Leuconostoc paramesenteroides* (sp. nov.) and *Leuconostoc oenos. Journal of General Microbiology*, **48**, 439–447.

Gits, J.J. and Grenson, M. (1967) Multiplicity of the amino acid permeases in *Saccharomyces cerevisiae*. III. Evidence for a specific methionine-transporting system. *Biochimica et Biophysica Acta*, **135**, 507–516.

Gits, J. J. and M. Grenson. (1969) Regulation of methyl amino acids in *Saccharomyces cerevisiae. Archives Internationales de Physiologie et du Biochimie*, **77**, 153–154.

Gockowiak, H. and Henschke, P.A. (1992) Nitrogen composition of grape juice and implications for fermentation: results of a survey made in NE Victoria. Submitted *Australian Grapegrower and Winemaker.*

Goldspink, B. and Gordon, C. (1991) Response of *Vitis vinifera* cv. Sauvignon blanc grapevines to timed applications of nitrogen fertilizers. In *Proceedings of the International Symposium on Nitrogen in Grapes and Wines* (Seattle, WA,1991), edited by J. Rantz, pp. 255–258. Davis, CA: American Society for Enology and Viticulture.

Greasham, R.L. and Moat, A.G. (1973) Amino acid transport in a polyaromatic amino acid auxotroph of *Saccharomyces cerevisiae. Journal of Bacteriology*, **115**, 975–981.

Grechko, N. Ya. and Emel'yanova, N.A. (1984) Change in the amino acid composition during thermal treatment of kvass must concentrate. *Izvestiya Vysshikh Uchebnykh Zavedenii Pishchevaya Tekhnologiya*, **3**, 52–53.

Gregory, M.E., Keenan, M.H.J. and Rose, A.H. (1982) Accumulation of L-Asparagine by *Saccharomyces cerevisiae* X-2180. *Journal of General Microbiology*, **128**, 2557–2562.

Grenson, M. (1983a) Inactivation-Reactivation process and repression of permease formation regulate several ammonia-sensitive permeases in the yeast *Saccharomyces cerevisiae*. *Journal of Biochemistry*, **133**, 135–139.

Grenson, M. (1983b) Study of the positive control of the general amino-acid permease and other ammonia-sensitive uptake systems by the product of the NPR1 gene in the yeast *Saccharomyces cerevisiae*. *Journal of Biochemistry*, **133**, 141–144.

Grenson, M. and Acheroy, B. (1982) Mutations affecting the activity and the regulation of the general amino-acid permease of *Saccharomyces cerevisiae*. Localisation of the cis acting dominant PGR regulatory mutation in the structural gene of this permease. *Molecular and General Genetics*, **188**, 261–265.

Grenson, M. and Dubois, E. (1982) Pleiotropic deficiency in nitrogen-uptake systems and derepression of nitrogen-catabolic enzymes in *npr-1* mutants of *Saccharomyces cerevisiae*. *European Journal of Biochemistry*, **121**, 643–647.

Grenson, M. and Hou, C. (1972) Ammonia inhibition of the general amino acid permease and its suppression in NADPH-specific glutamate dehydrogenaseless mutants of *Saccharomyces cerevisiae*. *Biochemical and Biophysical Research Communications*, **48**, 749–756.

Grenson, M., Hou, C. and Crabeel, M. (1970) Multiplicity of the amino acid permeases in *Saccharomyces cerevisiae*. IV Evidence for a General Amino Acid Permease. *Journal of Bacteriology*, **103**, 770–777.

Grenson, M., Mousset, M., Wiame, J.M. and Béchet. (1966) Multiplicity of the amino acid permeases in *Saccharomyces cerevisiae* I. Evidence for a specific arginine-transporting system. *Biochimica et Biophysica Acta*, **127**, 325–338.

Grenson, M., Muyldermans, F., Broman, K. and Vissers, S. (1987) 4-Amino butyric acid (GABA) uptake in baker's yeast *Saccharomyces cerevisiae* is mediated by the general amino acid permease, the proline permease and GABA-specific permease intergrated into the GABA-catabolic pathway. *Biochemistry (Life Science Adv.)*, **6**, 35–39.

Groat, M. and Ough, C.S. (1978) Effects of insoluble solids added to clarified musts on fermentation rate, wine composition and wine quality. *American Journal of Enology and Viticulture*, **29**, 112–119.

Guymon, J.F., Ingraham, J.L. and Crowell, E.A. (1961) Influence of aeration upon the formation of higher alcohols by yeasts. *American Journal of Enology and Viticulture*, **12**, 60–66.

Hardwick, W.A. (1983) Food and feed production with microorganisms. In *Biotechnology*, vol. 5, edited by G. Reed. pp. 169–229. Verlag Chemie: Weinheim.

Hartnell, P.C. and Spedding, D.J. (1979) Uptake and metabolism of S^{35}-sulphate by wine yeast. *Vitis*, **18**, 307–315.

Haukeli, A.D. and Lie, S. (1976) Effects of lipids and oxygen on yeast growth and the biosynthesis of acetoin during fermentation. *Journal of the Institute of Brewing*, **82**, 1611–1615.

Hawes, C.S. and Nicholas, D.J.D. (1973) Adenosine 5'-triphosphate sulphurylase from *Saccharomyces cerevisiae*. *Biochemical Journal*, **133**, 541–550.

Heinzel, M. and Trüper, H.G. (1976) Sulfite formation by wine yeasts. II Properties of ATP-sulfurylase. *Archives of Microbiology*, **107**, 293–297.

Heirholzer, G. and Holzer, H. (1963) Repression der synthese von DPN-abhangiger glutaminsauredehydrogenase in *Saccharomyces cerevisiae* durch ammoniumionen. *Biochemische Zeitschrift*, **339**, 175–185.

Henick-Kling, T. and Larsson, I.M. (1991) Nitrogen limitation and stuck fermentation *American Wine Society Journal*, Winter Issue; 124–126.

Henschke, P.A. (1990) Evaluating wine yeasts for improved wine quality. *Proceedings of the seventh Australian Wine Industry Technical Conference.* (Adelaide, SA,14–16 August 1989) edited by P.J. Williams, D.M. Davidson and T.H. Lee, pp. 157–165. Adelaide, SA: Australian Industrial Publishers.

Henschke, P.A. and Dixon, G.D. (1990) Effect of yeast strain on acetic acid accumulation during fermentation of *Botrytis* affected grape juice. *Proceedings of the seventh Australian wine industry technical conference* (Adelaide, SA, 14–16 August 1989) edited by P.J. Williams, D.M. Davidson and T.H. Lee. pp. 242–244. Adelaide, SA: Australian Industrial Publishers.

Henschke, P.A. and Jiranek, V. (1991) Hydrogen sulfide formation during fermentation: Effect of nitrogen composition in model grape must. In *Proceedings of the International Symposium on Nitrogen in Grapes and Wines* (Seattle, WA, June 1991), edited by J. Rantz, pp. 172–184. Davis, CA: American Society for Enology and Viticulture.

Henschke, P.A. and Ough, C.S. (1991) Urea accumulation in fermenting grape juice. *American Journal of Enology and Viticulture*, **42**, 317–321.

Henschke, P.A. and Rose, A.H. (1991) Plasma Membranes. In *The Yeasts*, vol.4, second edition. Edited by A.H. Rose. and J.S. Harrison, pp. 297–345. London: Academic Press.

Heresztyn, T. (1986) Formation of substituted tetrahydropyridines by species of Brettanomyces and Lactobacillus isolated from mousy wines. *American Journal of Enology and Viticulture*, **37**, 127–132.

Herraiz, T., Martin-Alvarez, P.J., Reglero, G., Herraiz, M. and Cabezudo, M.D. (1989) Differences between wines fermented with and without SO_2 using various selected yeasts. *Journal of the Science of Food and Agriculture*, **49**, 249–258.

Higgins, S.E. (1990) Revision and recodification of wine regulations; Final rule. Federal register, part III, **55**(118), 24974–25034. Department of Treasury, Bureau of Alcohol, Tobacco and Firearms.

Hinnesbusch, A.G. (1988) Mechanisms of gene regulation in the general control of amino acid biosynthesis in *Saccharomyces cerevisiae. Microbiological Reviews*, **52**, 248–273.

Holzberg, I., Finn, R.K. and Steinkraus, K.H. (1967) A kinetic study of the alcoholic fermentation of grape juice. *Biotechnology and Bioengineering*, **9**, 413–427.

Holzer, H. and Schneider, S. (1957) Anreicherung und trennung einer DPN-spezifischen und einer TPN-spezifischen glutaminsaure-dehydrogenase aus hefe. *Biochemische Zeitschrift,* **329**, 361–369.

Horák, J. (1986) Amino acid transport in eucaryotic microorganisms. *Biochimica et Biophysica Acta,* **864**, 223–256.

Horák, J. and Kotyk, A. (1977) Temperature effects in amino acid transport by *Saccharomyces cerevisiae. Experimental Mycology*, **1**, 63–68.

Horák, J. and Říhová, L. (1982) L-proline transport in *Saccharomyces cerevisiae. Biochimica et Biophysica Acta*, **691**, 144–150.

Houtman, A.C. and du Plessis, C.S. (1981) The effect of juice clarity and several conditions promoting yeast growth on fermentation rate, the production of aroma components and wine quality. *South African Journal for Enology and Viticulture*, **2**, 71–81.

Houtman, A.C. and du Plessis, C.S. (1986a) The effect of grape cultivar and yeast strain on fermentation rate and concentration of volatile components in wine. *South African Journal for Enology and Viticulture*, **7**, 14–20.

Houtman, A.C. and du Plessis, C.S. (1986b) Nutritional deficiencies of clarified white grape juices and their correction in relation to fermentation. *South African Journal for Enology and Viticulture,* **7**, 39–46.

Houtman, A.C., Morais and du Plessis, C.S. (1980a) Factors affecting the reproducibility of fermentation of grape juice and of the aroma composition of wines. I Grape maturity, sugar, inoculation concentration, aeration, juice turbidity and ergosterol. *Vitis*, **19**, 37–54.

Houtman, A.C., Morais and du Plessis, C.S. (1980b) The possibilities of applying present-day knowledge of wine aroma components: Influence of several juice factors on fermentation rate and ester production during fermentation. *South African Journal for Enology and Viticulture*, **1**, 27–33.

Huang, Z. and Ough, C.S. (1989) Effect of vineyard location, varieties and woodstocks on the juice amino acid composition of several cultivars. *American Journal of Enology and Viticulture*, **40**, 135–139.

Huang, Z. and Ough, C.S. (1991) Amino acid profiles of commercial grape juices and wines. *American Journal of Enology and Viticulture*, **42**, 261–267.

Iglesias, R., Ferreras, J.M., Arias, F.J., Munoz, R. and Girbes, T. (1990) Changes in the activity of the general amino acid permease from *Saccharomyces cerevisiae var. ellipsoideus* during fermentation. *Biotechnology and Bioengineering*, **36**, 808–810.

Iglesias, R., Ferreras, J.M., Arias, F.J., Munoz, R. and Girbes, T. (1991) Effect of continued exposition to ethanol on activity of the ammonium and fructose transport systems in *Saccharomyces cerevisiae var. ellipsoideus*. *Biotechnology and Bioengineering*, **37**, 389–391.

Ingledew, W.M. and Kunkee, R. (1985) Factors influencing sluggish fermentations of grape juice. *American Journal of Enology and Viticulture*, **36**, 65–76.

Ingledew, W.M., Magnus, C.A. and Patterson, J.R. (1987) Yeast foods and ethyl carbamate formation in wine. *American Journal of Enology and Viticulture*, **38**, 332–335.

Ingledew, W.M., Magnus, C.A. and Sosulski, F.W. (1987) Influence of oxygen on proline utilisation during the wine fermentation. *American Journal of Enology and Viticulture*, **38**, 246–248.

Ingledew, W.M., Sosulski, F.W. and Magnus, C.A. (1986) An assessment of yeast foods and their utility in brewing and enology. *Proceedings of the American Society of Brewing Chemists*, **44**, 166–170.

Iwashima, A., Nishino, H. and Nose, Y. (1973). Carrier-mediated transport of thiamine in baker's yeast. *Biochimica et Biophysica Acta*, **330**, 222–234.

Jiranek, V. (1992) PhD Thesis, University of Adelaide, Australia. In preparation.

Jiranek, V., Langridge, P. and Henschke, P.A. (1990) Nitrogen requirement of yeast during wine fermentation. In *Proceedings of the Seventh Australian Wine Industry Technical Conference* (Adelaide, 1989), edited by P. J. Williams, D. M. Davidson and T. H. Lee, pp. 166–171. Adelaide, S.A: Australian Industrial Publishers.

Jiranek, V., Langridge, P. and Henschke, P.A. (1991) Yeast nitrogen demand: Selection criterion for wine yeasts for fermenting low nitrogen musts. In *Proceedings of the International Symposium on Nitrogen in Grapes and Wine* (Seattle, 1991) Edited by J. Rantz. pp. 266–269. Davis, CA.

Johnson, B. and Brown, C.M. (1974) The enzymes of ammonia assimilation in *Schizosaccharomyces* spp. and in *Saccharomyces ludwigii*. *Journal of General Microbiology*, **85**, 169–172.

Joiris, C.R. and Grenson, M. (1969) Specificity and regulation of a permease of dicarboxylic amino-acids in *Saccharomyces cerevisiae*. *Archives Internationales de Physiologie et de Biochimie*, **77**, 154–156.

Jones, R.P. (1989) Biological principles for effects of ethanol. *Enzyme Microbiology Technology*, **11**, 130–153.

Jones, E.W. and Fink, G.R. (1982) Regulation of amino acid and nucleotide biosynthesis in yeast. In *The Molecular Biology of the Yeast Saccharomyces*. Metabolism and Gene Expression, edited by J. N. Strathern, E. W. Jones and J. B. Broach. pp. 181–300. Cold Spring Harbor Laboratory, New York .

Jones, M. and Pierce, J.S. (1964) Absorption of amino acids from wort by yeasts. *Journal of the Institute of Brewing*, **70**, 307–315.

Jones, M., Pragnell, M.J. and Pierce, J.S. (1969) Absorption of amino acids by yeast from a semi-defined medium simulating wort. *Journal of the Institute of Brewing*, **75**, 520–536.

Juroszek, J.-R., Feuillat, M. and Charpentier, C. (1987) Effect of the champagne method of a starter preparation on ethanol tollerance of yeast. *American Journal of Enology and Viticulture*, **38**, 194–198.

Keenan, M.H.J. and Rose, A.H. (1979) Plasma-membrane lipid unsaturation can effect the kinetics of solute transport in *Saccharomyces cerevisiae. FEMS Microbiology Letters*, **6**, 133–137.

Kitomoto, K., Yoshizawa, K., Ohsumi, Y. and Anraka, Y. (1988) Dynamic aspects of vacuolar and cytosolic amino acids pools of *Saccharomyces cerevisiae. Journal of Bacteriology*, **170**, 2683–2686.

Klein, H.P. (1955) Synthesis of lipids in resting cells of *Saccharomyces cerevisiae. Journal of Bacteriology*, **69**, 620–627.

Kliewer, W.M. (1968) Changes in the concentration of free amino acids in grape berries during maturation. *American Journal of Enology and Viticulture*, **19**, 166–174.

Kliewer, W.M. (1969) Free amino acids and other nitrogenous substances of table grape varieties. *Journal of Food Science*, **34**, 274–278.

Klingshirn, L.M., Liu, J.R. and Gallander, J.F. (1987) Higher alcohol formation in wines as related to the particle size profiles of juice insoluble solids. *American Journal of Enology and Viticulture*, **38**, 207–210.

Kluba, R.M., Mattick, L.R. and Hackler, L.R. (1978) Changes in the free and total amino acid composition of several *Vitis labruscana* varieties during maturation. *American Journal of Enology and Viticulture*, **29**, 102–111.

Knatchbull, F.B. and Slaughter, J.C. (1987) The effect of low CO_2 on the absorption of amino acids and production of flavour active volatiles by yeast. *Journal of the Institute of Brewing*, **93**, 420–424.

Koch, J. and Sajak, E. (1959) A review and some studies on grape protein. *American Journal of Enology and Viticulture*, **10**, 114–123.

Kotyk, A. and Dvoráková, M. (1990) Transport of L-Tryptophan in *Saccharomyces cerevisiae. Folia Microbiology*, **35**, 209–217.

Kumada, J., Nakajjma, S., Takahashi, T. and Narziss, L. (1975) Einfluss von Druck und Temperatur bei der Gärung auf den Stoffwechsel der Hefe und Bierqualität.*Proceedings of the 15th European Brewery Convention*, Nice, pp. 615–623.

Kunkee, R.E. (1991) Relationship between nitrogen content of must and sluggish fermentation. In *Proceedings of the International Symposium on Nitrogen in Grapes and Wines* (Seattle, WA,1991), edited by J. Rantz, pp. 148–155. Davis, CA: American Society for Enology and Viticulture.

Kunkee, R.E. and Amerine, M.A. (1970) Yeast in winemaking. In *The Yeasts*, volume 3,1st Edition, edited by A.H. Rose and J.S. Harrison, pp. 5–71, London: Academic Press.

Lafon, S. (1978) Les stériodes "facteurs de survie" des levures au cours de la vinification. *Annales de Technologie Agricole*, **27**, 215–217.

Lafon-Lafourcade, S. (1986) Applied microbiology. *Experientia*, **42**, 904–914.

Lafon-Lafourcade, S. and Guimberteau, G. (1962) Evolution des aminoacides au cours de la maturation des raisins. *Vitis*, **3**, 130–135.

Lafon-Lafourcade, S. and Peynaud, E. (1959) Dosage microbiologique des acides aminés des moûts de raisin et des vins. *Vitis*, **2**, 45–56.

Lafon-Lafourcade, S. and Ribéreau-Gayon, P. (1979) Quelques observations sur les problèmes microbiologiques de la vinification en blanc. *Connaissance de la Vigne et du Vin*, **13**, 51–76.

Lafon-Lafourcade, S., Geneix, S. and Ribéreau-Gayon, P. (1984) Inhibition of alcoholic fementation of grape must by fatty acids produced by yeasts and their elimination by yeast ghosts. *Applied and Environmental Microbiology*, **47**, 1246–1249.

Lagace, L.S. and Bisson, L.F. (1990) Survey of yeast acid proteases for effectiveness of wine haze reduction. *American Journal of Enology and Viticulture*, **41**, 147–155.

Lagunas, R., Dominguez, C., Busturia, A. and Saez, M.J. (1982) Mechanisms of appearance of the Pastuer effect in *Saccharomyces cerevisiae*: Inactivation of sugar transport systems. *Journal of Bacteriology*, **152**, 19–25.

Large, P.J. (1986) Degradation of organic nitrogen compounds by yeasts. *Yeast*, **2**,1–34.

Larue, F. and Lafon-Lafourcade, S. (1989) Survival factors in wine fermentation. In *Alcohol tolerance in yeasts and bacteria*, edited by N. van Uden. pp. 193–215. Boca Raton: CRC Press.

Larue, F., Lafon-Lafourcade, S. and Ribéreau-Gayon, P. (1978) Relationships entre la tenure cellulaire en sterols etl'activite fermentaire des levures dans le mout de raisin. Role fonctionnel des steroides. *Comptes Rendus de l'Académie des Sciences,* 287D, 1445–1448.

Larue, F., Lafon-Lafourcade, S. and Ribereau-Gayon, P. (1982) Inhibition de *Saccharomyces cerevisiae* dans le mout de raison. *Comptes Rendus de l'Académie des Sciences*, Paris, série D. 294, 587–590.

Lasko, P.F. and Brandriss, M.C. (1981) Proline transport in *Saccharomyces cerevisiae*: *Journal of Bacteriology*, **148**, 241–247.

Leão, C. and van Uden, N.J. (1983) Effects of ethanol and other alkanols on the ammonium transport system of *Sacccharomyces cerevisiae. Biotechnology and Bioengineering,* **25**, 2085–2090.

Leão, C. and van Uden, N.J. (1984a) Effects of ethanol and other alkanols on passive proton influx in the yeast *Sacccharomyces cerevisiae. Biochimica et Biophysica Acta,* **774**, 43–48.

Leão, C. and van Uden, N.J. (1984b) Effects of ethanol and other alkanols on the general amino acid permease of *Sacccharomyces cerevisiae. Biotechnology and Bioengineering,* **26**, 403–405.

Lee, W.J. (1987) Utilisation of nucleosides and nucleobases by the lager yeast, Saccharomyces carlsbergensis. *American Society of Brewing Chemists*, **45**, 128–131.

Lehmann, F.L. (1987) Free SO_2 and the petite mutation in wine yeasts. *Australian Grapegrower and Winemaker*, April, 1987, 68–69.

Lemperle, E. (1988) Fungicide residues in musts and wines. In *Proceedings of the second international symposium for cool climate viticulture and oenology*; 11–15 January 1988, Auckland, NZ, edited by R.E. Smart, R.J. Thornton, S.B. Rodriguez, J.E. Young, pp. 211–218. Auckland, NZ: NZSVO.

Lie, S. (1973) The EBC-Ninhydrin method for determination of free alpha amino nitrogen (FAN). *Journal of the Institute of Brewing*, **79**, 37–41.

Light, R.J., Lennarz, W.J. and Bloch, K. (1962) The metabolism of hydroxystearic acids in yeast. *Journal of Biological Chemistry*, **237**, 1793–1800.

Lloyd, D., Kristensen, B. and Degn, H. (1983) Glycolysis and respiration in yeasts: The effect of ammonium ions studied by mass spectroscopy. *Journal of General Microbiology*, **129**, 2125–2127.

MacDonald, J., Reeve, F.T.V., Ruddlesden, J.D. and White, F.H. (1984) Current approaches to brewery fermentations. *Progress in Industrial Microbiology*, **19**, 1–197.

Magaña-Schwencke, N. and Schwencke, J. (1969) A proline transport system in *Saccharomyces chevalieri. Biochimica et Biophysica Acta,* **173**, 313–323.

Magaña-Schwencke, N., Kuznar, J. and Schwencke, J. (1973) Imino acid transport in yeast: the uptake of sarcosine. *Biochimica et Biophysica Acta*, **318**, 281–288.

Mantachian, E.A. (1984) Synthesis of proline oxidase in *Sacccharomyces Vini*: Regulation by the introduction of repressing factor. *Biological Journal of Armenia*, **37**, 46–50.

Marder, R., Becker, J.M. and Naider, F. (1977) Peptide transport in yeast: Utilization of leucine- and lysine-containing peptides by *Saccharomyces cerevisiae. Journal of Bacteriology,* **131**, 906–916.

Maule, D.R., Pinnegar, M.A., Pronto, A.D. and Whitcar, A.L. (1966) Simultaneous assessment of changes occurring during batch fermentation. *Journal of the Institute of Brewing*, **72**, 488–494.

Mauricio, J.C., Guijo, S. and Ortega, J.M. (1991) Relationship between phospholipid and sterol content in *Saccharomyces cerevisiae* and *Torulaspora delbrueckii* and their fermentation activity in grape must. *American Journal of Enology and Viticulture,* **42**, 301–308.

Maw, G.A. (1965) The role of sulfur in yeast growth and in brewing. *Wallerstein Laboratory Communications*, **21**, 49–68.
Mayer, K., Pause, G. and Vetsch, U. (1973) Aminosauregehalte im verlauf der vinification einiger rotweine. *Mitteil Klosterneuburg*, **23**, 331.
McKelvey, J., Rai, R. and Cooper, T.G. (1990) GABA transport in *Saccharomyces cerevisiae. Yeast*, **6**, 263–270.
Messenguy, F., Colin, D. and Ten Have, J.-P. (1980) Regulation of compartmentation of amino acid pools in *Sacccharomyces cerevisiae* and its effects on metabolic control. *European Journal of Biochemistry*, **108**, 439–447.
Middelhoven, W.J. (1964) The pathway of arginine breakdown in *Saccharomyces cerevisiae. Biochimica et Biophysica Acta*, **93**, 650–652.
Miller, G.C., Amon, J.M., Gibson, R.L. and Simpson, R.F. (1985) Loss of wine aroma attributable to protein stabilisation with bentonite or ultrafiltration tangentielle. *Australian Grapegrower Winemaker*, **256**, 46, 49–50.
Moneton, P., Sarthou, P. and Le Goffie, F. (1986a) Role of the nitrogen source in peptide transport in *Saccharomyces cerevisiae. FEMS Microbiology Letters*, **36**, 95–98.
Moneton, P., Sarthou, P. and Le Goffie, F. (1986b) Transport and hydrolysis of peptides in *Saccharomyces cerevisiae. Journal of General Microbiology*, **132**(8), 2147–2153.
Monk, P.R. (1982) Effect of nitrogen and vitamin supplements on yeast growth and rate of fermentation of Rhine Riesling grape juice. *Food Technology in Australia*. **34**, 328–332.
Monk, P.R. (1984) Aspects of yeast technology. Advances in viticulture and oenology for economic gain: *Proceedings of the 5th Australian wine industry techical conference*, 29th Nov.–1 Dec. 1983, Perth, WA, edited by T.H. Lee and T.C. Sommers, pp. 403–418. Adelaide SA: The Australian Wine Research Institute.
Monk, P.R. (1986) Formation, utilization and excretion of hydrogen sulphide by wine yeast. *Australian and New Zealand Wine Industry Journal*, **1**, 10–16.
Monk, P.R. and Costello, P.J. (1984) Effect of ammonium phosphate and vitamin mixtures on yeast growth in preserved grape juice. *Food Technology in Australia*, **36**, 25–28.
Monk, P.R. and Storer, R.J. (1986) The kinetics of yeast growth and sugar utilization in tirage: the influence of different methods of starter culture preparation and inoculation levels. *American Journal of Enology and Viticulture*, **37**, 72–76.
Monk, P.R., Hook, D. and Freeman, B.M. (1987) Amino acid metabolism by yeast. In *Proceedings of the sixth Australian Wine Industry Technical Conference*, (Adelaide, S.A, 1986), edited by T.H. Lee, pp. 129–133. Adelaide, SA: Australian Industrial Publishers.
Monteiro, F. F. and Bisson, L. F. (1991a) Biological assay of nitrogen content of grape juice and prediction of sluggish fermentations. *American Journal of Enology and Viticulture*, **42**, 47–57.
Monteiro, F. F. and Bisson, L. F. (1991b) Amino acid utilisation and urea formation during vinification fermentations. *American Journal of Enology and Viticulture*, **42**, 199–208.
Monteiro, F.F., Trousdale, E.K. and Bisson, L.F. (1989) Ethyl carbamate formation in wine: Use of radioactively labeled precursors to demonstrate the involvement of urea. *American Journal of Enology and Viticulture*, **40**, 1–8.
Morozov, V.S. (1973) Effect of mineral fertilizers on the quality of wine materials. *Vinodelie i Vinogradarstvo SSSR*, **2**, 11–14.
Mounolou, J.-C. (1971) The properties and composition of yeast nucleic acids. In *The Yeasts*. Vol. 2, 1st edition, edited by A.H. Rose and J.S. Harrison. pp. 309–333. London: Academic Press.
Munoz, E. and Ingledew, W.M. (1989) Effect of yeast hulls on stuck and sluggish wine fermentations: Importance of the lipid component. *Applied and Environmental Microbiology*, **55**, 1560–1564.
Munoz, E. and Ingledew, W.M. (1990) Yeast hulls in wine fermentations-A review. *Journal of Wine Research*, **1**, 197–210.
Murphy, J.T. and Spence, K.D. (1972) Transport of *S*-adenosylmethionine in *Saccharomyces cerevisiae. Journal of Bacteriology*, **109**, 499–504.

Naider, F., Becker, J.M. and Katzir-Katchalski, E. (1974) Utilization of methionine-containing peptides and their derivatives by a methionine-requiring auxotroph of *Saccharomyces cerevisiae*. *Journal of Biological Chemistry*, **249**, 9–20.

Nelson, G. and Young, T.W. (1986) Yeast extracellular proteolytic enzymes for chill-proofing beer. *Journal of the Institute of Brewing*, **92**, 599–603.

Nes, W.R. (1973) Role of sterols in membranes. *Lipids*, **9**, 596–612.

Nordström, K. (1964) Formation of esters from acids by brewer's yeast. IV. Effect of higher fatty acids and toxicity of lower fatty acids. *Journal of the Institute of Brewing*, **70**, 233–242.

Nishino, H., Miyazaki, S. and Tohjo, K. (1985) Effect of osmotic pressure on the growth rate and fermentation activity on wine yeasts. *American Journal of Enology and Viticulture*, **36**, 170–174.

Nykänen, L. (1986) Formation and occurrence of flavour compounds in wine and distilled alcoholic beverages. *American Journal of Enology and Viticulture*, **37**, 84–96.

O'Conner-Cox, E.C. and Ingledew, W.M. (1989) Wort nitrogenous sources- Their use by brewing yeasts: A review. *American Society of Brewing Chemists Journal*, **47**, 102–108.

Ooghe, W. and Kastelijn, H. (1988) Aminozuurpatroon van druivenmost aangewend voor de bereiding van wijnen van gegarandeerde herkomst. *Belgian Journal of Food Chemistry and Biotechnology*, **43**(1), 15–21.

Ough, C.S. (1964) Fermentation rate of grape juice. I Effects of temperature and composition on white juice fermentation rates. *American Journal of Enology and Viticulture*, **15**, 167–177.

Ough, C.S. (1969) Ammonia content of California grapes. *American Journal of Enology and Viticulture*, **20**, 213–20.

Ough, C.S. (1976)Ethyl carbamate in fermentated beverages and foods. *Journal of Agricultural and Food Chemistry*, **24**, 323–328.

Ough, C.S. (1991) Influence of nitrogen compounds in grapes on ethyl carbamate formation in wines. In *Proceedings of the International Symposium on Nitrogen in Grapes and Wines* (Seattle, WA,1991), edited by J. Rantz, pp. 165–171. Davis, CA: American Society for Enology and Viticulture.

Ough, C.S. and Amerine, M.A. (1966) Fermentation rates of grape juice. IV Compositional changes affecting prediction equations. *American Journal of Enology and Viticulture*, **17**,163–173.

Ough, C.S. and Amerine, M.A. (1988) Nitrogen compounds. In *Methods for Analysis of Musts and Wines*, 2nd edn., pp. 172–195. New York: Wiley-Interscience.

Ough, C.S. and Bell, A.A. (1980) Effects of nitrogen fertilisation of grapevines on amino acid metabolism and higher alcohol formation during grape juice fermentation. *American Journal of Enology and Viticulture*, **31**, 122–123.

Ough, C.S. and Crowell, E.A. (1980) Nitrate determination in California USA musts and wines. *American Journal of Enology and Viticulture*, **31**, 344–346.

Ough, C.S. and Kriel, A. (1985) Ammonia concentrations of musts of different grape cultivars and vineyards in the Stellenbosh area. *South African Journal for Enology and Viticulture*, **6**, 7–11.

Ough, C.S. and Kunkee, R.E. (1968) Fermentation rates of grape juice. Biotin content of juice and its effect on alcoholic fermentation rate. *Applied Microbiology*, **16**, 572–576.

Ough, C.S. and Lee, T.H. (1981) Effect of vineyard nitrogen fertilisation level on the formation of some fermentation esters. *American Journal of Enology and Viticulture*, **32**, 125–127.

Ough, C.S. and Tabacman (1979) Gas chromatographic determinations of amino acid differences in Cabernet Sauvignan grapes and wines as affected by rootstocks. *American Journal of Enology and Viticulture*, **30**, 306–311.

Ough, C.S. and Trioli, G. (1988) Urea removal from wine by an acid urease. *American Journal of Enology and Viticulture*, **39**, 303–307.

Ough, C.S., Crowell, E.A. and Gutlove, B.R. (1988) Carbamyl compound reactions with ethanol. *American Journal of Enology and Viticulture*, **39**, 239–242.

Ough, C.S., Crowell, E.A. and Mooney, L.A. (1988) Formation of ethyl carbamate precursers during grape juice (Chardonnay) fermentation. I Addition of amino acid, urea and ammonia effects of fortification on nitrocellular and extracellular precursors. *American Journal of Enology and Viticulture*, **39**, 243–249.

Ough, C.S., Guymon, J.F. and Crowell, E.A. (1966) Formation of higher alcohols during the fermentation of grape juice at different temperatures. *Journal of Food Science*, **31**, 620–625.

Ough, C.S., Stevens, D. and Almy, J. (1989) Effects of vineyard and fertilization on urea and ethyl carbamate accumulation in wines. *American Journal of Enology and Viticulture*, **40**, 219–220.

Ough, C.S., Huang, Z., An, D. and Stevens, D. (1991) Amino acid uptake by four commercial yeasts at two different temperatures of growth and fermentation: Effects on urea excretion and readsorption. *American Journal of Enology and Viticulture*, **41**, 26–40.

Ough, C.S., Stevens, D., Sendovski, T., Huang, Z. and An, D. (1990) Factors contributing to urea formation in commercially fermented wines. *American Journal of Enology and Viticulture*, **41**, 68–73.

Palmqvist, U. and Äyräpää, T. (1969) Uptake of amino acids in bottom fermentations. *Journal of the Institute of Brewing*, **75**, 181–190.

Pampulha, M.E. and Loureiro, V. (1989) Interaction of the effects of acetic acid and ethanol on inhibitoin of fermentation in *Saccharomyces cerevisiae*. *Biotechnology Letters*, **11**, 269–274.

Papahadjopoulos, D., Cowden, M. and Kimelberg, H. (1973) Role of cholesterol in membranes: Effects on phospholipid-protein interactions, membrane permeability and enzymatic activity. *Biochimica et Biophysica Acta*, **330**, 8–26.

Peacock, W.L., Christensen, L.P. and Broadbent, F.E. (1989) Uptake, storage, and utilization of soil-applied nitrogen by Thompson seedless as affected by time of application. *American Journal of Enology and Viticulture*, **40**, 16–20.

Pekur, G.N., Bur'yan, N.I. and Pavlenko, N.M. (1981) Characteristics of nitrogen metabolism in wine yeasts under different fermentation conditions. *Applied Biochemistry and Microbiology*, **17**, 248–252.

Perl, M., Kearney, E.B. and Singer, T.P. (1976) Transport of riboflavin into yeast cells. *Journal of Biological Chemistry*, **251**, 3221–3228.

Petrov, V.V. and Okorokov, L.A. (1986) Ethanol increases the proton and anion permeability of *Saccharomyces carlsbergensis* yeast plasmalemma. Proceedings of the XII International Symposium on Yeasts, Lisbon, Abstracts, p. 127.

Peynaud, E. and Guimberteau, G. (1962) Sur la formation des alcools supérieurs pour les levures de vinification. *Annales de Technologie Argicole*, **11**, 85–105.

Pickerbell, A.T.W. (1986) The influence of free alpha-amino nitrogen in sorghum beer fermentation. *Journal of the Institute of Brewing*, **92**, 568–571.

Pierce, J.S. (1987) Horace Brown memorial lecture; the role of nitrogen in brewing. *Journal of the Institute of Brewing*, **93**, 378–381.

Pinto, I., Cardoso, H., Leao, C. and van Uden, N. (1989) High enthalpy and low enthalpy death in *Saccharomyces cerevisiae* induced by acetic acid. *Journal of Biotechnology and Bioengineering*, **33**, 1350–1352.

Prasad, R. and Rose, A.H. (1986) Involvement of lipids in solute transport in yeasts. *Yeast*, **2**, 205–220.

Proudlock, J.W., Haslam, J.M. and Linnane, A.W. (1969) Specific effect of unsaturated fatty acid depletion on mitochondrial oxidation phosphorylation in *Saccharomyces cerevisiae*. *Biochemical and Biophysical Research Communications*, **37**, 847–852.

Quain, D.E. (1988) Studies on yeast physiology-Impact of fermentation performance and product quality. *Journal of the Institute of Brewing*, **95**, 315–323.

Radler, F. (1978) Promotors of the yeast anaerobic growth. *Annales de Technologie Argicole.*, **27**, 203–213.
Radler, F. and Fäth, K.-P. (1991) Histamine and other biogenic amines in wines. In *Proceedings of the International Symposium on Nitrogen in Grapes and Wines* (Seattle, WA,1991), edited by J. Rantz, pp. 185–195. Davis, CA: American Society for Enology and Viticulture.
Rainbow, C. (1969) Brewer's Yeasts. In *The Yeasts*, 1st edition .Edited by A.H. Rose and J.S. Harrison, pp. 147–224. London: Academic Press.
Ramos, E.H., de Bongioanni, L.C. and Stoppani, A.O.M. (1980) Kinetics of L-[^{14}C] Leucine transport in *Saccharomyces cerevisiae. Biochimica et Biophysica Acta*, **599**, 214–231.
Rankine, B.C. (1963) Nature, origin and prevention of hydrogen sulphide aroma in wines. *Journal of the Science of Food and Agriculture*, **14**, 79–91.
Rankine, B.C. (1967) Formation of higher alcohols by wine yeasts, and relationship to taste thresholds. *Journal of the Science of Food and Agriculture*, **18**, 583–589.
Rankine, B.C. (1968) The importance of yeasts in determining the composition and quality of wines. *Vitis*, **7**, 22–49.
Rankine, B.C. (1977) Modern developments in selection and use of pure yeast cultures for winemaking. *Australian Wine and Brewing Spirit Review*, **96**(8), 31–33, **96**(9), 32–34.
Rankine, B.C. and Pocock, K.F. (1969) ß-Phenethanol and n-hexanol in wines: influence of yeast strain, grape variety and other factors; and taste thresholds. *Vitis*, **8**, 23–37.
Rankine, B.C., Fornachon, J.C.M., Boehm, E.W. and Cellier, K.M. (1971) Influence of grape variety, climate and soil on grape composition and on the composition and quality of table wines. *Vitis*, **10**, 33–50.
Rantz, J. (1991) *Proceedings of the International Symposium on Nitrogen in Grapes and Wines* (Seattle, WA, 1991). Davis, CA: American Society for Enology and Viticulture.
Rapp, A. and Franck, H. (1971) Über die bildung von äthanol und einiger aromastoffe bei modellgärversuchen in abhängigkeit von der aminosäurenkonzentration. *Vitis*, **9**, 299–311.
Rapp, A. and Versini, G. (1991) Influence of nitrogen compounds in grapes on aroma compounds of wine. In *Proceedings of the International Symposium on Nitrogen in Grapes and Wines* (Seattle, WA, 1991), edited by J. Rantz, pp. 156–164. Davis, CA: American Society for Enology and Viticulture.
Ratledge, C. and Evans, C.T. (1989) Lipids and their metabolism. In *The Yeasts* Vol. 3, 2nd edition, Edited by A.H. Rose and J.S. Harrison, pp. 367–455. Academic Press: London.
Rattray, J.B.M., Schibeci, A. and Kidby, D.K. (1975) Lipids of yeasts. *Bacteriological Reviews*, **39**(3), 197–231.
Reed, G. and Peppler, H.J. (1973) Yeast technology. Westport, Connecticut: AVI Publishing Company.
Ribéreau-Gayon, J. (1985) New developments in wine microbiology. *Amerian Journal of Enology and Viticulture*, **36**, 1–10.
Ribéreau-Gayon, J. and Peynaud, E. (1960) *Traité d'oenologie*. Paris: Lib. Polytechnique Béranger.
Ribéreau-Gayon, J. and Peynaud, E. (1966) Alimentation des levures et produits de leur métabolisme. *Annual de la Nutrition et de l'Alimentation*, **20**(1), 1–23.
Ribéreau-Gayon, J., Lafon-Lafourcade, S. and Bertrand, A. (1975) Le debourbage des mouts de vendange blanche. *Connaissance de la Vigne et du Vin*, **9**, 117–139.
Rogers, T.O. and Lichstein, H.C. (1969) Characterization of the biotin transport system in *Saccharomyces cerevisiae. Journal of Bacteriology*. **100**, 557–564.
Romano, P., Soli, M.G., Suzzi, G., Grazia, L. and Zambonelli, C. (1985) Improvement of wine *Saccharomyces cerevisiae. Applied and Environmental Microbiology*, **50**, 1064–1067.
Roon, R.J., Even, H. and Larimore, F. (1974) Glutamate synthase: Properties of the reduced nicotinamide adenine dinucleotide-dependent enzyme from *Saccharomyces cerevisiae. Journal of Bacteriology*, **118**, 89–95.

Roon, R.J., Larimore, F. and Levy, J.S. (1975) Inhibition of amino acid transport by ammonium ion in *Saccharomyces cerevisiae*. *Journal of Bacteriology*, **124**, 325–331.
Roon, R.J., Meyer, G.M. and Larimore, F.S. (1977) Evidence for a common component in kinetically distinct transport systems of *Saccharomyces cerevisiae*. *Molecular and General Genetics*, **158**, 185–191.
Roon, R.J., Even, H.L., Dunlop, P. and Larimore, F.L. (1975) Methylamine and ammonium transport in *Saccharomyces cerevisiae*. *Journal of Bacteriology*, **122**, 502–509.
Rose, A.H. (1987) Responses to the chemical environment. In *The Yeasts*, vol. 2, 2nd edn., edited by A.H. Rose and J.S. Harrison, pp. 5–40. London: Academic Press.
Rose, A.H. and Keenan, M.H.J. (1981) Amino-acid uptake by *Saccharomyces cererevisiae*. In *Proceedings of the European Brewing Convention Congress* (Copenhagen, 1981), pp. 207–216. London: IRL Press.
Rosén, K. (1989) Preparation of yeast for industrial use in production of beverages. In *Biotechnology applications in beverage production*. Edited by Cantarelli, C. and Lanzarini, G. pp. 169–187. London and New York: Elsevier Applied Science.
Rosi, I., Costamagna, L. and Bertuccioli, M. (1987) Screening for extracellular acid protease(s) production by wine yeasts. *Journal of the Institute of Brewing*, **93**, 322–324.
Rozes, N., Larue, F. and Ribéreau-Gayon, P. (1988) Effect of a variation of grape must temperature on the fermentative ability and the neutral lipid content of *Saccharomyces cerevisiae*. *Biotechnology Letters*, **10**, 821–824.
Rytka, J. (1975) Positive selection of general amino acid permease mutants in *Saccharomyces cerevisiae*. *Journal of Bacteriology*, **121**, 562–570.
Saita, M. and Slaughter, J.C. (1984) Acceleration of the rate of fermentation by *Saccharomyces cerevisiae* in the presence of ammonium ion. *Enzyme Microbiology Technology*, **6**, 375–378.
Salmon, J.M. (1989) Effect of sugar transport inactivation on sluggish and stuck oenological fermentations. *Applied and Environmental Microbiology*, **55**, 953–958.
Samvelyan, A.M., Khachartryon, A.L. and Martirosyan, K.B. (1972) Fertilization of vineyards and quality of sherry wine materials. *Sadovod. Vinograd, Vinodel. Mold*. **27**, 53–55.
Sato, T., Ohsumi, Y. and Anraku, Y. (1984) Substrate specificities of active transport systems for amino acids in vacuolar membrane vesicles of *Saccharomyces cerevisiae*. *Journal of Biology and Chemistry*, **259**, 11505–11508.
Schlenk, F., Dainko, J.L. and Svihla, G. (1970). The accumulation and intracellular distribution of biological sulfonium compounds in yeast. *Archives of Biochemistry and Biophysics*, **140**, 228–236.
Schreier, P. (1979) Flavour composition of wines: a review. *CRC Critical reviews in Food Science and Nutrition*. pp. 59–111.
Schwenke, J. (1991) Vacuoles, internal membranous systems and vesicles. In *The Yeasts*, Vol. 4 2nd. edition, edited by A.H. Rose and J.S. Harrison. pp. 347–432. London: Academic Press.
Sentheshanmuganathan, S. (1960) The mechanism of the formation of higher alcohols from amino acids by *Saccharomyces cerevisiae*. *Biochemistry Journal*, **74**, 568.
Shimazu, Y. and Watanabe, M. (1981) Effects of yeast strains and environmental conditions on forming of organic acids in must during fermentation (Japanese). *Journal of Fermentation Technology*, **59**, 27–32.
Shukla, H., Viswanathan, L. and Singh, M. (1982) A specific inducible serine transport system in distiller's yeast *Saccharomyces cerevisiae*. *Biochemistry International*, **5**, 253–262.
Simpson, R.F. (1979) Some important aroma components of white table wine. *Food Technology Australia*, **31**, 516, 518–522.
Sinton, T.H., Ough, C.S., Kissler, J.J. and Kasimatis, A.N. (1978) Grape juice indicators for prediction of potential wine quality. I. Relationship between crop level, juice and wine

composition, and wine sensory ratings and scores. *American Journal of Enology and Viticulture.* **29**, 267–271.

Slaughter, J.C. (1989) The effects of carbon dioxide on yeasts. In *Biotechnology applications in beverage production*, edited by C. Cantarelli, and G. Lanzarini, pp. 49–64. London and New York: Elsevier Applied Science.

Slaughter, J.C. and Housden, S.J. (1978) Effect of NH_4^+ on the production of ethanol and higher alcohols by *Saccharomyces cerevisiae. VIth International Specialised Symposium on Yeasts*, Montpellier, SII 7–8.

Spayd, S.E., Wample, R.L., Nagel, C.W., Stevens, R.G. and Evans, R.G. (1991) Vineyard nitrogen fertilization effects on must and wine composition and quality. *International Symposium on Nitrogen in Grapes and Wine*. (Seattle, WA, 1991) edited by J. Rantz, pp. 196–199. Davis, CA: American Society for Enology and Viticulture.

Sponholz, W.R. (1991) Nitrogen compounds in grapes, must, and wine. *International Symposium on Nitrogen in Grapes and Wine*. (Seattle, WA, 1991) edited by J. Rantz, pp. 67–77. Davis, CA: American Society for Enology and Viticulture.

Stratford, M. and Rose, A.H. (1985) Hydrogen sulphide production from sulphite by *Saccharomyces cerevisiae. Journal of General Microbiology*, **131**, 1417–1424.

Stratford, M. and Rose, A.H. (1986) Transport of sulphur dioxide by *Saccharomyces cerevisiae. Journal of General Microbiology*, **132**, 1–6.

Strydom, M., Kirschbaum, A.F. and Tromp, A. (1982) Ergosterol concentrations of several different *Saccharomyces cerevisiae* yeast strains. *South African Journal for Enology and Viticulture*, **3**, 23–28.

Stuckey, W., II., Henschke, P.A. and Gawel, R. (1991) The effect of lees contact time on Chardonnay wine composition. In *Proceedings of the International Symposium on Nitrogen in Grapes and Wines* (Seattle, WA, 1991), edited by J. Rantz, pp. 315–319. Davis, CA: American Society for Enology and Viticulture.

Sturley, S.L. and Young, T.W. (1988) Extracellular protease activity in a strain of *Saccharomyces cerevisiae. Journal of the Institute of Brewing*, **94**, 23–27.

Subden, R.E. (1991) Wine yeast: Selection and modification. In *Yeast Strain Selection.* Edited by C.H. Panchal. pp. 113–137. Xoma Corporation, Berkeley, California.

Suizu, T., Iimura, Y., Gomi, K., Tokahashi, K., Hara, S., Yoshizawa, K. and Tamura, G. (1990) Construction of urea non-producing yeast *Saccharomyces cerevisiae* by disruption of the CARI gene. *Agricultural and Biological Chemistry*, **54**, 537–539.

Sumrada, R. and Cooper, T.G. (1978) Control of vacuole permeability and protein degradation by the cell cycle arrest signal in *Saccharomyces cerevisiae. Journal of Bacteriology*, **136**, 234–246.

Sumrada, R. and Cooper, T.G. (1982) Isolation of the CARI gene from *Saccharomyces cerevisiae* and analysis of its expression. *Molecular and Cellular Biology*, **2**, 1514–1523.

Suomalainen, H. and Oura, E. (1971) Yeast nutrition and solute uptake. In *The Yeasts* Vol. 2, 1st Edition. edited by A.H. Rose and J.S. Harrison, pp. 3–74. London: Academic Press.

Surdin-Kerjan, Y., Cherest, H. and de Robichon-Szulmajster, H. (1973) Relationship between methionyl transfer ribonucleic acid cellular content and synthesis of methionine enzymes in *Saccharomyces cerevisiae. Journal of Bacteriology*, **113**, 1156–1160.

Suzzi, G., Romano, P. and Zambonelli, C. (1985) *Saccharomyces* strain selection in minimizing SO_2 requirement during vinification. *American Journal of Enology and Viticulture*, **36**, 199–202.

Tarantola, C. (1967) Vins "Moscato d'Asti" et "Asti Spumante". *Annals Instuto Sperimentale per l'Enologia*, Asti II **8**, 469 480 (1967–1969).

Tarantola, C. (1970) Nitrogenous matter in grapes and wine. *Vini d'Italia*, **12** (69), 457–459, 463.

Tegmo-Larsson, I.M. and Henick-Kling, T. (1990) Ethyl carbamate precursors in grape juice and the efficiency of acid urease on their removal. *American Journal of Enology and Viticulture*, **40**, 106–108.

Thomas, D.S. and Rose, A.H. (1979) Inhibitory effect of ethanol on growth and solute accumulation by *Saccharomyces cerevisiae* as affected by plasma-membrane lipid composition. *Archives of Microbiology*, **122**, 49–55.

Thornton, R.J. and Rodriguez, B. (1987) Genetics of wine microorganisms: potentials and problems. *Proceedings of the sixth Australian wine industry technical conference*; 14–17 July 1986; Adelaide, SA, edited by T.H. Lee, pp. 98–102. Adelaide, SA: Australian Wine Industry Technical Conference.

Tokuyama, T., Kuraishi, H., Aida, K. and Uemura, T. (1973) Hydrogen sulfide evolution due to a pantothenic acid deficiency in the yeast requiring this vitamin, with special reference to the effect of adenosine triphosphate on yeast cysteine desulfhyrase. *Journal of General and Applied Microbiology*, **19**, 439–466.

Tromp, A. (1984) The effect of yeast strain, grape solids, nitrogen and temperature on fermentation rate and wine quality. *South African Journal for Enology and Viticulture*, **5**, 1–6.

Usseglio-Tomasset, L. and Bosia, P.D. (1990) Amino acids and ogliopeptides development from the must to the wine. *Bulletin de l'Office International de la Vigne et du Vin*, **707–708**, 21–46.

Vandenbol, M., Jauniaux, J.-C., Vissers, S. and Grenson, M. (1987) Isolation of the NPR1 gene responsible for the reactivation of ammonia-sensitive amino-acid permeases in *Saccharomyces cerevisiae. European Journal of Biochemistry*, **164**, 607–612.

van Dijken, J.P. and Scheffers, W.A. (1986) Redox balances in the metabolism of sugars by yeasts. *FEMS Microbiology Reviews*, **32**, 199–224.

van Rooyen, P.C. and Tromp, A. (1982) The effect of fermentation time (as induced by fermentation and must conditions) on the chemical profile and quality of a chenin blanc wine. *South African Journal for Enology and Viticulture*, **3**, 75–80.

van Uden, N.J. (1989) Effects of alcohols on membrane transport in yeasts. In *Alcohol toxicity in yeasts and bacteria*, edited by N. J. van Uden, pp. 135–146. Boca Raton: CRC Press.

van Uden, N. (1989) Alcohol toxicity in yeasts and bacteria. Boca Raton: CRC Press

van Vuuren, H.J.J. and Wingfield, B.D. (1986) Killer yeasts-cause of stuck fermentations in a wine celler. *South African Journal for Enology and Viticulture*, **7**, 113–118.

van Wyk, C.J., Augustyn, P., de Wet, P. and Joubert, W.A. (1977) The role of isoamyl acetate in the typical fermentation bouquet of *Vitis vinifera* cultivar Pinotage. *Proceedings of the first congress and 20th annual general meeting of the South African Society for Enology and Viticulture*, Cape Town, SA, Nov. 21–22, 1977. pp. 155–171.

Vas, K. (1953) On the mechanism of antimicrobial action: Interference with the cytoplasmic membrane. *Agrokémia és Talajtan*, **2**, 1–16.

Verma, R.S., Rao, T.V.G. and Prasad, R. (1984) An inducible, specific and derepressible transport of L-serine in *Saccharomyces cerevisiae*. *Biochimica et Biophysica Acta*, **778**, 289–297.

Voilley, A., Farès, K., Lorient, D. and Simatos, D. (1988) Interaction between proteins and aroma compounds. In *Functional properties of food proteins*. Edited by Lasztity, R. and Ember-Karpal, M. Proceedings of International Seminar, MTESZ: Budapest.

Vos, P.J.A. (1981) Assimilable nitrogen-A factor influencing the quality of wines. In *International Association for modern winery technology and management. 6th International Oenological Symposium* (28–30th April 1981). Mainz, Germany. pp. 163–180.

Vos, P.J.A. and Gray, R.S. (1979) The origin and control of hydrogen sulfide during fermentation of grape must. *American Journal of Enology and Viticulture*, **30**, 187–197.

Vos, P.J.A., Crous, E. and Swart, L. (1980) Fermentation and the optimal nitrogen balance of musts. *Die Wynboer*, **582**, 58–59, 60, 63.

Vos, P.J.A., Zeeman, W. and Heyman, H. (1978) The effect on wine quality of di-ammonium phosphate addition to musts. *Proceedings of the South African Society for Enology and Viticulture*; 3–4 October 1978; Cape Town, South Africa. pp. 87–104. Stellenbosch: SASEV.

Wainwright, T. (1971) Production of H_2S by yeasts: Role of nutrients. *Journal of Applied Bacteriology*, **34**, 161–171.

Waters, E.J. (1991) Heat unstable wine proteins and their interactions with wine polysaccharides. PhD Thesis. University of Adelaide, S.A.

Watson, T.G. (1976) Amino–acid pool composition of *Saccharomyces cerevisiae* as a function of growth rate and amino-acid nitrogen source. *Journal of General Microbiology*, **96**, 263–268.

Weimken, A. and Durr, M. (1974) Characterisation of the amino acid pools in the vacuolar compartment of *Saccharomyces cerevisiae*. *Archives of Microbiology*, **101**, 45–57.

Wenzel, K. and Dittrich, H.H. (1978) Zur beeinflussung der schwefelwasserstoff-bildung der hefe durch trub, stickstoffgehalt, molekularen schwefel und kupfer bei der vergärung von traubenmost. *Wien-Wissenschaft*, **33**, 200–214.

Whitney, P.A. and Magasanik, B. (1973) The induction of arginase in *Saccharomyces cerevisiae*. *Journal of Biological Chemistry*, **248**, 6197–6202.

Wiame, J.-M., Grenson, M. and Arst, H.N., Jr. (1985) Nitrogen catabolite repression in yeast and filamentous fungi. *Advances in Microbial Physiology*, **26**, 2–88.

Wickner, R.B. (1975) Mutants of *Saccharomyces cerevisiae* that incorporate deoxythymidine 5'monophosphate into DNA in vivo. In *Methods in Cell Biology*. Vol.XI, edited by D.M. Prescott. pp. 295–302.

Williams P.J., Sefton, M.A. and Wilson, B. (1989) Nonvolatile conjugates of secondary metabolites as precursors of varietal grape flavour components. *Flavour chemistry trends and developments*; symposium at the third chemical congress of North America (195th national meeting of the American Chemical Society, Toronto, Ontario, Canada, 5–11 June 1988) edited by R. Teranishi, R.G. Buttery and F. Shahidi. pp. 35–48. Washington, DC: American Chemical Society.

Woodward, J.R. and Cirillo, P.V. (1977) Amino acid transport and metabolism in nitrogen-starved cells of *Saccharomyces cerevisiae*. *Journal of Bacteriology*, **130**, 714–723.

Wucherpfenning, K. and Bretthauer, G. (1970) Über die bildung von flüchtigen Aromastoffen in Traubenwein in Abhängigkeit von der Mostvorbehandlung sowie von der verwendeten Heferasse. *Mitteil Klosterneuburg*, **20**, 36–46.

Yamagata, S. (1987) Partial purification and some properties of homomserine O-acetyltransferase of a methionine auxotroph of *Saccharomyces cerevisiae*. *Journal of Bacteriology*, **169**, 3458–3463.

Yamagata, S. (1989) Roles of O-acetyl-L-homoserine sulfhydrylases in micro-organisms. *Biochimie*, **71**, 1125–1143.

Yoshimo, M. and Murakami, K.J. (1982) AMP deaminase reaction as a control system of glycolysis in yeast: activation of phosphofructokinase and pyruvate kinase by the AMP deaminase ammonia system. *Journal of Biology and Chemistry*, **257**, 2822–2828.

Zacharski, C.A. and Cooper, T.G. (1978) Metabolite compartmentation in *Saccharomyces cerevisiae*. *Journal of Bacteriology*, **135**, 490.

Zambonelli, C., Mutinelli, P. and Pacchetti, G.(1975) Biosynthesis of sulphur amino acids in *Saccharomyces cerevisiae*. I. Genetic analysis of leaky mutants of sulphite reductase. *Archives of Microbiology*, **102**, 247–251.

Zeeman, W., Snyman, J.P. and van Wyk, C.J. (1982) The influence of yeast strain and malolactic fermentation on some volatile bouquet substances and on quality of table wines. In: *Proceedings of the Grape and Wine Centennial Symposium*, pp. 79-90. University of California Press.

ACKNOWLEDGEMENTS

VJ gratefully acknowledges the receipt of a scholarship from The Australian Wine Research Institute. The Grape and Wine Research Council is thanked for financial support.

Chapter 5

YEASTS— METABOLISM OF ORGANIC ACIDS

FERDINAND RADLER

Institut für Mikrobiologie und Weinforschung der Johannes Gutenberg-Universität, Postfach 3980, D-6500 Mainz, Germany

Chapter Outline

INTRODUCTION

Yeasts can be used for the industrial production of organic acids, but in these processes paraffins are the substrates and they are oxidized by species of *Candida*, *Pichia*, *Brettanomyces* or others, as reviewed by Schindler *et al.* (1990). In winemaking, a completely different situation prevails. The main yeast is *Saccharomyces cerevisiae*, the conditions are anaerobic and the predominant substrates are sugars. The acids formed or metabolized by yeasts during the fermentation of grape must are never the main products. In fact, lactic acid bacteria have a more prominent role than yeasts in metabolizing organic acids during winemaking. The relative significance of yeasts and lactic acid bacteria in metabolizing organic acids during wine fermentation was the subject of much controversy in the early literature and this has been aptly reviewed by Kunkee (1967).

For quite a while, the metabolism of organic acids by yeasts during anaerobiosis has been neglected, but with the convenient and more sensitive analytical methods now available, considerable progress has been made in understanding the biochemistry of these reactions. For reviews of the metabolism of organic acids by yeasts see Whiting (1976), Radler (1981, 1984) and Dittrich (1987, 1989).

Grape must and wine contain a variety of organic acids. Their concentrations vary greatly and depend on the condition and maturity of the grapes as well as the biochemistry of fermentation. The main acids and some of the minor acids are listed in Table 5.1. The main acids of must are malic acid and tartaric acid which are produced by the grape. Many of the acids found in wine are of microbial origin. A comprehensive list of the minor acids has been compiled by Drawert *et al.* (1978).

FORMATION OF ORGANIC ACIDS

Succinic Acid

Succinate is the main carboxylic acid produced by yeasts during wine fermentation. It is generally present in amounts up to 2 g/l (Thoukis *et al.*, 1965). The formation of succinate is rapid; when sugar is added to resting yeast cells, an increase in acidity is observed after 10 min, and this is mainly caused by succinic acid (Sigler *et al.*, 1980). The formation of succinate depends on the yeast strain and is correlated with the formation of 2-oxoglutarate. Its production is particularly favored by the presence of glutamate (Heerde and Radler, 1978). When yeasts are grown anaerobically on glutamate, the oxidative pathway (via succinyl-CoA) is operative for the synthesis of succinate (Figure 5.1). Accordingly, the important citric acid cycle enzyme, 2-oxoglutarate dehydrogenase, has been found in fermenting yeast cells; this enables them to produce succinate during fermentation by the oxidative reactions. A related fact

TABLE 5.1 The acids of grape must and wine and their relation to yeast metabolism.

Acid	*Concentration in must or wine (g/l)*	*Action of yeast during fermentation*	
		Decomposition fermentation, or uptake	*Formation*
Major constituents			
Tartaric	1–4	–	–
L-Malic	0–8	+	+
Succinic	0.5–2	–	+
Acetic	0–3	(+)	+
L-Lactic	0.1–6	–	–
Amino acids	1–6	+	+
Gluconic acid[a]	0–3	–	–
Some minor constituents			
Citric	0–0.5	–	–
D-Lactic	0.1–0.5	–	+
Pyruvic	0–0.5	(+)	+
2-Oxoglutaric	0–0.2	–	+
Dimethyl glyceric	0.1–0.2	–	+
Citramalic	0.1–0.2	–	+
Galacturonic	0.1	–	–
Glucuronic	0.06	–	–
Mucic	0.05	–	–

[a] in moldy grapes.

is that fermenting yeasts contain succinyl-CoA synthetase, an enzyme that has been partially purified and characterized (Schwartz *et al.*, 1983). Malate and aspartate may be converted to small amounts of succinate by enzymes of the reductive pathway (via oxalacetate, malate, and fumarate). At least some of the oxalacetate is formed by a biotin-dependent pyruvate carboxylase (Schwartz and Radler, 1988). However, when fumarate is added to the medium, it is not reduced to succinate.

Malic Acid

Although reactions that decompose malate predominate in wine fermentations, some yeast strains are able to produce this acid (Whiting, 1976). Such strains

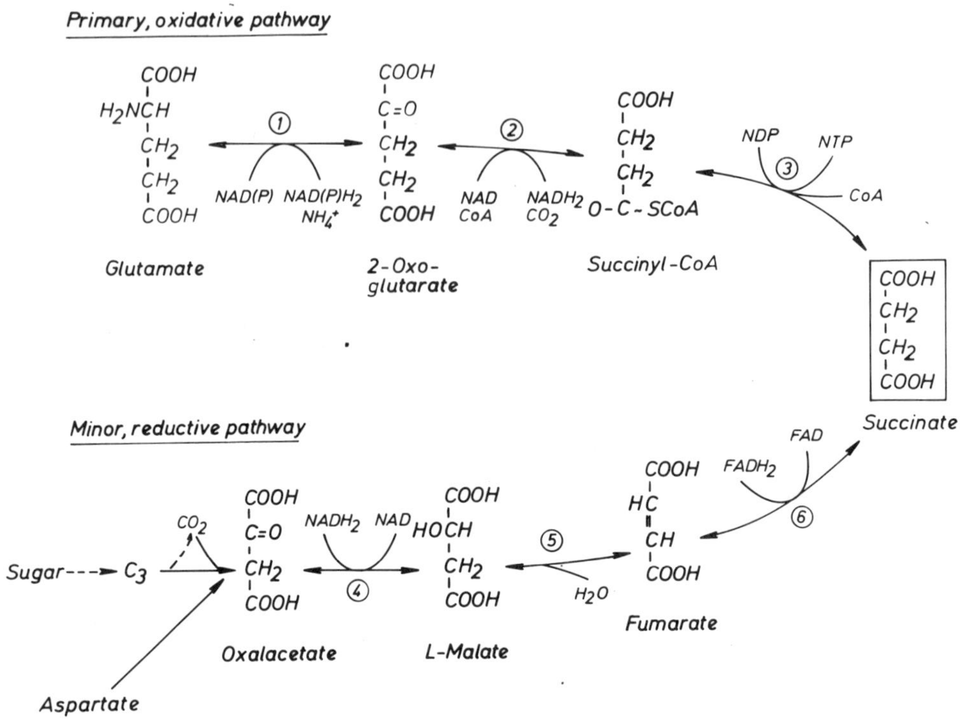

FIGURE 5.1 The main oxidative and the minor reductive reactions of the tricarboxylic acid cycle used by yeast for the formation of succinate during fermentation. 1. Glutamate dehydrogenase; 2. 2-oxoglutarate dehydrogenase; 3. succinyl-CoA synthetase; 4. malate dehydrogenase; 5. fumarase; 6. succinate dehydrogenase.

have been isolated from cider but they are also found among wine yeasts. Malate formation is greatly influenced by culture conditions: it is favored by high sugar concentrations (20–30%), pH-values of about pH 5, and limiting concentrations of nitrogen compounds (100–250 mg N/l). The presence of carbon dioxide is required (Radler and Lang, 1982). The known reactions that could potentially lead to malate formation are compiled in Figure 5.2. A backward reaction of malic enzyme is possible but not very likely.

When radioactive carbon dioxide is added to fermenting yeast cultures, it results in the formation of approximately equal quantities of labelled malate and succinate with one strain of *S. cerevisiae,* whereas about ten times more malate than succinate is formed by another strain formerly known as *Saccharomyces uvarum.* The synthesis of malate and succinate is not strictly linked. Under identical conditions, the ratio of malate to succinate varies from 2.1 to 0.2 with different yeast strains. Avidin, a specific inhibitor for biotin-dependent carboxylations, prevents the synthesis of malate, but does not significantly lower the formation of succinate. Therefore, it can be assumed that malate is

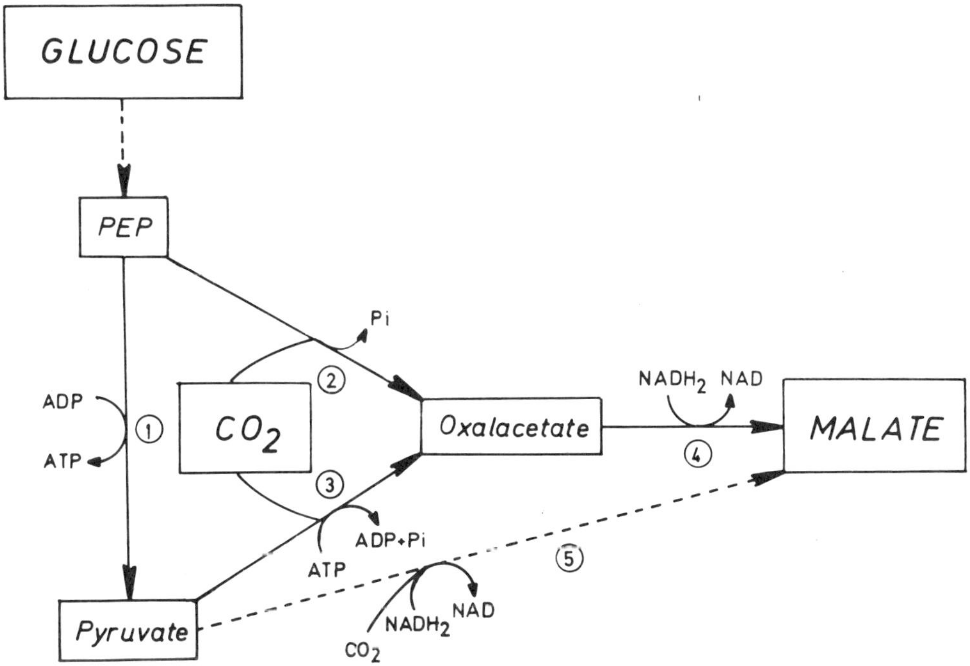

FIGURE 5.2 The most likely mechanism for the synthesis of malate by yeast during fermentation. 1. Pyruvate kinase; 2. phosphoenolpyruvate (PEP) carboxylase; 3. pyruvate carboxylase; 4. malate dehydrogenase; 5. malic enzyme (an unlikely reaction).

formed by reduction of oxalacetate, the synthesis of which is catalyzed by a biotin-dependent pyruvate carboxylase (Schwartz and Radler, 1988).

In wine, the usually high concentration of amino acids and the low pH may prevent the formation of malate and, therefore, its decomposition prevails. On the other hand, depending on the conditions, part of the malate found after fermentation might have been synthesized by yeasts.

Acetic Acid

Acetate is the main product of the oxidation of ethanol by acetic acid bacteria. But acetate can also be formed anaerobically by lactic acid bacteria from sugars, particularly pentoses, and from various carboxylic acids. Wine is legally regarded as spoiled when its acetate content is higher than about 20 meq/l. The exact value is different for various types of wines and varies in different countries. Usually, acetic acid is determined as volatile acid (Carr and Whiting, 1955). Acetate is also a normal by-product of the alcoholic fermentation by yeasts.

Shimazu and Watanabe (1981) surveyed the effect of 20 different yeast strains from seven genera on the content of various organic acids in wines after fermentation of Koshu grape must. The highest amount of acetic acid was formed by *Hansenula anomala*, *Kloeckera apiculata* and *Candida krusei*. They observed that acetic acid formation by strains of *S. cerevisiae* is affected by sugar concentration and pH. This was also reported by Monk and Cowley (1984) who, furthermore, found that addition of 5 mg/l nicotinic acid to synthetic medium enhanced the formation of acetic acid at low and high sugar concentrations.

In a thorough study, Hannemann (1985) demonstrated that various yeast strains show great differences in the anaerobic production of acetate (Figure 5.3). Among 100 strains of various genera, 13 strains of *S. cerevisiae* were found to produce more than 1 g/l of acetic acid in a synthetic medium. Such strains are not likely to be suitable for wine fermentation. Acetic acid is formed early during fermentation and its production stops as soon as the sugar is fermented. The amount of acetic acid formed depends on the sugar concentration, the nitrogen source and the pH. Generally, more acetic acid is produced under conditions that are growth limiting. The amount of acetic acid formed increases with sugar concentrations above about 20% (w/v). The nitrogen source has a variable effect. Some compounds (ammonium, glutamic acid, glutamine, asparagine) inhibit while others (methionine, valine) promote the formation of acetate when sufficient nitrogen compounds are available for growth. When the nitrogen

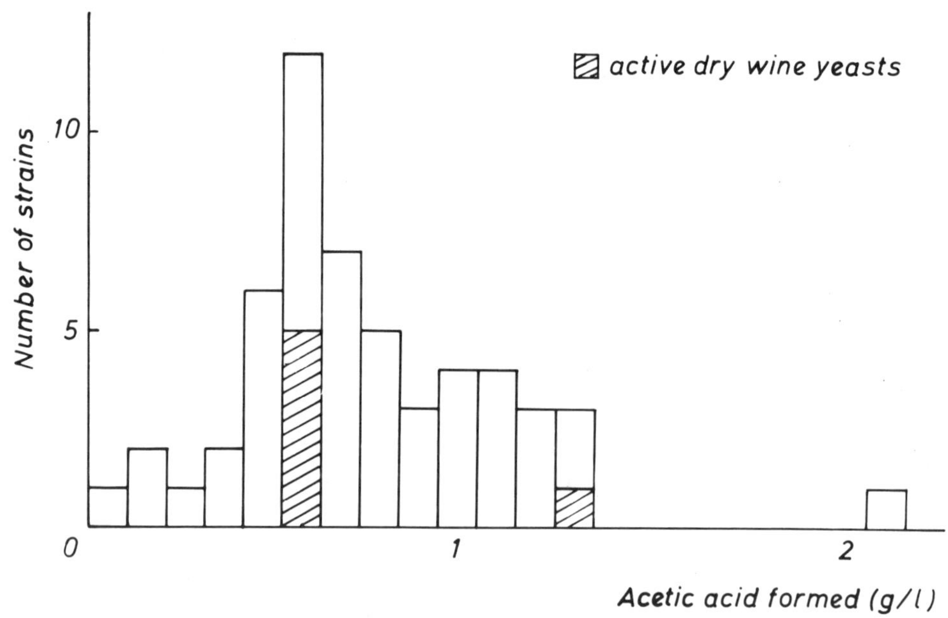

FIGURE 5.3 Distribution of the characteristic "formation of acetic acid" among yeast strains. The shaded area corresponds to strains isolated from preparations of active dry yeast (adapted from Hannemann, 1985).

supply is insufficient, an increase in sugar concentration favors the production of acetate. The pH-value has a great influence; increasing amounts of acetic acid are formed below pH 3.2 and at pH values more neutral than pH 4. Aeration has little influence on acetic acid formation by yeasts at high sugar concentrations. Resting cells of *S. cerevisiae* produce acetate from glucose, pyruvate or acetaldehyde. The latter two compounds yield acetate and ethanol in a molar ratio of 1:1. Of several enzyme activities, only the NADP-dependent aldehyde dehydrogenase (E.C.1.2.1.4.) showed a correlation with the ability of different yeast strains to form acetic acid. This correlation is shown in Figure 5.4 (Hannemann, 1985).

Millan and Ortega (1988) found that five yeasts (four strains of *S. cerevisiae* and one strain of *Torulaspora delbrueckii*) excreted different amounts of acetaldehyde and acetic acid during the fermentation of grape juice. Great differences in the activities of alcohol and acetaldehyde dehydrogenase were observed at different stages of fermentation and with different strains. It was proposed that the formation of acetic acid was not related to the activity of the enzymes alcohol dehydrogenase and aldehyde dehydrogenase but probably to the equilibrium between their oxidized and reduced coenzymes and to the rate of uptake of reducing sugars and acetate.

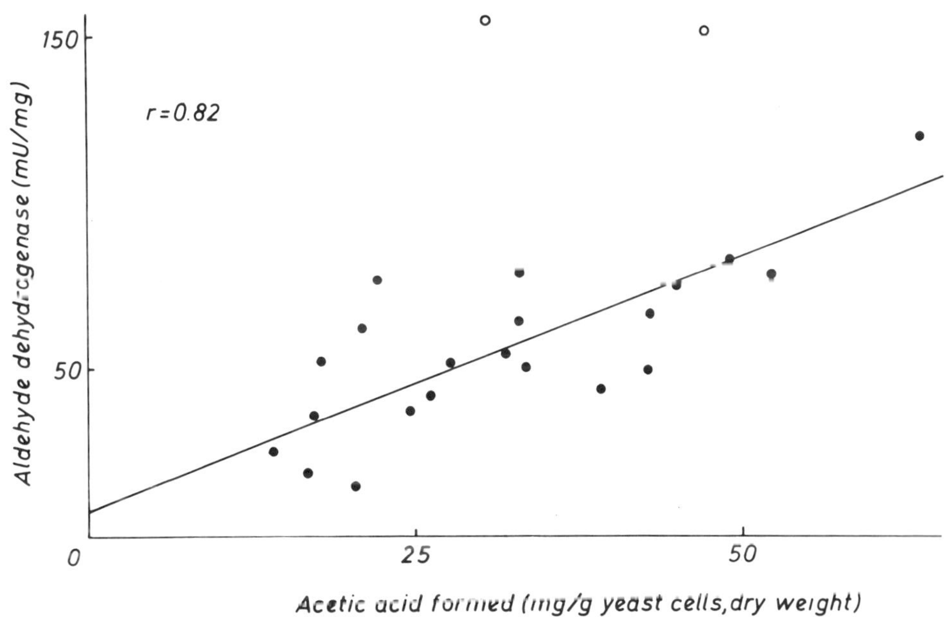

FIGURE 5.4 Correlation between the formation of acetic acid during fermentation and the activity of aldehyde dehydrogenase in cell extracts of various strains of *Saccharomyces cerevisiae* (adapted from Hannemann, 1985). o=strains forming ethylacetate that have not been included for the calculation of *r*.

Usually, less than 0.2 g/l of acetic acid are formed in beer. Aeration of pitching yeast increases the specific activity of pyruvate decarboxylase, resulting in higher acetic acid concentrations during wort fermentation as compared with fermentations conducted with anaerobic pitching yeast. A direct conversion of pyruvate to acetate is assumed to be catalyzed by Fe^{3+}-linked pyruvate decarboxylase of the pyruvate dehydrogenase complex (Amata and Germain, 1990).

Lactic Acid

Usually yeasts produce only small amounts of free carboxylic acids during fermentation (about 20–25 meq/l); of these acids only a small fraction is lactate. It has been observed that *Saccharomyces veronae* produces L-lactate (Peynaud *et al.*, 1967). Among several hundred yeast strains investigated, only a few strains were observed that produced comparatively large quantities of up to 100 meq or 9 g/l of L-lactate; see Figure 5.5 (Radler, 1972). These yeasts belong to the species *Torulopsis pretoriensis* (synonym *Saccharomyces pretoriensis*) which resembles *S. cerevisiae*. The acid is synthesized from sugar by reduction of pyruvate by L-lactate dehydrogenase which has been partially purified (Genitsariotis, 1979).

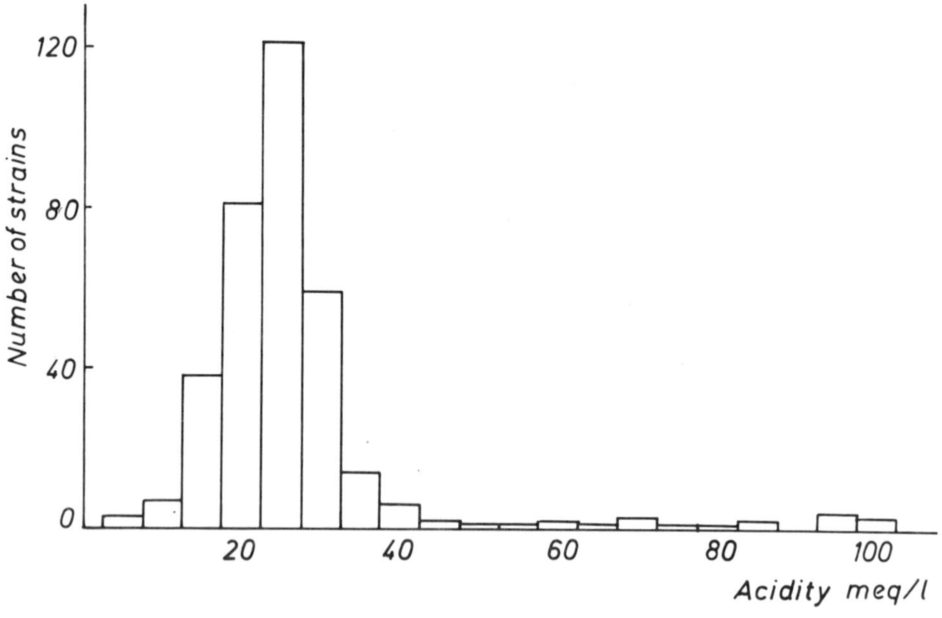

FIGURE 5.5 Distribution of the characteristic "acidification" among yeasts of various genera.

Oxo and Hydroxy Acids

Yeasts are able to produce several oxo acids: pyruvate, 2-oxoglutarate, 2-oxobutyrate, 2-oxoisovalerate, 2-oxo-3-methylvalerate and 2-oxoisocaproate. In brewing, these acids are regarded as products of amino acid metabolism (Dellweg and Kettler, 1975). The two most important oxo acids, pyruvate and 2-oxoglutarate, are present in grape must in low concentrations only. During fermentation, their amount increases, and pyruvate is later partially metabolized by the yeasts. Both acids may be formed from the corresponding amino acids alanine and glutamate but they are also excreted by yeast cells growing in the presence of low concentrations of nitrogen compounds (Ponader, 1973). At sub-optimum supply of nitrogen, more pyruvate is formed than at sufficient nitrogen supply. Furthermore, the pyruvate content does not decrease during the later stages of fermentation as is the case when sufficient nitrogen is available (Figure 5.6). The oxo acids are important in wine fermentation because they are able to bind to SO_2, thus lowering the content of

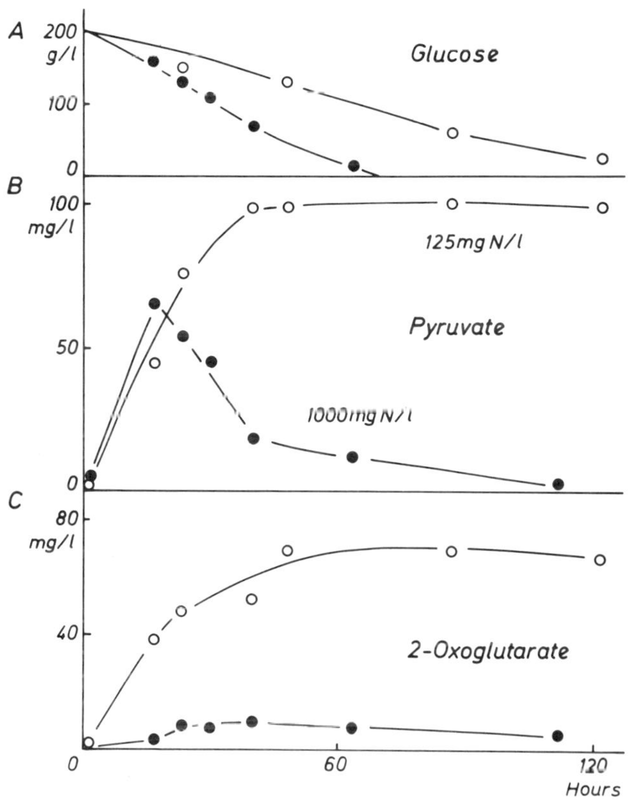

FIGURE 5.6 Influence of nitrogen (N) content of the medium on the rate of fermentation of glucose (A) and the formation of pyruvate (B) and 2-oxoglutarate (C) by wine yeast. o = Suboptimum N-supply (125 mg N/l); ● = sufficient N-supply (1000 mg N/l).

free SO_2 in wine which is regarded as the biologically active part and is necessary for the safe preservation of wine. Like the other acids mentioned in this paragraph, 2-methyl-2,3-dihydroxy butyric acid is a metabolite of amino acid biosynthesis. This acid is formed by reduction of 2-acetyl lactate, a precursor of valine. 2-Methyl-2,3-dihydroxy butyric acid is found in wine in amounts up to about 150 mg/l (Dittrich, 1989). It is formed during the later stages of fermentation.

Although citric acid is a regular constituent of grape must and wine (usually in amounts of less than 500 mg/l), yeasts do not contribute significantly to its content, which can increase or decrease during fermentation (Dittrich, 1989).

Fatty Acids

Although it is justified to regard the volatile acid of wine as acetic acid, a variety of other fatty acids are present, but usually more than 90% of the volatile acid consists of acetic acid. Sponholz and Dittrich (1986) have analyzed wines of different grape varieties, areas, year of harvest and sugar content for the presence of fatty acids (Table 5.2). Except for formic acid, that is assumed to originate from moldy grapes, only a few mg of the various volatile fatty acids were found. Although only minute amounts of these fatty acids are detected in wine, it has been claimed that some of them, in particular octanoic and decanoic acids that are formed by yeast, might be inhibitors of fermentation (Lafon-Lafourcade *et al.*, 1984). About 10 mg/l of octanoic and 5 mg/l (of each) hexanoic and decanoic may be formed by yeast in clarified juice without insoluble grape solids (Edwards *et al.*, 1990). Nagel *et al.* (1988) have observed that yeast strains differ in their sensitivity to decanoic and octanoic acids.

TABLE 5.2 The content of volatile fatty acids observed in 138 different German white wines and 46 various red wines. (From Sponholz and Dittrich, 1986.)

Acid	*Range of concentrations (mg/l)*	
	White wines	*Red wines*
Formic acid	2.4–89	3.3–22
Acetic acid	109–1150	129–860
Propionic acid	0.1–6.6	trace–13.9
2-Methylpropionic acid	0.3–6.4	trace–5.7
Butyric acid	0.3–3.1	0.3–2.9
2-Methylbutyric acid	0–0.7	0.1–1.5
3-Methylbutyric acid	0.1–1.8	0.2–1.4
Hexanoic acid	0.4–5.4	1.0–4.9
Octanoic acid	0–11.3	1.2–9.6
Decanoic acid	0–7.4	0.3–35

DECOMPOSITION OF ORGANIC ACIDS

Malic Acid

The metabolism of malate by wine yeasts has been studied by many authors (Peynaud *et al.*, 1964; Rankine, 1966; Fuck and Radler, 1972; Shimazu and Watanabe, 1981; Wenzel *et al.*, 1982; Carre *et al.*, 1983; Rodriguez and Thornton, 1990). Wine yeasts of *S. cerevisiae* are capable of metabolizing malate during fermentation but generally in small amounts, 3–45%, depending on the strain. Similar weak utilization is also exhibited by various wine species of *Kloeckera*, *Candida*, *Pichia* and *Hansenula*. Thus, these species as well as *S. cerevisiae* cause only minor changes to the total acidity of the wine. In contrast, however, strains of *Schizosaccharomyces pombe* and *Schizosacch. malidevorans* can completely degrade this acid. *Zygosaccharomyces bailii* is another yeast that gives strong (40–100%) degradation of L-malic acid.

The biochemical mechanism for the decomposition of malate is the same for yeasts of the genera *Saccharomyces* (Fuck *et al.*, 1973) and *Schizosaccharomyces* (Dittrich, 1964; Temperli *et al.*, 1965). Malate is oxidatively decarboxylated to pyruvate by NAD-dependent malic enzyme. Pyruvate is decarboxylated to acetaldehyde which is reduced to ethanol (Figure 5.7). These reactions are catalyzed by two enzymes that are involved in the alcoholic fermentation. The overall result is that one molecule of malate yields two molecules of carbon dioxide and one molecule of ethanol. In addition, part of the malate is probably transformed to succinate via fumarate (Kuczynski and Radler, 1982). So far, attempts to employ yeasts with an ability to ferment malate in winemaking have not been very satisfactory.

Yeasts of the genus *Schizosaccharomyces* that ferment malate completely grow best at elevated temperatures and are, therefore, easily outgrown by the less exacting wine yeasts of the genus *Saccharomyces*. In addition, a tendency to

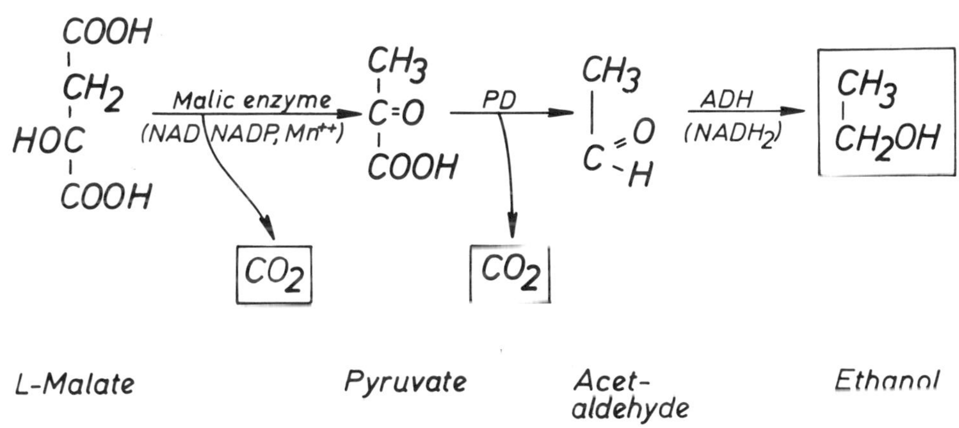

FIGURE 5.7 The biochemical reactions of the fermentative decomposition of malic acid by *Saccharomyces cerevisiae*. PD, pyruvate decarboxylase; ADH, alcohol dehydrogenase.

produce undesirable off-flavors has been observed with *Schizosaccharomyces* (Benda and Schmitt, 1969; Carre *et al.*, 1983). Red musts and partially fermented wines have been successfully deacidified with *Schizosacch. pombe* entrapped in Ca-alginate gel and no undesirable effects on wine quality were observed (Magyar and Panyik, 1989).

Zygosaccharomyces bailii (synonym *Saccharomyces bailii*) is also able to ferment malate completely (Baranowski and Radler, 1984). This species is not a usual wine yeast, although it is occasionally found in wine. It is resistant to preservative chemicals (sulfur dioxide and sorbic acid, among others) and high sugar concentrations. Therefore, this yeast is a well-known spoilage organism, especially of concentrates and beverages. The malic enzymes of *S. cerevisiae* and *Z. bailii* are somewhat different, in particular, with respect to their substrate affinities (expressed as K_m-values). However, this difference is insufficient to explain why *Z. bailii* and *Schizosaccharomyces* spp. can metabolize malate completely whereas the wine yeast (*S. cerevisiae*) cannot (Kuczynski and Radler, 1982). Malate enters the cells of *S. cerevisiae* by simple diffusion. In *Z. bailii,* the transport of malate is mediated by a carrier protein that is specific for the L(-)-isomer. As experiments have shown, this protein is not always present in the cells. Surprisingly, it is only synthesized when the yeast cells are grown on glucose. Cells grown on other carbon sources like fructose or glycerol (aerobically) are unable to transport malate rapidly, whereas glucose-grown cells transport malate by facilitated diffusion. Thus, the effective metabolism of malate by *Z. bailii* depends on the presence of a special transport system (a carrier protein for L-malate) and a malic enzyme with higher affinity for malate than in *S. cerevisiae.*

A constitutive system for malate transport has been demonstrated in *Schizosacch. pombe* (Osothsilp and Subden, 1986): inhibitors of energy metabolism strongly inhibited malate transport, suggesting that it may be energy dependent.

In *S. cerevisiae*, the transport of malic acid is not carrier mediated (Salmon 1987). Corte-Real and Leao (1990) demonstrated (by K_m measurements as a function of pH) that malic acid and some other acids were transported in *H. anomala* by an accumulative dicarboxylate proton symporter. It is repressed by glucose and therefore not operational when the yeast is growing in grape must (Corte-Real *et al.*, 1990), but a mutant has been obtained that does not show glucose repression.

Rodriguez and Thornton (1990) have shown that the concentration of glucose in the medium can significantly influence the utilization of L-malic acid by yeasts grown under semi-anaerobic conditions. The effect varies with the yeast species and seems related to catabolic repression of malate transport mechanisms. Varying glucose concentration did not affect malate utilization by *S. cerevisiae*, but malate utilization by *Schizosacch. malidevorans* and *Z. bailii* was reduced at high and low levels of glucose.

In wines produced from ripe grapes with a low content of malic acid, there is no need to transform this acid by microorganisms. In fact, it has been attempted to use yeast strains that do not decompose malic acid for the fermentation of grape must in order to avoid the usual loss of acidity (Fatichenti *et al.*, 1981). On the other hand, it may be beneficial to remove an excess of malate in wines from less mature grapes with high acidity. The use of malate-decomposing yeasts is possible but so far even selected strains of *Schizosaccharomyces* or *Z. bailii* have not been completely satisfactory because of the production of off-flavors (Benda and Schmitt, 1969; Radler, unpublished results). Attempts to hybridize wine yeasts with malate-decomposing strains of *Z. bailii* by conjugation or cell fusion have not been successful (Radler, unpublished results). However, Rodriguez and Thornton (1988) have reported successful trials of wine deacidification using a mutant of *Schizosacch. malidevorans* that gave rapid degradation of malic acid with minimal utilization of glucose. The grape juice was first fermented at 25°C with the mutant until nearly all the malic acid had been degraded, then the remaining sugar was fermented by inoculation with *S. cerevisiae*.

An interesting new approach is transfer of the gene for the malolactic enzyme from a lactic acid bacterium (*Lactobacillus delbrueckii*) into yeast by genetic engineering using a shuttle plasmid that is replicated in a bacterial cell or in a yeast cell (Williams *et al.*, 1984). The genetic manipulation has been successful and the resulting yeast strains synthesized the bacterial malolactic enzyme. However, only small amounts of malate were decarboxylated to lactate. I assume that transfer of the malolactic enzyme into a yeast cell is not sufficient for rapid metabolism of malate, for it is likely that an additional specific transport system for malate is required, as has been demonstrated in *Z. bailii*.

Various Acids

Only traces of fumaric acid are present in grape must, but in musts from moldy grapes up to 130 mg/l have been observed (Dittrich, 1989). It is not known whether these small amounts of fumaric acid are affected by yeasts. However, if fumarate is added to must prior to fermentation, it is metabolized by fumarase to malate which is further metabolized by malic enzyme (Wagener *et al.*, 1971; Radler, unpublished results). Thus fumarate, that is known to inhibit the bacterial malolactic fermentation (Cofran and Meyer, 1970), cannot be used as an acidulant for grape must before the alcoholic fermentation is completed (Wagener *et al.*, 1971; Pilone *et al.*, 1973). However, if the gene FUM1 that encodes fumarase is overexpressed in *S. cerevisiae*, L-malate can be produced from fumarate at a higher conversion rate than with industrial strains of *Brevibacterium* (Peleg *et al.*, 1990).

Polygalacturonic acid was degraded by 9 of 33 strains of *S. cerevisiae* examined, although the yeasts were unable to use this substance as sole carbon source (McKay, 1990). However, it is not likely that yeasts hydrolyze pectins of grape must to a significant extent.

Gluconic acid is found in musts that are prepared from moldy grapes. This acid is the product of fungi and acetic acid bacteria. During fermentation this acid is not significantly metabolized by yeasts (Holbach and Woller, 1979).

ACIDULANTS AND INHIBITORS

Di- and Tricarboxylic Acids

A traditional procedure that has been used in certain viticultural areas to increase the acidity of wine has been the addition of gypsum ($CaSO_4$) to grape must. Besides the reduction in pH, an improvement in the color of red wines is achieved. At the same time, the process of pressing the mash is facilitated. Gypsum has the effect of converting the partly soluble potassium hydrogen tartrate into the insoluble calcium tartrate, potassium sulphate and free tartaric acid (Amerine and Cruess, 1960).

Depending on the legal requirements of the country, the following acids may be added to the must to increase its acidity: tartaric acid, D,L-malic acid, citric acid, fumaric acid, and lactic acid. Such additions are done to reduce pH and allow for cleaner fermentations with less danger of bacterial growth. Tartaric acid may be used in those areas of the European Community where the addition of acid is authorized (EC Order 822/87 from 16.3.1987). Of these acids, tartaric acid and lactic acid are regarded as stable. Depending on the concentration of potassium and calcium ions, some tartaric acid will be lost by precipitation of the corresponding tartrates that are not very soluble, particularly in the presence of alcohol. There appears to be no reports that tartrate is metabolized by yeasts during fermentation. A few species of lactic acid bacteria are able to decompose tartrates. Lactic acid is similarly stable under anaerobic conditions.

The fate of some organic acids when added to grape must prior to fermentation was investigated by Wagener *et al.* (1971). Some of the malic acid was decomposed and citric acid showed a steady decrease from the start to the finish of the fermentation. The decrease was generally about 0.5 to 1 g per liter. Fumaric acid was almost completely metabolized, presumably by fumarase to malate. In the presence of excess fumaric acid, no additional succinate was produced.

As the taste of the various acids is not very different (Amerine *et al.*, 1965), their use will mainly depend on the cost and the state of authorization that is different in various countries. There is one puzzling aspect. In several viticultural areas, acids are added to the must before fermentation, but after fermentation the bacterial malolactic fermentation is induced causing a loss in malic acid that is decarboxylated to lactic acid.

Sorbic Acid

Sorbic acid or 2,4-hexadienoic acid is an unsaturated fatty acid that is widely used as an antimicrobial agent. Although different yeast species differ in their sensitivity for sorbic acid, it tends to prevent secondary fermentation at very low cell densities. When added at the bottling stage, unwanted fermentations are suppressed even in the presence of residual sugar. The exact mode of action of sorbic acid is not known, probably the effect is unspecific. Due to its lipophilic nature sorbic acid is adsorbed by the yeast cell (Reinhard and Radler, 1981). It is ineffective against lactic acid bacteria that may transform it and produce an off-flavor (Edinger and Splittstoesser, 1986).

OUTLOOK

Organic acids contribute significantly to the taste of wine. Due to the interaction of various factors affecting taste, an optimum acidity is required for each different type of wine. Although the acid composition of the grape is the main factor controlling the acidity of must and wine, this property can be significantly affected by microbial metabolism. In this respect, lactic acid bacteria are more important than yeasts. However, almost all organic acids present in wine are affected by yeasts. Most acids are formed during fermentation and some are degraded, although, like malic acid, only to a limited extent. So far, only wild type yeast strains have been used for winemaking. Since the biochemical basis of yeast metabolism of organic acids is now largely understood, it can be assumed that either breeding or genetic engineering will result in yeast strains with more efficient acid metabolism.

REFERENCES

Amata, I.A. and Germain, P. (1990) The effect of pitching yeast aeration on the production of acetic acid during fermentations with brewers' yeast: an enzymatic approach. *Journal of the Institute of Brewing*, **96**, 131–134.

Amerine, M.A. and Cruess, W.V. (1960) The technology of wine making. Westport, Connecticut: AVI Publ. Co. Inc.

Amerine, M.A., Roessier, E.B. and Ough, C.S. (1965) Acids and the acid taste. I. The effect of pH and titratable acidity. *American Journal of Enology and Viticulture*, **16**, 29–37.

Baranowski, K. and Radler, F. (1984) The glucose-dependent transport of L-malate in *Zygosaccharomyces bailii*. *Antonie van Leeuwenhoek*, **50**, 329–340.

Benda, I. and Schmitt, A. (1969) Untersuchungen zum Säureabbau im Most durch verschiedene Hefestämme aus der Gattung *Schizosaccharomyces*. *Weinberg und Keller*, **16**, 71–83.

Carr, J.G. and Whiting, G.C. (1955) Acid change in cider fermentation. II. The effect of juice composition. *Report of the Agricultural and Horticultural Research Station, Long Ashton*, pp. 163–168.

Carre, E., Lafon-Lafourcade, S. and Bertrand, A. (1983) Desacidification biologique des vins blancs secs par fermentation de l'acide malique par les levures. *Connaissance de la Vigne et du Vin*, **17**, 43–53.

Cofran, D.R. and Meyer, B.J. (1970) The effect of fumaric acid on malolactic fermentation. *American Journal of Enology and Viticulture*, **21**, 189–192.

Corte-Real, M. and Leão, C. (1990) Transport of malic acid and other dicarboxylic acids in the yeast *Hansenula anomala*. *Applied and Environmental Microbiology*, **56**, 1109–1113.

Corte-Real, M., Leão, C. and van Uden, N. (1990) Inverse diauxy in the yeast *Hansenula anomala*: mutants derepressed for malic acid utilization in the presence of glucose. *Applied and Environmental Microbiology*, **56**, 3402–3404.

Dellweg, H. and Kettler, R. (1975) Über das Auftreten von Alpha-Ketosäure und anderen organischen Säuren bei der Biergärung. *Monatsschrift für Brauerei,* **28**, 169–178.

Dittrich, H.H. (1964) Die alkoholische Vergärung der L-Äpfelsäure durch *Schizosaccharomyces pombe* var. acidodevoratus. *Zentralblatt für Bakteriologie, Parasitenkunde, Infektionskrankheiten und Hygiene,* II. Abt. 118, 406–421.

Dittrich, H.H. (1987) *Mikrobiologie des Weines.* Stuttgart: Ulmer.

Dittrich, H.H. (1989) Die Gärung in *Chemie des Weines* edited by G. Würdig and R. Woller, pp. 184–221. Stuttgart: Ulmer.

Drawert, F., Leupold, G. and Lessing, V. (1978) Gaschromatographische Bestimmung der Inhaltsstoffe von Gärungsgetränken. 12. Mitt. Gaschromatograpische Bestimmung von organischen Säuren, Neutralstoffen (Kohlenhydraten), Aminosäuren und phenolischen Verbindungen in Weinen. *Wein-Wissenschaft,* **33**, 54–70.

Edinger, W.D. and Splittstoesser, D.F. (1986) Production by lactic acid bacteria of sorbic alcohol, the precursor of the geranium odor compound. *American Journal of Enology and Viticulture,* **37**, 34–38.

Edwards, C.G., Beelman, R.B., Bartley, C.E. and McConnell, A.L. (1990) Production of decanoic acid and other volatile compounds and the growth of yeast and malolactic bacteria during vinification. *American Journal of Enology and Viticulture,* **41**, 48–56.

Fatichenti, F., Farris, G.A., Deiana, P., Ceccarelli, S. and Serra, A. (1981) Commercial trial of winemaking using two selected starters of *Saccharomyces cerevisiae* which do not reduce malic acid content. *American Journal of Enology and Viticulture,* **32**, 236–240.

Fuck, E. and Radler, F. (1972) Äpfelsäurestoffwechsel bei *Saccharomyces.* I. Der anaerobe Äpfelsäureabbau bei *Saccharomyces cerevisiae. Archiv für Mikrobiologie,* **87**, 149–164.

Fuck, E., Stärk, G. and Radler, F. (1973) Äpfelsäurestoffwechsel bei *Saccharomyces.* II. Anreicherung und Eigenschaften eines Malatenzyms. *Archiv für Mikrobiologie,* **89**, 223–231.

Genitsariotis, R. (1979) *Über die Bildung von L(+)-Lactat bei Saccharomyces pretoriensis und die NAD-abhängige Lactat-Dehydrogenase.* Dissertation Mainz.

Hannemann, W. (1985) *Ausscheidung von Essigsäure durch gärende Hefen und die Reinigung und Charakterisierung einer NADP+-spezifischen Aldehyd-Dehydrogenase aus Saccharomyces cerevisiae.* Dissertation, Mainz.

Heerde, E. and Radler, F. (1978) Metabolism of the anaerobic formation of succinic acid by *Saccharomyces cerevisiae. Archives of Microbiology,* **117**, 269–276.

Holbach, B. and Woller, R. (1979) Zum Einfluß von Hefen auf Gluconsäure während der Gärung. *Wein und Rebe,* **60**, 1358–1360.

Kuczynski, J.T. and Radler, F. (1982) The anaerobic metabolism of malate of *Saccharomyces bailii* and the partial purification and characterization of malic enzyme. *Archives of Microbiology,* **131**, 266–270.

Kunkee, R.E. (1967) Malo-lactic fermentation. *Advances in Applied Microbiology,* **9**, 235–279.

Lafon-Lafourcade, S., Geneix, C. and Ribéreau-Gayon, P. (1984) Inhibition of alcoholic fermentation of grape must by fatty acids produced by yeasts and their elimination by yeast ghosts. *Applied and Environmental Microbiology,* **47**, 1246–1249.

Magyar, I. and Panyik, I. (1989) Biological deacidification of wine with *Schizosaccharomyces pombe* entrapped in Ca-alginate gel. *American Journal of Enology and Viticulture,* **40**, 233–240.

McKay, A.M. (1990) Degradation of polygalacturonic acid by *Saccharomyces cerevisiae. Letters of Applied Microbiology,* **11**, 41–44.

Millan, C. and Ortega, J.H. (1988) Production of ethanol, acetaldehyde, and acetic acid in wine by various yeast races: role of alcohol and aldehyde dehydrogenase. *American Journal of Enology and Viticulture,* **39**, 107–112.

Monk, P. R. and Cowley, P. J. (1984) Effect of nicotinic acid and sugar concentration of grape juice and temperature on accumulation of acetic acid during yeast fermentation. *Journal of Fermention Technology,* **62**, 515–525.

Nagel, C.W., Anderson, J.D. and Weller, K.M. (1988) A comparison of the fermentation patterns of six commercial wine yeasts. *Vitis*, **27**, 173–182.

Osothsilp, C. and Subden, R.E. (1986) Malate transport in *Schizosaccharomyces pombe*. *Journal of Bacteriology*, **168**, 1439–1443.

Peleg, Y., Rokem, J.S., Goldberg, I. and Pines, O. (1990) Inducible overexpression of the FUM1 gene in *Saccharomyces cerevisiae*: localization of fumarase and efficient fumaric acid bioconversion to L-malic acid. *Applied and Environmental Microbiology*, **56**, 2777–2783.

Peynaud, E. (1968) Études récentes sur les bactéries lactiques du vin. Fermentation et vinification. 2e Symposium international d'Oenologie, vol. 1, pp. 219–255. Paris: Institut National de la Recherche agronomique.

Peynaud, E., Domercq, S., Boidron, A., Lafon-Lafourcade, S. and Guimberteau, G. (1964) Etude der levures *Schizosaccharomyces* metabolisant l'acide L-malique. *Archiv für Mikrobiologie*, **48**, 150–165.

Peynaud, E., Lafon-Lafourcade, S. and Guimberteau, G. (1967) Nature de l'acide lactique formé par les levures un caractére spécifique de *Saccharomyces veronae* Lodder et Kreger van Rij. *Antonie van Leeuwenhoek*, **33**, 49–55.

Pilone, D.A., Pilone, G.J. and Rankine, B.C. (1973) Influence of yeast strain, pH, and temperature on degradation of fumaric acid in grape juice fermentation. *American Journal of Enology and Viticulture*, **24**, 97–102.

Ponader, W. (1973) *Einfluß von Stickstoffverbindungen auf die Bildung von Carbonylverbindungen durch Hefen*. Dissertation, Mainz.

Radler, F. and Lang, E. (1982) Malatbildung bei Hefen. *Wein-Wissenschaft*, **37**, 391–399.

Radler, F. (1966) Die mikrobiologischen Grundlagen des Säureabbaus im Wein. *Zentralblatt für Bakteriologie, Parasitenkunde, Infektionskrankheiten und Hygiene*, II. Abt. 120, 237–287.

Radler, F. (1972) The formation of non volatile acids by strains of Saccharomyces during fermentation. In *Yeasts. Models in Sciences and Technics* edited by A. Kocková-Kratochvilová and E. Minarik, pp. 519–529. Bratislava: Slovac Academy of Sciences.

Radler, F. (1981) The metabolism of organic acids by Saccharomyces. Symposium Proceedings, pp. 103–108. Davis: University of California.

Radler, F. (1984) Über den anaeroben Stoffwechsel von Carbonsäuren bei Hefen. *Forum Mikrobiologie*, **7**, 286–291.

Rankine, B.C. (1966) Decomposition of L-malic and by wine yeasts. *Journal of the Science of Food and Agriculture*, **17**, 312–316.

Reinhard, L.K. and Radler, F. (1981) Die Wirkung von Sorbinsäure auf *Saccharomyces cerevisiae* II. Adsorption, Desorption und Abbau von Sorbinsäure. *Zeitschrift für Lebensmittel-Untersuchung und -Forschung*, **172**, 382–388.

Rodriquez, S. and Thornton, R. (1988) Rapid utilization of malic acid by a mutant of *Schizosaccharomyces malidevorans*. In *Proceedings of the Second Cool Climate Viticulture and Oenology Symposium*, pp. 313–315. New Zealand.

Rodriguez, S. and Thornton, R. (1990) Factors influencing the utilization of L-malate by yeasts. *FEMS Microbiology Letters*, **72**, 17–22.

Salmon, J.M. (1987) L-Malic acid permeation in resting cells of anaerobically grown *Saccharomyces cerevisiae*. *Biochimica et Biophysica Acta*, **901**, 30–34.

Schindler, J., Meussdoerffer and Giesel-Bühler, H. (1990) Mikrobiologische Herstellung von Industriechemikalien: Grundlagen der Dicarbonsäurebildung mit Hefen. *Forum Mikrobiologie*, **5**, 274–281

Schwartz, H. and Radler, F. (1988) Formation of L(-)malate by *Saccharomyces cerevisiae* during fermentation. *Applied Microbiology and Biotechnology*, **27**, 553–560.

Schwartz, H., Steitz, H.-O. and Radler, F. (1983) Partial purification and characterization of succinyl-CoA synthetase from *Saccharomyces cerevisiae*. *Antonie van Leeuwenhoek*, **49**, 69–78.

Shimazu, Y. and Watanabe, M. (1981) Effects of yeast strains and environmental conditions on formation of organic acids in must during fermentation. *Journal of Fermention Technology,* **59**, 27–32.

Sigler, K., Knotková, Paca, J. and Wurst, M. (1980) Extrusion of metabolites from baker's yeast during glucose-induced acidification. *Folia Microbiologica,* **25**, 311–317.

Sponholz, W. R. and Dittrich, H. H. (1986) Flüchtige Fettsäuren in Weinen verschiedener Qualitätsstufen. *Zeitschrift für Lebensmittel-Untersuchung und -Forschung,* **183**, 344–347

Sponholz, W.R., Dittrich, H.H., Haas, F. and Wünsch, B. (1981) Die Bildung von flüchtigen Fettsäuren durch Saccharomyceshefen während der Gärung von Traubenmost. *Zeitschrift für Lebensmittel-Untersuchung und -Forschung,* **173**, 297–300.

Temperli, A., Künsch, U., Mayer, K. and Busch, I. (1965) Reinigung und Eigenschaften der Malatdehydrogenase (decarboxylierend) aus Hefe. *Biochimica et Biophysica Acta,* **110**, 630–632.

Thoukis, G., Ueda, M. and Wright, D. (1965) The formation of succinic acid during alcoholic fermentation. *American Journal of Enology and Viticulture,* **16**, 1–8.

Wagener, W.W.D., Ough, C.S. and Amerine, M.A. (1971) The fate of some organic acids in grape juice prior to fermentation. *American Journal of Enology and Viticulture,* **22**, 167–171.

Wenzel, K., Dittrich, H.H. and Pietzonka, B. (1982) Untersuchungen zur Beteiligung von Hefen am Äpfelsäureabbau bei der Weinbereitung. *Wein-Wissenschaft,* **37**, 133–138.

Whiting, G.C. (1976) Organic acid metabolism of yeasts during fermentation of alcoholic beverages. A review. *Journal of the Institute of Brewing,* **82**, 84–92.

Williams, A.S., Hodges, R.A., Strike, T.L., Snow, R. and Kunkee, R.L.(1984) Cloning the gene for the malolactic fermentation of wine from *Lactobacillus delbrueckii* in *Escherichia coli* and yeasts. *Applied and Environmental Microbiology,* **47**, 288–292.

Chapter 6

YEASTS—
PRODUCTION OF SULFUR COMPOUNDS

DORIS RAUHUT

Forschungsanstalt Geisenheim, Institut für Oenologie und Getränkeforschung, Fachgebiet Mikrobiologie und Biochemie, von-Lade-Straße 1, D-6222 Geisenheim, Germany

Chapter Outline

Chapter Outline (Cont.)

INTRODUCTION

Sulfur (S)-containing compounds occur in a range of chemical combinations in the natural environment. Sulfur is present in essential vitamins, coenzymes and amino acids such as methionine and cysteine (Richmond, 1973). Thiol-groups have important roles in the active sites of several enzymes, in electron transport, and in the processes of growth and cell division. The oxidation state of sulfur ranges from –2 to +6. S-substances with intermediate oxidation states are chemically and biologically active, because sulfur is unstable below the oxidation state of +6. The free energy for oxidation is relatively high (Huxtable, 1986; Kelly, 1988). Yeasts are able to reduce sulfate, a compound with the highest oxidation state of sulfur, to sulfide which is then used for further metabolism.

The first part of this chapter reviews yeast metabolism of sulfate and biosynthesis of S-containing amino acids and other significant S-substances. The second part covers the volatile S-compounds produced by yeasts during winemaking. These substances make a significant contribution to the flavor of wine because of their extremely low threshold values. In some circumstances, they are responsible for undesirable sulfurous off-flavors (Schutte, 1975; Peppard, 1988).

CONTENT OF SULFUR AND S-COMPOUNDS IN YEASTS

The total amount of sulfur in yeast cells is low compared with its importance to yeast metabolism. In *Saccharomyces cerevisiae*, the sulfur content ranges from 0.2–0.9% of the dry weight (Maw, 1963a, 1965). In baker's yeast, 0.14% sulfur occurs in the cell wall (cited in Phaff, 1971). The most abundant S-compounds are the amino acids methionine and cysteine, which are present in peptides and proteins, and the tripeptide, glutathione (Hartnell and Spedding, 1979). Glutathione acts as coenzyme, provides a reservoir of thiol groups and plays an essential role as a biological redox system (Karlson, 1988). Values for the content of these compounds in strains of *S. cerevisiae*, expressed as dry weight of sulfur are: 0.13–0.27 % for cystine; 0.15–0.28 % for methionine; and 0.05–0.10 % for glutathione (Maw, 1965). Glutathione-sulfur is about 20 % of the total sulfur, or about 40 % of the cystine-sulfur and is mostly (90%) present in the reduced form. Other S-containing compounds, such as thiamine, acetyl-CoA, biotin and lipoic acid, represent only a small proportion of the total sulfur. In *Saccharomyces* spp., thiamine ranges from 29–90 μg/g dry weight of the cells. Coenzyme A was found in a brewer's strain of *S. cerevisiae* at 16.9 μg/g dry weight. In a baker's strain of *S. cerevisiae* and *Candida utilis*, S-adenosylmethionine (SAM) was detected at 0.5 and 2.5 μmoles/g dry weight, respectively. The biotin contents of four *Saccharomyces* spp. were in the range 0.5–4.5 μg/g dry weight of cells, depending on the culture medium (Maw, 1965).

Inorganic sulfur, mainly as sulfate, is present in minor quantities, but the available data vary. Studies with S^{35}-labelled sulfate in a brewer's yeast yielded up to 1.3 % of total sulfur for inorganic sulfate and up to 3.0 % for bound sulfate (Maw, 1965). Maw (1963b) also showed with a brewer's yeast that, after one hour, 60 % of the sulfate-sulfur taken up from the medium occurred as protein sulfur, while only 5 % could be measured as sulfate ions. This illustrates that free inorganic sulfate has a short intracellular residence time before its sulfur is transferred into organic substances.

UPTAKE OF SULFUR COMPOUNDS

A source of assimilable sulfur is essential for yeast growth. Almost all yeasts utilize inorganic sulfate. Numerous studies deal with the mechanism of sulfate uptake, utilization of S-containing amino acids and their partial or total replacement by other sources (Maw, 1960, 1961, 1963 a, b, c, 1965; Hysert and Morrison, 1976; Breton and Surdin-Kerjan, 1977).

It is not fully elucidated whether one or two transport systems exist in *S. cerevisiae* for accumulation of sulfate (Cartwright *et al.*, 1989). McCready and Din (1974) demonstrated the activity of a sulfate permease in *S. cerevisiae*. Kotyk (1959), Maw (1963 b) and Roomans *et al.* (1979) described a single sulfate transport system by Michaelis-Menten-kinetics. The K_m and V_{max} values depended on the intracellular and extracellular pH values. However, Breton and Surdin-Kerjan (1977) proposed that sulfate uptake by *S. cerevisiae* probably involved two permeases because it followed biphasic Lineweaver-Burke kinetics. The activity of these two permeases was inhibited by endogenous sulfate and adenosine-5'-phosphosulfate (APS), an intermediate in sulfate assimilation. Biosynthesis of the permeases was coordinated with the production of enzymes for methionine synthesis.

Maw (1963 b) showed that optimum uptake of sulfate by cell suspensions of a brewer's yeast depended on the presence of an energy source, such as glucose, in combination with ammonium and citrate in the medium. Sulfate incorporation paralleled growth and protein synthesis, and depended on temperature. Omission of phosphate had little or no effect. Ammonium ions could be substituted almost entirely by L-glutamine. L-methionine, L-ethionine, L-cysteine and DL-homocysteine partly suppressed sulfate uptake. This effect was not observed with glutathione or cystine. Horák *et al.* (1981) reported that sulfate transport was increased by preincubation of cells with energy-providing substances. A further increase was obtained on addition of arginine, lysine or potassium ions. Decreased transport occurred in the presence of neutral or anionic amino acids and ammonium. Sulfate uptake was inhibited by azide, 2,4-dinitrophenol, selenate and chromate (Maw, 1963 b; McCready and Din, 1974; Cooper, 1982).

Most *Saccharomyces* spp. grow when sulfate is substituted by sulfite or thiosulfate. Growth rate is increased when methionine and glutathione are included in the medium, in addition to sulfate. Sometimes methionine can be utilized instead of sulfite. The utilization of glutathione is generally similar or less than that of methionine. Inorganic sulfate and methionine are entirely taken up from the medium (Maw, 1963 b, c; 1965; Suomalainen and Oura, 1971). Two different systems are demonstrated for methionine uptake by *S. cerevisiae* (Cartwright *et al.*, 1989). Cystine is not removed at all, and cysteine is taken up only slowly. The majority of *Saccharomyces* yeasts cannot use the sulfur of cysteine and cystine effectively, in contrast to *Candida utilis*, which can grow on a larger number of S-sources than *S. cerevisiae* (Suomalainen and Oura, 1971). Cystine shows no effect, because it cannot permeate the cell envelope of most *Saccharomyces* yeasts. Cysteine inhibits growth and depresses respiration probably by complexing with metal ions, which are necessary for the activity of several enzymes (Maw, 1965). Similar inhibition occurs with other compounds, which carry a sulfhydryl group, e.g. mercaptoacetate, cysteamine, 2-mercapto-ethanol and DL-homocysteine (Maw, 1965). However, reduced glutathione and mercaptosuccinate are not growth inhibitors (Maw, 1961). Also, yeasts can grow in the presence of high concentrations of S-methyl-L-cysteine, which is thought to be metabolized to methionine (Maw, 1965, 1982).

SULFUR REQUIREMENTS FOR GROWTH

The effect of sulfur deficiency on the composition and morphology of yeasts has been studied insufficiently. It is known that growth is poor or negligible under these conditions and that the cells are largely depleted of glutathione (Maw, 1965). A sulfate content giving 10 mg/l of total sulfur is sufficient for growth of a brewer's strain of *S. cerevisiae* (Maw, 1965).

Yeasts differ in their requirements for the sulfur containing vitamins, biotin and thiamine. *Candida utilis* is one of the few yeasts which can synthesize biotin and thiamine. *Saccharomyces* yeasts require biotin, but can grow without thiamine. For optimum growth, however, a supply of this vitamin is necessary. Maximum growth is obtained with 100 μg/l of biotin, but for many yeasts 1 μg/l is sufficient. Requirements for thiamine vary between 50–500 μg/l (Maw, 1965). Biotin deficiency causes an inhibition in fermentation of glucose and fructose and a reduction of C_{18} fatty acids synthesis. DNA, RNA and total protein contents are reduced. Furthermore, a change in polysaccharide composition is noticeable. The content of wall glucan is increased and the mannan content decreased, giving increased cell fragility (Maw, 1965).

METABOLISM AND FORMATION OF SULFUR COMPOUNDS

Assimilatory Sulfate Reduction Pathway

The most common S-source, sulfate, is reduced to sulfide by an assimilatory reduction pathway (see reviews of De Robichon-Szulmajster and Surdin-Kerjan, 1971; Jones and Fink, 1982). The individual steps of the pathway are shown in Figure 6.1 and have been assigned numbers which are used in the following discussion.

The free energy drop ($\Delta G°$) between sulfate and sulfide is about 800 kJ/mol (Huxtable, 1986) so that considerable energy is needed for this reduction. After transport of extracellular sulfate with sulfate permease [1], an activation of sulfate takes place. Sulfate displaces the pyrophosphate of adenosine-5'-triphosphate (ATP) and adenosine-5'phosphosulfate (APS) is formed [2]. This reaction is catalyzed by ATP-sulfurylase. The reaction equilibrium is in favor of sulfate and ATP. A reaction is only explainable because pyrophosphate is taken away through the action of inorganic pyrophosphatase, and because the K_m value for APS of the next enzyme in the pathway is very low (Slaughter, 1989). Phosphorylation of APS to 3'-phosphoadenosine-5'-phosphosulfate (PAPS) occurs in the next reaction [3] and is catalyzed by APS-kinase (Robbins and Lipmann, 1958 a, b). The sulfonic group of PAPS can be used by sulfotransferases to form sulfate esters, like the sulfation of phenols and steroids (Baumgarten, 1981; Bidan and Collon, 1988).

The reduction of PAPS to sulfite and 3'–5' diphosphoadenosine (PAP) involves PAPS-reductase and NADPH (NADPH+H^+) as electron donor [4] (Peck and Lissolo, 1988). This reaction is carried out by a complex electron transport system (Asahi *et al.*, 1961; Wilson *et al.*, 1961; Wilson and Bierer, 1976). *In vitro*, the reduction of PAPS requires three isolated protein fractions. Two of them are heat labile, high molecular weight enzymes which are designated as fractions A and B. The third, fraction C, is a heat resistant, non-dialyzable protein of low molecular weight (Wilson *et al.*, 1961; Maw, 1965; Lawrence and Cole, 1968). Evidence suggests that sulfite is not always free in the medium (Hilz *et al.*, 1959; Hilz and Kittler, 1960), but remains bound to the sulfhydryl group of fraction C. After releasing the sulfite anion, a disulfide in fraction C is formed, which is regenerated by NADPH. It is assumed that enzyme A (fraction A), is a flavoprotein and couples reduction of the disulfide group in fraction C with the oxidation of NADPH (Figure 6.2). The enzyme of fraction B, generally called PAPS-reductase, is less studied and difficult to purify (Lawrence and Cole, 1968; Baumgarten, 1981, Bidan and Collon, 1988). During transfer of the sulfonic group to fraction C, the oxidation status of the sulfur atom is not directly reduced from +6 to +4. First, a change in the valency of the sulfur from +6 to +5 takes place, then subsequently a spontaneous intramolecular redox reaction occurs, whereby a disulfide is formed and a sulfite anion with the oxidation state +4 is released. Therefore, PAPS-reductase is called PAPS-thiol-sulfotransferase (Baumgarten, 1981; Peck and Lissolo, 1988). Porqué *et al.* (1970) proved that the three proteins

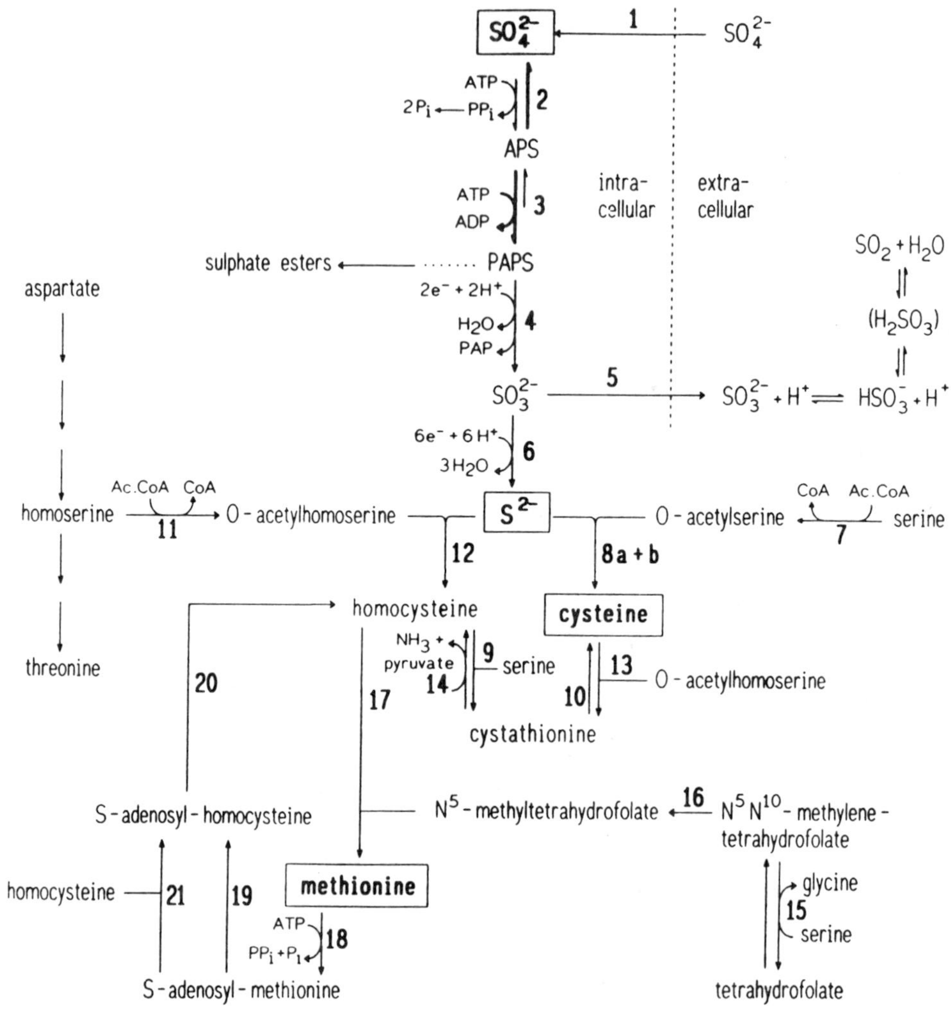

FIGURE 6.1 Assimilatory reduction pathway of sulfate and biosynthesis of methionine and cysteine by yeasts (modified from Jones and Fink, 1982) [1. sulfate permease (two enzymes); 2. ATP sulfurylase (EC* 2.7.7.4); 3. APS kinase (EC 2.7.1.25); 4. PAPS reductase; 5. sulfite permease; 6. sulfite reductase (EC 1.8.1.2); 7. serine acetyltransferase (EC 2.3.1.30); 8a. cysteine synthase (EC 4.2.99.8); 8b = 12. homocysteine-cysteine synthase (EC 4.2.99.10); 9. β-cystathionine synthase (EC 4.2.1.22); 10. γ-cystathionase (EC 4.4.1.1); 11. homoserine acetyltransferase (EC 2.3.1.31); 12 = 8b. homocysteine synthase (EC 4.2.99.10); 13. γ-cystathionine synthase (EC 4.2.99.9); 14. β-cystathionase (EC 4.4.1.8); 15. serine hydroxymethyltransferase (EC 2.1.2.1); 16. 5,10-methylene THF reductase (THF = tetrahydrofolate; EC 1.1.99.15); 17. homocysteine methyltransferase (EC 2.1.1.14); 18. SAM synthetase (two isozymes) (EC 2.5.1.6); 19. SAM demethylase; 20. adenosylhomocysteinase (EC 3.3.1.1)].
*EC = Enzyme Commission number.

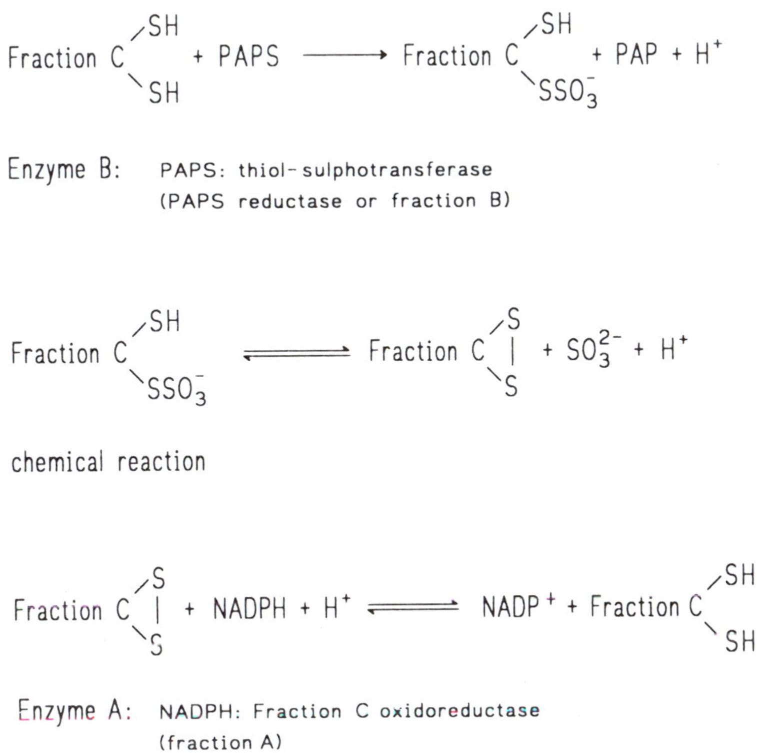

FIGURE 6.2 Proposed scheme for the reduction of 3-phosphoadenosine-5-phosphosulfate (PAPS). Fraction C represents a low molecular weight protein, e.g. lipoic acid or thioredoxin (modified from Maw, 1965; Lawrence and Cole, 1968; Peck and Lissolo, 1988).

needed for *in vitro* reduction of PAPS are thioredoxin, thioredoxin reductase and PAPS-thiol-sulfotransferase, corresponding to the enzyme fractions C, A, and B (Wilson *et al.*, 1961; De Robichon-Szulmajster and Surdin-Kerjan, 1982; Bidan and Collon, 1988).

The intermediate sulfite can be excreted to the medium either actively with sulfite permease [5] by an active carrier mediated process or passively by diffusion (Dott and Trüper, 1978 c; Stratford and Rose, 1986). It is also possible that the reduction of PAPS does not occur by release of the sulfite anion, and that sulfate is transferred to a carrier protein, containing a thiol group (see later) (Torii and Bandurski, 1967).

Sulfite is reduced to sulfide by sulfite reductase [6]. Yoshimoto and Sato (1968 a, b, 1970) demonstrated that, in yeast sulfite reductase, the electron transfer is provided by NADPH via flavin and a chromophore to sulfite. Purified sulfite reductase of a wild type yeast is a complex protein, which contains FMN, FAD, a chromophore adsorbing at 587 nm, essential sulfhydryl group(s), iron and acid-labile sulfide. Prabhakararao and Nicholas (1969, 1970) designated sulfite reductase as an hemoflavoprotein, since they noticed the participation of a flavoprotein and hemoprotein moiety in the electron transport. In textbooks of biochemistry, sulfite reductase is indicated as having a special prosthetic-group, the sirohaem, an iron-porphyrin ring, in which two of the cyclic pyrrols are reduced. NADPH reduces the sirohaem-protein with the assistance of a flavoprotein (Lehninger, 1977; Karlson, 1988). Most microbial sulfite reductases contain an exchange-coupled sirohaem [4Fe-4S] unit at the catalytic centre (Peck and Lissolo, 1988). Cooper (1983) and Slaughter (1989) described that, for sulfite reduction in yeast, thioredoxin is needed in addition to NADPH, and that lipoic acid seems to be a cofactor. The function of thioredoxin is not completely elucidated.

In *S. cerevisiae*, there are two distinct sulfite reductases. One of them uses carrier-bound sulfite and the other uses free sulfite (Umbarger, 1978). In the case of bound sulfite (Figure 6.3), the sulfonic group of PAPS is given up directly by PAPS-thiol-sulfotransferase to thioredoxin to yield a thiosulfonate (thioredoxin-

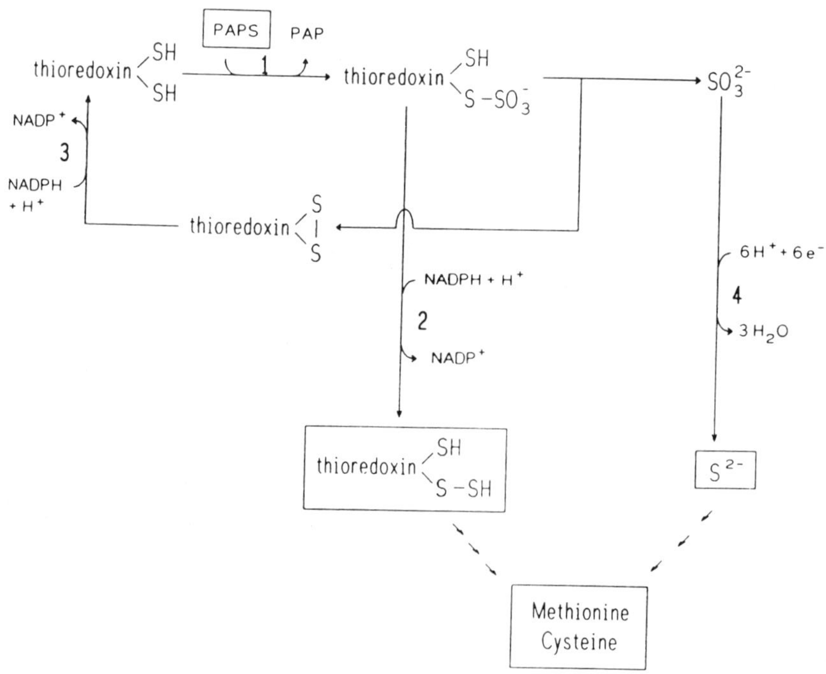

FIGURE 6.3 Supposed reduction of PAPS via a free or a bound sulfite anion (modified from Wilson and Bierer, 1976; Baumgarten, 1981; Jones and Fink, 1982) 1 = PAPS: thiol-sulfotransferase; 2 = thiol-sulfonate reductase; 3 = thioredoxin reductase (NADP: oxidized thioredoxin; oxidoreductase, E.C. 1.6.4.5); 4 = sulfite reductase (hydrogen sulfide: NADP oxidoreductase, E.C. 1.8.1.2).

S-SC_3^-), which is then reduced by NADPH dependent thiosulfonate reductase to form a persulfide (thioredoxin-S-S^-). The persulfide transfers sulfide to the acetylated amino acid acceptors, O-acetylhomoserine or O-acetylserine (Jones and Fink, 1982; Figure 6.1). The reactions involving carrier-bound sulfite are only conjecture and have not been discovered in yeasts.

The enzyme ADP-sulfurylase, which has been partially purified and characterized from yeasts, catalyzes the reaction of APS with inorganic phosphate, whereby ADP and sulfate are formed. Its function in sulfate assimilation is still unknown (Heinzel and Trüper, 1978). With 3'phosphonucleotidase, a reaction from PAPS to APS is possible. This reaction has been only observed *in vitro* (Bidan and Collon, 1988).

Biosynthesis of Cysteine and Methionine

Reduced sulfur is incorporated into carbon compounds by reaction of sulfide with several precursors to synthesize the amino acids cysteine and methionine (Figure 6.1).

Two pathways exist to form cysteine (Jones and Fink, 1982; Ono *et al.*, 1988). The first pathway involves acetylation of serine by serine acetyltransferase [7] to form O-acetylserine. This reaction is followed by sulfhydrylation to form cysteine [8 a, b] using either free sulfide or bound sulfide (Tsang and Schiff, 1976). Two enzymes catalyze the synthesis of cysteine from O-acetylserine. Cysteine synthase [8 a] is responsible for the production of cysteine (Jones and Fink, 1982), whereas the bifunctional homocysteine synthase (also called bifunctional sulfhydrylase or O-acetylhomoserine-O-acetylserine sulfhydrylase) produces both cysteine [8 b] and homocysteine [12] (Yamagata and Takeshima, 1976; Yamagata, 1989). The second pathway begins with condensation of homocysteine and serine by ß-cystathionine synthase [9]. Subsequently, cleavage occurs by γ-cystathionase and cysteine is formed [10]. Schlossmann and Lynen (1957) showed that cysteine is also formed from serine and H_2S by the enzyme serine sulfhydrylase. Dott and Trüper (1978 c) suggested that this enzyme has more of a degradative function and splits cysteine to the initial products. There is no evidence that this enzyme functions as a cysteine synthase in *S. cerevisiae* (Yamagata, 1989).

The formation of methionine begins with an acetyl-CoA-dependent acetylation of homoserine by homoserine acetyltransferase to form O-acetylhomoserine [11] (Naiki and Yamagata, 1973). There are two possible ways to transform O-acetylhomoserine to homocysteine. One is direct sulfhydrylation of O-acetylhomoserine to form homocysteine [12]. The sulfhydryl group can be transferred as free sulfide or by a carrier such as thioredoxin. The second possibility is the condensation of O-acetylhomoserine with cysteine by γ-cystathionine synthase to form cystathionine [13]. Then, ß-cystathionase splits cystathionine, whereby homocysteine, pyruvate and ammonia are obtained [14]. It is not clear which of the two pathways is predominant (Yamagata, 1989). Methionine is synthesized from homocysteine by methylation of the sulfur atom with 5-tetrahydropteroyl-triglutamate [17]. The catalyzing enzyme is homocysteine methyltransferase, which needs the diglutamyl or triglutamyl derivative of 5-methyl-tetrahydrofolate (THF = tetrahydrofolate) for synthesis (Jones and Fink, 1982). The enzymes involved are serine hydroxymethyltransferase [15] and 5, 10 methylene-THF-reductase [16].

Methionine can also be a precursor for cysteine (Ono *et al.*, 1988). The biosynthesis by a trans-sulfuration pathway starts with the formation of S-adenosylmethionine (SAM) through SAM synthetase [18]. The next reaction is demethylation of SAM to S-adenosylhomocysteine with SAM demethylase [19]. S-adenosyl-homocysteine is transformed to homocysteine [20] with adenosyl-homocysteinase. Homocysteine is changed to cystathionine with ß-cystathionine synthase [9] (Delavier-Klutchka and Flavin, 1965). The last step to cysteine and 2-keto-butyric acid is obtained with γ-cystathionase [10] (De Robichon-Szulmajster and Surdin-Kerjan, 1971; Jones and Fink, 1982).

The biosynthesis of methionine and cysteine is regulated by feedback inhibition and control of enzyme formation. This topic is well-covered in the literature and only an outline is presented here (Jones and Fink, 1982; Gyllang *et al.*, 1989). The enzymes of the sulfate reduction pathway and the acetylation and sulfhydrylation reactions have been called the 'met-group-I-enzymes'. A coordinated regulation of this group was postulated (Cherest *et al.*, 1971; Surdin-Kerjan *et al.*, 1976; Breton and Surdin-Kerjan, 1977), but the results of Dott (1977), Heinzel (1978) and Baumgarten (1981) indicated that such regulation was not coordinated for the enzymes ATP-sulfurylase, sulfate-permease, sulfite-reductase and PAPS-sulfotransferase. The biosynthesis of APS-kinase was not influenced by methionine.

Investigations of methionine and SAM pools indicated that repression of 'met-group-I-enzymes' by methionine is caused by the transformation of methionine to SAM, which then functions as a repressor. Addition of methionine to wild-type yeast cells in a minimal medium gave an increase in the pools of SAM and methionine whereas addition of SAM caused an increase only in the SAM pools. Evidence has been presented for the occurrence of two pools of SAM because there are two isozymes for its synthesis (Cherest *et al.*, 1973 a, b; Jones and Fink, 1982).

Methionine biosynthesis is repressed by methionyl-tRNA at the transcription level, whereas SAM operates at the translation level (Masselot and de Robichon-Szulmajster, 1972; Surdin-Kerjan *et al.*, 1976). Both processes function independent of each other. Methionine inhibits the carbon flow from O-acetylserine to homocysteine. The bifunctional enzyme, homocysteine-cysteine synthase is inhibited by methionine whether O-acetylhomoserine (Cherest *et al.*, 1969) or O-acetylserine (Yamagata *et al.*, 1975) is the substrate. Methionine and SAM influence the carbon flow of single carbons from serine to methionine.

In some degree, methionine inhibits 5,10-methylene-THF reductase. The 5-methyl-THF-dependent homocysteine methyltransferase is repressed by methionine. The inhibition and repression of SAM has not been tested (Jones and Fink, 1982). Inhibition of serine hydroxymethyltransferase is caused by methionine and SAM (Botsford and Parks, 1969; Jones and Fink, 1982).

The synthesis of ATP-sulfurylase is induced by cysteine and the synthesis of sulfite reductase is repressed when cysteine is the only sulfur source (De Vito and Dreyfuss, 1964). The repressive function of cysteine and sulfite has to be seen as indirect, since they can be transformed to methionine and other repressors (Baumgarten, 1981).

In *S. cerevisiae*, methionine biosynthesis can be separated into three sequences; synthesis of the four carbon skeleton, reduction of sulfate to sulfide (free or bound), and synthesis and addition of the methyl group (Cherest *et al.*, 1969). Fundamental investigations on the regulation of methionine and cysteine biosynthesis in *S. cerevisiae* were conducted by Cherest *et al.* (1969, 1971, 1973 a, b), Masselot and de Robichon-Szulmajster (1972), Antoniewski and De Robichon-Szulmajster (1973), Surdin-Kerjan *et al.* (1976) and Masselot and Surdin-Kerjan (1977). Methionine is preferably synthesized than cysteine (Eschenbruch *et al.*, 1973) and, also, it seems to be involved in the formation of repressors.

Additional Reactions

Cysteine and methionine are necessary for the synthesis of peptides (glutathione) and proteins. Nwoguh and Berry (1990) suggested that methionine is necessary for biomass and pseudomycelial formation by *Saccharomycopsis fibuligera* in defined medium. The sulfhydryl group of cysteine is important in the catalytic centre of numerous enzymes and is involved in the binding of metal ions. Cysteine is a part of the tripeptide glutathione and of the amino acid cystine. SAM is important for methylation and delivers the propylamino groups for the synthesis of polyamines (Slaughter, 1989).

Bivalent sulfur is incorporated in some important vitamins and growth factors. The sulfhydryl group of cysteamine, the decarboxylation product of cysteine, is the functional unit of coenzyme A. The sulfur source for the vitamins, biotin and thiamine, which have thioether linkages, is not fully elucidated. The bivalent sulfur for biotin synthesis from dethiobiotin can be supplied from methionine sulfoxide and methionine, and also from Na_2SO_3, Na_2S, Na_2SO_4, homocysteine and S-adenosylmethionine (Umezawa and Kishi, 1989). Sulfate, sulfite and cysteine cannot be used as sulfur source for biotin synthesis from *Rhodotorula glutinis*, whereas methionine was utilized. The sulfur source for the thiazole moiety of thiamine in yeast can be methionine or glutathione (Maw, 1985; Umezawa and Kishi, 1989).

Lipoic acid has a disulfide group and can be reversibly reduced or oxidized. It is located in the multienzyme complex required for the decarboxylation of pyruvic acid and α-ketoglutaric acid. Lipoic acid also has a function in the assimilatory reduction pathway of sulfate (Slaughter, 1989).

SULFUR COMPOUNDS IN WINE

Origin

Sulfur containing substances are formed in grapes during ripening and, later during the process of winemaking. Both enzymatic and non-enzymatic reactions are involved in their production (Schreier and Drawert, 1975). Enzymatic formation occurs in both intact and crushed grapes, and as a consequence of yeast metabolism. In addition, S-containing pesticides or their breakdown products can be used as precursors in biochemical reactions arising from grape juice enzymes or yeast metabolism. Non-enzymatic formation of S-compounds occurs through the influence of temperature, light (sun-struck flavor), and chemical transformations such as the Maillard-reaction or Strecker-degradation. Other factors which cause non-enzymatic reactions are vinification techniques, hydrolysis or chemical degradation of pesticides, and artificial reactions with reactive compounds such as hydrogen sulfide or mercaptans.

In must, sulfur is naturally available to yeasts as sulfate, which is utilized by the pathway outlined in Figure 6.1. The concentration of sulfate in the must is about 200 mg/l (Dittrich, 1987), but higher concentrations occur. The sulfate content in wine varies widely, between 30–2200 mg/l (Zee, 1983). In German wines, sulfate concentrations of 160–400 mg/l are detected, (Lemperle and Lay, 1989).

Sulfite is present in must and wine at 50–200 mg/l through its deliberate addition as an antioxidant and antimicrobial agent. Normally, yeasts produce between 10–30 mg/l of sulfite, but some strains are able to form more than 100 mg/l of sulfite (Eschenbruch, 1974). At the pH of wine, sulfite dissociates into a mixture of sulfite anions (SO_3^{2-}), bisulfite anions (HSO_3^-) and sulfur dioxide (SO_2) (Macris and Markakis, 1974). It also reacts with compounds containing carbonyl groups to form α-hydroxysulfonates (Wucherpfennig, 1984).

Methionine is the major S-containing amino acid in must. In must from healthy grapes, its content varies from 17–76 mg/l while in grapes infected with *Botrytis cinerea*, its content is 5–22 mg/l (Rapp and Reuther, 1971; Dittrich and Sponholz, 1975; Dittrich, 1987). As expected, the concentrations of methionine are decreased after fermentation. In wine, its concentration ranges between 0–15 mg/l (Rapp, 1965; Dittrich and Sponholz, 1975).

Only little information is available about the concentrations of cysteine and cystine in musts. Either it was not detected or was found only in trace amounts (Eschenbruch, 1974). The glutathione content of must ranges from 14–102 mg/l (Cheynier *et al.*, 1989). However, it is important to consider that the amino acid content of musts, generally, is affected by different grape varieties, vintages, climatic conditions, soil types, fertilization, *B. cinerea* infection and the winemaking process (e.g. treatment with bentonite or enzymes).

Sulfur-containing vitamins such as thiamine (vitamin B_1) and biotin, occur in musts, but they can be lacking after special treatments (e.g. heating). The thiamine content is also reduced in musts obtained from *B. cinerea* infected grapes. Normally, thiamine is present in must at 160–330 μg/l (Radler, 1989).

During fermentation, its concentration is decreased to about 50 μg/l, and biotin is decreased from 9–11 μg/l to 2–6 μg/l (Lemperle and Lay, 1989).

Non-natural S-sources can be elemental sulfur, which is used as a grape fungicide, and other S-containing pesticides (see later Sections).

Production by Yeasts

Diversity and sensory thresholds

Sulfur-containing compounds have a significant role in the flavor of wines. This is related to their high volatility, reactivity and potency at very low concentrations (Schutte, 1975).

The potency of a flavor substance is generally expressed as its threshold concentration, that is the value below which its sensory effect is not detected. Many S-compounds, especially those of low molecular weight, are extremely flavor active and have thresholds far below 0.002 ppb (Soltoft, 1988). However, these values vary depending on the medium (e.g. wine variety) in which they are measured, as well as the taste panel member. Table 6.1 lists threshold values determined for some of the main sulfur compounds found in wine. Table 6.2 represents a collation of S-compounds detected in wines along with the range of concentrations reported. Both tables serve as guides to key publications on these topics and demonstrate the diversity in chemical form of these substances. Some of these substances have a desirable impact on wine quality, while others contribute strong objectionable flavors, even at extremely low concentrations. Peppard (1988) discusses analytical techniques for the detection and identification of volatile S-compounds in alcoholic beverages.

Sulfite

It is well known that yeasts are able to form sulfite during fermentation. Normally, strains of *S. cerevisiae* produce sulfite in the range 10–30 mg/l (Eschenbruch, 1974; Dott *et al.*, 1976; Dittrich, 1987). However, some strains, called 'SO_2-forming yeasts' or 'yeasts with high-sulfite formation', produce sulfite in amounts exceeding 100 mg/l (Eschenbruch, 1974). Such sulfite is of considerable enological interest, because it binds to acetaldehyde and other compounds, and it can contribute to unacceptably high levels of SO_2 in wines (Rankine and Pocock, 1969; Eschenbruch, 1987). Thus, careful selection of strains for conducting the alcoholic fermentation can overcome problems of excessive SO_2 production, especially by wild or indigenous strains (Bidan and Salgues, 1981; Dittrich, 1987).

High-sulfite formation by yeasts is influenced by many vinification factors (Minarik, 1977; Larue *et al.*, 1985). Initial studies showed that sulfite production increases during fermentation, depends on the yeast strain and varies with the degree of must clarification (Zang and Franzen, 1967; Mayer and Pause, 1968; Eschenbruch, 1972c; Fetterroll and Würdig, 1977; Dott and Trüper, 1978 b). Strong clarification leads to more sulfite formation. In unclarified musts there is, in general, lower formation of sulfite, possibly because of the increased surface

TABLE 6.1. Flavor threshold values of volatile sulfur-substances.

S-substance		*Threshold Concentration* [μg/l]	*References*
Hydrogen sulfide	O[a]	50–80 (wine)	Wenzel *et al.*, 1980
Dimethyl sulfide	O	25 (white wine)	Goniak and Noble, 1987
	O	60 (red wine)	De Mora *et al.*, 1987
Diethyl sulfide	O	0.92 (white wine)	Goniak and Noble, 1987
Dimethyl disulfide	O	29 (white wine)	Goniak and Noble, 1987
Methyl mercaptan	O/F	0.02/2 (water)[b]	Fors, 1983
	F[c]	2–12 (beer)[b]	Soltoft, 1988
Ethyl mercaptan	O	1.1 (white wine)	Goniak and Noble, 1987
S-Methyl thioacetate	F	300 (beer)[b]	Soltoft, 1988
S-Ethyl thioacetate	F	40 (beer)[b]	Soltoft, 1988
Carbon disulfide	F	>38 (white wine)	Spedding and Raut, 1982
3-(Methylthio)-1-propanol	O	2 000 (beer)	Meilgaard, 1981
3-(Methylthio)-1-propanal	F	10-50 (water)[b]	Fors, 1983
	F	250 (beer)[b]	Fors, 1983
3-(Methylthio)-propylacetate	O	600 (beer)	Meilgaard, 1981
Benzothiazole	O	24 (wine)	Keck, 1989
4-(Methylthio)-butanol-1	O	100–1 000 (15% (v/v) ethanol/water)	Rapp *et al.*, 1985
2-[[(Methylthio)methyl]-thio]-ethanol	F	>100 (estimated in wine)	Keck, 1989
2-Mercaptoethanol	O	10^3–10^4 (15% (v/v) ethanol/water)	Rapp *et al.*, 1985

[a]odor threshold
[b]ppb
[c]flavor threshold, odor and taste together or no differentiation.

area of particulate material and higher quantity of amino acids (Fetterroll and Würdig, 1977; Würdig, 1985). Both pH and temperature affect SO_2 formation by yeasts but the response varies with the strain (Rankine and Pocock, 1969; Würdig and Schlotter, 1967, 1971; Eschenbruch and Bonish, 1976 a).

Dittrich and Staudenmayer (1968, 1970) proposed that sulfite-producing yeasts are mutants which are unable to reduce sulfite to sulfide (reaction 6, Figure 6.1). Contrary to this hypothesis, Eschenbruch (1972c) and Eschenbruch *et al.* (1973) proved that both normal and sulfite-producing strains form sulfide in similar concentrations.

TABLE 6.2 Concentrations of volatile sulfur-substances in wine.

S-substance	*Concentration* [µg/l]		*References*
Hydrogen sulfide	<1.5	(normal wine)	Spedding *et al.*, 1983
	20–30	(young wine)	Dittrich and Staudenmayer, 1970
	>80	(off–flavor wine)	Wenzel *et al.*, 1980
Dimethyl sulfide	0–474		Loubser and Du Plessis, 1976
	5–44	(white wine)	Leppänen *et al.*, 1979
	10–32	(red wine)	
	5–20	(rosé wine)	
	<910		De Mora *et al.*, 1987
Diethyl sulfide	qualitative		Schreier, 1979; Januik, 1984
Dimethyl disulfide	traces–1.1	(white wine)	Leppänen *et al.*, 1979, 1980
	0.3–1.6	(red wine)	
	traces–0.5	(rosé wine)	
Methyl mercaptan	qualitative	(fermentation experiments)	De Mora *et al.*, 1986
		(off-flavor wines)	Rauhut *et al.*, 1986 Rauhut and Sponholz, 1990
Ethyl mercaptan	qualitative	(suspected)	Rankine, 1963
S-Methyl thioacetate	2–13 (red wine)	(white wine)	Leppänen *et al.*, 1979, 1980
	7–16	(red wine)	
	11–15	(rosé wine)	
S-Ethyl thioacetate	traces–2	(white wine)	Leppänen *et al.*, 1979, 1980
	traces–4	(red wine)	
	traces	(rosé wine)	
Carbon disulfide	traces–10	(white wine)	Leppänen *et al.*, 1979, 1980
	traces–2.1	(red wine)	
	0.7–1.4	(rosé wine)	
Carbonyl sulfide	traces–2.5	(2.4 nl/l)	Spedding *et al.*, 1983
3-(Methylthio)-1-propanol	145–520		Schreier *et al.*, 1974
	507–998	(white wine)	Baumes *et al.*, 1986

TABLE 6.2 (Cont.) Concentrations of volatile sulfur-substances in wine.

S-substance	*Concentration* [μg/l]		*References*
	1363–2314	(red wine)	
	500–6300		Keck, 1989
3-(Methylthio)-1-propanal	traces		Schreier *et al.*, 1976
3-(Ethylthio)-1-propanol	11-68		Schreier *et al.*, 1974
	6–23	(white wine)	Baumes *et al.*, 1986
	21–39	(red wine)	
3-(Methylthio)-propylacetate	traces–15		Schreier *et al.*, 1974
	0.4–20		
Methyl 3-(methylthio)-propanoate	qualitative		Schreier *et al.*, 1974
Ethyl 3-(methylthio)-propanoate	traces–5.7		Schreier *et al.*, 1974
3-(Methylthio) propanoic acid	qualitative		Schreier *et al.*, 1974
N-3-(Methylthio)-propylamine	qualitative		Schreier *et al.*, 1976
N-[3-Methylthio)propyl]-acetamide	qualitative		Schreier and Drawert, 1974
	<10–340	(normal wine)	Keck, 1989
N-(1-Acetyl)methionineamide	qualitative	(white wine)	Rapp and Güntert, 1986
Benzothiazole	qualitative	(white wine)	Etievant and Bayonove, 1983
	qualitative	(red wine)	Vernin *et al.*, 1988
	0.4–7		Keck, 1989
5-(2-Hydroxyethyl)-	qualitative		Rapp *et al.*, 1985
4-methylthiazole	15–384		Keck, 1989
4-(Methylthio)-butanol-1	qualitative		Rapp *et al.*, 1985
	9–50		Keck, 1989
2-[[Methylthio)methyl]thio]-ethanol	qualitative		Rapp *et al.*, 1985
	0.4–47		Keck, 1989
2-Methyltetrahydro-thiophene-3-one	qualitative		Schreier, 1976
	7.–45	(white wine)	Baumes *et al.*, 1986
	92–167	(red wine)	
	<60–1040		Keck, 1989

TABLE 6.2 (Cont.) Concentrations of volatile sulfur-substances in wine.

S-substance	*Concentration* [μg/l]		*References*
cis-2-Methyltetrahydro-thiophene-3-ol	qualitative		Rapp *et al.*, 1984
	<5–8	(white wine)	Baumes *et al.*, 1986
	13–25	(red wine)	
	106–657		Keck, 1989
trans-2-Methyltetrahydro-thiophene-3-ol	qualitative		Rapp *et al.*, 1984
	<5–12	(white wine)	Baumes *et al.*, 1986
	14–23	(red wine)	
	2–160		Keck, 1989
2-Mercaptoethanol	$<10^{-2}$	(estimated)	Rapp *et al.*, 1985
Ethyl methane disulphonate	qualitative		Yunome *et al.*, 1981
2,4-Di-tert-butylthiophene	qualitative		Shimizu and Watanabe, 1982a
2-(1-Butyl)-5-(2-methyl-propyl)-thiophene	qualitative		Shimizu and Watanabe, 1982b
2,5-Di-2-methylpropyl-thiophene	qualitative		Shimizu and Watanabe, 1982b
N-Acetylmethionine ethyl ester	14–215		Keck, 1989

Altered properties and activities of enzymes involved in the sulfate reduction pathway as well as variations in the biosynthesis of these enzymes could explain the high SO_2 production by some yeast strains (Dott and Trüper, 1978 c). Thus, accumulation of sulfite could be due to abnormally high activities of sulfite generating enzymes or to decreased or missing activities of sulfite-decomposing enzymes (Eschenbruch, 1974; Heinzel *et al.*, 1979). Higher specific activity of sulfate permease has been observed in such yeasts along with a partial or total lack in repression of this enzyme by methionine (Dott *et al.*, 1977; Bonish and Eschenbruch, 1976). Intracellular sulfate concentration is increased in high SO_2-producing strains, compared with normal strains (Jorzig, 1985). Also, sulfite causes an inhibition of sulfate transport. The activity of ATP-sulfurylase may be higher (Heinzel and Trüper, 1976; Heinzel *et al.*, 1979) and may not be subject to sulfite or methionine repression (Heinzel and Trüper, 1978). Some SO_2-producing yeasts have a defective feedback inhibition of ATP-sulfurylase (Heinzel and Trüper, 1976).

In all SO_2-forming yeast strains examined, a very low specific activity of NADPH-sulfite reductase (reaction 6, Figure 6.1) was observed, along with a decreased affinity of their enzyme for its substrate, sulfite (Dott and Trüper, 1976, 1978 a; Heinzel *et al.*, 1979).

The nutrient composition of the must influences sulfite formation. Increased additions of sulfate causes increased sulfite production by high sulfite-forming strains. Abundant amounts of fermentable substrates, such as glucose, gives higher SO_2-formation (Gyllang *et al.*, 1989). A decrease in the concentration of pantothenic acid results in reduced sulfite formation by high-sulfite producers. (Wainwright, 1970, 1971; Beech *et al.*, 1979; Hartnell and Spedding, 1979).

In low-sulfite forming strains, sulfite production can be suppressed by the addition of sulfite. This, presumably, represses the ATP-sulfurylase (Heinzel and Trüper, 1976; Dott and Trüper, 1978 a). Bonish and Eschenbruch (1976) and Eschenbruch and Bonish (1976 a, b) demonstrated a linear correlation between ATP-sulfurylase activity and SO_2-formation. Further regulation of SO_2-formation occurs via the main products of sulfur utilization, methionine and cysteine. The addition of increasing amounts of methionine and/or cysteine reduces the uptake of sulfate and the formation of sulfite and sulfide by low as well as by high-sulfite-forming strains. Eschenbruch (1972 a, b) assumed that these effects are caused by the repression of ATP-sulfurylase by methionione and a repression of sulfite reductase by cysteine. Cysteine may act via methionine (Heinzel *et al.*, 1979) and probably has a repressive influence on sulfate-permease (Eschenbruch 1972 a, b, 1974).

Sulfite production is very energy dependant. Therefore, SO_2-forming yeasts show a distinct loss in cellular metabolism, which explains the reduced production of biomass and, in most cases, a slow fermentation. An intracellular inhibition by the produced sulfite is also conceivable.

Hydrogen sulfide

Hydrogen sulfide (H_2S), the end-product of the sulfate reduction pathway, is formed in small quantities during fermentation (Rankine, 1963, 1964; Dittrich, 1987, reviewed by Monk, 1986). The odor threshold value of H_2S in wine varies from 50–80 μg/l (Wenzel *et al.*, 1980), with concentrations exceeding these values causing an undesirable sulfurous off-flavor that is reminiscent of rotten eggs. In young wines without SO_2-treatment, the production of 20–30 μg/l of H_2S may have a positive contribution to 'yeasty' flavor (Dittrich and Staudenmayer, 1970).

Many chemical and biological factors affect the production of H_2S in wines. Most significantly, its production varies with the strain of *S. cerevisiae*, with some strains producing amounts exceeding 1 mg/l (Acree *et al.*, 1972; Eschenbruch, 1974, 1978; Eschenbruch *et al.*, 1978; Vos and Gray, 1979). Such variation is genetically based, but is also influenced by environmental factors such as the composition of the must and fermentation conditions (Monk, 1986). As noted already, H_2S is formed as an intermediate in the biosynthesis of S-containing amino acids. Consequently, its formation in wine is related to both sulfur and nitrogen metabolism by yeasts (Henschke and Jiranek, 1991).

Increased H_2S production can occur during fermentation at higher temperatures and at higher pH values (Rankine, 1968). Increased fermentation intensity, which is often associated with higher starting temperatures of the must, high must volume and large vessel capacity, can also lead to increased H_2S production (Wenzel and Dittrich, 1978). More H_2S is produced during fermentation of musts that have a higher content of solids. Grape must solids consist partly of protein which, possibly, are metabolized by yeasts to release S-containing amino acids that lead to H_2S formation (Wenzel and Dittrich, 1978, 1983).

Deficiency in a readily available nitrogen supply (e.g. ammonium ions or glutamate) in the must is an important cause of H_2S formation by yeasts (Monk, 1986; Henschke and Jiranek, 1991). Vos and Gray (1979) and Vos (1981) suggested that musts with low concentrations of free amino nitrogen might stimulate extracellular protease production by yeasts, thereby leading to protein degradation and significantly increased H_2S formation from the S-containing amino acids released. However, Henschke and Jiranek (1991), in concurrence with the studies of Stratford and Rose (1985) consider that sulfite and not protein is the principal source of excessive H_2S production in nitrogen deficient musts. The addition of ammonium nitrogen to the must provides an easily assimilated nitrogen source for yeast, and reduces H_2S formation. This practice is used in warmer grape growing regions to control H_2S production, since musts of such grapes contain low assimilable nitrogen (Monk, 1986).

An abundant quantity of organic sulfur compounds, like cysteine, in the must can give high H_2S values. Desulfuration of cysteine to H_2S is possible with cysteine desulfhydrase. It has been proposed that α-aminoacrylic acid (2-amino-propenoic acid) and H_2S are formed simultaneously (Figure 6.4). After this reaction, α-aminoacrylic acid isomerizes and decomposes to pyruvate and ammonia (Maw, 1965). Splitting of cysteine by the enzyme serine sulfhydrase also gives H_2S as well as serine (Dott and Trüper, 1978c).

Deficiencies in concentrations of the vitamins pantothenate or vitamin B_6 in the medium cause increased H_2S production by yeast (Wainwright, 1970, 1971). A deficiency of pantothenate leads to deficiency of coenzyme A. This results in a decrease of O-acetylhomoserine. Lack of pantothenate could also result in

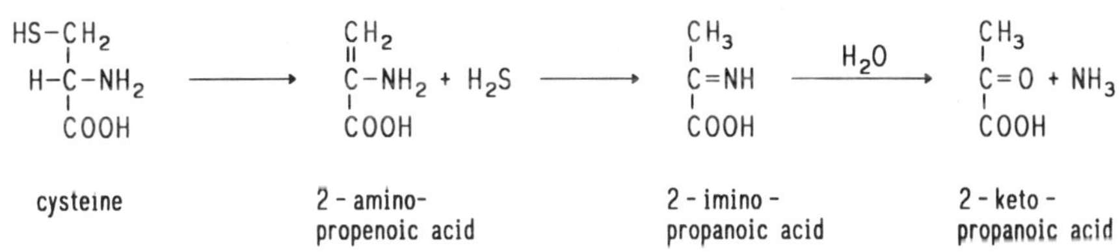

FIGURE 6.4 Suggested mechanism for H_2S-formation from cysteine by yeasts (modified from Maw, 1965).

accumulation of cysteine. It is not elucidated if H_2S is formed by the splitting of cysteine by cysteine desulfhydrase or if it arises from the biosynthetic pool (Slaughter, 1989). However, in must the pantothenate concentration is sufficient and the cysteine content very low; therefore it is unlikely that they would influence H_2S formation during fermentation. Pyridoxin is necessary for several reactions in the pathway of methionine synthesis. Large amounts of H_2S are produced if yeast strains are not able to synthesize pyridoxin or if the available concentration in the must is below 2 mg/l.

As noted before, sulfite (or SO_2) added to grapes at the stage of crushing can be another source of sulfur for H_2S formation by yeasts (Eschenbruch, 1974; Monk, 1986). The biochemical bases of this reaction have been studied by Stratford and Rose (1985, 1986).

Elemental or molecular sulfur is used as a grape fungicide. About 60% of this sulfur can be found as a residue in unclarified must, while about 10% occurs in clarified musts. Elemental sulfur residues in unclarified musts range from 0.3–8.9 mg/l. Wines produced from grapes containing more than 1 mg/l of residual elemental sulfur had H_2S levels far beyond the threshold value (Wenzel *et al.*, 1980; Wenzel and Dittrich, 1983). The production of H_2S from this source depends on the kind and amount of elemental sulfur, fermentation temperature, ethanol concentration, the oxidation-reduction state of the medium and the physiological condition of the yeast (Schütz and Kunkee, 1977). Residues of elemental sulfur could originate also from incompletely burned sulfur wicks, used to disinfect casks. Elemental sulfur may undergo non-enzymatic reduction to H_2S by compounds (e.g. glutathione) that possess a sulfhydryl group (Maw 1965; Wainwright 1971). Thus,

$$2\,RSH + S \rightarrow R\text{-}S\text{-}S\text{-}R + H_2S$$

Normally, most of the H_2S formed during fermentation is carried out with the carbon dioxide. Therefore, in normal, faultless wines H_2S is not detectable, despite a detection limit of 1.5 μg/l (Spedding *et al.*, 1983). In most cases, H_2S which remains after fermentation will disappear slowly. Increased amounts of H_2S can be purged out or oxidized by aeration or oxidized by treatment with sulfite (Tanner, 1969).

$$2H_2S + O_2 \rightarrow 2H_2O + S$$

$$2H_2S + SO_2 \rightarrow 2H_2O + 3S$$

If the H_2S is not removed by these steps, unpleasant stagnant or rubbery aromas develop with time and storage (Rankine, 1963, 1968). This is attributed to the formation of other volatile S-compounds, which probably arise from reactions of H_2S. It has been proposed that ethyl mercaptan is formed by the reaction of H_2S with acetaldehyde (ethanal) *via* a cyclic trithioacetaldehyde (trithioethanal) intermediate (Rankine, 1963, 1968; Tanner, 1969) but this reaction has not been demonstrated in model solutions (Bobet *et al.*, 1990).

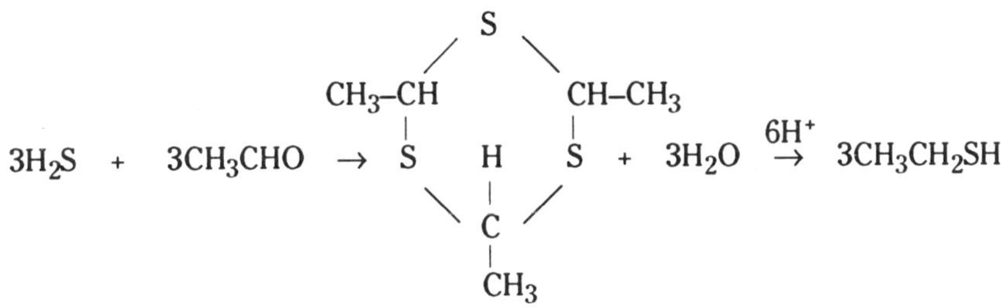

The reaction of hydrogen sulfide and acetaldehyde to 1,1-ethanedithiol was determined in wine-like model solutions and wine (Rauhut *et al.*, 1992). Moreover, its oxidation product 3,6-dimethyl-1,2,4,5-tetrathiane, the reaction products 3,5-dimethyl-1,2,4-trithiolane, 4,7-dimethyl-1,2,3,5,6-pentathiepane and their isomers were identified. After the addition of acetaldehyde and hydrogen sulfide to wine further S-compounds were detected. 1,1-ethanedithiol causes a sulfury and rubbery flavor note. This S-substance and some of its reaction products were also identified in yeast extract (Werkhoff *et al.*, 1991). There, they are responsible for the main impact of the flavor. A contribution of these S-substances to off-flavors in wine can be expected.

It has also been suggested that ethanol reacts with H_2S to form ethyl mercaptan (EtSH), but this has not been proven (Rankine, 1963, 1968).

$$C_2H_5OH + H_2S \rightarrow C_2H_5SH + H_2O$$

An unacceptably high amount of H_2S can be removed by treatment with copper or silver ions, according to the following reaction, but copper or silver sulfide is formed (Tanner, 1969), and such treatments must conform to prevailing legislation.

$$H_2S + CuSO_4 \rightarrow CuS + H_2SO_4$$

The formation of hydrogen sulfide in sterile wine can be explained from reduction of SO_2 caused by the contact with metal closures, metal surfaces or unidentified chemical reactions in which organic S-containing compounds are involved.

Dimethyl sulfide

The thioether, dimethyl sulfide (DMS), has a relatively low boiling point (38°C) and contributes beneficially to the aroma and flavor of many foods and beverages.

The flavor note of DMS is described as reminiscent of asparagus, corn and molasses. The odor threshold of DMS for white wine is 25 μg/l (Goniak and Noble, 1987) and for red wine it is 60 μg/l (De Mora *et al.*, 1987).

The concentration of DMS in wines ranges from 0–474 μg/l (Loubser and Du Plessis, 1976) or, in special cases, even higher. Leppänen *et al.* (1979) reported concentrations of 5–44 μg/l, in white wine, 10–32 μg/l in red wine, and 5–20 μg/l in rosé wine. In high concentrations, it may contribute to the development of bottle age bouquet and can be important in the aroma of late harvest wine (Du Plessis and Loubser, 1974; Simpson, 1979).

Marais (1979) observed that DMS concentrations increase with storage time and temperature. The DMS concentration in white wines aged one to two years was 0 to trace, and in red wines aged one year was 3.2 to 6.0 μg/l (Goto and Takamuro, 1987). It appeared in red wines earlier than in white wines and increased in concentration during bottle ageing. A maximum concentration (7.6–68.4 μg/l) occurred in wines aged 5–10 years and, thereafter, its concentration decreased. It was suggested that the presence of DMS contributes to bottle bouquet.

Eschenbruch *et al.* (1986) found that formation of DMS starts at the beginning of fermentation. The concentration of DMS in New Zealand wines showed large variation, which was considered to depend on the process of winemaking. Older wines generally had more DMS (Spedding *et al.*, 1980, 1983). However, no consistent relationship between storage time and DMS concentrations was found in some Australian wines, especially in Cabernet Sauvignon wines from the Coonawara district in South Australia (De Mora *et al.*, 1987). In these wines, DMS levels were sometimes extremely high (up to 910 μg/l), and this variation was related to vintage, viticultural practice and vinification techniques. The tolerance of DMS, even in high concentrations, depended on the character of the wine and was mainly observed as a part of the bouquet. Low concentrations had a positive effect on flavor (Spedding and Raut, 1982) and could enhance fruit character (De Mora *et al.*, 1987).

Only fragmentary information is available about the formation and precursors of DMS in wine. In beer, the main precursor of DMS is considered to be S-methylmethionine (SMM), which is formed during malting of barley (Bamforth, 1980; Anness and Bamforth, 1982). A part of DMS found in beer is released by decomposition of SMM during malt kilning. To some extent, DMS is transformed to dimethyl sulfoxide (DMSO). This reaction is catalyzed by heating during the brewing process. The production of DMS in fermenting worts is attributed to the ability of yeast to reduce DMSO to DMS (Figure 6.5). There are indications that there is no special enzyme, but that the synthesis is carried out by the NADPH linked methionine sulfoxide reductase. The functional system of this enzyme consists of three parts: a methionine sulfoxide reducing protein, thioredoxin and thioredoxin reductase (Bamforth and Anness, 1981).

There is one view that yeasts cannot synthesize DMS directly from SMM (Anness and Bamforth, 1982), but another suggestion that, in particular circumstances, SMM can be enzymatically hydrolyzed to DMS and homoserine (Figure 6.6; Booer and Wilson, 1979). Under special conditions yeasts may produce SMM via transmethylation between S-adenosyl-L-methionine and methionine (Booer and Wilson, 1979). However, Schreier *et al.* (1976) postulated

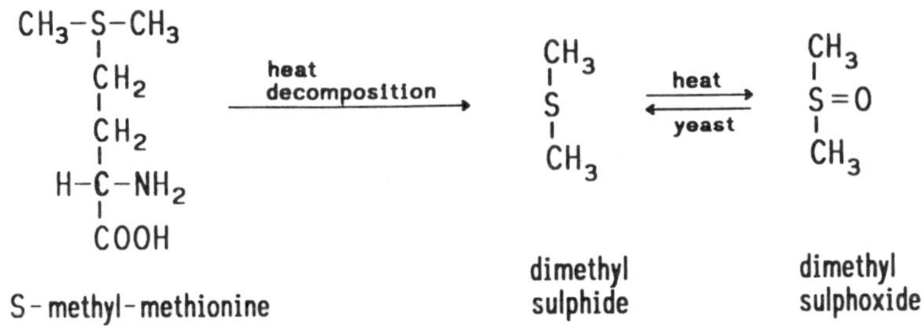

FIGURE 6.5 Mechanisms of dimethyl sulfide formation during the various heating stages of the brewing process (modified from Anness and Bamforth, 1982; Slaughter, 1989).

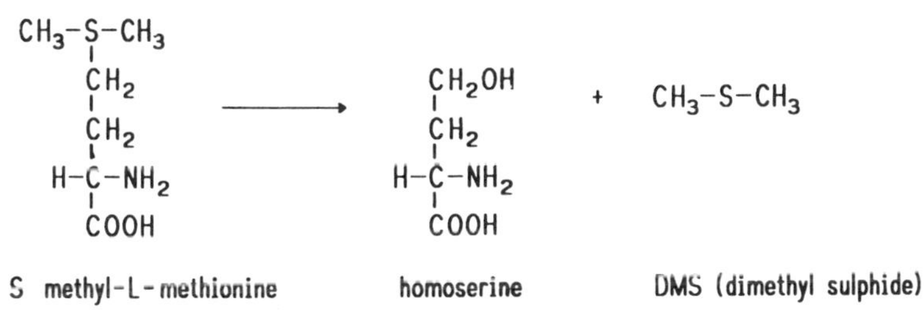

FIGURE 6.6 Suggested pathway for the formation of dimethyl sulfide from S-methyl-L-methionine by yeasts (modified from Booer and Wilson, 1979).

that yeasts are not capable of forming SMM. They indicated that DMS can be synthesized by yeasts from cystine, cysteine or glutathione. De Mora *et al.* (1986) noted that cysteine can be a precursor for DMS especially when the wine is racked late, suggesting that lees are involved in DMS formation before the wines are bottled.

In wine, it is also assumed that DMSO is a precursor to the DMS which develops during bottle storage and ageing (Spedding *et al.*, 1980). This has not been proved analytically.

There are no data regarding the concentrations of SMM or DMSO in must or wine and systematic investigation to prove that these substances can be precursors for DMS in wine is absent. The biosynthesis of DMS in wine has not been sufficiently elucidated and further research is necessary.

Other sulfides and disulfides

Only few qualitative and quantitative results are available on the occurrence of other sulfides and disulfides in wine. Diethyl sulfide (DES) occurs in wine and has a flavor reminiscent of cooked vegetables, onion and garlic (Fors, 1983; Januik, 1984; Goniak and Noble, 1987). Its flavor threshold is 0.92 $\mu g/l$ (Goniak and Noble, 1987).

Dimethyldisulfide (DMDS) was determined in white wines at concentrations from traces to 1.1 μg/l, in rosé wines from traces to 0.5 μg/l and in red wines from 0.3–1.6 μg/l (Leppänen *et al.*, 1979; 1980). These small concentrations of DMDS can be formed from methyl mercaptan (MeSH) by oxidation. Higher amounts arising from pesticides can occur in sulfurous off-flavor wines. Yeasts have the ability to reduce disulfides, such as DMDS, to mercaptans (Neuberg and Schwenk, 1915). The odor of DMDS is characterized as cooked cabbage and intensively onion-like (Fors, 1983) and has a threshold value in wine of 29 μg/l (Goniak and Noble, 1987).

Diethyldisulfide (DEDS) can be produced by the oxidation of two molecules of ethyl mercaptan (EtSH). The flavor of DEDS was described as reminiscent of garlic and burnt rubber and has a threshold of 4.3 μg/l in white wine (Fors, 1983; Goniak and Noble, 1987).

Mercaptans

Mercaptans, like methyl mercaptan (MeSH) and ethyl mercaptan (EtSH), are highly reactive substances with very low boiling points (MeSH 5.95°C; EtSH, 35°C). The odor threshold of MeSH is extremely low at 0.02–2 ppb in water (Fors, 1983) and 2–12 ppb in beer (Soltoft, 1988). The flavor of MeSH has been described as being similar to rotten eggs or cabbage (Fors, 1983; Soltoft, 1988).

Niefind and Späth (1971) reported the formation of MeSH during fermentation of synthetic media containing the sulfur sources cysteine, methionine, S-methylmethionine and sulfate. More recent fermentation studies by De Mora *et al.* (1986) suggested that it can be synthesized by yeast metabolism of methionine. It can also arise in wine through the hydrolysis of pesticides and from methionine after exposure to light during storage (Thoukis and Stern, 1962; Haye *et al.*, 1977; Maujean, 1984; Laparra, 1990). A pH-dependent formation of MeSH from methionine in model solutions was demonstrated by Wainwright *et al.* (1972). In normal faultless wines, MeSH has not been investigated neither qualitatively nor quantitatively.

Ethyl mercaptan can probably arise by the reaction of H_2S with ethanal or ethanol. The aroma of EtSH is described as similar to onion or rubber and the odor threshold value in wine was 1.1 μg/l (Goniak and Noble, 1987). There are no data on the qualitative and quantitative determination of EtSH in wine.

Mercaptans are very reactive substances and can be easily oxidized to disulfides. This reaction takes place in the presence of oxygen, mild oxidizing agents and is catalyzed by heating processes and radiation. Thus,

$$4C_2H_5SH + O_2 \leftrightarrow 2C_2H_5SSC_2H_5 + 2H_20$$

The disulfides can be reduced to mercaptans again by the addition of sulfite ions (Tanner, 1969; Bobet *et al.*, 1990). In wines without SO_2-treatment, the reaction of ethyl mercaptan and ethanal to 1,1-bis-(ethylthio)-ethane is supposed (Tanner, 1969).

$$2C_2H_5SH + CH_3CHO \rightarrow \begin{array}{c} H \\ | \\ CH_3\text{–}C\text{–}SC_2H_5 \\ | \\ SC_2H_5 \end{array} + H_2O$$

High amounts of mercaptans can be removed by treatment with copper or silver ions. The reaction with copper ions leads to copper mercaptides and to the corresponding disulfides (Tanner, 1969). Disulfides cannot react with copper ions. Therefore, in some cases an off-flavor is not removable, because disulfides can cause an unpleasant smell.

Thioesters

S-Methyl thioacetate (S-methyl acetothioate; MeSAc) and S-ethyl thioacetate (S-ethyl acetothioate; EtSAc) have been found in wine by Leppänen *et al.* (1979, 1980). MeSAc was determined in white wine in concentrations from 2–13 μg/l, in red wine from 7–16 μg/l and in rosé wines from 11–15 μg/l (Leppänen *et al.*, 1979). It has been postulated that MeSAc is synthesized from MeSH, which is formed by the enzymatic degradation of methionine by yeasts (Matsui and Amaha, 1981). EtSAc appears in smaller amounts, at a content up to 2 μg/l in white wine, up to 4 μg/l in red wine, and only traces in rosé wines (Leppänen *et al.*, 1979)

Both thioesters have relatively high sensory thresholds, (e.g. MeSAc 300 ppb in beer and EtSAc 40 ppb in beer). MeSAc has an objectionable odor of rotten vegetables. The flavor notes of thioesters are often described as cheesy, onion-like, burnt and sulfury (Soltoft, 1988). They cause sulfurous off-flavors in wine because they can be hydrolyzed to the corresponding mercaptans (Leppänen *et al.*, 1980).

Carbonyl sulfide and carbon disulfide

Leppänen *et al.* (1980) found carbon disulfide (CS_2) from traces to 10 μg/l in white wines, from 0.7–1.4 μg/l in rosé wines and from traces to 2.1 μg/l in red wines. In the headspace of some New Zealand wines, levels up to 15–18.9 μg/l of CS_2 were reported by Spedding *et al.* (1983). Carbonyl sulfide (COS) was also identified from traces to 2.5 µg/l.

Eschenbruch *et al.* (1986) observed the development of COS and CS_2 during clarification of fresh musts of the varieties Müller-Thurgau and Riesling. An increase of both substances was measured immediately after crushing. The formation was accelerated with elevated temperatures. A decrease of COS and CS_2 levels occurred during fermentation. The addition of SO_2 had no definite effect on the development of COS and CS_2.

CS_2 has a sweet, ethereal odor and a boiling point of 46.3°C. Commercially available CS_2 often has unknown impurities. These impurities are responsible for the foul smell rather than CS_2 (Neumüller, 1987; Eschenbruch *et al.*, 1986). Spedding and Raut (1982) tested the influence of CS_2 on wine bouquet. It seems that the flavor threshold value is above 37.8 μg/l, which is higher than the concentrations detected in wines; therefore no significant influence on the flavor of wine is supposed. COS (b.p. –50°C) is also unlikely to have an influence on the sensory quality of wines, because it is a colorless and odorless gas (Spedding *et al.*, 1980).

Volatile S-compounds, like COS and CS_2, are produced by phytoplankton in the photosynthetic zone of the oceans. Subsequently, they enter the atmosphere by turbulence-induced gas exchange. Another mode of formation of COS and CS_2 is the decomposition of plant material by microorganisms (Bingemer *et al.*, 1987). There is evidence that, in wine, both substances have a metabolic origin (Eschenbruch *et al.*, 1973). It is also thought that COS and CS_2 can result from the degradation of fungicides, such as ethylenebisdithiocarbamate and tetramethylthiuramdisulfide (Burkhardt, 1978; Schmitt, 1987; Maujean, 1989). Additionally, it was postulated that COS can be formed by the reaction of SO_2 and CO_2 (Shaw and Nagy, 1981).

Miscellaneous volatile sulfur compounds

Several other S-compounds, which have been determined in wine include thioethers, cyclic and heterocyclic compounds (Table 6.2).

The substance, 3-(methylthio)-1-propanol, which has a raw potato odor, was first identified in wine by Muller *et al.* (1971). This compound is mainly produced during fermentation, but Mandery (1983) showed that traces are also in the must. It is formed from methionine by the Ehrlich-mechanism. According to this mechanism, amino acids are transformed to keto acids, which are then decarboxylated to aldehydes. Subsequently, the aldehydes are enzymatically reduced to alcohols (Bärwald and Kliem, 1971; Figure 6.7).

Schreier *et al.* (1974) reported concentrations of 3-(methylthio)-1-propanol from 145–520 μg/l in wines. Di Stefano (1985) obtained similar results for German Riesling wines. It has been suggested that there is a correlation between methionine content in the must and 3-(methylthio)-1-propanol content in wine. An increase in 3-(methylthio)-1-propanol is probably an indicator of the high degree of ripeness of the grapes (Schreier *et al.*, 1974). As investigated by Baumes *et al.* (1986), white wines contain 507–998 μg/l of 3-(methylthio)-1-propanol and red wines 1363–2314 μg/l. A negative contribution to wine aroma was presumed, and the flavor note of this S-substance was described as cooked

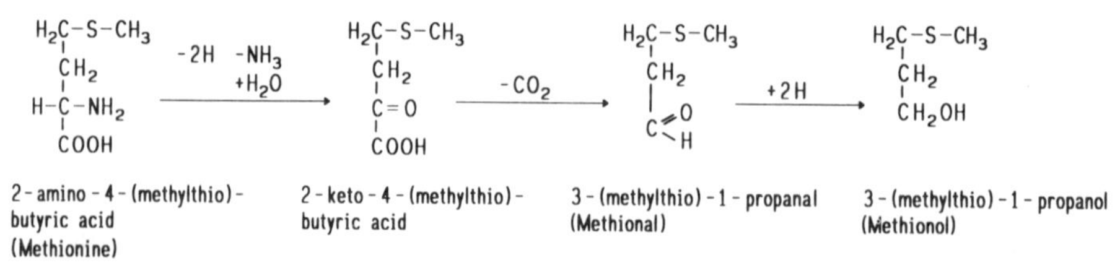

FIGURE 6.7 Pathway for the biosynthesis of 3-(methylthio)-1-propanol (modified from Bärwald and Kliem, 1971).

potatoes. More recent studies showed a variation in the content of 3-(methylthio)-1-propanol from 500–630 μg/l. Wines which were judged negatively tended to have higher amounts of 3-(methylthio)-1-propanol. Therefore, high concentrations of this S-compound can probably be an indicator of inferior wine quality (Keck, 1989). Although the contribution of 3-(methylthio)-1-propanol to the overall flavor of wine should be taken into account, its odor threshold value is probably relatively high (2 mg/l in beer; Meilgaard, 1981).

The precursor of 3-(methylthio)-1-propanol, namely, 3-(methylthio)-1-propanal, has only been detected in trace amounts (Schreier *et al.*, 1976). Its flavor threshold is 10–50 ppb in water and 250 ppb in beer. Its aroma is associated with potatoes (Fors, 1983).

The concentration of 3-(ethylthio)-1-propanol in wines has varied from 6–68 μg/l (Schreier *et al.*, 1974; Baumes *et al.*, 1986). Keck (1989) observed this S-compound in every wine at amounts between 3–27 μg/l, and sometimes higher. However, as 3-(ethylthio)-1-propanol has a high threshold value similar to 3-(methylthio)-1-propanol, it probably contributes very little to wine aroma (Keck, 1989).

Schreier *et al.* (1974) detected 3-(methylthio)-propylacetate in wines in amounts from traces to 15 μg/l. Keck (1989) obtained concentrations of 0.4–20 μg/l. This S-substance is formed by the reaction of 3-(methylthio)-1-propanol and acetic acid during fermentation. Its flavor threshold was estimated to be 600 μg/l (Meilgaard, 1981). Since this value is higher than the concentrations found in wine, it seems that this S-compound has little influence on wine quality.

The thioether ester, methyl 3-(methylthio)-propanoate, and the corresponding ethyl ester were determined in wine by Schreier *et al.* (1974). The former was identified only qualitatively and the latter in concentrations of 5.7 μg/l. Methyl 3-(methylthio)-propanoate has a sweet, pineapple flavor in high dilutions (Shankaranarayana *et al.*, 1982; Keck, 1989). The addition of ethyl 3-(methylthio)-propanoate to wine at 1 mg/l had no distinct negative effect on aroma (Keck, 1989). The precursor of these thioether esters, 3-(methylthio)-propanoic acid, was only detected qualitatively (Schreier *et al.*, 1974).

The S-containing amine, N-3-(methylthio)-propylamine, has been identified in flavor extracts of beer and wine (Schreier *et al.*, 1976) but its importance to the aroma of wine has not been investigated. Amines are already present in grapes. In plants, their mode of formation is proposed to be a transamination reaction between aldehydes and amino acids to yield an amine and the corresponding keto acid (Keck, 1989). Schreier *et al.* (1976) has discussed their formation by bacterial metabolism.

Schreier *et al.* (1975) reported that yeasts are able to acetylate primary amines to secondary acetamides. This would explain why the concentration of amines decreases during fermentation. Sometimes, an increase in the concentrations of these substances can occur through bacterial biosynthesis during the malolactic fermentation (Keck, 1989). The acetamide, N-[3-(methylthio)propyl]-acetamide, is derived from its corresponding amine and has been identified in wines by Schreier and Drawert (1974) and Keck (1989) between 10–340 μg/l. Five wines, two with a mousy taste, two with a medicinal note and one with a rubber aroma, had concentrations of the acetamide exceeding 900 μg/l. One of the wines with a mousy taint had even 4060 μg/l. At normal concentrations, however, it was presumed that N-[3-(methylthio)-propyl]-acetamide did not contribute to the taste or aroma of wines (Keck, 1989). Very good wines, from the Riesling variety, had the lowest concentrations of the acetamides determined. Wines which were judged negatively tended to have higher amounts of these compounds. It was also noted that increased amounts of acetamides occurred after skin-contact (Keck, 1989). N-(1-acetyl)methionineamide was qualitatively determined in white wine by Rapp and Güntert (1986). Adverse microbial activity probably accounts for the very high concentrations of acetamides in wines.

Benzothiazole was identified in white (Etievant and Bayonove, 1983) and red wine (Vernin *et al.*, 1988) and has been observed in grapes (Schreier *et al.*, 1976). Generally, thiazoles are found in foods that have undergone heat treatment or Maillard-reactions. Kato *et al.* (1973) showed that thiazole derivatives can be formed by the thermal interaction of cysteine and cystine with reactive compounds like carbohydrates or carbonyls. The degradation of thiamine is another possibility for their formation (Belitz and Grosch, 1985). Benzothiazole can be also formed as an artifact during the preparation of flavor extracts (Shankaranarayana, 1975). The concentrations of benzothiazole in wines vary from 0.4–7 μg/l. A negative influence on the aroma of wine was observed by the addition of 24 μg/l of benzothiazole. A rubber-like odor became apparent, when somewhat more than 70 μg/l was added (Keck, 1989). The odor of benzothiazole is also described as being similar to quinoline (Belitz and Grosch, 1985). The investigations of Keck (1989) showed that only one wine with a mousy taint had a very high value of benzothiazole. This wine, however, had no rubbery aroma. Every wine contains benzothiazole in small amounts. An increase in its concentration can occur in wines spoiled by yeasts or bacteria, but its biochemical production needs proof.

Rapp *et al.* (1985) identified 5-(2-hydroxyethyl)-4-methylthiazole in wine. Keck (1989) reported concentrations between 15–384 μg/l for this S-compound, but since it is odorless, it probably has no influence on wine aroma.

Rapp *et al.* (1985) and Keck (1989) reported the presence of 4-(methylthio)-butanol-1 in wine at 9–50 μg/l. It has a sharp chive-garlic aroma with a threshold value of 0.1–1 mg/l in a 15 % (v/v) solution of ethanol in water. In wine, it produces a metallic-bitter or grassy note depending on concentration, but is considered not to have a negative influence. The biosynthesis of this S-substance from homomethionine was proposed to take place through the Ehrlich-mechanism (Rapp *et al.*, 1985).

2-[[(Methylthio)methyl]thio]-ethanol has been identified in wines at 0.4–47 μg/l by Rapp *et al.* (1985) and Keck (1989). The investigations included wines with a mousy taint or strong rubber aroma. Some of these faulty wines had increased concentrations of this S-substance. However, information about its flavor note is contradictory. Gas-chromatography sniff tests performed by Rapp *et al.* (1985) showed that the compound has a very strong cauliflower aroma, whereas Keck (1989) observed a rubbery aroma. According to Keck (1989), the presence of this substance is noticeable only at concentrations exceeding 100 μg/l and may not cause the mousy or rubbery taints just mentioned. A proposed precursor of this S-compound is djenkolic acid, but there is no information on the presence this amino acid in must or wine.

2-Methyltetrahydrothiophene-3-one was identified by Schreier *et al.* (1975, 1976) in model fermentation systems with methionine as the only nitrogen source. Baumes *et al.* (1986) detected concentrations of 7–45 μg/l in white wines and 92–167 μg/l in red wines. Sniff tests indicated a butane-like odor. Keck (1989) reported up to 1040 μg/l of this compound in wine. Its mode of formation in wine is unknown. It is produced in food by roasting, which involves the reaction of cysteine with α-dicarbonyl compounds. It can also be formed in aqueous solution, during heating of thiamine, of D-glucose and H_2S, or by the heating of S-containing amino acids with sugar (Ledl and Severin, 1973).

Cis- and *trans*-2-methyltetrahydrothiophene-3-ol were detected by Rapp *et al.* (1984) and Baumes *et al.* (1986) at concentrations less than 5–12 μg/l in white wines and between 13–25 μg/l in red wines. Both of these cyclic compounds are by-products of fermentation. They are not detectable in grapes (Mandery, 1983). Keck (1989) observed the *cis*-isomer in every wine at concentrations of 106–657 μg/l. The *trans*-isomer seemed to be present in lower concentrations, of 2–160 μg/l. Little is known about the sensory properties of these substances but it appears that both have low odor activity (Keck, 1989).

2-Mercaptoethanol was identified in wine by Rapp *et al.* (1985). Cysteine was proposed as the source of its biochemical formation. It has an odor threshold between 1.0–10.0 mg/l (15 % (v/v) solution of ethanol in water) and can probably cause sulfurous off-flavors, perhaps synergistically with other S-substances (Rapp *et al.*, 1985). However, its influence on wine quality seems unlikely, because the estimated concentrations in wine, 10^{-2} μg/l or less, are far below its threshold value.

Ethyl methanedisulfonate (Yunome *et al.*, 1981), 2,4-di-tert-butylthiophene (Shimizu and Watanabe, 1982 a), 2-(1-butyl)-5-(2-methylpropyl)thiophene and 2,5-di-2-methylpropylthiophene (Shimizu and Watanabe, 1982 b) were qualitatively investigated. N-Acetylmethionine ethyl ester was first identified in wine at concentrations between 14–215 μg/l by Keck (1989). The addition of 200 μg/l of this substance to wine has shown no influence on the flavor. Wildenradt *et al.* (1975) determined 1,3,5,7-tetrathiocane and *cis*- or *trans*-3,5-dimethyl-1.2,4-trithiolane in steam distillates of grape leaves.

Degradation of Pesticide Residues

For some years now, it has been suspected that residues of sulfur containing pesticides in musts could be the cause of some of the sulfurous off-odors and off-flavors in wines (Burkhardt, 1978; Schmitt *et al.*, 1986 a, b). The fate of such residues can be affected by the winemaking process, adsorption by solids or lees, chemical degradation (acid hydrolysis) and metabolism by yeasts during fermentation (Cabras *et al.*, 1983, 1987, 1988).

Jakob (1961) postulated that the fungicide, tetramethylthiuramdisulfide (TMTD), causes sulfurous off-flavors through its reduction to two molecules of dimethyldithiocarbamate during fermentation. Schmitt (1987) indicated that, at the pH of wine, TMTD can be hydrolyzed to volatile S-containing substances such as dimethyldithiocarbamate, dimethylthiocarbamate, CS_2 and COS.

Significant research has been done on the degradation and metabolism of the ethylenebisdithiocarbamate (EBDC) group of fungicides (Cantarelli *et al.*, 1964; Casanova and Guichon, 1988). According to Cantarelli *et al.* (1964), yeasts can utilize EBDCs to give volatile sulfur compounds such as mercaptans, H_2S and CS_2. Marshall (1977) showed that commercial EBDCs are degraded via ethylenethiurammonosulfide (ETM). At high temperatures, ETM is transformed into ethylenethiourea (ETU) by splitting of CS_2 (Figure 6.8). The same reaction has been demonstrated between baker's yeast and reducing substances like glutathione (Vonk and Sijpesteijn, 1976).

The development of sulfurous off-flavors in bottled wine from residues of acephate, the active ingredient of an insecticide with the trade name Orthene, has resulted in severe economic loss in some of the Western European winemaking countries since the end of the 1970s. Acephate (O,S-dimethyl-N-acetyl-amido-thiophosphate) is hydrolyzed via methamidophos (O,S-dimethylphosphoramidothioate) and/or other S-containing intermediates to methyl mercaptan (MeSH) and additional metabolites (Chukwudebe *et al.*, 1984; Rauhut, 1985; Rauhut *et al.*, 1986; Rauhut 1990) (Figure 6.9). These reactions depend upon the concentration of residue, wine pH and the time and temperature of wine storage. MeSH and its oxidation product, dimethyl disulfide (DMDS), have an unpleasant smell and are responsible for the sulfurous off-flavor caused by acephate.

$$4CH_3SH + O_2 \leftrightarrow 2CH_3SSCH_3 + 2H_2O$$

Wines with acephate residues can yield high amounts of MeSH and DMDS, far exceeding their threshold values (Tables 6.1 and 6.2) (Rauhut, 1990; Rauhut and Sponholz, 1990).

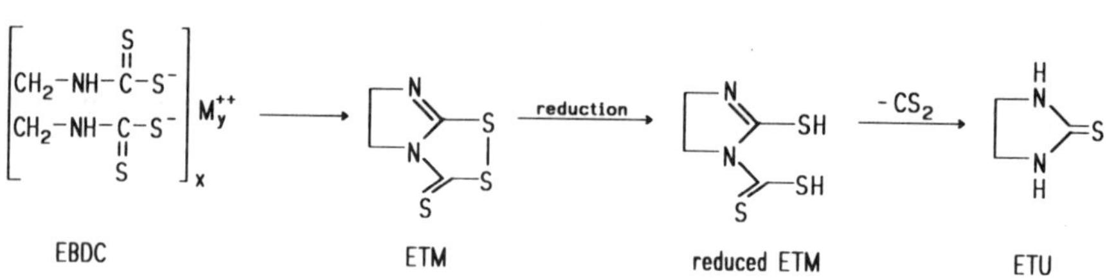

FIGURE 6.8 Ethylenebisdithiocarbamates (EBDCs) and some of their main metabolites. ETM = ethylenethiurammonosulfide; ETU = ethylenethiourea (M^{++} = bivalent metal; modified from Cabras *et al.*, 1987).

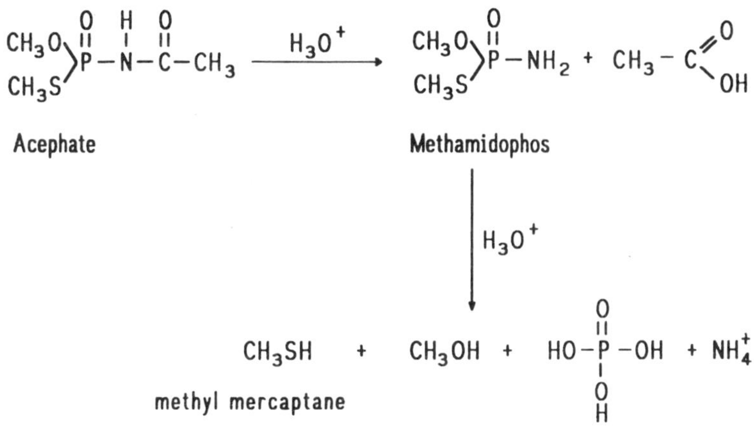

FIGURE 6.9 Hydrolysis of acephate (O,S-dimethyl-N-acetyl-amido-thiophosphate).

CONCLUSION

This chapter has indicated the range of volatile S-compounds that occur in wine, their possible origin and how they contribute to wine quality. Despite significant progress in this field during the last 10 years, many questions remain unanswered, especially in relation to the production of these substances by yeast metabolism. In this context, it should be noted that nearly all the studies to date concern the biochemical activities of *S. cerevisiae*, and virtually no attention has been given to the production of S-compounds by other species, some of which (e.g. *Kloeckera*, *Hanseniaspora* and *Candida* spp.) can make significant contributions to wine fermentations.

REFERENCES

Acree, T.E., Sonoff, E.P. and Splittstoesser, D.F. (1972) Effect of yeast strain and type of sulfur compounds on hydrogen sulfide production. *American Journal of Enology and Viticulture*, **23**, 6–9.

Anness, B.J. and Bamforth, C.W. (1982) Dimethyl sulfide—a review. *Journal of the Institute of Brewing*, **88**, 244–252.

Antoniewski, J. and De Robichon-Szulmajster, H. (1973) Biosynthesis of methionine and its control in wild type and regulatory mutants of *Saccharomyces cerevisiae*. *Biochimie*, **55**, 529–539.

Asahi, T., Bandurski, R.S. and Wilson, L.G. (1961) Yeast sulfate-reducing system II. Enzymatic reduction of protein disulfide. *Journal of Biological Chemistry*, **236**, 1830–1835.

Bärwald, G. and Kliem, D. (1971) Ein Beitrag zum Methioninstoffwechsel der Hefe (*S. cerevisiae*). *Chemie Mikrobiologie Technologie der Lebensmittel*, **1**, 27–32

Bamforth, C.W. (1980) Dimethyl sulfoxide reductase of *Saccharomyces* spp. *FEMS Microbiology Letters*, **7**, 55–59

Bamforth, C.W. and Anness, B.J. (1981) The role of dimethyl sulfoxide to dimethyl sulfide during fermentations. *Journal of the Institute of Brewing*, **87**, 244–252.
Baumes, R., Cordonnier, R., Nitz, S. and Drawert, F. (1986) Identification and Determination of volatile constituents in wines from different vine cultivars. *Journal of the Science of Food and Agriculture*, **37**, 927–943.
Baumgarten, F. (1981) Aspekte der assimilatorischen Sulfatreduktion und Sulfitresistenz bei Weinhefen mit und ohne Sulfitproduktion. Thesis, Universität Bonn, Germany.
Beech, F.W., Burroughs, L.F., Timberlake, C.F. and Whiting, G.C. (1979) Progrès récents sur l'aspect chimique et l'action antimicrobienne de l'anhydride sulfureux (SO_2). *Bulletin de l'Office International du Vin*, **52**, 1001–1022.
Belitz, H.-D., Grosch, W. (1985) Aromastoffe. In *Lehrbuch der Lebensmittelchemie*. 2. Aufl., pp. 264–313. Berlin: Springer-Verlag.
Bidan, P. and Collon, Y. (1988) Le métabolisme du soufre chez la levure. In *Application à l'oenologie des progrès récents en microbiologie et en fermentation* (Paris, 1988), ENSSAA-Dijon and ENSA-Montpellier, pp. 203–233. Paris: Office International du Vin.
Bidan, P. and Salgues, M. (1981) Der Schwefelstoffwechsel der Hefe. Sixth Internationales Önologisches Symposium, 28–30 April 1981, Mainz, edited by E. Lemperle and J. Frank, pp. 247–260. Breisach: Eigenverlag der Internationalen Interessengemeinschaft für moderne Kellertechnik und Betriebsführung.
Bingemer, H., Georgii, H.W., Bürgermeister, S. and Ockelmann, G. (1987) Die Bedeutung biogener Schwefelemissionen für den troposphärischen Schwefelkreislauf. In *Atmosphärische Spurenstoffe*, Deutsche Forschungsgemeinschaft (DFG), edited by Jaenicke, R. pp. 145–171. Weinheim: VCH Verlagsgesellschaft.
Bobet, R.A. (1987) Interconversion reactions of sulfides in model solutions. M.S. Thesis, University of California, Davis, USA
Bobet, R.A., Noble, A.C. and Boulton, R.B. (1990) Kinetics of the ethanethiol and diethyl disulfide interconversion in wine-like solutions. *Journal of Agricultural and Food Chemistry*, **38**, 449–452.
Bonish, P. and Eschenbruch, R. (1976) Sulfite reductase and ATP sulfurylase in low- and high-sulfite forming wine yeasts. *Archives of Microbiology*, **109**, 85–88.
Booer, C.D. and Wilson, R.J.M. (1979) Synthesis of dimethyl sulfide during fermentation by a route not involving the heat- labile DMS precursor of malt. *Journal of the Institute of Brewing*, **85**, 35–37.
Botsford, J.L. and Parks, L.W. (1969) Serine transhydroxymethylase in methionine biosynthesis in Saccharomyces cerevisiae. *Journal of Bacteriology*, **113**, 1176–1183.
Breton, A. and Surdin-Kerjan, Y. (1977) Sulfate uptake in *Saccharomyces cerevisiae*: Biochemical and genetic study. *Journal of Bacteriology*, **132**, 224–232.
Burkhardt, R. (1978) Verschiedene geruchswirksame Verbindungen des Schwefels in Mosten und Weinen. *Der Deutsche Weinbau*, **28**, 1281–1283.
Cabras, P., Meloni, M. and Pirisi, F.M. (1983) The effect of clarifying substances on the content of some insecticides and fungicides in white wine. *American Journal of Enology and Viticulture*, **34**, 103–107.
Cabras, P. Meloni, M. and Pirisi, F.M. (1987) Pesticide fate from vine to wine. *Reviews of Environmental Contamination and Toxicology*, **99**, 83–117.
Cabras, P. Meloni, M., Pirisi, F.M., Farris, G.A. and Fatichenti, F. (1988) Yeast and pesticide interaction during aerobic fermentation. *Applied Microbiology and Biotechnology*, **29**, 298–301.
Cantarelli, D., Tarufi, F., Martini, A. (1964) Chemical and microbiological surveys on the effects of dithiocarbamate fungicides on wine-making. *Journal of the Science of Food and Agriculture*, **15**, 186–196.
Cartwright, C.P., Rose, A.H., Calderbank, J. and Keenan, M.H.J. (1989) Solute Transport. In *The Yeasts*, Volume 3, 2nd edn, *Metabolism and Physiology of Yeasts*, edited by A.H. Rose and J.S. Harrison, pp. 5–56. London: Academic Press.

Casanova, M. and Guichon, R. (1988) Residues of EBDC Fungicides and ETU in experimental and commercial beverages (beer and wine). *Journal of Environmental Science and Health*, **B23**, 179–188.

Cherest, H., Eichler, F. and De Robichon-Szulmajster, H. (1969) Genetic and regulatory aspects of methionine biosynthesis in *Saccharomyces cerevisiae*. *Journal of Bacteriology*, **97**, 328–336.

Cherest, H., Surdin-Kerjan, Y., Antoniewski, J. and De Robichon-Szulmajster, H. (1973a) S-adenosylmethionine-mediated repression of methionine biosynthetic enzymes in *Saccharomyces cerevisiae*. *Journal of Bacteriology*, **114**, 928–933.

Cherest, H., Surdin-Kerjan, Y., Antoniewski, J. and De Robichon-Szulmajster, H. (1973b) Effects of regulatory mutations upon methionine biosynthesis in *Saccharomyces cerevisiae*: Loci eth2-eth3-eth10. *Journal of Bacteriology*, **115**, 1084–1093.

Cherest, H., Surdin-Kerjan, Y. and De Robichon-Szulmajster, H. (1971) Methionine-mediated repression in *Saccharomyces cerevisiae*: a pleiotrophic regulatory system involving methionyltransfer ribonucleic acid and the product of gene eth2. *Journal of Bacteriology*, **106**, 758–772.

Cheynier, V., Souquet, J.M. and Moutounet, M. (1989) Glutathione content and glutathione to hydroxycinnamic acid ratio in vitis vinifera grapes and musts. *American Journal of Enology and Viticulture*, **40**, 320–324.

Chukwudebe, A.C., Hussain, M.A. and Oloffs, P.C. (1984) Hydrolytic and metabolic products of Acephate in water and mouse liver. *Journal of Environmental Science and Health*, **B19**, 501–522.

Cooper, T.G. (1982) Transport in *Saccharomyces cerevisiae*. In *The Molecular Biology of the Yeast Saccharomyces, Metabolism and Gene Expression*, edited by J.N. Strathern, E.W. Jones and J.R. Broach, pp. 399–461. New York: Cold Spring Harbor Laboratory.

Cooper, A.J.L. (1983) Biochemistry of sulfur-containing amino acids. *Annual Reviews of Biochemistry*, **52**, 187–222.

Delavier-Klutchko, C. and Flavin, M. (1965) Enzymatic synthesis and cleavage of cystathionine in fungi and bacteria. *Journal of Biological Chemistry*, **240**, 2537–2549.

De Mora, S.J., Eschenbruch, R., Knowles, S.J. and (the late) Spedding, D.J. (1986) The formation of dimethyl sulfide during fermentation using a wine yeast. *Food Microbiology*, **3**, 27–32.

De Mora, S.J., Knowles, S.J., Eschenbruch, R. and Torrey, W.J. (1987) Dimethyl sulfide in some Australian red wines. *Vitis*, **26**, 79–84.

De Robichon-Szulmajster, H. and Surdin-Kerjan,Y. (1971) Nucleic acid and protein synthesis in yeasts: regulation of synthesis and activity. In *The Yeasts*, edited by A.H. Rose and J.S. Harrison, Vol. 2, *Physiology and Biochemistry of Yeasts*, pp. 335–418. London: Academic Press.

De Vito, P.C. and Dreyfuss, J. (1964) Metabolic regulation of adenosine triphosphate sulfurylase in yeast. *Journal of Bacteriology*, **88**, 1341–1348.

Di Stefano, R. (1985) Presenza di caratteri organolettici favorevoli in vini bianchi lungamente invecchiati. Indagine sui composti volatili e su alcuni parametri chimici e fisici di Riesling prodotti in Germania. *Rivista di Viticultura e di Enologia Conegliano*, **4**, 228–241.

Dittrich, H.H. (1987) *Mikrobiologie des Weines*, 2nd edn, pp. 138–147; 111–118. Stuttgart: Verlag Eugen Ulmer.

Dittrich, H.H. and Sponholz, W.-R. (1975) Die Aminosäureabnahme in Botrytis-infizierten Traubenbeeren und die Bildung höherer Alkohole in diesen Mosten bei ihrer Vergärung. *Wein-Wissenschaft*, **30**, 188–210.

Dittrich, H.H. and Staudenmayer, T. (1968) SO_2-Bildung, Böckserbildung und Böckserbeseitigung. *Deutsche Wein-Zeitung*, **24**, 707–709.

Dittrich, H.H. and Staudenmayer, T. (1970) Über die Zusammenhänge zwischen der Sulfit-Bildung und der Schwefelwasserstoff-Bildung bei *Saccharomyces cerevisiae*.

Zentralblatt für Bakteriologie, Parasitenkunde, Infektionshrankhbeiten und Hygiene, II. Abt. Bd. **124**, 113–118.
Dott, W. (1977) Enzymatische und regulatorische Untersuchungen der assimilatorischen Sulfatreduktion von Weinhefen der Gattung Saccharomyces im Hinblick auf eine erhöhte Sulfitproduktion während der Gärung. Thesis, Universität Bonn, Germany.
Dott, W., Heinzel, M. and Trüper, H.G. (1977) Sulfite Formation by wine yeasts. IV. Active uptake of sulfate by "low" and "high" sulfite producing wine yeasts. *Archives of Microbiology*, **112**, 283–285.
Dott, W., Heinzel, M. and Trüper, H.G. (1976) Sulfite formation by wine yeasts. I. Relationships between growth, fermentation and sulfite formation. *Archives of Microbiology*, **107**, 289–292.
Dott, W. and Trüper, H.G. (1976) Sulfite formation by wine yeasts. III. Properties of sulfite reductase. *Archives of Microbiology*, **108**, 99–104.
Dott, W. and Trüper, H.G. (1978a) Sulfite formation by wine yeast. VI. Regulation of biosynthesis of NADPH- and BV-dependent sulfite reductases. *Archives of Microbiology*, **118**, 249–251.
Dott, W. and Trüper, H.G. (1978b) Über die Notwendigkeit standardisierter Bedingungen zur Charakterisierung sulfitbildender Hefen. 1. Korrelation von Sulfitbildung mit Wachstumsparametern der Hefen. *Wein- Wissenschaft*, **33**, 159–169.
Dott, W. and Trüper, H.G. (1978c) Über die Notwendigkeit standardisierter Bedingungen zur Charakterisierung sulfitbildender Hefen. 2. Vergleich zwischen drei schwach und zwei sehr stark sulfitbildenden Hefestämmen der Arten *Saccharomyces cerevisiae var. ellipsoideus und S. bayanus. Die Wein-Wissenschaft*, **33**, 277–288.
Du Plessis, C.S. and Loubser, G.J. (1974) The bouquet of "late harvest" wine. *Agrochemophysica*, **6**, 49–52.
Eschenbruch, R. (1972a) Der Einfluß von Methionin und Cystein auf die SO_2-Bildung einiger Stämme von *Saccharomyces cerevisiae* bei der Vergärung von Traubenmost. *Vitis*, **11**, 53–57.
Eschenbruch, R. (1972b) Sulfate uptake and sulfite formation related to the methionine and/or cysteine content of grape must during the fermentation by strains of *Saccharomyces cerevisiae. Vitis*, **11**, 222–227.
Eschenbruch, R. (1972c) Zur Substratabhängigkeit der H_2S- und SO_2-Bildung bei *Saccharomyces cerevisiae* Stämmen. *Wein-Wissenschaft*, **27**, 40–44.
Eschenbruch, R. (1974) Sulfite and sulfide formation during wine making. A review. *American Journal of Enology and Viticulture*, **25**, 157–161.
Eschenbruch, R. (1978) Sulfite and sulfide formation by wine yeasts. In *5th International Oenological Symposium*, 13.–15. February 1978, Auckland, New Zealand, edited by E. Lemperle and J. Frank, pp. 267–274. Breisach: Eigenverlag der Internationalen Interessengemeinschaft für moderne Kellertechnik und Betriebsführung.
Eschenbruch, R. (1987) Can the grapegrower contribute towards lower SO_2 levels in wine? *American Journal of Enology and Viticulture*, **38**, 78–79.
Eschenbruch, R. and Bonish, P. (1976a) The influence of pH on sulfite formation by yeasts. *Archives of Microbiology*, **107**, 229–231.
Eschenbruch, R. and Bonish, P. (1976b) Production of sulfite and sulfide by low- and high-sulfite forming wine yeasts. *Archives of Microbiology*, **107**, 299–302.
Eschenbruch, R., Bonish, P. and Fisher, B.M. (1978) The production of H_2S by pure culture wine yeasts. *Vitis*, **17**, 67–74.
Eschenbruch, R., De Mora, S.J., Knowles, S.J., Leonard, W.K., Forrester, T. and Spedding, D.J. (1986) The formation of volatile sulfur compounds in unclarified grape juice. *Vitis*, **25**, 53–57.
Eschenbruch, R., Hassbrock, F.S. and De Villiers, J.F. (1973) On the metabolism of sulfate and sulfite during the fermentation of grape must by *Saccharomyces cerevisiae. Archives of Microbiology*, **93**, 259–266.

Etievant, P.X. and Bayonove, C.L. (1983) Aroma components of pomaces and wine from the variety Muscat de Frontignan. *Journal of the Science of Food and Agriculture*, **34**, 393–403.

Fetterroll, B.M. and Würdig, G. (1977) Die Vergrößerung der "inneren Oberfläche" des Mostes und deren Wirkung auf sulfitbildende Hefen. *Wein-Wissenschaft*, **32**, 25–33.

Fors, S. (1983) Sensory properties of volatile Maillard reaction products and related compounds: a literature review. In *The Maillard reaction in food and nutrition*, edited by G.R. Waller and M.S. Feather, pp. 185–286. Washington, DC: American Chemical Society (ACS symposium series, 215).

Goniak, O.J. and Noble, A.C. (1987) Sensory study of selected volatile sulfur compounds in white wine. *American Journal of Enology and Viticulture*, **38**, 223–227.

Goto, S. and Takamuro, R. (1987) Concentration of dimethyl sulfide in wine after different storage times. *Hakkokogaku*, **65**, 53–57.

Gyllang, H., Winge, M. and Korch, C. (1989) Regulation of SO_2 formation during fermentation. European Brewery Convention. In *Proceedings of the 22nd Congress*, Zürich 1989, 347–354.

Hartnell, P.C. and Spedding, D.J. (1979) Uptake and metabolism of ^{35}S-sulfate by wine yeast. *Vitis*, **18**, 307–315.

Haye, B., Maujean, A., Jacquemin, C. and Feuillat, M. (1977) Contribution a l'étude des "Goûts de lumière" dans le vin de Champagne. *Connaissance Vigne Vin*, **11**, 243–254.

Heinzel, M.A. (1978) Regulation der Sulfat-Assimilation bei Weinhefen, Ursachen einer SO_2-Bildung. Thesis, Universität Bonn, Germany.

Heinzel, M.A., Dott, W. and Trüper, H.G. (1979) Ursachen einer biologischen SO_2-Bildung bei der Weingärung. *Wein-Wissenschaft*, **34**, 192–211.

Heinzel, M. and Trüper, H.G. (1976) Sulfite formation by wine yeasts. II. Properties of ATP-sulfurylase. *Archives of Microbiology*, **107**, 293–297.

Heinzel, M.A. and Trüper, H.G. (1978) Sulfite Formation by wine yeasts. V. Regulation of Biosynthesis of ATP-and ADP-sulfurylase by sulfur- and selenium-compounds. *Archives of Microbiology*, **118**, 243–247.

Henschke, P.A. and Jiranek, V. (1991) Hydrogen sulfide formation during fermentation: effect of nitrogen composition in model grape musts. "International Symposium on Nitrogen in Grapes and Wine", *American Society for Enology and Viticulture*, 18–19 June, Seattle.

Hilz, H. and Kittler, M. (1960) Reduction of active sulfate (PAPS) by dihydrolipoic acid as substrate. *Biochemical Biophysics Research Communications*, **3**, 140–142.

Hilz, H., Kittler, M. and Knappe, G. (1959) Die Reduktion von Sulfat in der Hefe. *Biochemische Zeitung*, **332**, 1561–166.

Horák, J., Rihová, L. and Kotyk, A. (1981) Energetization of sulfate transport in yeast. *Biochimica et Biophysica Acta*, **649**, 436–440.

Huxtable, R.J. (1986) *Biochemistry of Sulfur*, pp. 1–9. New York: Plenum Press.

Hysert, D.W. and Morrison, N.M. (1976) Sulfate metabolism during fermentation. *American Society of Brewing Chemists Proceedings*, **34**, 25–31.

Jakob, L. (1961) Veränderungen von Thiuramen bei der alkoholischen Gärung. *Wein-Wissenschaft*, **16**, 97–104.

Januik, M.T. (1984) The development of a technique to quantify and identify wine sulfur volatiles by gas chromatography. Thesis, University of California, Davis.

Jones, E.W. and Fink, G.R. (1982) Regulation of amino acid and nucleotide biosynthesis in yeast. In *The Molecular Biology of the Yeast Saccharomyces*, edited by J.N. Strathern, E.W. Jones, J.R. Broach, pp. 181–299. New York: Cold Spring Harbor Laboratory.

Jorzik, E.B. (1985) Sulfitbildung durch Weinhefen der Gattung *Saccharomyces*: Untersuchbungen zum Sulfit-Transport. Thesis, Universität Bonn, Germany.

Kato, S., Kurata, T. and Fujimaki, M. (1973) Volatile compounds produced by the reaction of L-cysteine or L-cystine with carbonyl compounds. *Agricultural and Biological Chemistry*, **37**, 539–544.

Karlson, P. (1988) *Kurzes Lehrbuch der Biochemie für Mediziner und Naturwissenschaftler.* 13. neubearbeitete Auflage, pp. 184–185. Stuttgart: Georg Thieme Verlag.

Keck, S. (1989) Untersuchungen zur Bedeutung flüchtiger phenolischer, schwefelhaltiger und stickstoffhaltiger Verbindungen für unerwünschte Aromanoten des Weines mittels Gaschromatographie/Massenspektrometrie. Thesis, Universität Karlsruhe, Germany.

Kelly, D.P. (1988) Oxidation of sulfur compounds. In *The Nitrogen and Sulfur Cycles*, edited by J.A. Cole and S.J. Ferguson, pp. 65–98. Cambridge: Cambridge University Press.

Kotyk, A. (1959) Uptake of inorganic sulfate by the yeast cell. *Folia Microbiologica.*, **4**, 363–372.

Laparra, J. (1990) Les procyanidines du vin action inhibitrice sur les radicaux libres oxygénés. In *Actualités Oenologiques 89* comptes rendus du 4e Symposium International d'Oenologie, Bordeaux 15–17 June 1989, edited by P. Ribéreau-Gayon and A. Lonvaud, pp. 392–397. Paris: Dunod.

Larue, F., Park, M.K. and Caruana, C. (1985) Quelques observations sur les conditions de la formation d'anhydride sulfureux en vinification. *Connaissance Vigne Vin*, **19**, 241–248.

Lawrence, W.C. and Cole, E.R. (1968) Yeast sulfur metabolism and the formation of hydrogen sulfide in brewery fermentations. *Wallerstein Laboratory Communications*, **31**, 95–115.

Ledl, F. and Severin, T. (1973) Sulfur compounds from cysteine and xylose. *Zeitschrift für Lebensmittel Untersuchung und Forschung*, **154**, 29–31

Lehninger, A.L. (1977) *Biochemie*, 2. neubearbeitete und erweiterte Auflage, pp. 593–594. Weinheim: Verlag Chemie.

Lemperle, E. and Lay, H. (1989) Zusammensetzung und Beurteilung der Weine. In *Chemie des Weines*, edited by G. Würdig and R. Woller, pp.(Anionen) 568–571; (Vitamine) 615–621. Stuttgart: Verlag Eugen Ulmer.

Leppänen, O.A., Denslow, J. and Ronkainen, P.P. (1979) A gas chromatographic method for the accurate determination of low concentrations of volatile sulfur compounds in alcoholic beverages. *Journal of the Institute of Brewing*, **85**, 350–353.

Leppänen, O.A., Denslow, J. and Ronkainen, P.P. (1980) Determination of thiolacetates and some other volatile sulfur compounds in alcoholic beverages. *Journal of Agricultural and Food Chemistry*, **28**, 359–362.

Loubser, G.J. and Du Plessis, C.S. (1976) The quantitative determination and some values of dimethyl sulfide in white table wines. *Vitis*, **15**, 248–252.

Mandery, H. (1983) Gaschromatograghisch-massenspektrometrische Untersuchungen flüchtiger Inhaltsstoffe des Traubenmost- und Weinaromas: Auswirkung der Süßung auf die Aromazusammensetzung. Thesis, Universität Karlsruhe, Germany.

Marais, J. (1979) Effect of storage time and temperature on the formation of dimethyl sulfide and on white wine quality. *Vitis*, **18**, 254–260.

Marcis, B.J. and Markakis, P. (1974) Transport and toxicity of sulfur dioxide in *Saccharomyces cerevisiae* var. *ellipsoideus*. *Journal of the Science of Food and Agriculture*, **25**, 21–29.

Marshall, W. (1977) Thermal decomposition of ethylenbisdithiocarbamate fungicide to ethylenthiourea in aqueous media. *Journal of Agricultural and Food Chemistry*, **25**, 357–361.

Masselot, M. and De Robichon-Szulmajster, H. (1972) Nonsense mutation in the regulatory gene ETH2 involved in methionine biosynthesis in *Saccharomyces cerevisiae*. *Genetics*, **71**, 535–550.

Masselot, M. and Surdin-Kerjan, Y. (1977) Methionine biosynthesis in *Saccharomyces cerevisiae* II. Gene-enzyme relationships in the sulfate assimilation pathway. *Molecular and General Genetics*, **154**, 23–30.

Matsui, S. and Amaha, M. (1981) Production of S-methyl thioacetate from methyl mercaptan by brewer's yeast. *Agricultural and Biological Chemistry*, **45**, 1341–1349.

Maujean, A. (1984) L'influence de la lumière sur les vins. Symposium international de montréal sur le transport de vins. *Revue Francaise d'Oenologie*, **96**, 3–10.
Maujean, A. (1989) Goûts anormaux. *Revue des Oenologues*, No. 53S (November), 31–32.
Maw, G.A. (1960) Utilization of sulfur compounds by a brewer's yeast. *Journal of the Institute of Brewing*, **66**, 162–167.
Maw, G.A. (1961) Effects of cysteine and other thiols on the growth of a brewer's yeast. *Journal of the Institute of Brewing*, **67**, 57–63.
Maw, G.A. (1963a) Sulfur utilization by yeast. *Pure and Applied Chemistry*, **7**, 655–668.
Maw, G.A. (1963b) The uptake of inorganic sulfate by a brewer's yeast. *Folia Microbiologica*, Praha **8**, 325–332.
Maw, G.A. (1963c) The uptake of some sulfur-containing amino acids by a brewer's yeast. *Journal of General Microbiology*, **31**, 247–259.
Maw, G.A. (1965) The role of sulfur in yeast growth and in brewing. *Wallerstein Laboratory Communications*, Vol. XXVIII, No. 95, April, 49–70.
Maw, G.A. (1982) Biochemistry of S-methyl-L-cysteine and its principal derivatives. *Sulfur Reports*, **2**(1), Feb., pp. 1–32. New York: Harwood Academic Publishers.
Mayer, K. and Pause, G. (1968) Über die Bildung von schwefliger Säure und Schwefelwasserstoff während der Weingärung. *Mitteilungen aus dem Gebiet der Lebensmitteluntersuchung und Hygiene*, **59**, 387–392.
McCready, R.G.L. and Din, G.A. (1974) Active sulfate transport in *Saccharomyces cerevisiae*. *FEBS Letters*, **38**, 361–363.
Meilgaard, M.C. (1981) Beer Flavour, Thesis, Technical University of Denmark, Ann Arbor, Michigan, USA, University Microfilms International.
Minarik, E. (1977) Métabolisme et production de composés soufrés par la levure. *Bulletin de l'Office International du Vin*, **50**, 559, 641–648.
Monk, P.R. (1986) Formation, utilization and excretion of hydrogen sulphide by wine yeast. *Wine Industry Journal*, November, 10–16.
Muller, C.J., Kepner, R.E. and Webb, A.D. (1971) Identification of 3-(methylthio)-propanol as an aroma constituent in 'Cabernet Sauvignon' and 'Ruby Cabernet' wines. *American Journal of Enology and Viticulture*, **22**, 156–160.
Naiki, N. and Yamagata, S. (1973) Isolation of O-acetylhomoserine from culture broth of a methionine auxotroph of *Saccharomyces cerevisiae*. *Plant Cell Physiology*, **14**, 1193–1197.
Neuberg, C. and Schwenk, E. (1915) Phytochemische Reduktionen. XI. Die Umwandlung von Äthyldisulfid in Äthylmercaptan. *Biochemische Zeitschrift*, **71**, 118–121.
Neumüller, O.-A. (1987) *Römpps Chemie-Lexikon*, 8. neubearbeitete und erweiterte Auflage, Band 5: PL-S, Stuttgart: Franckh'sche Verlagshandlung.
Niefind, H.J. and Späth, G. (1971) Die Bildung flüchtiger Aromastoffe durch Mikroorganismen. *European Brewing Convention Proceedings 13th Congress*, Estoril, 459–468.
Nwoguh, C.E. and Berry, D.R. (1990) Effect of methionine on the growth and morphogenesis of *Saccharomycopsis fibuligera* in defined medium. *Biotechnology Letters*, **12**, 271–276.
Ono, B.-I., Shirahige, Y.-I., Nanjoh, A., Andou, N., Ohue, H. and Ishino-Arao, Y. (1988) Cysteine biosynthesis in Saccharomyces cerevisiae: Mutation that confers cystathionine β-synthase deficiency. *Journal of Bacteriology*, **170**, 5883–5889.
Peck, H.D. and Lissolo, T. (1988) Assimilatory and dissimilatory sulfate reduction: enzymology and bioenergetics. In *The Nitrogen and Sulfur Cycles*, edited by J.A. Cole and S.J. Ferguson, pp. 99–132. Cambridge: Cambridge University Press.
Peppard, T. (1988) Analytical Measurement of Volatile Sulfur Compounds in Beer. In *Beer Analysis*, edited by H.F. Linskens and J.F. Jackson, pp. 241–253. Berlin: Springer-Verlag.
Phaff, H.J. (1971) Structure and biosynthesis of the yeast cell envelope. In *The Yeasts*, Vol. 2 Physiology and Biochemistry of Yeasts, edited by A.H. Rose and J.S. Harrison, pp. 135–210. London: Academic Press.

Porqué, P.G., Baldesten, A. and Reichard, P. (1970) The involvement of the thioredoxin system in the reduction of methionine sulfoxide and sulfate. *Journal of Biological Chemistry*, **245**, 2371–2374.

Prabhakararao, K. and Nicholas, D.J.D. (1969) Sulfite reductase from baker's yeast: a haemoflavoprotein. *Biochimica et Biophysica Acta*, **180**, 253–263.

Prabhakararao, K. and Nicholas, D.J.D. (1970) The reduction of sulfite, nitrite and hydroxylamine by an enzyme from baker's yeast. *Biochimica et Biophysica Acta*, **216**, 122–129.

Radler, F. (1989) Vitamine und Wuchsstoffe. In *Chemie des Weines*, edited by G. Würdig and R. Woller, pp. 119–120. Stuttgart: Verlag Eugen Ulmer.

Rankine, B.C. (1963) Nature, origin and prevention of hydrogen sulfide aroma. *Journal of the Science of Food and Agriculture*, **14**, 79–91.

Rankine, B.C. (1964) Hydrogen sulphide production by yeasts. *Journal of the Science of Food and Agriculture*, **12**, 872–877.

Rankine, B.C. (1968) The importance of yeasts in determining the composition and quality of wines. *Vitis*, **7**, 22–49.

Rankine, B.C. and Pocock, K.F. (1969) Influence of yeast strain on binding of sulfur dioxide in wines, and on its formation during fermentation. *Journal of the Science of Food and Agriculture*, **20**, 104–109.

Rapp, A. (1965) Über die Inhaltsstoffe von Traubenmosten und Weinen unter besonderer Berücksichtigung der flüchtigen Verbindungen und des stofflichen Geschehens während der Hefegärung. Thesis, Mainz, Germany.

Rapp, A. and Reuther, K.H. (1971) Der Gehalt an freien Aminosäuren in Traubenmosten von gesunden und edelfaulen Beeren verschiedener Rebsorten, *Vitis*, **10**, 51–58.

Rapp, A. and Güntert, M. (1986) Changes in aroma substances during the storage of white wines in bottles. In *The Shelf Life of Foods and Beverages*. Proceedings of the Fourth International Flavour Conference, Rhodes, Greece 23.–26. July, 1985, edited by G. Charalambous, pp. 141–167. Amsterdam: Elsevier Science Publishers.

Rapp, A., Güntert, M. and Almy, J. (1984) Neue schwefelhaltige Aromastoffe in Wein - cis- und trans-2-methyl-thiolan-3-ol. *Vitis*, **23**, 66–72.

Rapp, A., Güntert, M. and Almy, J. (1985) Identification and significance of several sulfur-containing compounds in wine. *American Journal of Enology and Viticulture*, **36**, 219–221.

Rauhut, D. (1985) Versuche zum Nachweis extrem flüchtiger Schwefelverbindungen mit "Böckser-Wirkung". Diplomarbeit, Diplom- Oenologie, Universität Gießen, Germany.

Rauhut, D. (1990) Trace analysis of sulfurous off-flavours in wine caused by extremely volatile S-containing metabolites of pesticides e.g. Orthene. In *Actualités Oenologiques 89* comptes rendus du 4e Symposium International d'Oenologie, Bordeaux 15–17 June 1989, edited by P. Ribéreau-Gayon and A. Lonvaud, pp. 482–487. Paris: Dunod.

Rauhut, D., Kürbel, H. and Dittrich, H.H. (1992) Sulfur compounds and their influence on wine quality. 3rd International Cool Climate Symposium, Forschungsanstalt Geisenheim-Universität Mainz, Mainz 8–12 June 1992.

Rauhut, D. and Sponholz, W.-R. (1990) Sulfur compounds in wine. In Proceedings of the Seventh Australian Wine Industry Technical Conference, Adelaide, South Australia, 13–17 August 1989, edited by P.J. Williams, D.M. Davidson and T. Lee, pp. 196–202. Adelaide: Winetitles.

Rauhut, D., Sponholz, W.-R. and Dittrich, H.H. (1986) Beeinflussen Pflanzenschutzmittel die Weinqualität? *Der Deutsche Weinbau*, **16**, 872–874.

Richmond, D.V. (1973) Sulfur Compounds. *Phytochemistry*, 3, pp. 41–73. New York: Van Nostrand Reinhold.

Robbins, P.W. and Lipmann, F. (1958a) Separation of the two enzymatic phases in active sulfate synthesis. *Journal of Biological Chemistry*, **233**, 681–685.

Robbins, P.W. and Lipmann, F. (1958b) Enzymatic synthesis of adenosine phosphosulfate. *Journal of Biological Chemistry*, **233**, 686–690.

Roomans, G.M., Kuypers, G.A.J., Theuvenet, A.P.R. and Borst-Pauwels, G.W.F.H. (1979) Kinetics of sulfate uptake by yeast. *Biochimica et Biophysica Acta*, **551**, 197–206.

Schlossmann, K. and Lynen, F. (1957) Biosynthese des Cysteins aus Serin und Schwefelwasserstoff. *Biochemische Zeitung*, **328**, 591–594.

Schmitt, A. (1987) Untersuchung des chemischen Verhaltrens von Tetramethylthiuramdisulfid in Lebensmitteln. Thesis, Universität Karlsruhe.

Schmitt, A., Köhler, H., Miltenberger, R. and Curschmann, K. (1986 a+b) Einfluß von Pflanzenschutzmitteln auf Most und Wein. I und II. Der Deutsche Weinbau, **41**, 323–329 (a), 365–369 (b).

Schreier, P. (1979) Flavour composition of wines: A review. CRC *Critical Reviews in Food Science and Nutrition*, **12**, 59–111.

Schreier, P. and Drawert, F. (1974) Gaschromatographisch-massenspektrometrische Untersuchung flüchtiger Inhaltsstoffe des Weines. V. Mitteilung, Alkohole, Hydroxy-Ester, Lactone und andere polare Komponenten des Weinaromas. *Chemie Mikrobiologie Technologie der Lebensmittel*, **3**, 154–160.

Schreier, P. and Drawert, F. (1975) Schwefelhaltige Aromastoffe. In *Geruch- und Geschmackstoffe*, edited by F. Drawert, pp. 113–131. Nürnberg: Verlag Hans Karl.

Schreier, P., Drawert, F. and Junker, A. (1974) Gaschromatographisch-massenspektrometrische Untersuchung flüchtiger Inhaltsstoffe des Weines II. Thioäther-Verbindungen des Weinaromas. *Zeitschrift für Lebensmittel Untersuchung und Forschung* **154**, 279–284.

Schreier, P., Drawert, F. and Junker, A. (1975) Über die Biosynthese von Aromastoffen durch Mikroorganismen. I. Bildung von N-Acetyl-aminen durch *Saccharomyces cerevisiae*. *Zeitschrift für Lebensmittel Untersuchung und Forschung* **158**, 351–360.

Schreier, P., Drawert, F., Junker, A., Barton, H. and Leupold, G. (1976) Über die Biosynthese von Aromastoffen durch Mikroorganismen. II. Bildung von Schwefelverbindungen aus Methionin durch *Saccharomyces cerevisiae*. *Zeitschrift für Lebensmittel Untersuchung und Forschung*, **162**, 279–284.

Schütz, M. and Kunkee, E. (1977) Formation of hydrogen sulfide from elemental sulfur during fermentation by wine yeast. *American Journal of Enology and Viticulture*, **28**, 137–144.

Schutte, L. (1975) Precursors of sulfur-containing flavor compounds. In *Fenaroli's Handbook of Flavor Ingredients*, edited by T.E. Furia and N. Bellanca, pp. 132–183. Cleveland, Ohio: CRC Press.

Shankaranarayana, M.L. (1975) Volatile sulfur compounds in food flavors. In *Fenaroli's Handbook of Flavor Ingredients*, edited by T.E. Furia and N. Bellanca, pp. 184–227. Cleveland, Ohio: CRC Press.

Shankaranarayana, M.L., Raghavan, B., Abraham, K.O. and Natarajan, C.P. (1982) Sulfur compounds in flavours. In *Food Flavours*, Part A, edited by I.D. Morton and A.J. MacLeod, pp. 169–281, Amsterdam: Elesevier Science Publishers.

Shaw, P.E. and Nagy, S. (1981) Analysis and flavour effects of volatile sulfur compounds in citrus juices. In *The Quality of Foods and Beverages. Chemistry and Technology* edited by G. Charalambous and G. Inglett, pp. 361–376, New York: Academic Press.

Shimizu, J. and Watanabe, M. (1982a) Volatile components identified in the acidic fractions of wines from Koshu and Zenkoji grapes. *Agricultural and Biological Chemistry*, **46**, 5, 1377–1380

Shimizu, J. and Watanabe, M. (1982b) Investigations of volatile components in wines from Koshu and Zenkoji grapes. *Agricultural and Biological Chemistry*, **46**, 2353–2355.

Simpson, R.F. (1979) Aroma composition of bottle aged white wine. *Vitis*, **18**, 148–154.

Slaughter, J.C. (1989) Sulfur compounds in fungi. In *Nitrogen, Phosphorous and Sulfur Utilization by Fungi*, edited by L. Boddy, R. Marchant and D.J. Read, pp. 91–105, Cambridge: Cambridge University Press.

Soltoft, M. (1988) Flavour active sulfur compounds in beer. *Brygmesteren*, **2**, 18–248.

Spedding, D.J., Eschenbruch, R. and McGregor, P.J. (1983) Sulfur compounds in the headspace of some New Zealand commercial wines. *Food Technology in Australia*, **35**, 22–23.

Spedding, D.J., Eschenbruch, R. and Purdie, A. (1980) The distribution of dimethyl sulfide in some New Zealand wines. *Vitis*, **19**, 241–245.

Spedding, D.J. and Raut, P. (1982) The influence of dimethyl sulfide and carbon disulfide in the bouquet of wines. *Vitis*, **21**, 240–246.

Stratford, M. and Rose, A.H. (1985) Hydrogen sulphide production from sulphite by *Saccharomyces cerevisiae*. *Journal of General Microbiology*, **131**, 1417–1424.

Stratford, M. and Rose, A.H. (1986) Transport of sulphur dioxide by *Saccharomyces cerevisiae*. *Journal of General Microbiology*, **132**, 1–6.

Suomalainen, H. and Oura, E. (1971) Yeast nutrition and solute uptake. In *The Yeasts*, Volume 2, *Physiology and Biochemistry of Yeasts*, edited by A.H. Rose and J.S. Harrison, pp. 3–74. London: Academic Press.

Surdin-Kerjan, Y., Cherest, H. and De Robichon-Szulmajster, H. (1976) Regulation of methionine synthesis in *Saccharomyces cerevisiae* operates through independent signals: methionyl-tRNA^{met} and S-adenosylmethionine. *Acta Microbiologica Academiae Scintarum Hungaricae, Budapest*, **23**, 109–120.

Tanner, H. (1969) Der Weinböckser, Entstehung und Beseitigung. *Zeitschrift für Obst- und Weinbau*, **105**, 78. Jhg., 252–258.

Thoukis, G. and Stern, L.A. (1962) A review and some studies of the effect of sulfur on the formation of off-odours in wine. *American Journal of Enology and Viticulture*, **13**, 133–140.

Torii, K. and Bandurski, R.S. (1967) Yeast sulfate-reducing system III. An intermediate in the reduction of 3'-phosphoryl-5'-adenosinephosphosulfate to sulfite. *Biochimica et Biophysica Acta*, **136**, 286–295.

Tsang, M.L.-S. and Schiff, J.A. (1976) Sulfate-reducing pathway in *Escherichia coli* involving bound intermediates. *Journal of Bacteriology*, **125**, 923–933.

Umbarger, H.E. (1978) Amino acid biosynthesis and its regulation. In *Annual Reviews of Biochemistry*, **47**, 533–606.

Umezawa, C. and Kishi, T. (1989) Vitamin Metabolism. In *The Yeasts*, Volume 3, 2nd edn, *Metabolism and Physiology of Yeasts*, edited by A. H. Rose and J.S. Harrison, pp. 457–488. London: Academic Press.

Vernin, G., *et al.* (1988) Aromas of Syrah wines: Identification of volatile compounds by GC-MS-spectra data bank and classification by statistical methods. Frontiers of Flavour, In *Proceedings of the 5th International Flavor Conference*, Porto Karras, Chalkidiki, Greece, edited by G. Charalambous, pp. 655–685. Amsterdam: Elsevier Science Publishers.

Vonk, J.W. and Sijpesteijn, A.K. (1976) Formation of ethylenethiourea from 5,6-dihydro-3*H*-imidazo[2,1-c]-1,2,4-dithiazole-3-thione by microorganisms and reducing agents. *Journal of Environmental Science and Health, New York*, **B11** (1), 33–47.

Vos, P.J.A. (1981) Assimilierbarer Stickstoff, ein Faktor, der die Weinqualität beeinflußt. In *6. Internationales Önologisches Symposium.*, 28–30 April 1981, Mainz, Germany, edited by E. Lemperle and J. Frank, pp. 163–180. Breisach: Eigenverlag der Internationalen Interessengemeinschaft für moderne Kellertechnik und Betriebsführung.

Vos, P.J.A. and Gray, R.S. (1979) The origin and control of hydrogen sulfide during fermentation of grape must. *American Journal of Enology and Viticulture*, **30**, 187–197.

Wainwright, T. (1970) Hydrogen sulfide production by yeast under conditions of methionine, pantothenate or vitamin B_6 deficiency. *Journal of General Microbiology*, **61**, 107–119.

Wainwright, T. (1971) Production of H_2S by yeasts: role of nutrients. *Journal of Applied Bacteriology*, **34**, 161–171.

Wainwright, T. McMahon, J.F. and McDowell, J. (1972) Formation of methional and methanethiol from methionine. *Journal of the Science of Food and Agriculture*, **23**, 911–914.

Wenzel, K. and Dittrich, H.H. (1978) Zur Beeinflussung der Schwefelwasserstoff-Bildung der Hefe durch Trub, Stickstoffgehalt, molekularen Schwefel und Kupfer bei der Vergärung von Traubenmost. *Wein-Wissenschaft*, **33**, 200–213.

Wenzel, K. and Dittrich, H.H. (1983) Hydrogen sulfide (H_2S) formation by yeasts during wine making. In *The 3rd. International Mycology Congress*, 1983, Tokyo, Japan, Abstract, 339.

Wenzel, K., Dittrich, H.H., Seyffardt, H.P. and Bohnert, J. (1980) Schwefelrückstände auf Trauben und im Most und ihr Einfluβ auf die H_2S-Bildung. Wein-Wissenschaft, **35**, 414–420.

Werkhoff, P., Bretschneider, W., Emberger, R., Güntert, M., Hopp, R. and Köpsel, M. (1991) Recent developments in the sulfur flavour chemistry of yeast extracts. *Chemie Mikrobiologie Technology der Lebensmittel*, **13**, 30–57.

Wildenradt, H.L., Christensen, E.N., Stackler, B., Caputi, A., Slinkard, K. and Scutt, K. (1975) Volatile constituents of grape leaves. I. Vitis vinifera variety 'Chenin Blanc'. *American Journal of Enology and Viticulture*, **26**, 148–153..

Wilson, L.G., Asahi, T. and Bandurski, R.S. (1961) Yeast sulfate-reducing system I. Reduction of sulfate to sulfite. *Journal of Biological Chemistry*, **236**, 1822–1829.

Wilson, L.G. and Bierer, D. (1976) The formation of exchangeable sulfite from adenosine 3'-phosphate-5'-sulfatophosphate in yeast. *Biochemical Journal*, **158**, 255–270.

Wucherpfennig, K. (1984) Die schweflige Säure im Wein—önologische und toxikologische Aspekte. In *Deutsches Weinbau-Jahrbuch*, edited by B. Götz and W. Madel, **35**, pp. 213–241. Waldkirch: Waldkircher Verlagsgesellschaft, Germany.

Würdig, G. (1985) Levures produisant du SO_2. *Bulletin de l'Office International du Vin*, 652–653, 582–589.

Würdig, G. and Schlotter, H.A. (1967) SO_2-Bildung in gärenden Traubenmosten. *Zeitschrift für Lebensmittel Untersuchung und Forschung*, **134**, 1, 7–13.

Würdig, G. and Schlotter, H.A. (1971) Über das Vorkommen SO_2-bildender Hefen im natürlichen Hefegemisch des Traubenmostes. *Deutsche Lebensmittel-Rundschau*, **67**, 86–91.

Yamagata, S. (1989) Roles of O-acetyl-L-homoserine sulfhydrylases in microorganisms. *Biochimie*, **71**, 1125–1143.

Yamagata, S. and Takeshima, K. (1976) O-acetylserine and O-acetylhomoserine sulfhydrylase of yeast. Further purification and characterization as a pyridoxal enzyme. *Journal of Biochemistry*, **80**, 777–785.

Yamagata, S., Takeshima, K. and Naiki, N. (1975) O-acetylserine and O-acetylhomoserine sulfhydrylase of yeast: Studies with methionine auxotrophs. *Journal of Biochemistry*, **77**, 1029–1036.

Yoshimoto, A. and Sato, R. (1968a) Studies on yeast sulfite reductase. I. Purification and characterization. *Biochimica et Biophysica Acta*, **153**, 555–575.

Yoshimoto, A. and Sato, R. (1968b) Studies on yeast sulfite reductase. II. Partial purification and properties of genetically incomplete sulfite reductases. *Biochimica et Biophysica Acta*, **153**, 576–588.

Yoshimoto, A. and Sato, R. (1970) Studies on yeast sulfite reductase. III. Further characterisation. *Biochimica et Biophysica Acta*, **220**, 190–205.

Yunome, H., Nishimura, K., Masuda, M., Zenibayashi, Y. and Ohkawa, E. (1981) Some neutral volatile compounds of botrytized wine. *Journal of Fermentation Technology*, **59**, 177–184.

Zang, K. and Franzen, K. (1967) Bildung von schwefliger Säure bei der Weinbereitung und ihre möglichen Ursachen. *Deutsche Wein-Zeitung*, **5**, 88–89.

Zee, J.A., Szöghy, M., Simard, R.B. and Temblay, J. (1983) Elemental analysis of Canadian, European and American Wines by photon induced X-Ray fluorescence. *American Journal of Enology and Viticulture*, **34**, 152–156.

Chapter 7

YEAST AUTOLYSIS

CLAUDINE CHARPENTIER and MICHEL FEUILLAT

Laboratoire d'Oenologie, Université de Bourgogne, Dijon, France

Chapter Outline

INTRODUCTION

The term autolysis was introduced into the literature by Salkowsky in 1875 (Farrer, 1956). Essentially, it represents an enzymatic self-degradation of cellular constituents that commences immediately after cell death. Yeast autolysis has been the subject of numerous studies because of its applications in the extraction and purification of industrial enzymes and in the production of yeast extracts for the food industry (Peppler, 1982). The biochemistry and general significance of yeast autolysis have been reviewed by Joslyn (1955), Joslyn and Vosti (1955), Farrer (1956) and Arnold (1981).

Yeast autolysis has several implications in winemaking. It occurs during storage of wine on lees, releasing into the wine yeast constituents that could influence wine sensory properties and encourage the growth of malolactic and spoilage bacteria. It has an unique occurrence and significance during production of sparkling wine by the "methode champenoise".

The first part of this chapter considers fundamental information about the chemistry and biochemistry of yeast autolysis. This is followed by a discussion of the occurrence and significance of yeast autolysis in winemaking.

AUTOLYSIS—GENERAL CONSIDERATIONS

Early studies, reported and reviewed by Vosti and Joslyn (1954), clearly established that cellular proteins and nucleic acids were extensively degraded during autolysis and that the kinetics of these reactions were influenced by environmental conditions. When yeast (*Saccharomyces cerevisiae*) cell suspensions were incubated for 72 h at 45°C and a pH near 5.0, more than 75% of the cell nitrogen and 85% of the cell phosphorus were solubilized and could be recovered from the extracellular medium. These authors distinguished two phases of autolysis: (i) a latent period of about four hours during which yeast glycogen was degraded and there was only weak solubilization of nitrogen and phosphorus; and (ii) a subsequent phase of about 20 h during which there was intense liberation of cell nitrogen and phosphorus. Similar autolytic degradation and solubilization of intracellular components were reported by Hough and Maddox (1970) for brewing strains of *S. cerevisiae*.

Subsequent work, reported by Arnold (1981) and Babayan and Bezrukov (1985), indicated that the yeast cell envelope, consisting of the cytoplasmic membrane and the cell wall, played an important role during autolysis. Essentially, break down of the membranous systems within the cell either naturally or through physical (e.g. temperature, pressure) and chemical (e.g. detergents, organic solvents) influences, results in loss of cellular organization and causes hydrolytic enzymes to come in contact with their substrates. At the same time, degradation of cell wall components, such as the glucans and mannoproteins, increases the porosity of the cell, facilitating the release of autolytic products to the extracellular environment. Thus, from a biochemical perspective, autolysis could be considered to consist of five stages:

1. Disorganization of the membranous systems of the cell (e.g. lysosomes, cytoplasmic membrane, other organelles) permitting the liberation of hydrolytic enzymes;

2. Inactivation of specific inhibitors of these enzymes by the action of proteases or, alternatively, the proteolytic activation of these enzymes;
3. Enzymatic degradation of intracellular macromolecules, leading to an accumulation of the products of degradation;
4. An increase in the porosity of the cell wall, allowing release of the autolytic products into the extracellular environment; and
5. Further autolytic degradation of substances in the extracellular environment.

Wines impose particular conditions on the autolytic process, namely, low pH (3.0–3.5), low temperatures (often less than 15°C) and the presence of ethanol. Compared with the conditions reported earlier (45°C, pH 5.0), these factors might be expected to significantly decrease the rate of the autolytic reactions. Nevertheless, wines are often left in contact with yeasts for periods of months and years.

Bidan (1975) has reviewed early studies on the autolysis of yeasts in wines, notably sparkling wines. These studies demonstrated the influence of autolysis on the composition of wine, in particular, on the nature and concentration of volatiles and nitrogenous substances. More recent studies on these topics have been described by Feuillat (1980), Loyaux (1981) and Feuillat and Charpentier (1982). Extensive studies in our laboratory have examined the chemical and biochemical mechanisms of yeast autolysis under acid conditions, as would occur in wines (Nguyen van Long, 1985; Chung, 1986; Lurton, 1987; Freyssinet, 1988). Such studies have enabled a better understanding of the autolytic changes that occur during the maturation of sparkling wines (Feuillat *et al.*, 1988) and of certain white wines kept on lees (Ferrari and Feuillat, 1988; Feuillat, Freyssinet and Charpentier, 1989).

AUTOLYSIS OF YEASTS IN ACID MEDIA

Studies on yeast autolysis under acid conditions can be considered under two categories: (i) proteolysis, and (ii) degradation of the cell wall.

Proteolysis

Saccharomyces cerevisiae possesses a diverse array of proteases, of which the principal enzymes are involved in hydrolyzing cell constituents after cell death (Achstetter and Wolf, 1985). However, the pH (3.0–3.5) of wine is unfavorable for most of these proteases, so that autolysis in an acid environment becomes a specific case.

Lurton (1987) and Lurton, Seguin and Feuillat (1989) have studied the role of proteases in yeast autolysis at different pH values by using inhibitors specific to the four classes of proteases present in *S. cerevisiae*. Phenylmethylsulfonylfluoride (PMSF) inhibits the serine proteases, mercuric chloride acts against the thiol proteases containing functional SH groups, pepstatine is an inhibitor of protease A, (the only acid protease in *S. cerevisiae*) and ethylene-diamino-tetra-acetic acid (EDTA) inactivates the metalloproteases. The combined addition of the four inhibitors gives a strong decrease in the liberation of soluble nitrogen from *S. cerevisiae* during autolysis for 7 days at 40°C, but it is not completely inhibited (Lurton *et al.*, 1989). Some 25–30% of the cell nitrogen is still released and, consequently, seems not to arise from enzymatic degradation. Essentially, it comes from an intracellular pool of free amino acids, including more than 50% of glutamic acid and alanine, which are liberated by the passive mechanism of exsorption, as generally observed at the end of the alcoholic fermentation (Dupuy *et al.*, 1967). Compared with the action of the combined inhibitors, pepstatine alone, at pH 3.0, gives significant inhibition of the release of the different forms of proteolytic nitrogen. Thus, at the time of proteolysis, pepstatine inhibits the release of 85% of total nitrogen, 77% of amino nitrogen and 84% of amino acids. This observation demonstrates the key role of protease A in autolysis at acid pH. It acts as an endoprotease because most of the liberated nitrogen consists of peptides of molecular weight between 1–10 kDa (Lurton *et al.*, 1989). As shown already by Lenney *et al.* (1974), free amino acids represent only 13% of the total nitrogen during autolysis at pH 3.0 but, at pH 5.0, they represent 60%. It seems that, at the higher pH, protease A plays a role of activator of other proteases, having little lytic activity (Bẹhalova and Beran, 1979). Carboxypeptidase Y, which liberates free amino acids at pH 3.0, is a likely target of protease A activation.

Freyssinet (1988) examined the concentrations of different nitrogen components released during yeast autolysis at pH 3.0 and 40°C over a period of 14 days (Figure 7.1). Strongest release of total nitrogen occurred during the first 24 hours; 42% of the total nitrogen was released during this period compared with 58% after 14 days. Initially, this nitrogen consisted mainly of proteins with

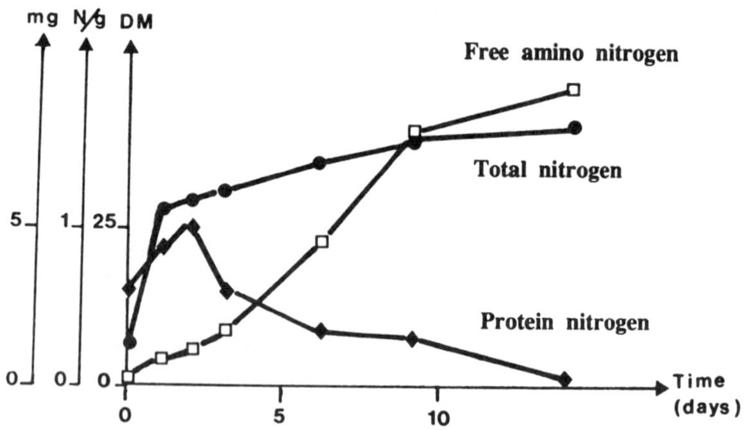

FIGURE 7.1 Evolution of the different forms of nitrogen compounds during yeast autolysis.

molecular weights in the range of 12.8 kDa. However, after 48 h, these were hydrolyzed to smaller molecular weight fractions, in particular, to amino acids whose concentrations now increased significantly. This kinetic behavior could be explained by the release of protease A from the intracellular location to the extracellular environment. The molecular weight of protease A is approximately 42 kDa, (Achstetter and Wolf, 1985), so that some degradation of the cell wall would seem necessary to permit its transfer to the extracellular location.

Degradation of the Cell Wall

Composition and structure

The chemical composition and structure of the wall has been reviewed by Fleet (1991). Using transmission electron microscopy, Avakyantz (1982) examined alterations in the structure and organization of wine yeast during storage in wine at 20°C. The following changes were noted: (i) progressive disorganization of the cytoplasm and its organelles; (ii) an increase in the number and size of vacuoles; (iii) an accumulation of chromatin and phospholipids; (iv) progressive separation of the cytoplasmic membrane from the cell wall, involving rupture of the membrane; and (v) a thickening of the wall, without its disruption. Avakyantz (1982) also showed, with scanning electron microscopy, that the external surface of the cell became wrinkled, which he attributed to plasmolysis. However, Silva *et al.* (1987) observed that the wall and membrane were only little modified during storage of yeasts in sparkling wine. The main change was detachment of the membrane from the cell wall, that appeared only after storage for five years.

Subsequent studies by Charpentier *et al.* (1986) have confirmed changes to the cell wall during autolysis of *S. cerevisiae* in buffer pH 3.0, containing 10% ethanol. Scanning electron micrographs (Figure 7.2) showed the development of ridges on the wall surface while transmission electron micrographs (Figure 7.3) demonstrated about 10% increase in wall thickness. These changes were associated with an increase (about 26%) in the content of wall polysaccharide (demonstrated by cytological staining), a decrease in the concentration of wall amino acids, an enrichment in the concentration of wall hexose, and a strong increase in the ratio of wall mannose:glucose (Table 7.1).

Role of enzymes

Fleet and Phaff (1974) and Hien and Fleet (1983) have shown that isolated cell walls of yeasts undergo self-degradation, or autolysis. The products of this autolysis were mainly oligosaccharides containing β-(1,3)-glucosidic linkages. The walls contained strong β-(1,3)-glucanase activity which was considered to cause their degradation. These authors also noted the association of protease activity with isolated cell walls. The role of enzymes in yeast wall degradation has been reviewed (Fleet, 1991).

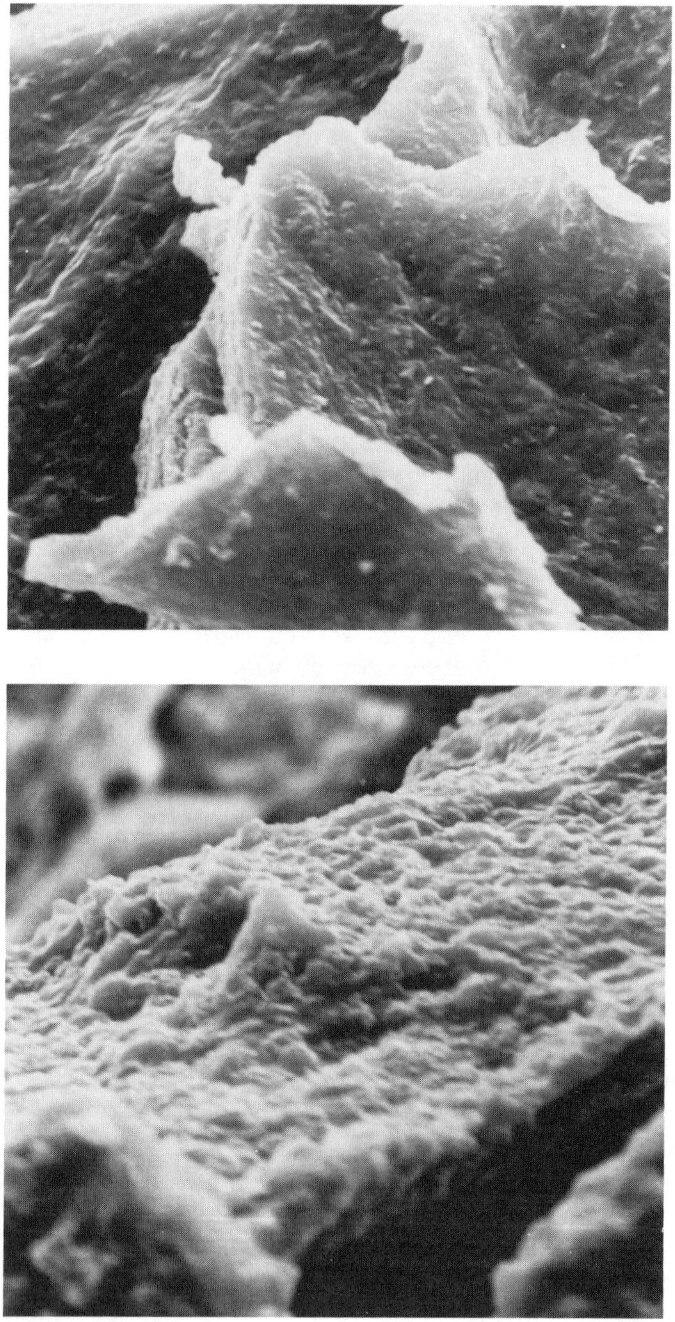

FIGURE 7.2 Scanning electron micrographs of the cell wall of *Saccharomyces cerevisiae* before (A) and after (B) autolysis for 14 days. (Magnification 6800×).

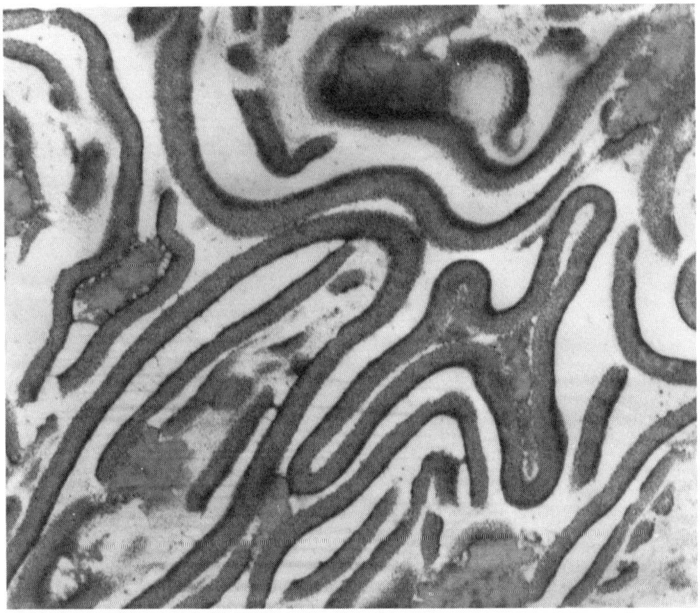

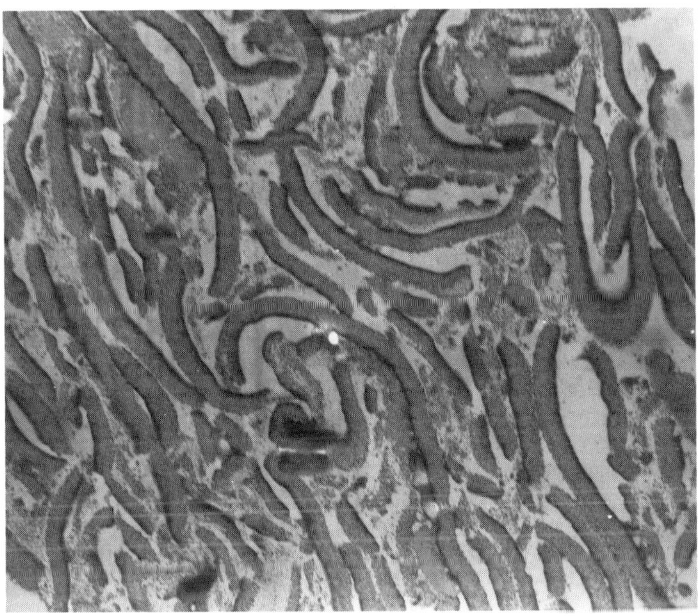

FIGURE 7.3 Transmission electron micrographs of sections of the cell wall of *Saccharomyces cerevisiae* before (A) and after (B) autolysis for 14 days. (Magnification 6800×).

TABLE 7.1 Changes in the composition of the cell wall of *Saccharomyces cerevisiae* during autolysis.

	Saccharomyces cerevisiae			*Saccharomyces cerevisiae (bayanus)*		
	Time of autolysis (days)					
	0	7	14	0	7	14
Hexoses (μg/mg cell wall)	656	841	809	780	790	826
Mannose/glucose	1.0	1.7	2.0	1.6	1.2	3.8
Amino acids (μg/mg cell wall)	227	122	77	134	106	72

The macromolecules liberated during the autolysis of isolated cell walls have been precipitated by ethanol and separated by gel filtration (Figure 7.4). The polysaccharides released after the first day are heterogeneous in size, having molecular weights between 2–400 kDa. During autolysis over the next 14 days, the smaller size components are hydrolyzed, resulting in an autolysate enriched in polysaccharides with molecular weights of 300–400 kDa, which are identical to those found in autolysates of intact cells (Figure 7.5). The high molecular weight colloids are characterized by a higher content of mannose, suggesting a predominance of mannan polysaccharide (Freyssinet, 1988; Charpentier and Freyssinet, 1989). These findings suggest the following mechanism of wall autolysis in *S. cerevisiae*:

1. In the early stages, exo- and endo-β-(1,3)-glucanases act on the wall to release a mixture of polysaccharides and short chain oligosaccharides. Part of this polysaccharide is mannoprotein that was probably covalently linked to the glucan in the intact wall (Fleet, 1991).
2. Continued hydrolysis of the glucan components destabilizes the wall structure, allowing further liberation of high molecular weight mannoproteins that contain only a small percentage of glucose, and most likely originate from the periplasmic zone.
3. In the later stages of autolysis, wall glucans continue to be attacked by β-(1,3)-glucanases still remaining associated with the wall but, also, they will be degraded from the exterior by glucanases that have been solubilized in the autolysate.
4. Exo-β-(1,3)-glucanases solubilized in the autolysate will degrade any glucans linked to the mannoprotein and, in addition, mannoproteins will be degraded by α-mannosidases. Proteolytic enzymes will cleave the protein

component of the mannoprotein as well as other proteins released from the wall.

This model is supported by the findings of Piton, Charpentier and Troton (1988) who showed that, during storage of champagne on lees of *S. cerevisiae*, the inner layers of the wall disappeared within the first six months, followed by degradation of the mannan after eight to eleven years.

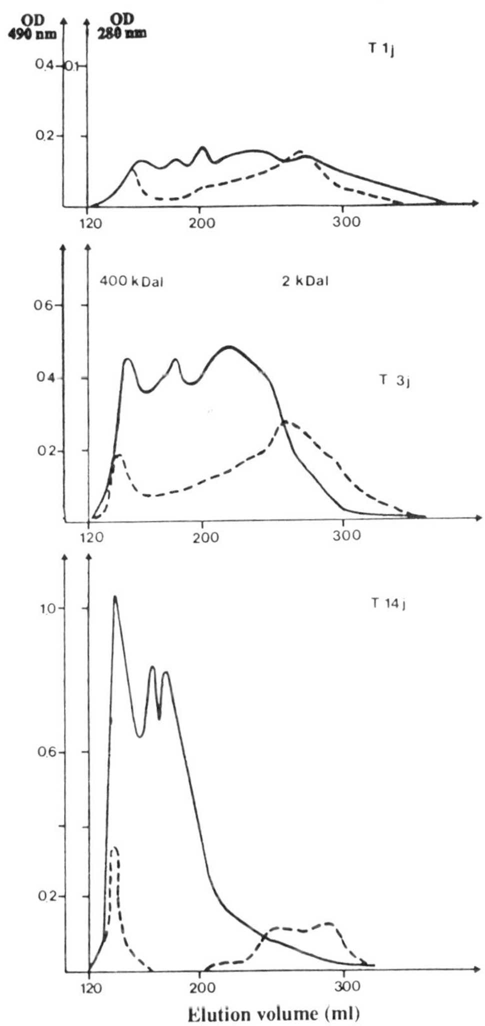

FIGURE 7.4 Fractionation on Sephacryl S-300 of glucidic and proteic colloids released during autolysis of cell walls of *Saccharomyces cerevisiae*.

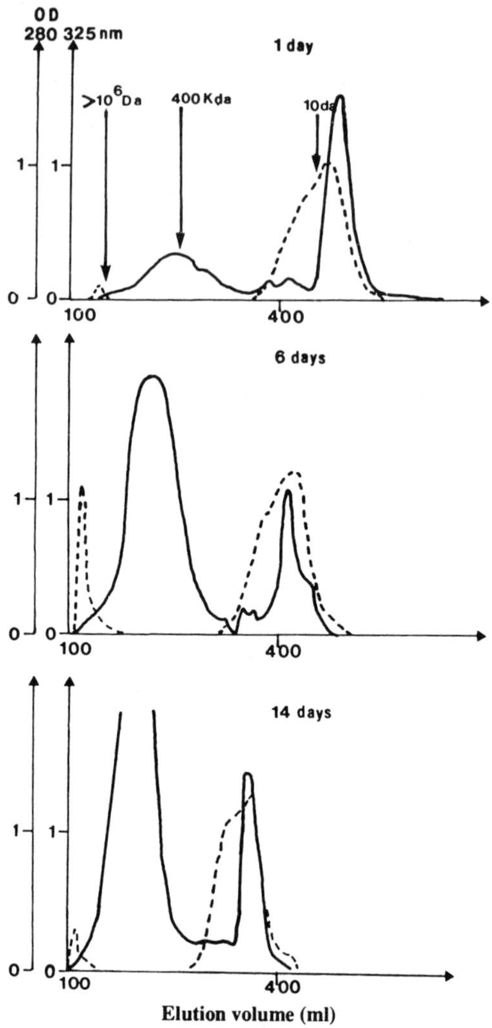

FIGURE 7.5 Fractionation on Sepharose C1-6B of glucidic and proteic colloids released during autolysis of cells of *Saccharomyces cerevisiae.*

ENOLOGICAL SIGNIFICANCE OF YEAST AUTOLYSIS

Sparkling Wine Produced by the Methode Champenoise

Nitrogen components

Because of the abundance of protein in yeasts and its susceptibility to enzymatic degradation, many authors have characterized autolysis during champagne production by the release of amino acids into the wine (Bidan, 1975; Colagrande and Mazzoleni, 1977; Suarez, Polo and Llaguno, 1979; Feuillat, 1980; Margheri *et al.*, 1982, 1983; Feuillat and Charpentier, 1982; Colagrande, Silva and Casoli, 1984; Silva *et al.*, 1987).

In reality, however, the increase in concentration of amino acids in wine is not always significant, and depends on the stage of production. Moreover, there are some changes to amino acid concentration, that are independent of autolysis. Immediately after tirage, the wine undergoes a secondary fermentation, which is accompanied by the utilization of sugar and the production of carbon dioxide. During this stage, there is a decrease in the concentration of amino acids, corresponding to a partial assimilation by the yeasts. This is quickly followed by a stage where the yeast cells release amino acids from their intracellular pool into the wine by a passive process of desorption. This appears to be a physiological response of the yeast to the exhaustion of sugars (Bidan, Feuillat and Moulin, 1985). These early changes in amino acids depend upon the concentration of ammonia nitrogen in the wine (Silva *et al.*, 1987).

After the secondary fermentation has been completed, the concentration of total and amino nitrogen in the wine remains stable, generally for a period of three to four months. It is believed that this period corresponds to a phase of activation, preceding autolysis. The slowness of this phase is probably due to the low pH (3.0–3.5) of the wine and the low temperature (< 15°C) of storage (Bergner, 1968; Suarez *et al.*, 1979; Feuillat, 1981). Subsequently, there is an increase in the concentration of amino acids in the wine. This development corresponds to the activity of protease A, which is now detected in the cells and in the wine. Its activity actually increases during the subsequent four to five years, provided the yeast is not separated from the wine (Leroy *et al.*, 1990).

Quantitatively, the total increase in concentration of amino acids remains relatively weak and is of the order of only a few mg/l. There are two possible explanations for this observation. First, protease A is an endo-protease that produces mostly peptide products, rather than free amino acids. Second, some amino acids already released (e.g. glutamic, alanine, arginine) are further transformed by decarboxylation and deamination reactions, causing a decrease in their concentrations (Loyaux, 1981).

Because of the relatively weak release of amino acids into the wine, some authors have concluded that yeast autolysis may have only a minimal effect on the sensory quality of sparkling wines produced by the "methode champenoise" (Usseglio-Tomasset *et al.*, 1983; Usseglio-Tomasset, 1985). However, apart from amino acids, there is a major transfer of cell extract from the yeast into the wine. The bulk of this transfer occurs by the end of the secondary fermentation, but continues during subsequent ageing in the bottle. Both proteins and nucleic acids represent a significant part of this extract (Leroy *et al.*, 1990).

Polysaccharides

The yeast polysaccharides found in wine stored on lees originate from degradation of the cell wall. These macromolecules are, essentially, glycoproteins consisting of about 85–90% carbohydrate and 10–15% protein.

The cell wall of *S. cerevisiae* comprises mainly glucans and mannoproteins whose composition and structure have been reviewed by Fleet (1991).

The polysaccharides of wine base are composed mostly of arabinose (66%) but, after secondary fermentation, polysaccharides consisting of mannose (43%) and glucose (31%) are also found (Gizzy, 1984; Feuillat, 1986). Such observations have been confirmed by studies in synthetic media with several strains. The concentrations of these glucan-mannan polysaccharides increase by about 3–4 fold during sparkling wine production, their liberation occurring during the secondary fermentation as well as during the subsequent eight months (Feuillat *et al.*, 1988). Thereafter, their concentrations remain stable, and could even decrease due to partial hydrolysis by β-(1,3)-glucanases also released by the yeast (Freyssinet, 1988; Freyssinet, Feuillat and Charpentier, 1989). The mannoproteins or peptidomannans that are released contribute significantly to the sensory quality of the wine, especially to the finesse and persistence of the bubbles that are so characteristic of sparkling wines. This is one of the reasons why sparkling wines produced by the "methode champenoise" are superior to those produced by other methods where the influence of yeast autolysis is very limited (Brissonet and Maujean, 1991).

Fatty acids

The concentration of fatty acids in sparkling wines made by the "methode champenoise" exceed 2 μ moles/l at the end of the secondary fermentation and 4μ moles/l after bottle ageing in contact with yeasts (Ferrari, 1984; Ferrari, Meunier and Feuillat, 1987). In experiments with synthetic media, these authors showed that the main fatty acids released during autolysis were saturated and were composed of either 8, 10, 12 or 16 carbon atoms. Similar findings were reported by Chen, Jamieson and van Gheluwe (1980) during the study of yeast autolysis in beer. However, once released the fatty acids may decrease in concentration, because they could be involved in the formation of esters, ketones and aldehydes. Such substances are volatile with low sensory thresholds and, consequently, could influence the flavor of the wine (Chung, 1986).

Volatile constituents

Early studies on the release of volatile constituents during yeast autolysis were reviewed by Bidan (1975). Rodopulo *et al.* (1975) compared the aromatic profiles of nine sparkling wines, linking their quality to the concentration of heavy esters such as isoamyl caproate, octylacetate, phenylethylacetate, phenylethylcaproate, ethyllinoleate and diethylsuccinate. Avakyantz and Schakarova (1973) had previously shown the release of such esters during the autolysis of yeasts at 20°C for one month or at 45–60°C for two months in a model medium of similar composition to wine. These findings have been confirmed by Molnar, Oura and Suomalainen (1981) and Chung (1986).

Molnar *et al.* (1981) and Chung (1986) also reported the release of terpenoid substances such as linalol and α-terpenol during autolysis. In a study of champagnes aged for different times on yeasts, Loyaux (1981) observed an enrichment of nerolidol in the wine during the first two years, followed by a slow disappearance of this substance.

Farnesol has been detected in sparkling wines and is considered to contribute to their aromatic qualities (Rodopulo *et al.*, 1975; Molnar *et al.*, 1981; Chung, 1986).

Concerning higher alcohols, such as isoamylalcohol, Konovalova (1977) observed a decrease in the concentration of these substances after three months of bottle storage and considered this change to improve the quality of the wine. In conducting yeast autolysis experiments in model media, Chung (1986) confirmed these findings but showed that the changes were more complex and depended on the nature of the alcohol under consideration. Phenylethylalcohol was identified in the products of yeast autolysis and its concentration increased strongly when the yeast cells were broken.

By comparing champagnes analyzed immediately after secondary fermentation and after ageing on yeasts for six years, Loyaux (1981) identified several other volatile substances that seemed to originate from yeast autolysis. These included methyl-2-ethoxy-2-furane that has a lightly phenolic but particularly strong fruity odor, and the lactone, dimethyl-4, 5-tetrahydrofurane-2, 3-dione, which has an odor of green nuts. Also identified in older champagnes was the sulfur volatile, vitispirane, which has a floral odor.

White Wine Left on Lees

In the Burgundy region of France, some white wines of the Chardonnay variety are fermented in wooden barrels of 228 liter capacity and then allowed to remain on the yeast lees for several months. During this period, the yeast sediment is occasionally brought into suspension ("batonnés") by mixing the wine with a baton. Analyses have been made on samples of such wines racked immediately after alcoholic fermentation and not allowed to remain on lees, and samples allowed to remain on lees with and without periodic resuspension (Ferrari and Feuillat, 1988). The wines left on lees have a higher concentration of total nitrogen and free amino acids, suggesting significant autolytic release from yeasts. This release is greater in the "batonnés" wines where, after six months, the concentration of amino acids was increased by 60%. For "non-batonnés" wines the corresponding increase was only 20%. The "batonnés" wines also become enriched in ethanol precipitable macromolecules during the first few months of storage on yeasts. These are mostly polysaccharides, principally mannoproteins, that originate from degradation of the yeast cell wall. On long term storage, however, partial hydrolysis of these polysaccharides occurs (Llauberes, 1988; Feuillat *et al.*, 1989). Some degradation of proteins by acid protease liberated during yeast autolysis is also evident.

Despite the autolytic changes that occur in wines left on yeasts, there appears to be no significant sensory preference for such wines over wines not left on yeasts. However, the wines kept on lees were considered to have better potential for long-term bottle storage (Ferrari and Feuillat, 1988).

Use of Yeast Autolysates

Given the fact that yeast autolysis can have a favorable effect on wines, but that the natural process is slow and uneconomic, some consideration has been given to the addition of *in vitro* preparations of yeast autolysates to wines. Yeast extracts or yeast autolysates generally used in the food industry (Peppler, 1982) have undergone excessive proteolysis, and give off-flavors when added to wines. For this reason, a specific procedure has been developed for preparing autolysates of wine yeasts for addition to wines (Feuillat, 1986). This procedure involved the selection of a strongly proteolytic strain of *S. cerevisiae* (variety *bayanus*) and conducting its autolysis in a hydro-alcoholic medium at pH 3.0 and temperatures of 30–50°C. This gives an autolysate that is little degraded, where only 20–30% of the total nitrogen is in the form of amino-nitrogen (Feuillat, Brillant and Rochard, 1980). At the end of autolysis, the cell envelopes are removed by centrifugation and the supernatant (autolysate) dried by heating. These autolysates have been used at concentrations of 1–1.5 g/l in experiments with sparkling wines produced by the "methode champenoise" and with white wines allowed to remain on lees (Feuillat, 1987; Ferrari and Feuillat, 1988).

As expected, wines containing added autolysates are enriched in their concentrations of total nitrogen, protein and amino acids (notably lysine, arginine, serine, glutamic acid and alanine) as well as ethanol precipitable macromolecules such as polysaccharides (Table 7.2; Feuillat, 1987). Sparkling wines to which autolysates had been added showed accelerated maturation and, after 7–8 months of ageing, had significantly better aromatic and bubble qualities than untreated control wines. White wines containing added autolysate had sensory qualities comparable to those allowed to remain on lees for 7–8 months, thus demonstrating an acceleration of the maturation-autolytic process (Ferrari and Feuillat, 1988).

TABLE 7.2 Influence of addition of yeast autolysate on the nitrogen and colloidal content of Champagne wine.

Sample	*Total colloids (mg/l)*	*Total nitrogen (mg/l)*	*Protein (mg/l)*	*Amino acids (mg/l)*
Wine (control)	357	350	15	116
Wine + autolysate	529	434	24	175
Wine + autolysate	648	469	32	129

CONCLUSION

Studies with model solutions have enabled significant advances in understanding the autolysis of yeasts at acid pH. At pH 3.0–3.5, proteolysis is essentially conducted by protease A, which is an endo-protease producing mostly peptides of 1–10 kDa. Consequently, within this pH range amino acids represent only about 15% of the total nitrogen released by autolysis, whereas at pH 5.0 they represent about 60% of this nitrogen. During autolysis, protease A passes into the extracellular environment where it continues to be active.

The cell wall undergoes partial degradation by *β*-(1,3)-glucanases during autolysis, releasing large molecular weight mannoproteins to the extracellular location. However, the integrity of the wall is not ruptured and, in fact, its thickness increases slightly, in conjunction with a significant increase in its content of mannose with respect to glucose.

There is a significant effect of yeast autolysis on the sensory character of wines, including sparkling wines, allowed to rest on lees. This influence is due to changes in the composition of free amino acids, volatile constituents and an enrichment in mannoprotein colloids arising from cell wall degradation. Recent studies (Voilley *et al.*, 1990) suggest that these mannoproteins can interact with aromatic wine constituents, such as *β*-ionone, and processes such as bentonite addition and membrane filtration affect these interactions.

Yeast autolysates, produced under certain conditions, can accelerate the process of maturation when added to sparkling wines and some white wines.

REFERENCES

Achstetter, T. and Wolf, D.H. (1985) Proteinases, proteolysis and biological control in the yeast *S. cerevisiae*. *Yeast*, **1**, 139–157.

Arnold, W.N. (1980) Autolysis. In *Yeast Cell Envelopes. Biochemistry, Biophysics and Ultrastructure*, Volume 2, edited by W.N. Arnold, pp. 93–103. Boca Raton: CRC Press.

Avakyantz, S.P. (1982) Etude au microscope électronique de l'autolyse des levures de vin. *Vinodelie i Vinogradstro*, URSS, **2**, 55–59.

Avakyantz, S.P. and Schakarova, F.I. (1973) Influence de l'autolyse des levures sur la composition des substances du bouquet du vin. *Izvestija Vyssikh Uchebnykh Zavedessij Piscevaja Tekhnologija*, **4**, 95–98.

Babayan, T.L. and Bezrukov, M.G. (1985) Autolysis in yeasts. *Acta Biotechnology*, **5**, 129–136.

Behalova, B. and Beran, K. (1979) Activation of proteolytic enzyme during autolysis of disintegrated baker's yeast. *Folia Microbiologia*, **24**, 455–461.

Bergner, K. (1968) Les acides aminés dans les vins mousseux et leurs variations en fonction des procédés d'élaboration. *Bulletin de l'Office International du Vin*, **41**, 446, 460–467.

Bidan, P. (1975) Les vins mousseux. Synthése bibliographique. CDIUPA, 7, Paris, France.

Bidan, P., Feuillat, M. and Moulin, J.P. (1985) Les vins mousseux. Rapport français. 65ème Assemblée Générale de l'OIV. *Bulletin de l'Office International du Vin*, **59**, no. 663–64, 563–626.

Brissonet, F. and Maujean, A. (1991) Identification of some foam-active compounds in champagne base wines. *American Journal of Enology and Viticulture*, **42**, 97–102.

Charpentier, C., Nguyen van Long, T., Bonaly, R. and Feuillat, M. (1986) Alteration of cell wall structure in *Saccharomyces cerevisiae* and *Saccharomyces bayanus* during autolysis. *Applied Microbiology and Biotechnology*, **24**, 405–413.

Charpentier, C. and Freyssinet, M. (1989) The mechanism of yeast autolysis in wine. *Yeasts*, **5**, S181–S186.

Chen, C.H., Jamieson, A.M. and Van Gheluwe, G. (1980) The release of fatty acids as a consequence of yeast autolysis. *American Society of Brewing Chemists Journal*, **38**, 13–18.

Chung, S.H. (1986) Contribution à l'étude de la formation des composés volatils au cours de l'autolyse de levures de vinification. Thèse Doctorat 3ème cycle, Université de Bourgogne, Dijon, France.

Colagrande, O. and Mazzoleni, F. (1977) Etude des caractères analytiques des vins mousseux italiens. *Connaissance de la Vigne et du Vin*, **11**, 325–337.

Colagrande, O., Silva, A. and Casoli, A. (1984) Acides aminés dans les vins mousseux. *Connaissance de la Vigne et du Vin*, **18**, 27–48.

Cordonnier, R. (1966) Recherches sur l'aminocarboxypeptidase de *Saccharomyces cerevisiae* HANSEN var. ellipsoideus. Thèse Doc. Sci., Université Aix Marseille, France.

Dupuy, P., Morfaux, J.N. and Usciati, M. (1967) Métabolisme azoté des levures en oenologie. IIème Symposium International d'Oenologie de Bordeaux Cognac, Bordeaux, France.

Farrer, K.T.H. (1956) The autolysis of yeast. *Food Science Abstracts*, **28**, 1–12.

Ferrari, G. (1985) Etude des lipides du vin. Thèse Doc. Ing., Université de Bourgogne, Dijon, France.

Ferrari, G. and Feuillat, M. (1988) L'élevage sur lies des vins blancs de Bourgogne. 1ère Partie: Etude des composés azotés, des acides gras et analyse sensorielle des vins. *Vitis*, **27**, 183–193.

Ferrari, G., Meunier, Y.M. and Feuillat, M. (1987) Dosage des acides gras totaux du vin et des levures de vinification. *Sciences des Aliments*, **7**, 61–76.

Feuillat, M. (1980) Vieillissement du vin de Champagne sur levures: phénomènes d'autolyse. Relations avec le développement des arômes. *Revue Francaise d'Oenologie*, **79**, 35–46.

Feuillat, M. (1986) Autolysats de levures à usage oenologique et leur procédé de fabrication. Brevet ANVAR no. 86–16318. Université de Bourgogne, Dijon, France.

Feuillat, M. (1987) Préparation d'autolysats de levures et addition dans les vins effervescents élaborés selon la méthode champenoise. *Revue Francaise d'Oenologie*, **109**, 17–27.

Feuillat, M., Brillant, G. and Rochard, J. (1980) Mise en évidence d'une production de protéases exocellulaires par les levures au cours de la fermentation alcoolique. *Connaissance de la Vigne et du Vin*, **1**, 37–52.

Feuillat, M. and Charpentier, C. (1982) Autolysis of yeasts in Champagne. *American Journal of Enology and Viticulture*, **38**, 6–13.

Feuillat, M., Charpentier, C., Picca, G. and Bernard, P. (1988) Production de colloïdes par les levures dans les vins mousseux élaborés selon la méthode champenoise. *Revue Francaise d'Oenologie*, **111**, 36–45.

Feuillat, M., Freyssinet, M. and Charpentier, C. (1989) L'élevage sur lies des vins blancs de Bourgogne. 2ème partie: Evolution des macromolécules (polysaccharides et protéines). *Vitis*, **28**, 161–176.

Fleet, G.H. (1991) Cell Walls. In *The Yeasts. Yeast Organelles*. Volume 4, edited by A.H. Rose and J.S. Harrison, pp. 199–277. London: Academic Press.

Fleet, G.H. and Phaff, H.J. (1974) Glucanases in *Schizosaccharomyces*—Isolation and properties of the cell wall-associated β-(1,3) glucanases. *Journal of Biological Chemistry*, **249**, 1717–1728.

Freyssinet, M. (1988) Etude des mécanismes de libération des constituents intracellulaires et pariétaux au cours du processus d'autolyse chez *Saccharomyces cerevisiae*. Nouvelle Thèse d'Université, Université de Bourgogne, Dijon, France.

Freyssinet, M., Feuillat, M. and Charpentier, C. (1989) Rôle de la paroi cellulaire dans l'autolyse des levures. Applications oenologiques. In *Actualités oenologiques 89. Comptes Rendus 4ème Symposium International d'Oenologie*, Bordeaux, edited by Masson, pp. 160–168, Paris, France.

Gizzy, A. (1984) Contribution à l'étude des colloïdes des vins effervescents. Thèse de Doctorat de 3ème cycle, Université de Bourgogne, Dijon, France.

Hayashi, R. (1976) Carboxypeptidase Y. In *Methods in Enzymology*, 45B, pp. 568–587, New York: Academic Press.

Hien, H.G.H. and Fleet, G.M. (1983) Separation and characterization of six (1,3) β-glucanases from *Saccharomyces cerevisiae*. *Journal of Bacteriology*, **156**, 1204–1213.

Hough, J.S. and Maddox, I.S. (1970) Yeast autolysis. *Process Biochemistry*, **5**, 50–53.

Joslyn, M.A. (1955) Yeast autolysis. Chemical and cytological changes involved in autolysis. *Wallerstein Laboratories Communications*, **18**, 107–119.

Joslyn, M.A. and Vosti, D.C. (1955) Factors influencing the rate and extent of autolysis. *Wallerstein Laboratories Communications*, **18**, 191–201.

Konovalova, L.A. (1977) Formation d'alcools supérieurs dans le vin. *Vinodelie i Vinogradarstvo*, **2**, 14 [cité par CHUNG (1986)].

Lenney, J.F., Matile, P., Wiemken, A., Schellenberg, M. and Meyer, J. (1974) Activities and cellular localization of yeast proteases and their inhibitors. *Biochemistry and Biophysics Research Communications*, **60**, 1378–1383.

Leroy, M.J., Charpentier, M., Duteurtre, B., Feuillat, M. and Charpentier, C. (1990) Yeast autolysis during champagne ageing. *American Journal of Enology and Viticulture*, **41**, 21–28.

Llauberes, R.M. (1988) Les polysaccharides decrétés dans les vins par *Saccharomyces cerevisiae* et *Pediococcus*. Nouvelle thèse d'Université, Université de Bordeaux II, Bordeaux, France.

Loyaux, D. (1981) Analyse des composés volatils du champagne. Etude de leur évolution au cours du vieillissement en présence des levures. Thèse Doc. Ing., Université de Dijon, France.

Lurton, L. (1987) Etude de la protéolyse intervenant au cours du processus d'autolyse chez *Saccharomyces cerevisiae*. Applications oenologiques. Nouvelle Thèse d'Université, Université de Bourgogne, Dijon, France.

Lurton, L., Seguin, Y.P. and Feuillat, M. (1989) Etude de la protéolyse au cours de l'autolyse de levures en milieu acide. *Sciences des Aliments*, **9**, 111–124.

Margheri, G., Gianotti, L., Mattarei, C. and Pellegrini, R. (1982) Evoluzione degli aminoacidi liberi nel corso della ellaborazione dei vini spumanti. *Vignevini*, **11**, 19–30.

Molnar, I., Oura, E. and Suomalainen, H. (1981) Study of volatile substances produced during the autolysis of champagne yeasts. *Acta Alimentaria*, **10**, 27–36.

Nguyen Van Long, T. (1985) Etude des processus fermentaires en champagnisation. Thèse Doc. Ing., Université de Bourgogne, France.

Peppler, H.J. (1982) Yeast extracts. In *Economic Microbiology*, Volume 7, *Fermented Foods*, edited by A.H. Rose, pp. 293–312. London: Academic Press.

Piton, F., Charpentier, M. and Troton, D. (1988) Cell wall and lipid changes in *Saccharomyces cerevisiae* during ageing of Champagne wine. *American Journal of Enology and Viticulture*, **33**, 221–226.

Rodopulo, A., Egeron, I.A., Kormatova, T.A. and Bezzoubou, R.A. (1975) Nature chimique des substances déterminant le bouquet du champagne. *Vinodelie i Vinogradarstvo*, **3**, 319, 14–18.

Sanz, P., Herrero, E. and Sentandreu, R. (1985) Autolytic release of mannoproteins from cell walls of *Saccharomyces cerevisiae*. *Journal of General Microbiology*, **181**, 2925–2932.

Silva, A., Fumi, M.D., Monterissa, G., Colombi, M.G. and Colagrande, O. (1987) Incidence de la conservation en présence de levures sur la composition des vins mousseux. *Connaissance de la Vigne et du Vin*, **21**, 141–162.

Suarez, M.A., Polo, M.C. and Llaguno, C. (1979) Etude de la composition de vins mousseux pendant la prise de mousse et au cours du vieillissement en bouteilles. I. Etude des acides aminés libres, du glycérol, des sucres et des acides organiques. *Connaissance Vigne et Vin*, **13**, 199–217.

Usseglio-Tomasset, L., Bosia, P.D., Di Stefabi, R. and Castubi, M. (1983) Oggettiva influenza del contatto con i lieviti sulle caratteristiche degli spumanti preparati con il metodo classico. *Vini d'Italia*, **25**, 142, 3–9.

Voilley, A., Lamer, C., Dubois, P. and Feuillat, M. (1990) Influence of macromolecules and treatments on the behaviour of aroma compounds in a model wine. *Journal of Agricultural Food Chemistry*, **38**, 248–251.

Vosti, D.C. and Joslyn, M.A. (1954) Autolysis of baker's yeast. *Applied Microbiology*, **2**, 70–78.

Chapter 8

KILLER YEASTS

KENICHI SHIMIZU

Research Laboratory, Sainte Neige Wine Co., Ltd., 107-1 Kamikanogawa, Yamanashi-shi, Yamanashi 405, Japan

Chapter Outline

INTRODUCTION

The killer character in yeasts was first reported by Bevan and Makower (1963). The killing action is due to toxins which are produced and secreted by the killer strains. Killer toxins are either proteins (Palfree and Bussey, 1979; Ashida *et al.*, 1983; Bostian *et al.*, 1984) or glycoproteins (Bussey and Skipper, 1975; Middelbeek *et al.*, 1979; Pfeiffer and Radler, 1982; Sugisaki *et al.*, 1984; Yokomori *et al.*, 1988a) which are lethal to sensitive yeasts. However, killer strains are immune to the toxin they produce. The killer phenomenon has been observed in many yeast genera from various sources including laboratory and industrial yeasts, and yeasts isolated from clinical and natural habitats. The killer property in *Saccharomyces cerevisiae* is associated with cytoplasmically inherited double-stranded RNA(dsRNA) plasmids (Mitchell *et al.*, 1973) which are encapsulated in virus-like particles (Herring and Bevan, 1974) and encode toxin precursors (Bostian *et al.*, 1980). The killer character in most other yeast genera appears to be determined by chromosomal genes (Starmer *et al.*, 1987; Yokomori *et al.*, 1988b), but there are some strains where it is determined by either linear DNA plasmids (Gunge *et al.*, 1981) or dsRNA plasmids (Kitano *et al.*, 1987; Zorg *et al.*, 1988).

As for the practical application of the killer phenomenon, utilization of killer yeasts for commercial fermentation of alcoholic beverages (Ouchi *et al.*, 1979; Hara *et al.*, 1980; Young, 1981; Seki *et al.*, 1985; Shimizu *et al.*, 1986; Tredoux *et al.*, 1986; Yokomori *et al.*, 1989) and clinical purposes (Morace *et al.*, 1984; Polonelli and Morace, 1986) has been investigated. Killer yeasts are closely associated with the winemaking process. A number of wild killer yeasts from various genera have been isolated from grape must, and some of these strains have caused stuck or slow fermentations (Kitano *et al.*, 1984; Shimizu *et al.*, 1985a; van Vuuren and Wingfield, 1986; Heard and Fleet, 1987; Benda, 1989). Occurrence of killer activity has been observed also among strains which are employed in the commercial fermentation of wine (Barre and Biron, 1982; Shimizu *et al.*, 1985b; Thornton, 1986).

This chapter is concerned with killer yeasts and their relation to the winemaking process. A general review of killer yeasts is given by Young (1987).

KILLER PHENOMENON

Distribution among Yeast Communities

Since the first demonstration of killer character in a laboratory strain of *S. cerevisiae*, killer strains have been isolated from a number of different sources and shown to be wide-spread among yeast genera, including *Saccharomyces*, *Hansenula*, *Pichia*, *Kluyveromyces*, *Candida*, *Kloeckera*, *Hanseniaspora*, *Rhodotorula*, *Trichosporon*, *Debaryomyces* and *Cryptococcus*.

A survey of 964 strains from the British National Collection of Yeast Cultures revealed the presence of killer character in 59 strains, of which 38 strains occurred in the genus *Saccharomyces* and the rest in *Candida*, *Pichia*, *Hansenula*, *Kluyveromyces* and *Debaryomyces* (Philliskirk and Young, 1975). Furthermore, the killer yeasts were classified into 10 groups (Kl to K10) based on cross-reactions between the killer strains of various genera and species (Young and Yagiu, 1978). A killer strain of *Candida glabrata* was detected in a laboratory culture collection (Bussey and Skipper, 1975) and appended to the classification as the eleventh category, K11 (Wickner, 1979).

Killer toxin producing strains have been isolated from natural habitats and many fermented food and beverage products. Seventeen percent of isolates from fruits, mushrooms, decaying plants and soils exhibited killer activity, and these strains were distributed among six yeast genera, *Saccharomyces*, *Pichia*, *Hansenula*, *Kluyveromyces*, *Debaryomyces* and *Cryptococcus* (Stumm *et al.*, 1977). Starmer *et al.* (1987) isolated killer strains of various genera and species from yeast communities associated with plants. They occur as contaminants in commercial fermentations of beer (Maule and Thomas, 1973), wine (Heard and Fleet, 1987) and sake (Imamura *et al.*, 1974). In some cases, the killer strains caused serious problems in the fermentations and/or off-flavors in the products. In a survey of 1294 yeast strains isolated from fermented foods in Italy, Rosini (1983) detected the killer property in 31 of 782 *S. cerevisiae* strains and observed high incidence of killer activity among isolates of *Hansenula anomala*. Bortol *et al.* (1986) isolated four killer strains of *S. cerevisiae* from bread, apple, asparagus and vinegar, respectively. Killer yeasts were found in fermented foods such as miso, soy sauce and salted vegetables, and were identified as *H. anomala*, *Pichia farinosa*, *Debaryomyces hansenii* and *Candida naeodendra* (Kagiyama *et al.*, 1988; Suzuki *et al.*, 1989). Suzuki *et al.* (1989) reported killer yeasts, mainly *Debaryomyces*, in 76% of the salted vegetables surveyed.

Killer strains have been also detected among pathogenic yeasts from the genera *Cryptococcus*, *Candida* and *Trichosporon* (Kandel and Stern, 1979; Middlebeek *et al.*, 1980a).

Biology and Genetics

The genetic determinants of the killer property are dsRNA and linear DNA plasmids, and nuclear genomes of yeasts.

Killer systems associated with dsRNA plasmids

The killer systems controlled by two dsRNA plasmids occur in *S. cerevisiae*, *Hanseniaspora uvarum* and its anamorph, *Kloeckera apiculata*. In *S. cerevisiae*, five killer types associated with dsRNA plasmids have been identified. The types K1, K2 and K3 were classified by Young and Yagiu (1978) and two additional types (KT28 and "K3") were found in yeasts from a culture collection (Pfeiffer and Radler, 1982) and wine yeasts (Extremera *et al.*, 1982), respectively. The *S. cerevisiae* killer systems are essentially comprised of two segments of dsRNA distinguished as L and M by size. The sizes of the M dsRNAs differ among the five different killer types of *S. cerevisiae* (K1, 1.9Kb; K2, 1.7Kb; K3, 1.5Kb; KT-28,

1.99Kb; "K3", 1.5Kb and 2.0Kb), whereas the sizes of the L dsRNAs show little difference among the types (4.7Kb). The killer strains of *H'spora uvarum* and *K. apiculata* contain L dsRNAs of almost the same size as those found in *S. cerevisiae*, and M dsRNAs with sizes that differ significantly among the strains (Kitano *et al.*, 1987; Zorg *et al.*, 1988). The L and M dsRNAs are separately packaged by a common major capsid protein into virus-like particles (VLPs) which are transmitted cytoplasmically (Herring and Bevan, 1974; Hopper *et al.*, 1977).

Of the two dsRNA species, the M genome encodes the production of killer toxin precursor while the L genome codes for polymerase and capsid protein for both genomes (Tipper and Bostian, 1984). Among the *S. cerevisiae* killer systems, the K1 system has been most extensively investigated. Bostian *et al.* (1984) identified the K1 toxin precursor gene on the M1 dsRNA genome (M dsRNA of K1 killer) and elucidated its structure. Furthermore, plasmids containing the cDNA copy of the toxin precursor gene with an appropriate promoter proved to confer both K1 killer and immunity phenotypes in sensitive host yeast strains (Lolle *et al.*, 1984). The K2 toxin gene has been cloned and its sequence shows no identity with that encoding for the K1 toxin, in spite of the apparent similarity in action of the two toxins (Bussey *et al.*, 1990). As shown in Figure 8.1, the M1 dsRNA consists of two segments, M1-1(1Kb) and M1-2(0.6Kb), separated by a highly adenine-uracil (AU) rich sequence (0.2Kb), of which M1-1 encodes the K1 toxin precursor, preprotoxin (Bostian *et al.*, 1984).

The expression of plasmid functions depends upon numerous chromosomal genes (Wickner, 1978; Tipper and Bostian, 1984). The *MAK* and *SKI* genes affect the maintenance of dsRNA plasmids and its replication (Wickner, 1986). The *KEX1* and *KEX2* genes encode for carboxypeptidase and endopeptidase, respectively, which are involved in the processing of killer toxins (Bussey *et al.*, 1983). Furthermore, several additional genes including *SEC* (toxin secretion), *END* (endocytosis), *KRE* (toxin binding), *VPL* (vacuolar protein localization) and *REX* (resistance expression) have been reported (Wickner and Leibowitz, 1976; Al-Aidroos and Bussey, 1978; Bussey *et al.*, 1983; Chvatchko *et al.*, 1986; Rothman and Stevens, 1986).

In addition, a novel killer system associated with L and M dsRNAs was recently found in *Candida stellata* (Shimizu *et al.*, unpublished work). Moreover, so-called neutral strains of *S. cerevisiae* with M dsRNAs that do not encode for

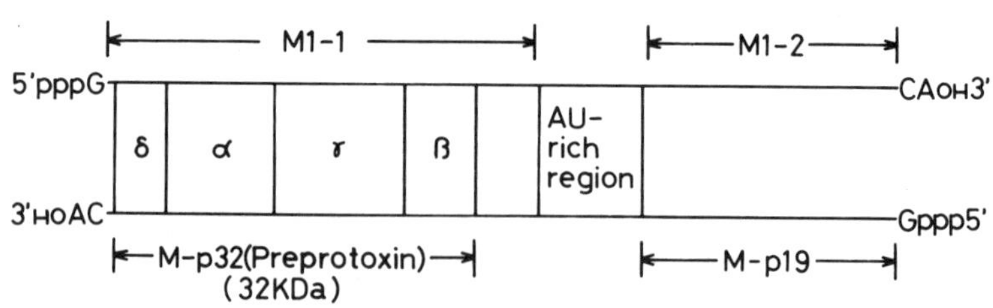

FIGURE 8.1 The structure of M1 dsRNA plasmid (Wickner, 1986).

killer toxins but confer only immunity to either K1 (Bevan and Somers, 1969) or K2 toxin (Wingfield *et al.*, 1990; Shimizu *et al.*, unpublished work) have been identified.

Killer system associated with linear DNA plasmids

The killer system associated with DNA plasmids has been found only in the yeast, *Kluyveromyces lactis* (Gunge *et al.*, 1981). The killer strain, *Kluy. lactis* IFO 1267, harbors two cytoplasmically inherited double-stranded linear DNA plasmids, pGKL1 (8.9Kb) and pGKL2(13.4Kb), of which pGKL1 codes for the production of the killer toxin (α, β and γ subunits) and the immunity determinant to the toxin, while pGKL2 is required for the maintenance of pGKL1 (Gunge *et al.*, 1981; Hishinuma *et al.*, 1984; Stark and Boyd, 1986). Nucleotide sequencing of pGKL1 (887bp) and pGKL2 (13457bp) has been carried out by Hishinuma *et al.* (1984) and Tommasino and Galeotti (1988), respectively. These pGKL plasmids which possess terminal proteins attached at their 5' ends (Kikuchi *et al.*, 1984), can be transferred into *S. cerevisiae* by protoplast fusion (Gunge and Sakaguchi, 1981) or transformation (Gunge *et al.*, 1982) and are stably maintained in *S. cerevisiae*, where the plasmids express the killer character. Unlike the dsRNA killer systems, no viral nature has been known in the pGKL plasmids.

As found with dsRNA killer systems, chromosomal gene function is required for the expression of killer character in *Kluy. lactis*. The *Kluy. lactis* KEX1 gene, which has been cloned and shown to be functionally homologous to *S. cerevisiae* KEX2 endopeptidase, is thought to perform the processing of the $\alpha\beta$ precursor into the mature α and β subunits of the killer toxin (Wesolowski *et al.*, 1988).

Other killer systems

In addition to the killer systems encoded by cytoplasmic plasmids, some killer phenotypes may be associated with determinants in chromosomal genes. Kitano *et al.* (1988) identified novel killer systems, designated *KHR* and *KHS* killers, in *S. cerevisiae* and showed that both killer characters were encoded by nuclear genes. Furthermore, it was shown that *KHR* and *KHS* genes were located on the left arm of chromosome IX and the right arm of chromosome V, respectively. Possible chromosomal location of killer determinants has been suggested in strains of *Candida glabrata* (Sriprakash and Batum, 1984), *Candida* spp. SW-55 (Yokomori *et al.*, 1988b), *Pichia kluyveri* (Starmer *et al.*, 1987), *P. farinosa* (Suzuki and Nikkuni, 1989), *H. anomala* (Kagiyama *et al.*, 1988) and *Hanseniaspora* spp. (Shimizu *et al.*, unpublished work).

Killer Toxins

Characterization

The killer toxins so far isolated and characterized are listed in Table 8.1. They are either proteins or glycoproteins, some of which consist of subunits. Most of the toxins are unstable at high temperature and at high pH; the toxins produced

TABLE 8.1 Comparison of characteristics of killer toxins from different yeast strains.

Killer strain	*Nature*	Molecular mass (kD[a])	*Optimum pH*	*Reference*
Saccharomyces cerevisiae (K1)	Protein	20	4.6–4.8	Bostian *et al.* (1984)
Saccharomyces cerevisiae (K2)	Glycoprotein	16	4.2	Pfeiffer and Radler (1984)
Saccharomyces cerevisiae (KT28)	Glycoprotein	16	5.8	Pfeiffer and Radler (1984)
Saccharomyces cerevisiae (KHR)	Protein	20	5.7	Goto *et al.* (1990)
Kluyveromyces lactis	Glycoprotein	180	4.0–8.0	Stark *et al.* (1990)
Pichia kluyveri	Glycoprotein	19	3.8–4.0	Middelbeek *et al.* (1970)
Hansenula mrakii	Protein	10.7	4.0–9.0	Ashida *et al.* (1983)
Hansenula saturnus	Protein	11–12	3.5–7.0	Henschke (1979)
Candida sp. SW–55	Glycoprotein	>200	3.8–5.8	Yokomori *et al.* (1988a)
Pichia farinosa	Protein	25	2.0–4.0	Suzuki and Nikkuni (1989)
Hansenula anomala	Glycoprotein	300	–	Kagiyama *et al.* (1988)
Hanseniaspora uvarum	Protein	18	4.0–4.5	Radler *et al.* (1990)

[a]monomer

by the genus *Hansenula* show relatively higher stability (Henschke, 1979; Ashida *et al.*, 1983). It is interesting that the two halophillic killer yeast strains, *P. farinosa* and *H. anomala*, produce toxins whose killer activities increase with elevation of NaCl concentration (Kagiyama *et al.*, 1988; Suzuki and Nikkuni, 1989). The killer toxin from the K3 strain of *S. cerevisiae (capensis)* was found to be a 18 kDa glycoprotein. Bussey and Skipper (1975) revealed that the toxin of *C. glabrata* consisted of several molecular species, at least one of which was a glycoprotein.

The killer toxins of *S. cerevisiae*, except the *KHR* toxin, are active only against sensitive strains of a few yeast species including *S. cerevisiae* and *C. glabrata*, while those of *H. mrakii*, *Kluy. lactis*, *Kluy. drosphilarum* and *Candida* spp. SW-55 show much broader anti-yeast spectra.

Biosynthesis and secretion

The biogenesis and secretion of killer toxins have been extensively investigated for *S. cerevisiae* K1 and *Kluy. lactis*.

As mentioned already (see Figure 8.1), the K1 preprotoxin (32 kDa) encoded by the M1-1 segment of the M1 dsRNA plasmid is composed of δ (leader sequence), α (α subunit of toxin), γ (γ peptide, immunity determinant) and β (β subunit of toxin) components. Among the components, the γ-component (44 amino acid residues) contains the amino-terminal signal peptide of 26 amino acid residues which is necessary for entry into the endoplasmic reticulum (Lolle *et al.*, 1984). The γ-component has 3 Asn-X-(Thr or Ser) glycosylation sites which appear to be glycosylated with $(Glu)_3(Man)_9(GlcNAc)_2$ cores, whereas other components are not glycosylated. After glycosylation in the endoplasmic reticulum, the 43 kDa glycosylated K1 preprotoxin (K1 protoxin) is transported into the Golgi apparatus and converted to the mature K1 toxin mostly in the secretion vesicles (Bostian *et al.*, 1983; Bussey *et al.*, 1983; Lolle *et al.*, 1984). The hydrophobic amino-terminal leader peptide is not removed in the endoplasmic reticulum and appears to be the attachment site of the K1 protoxin to the cytoplasmic membrane. The maturation of the membrane-associated K1 protoxin depends on proteolytic enzymes encoded by two chromosomal genes of the K1 killer yeasts, KEX1 and KEX2. Endoproteolytic cleavage by enzymes produced by the KEX2 gene generates the α (9.5 kDa) and β (9 kDa) toxin subunits (Julius *et al.*, 1984; Wickner and Leibowitz, 1976), while the COOH-terminal basic residues of the α subunit are removed by the carboxypeptidase coded by the KEX1 gene (Dmochowska *et al.*, 1987). After this processing, the matured K1 toxin is secreted as a disulfide-bonded α-β dimer via the constitutive secretory pathway (Tipper and Bostian, 1984).

The K2 preprotoxin is thought to have an overall structure similar to that of the K1 preprotoxin. The presence of 43 kDa intracellular and 21 kDa extracellular species have been observed, which may correspond to the glycosylated K2 protoxin and a glycosylated subunit derived from the protoxin, respectively (Bussey *et al.*, 1990).

The secreted killer toxin of *Kluy. lactis* consists of three subunits, α (99 kDa), β (30 kDa) and γ (27.5 kDa), of which only the α subunit is glycosylated (Stark and Boyd, 1986). As mentioned already, the $\alpha\beta$ and γ subunits of the toxin are encoded by the pGKL1 plasmid. Furthermore, it has been shown that processing of the $\alpha\beta$ precursor into mature α and β subunits is an essential step for secretion of the mature toxin and that processing is performed by the *Kluy. lactis* KEX1 endopeptidase. Although the γ subunit has its own secretion signal, as is the case with the $\alpha\beta$ precursor, its secretion is blocked in the absence of the $\alpha\beta$ precursor, which suggests that specific assembly steps, like formation of the β-γ disulfide bond, are required for the biogenesis of this toxin (Stark *et al.*, 1990). Since all the proteins secreted by yeasts are either glycosylated or cleaved from glycosylated precursors, it would be necessary for non-glycosylated species, such as the γ-subunit, to associate with a glycosylated species during secretion.

The *Kluy. lactis* immunity determinant, also encoded by pKGL1, appears to be a cytoplasmic protein because it has neither an amino-terminal signal peptide nor any other strongly hydrophobic regions (Stark *et al*., 1990).

Mode of action

The mechanism of killer toxin action has been investigated extensively in the K1 killer system. The primary event in the action of K1 toxin is an energy-independent binding of the toxin to a cell wall receptor (Al-Aidroos and Bussey, 1978; Bussey *et al*., 1979) identified as a (1-6)-β-D-glucan (Hutchins and Bussey, 1983). This is followed by an energy-dependent insertion of the toxin into its action site, the cytoplasmic membrane (Skipper and Bussey, 1977). The toxin-induced killing of sensitive cells also requires metabolic energy (Bussey and Sherman, 1973). It is thought that the toxin alters the proton permeability of the cytoplasmic membrane, leading to potassium efflux, acidification of cell contents, metabolic inhibition and leakage of ATP (de la Pena *et al*., 1981). According to Bussey *et al*. (1990), the α subunit of the K1 toxin is involved in reaction with the membrane. Similar physiological effects have been observed with the toxin of *P. kluyveri* (Middelbeek *et al*., 1980b). This toxin is thought to perturb membrane permeability by non-selective channel formation (Kagan, 1983). Schmitt *et al*. (1989) proposed that the KT28 killer toxin of *S. cerevisiae* inhibited DNA synthesis. Moreover, the initial receptor site of this toxin may be the cell wall mannoprotein, rather than glucan (Radler and Schmitt, 1987).

It has been suggested that the *Kluy. lactis* toxin causes specific inhibition of cell division at the G1 stage of the cell cycle by affecting adenylate cyclase of sensitive cells (Sugisaki *et al*., 1983; Stark *et al*., 1990). However, a strong argument against the involvement of adenylate cyclase in toxin action has been advanced (White *et al*., 1989). Among the three subunits of the toxin, the γ subunit is known to be responsible for toxin activity (Tokunaga *et al*., 1989). In contrast, the functions of the other two subunits are still uncertain.

The killer toxin of *H. mrakii* is thought to act by selective inhibition of cell wall glucan synthesis in sensitive yeast cells (Yamamoto *et al*., 1986).

In the K1 killer system (see Figure 8.1), the precise immunity determinant has not been determined. Boone *et al*. (1986) suggested that immunity was conferred by either the intact or incompletely processed toxin precursor that contained the α subunit. In this case, the immunity determinant may interfere with toxin action by binding to the α toxin receptor on the membrane or to the target of α toxin action. The glycosylated γ peptide may be also a candidate for the immunity determinant because of possible interaction with the carbohydrate receptor on the cytoplasmic membrane (Tipper and Bostian, 1984).

Since the immunity determinant of the *Kluy. lactis* killer system appears to be a cytoplasmic protein, its mode of action is suggested to be interaction with either the γ subunit or the intracellular target of toxin action (Tokunaga *et al*., 1989).

KILLER YEASTS IN WINEMAKING

Occurrence of Wild Killer Yeasts

Killer yeasts have been isolated from various enological sources including grape berries, grape musts, wine fermentations and wines. A survey performed in France in 1981 revealed that the incidence of killer yeasts in native enological flora varied from 0–90% according to the region. Very high incidences were observed in Mediterranean (65–90%) and Beaujolais (83%) vineyards, while lower frequencies (0–16%) occurred in other regions (Cuinier and Gros, 1983). Rosini *et al.* (1988) reported high incidence of K2 killer (12–83%) and neutral strains (17–87%) of *S. cerevisiae* in Segrantino wines and wineries in Italy. Ciolfi (1985) isolated killer strains of *Candida* sp. and *S. cerevisiae* from spontaneous fermentations of Barbera, Grignolino, Nebbiolo, Chardonnay, Malvasia and Freisa wines. In Germany, Benda (1985) isolated 209 killer strains of *Saccharomyces* from 30 wine fermentations in the Frankonia region. In a survey of natural enological microflora in Geisenheim, Germany, Shimizu *et al.* (1985a) isolated four killer strains of *S. cerevisiae* from natural wine fermentations and two *Candida* killer strains from grape musts. All the killer strains of *S. cerevisiae* showed the K2 killer character. The high incidence (21%) of K2 killer strains of *S. cerevisiae* has been also reported in spontaneous fermentations of Ribeiro wines in Spain (Cansado *et al.*, 1989). In contrast, a survey performed by Kitano *et al.* (1984) in Japan revealed that all the killer strains of *S. cerevisiae* isolated from several enological sources had either the K1 or novel killer properties. Similar results were also obtained in vineyards located in Yamanashi, Japan. Most of the killer yeasts isolated from these vineyards belonged to either *S. cerevisiae* or the genus *Hansenula*. Furthermore, among the 31 killer isolates of *S. cerevisiae*, only one showed the K2 phenotype, while the rest, but one, exhibited K1 killer character (Shimizu *et al.*, unpublished work).

The presence of killer yeasts in natural enological sources has been also observed in the Southern Hemisphere. Heard and Fleet (1987) examined 16 wine fermentations in four different Australian wineries for the presence of killer yeasts. Killer activity was detected in nine isolates of *S. cerevisiae* but not in isolates of *Kloeckera*, *Candida*, *Hansenula* and *Torulaspora*. In South Africa, 230 killer strains were isolated from Chenin blanc grapes and grape skins collected from six wineries (Jacobs *et al.*, 1988). Furthermore, high incidence (>27%) of killer yeasts was observed in grape musts from local wineries in Brazil (Pasqual *et al.*, 1990).

Killer Property in Wine Strains of *Saccharomyces cerevisiae*

The killer property has been detected in many wine strains of *S. cerevisiae* that have been deposited in culture collections throughout the world. These strains, as well as the strains now commercially available, are frequently used to initiate wine fermentations. Naumov and Naumova (1973) reported the presence of 17 killer strains and 19 sensitive strains in a collection of winemaking *S. cerevisiae* held in the USSR. In a survey of almost 1000 wine strains of *S. cerevisiae* in collections in France, approximately 60% were killer positive, 30% were sensitive and 10% were neutral (Barre, 1980; Barre and Biron, 1982). Of 454 cultures of wine *S. cerevisiae* from a collection in Yalta, USSR, only 9% were killers while 90% were sensitive and 1% neutral (Tiourina *et al.*, 1980). Similar low incidences (<10%) of killer strains have been reported in surveys of collections in Germany (Radler, 1980), Italy (Gaia, 1984) and South Africa (Tredoux *et al.*, 1986). In a survey of 25 culture *S. cerevisiae* from New Zealand and Australia, eight strains were found to have K2 killer character (Thornton, 1986). Shimizu *et al.* (1985b) surveyed wine yeasts from the culture collection in Geisenheim, Germany. They found two South African strains (WE14 and WE372) and four commercial strains (Montpellier 1116, and Prise de Mousse 1118, 71B 1122, and CAW) to be K2 killer yeasts. The strain Geisenheim74 (G74) was a mixed culture of a K2 killer strain (G74K2), a K2 neutral strain (G74Sen$^-$) and a K2 sensitive strain (G74Sen$^+$), of which G74 Sen$^+$ was recently identified as a *KHS* killer strain (see below). Moreover, a K2 killer strain was isolated from the stock culture of Schloss Boeckelheim. Recently, some killer strains, namely, K-1 (Lallemand), 796, 729A, R2 (Intek), have become commercially available.

Kitano *et al.* (1988) reported the wide distribution (75%) of a novel killer property among wine yeasts. This killer phenotype, encoded by a yeast chromosomal gene designated *KHS*, was detected by using *C. glabrata* IFO 0622 as a sensitive strain. The *KHS* killer property was also detected in well-known wine strains including Winningen, Geisenheim74 (G74 Sen$^+$), Epernay, Montrachet, Tokay, Liebfraumilch, Champagne, OC-2, W-3 and Ruedesheimerberg (Yokomori, personal communication). Although the enological importance of the *KHS* killer property is negligible because it does not affect *S. cerevisiae*, the wide distribution of this killer phenomenon among wine yeasts may have ecological interest.

Killer Property in Wild Yeasts Associated with Winemaking

Species of *Kloeckera*, *Hanseniaspora*, *Candida*, *Hansenula* and *Pichia* may establish significant growth during some wine fermentations (Fleet, 1990). Consequently, the potential of these species to produce killer toxins that affect *S. cerevisiae* becomes important, as does the susceptibility of these yeasts to toxins produced by *S. cerevisiae*.

Radler and co-workers (Radler, 1980; Pfeiffer and Radler, 1984; Zorg *et al.*, 1988; Radler and Knoll, 1988) have identified strains of *Candida krusei*, *Pichia fermentans*, *P. kluyveri* and *H'spora uvarum* that exhibit killer action against strains of *S. cerevisiae*. Rosini (1983, 1985) and Nomoto *et al.* (1984) reported the ability of some strains of *H. anomala* to kill strains of *S. cerevisiae*. As yet, the enological significance of such interactions requires analysis. Heard and Fleet (1987) could not demonstrate any effect of a killer strain of *S. cerevisiae* on the growth of wine isolates of *C. krusei*, *C. pulcherrima*, *K. apiculata* and *H. anomala* during experimental grape juice fermentations.

Influence of Wild Killer Yeasts on Wine Fermentation

Wild killer strains, regardless of the genus and species to which they belong, may interfere with wine fermentation and subsequent wine quality. Their presence and activity become particularly significant in wine fermentations conducted by inoculation with selected strains of *S. cerevisiae*. Their growth and activity have the potential to delay the onset of fermentation, cause stuck fermentations, and give wines with increased levels of acetaldehyde, lactic acid, acetic acid, and other undesirable sensory qualities (Benda, 1985; van Vuuren and Wingfield, 1986). The ability of killer strains of *S. cerevisiae* to eliminate sensitive strains of this species and, therefore, dominate the wine fermentation has been observed in experimental cultures (Barre, 1984; Tredoux *et al.*, 1986; Heard and Fleet, 1987), and in commercial fermentations (Cuinier and Gros, 1983; Benda, 1986). Killer strains of genera other than *Saccharomyces* may also cause retardation of wine fermentations. Radler and Knoll (1988) have convincingly demonstrated that a killer strain of *H'spora uvarum* can delay the fermentation of grape juice by a sensitive strain of *S. cerevisiae* for up to 10–20 days. Benda (1989) noted *P. fermentans* and *H. anomala* to interfere with wine fermentations. In Japanese wineries, there have been instances where killer strains of *S. cerevisiae* that produced cells with a hydrophobic surface, caused film formation on the surface of wines (Shimizu, unpublished data). These facts indicate a marked technological significance of the killer phenomenon in the winemaking process.

Specific use of Killer Yeasts in Wine Fermentation

Although the wild occurrence of killer yeasts can negatively affect wine fermentations, there may be distinct advantages in deliberately exploiting yeasts with this property.

Killer wine yeasts may offer three advantages over conventional wine yeasts when employed for fermentation. Firstly, they could be selected to kill certain wild yeasts which cause problems such as delay of fermentation, stuck fermentation and production of off-flavors. Secondly, they could be selected to have immunity against any killing action of wild yeasts and, therefore, have a greater chance of dominating the fermentation. Finally, they could be selected to produce stable killer toxins that would protect the wine from infection by spoilage yeasts.

Killer strains of *S. cerevisiae* must have several properties for proper exploitation in wine fermentations. These properties include good fermentation kinetics, the production of good quality wine and the secretion of killer toxins that sustain their activity under the extreme environment of the wine. Specific wine properties that affect the activity and stability of killer toxins include pH, ethanol concentration, sulfur dioxide concentration and complexing with tannins and fining agents (Ciolfi, 1984; Shimizu *et al.*, 1986; Radler and Schmitt, 1986; Heard and Fleet, 1987).

Generally, the killer toxins of *S. cerevisiae* become unstable at high ethanol concentrations and high temperatures (Shimizu unpublished data) and at very low pH value (Ciofli, 1984; Shimizu *et al.*, 1986). The K1, K2 and K3 killer activities of *S. cerevisiae* have been compared at different pH values (Figure 8.2). Activity of the K1 toxin is not observed at pH values less than 3.5 whereas the K2 and K3 toxins show significant activity at pH 2.9. The K2 activity is superior to the K3 activity under acidic conditions. Heard and Fleet (1987) also noted the activity of K2 toxin during experimental wine fermentations at pH 3.0. With respect to stability, K2 killer activity is most stable at pH values less than 4.0 (Shimizu *et al.*, 1986). The K2 killer toxin produced by *S. cerevisiae*, "Prise de Mousse" strain during the fermentation of Koshu white wine was active for more than 30 days during storage at pH 3.1 and 15°C (Shimizu, unpublished data). These facts suggest that a K2 killer strain of *S. cerevisiae* should be selected when aiming to exploit the killer phenomenon during commercial wine fermentation.

Figure 8.3 shows the killing of FY-3, a film-forming strain of *S. cerevisiae*, by a respiratory deficient mutant of a K2 wine strain, "Prise de Mousse", in the course of wine fermentation. The cells of FY-3 are killed completely during fermentation as long as their initial population is not more than 1×10^3 cells/ml. Generally, the population of film forming strains of *S. cerevisiae* in clarified grape musts is less than 5×10^2 cells/ml (Shimizu, unpublished work). Thus, the growth of these undesirable yeasts during commercial fermentations could be prevented by inoculating the must with a strain of *S. cerevisiae* possessing the K2 killer property.

Grape must inoculated with *S. cerevisiae* strain "Prise de Mousse" (K2 killer) exhibits maximum killer activity after three days. Thereafter, the activity declines gradually but significant activity still remains even at the end of

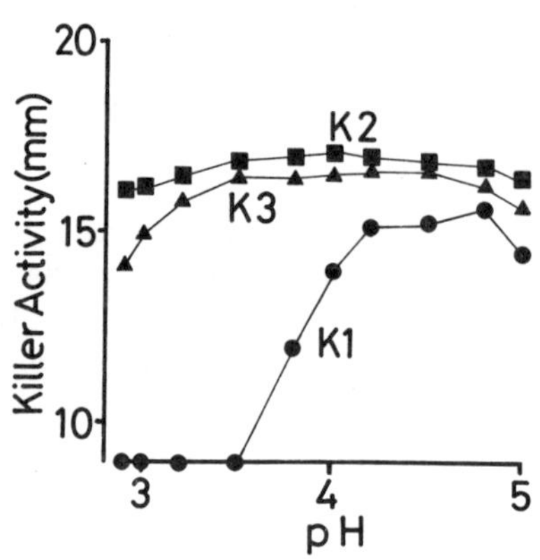

FIGURE 8.2 Optimum pH for killer activity. Killer yeast cultures in YPD medium containing 0.1M citrate-phosphate buffer (pH 4.0) were filtered through 0.25 μm membranes. The filtrates were examined for killer activity by an agar diffusion method on YPD - methylene blue agar adjusted to different pH values. *S. cerevisiae* (Epernay) was used as a sensitive strain.

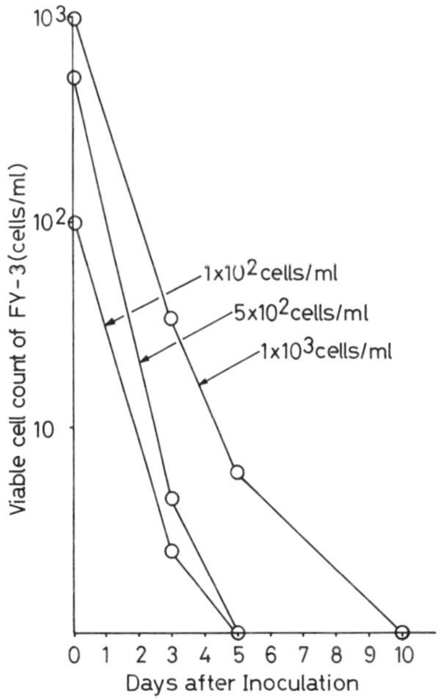

FIGURE 8.3 Killing of spoilage yeast, *S. cerevisiae* FY-3, by a respiration-deficient mutant of a killer wine yeast, "Prise de Mousse" during fermentation of Koshu grape juice (pH 3.1) at 20°C. The killer yeast was inoculated at 5×10^5 cells/ml.

fermentation (Figure 8.4). This killer activity can be removed by bentonite treatment at a concentration of 1 g/1 (Radler and Schmitt, 1987) but a bentonite-untreated wine keeps the killer activity until after filtration. As shown in Figure 8.5, the wine produced by the K2 killer strain, "Prise de Mousse", is more resistant against infection by a spoilage yeast, strain RE-2, than wine produced by a conventional non-killer yeast strain, Epernay.

The experiments described in Figures 8.2 to 8.5 provide evidence that a strain of *S. cerevisiae* possessing the K2 killer property could be used to control the growth of an undesirable wild yeast during fermentation and protect the fermented wine against the growth of a spoilage yeast. The wines produced with this K2 killer strain did not have any negative sensory character (Shimizu unpublished work) as reported, also, with other K2 killer strains (Tyurina *et al.*, 1980). However, these data apply only to the production of white wines. In red wine fermentations, K2 killer activity disappeared within three days after inoculation of the yeast into the musts and, presumably, this is related to complexing of the toxin proteins with tannins.

Although the employment of K2 killer yeasts for commercial wine fermentations may be advantageous, two factors must be taken into consideration for their successful use. One is the temperatures for fermentation and storage of wine and the other is the ratio of the killer strain, on a population

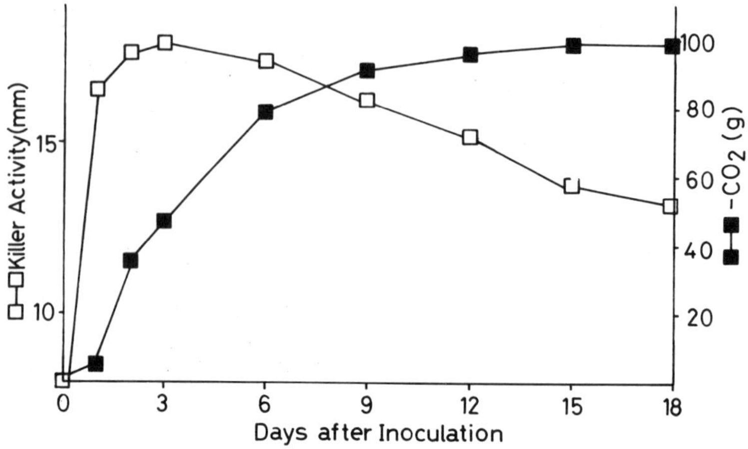

FIGURE 8.4 Change of killer activity during wine fermentation. Koshu grape must (pH 3.1) was fermented at 20°C with a K2 killer wine strain, *S. cerevisiae* "Prise de Mousse". Samples were taken at intervals and examined for killer activity at pH 4.5 as described in the legend for Figure 8.2. Time course of the fermentation was monitored by production of carbon dioxide.

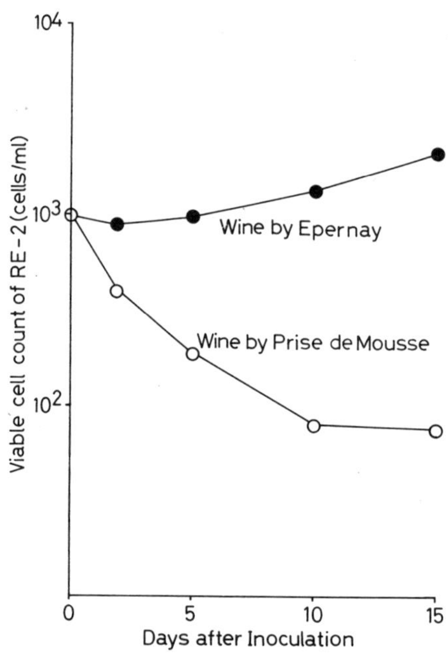

FIGURE 8.5 Growth of a spoilage yeast, *S. cerevisiae* RE-2, in wines produced by a killer (Prise de Mousse) and a non-killer (Epernay) wine yeast. Koshu white wines produced by either the non-killer or killer wine strains, were sterile-filtered and inoculated with *S. cerevisiae* RE-2 followed by incubation at 20°C. Samples were taken at intervals and examined for viable cell counts of RE-2.

basis, to the sensitive species. The former affects the stability of killer activity, and temperatures above 25°C can lead to the rapid inactivation (within 5 days) of activity (Shimizu, unpublished work). As for the population ratios, some authors have obtained good expression of the killer effect and dominance of the killer strains when they are inoculated at levels as low as 4–10% of the population (Hara *et al.*, 1980; Barre, 1984), whereas others only noted killer dominance when the killer strains were inoculated at much higher proportions (Seki *et al.*, 1985; Shimizu *et al.*, 1985; Rosini, 1985; Heard and Fleet, 1987; Petering *et al.*, 1991). This discrepancy may be attributed to properties and quantities of killer toxins produced by the killer strains but further studies are required.

Genetic Construction of Killer Yeasts

Since K2 killer wine yeasts are now commercially available, there is an increasing number of the wineries using them. However, it is becoming increasingly significant that strains with desirable killer properties must also possess excellent enological characteristics. To obtain this match, genetic construction of strains offers much promise. Particularly appealing is the prospect of constructing wine strains with broad spectrum killer activity.

So far, the introduction of killer characters into non-killer yeast strains has been carried out utilizing donor strains whose killer phenotypes are encoded by linear DNA plasmids (Gunge and Sakaguchi, 1981; Gunge *et al.*, 1982) or dsRNA plasmids. Construction of industrial yeast strains with killer activity has been reported for baker's (Bortol *et al.*, 1986), brewer's (Janderova *et al.*, 1986; Yong, 1981), sake (Ouchi *et al.*, 1979) and wine yeasts. The killer wine yeasts have been constructed by either sexual hybridization (Hara *et al.*, 1980) or protoplast fusion between a killer strain and a wine strain, and are classified into two groups. One group comprises the cytoductants in which only the cytoplasmic factors, mainly killer plasmids and mitochondria, are transferred without altering the karyotype of the wine strain (Seki *et al.*, 1985; Yokomori *et al.*, 1989). The other group comprises true hybrids that not only have the killer plasmids from the killer strain but possess the intermediate karyotype between the wine and killer strains (Hara *et al.*, 1980; Shimizu *et al.*, 1986; Yokomori *et al.*, 1989, 1990). Boone *et al.* (1990) have constructed K1/K2 double killer strains by introducing the K1 killer gene into K2 killer strains through chromosomal integration of the toxin gene. As for the practical employment of genetically constructed killer wine yeasts, three strains, constructed through protoplast fusion and designated GS-31-16, SWW-12 and BW-1019, respectively, have been applied successfully for commercial fermentation in a Japanese winery (Yokomori *et al.*, 1990).

FUTURE PROSPECTS

There is little doubt that killer yeasts can be used to improve the winemaking process and wine quality but further developments are necessary. First, although the K2 killer toxin is the most stable of the toxins in grape must and wine, it does not retain its activity throughout the wine storage period. Second, the anti-yeast spectrum of the K2 toxin is so restricted that the K2 killer wine yeast is not capable of killing wild yeasts other than strains of *S. cerevisiae* and *C. glabrata*. Thus, wild yeasts such as species of *Hansenula*, *Pichia* and *Kloeckera*, grow unaffected by this killer activity. The screening for novel killer yeasts which produce highly stable killer toxins with a broad spectrum of anti-yeast activity must continue. Considering the processes of genetic construction, it would be desirable that the novel killer character be coded by cytoplasmic factors. However, since the plasmid-encoded killer toxins so far identified either show restricted anti-yeast spectra (dsRNA-encoded toxins) or are ineffective at low pH (linear DNA-encoded toxin), transformation of nuclear killer genes may be required for the construction of wine yeasts with improved killer property. Finally, further research to understand the molecular structure of killer toxins may lay the foundations for protein engineering of toxins with improved stability, and a broader spectrum of action.

REFERENCES

Al-Aidroos, K. and Bussey, H. (1978) Chromosomal mutants of *Saccharomyces cerevisiae* affecting the cell wall binding site for killer factor. *Canadian Journal of Microbiology*, **24**, 228–237.

Ashida, S., Shimazaki, T., Kitano, K. and Hara, S. (1983) New killer toxin of *Hansenula mrakii*. *Agricultural and Biological Chemistry*, **47**, 2953–2955.

Barre, P. (1980) Rôle de facteur "killer" dans la concurrence entre souches de levures. *Bulletin de l'Office International de la Vigne et du Vin*, **53**, 560–567.

Barre, P. (1984) The killer mechanism in mixed strains of yeasts. Studies and measurement. *Bulletin de l'Office International de la Vigne et du Vin*, **57**, 635–643.

Barre, P. and Biron, M.J. (1982) Identification of 'Killer' group in 55 strains of *Saccharomyces* isolated from wine *Science des Aliments*, **2**, 125–130.

Benda, I. (1985) Yeasts in winemaking—investigating so-called killer yeasts during fermentation. *Der Deutscher Weinbau*, **40**, 1166–1171.

Benda, I. (1986) The effect of killer yeasts on the structure of yeast populations during must fermentation. *Wein Wissenschaft*, **41**, 345–355.

Benda, I. (1989) Occurrence of killer toxin-producing strains of film-forming yeasts in grape must and their influence on alcoholic fermentations. *Wein Wissenschaft*, **44**, 56–60.

Bevan, E.A. and Makower, M. (1963) The physiological basis of the killer character in yeast. In *Proceedings of the Eleventh International Congress of Genetics* (Hague, 1963), edited by S.J. Goerts, 202 pp.

Bevan, E.A. and Somers, J.M. (1969) Somotic segregation of the killer(k) and neutral(n) cytoplasmic genetic deteminants in yeast. *Genetic Research*, **14**, 71–77.

Boone, C., Bussey, H., Greene, D., Thomas, D.Y. and Vernet, T. (1986) Yeast killer toxins:site-directed mutations implicate the precursor protein as the immunity component. *Cell*, **46**, 105–113.

Boone, C., Sdicu, A.M., Wagner, J., Degrî, R., Sanchez, C. and Bussey, H. (1990) Integration of the yeast K1 killer toxin gene into the genome of marked wine yeasts and its effect on vinification. *American Journal of Enology and Viticulture*, **41**, 37–42.

Bortol, A., Nudel, C., Fragile, E., de Torres, R., Giulietti, A., Spencer, J.F.T. and Spencer, D. (1986) Isolation of yeast with killer activity and its breeding with an industrial baking strain by protoplast fusion. *Applied Microbiology and Biotechnology*, **24**, 414–416.

Bostian, K.A., Elliot, Q., Bussey, H., Burn, V., Smith, A. and Tipper, D.J. (1984) Sequence of the preprotoxin dsRNA gene of type 1 killer yeast:multiple processing events produce a two component toxin. *Cell*, **36**, 741–751.

Bostian, K.A., Hopper, J.E., Rogers, D.J. and Tipper, D.J. (1980) Translational analysis of the killer-associated virus-like particle dsRNA genome of *S. cerevisiae*:M-dsRNA encodes toxin. *Cell*, **19**, 403–414.

Bostian, K.A., Jayachandran, S. and Tipper, D.J. (1983) A glycosylated protoxin in killer yeast: models for its structure and maturation. *Cell*, **32**, 169–180.

Bussey, H., Boone, C., Zhu, H., Vernet, T., Whiteway, M. and Thomas, D.Y. (1990) Genetic and molecular approaches to synthesis and action of the yeast killer toxin. *Experientia*, **46**, 193–201.

Bussey, H., Saville, D., Greene, D. and Tipper, D.J. (1983) Secretion of the yeast killer toxin:processing of the glycosylated precursor. *Molecular and Cellular Biology*, **3**, 1362–1370.

Bussey, H., Saville, D., Hutchins, K. and Palfree, R.G.E. (1979) Binding of yeast killer toxin to a cell wall receptor on sensitive *Saccharomyces cerevisiae*. *Journal of Bacteriology*, **140**, 888–892.

Bussey, H. and Sherman, D. (1973) Yeast killer factor:ATP leakage and coordinate inhibition of macromolecular synthesis in sensitive cells. *Biochimica et Biophysica Acta*, **298**, 868–875.

Bussey, H. and Skipper, N. (1975) Membrane-mediated killing of *Saccharomyces cerevisiae* by glycoproteins from *Torulopsis glabrata*. *Journal of Bacteriology*, **124**, 476–483.

Cansado, J., Longo, E., Agrelo, D. and Villa, T.G. (1989) Yeast species involved in spontaneous fermentation of Ribeiro wine. *Microbiologia*, **5**, 79–88.

Chvatchko, Y., Howald, I. and Riezman, H. (1986) Two yeast mutants defective in endocytosis are defective in pheromone response. *Cell*, **46**, 355–364.

Ciolfi, G. (1984) Effect of pH on the killer phenotype in *Saccharomyces cerevisiae*. *Rivista di Viticoltura e di Enologia*, **37**, 76–79.

Ciolfi, G. (1985) Distribution of killer, neutral and sensitive phenotypes during spontaneous fermentations. *Rivista di Viticoltura e di Enologia*, **38**, 242–253.

Conde, J. and Fink, G.R. (1976) A mutant of *Saccharomyces cerevisiae* defective for nuclear fusion. *Proceedings of the National Academy of Science*, **73**, 3651–3655.

Cuinier, N.C. and Gros, C. (1983) Inquiry on "Killer" yeasts in France. *Vignes et Vins*, **318**, 25–27.

de la Pena, P., Barros, F., Gaccon, S., Lazo, P.S. and Romos, S. (1981) Effects of yeast killer toxin on sensitive cells of *Saccharomyces cerevisiae*. *Journal of Biological Chemistry*, **256**, 10420–10425.

Dmochowska, A., Dignard, D., Henning, D., Thomas, D.Y. and Bussey, H. (1987) Yeast KEX1 gene encodes a putative protease with a carboxypeptidase B-like function involved in killer toxin and α-factor precursor processing. *Cell*, **50**, 593–584.

Extremera, A.L., Martin, I. and Montoya, E. (1982) A new killer toxin produced by *Saccharomyces cerevisiae*. *Current Genetics*, **5**, 17–19.

Fleet, G.H. (1990) Growth of yeasts during wine fermentation. *Journal of Wine Research*, **1**, 211–224.

Gaia, P. (1984) Lieviti in associazione contemporanea. *Vin. d'Italia*, **26**, 47–56.

Goto, K., Iwase, T., Kichise, K., Kitano, K., Totsuka, A., Obata, T. and Hara, S. (1990) Isolation and properties of a chromosome-dependent *KHR* killer toxin in *Saccharomyces cerevisiae*. *Agricultural and Biological Chemistry*, **54**, 505–509.

Gunge, N., Murata, K. and Sakaguchi, K. (1982) Transformation of *Saccharomyces cerevisiae* with linear DNA killer plasmids from *Kluyveromyces lactis*. *Journal of Bacteriology*, **151**, 462–464.

Gunge, N. and Sakaguchi, K. (1981) Intergeneric transfer of deoxyribonucleic acid killer plasmids pGKL1 and pGKL2, from *Kluyveromyces lactis* into *Saccharomyces cerevisiae* by cell fusion. *Journal of Bacteriology*, **147**, 155–160.

Gunge, N., Tamura, A., Ozawa, F. and Sakaguchi, K. (1981) Isolation and characterization of linear deoxyribonucleic acid plasmids from *Kluyveromyces lactis* and the plasmid-associated killer character. *Journal of Bacteriology*, **145**, 382–390.

Hara, S., Iimura, Y. and Otsuka, K. (1980) Breeding of useful killer wine yeasts. *American Journal of Enology and Viticulture*, **31**, 28–33.

Heard, G.M. and Fleet, G.H. (1987) Occurrence and growth of killer yeasts during wine fermentation. *Applied and Environmental Microbiology*, **53**, 2171–2174.

Henschke, P.A. (1979) Killer factors of the genus *Hansenula*, in particular *Hansenula saturnus*. PhD thesis, The University of Adelaide, Australia.

Herring, A.J. and Bevan, E.A. (1974) Virus-like particles associated with the double-stranded RNA species found in killer and sensitive strains of the yeast *Saccharomyces cerevisiae*. *Journal of General Virology*, **22**, 387–394.

Hishinuma, F., Nakamura, K., Hiral, K., Nishizawa, R., Gunge, N. and Maeda, T. (1984) Cloning and nucleic acid sequences of the linear DNA plasmids from yeast. *Nucleic Acids Research*, **12**, 7581–7597.

Hopper, J.E., Bostian, K.A., Rowe, L.B. and Tipper, D.J. (1977) Translation of the L-species dsRNA genome of the killer-associated virus-like particles of *Saccharomyces cerevisiae*. *Journal of Biological Chemistry*, **252**, 9010–9017.

Hutchins, K. and Bussey, H. (1983) Cell wall receptor for yeast killer toxin:involvement of (1-6)-β-D-glucan. *Journal of Bacteriology*, **154**, 161–169.

Imamura, T., Kawamoto, T. and Takaoka, Y. (1974) Characteristics of main mash infected by killer yeast in sake breing and the nature of its toxic factor, *Journal of Fermentation Technology, (Hakko Kogaku Zasshi)*, **52**, 293–299.

Jacobs, C.J., Fourie, I. and van Vuuren, H.J. (1988) Occurrence and detection of killer yeasts on Chenin blanc grape and grape skins. *South African Journal of Enology and Viticulture*, **9**, 28–31.

Janderova, B., Davaasurengijn, T., Vondrejs, V. and Bendova, O. (1986) A new killer brewing yeast capable of degrading dextrin and starch. *Journal of Basic Microbiology*, **26**, 627–631.

Julius, D., Brake, A., Blair, L., Kunisawa, R. and Thorner, J. (1984) Isolation of the putative structural gene for the lysine-arginine cleaving endopeptidase required for processing of yeast prepro-α-factor. *Cell*, **37**, 1075–1089.

Kagan, B.L. (1983) Mode of action of yeast killer toxins:channel formation in lipid bilayer membranes. *Nature*, **302**, 709–711.

Kagiyama, S., Aiba, T., Kadowaki, K. and Mogi, K. (1988) New killer toxins of halophilic *Hansenula anomala*. *Agricultural and Biological Chemistry*, **52**, 1–8.

Kandel, J.S. and Stern, T.A. (1979) Killer phenomenon in pathogenic yeast. *Antimicrobial Agents and Chemotherapy*, **15**, 568–571.

Kikuchi, Y., Hirai, K. and Hishinuma, F. (1984) The yeast linear DNA killer plasmids, pGKL1 and pGKL2, possess terminally attached proteins. *Nucleic Acids Research*, **12**, 5685–5692.

Kitano, K., Sato, M., Shimazaki, T. and Hara, S. (1984) Occurrence of wild killer yeasts in Japanese wineries and their characteristics. *Journal of Fermentation Technology*, **62**, 1–6.

Kitano, K., Totsuka, A. and Goto, S. (1987) dsRNA plasmid-encoded killer phenomenon in the genus *Kloeckera*. In *Proceedings of the Annual meeting of the Japanese Society of Fermentation Technology* (Osaka, 1987), 341 pp.

Kitano, K., Totsuka, A. and Hara, S. (1988) Application of killer yeasts in winemaking. In *Proceedings of the Annual Meeting of the Japanese Society of Fermentation Technology*, 96 pp.

Lolle, S.J., Skipper, N., Bussey, H. and Thomas, D.Y. (1984) The expression of cDNA clones of yeast M1 double-stranded RNA in yeast confers both killer and immunity phenotypes. *EMBO Journal*, **3**, 1383–1387.

Maule, A.P. and Thomas, P.D. (1973) Strains of yeast lethal to brewery yeasts. *Journal of the Institute of Brewing*, **79**, 137–141.

Middelbeek, E.J., Hermans, J.M.H. and Stumm, C. (1979) Production, Purification and Properties of a *Pichia kluyveri* killer toxin. *Antonie van Leeuwenhoek*, **45**, 437–450.

Middelbeek, E.J., Hermans, J.M.H., Stumm, C. and Muytjens, H.L. (1980a) High incidence of sensitivity to killer toxins using *Candida* and *Torulopsis* isolates of human origin. *Antimicrobial Agents and Chemotherapy*, **17**, 350–354.

Middelbeek, E.J., Stumm, C. and Vogels, G.D. (1980b) Effects of *Pichia kluyveri* killer toxin on sensitive cells. *Antoinie van Leeuwenhoek*, **46**, 205–220.

Mitchel, D.J., Bevan, E.A. and Herring, A.J. (1973) The correlation between dsRNA in yeast and 'killer character'. *Heredity*, **31**, 133–134.

Morace, G., Archibusacci, C., Sestito, M. and Polnelli, L. (1984) Strain differentiation of pathogenic yeasts by the killer system. *Mycopathologia*, **84**, 81–85.

Naumov, G.I. and Naumova, T.I. (1973) Comparative genetics of yeast communication. XIII. Comparative study of killer strains of *Saccharomyces* from different collections. *Genetika*, **9**, 140–145.

Nomoto, H., Kitano, K., Shimizaki, T., Kodoma, K. and Hara, S. (1984) Distruction of killer yeasts in the genus *Hansenula*. *Agricultural and Biological Chemistry*, **48**, 807–809.

Ouchi, K., Wickner, R.B., Toh-e, A. and Akiyama, H. (1979) Breeding of killer yeasts for sake brewing by cytoduction. *Journal of Fermentation Technology*, **57**, 483–487.

Palfree, R.G.E. and Bussey, H. (1979) Yeast killer toxin: purification and characterization of the protein toxin from *Saccharomyces cerevisiae*. *European Journal of Biochemistry*, **93**, 487–493.

Pasqual, M.S., Carrau, J.L., Serfini, L.A. and Dillion, A.J.P. (1990) A simple method to detect killer yeasts in industrial systems. *Journal of Fermentation and Bioengineering*, **70**, 180–181.

Petering, J.E., Symons, M.R., Langridge, P. and Henschke, P.A. (1991) Determination of killer yeast activity in fermenting grape juice by using a marked *Saccharomyces* wine yeast strain. *Applied and Environmental Microbiology*, **57**, 3232–3236.

Pfeiffer, P. and Radler, F. (1982) Purification and characterization of extracellular and intracellular killer toxin of *Saccharomyces cerevisiae* Strain 28. *Journal of General Microbiology*, **128**, 2699–2706.

Pfeiffer, P. and Radler, F. (1984) Comparison of the killer toxin of several yeasts and the purification of a toxin of type K2. *Archives of Microbiology*, **137**, 357–361.

Philliskirk, G. and Young, T.W. (1975) The occurrence of killer character in yeasts of various genera. *Antonie van Leeuwenhoek*, **41**, 147–151.

Polonelli, L. and Morace, G. (1986) Reevaluation of the yeast killer phenomenon. *Journal of Clinical Microbiology*, **24**, 866–869.

Radler, F. (1980) Les facteurs "killer" des levures. *Bulletin de l'Office International de la Vigne et du Vin*, **53**, 568–572.

Radler, F. and Knoll, C. (1988) Formation of killer toxin by apiculate yeasts and interference with fermentation. *Vitis*, **27**, 111–132.

Radler, F. and Schmitt, M. (1987) Killer toxins of yeasts: inhibitors of fermentation and their adsorption. *Journal of Food Protection*, **50**, 234–238.

Radler, F., Schmitt, M.J. and Meyer, B. (1990) Killer toxin of *Hanseniaspora uvarum*. *Archives of Microbiology*, **154**, 175–178.

Rosini, G. (1983) The occurrence of killer characters in yeasts. *Canadian Journal of Microbiology*, **29**, 1462–1464.

Rosini, G. (1985) Effect of a 'killer' yeast *Saccharomyces cerevisiae* on a sensitive strain of the same species, producing no H_2S and selected for vinification in a mixed culture. *Bulletin de l'Office International de la Vigne et du Vin*, **58**, 214–217.

Rosini, G. (1985) Interaction between killer strains of *Hansenula anomala* var *anomala* and *Saccharomyces cerevisiae* yeasts species. *Canadian Journal of Microbiology*, **31**, 300–302.

Rosini, G., Cianti, M. and Vaughan, A.E. (1988) Segrantino DIC wine correlation between the cultures of *Saccharomyces cerevisiae* isolated from the wines and those found in the winemaking facilities. *Annali di microbiologia ed Enzimologia*, **38**, 171–180.

Rothman, J.H. and Stevens, T.H. (1986) Protein sorting in yeast: mutants defective in vacuole biogenesis mislocalize vacuolar proteins into the late secretory pathway. *Cell*, **47**, 1041–1051.

Schmitt, M., Brendel, M., Schwarz, R. and Radler, F. (1989) Inhibition of DNA synthesis in *Saccharomyces cerevisiae* by yeast killer toxins KT28. *Journal of General Microbiology*, **135**, 1529–1536.

Seki, T., Choi, E.H. and Ruy, D. (1985) Construction of killer wine yeast strain. *Applied and Environmental Microbiology*, **49**, 1211–1215.

Shimizu, K., Adachi, T., Kitano, K., Shimazaki, T., Totsuka, A., Hara, S. and Dittrich, H.H. (1985a) Occurrence and properties of 'wild' yeasts during winemaking. *Wein-Wissenschaft*, **40**, 258–270.

Shimizu, K., Adachi, T., Kitano, K., Shimazaki, T., Totsuka, A., Hara, S. and Dittrich, H.H. (1985b) Killer properties of wine yeasts and characterization of killer wine yeasts. *Journal of Fermentation Technology*, **63**, 421–429.

Shimizu, K., Hara, T., Yokomori, Y., Adachi, T. and Akiyama, H. (1986) The breeding and use of killer wine yeasts in winemaking. In *Proceedings of XIXe congres International de la Vigne et du Vin* (Santiago, 1986), pp. 131–155.

Skipper, N. and Bussey, H. (1977) Mode of action of yeast toxins: energy requirement for *Saccharomyces cerevisiae* killer toxin. *Journal of Bacteriology*, **129**, 668–677.

Sriprakash, K.S. and Batum, C. (1984) Possible chromosomal location for the killer determinant in *Torulopsis glabrata*. *Current Genetics*, **8**, 115–119.

Stark, M.J.R. and Boyd, A. (1986) The killer toxin of *Kluyveromyces lactis*: characterization of the toxin subunits and identification of the genes which encode them. *EMBO Journal*, **5**, 1995–2002.

Stark, M.J.R., Boyd, A., Mileham, A.J. and Romanos, M.A. (1990) The plasmid-encoded killer system in *Kluyveromyce lactis*: a review. *Yeast*, **6**, 1–29.

Starmer, W.T., Ganter, P., Aberdeen, V., Lachance, M.A. and Phaff, H.J. (1987) The ecological role of killer yeasts in natural communities of yeasts. *Canadian Journal of Microbiology*, **33**, 783–796.

Stumm, C., Hermans, J.M., Middelbeek, E.J., Croes, A.F. and de Vries, G.J.M.L. (1977) Killer-sensitive relationships in yeast from natural habitats. *Antonie van Leeuwenhoek*, **43**, 125–128.

Sugisaki, Y., Gunge, N., Sakaguchi, K., Yamasaki, M. and Tamura, G. (1983) *Kluyveromyces lactis* killer toxin inhibits adenylate cyclase of sensitive yeast cells. *Nature*, **304**, 464–466.

Sugisaki, Y., Gunge, N., Sakaguchi, K., Yamasaki, M. and Tamura, G., (1984) Characterization of a novel killer toxin encoded by a double-stranded linear DNA plasmid of *Kluyveromyces lactis*. *European Journal of Biochemistry*, **141**, 241–245.

Suzuki, C. and Nikkuni, S. (1989) Purification and properties of the killer toxin produced by a halotolerant yeast, *Pichia farinosa*. *Agricultural Biological Chemistry*, **53**, 2599–2604.

Suzuki, C., Yamada, K., Okada, N. and Nikkuni, S. (1989) Isolation and characterization of halotolerant killer yeasts from fermented foods. *Agricultural Biological Chemistry*, **53**, 2593–2598.

Thornton, R.J. (1986) Genetic characterization of New Zealand and Australian wine yeasts. *Antonie van Leeuwenhoek*, **52**, 97–103.

Tiourina, L.V., Bourjan, N.I. and Skarikova, T.K. (1980) Emploi des cultures pures du phenotype "killer" dans la fermentation des mouts de raisin. *Bulletin de l'Office de la Vigne et du Vin*, **53**, 573–576.

Tipper, D.J. and Bostian, K.A. (1984) Double-stranded ribonucleic acid killer systems in yeasts. *Microbiological Reviews*, **48**, 125–156.

Tokunaga, M., Kawamura, A. and Hishinuma, F. (1989) Expression of pGKL killer 28K subunit in *Saccharomyces cerevisiae*: identification of 28K subunit as a killer protein. *Nucleic Acids Research*, **17**, 3435–3446.

Tommashino, M. and Galeotti, C.L. (1988) Genome organisation of the killer plasmid pGKL2 from *Kluyveromyces lactis*. *Nucleic Acids Research*, **16**, 5863–5878.

Tredoux, H.G., Tracey, R.P. and Tromp, A. (1986) Killer factor in wine yeasts and its effect on fermentation. *South African Journal for Enology and Viticulture*, **7**, 105–112.

van Vuuren, H.J.J. and Wingfield, B.D. (1986) Killer yeasts—cause of stuck fermentations in a wine cellar. *South African Journal for Enology and Viticulture*, **7**, 113–118.

Wesolowski, M., Tanguy-Rougeau, C. and Fukuhara, H. (1988) A nuclear gene required for the expression of the linear DNA-associated killer system in the yeast *Kluyveromyces lactis*. *Yeast*, **4**, 71–81.

White, J.H., Butler, A.R. and Stark, M.J.R. (1989) *Kluyveromyces lactis* toxin does not inhibit yeast adenyl cyclase. *Nature*, **341**, 66–668.

Wickner, R.B. (1978) Twenty-six chromosomal genes needed to maintain the killer double-stranded RNA plasmid of *Saccharomyces cerevisiae*. *Genetics*, **88**, 419–425.

Wickner, R.B. (1979) The killer double-stranded RNA plasmids of yeasts. *Plasmid*, **2**, 303–322.

Wickner, R.B. (1986) Double-stranded RNA replication in yeast: the killer system. *Annual Review of Biochemistry*, **55**, 373–395.

Wickner, R.B. and Leibowitz, M.J. (1976) Two chromosomal genes required for killer expression in killer strains of *Saccharomyces cerevisiae*. *Genetics*, **82**, 429–442.

Wingfield, B.D., Southgate, V.J., Pretorius, I.S. and van Vuuren, H.J.J. (1990) A K2 Neutral *Saccharomyces cerevisiae* strain contains a variant K2 M. Genome. *Yeast*, **6**, 159–169.

Yamamoto, T., Hiratani, T., Hirata, H., Imai, M. and Yamaguchi, H. (1986) Killer toxin from *Hansenula mrakii* selectively inhibits cell wall synthesis yeast *Saccharomyces cerevisiae*. *FEBS Letters*, **197**, 50–54.

Yokomori, Y., Akiyama, H. and Shimizu, K. (1988a) Toxins of wild *Candida* killer yeast with a novel killer property. *Agricultural and Biological Chemistry*, **52**, 2797–2801.

Yokomori, Y., Akimyama, H. and Shimizu, K. (1988b) Isolation of a wild *Candida* killer yeast with a novel killer property. *Agricultural and Biological Chemistry*, **52**, 2791–2796.

Yokomori, Y., Akiyama, H. and Shimizu, K. (1989) Breeding of wine yeasts through protoplast fusion. *Yeast*, **5** (special issue), 145–150.

Yokomori, Y., Akiyama, H. and Shimizu, K. (1990) Breeding of wine yeasts through proptoplast fusion between champagne and wine yeasts. In *Fermentation Technologies, Industrial Applications*, edited by Pak-Lam Yu, pp. 258–263. London and New York: Elsevier Applied Science.

Young, T.W. (1981) The genetic manipulation of killer character into brewing yeast. *Journal of the Institute of Brewing*, **87**, 292–295.

Young, T.W. (1987) Killer yeasts. In *The Yeasts* Volume 2, Second Edition, edited by A.H. Rose and J.S. Harrison, pp. 131–164. London: Academic Press.

Young, T.W. and Yagiu, M. (1978) A comparison of the killer character in different yeasts and its classification. *Antonie van Leeuwenhoek*, **44**, 59–77.

Zorg, J., Killian, S. and Radler, F. (1988) Killer toxin producing strains of the yeasts *Hanseniaspora uvarum* and *Pichia kluyveri*. *Archives of Microbiology*, **149**, 261–267.

ACKNOWLEDGEMENT

The author is grateful to President Toshiaki Kinoshita, Sainte Neige Wine Co. Ltd, for permission to publish this manuscript and also to Mr Hideki Yamanaka and Miss Noriko Mukoyama, Sainte Neige Wine Co. Ltd, for typing the manuscript.

Chapter 9

GENETIC IMPROVEMENT OF WINE YEASTS

PIERRE BARRE*, FRANCOISE VEZINHET*, SYLVIE DEQUIN* and BRUNO BLONDIN**

**Institut des Produits de la Vigne, Institut National de la Recherche Agronomique, 34060 Montpellier Cedex 1, France*
***Ecole Nationale Supérieure Agronomique, 34060 Montpellier Cedex 1, France*

Chapter Outline

Chapter Outline (Continued)

INTRODUCTION

Alcoholic fermentation is one of the main phases in wine production. It is usually conducted with yeasts belonging to the species *Saccharomyces cerevisiae*. Problems encountered during this fermentation can be broadly classified into two categories: (i) kinetic phenomena such as length of fermentation, sluggish or stuck fermentation, excessive momentary demands for cooling due to very high fermentation rates, making it impossible to maintain an adequate temperature, etc.; and (ii) analytic and especially organoleptic characteristics of the product (particularly off-flavors).

There are three general ways by which fermentation can be modified: (i) by changing the composition of the medium (e.g. by adding assimilable nitrogen), which is obviously limited by the alimentary status of the product and its related legal and hygienic requirements; (ii) by altering external parameters (temperature, aeration, etc.); and (iii) by modifying the composition of the yeast flora conducting fermentation (Rankine, 1968; Radler, 1973), which, if not changed, consists of a mixture of wild strains that may or may not succeed one another sequentially (Fleet 1990). The last approach calls for expertise in obtaining fermentation by pure strains, involving technological restrictions and control methods that are not always respected.

Historically, strain selection from wild populations has provided the biological material for controlling alcoholic fermentation by inoculation and fermentation with pure strains. It is not within the scope of the present article to describe the development of this research, which started at the end of the nineteenth century (Ventre, 1911). It gradually led to the selection of about thirty strains that are now distributed throughout the world in the form of active dry yeast, not including strains whose development has remained confined to particular winemaking regions. This array of strains does not completely satisfy winemakers who sometimes need an additional property or, inversely, they want a property removed from a strain that otherwise gives full satisfaction. Moreover, they may desire a property that has not yet been described in *Saccharomyces*. These demands are met using the technologies of genetic manipulation. These procedures are not at all in conflict with the conventional techniques of selection from wild populations, but rather the two approaches should be used in parallel and should complement each other. Information concerning the genetic approach have been already given by Snow (1983) and Johnston (1990).

In this chapter, we describe the genomic context of commercial strains, which differs quite distinctly from that of laboratory haploid strains used to determine the genetics of *Saccharomyces cerevisiae*. We then deal with the manner of analyzing a problem before deciding on a particular manipulation: each technological problem requires specific analysis, but a whole series of questions must be asked before any decision can be made. We examine different manipulation methods, which are illustrated with representative examples of genetic procedures carried out on commercially important strains. Lastly, we suggest several possibilities for the future, taking into account regulations and consumer needs. The reader is referred to the recent review by Preterious and van der Westhuizen (1991) which also considers the impact of yeast genetics on the wine industry.

GENETIC BACKGROUND OF WINE YEASTS

The genetic material of yeasts consists of chromosomal, mitochondrial, and plasmid DNA. Genetic information is carried in some strains by double stranded RNA ("killer" yeasts).

Chromosomal DNA

Chromosomal DNA represents 80–85% of the total DNA in haploid strains of *S. cerevisiae*. In genetically defined laboratory strains, DNA is distributed among 16 chromosomes whose sizes vary from 250 to 2000 kb. Commercial strains are never haploid but, rather, diploid or polyploid, with the possibility of aneuploidy. Bakalinsky and Snow (1990) have shown that a single strain can be diploid for some chromosomes and triploid or quadruploid for others. This property affects sporulating ability; 0–75% ascus formation can be observed on a sporulation medium, and spore viability is also highly variable, ranging from 0–98%.

Ploidy

Ploidy higher than n implies that there are at least two copies of a given gene (two alleles) in the nuclear genome. This confers greater genetic stability on strains in the natural environment, but it also reduces possibilities for genetic improvement. For example, direct mutagenesis in a commercial strain does not allow expression of a recessive mutation unless a highly effective screening procedure makes it possible to isolate homozygotes for the gene of interest; these homozygotes are generated by mitotic gene conversion or mitotic recombination.

Separation of chromosomal DNA by pulsed-field electrophoresis (Carle and Olson, 1985) has revealed considerable polymorphism among strains of enological interest (Blondin and Vezinhet, 1988; Petering, Langridge and Henschke, 1988). The number of bands and their position and relative density vary from one strain to another (Figure 9.1). These differences can be used as a tool for characterizing strains and can be applied to: show that a strain is

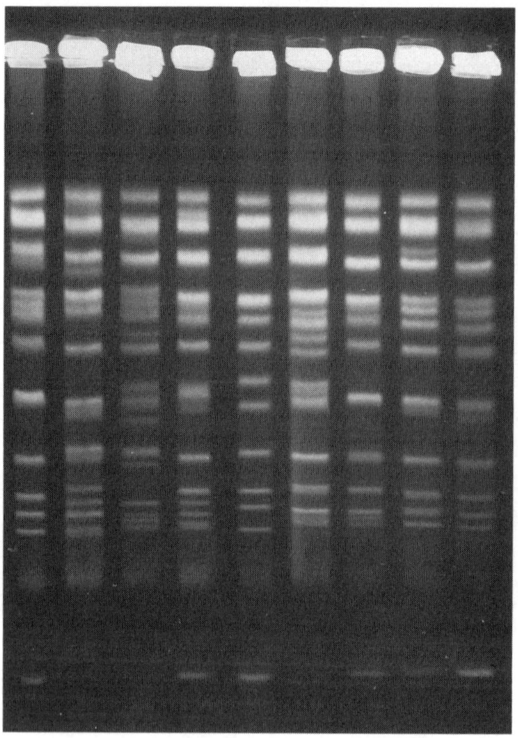

FIGURE 9.1 Electrophoretic separation of chromosomal DNA of *Saccharomyces cerevisiae* (Transverse alternating field electrophoresis - TAFE) Lanes 1 to 4 and 6 to 9: commercial wine yeasts. Lane 5: haploid laboratory strain.

identical to a recorded strain; monitor the manufacture of industrial biomass; analyze indigenous yeast populations in a given region; study population dynamics during vinification; and, particularly, monitor the efficacy of yeast addition. Chromosomal polymorphism can be attributed partly to size differences between homologous chromosomes because several probes specific for a given chromosome in a laboratory strain can hybridize with two distinct bands of a commercial strain. Analysis using specific probes has revealed complex situations; for example a probe from a laboratory strain may not hybridize with any of the bands of a commercial strain. In contrast, a chromosomal band may hybridize with two probes from two different chromosomes of a laboratory strain (Bidenne *et al.*, 1992). This polymorphism is the sign of great genetic variability in *S. cerevisiae*, especially among the wild strains obtained from wine fermentation. Less variation occurs in laboratory strains which have originated from a small number of common ancestors. Wine strains also differ from bottom-fermentation brewing strains, which are much more homogeneous (Pedersen, 1986). Thus, *S. cerevisiae* represents a natural genetic pool that has, perhaps, been underestimated and whose systematic exploitation could be of great interest.

Homothallism and heterothallism

Meiotic segregants from commercial strains diploidize with high frequency, which indicates a high rate of homothallism (Thornton and Eschenbruch, 1976). The only haploid stage is the ascospore before cell multiplication. This fact must be taken into account when applying techniques such as hybridization or rare mating to commercial strains. With some strains it is possible to obtain stable meiotic haploid segregants that can, therefore, be crossed with laboratory strains carrying auxotrophic markers, yielding 2/2 segregation (Vezinhet, unpublished results). This property indicates that these strains are heterothallic.

Heterozygosity

Considerable heterozygosity (different alleles of the same gene) has been observed in both homothallic and heterothallic strains. Meiotic segregants of these strains show polymorphism with regard to phenotype (Romano *et al.*, 1985) and karyotype (Pardo, Blondin and Vezinhet, 1989). Heterozygosity is a disadvantage if a haploid stage is essential for genetic improvement. Large changes often occur in strains merely as a result of getting meiotic segregants (forming haploids from diploids) and crossing them back (forming diploids from haploids). On the other hand, heterozygosity can be used to advantage when removing a defect from a strain, such as excessive foaming. The genome of a strain was changed by crossing meiotic segregants to obtain clones that reduce foam production while retaining the winemaking properties of the commercial strain (Vezinhet, 1989).

Mitochondrial DNA

In laboratory strains, mitochondrial DNA consists of circular molecules 75 000 base-pairs long. These molecules can be extracted from a cell population, purified, and analyzed by classical electrophoresis after cleavage with restriction enzymes. Such analyses show substantial polymorphism in the restriction patterns of commercial strains (Dubourdieu *et al.*, 1987; Hallet *et al.*, 1988), and constitutes another technique for identifying strains. Vezinhet, Blondin and Hallet, (1990) have comparatively evaluated chromosomal DNA profiles and mitochondrial DNA polymorphism as tools for the identification of enological strains of *S. cerevisiae*.

Mitochondrial DNA encodes proteins of the respiratory chain and theoretically should have no effect on alcoholic fermentation characteristics. However, an effect of this kind has been demonstrated in brewing strains (Hammond and Eckersley, 1984). Consequently, this possibility should be considered when transferring genetic material from one strain to another by hybridization or cytoduction techniques.

Plasmid DNA

The nucleus of *S. cerevisiae* contains a 2μm plasmid, which has scarcely been studied in wine strains. Neuville, Bonneu and Aigle, (1990) found it in 21 out of 22 wine strains analyzed. The restriction patterns of the plasmid isolated from these strains showed very little polymorphism, in contrast with chromosomal and mitochondrial DNA. Use of the 2 μm plasmid as a cloning vector, which is very frequent in laboratory strains, could be applied to commercial wine strains.

Killer RNA

The genetic determinants of the killer property consist of nuclear genes and two double-stranded cytoplasmic RNAs (L and M) (Young, 1987). The nuclear genes and L RNA are generally present in strains, and only M RNA is specific to neutral and killer strains. These strains have been characterized among indigenous wine strains. The most widely distributed toxin is the K2 form. The cDNA of the gene encoding the toxin can be used as an autoselection system in recombinant DNA technology.

ANALYSIS OF THE PROBLEM

The first step is to gather all biochemical and physiological data related to the desired property. The more information there is, the more feasible genetic alteration becomes. If this information is insufficient, it is usually preferable to abandon the genetic approach and carry out selection from wild flora. This would be the case, for example, when seeking to improve complex aroma characteristics whose chemical bases and, consequently, corresponding metabolic pathways and regulation, are very poorly known. In contrast, for example, the right kind of information is available to genetically introduce the property of malolactic fermentation into yeasts. Thus, the monoenzymatic character of the reaction and the single polypeptide chain forming the enzyme (Schutz and Radler, 1973) indicated that there is only one structural gene; sequencing the N terminus of the enzyme led to the development of a family of specific probes for the gene (Lonvaud, 1990; Renault, 1990); possibilities for malic acid permeation have been demonstrated (Salmon, 1987), implying that entry of substrate into the cell does not impede intracellular degradation.

Another group of essential data concerns genetic control of the desired property. Is the character determined by a single gene or several genes? Will the modification be dominant or recessive? Does it involve removal or amplification of a gene or modification of its product? What is the availability of the gene(s) to be inserted. Are they present in an haploid laboratory strain, in another species of *Saccharomyces*, in another kind of yeast, or even in a non-yeast organism?

Another important criterion is the existence of an easy-to-use method for screening for the desired property. Often, as many as several thousand clones must be tested to isolate those that have acquired the desired property. An adequate screen can vary from highly feasible to practically impossible depending on whether it involves, for instance, cell growth or the presence of a complex aroma characteristic.

Lastly, it should be noted that even if genetic alteration is successful, there is still the problem of technological validation of the new strain and also of its conformity with the original strain (except for the modified property, of course). Validation requires large-scale trials and is always long and painstaking, as when validating newly isolated wild strains. This constraint continually impels the geneticist to seek manipulation methods that are increasingly precise in the genome so that determinants unrelated to the desired property are not altered and, consequently, other winemaking abilities are not changed.

MAIN TYPES OF GENETIC MANIPULATION AND THEIR APPLICATION TO WINE YEASTS

We describe here several concrete cases illustrating the main types of genetic manipulation (Table 9.1). It is not within the scope of this chapter to give a detailed description of the methods used. For further information, the reader can consult Guthrie and Fink (1991) as well as Spencer, Spencer, and Bruce (1989) on classical genetics and Rose (1987) on DNA transfer, and the literature cited in the text. With regard to technical matters, we restrict ourselves to several details that must be indicated because of the nature of wine yeast.

TABLE 9.1 Some applications of genetic manipulation to wine yeasts.

Modified character	*Type of manipulation*	*Reference*
Leucine auxotrophy (lower production of isoamyl alcohol)	Chromosomal DNA mutagenesis	Rous and Snow (1983)
Introduction of markers	Mitochondrial DNA mutagenesis	Vezinhet (1985)
Introduction of flocculation	Hybridization	Thornton (1985)
Introduction of flocculation and elimination of H_2S formation	Hybridization	Romano *et al.* (1985)
Introduction of markers	Rare mating and cytoduction	Loiseau (1986)
Schizosaccharomyces-like metabolism of malic acid	Protoplast fusion	Carrau (1982)
Introduction of killer character	Protoplast fusion + cytoduction	Seki, Choi and Ryu (1985)
Introduction of flocculation	Cytoduction	Vezinhet and Barre (1988)
Introduction of a malolactic activity	Gene transfer	Williams *et al.* (1984)
Construction of a K1/K2 killer strain	Gene transfer	Boone *et al.* (1990)
Introduction of markers	Gene transfer	Petering *et al.* (1991)

Mutagenesis

This approach is often used when the biochemical and genetic bases of the alteration are poorly known or, inversely, when the product of a particular gene is to be eliminated, amplified, or altered. The most frequently used mutagens are physical (UV radiation at 254 nm) or chemical (ethylmethanesulfonate and nitrosoguanidine). Once mutations are induced, many problems arise with attempts to exploit them. First, since most mutations are recessive, they are not expressed in a polyploid genome. Direct application of mutagenesis to such genomes has little chance of success because different alleles have to be affected simultaneously. Consequently, stable haploid lineages must be used (as in the case of an initial heterothallic diploid strain), or else mutagenesis must be carried out directly on spores. In any case, it is advisable to perform rediploidization so that the strain is homozygous for the mutation, and expression is assured. Second, screening methods must be simple enough to detect mutants that are present at a frequency of 10^{-4} to 10^{-5} in the surviving population. Third, because of the imprecision of the procedure, unrelated genes can be affected and this can cause undesirable changes in winemaking abilities.

They can be affected also by the haploidization-rediploidization process which, as mentioned above, can cause significant genomic variations. Lastly, reversions are possible, which can jeopardize the stability of the mutation.

Mutagenesis has produced auxotrophs for valine, leucine and isoleucine (Rous and Snow, 1983), which give reduced production of isobutyl, active amyl, and isoamyl alcohols during fermentation. Mutagenesis with ethylmethanesulfonate was performed on sporulated cultures of a homothallic strain, yielding diploids that were homozygous for the auxotrophy after diploidization of mutant spores. An auxotroph for leucine (lacking α-isopropylamate dehydratase) has been constructed, which reduces isoamyl alcohol production by 50% and results in wines of satisfactory quality. The strain seems to be suitable for producing wines for distillation.

Mutagenesis with U.V. radiation has been performed on cryophilic laboratory strains to improve their fermentation performance (Kuzewicz, 1976). Genetic analysis showed the strains to be diploid or polyploid (Kuzewicz and Johnston, 1980). Mutants were obtained whose low-temperature fermentation performance was increased by 10–15%.

Knowledge of the biosynthetic pathway of ergosterol (a component of the cytoplasmic membrane of *Saccharomyces*) made it possible to obtain mutations which conferred the ability to secrete prenyl alcohols (farnesol and geraniol) (Chambon *et al.*, 1990). Complementary mutations aimed at preventing reversion and partially removing ergosterol auxotrophy were also obtained. Alcoholic fermentation with a diploid strain homozygous for all these mutations gives a product with a strong muscatel-like aroma due to the formation of geraniol and linalol (Javelot *et al.*, 1990).

The last example of an application of mutagenesis concerns the use of a mutagen (manganese chloride) that is specific for mitochondrial DNA. The objective was to introduce resistance to mitochondrion-specific inhibitors so that strains marked in this manner could easily be differentiated and monitored among indigenous strains (Vezinhet, 1985; Loiseau *et al.*, 1987). Since some of these inhibitors are antibiotics, it must be said that the mechanism of resistance is totally different from that generally introduced with similar markers. The site of action of the inhibitor was modified by altering mitochondrial DNA. The modification can only be expressed in the organelle *in situ* and is not subject to natural dissemination. It should also be mentioned that spontaneous mutants showing the same combination of resistance have regularly been isolated. The modification confers no selective advantage in natural habitats, since the resistance only functions in laboratory studies where the inhibitor is added to the medium for growing the yeasts. Several studies have successfully used strains of *S. cerevisiae* carrying mitochondrial markers to examine the effect of vinification factors upon the dominance or non-dominance of inoculated *S. cerevisiae* during grape juice fermentation, in relation to indigenous strains (Loiseau *et al.*, 1987; Delteil and Aizac, 1988).

Hybridization

This is the first method to consider when the desired property is discovered in another strain, especially when the property is under multigenic control. Its main disadvantage is its imprecision and consequent disregard for the genome of the original strain. There are of course procedures, such as backcrossing, that can potentially regenerate the original genome, as shown for example by Thornton (1985). The regeneration is often only partial due to meiotic and mitotic recombination phenomena.

Hybridization is usually only feasible with two haploid cells of opposite mating types belonging to the same species. However, in addition to true hybridization, there are procedures that can overcome these restrictions at least partially, i.e. rare mating and protoplast fusion. Lastly, cytoduction is a very particular case of hybridization in that it allows transfer of cytoplasmic material alone, and sometimes a limited quantity of nuclear material, from the donor strain.

True hybridization

Ideally, true hybridization involves a cross between two heterothallic diploid strains of opposite mating types with high spore viability, which can thus produce stable haploid lineages. Two homothallic strains can also be used, on condition that the cultures are sporulating and fusion can occur between spore germination and diploidization. However, hybrids are obtained with lower frequency in this case.

Hybridization has often been used blindly to eliminate an undesirable property from a strain or to confer a particular ability. Eschenbruch *et al.* (1982) achieved this objective by crossing a commercial strain with a laboratory strain and screening the hybrids. On the same basis, Thornton (1982) crossed a diploid commercial strain with a heterothallic haploid laboratory strain. The segregants were then crossed among themselves for three generations, in parallel with a screening procedure. The author thereby managed to isolate clones with higher ethanol yields and sulfur dioxide resistance, without other apparent significant differences from the original strain. This type of procedure tends to produce a whole range of variation from the original strain, within which screening is carried out. However, it cannot be excluded that equivalent results could be obtained by conducting similar screening on indigenous populations.

Generally, hybridization provides a solution for a precise objective, based on a correctly defined genetic context. Romano *et al.* (1985) constructed non-H_2S producing flocculent hybrids by conjugating spores of two homothallic commercial strains that analysis had shown to be homozygous for the desired characters: this combination is rarely found in indigenous populations. By conjugating spores, Thornton (1985) introduced the flocculent property of a heterothallic haploid laboratory strain into a non-flocculent homothallic commercial strain. He then attempted to regenerate the original genome of the commercial strain by five successive backcrosses, thereby obtaining clones that did not differ from the original strain with regard to fermentation kinetics and organoleptic properties. The modified strain containing the flocculent property should allow better separation of yeast cells from the wine, thereby facilitating the process of clarification.

Rare mating

This method is based on the fact that in a population of diploid or polyploid cells, some individuals become homozygous for the mating type and can, therefore, be crossed with a haploid cell of the opposite type. This cross is a rare event that can only be detected with a highly effective screen. The procedure has the great advantage of allowing hybridization between strains of different ploidy and avoids the need for sporulation, which often does not work with commercial strains and runs the risk of genome alteration mentioned above. A very detailed description of its use is given by Hammond and Eckersley (1984), who transferred the killer property to a brewing strain. Rare mating has been used in our laboratory and was found to be effective with wine strains (Loiseau, 1986).

Protoplast fusion

Protoplast fusion can be used to obtain interspecific and even intergeneric hybrids via a mechanism that has not been completely elucidated, but is similar to somatic fusion. The final result is often a large predominance of the genome of one of the strains plus several genes of the other strain. However, considerable difficulties have been encountered in developing a coherent screen and in stabilizing the fusion products. Seki *et al.* (1985) have used this method in combination with cytoduction to introduce the killer property into a sensitive commercial strain. Carrau *et al.* (1982) have described the fusion of *Schizosaccharomyces pombe* with *S. cerevisiae* to obtain a *Saccharomyces*-like strain that expresses *Schizosaccharomyces*-like metabolism of malic acid.

Cytoduction

Conde and Fink (1976) isolated *kar* mutants in which nuclear fusion does not occur after cell fusion (Figure 9.2). These mutants can be used in the following procedures: (i) transfer of cytoplasmic material without transfer of nuclear material; and (ii) simultaneous transfer of one or two chromosomes at a frequency depending on the size of the chromosome involved (Dutcher, 1981). The latter characteristic can be used for partial and selective transfer of nuclear material.

We have used the first procedure for marking purposes, introducing mitochondria from a haploid *kar* laboratory strain, with mutations conferring resistance to specific inhibitors, into a commercial strain (Loiseau, 1986). However, the results in technological validation trials were less favorable than with strains obtained by direct mutagenesis with $MnCl_2$ or by selection of spontaneous mutants. We have also used the second procedure to introduce a single chromosome 1, containing gene FLO5, which confers the flocculent

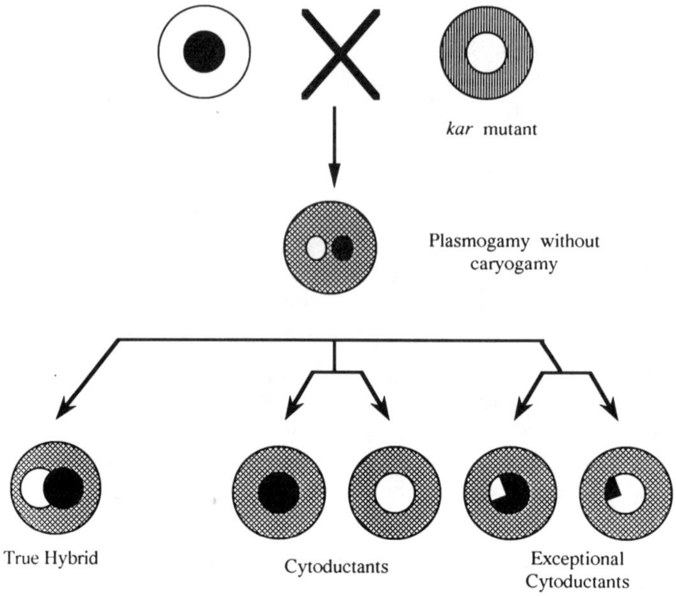

FIGURE 9.2 Schema of cytoduction. Progeny of a cross between a strain and a *kar* mutant is composed of true hybrids (plasmogamy and caryogamy), cytoductants (plasmogamy without caryogamy), exceptional cytoductants (plasmogamy and partial caryogamy).

characteristic into a strain used in producing sparkling wine, thereby facilitating champagnization (Vezinhet and Barre, 1988; Vezinhet, Blondin and Barre, 1991).

Gene Transfer

Since the first studies on yeast transformation (Hinnen, Hicks and Fink, 1978), tools and methods for cloning microorganisms have steadily been improved until it is now relatively easy to manipulate genes in *S. cerevisiae*. The main advantage of molecular cloning techniques is their genetic precision, which allows discrete transfer of clearly defined DNA sequences. This precision can be used to obtain mutants: a cloned and disrupted wild-type gene can replace a host gene by homologous recombination. Molecular biology also makes it possible to affect the function of existing genes. *In vitro* modifications of sequences can modulate gene expression as well as the function of its product. Recombinant DNA methods can also be used to introduce and express genes from organisms other than yeast, which considerably widens the possibilities for strain improvement. However, these methods are limited to step-by-step or gene-by-gene procedures. Theoretically, there is no obstacle to introducing several genes into yeast, but it is not feasible in practical terms to transfer a character that depends on a large number of genes.

Vectors

The first step consists in inserting the gene of interest into a vector, which is usually a shuttle plasmid that can replicate in *Escherichia coli* and in the yeast. Depending on the structural components contained in the vector, it can either replicate autonomously in the yeast or integrate into the chromosomal genome. Table 9.2 shows the main types of vectors available. Yeast integrating plasmids (YIp) cannot replicate in yeast and are stabilized when integrated into chromosomal DNA by homologous recombination. The ability to replicate autonomously can be conferred by adding autonomous replicating sequences (ARS) yielding yeast replicating plasmids (YRp), or by adding 2 μm plasmid sequences to produce yeast episomal plasmids (YEp). However, these high copy number vectors are very unstable (especially YRp) and are rapidly lost in the absence of selection pressure. Addition of a centromeric sequence to a replicating vector confers high stability and reduces the copy number per cell to a level near one (yeast centromeric plasmid, YCp). Various reviews (Struhl, 1983; Parent, Fenimore and Bostian, 1985; Boguslawski, 1985) give detailed descriptions of the properties of these yeast cloning vectors.

Transformation

Transformation involves placing the above-mentioned vectors within the yeast. The oldest methods require the formation of spheroplasts, which are treated with calcium chloride and polyethylene glycol to enable them to absorb exogenous DNA (Hinnen, *et al.*, 1978). More rapid techniques based on the treatment of whole cells with lithium salts are also used (Ito *et al.*, 1983). More recently, electroporation has been used to transform intact cells (Delorme, 1989), including baker's yeast (Gysler, Kneuss and Niederberger, 1990). Transformation efficiency depends on the method used, the strain, and the transformant selection system. Brewery strains are known to transform with low efficiency and the scarce data in the literature on wine strains indicate that their transformation efficiency is also low (Hadfield, Cashmore and Meacock, 1986; Casey, Xiao and Rank, 1988).

TABLE 9.2 Main types of vectors available for yeast transformation.

Plasmid type	*Maintenance mode*	*Yeast sequence*	*Copy number*	*Mitotic stability*
YIp integrating	Integrative	chromosomal	1[a]	+
YRp replicating	Autonomous	ARS	3–30	–
YEp episomal	Autonomous	2 μ	5–40	±
YCp centromeric	Autonomous	ARS,CEN	1	+

[a]The copy number depends on the number of integration sites.

Selection systems

Selection systems make it possible to isolate transformants present in very low proportions in a cell suspension. Systems used for laboratory strains such as the markers LEU2, URA3, or TRP1 which complement corresponding auxotrophic mutations in the recipient cell, are practically unusable in wine strains because of their ploidy. Consequently, dominant resistance marker systems had to be developed for industrial strains. These markers are designed to identify clones that have received exogenous DNA, to eliminate non-transformed cells, and also to exert selection pressure when using high copy number plasmids, which are often unstable. Recombinant selection systems should be usable under medium conditions resembling winemaking to allow analysis of the physiological consequences of gene insertion. Under these criteria, the choice of markers is one of the key elements in gene manipulation. Twenty dominant resistance markers have been described (Rank, Casey and Xiao, 1988), some of which have been tested on industrial strains. The most frequently used are the following:

1. Genes from bacterial transposons conferring resistance to a drug, i.e. resistance to chloramphenicol (Hadfield, *et al.*, 1986), to G418 (Jimenez and Davies, 1980; Webster and Dikson, 1983), to hygromycin B (Gritz and Davies, 1983; Kaster, Burgett and Ingolia, 1984), and to phleomycin (Gatignol, Baron and Tiraby, 1987). To achieve maximum transformation efficiency, many copies of the genes must be present or else they must be under the control of a strong yeast promoter.
2. Yeast genes conferring resistance to different agents. Gene SMRI-410, a modified allele of ILV2 encoding resistance to the herbicide sulfometuron methyl, functions as a single copy and has been tested in wine strains (Casey *et al.*, 1988). Gene CUP1, conferring resistance to copper (Henderson, Coy and Tubb, 1985), which is frequently used in brewery strains, cannot always be used in wine strains, since some of them are copper resistant (Thornton and Bunker, 1989). Immunity to killer toxin K1 has also been used for selectable markers in industrial strains (Bussey and Meaden, 1985). Cells transformed by a plasmid containing the K1 killer/immunity precursor gene can be selected in the presence of active toxin. Moreover, the system is autoselective, since cells that have lost the plasmid are eliminated by the toxin produced by recombinant cells.

Stabilization of transformants

Stabilization is an essential step in the construction of recombinant industrial strains. In addition, regulations require that all DNA sequences unnecessary for gene expression be removed, notably those from *E. coli*.

The most frequently used stabilization system involves integrating the sequence of interest into the genome by homologous recombination. Different strategies are possible. For example, during cotransformation (Rudolph, Koening-Rauseo and Hinnen, 1985) the gene is introduced in linear form and transformants are selected using a replicating plasmid containing a dominant resistance marker that can easily be removed by segregation. Another possibility is two-step integration using an integrating vector containing the selectable marker and the gene of interest (Hinnen, *et al.*, 1978; Orr-Weaver, Szostak and Rothstein, 1981). This strategy makes it possible to remove the vector sequences after integration (Figure 9.3). The integration target can be the homologous sequence of a modified gene, in which case it is often necessary to replace all the alleles of a polyploid strain by successive transformations. With a heterologous gene, the integration target must be a sequence whose inactivation in no way changes the industrial performance of the strain. Lastly, expression of the inserted gene can be enhanced by integration into repeated sequences. This method was used to integrate more than 20 copies of the acetolactate decarboxylase gene into the rDNA of a brewery strain (Fuji *et al.*, 1990).

The use of replicating vectors in industrial strains is restricted to vectors that lack bacterial DNA and are stable in the absence of selection agents. Use of such agents is inconceivable in beverage production. All yeast plasmids constructed *in vitro* have produced stable transformants in brewery strains (Panchal *et al.*, 1987). By means of specific sequences, hybrid *E. coli*-yeast plasmids can

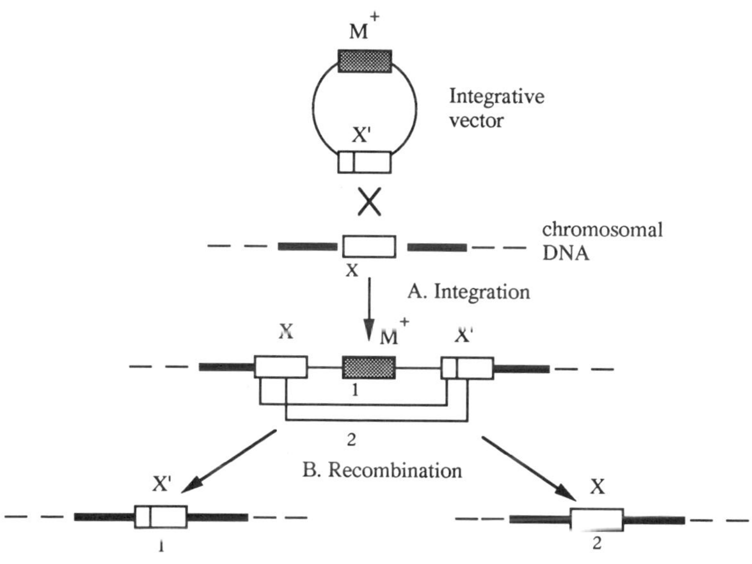

FIGURE 9.3 An example of stabilisation by the two step integration mechanism. A: Recombination between homologous regions on the plasmid and the chromosome results in integration of the plasmid sequences flanked by direct repeats of the homologous DNA. B: Recombination events within the flanked sequences result in excision of the vector sequences. If an alteration (deletion, insertion...) is made in the plasmidic homologous gene, each allele can be recovered, depending on the site of the recombination event 1 or 2. M^+ Dominant marker X Yeast target gene X' *In vitro* altered copy.

recombine *in vivo* with endogenous 2 μm plasmid of *S. cerevisiae* allowing removal of bacterial sequences and integration of the gene of interest into the 2 μm genome of the recipient strain. These "disintegration vectors" constitute a very stable system (Chinery and Hinchliffe, 1989).

Lastly, the advantages of certain yeast dominant resistance markers should be emphasized, since they can be retained to stabilize the transformants. This is the case of the K1 killer/immunity system, which confers an autoselection ability, and the SMRI-410 marker, which can direct integration into its ILV2 allele.

Control of gene expression

When introducing and stabilizing a gene to obtain a given technical property, it is often useful, and sometimes indispensable, to control its expression. This can be done by changing the copy number, but more accurate control can be attained on the transcriptional level, by changing the strength of the promoter. If expression is to be increased, which is often the case with expression of a heterologous gene, the original promoter can be replaced by a strong yeast promoter. The most frequently used promoters are those related to genes encoding abundantly expressed glycolytic enzymes, such as the genes for alcohol dehydrogenase (ADC1), phosphoglycerate kinase (PGK), or glyceraldehyde 3 phosphate dehydrogenase (GPD) (Ammerer, 1983; Tuite *et al.*, 1982; Bitter and Egan, 1984).

In some cases, overexpression of a gene can have toxic effects on the host cell or negative effects on the characteristics of the transformed products, at least in certain stages of the fermentation cycle. It is then feasible to use an inducible promoter allowing gene expression only during a phase of weak protein synthesis. We use various inducible promoters in our laboratory, such as PHO5, GAL, and CUP1. However, none of these systems can be used in commercial winemaking because of the high sugar concentration in the media and restrictions concerning product composition. Just recently, Kobayashi *et al.* (1990) described an ingenious system of thermoinducible expression using genes controlling the mating type, which allows induction by a mere heat shock of 25 to 30°C. This is one of the rare systems that is potentially applicable to wine yeasts.

Protein secretion

One possible objective of genetic manipulation of wine strains is the secretion of an enzymatic protein into the medium. *Saccharomyces cerevisiae* has a limited but exploitable ability for such secretions. The secreted proteins are synthesized in the form of precursors containing a peptide signal of twenty amino acids at the N-terminus, which is removed by proteolytic cleavage at the beginning of the secretion process. In some cases, secretion of heterologous proteins by *S. cerevisiae* has been achieved under the control of their own signal sequence, α-amylase and mold cellulase (Innis *et al.*, 1985; Penttila *et al.*, 1987), and lysozyme (Oberto and Davison, 1985). For other proteins, effective secretion is only obtained using yeast signal sequences such as MATα, invertase, MEL1, and K1 killer toxin (Shuster, 1989). It should be noted that the secretion of heterologous proteins achieved until now is only that of proteins which are themselves secreted.

Applications to winemaking

There are still very few concrete examples of wine yeast modification or improvement by transfer of cloned DNA. Williams *et al.* (1984) have cloned the malolactic enzyme gene of *Lactobacillus delbrueckii*, but attempts to induce expression of the cloned DNA fragment in wine yeast have not produced significant malolactic activity. Boone *et al.* (1990) introduced the cDNA of RNA encoding the preprotoxin K1 into a K2 killer wine strain. The cDNA was integrated, under the control of a strong ADH promoter, into the non-transcribed 3' sequence of the LEU2 gene. The transformants obtained were stable and showed double K1/K2 activity. Consequently, the modified strain has a broader competitive advantage over indigenous populations, while its industrial performances in laboratory tests are comparable to those of the original strain. In our laboratory we have obtained disrupted mutants for pyruvate carboxylase (Stucka *et al.*, 1991), which we are using in studies of yeast carbon metabolism. Such mutants could never have been obtained by classical genetics. Lastly Petering, Henschke and Langridge (1991) have recently introduced the *E. coli* β-glucuronidase gene into *S. cerevisiae*. They intend to use this gene as a marker for identification of the manipulated strain among wine yeast strains.

FUTURE STUDY

The use of genetic methods for yeast improvement largely depends on available data and methods. It also depends on social aspects such as regulations and consumer demands. Taking these aspects into account, several prospects for the future can be suggested.

Acquisition of New Data and Methods

The future of genetic manipulation of yeasts closely depends on progress in determining the biochemical, physiological, and genetic bases of winemaking properties related to alcoholic fermentation. Consequently, the rate of progress in these fields will largely decide the rate at which genetic manipulation programs are undertaken. Unfortunately, this implies, for example, that the improvement of wine aromas, which are usually controlled by complex factors, will remain beyond the scope of genetic manipulation for some time. Further progress can be expected in genetic methods. This concerns the regulation of gene expression (both native and transferred genes), the stability of their products, and the physiological consequences of their expression. The development of modules for gene regulation, that depend on parameters compatible with enology (such as temperature, the concentration of NH_4^+ ions, the cell growth stage) will probably open new horizons.

Regulation and Consumer Demands

There is always a question as to whether genetically altered strains whose winemaking ability is entirely satisfying can be used on an industrial scale. It is difficult to provide a universally valid answer to the question, because of differences in legislation in different countries. It can be said, however, that any strain obtained by clonal DNA transfer will face more difficulties in gaining approval than strains obtained by mutagenesis or protoplast fusion. As an example, let us consider the administrative guidelines laid down by the European Community (Anonymous, 1990a and b) for authorizing the restricted use and deliberate dissemination of manipulated organisms. These procedures comprise the opinions of committees of scientific experts whose criteria differ considerably. It does, however, appear that a certain number of guarantees will often be required, starting of course with assurance of the non-pathogenic character of the target and donor strains. These guarantees entail the following: (i) complete definition of the DNA sequence introduced as well as elimination of any sequence that is not indispensable for expression of the desired property; (ii) absence of any selective advantages conferred on the strain that could allow it to become dominant in natural habitats; and (iii) thorough analysis of the possibility for the new sequence to be introduced into other microorganisms, even if the sequence can in no way endanger human health. The risk of transfer must remain negligible and all efforts must be made to reduce this possibility; for example, by rendering the strain incapable of sporulation, thereby considerably diminishing the chances of cell fusion in the natural habitat. Further requirements may emerge as new cases are studied, such as the rejection of any manipulation that introduces auxotrophy (either as a marker or as a result of insertion) into a haploid form, even if it is compensated by the corresponding dominant allele in the final diploid.

Another factor influencing the acceptability of manipulated strains is the reaction of consumers to the resulting transformed products. It is possible that the public could develop some fear of these manipulated strains. Geneticists must therefore make an effort to inform the public and be open about their work. It would be preferable if the first products in this field could show organoleptic, hygienic and economic advantages that are unquestionable and easily perceptible to everyone from producers to consumers.

Prospects for the Future

In spite of the reservations mentioned above, we consider that clonal DNA transfer methods will predominate in the future construction of wine strains. Classical genetics clearly provides a means of increasing and directing the pool of diversity that can be exploited using indigenous populations, but it inevitably lacks the precision and efficacy of direct manipulation.

It would appear that three main areas will benefit from yeast genetics in the medium and long terms. The first concerns improvements in the control and reliability of fermentation processes, i.e. expression of malolactic fermentation by yeast, control of flocculation and foaming, control of the composition of the yeast population, and regulation of certain metabolic activities by external signals. The second involves the prevention of organoleptic (e.g. sulfide compounds) and hygienic (e.g. ethylcarbamate) defects. Third, a large field for the future, whose applications will appear in the long term, concerns the comprehensively monitored modification of alcoholic fermentation by replacing part of the ethanol with metabolites that are less deleterious to consumer health. On the other hand, it seems less likely that the construction of strains secreting enologically advantageous enzymes (pectolytic and proteolytic enzymes, etc.) or biologically active substances (specific inhibitors of microorganisms) can offer real advantages over direct addition of such substances produced elsewhere. Lastly, it should be expected that genetic solutions to problems will sometimes be in competition with physical solutions (e.g. membrane techniques). A typical example of this situation is the improvement of champagnization conditions: should flocculent activity be conferred on a given strain by genetic manipulation or should this strain be included in alginate beads?

CONCLUSION

Yeast genetics remains an important field for the future of enological research, despite the fact that the improvement of overall wine quality is a much more diffuse and less measurable objective than those addressed in other areas of industrial microbiology, such as high yields of biomass or antibiotics. It must be admitted that, until now, genetically altered strains (apart from labeled stains) are not among the most widely used strains in the world. Nevertheless, the proliferation of methods available to modern genetics and the increasing documentation of yeast physiology in areas of crucial interest to enology should change this situation in the future, with respect for winemaking traditions, product quality and consistency, and consumer health.

REFERENCES

Anonymous (1990a) Directive du conseil du 23 avril 1990 relative à l'utilisation confinée de microorganismes génétiquement modifiés. *Journal Officiel des Communautés Européennes*, L 117 du 8 mai 1990, 1–14.

Anonymous (1990b) Directive du conseil du 23 avril 1990 relative à la dissémination volontaire d'organismes génétiquement modifiés dans l'environnement. *Journal Officiel des Communautés Européennes*, L 117 du 8 mai 1990, 15–27.

Ammerer, G. (1983) Expression of genes in yeast using the ADC1 promoter. In *Methods in Enzymology*, vol 10, edited by Wu, R., Grossman, L. and Moldave, K., pp. 192–201. New York: Academic Press.
Bakalinsky, A.T. and Snow, R. (1990) The chromosomal constitution of wine strains of *Saccharomyces cerevisiae*. *Yeast*, **6**, 367–382.
Bidenne, C., Blondin, B., Dequin, S. and Vezinhet, F. (1992) Analysis of the chromosomal DNA polymorphism of wine strains of *Saccharomyces cerevisiae*. *Current Genetics*, **21**, (In press).
Bitter, G.A. and Egan, K.M. (1984) Expression of heterologous genes in *Saccharomyces cerevisiae* from vectors utilizing the glyceraldehyde-3-phosphate dehydrogenase gene promoter. *Gene*, **32**, 263–274.
Blondin, B. and Vezinhet, F. (1988) Identification de souches de levures oenologiques par leurs caryotypes obtenus en électrophorèse en champ pulsé. *Revue Française d'Oenologie*, **28**, 7–11.
Boguslawski, G. (1985) Yeast transformation. In *Gene manipulation in Fungi*, edited by J.W. Bennet and L.L. Lasure, pp. 161–195. London: Academic Press.
Boone, C., Sdicu, A.M., Wagner, J., Degré, R., Sanchez, C. and Bussey, H. (1990) Integration of the yeast K1 killer toxin into the genome of marked wine yeasts and its effect on vinification. *American Journal of Enology and Viticulture*, **41**, 37–42.
Bussey, H. and Meaden, P. (1985) Selection and stability of yeast transformants expressing cDNA of an M1 killer toxin-immunity gene. *Current Genetics*, **9**, 285–291.
Carle, G.F. and Olson, M.V. (1985) An electrophoretic karyotype for yeast. *Proceedings of the National Academy of Science, USA*, **82**, 3756–3760.
Carrau, J.L., De Avezedo, J.L., Subdery, P. and Campbell, D. (1982) Methods for recovering fusion products among enological strains of *Saccharomyces cerevisiae* and *Schizosaccharomyces pombe*. *Brazilian Journal of Genetics*, **5**, 221–226.
Casey, G.P., Xiao, W. and Rank, G.H. (1988) A convenient dominant selection marker for transfer in industrial strains of *Saccharomyces yeast*: SMR1 encoded the resistance for the herbicide sulfometuron methyl. *Journal of the Institute of Brewing*, **94**, 93–97.
Chambon, C., Ladeveze, V., Oulmouden, A., Servouze, M. and Karst, F. (1990) Isolation and properties of yeast mutants affected in farnesyl diphosphate synthetase. *Current Genetics*, **18**, 41–46.
Chinery, S.A. and Hinchliffe, E. (1989) A novel class of vector for yeast transformation. *Current Genetics*, **16**, 21–25.
Conde, J. and Fink, G.R. (1976) A mutant of *Saccharomyces cerevisiae* defective for nuclear fusion. *Proceedings of the National Academy of Science*, USA, **73**, 3651–3655.
Delorme, E. (1989) Transformation of *Saccharomyces cerevisiae* by electroporation. *Applied and Environmental Microbiology*, **55**, 2242–2246.
Delteil, D. and Aizac, T. (1988) Comparaison de différentes techniques de levurage par suivi de l'implantation d'une souche de levure oenologique marquée. *Revue Française d'Oenologie*, **28**, 11–18.
Dubourdieu, D., Sokol, A., Zucca, J., Thalouarn, P., Dattee, A. and Aigle, M. (1987) Identification des souches de levures isolées de vins par l'analyse de leur ADN mitochondrial. *Connaissance de la Vigne et du Vin*, **21**, 267–278.
Dutcher, S.K. (1981) Internuclear transfer of genetic information in kar 1-1/KAR 1 heterokaryons in *Saccharomyces cerevisiae*. *Molecular and Cellular Biology*, **1**, 245–253.
Eschenbruch, R., Cresswell, K.J., Fisher, B.M. and Thornton, R.J. (1982) Selective hybridization of pure culture wine yeasts. I Elimination of undesirable winemaking properties. *European Journal of Applied Microbiology and Biotechnology*, **14**, 155–158.
Fleet, G.H. (1990) Growth of yeasts during wine fermentation. *Journal of Wine Research*, **1**, 211–223.
Fuji, T., Kondo, K., Shimizu, F., Sone, H., Tanaka, J.I. and Inoue, T. (1990) Application of ribosomal DNA integration vector in the construction of a brewer's yeast having α-

acetolactate decarboxylase activity. *Applied and Environmental Microbiology*, **56**, 997–1003.

Gatignol, A., Baron, M. and Tiraby, G. (1987) Phleomycin resistance encoded by the *ble* gene transposon Tn5 as a dominant selectable marker in *Saccharomyces cerevisiae*. *Molecular and General Genetics*, **207**, 342–348.

Gritz, L. and Davies, J. (1983) Plamid encoded hygromycin B resistance: the sequence of hygromycin B phosphotransferase gene and its expression in *Escherichia coli* and *Saccharomyces cerevisiae*. *Gene*, **25**, 179–188.

Guthrie, C. and Fink, G.R. (1991) Guide to yeast genetics and molecular biology. *Methods in Enzymology*, vol. 194. New York: Academic Press.

Gysler, C.H., Kneuss, P. and Niederberger, P. (1990) Transformation of commercial baker's yeast strains by electroporation. *Biotechnology Techniques*, **4**, 285–290.

Hadfield, C., Cahsmore, A.M. and Meacock, P.A. (1986) An efficient chloramphenicol resistance marker for *Saccharomyces cerevisiae* and *Escherichia coli*. *Gene*, **45**, 149–158.

Hallet, J.N., Craneguy, B., Zucca, J. and Poulard, A. (1988) Caractérisation de différentes souches industrielles de levures oenologiques par les profils de restriction de leur ADN mitochondrial. *Le Progrès Agricole et Viticole*, **105**, 328–333.

Hammond, J.R.M. and Eckersley, K.W. (1984) Fermentation properties of brewing yeast with killer character. *Journal of the Institute of Brewing*, **90**, 167–177.

Henderson, R.C.A., Cox, B.C. and Tubb, R. (1985) The transformation of brewing yeasts with a plasmid containing the gene for copper resistance. *Current Genetics*, **9**, 133–138.

Hinnen, A., Hicks, J.B. and Fink, G. (1978) Transformation of yeast. *Proceedings of the National Academy of Science*, USA, **75**, 1929–1933.

Innis, M.A., Holland, M.I., McCabe, P.C., Cole, G. E., Wittman, V.P., Tal, R., Watt, K.W.K., Gelfand, D.H., Holland, J.P. and Meade, J.H. (1985) Expression, glycosylation and secretion of an *Aspergillus glucoamylase* by *Saccharomyces cerevisiae*. *Science*, **228**, 21–25.

Ito, H., Fukuda, Y., Murata, K. and Kimura, A. (1983) Transformation of intact yeast cells treated with alcali cations. *Journal of Bacteriology*, **153**, 163–168.

Javelot, C., Karst, F., Ladeveze, V., Chambon, C. and Vladescu, B. (1990) Production of monoterpenes by yeast mutants, defective in sterol biosynthesis. In *Microbiology Applications in Food Biotechnology*, edited by B.H. Nga and Y.K. Lee, pp. 101–122. London : Elsevier Applied Science.

Jimenez, A. and Davies, J. (1980) Expression of a transposable antibiotic resistance element in *Saccharomyces*. *Nature*, **287**, 869–871.

Johnston, J.R. (1990) Brewing and distilling yeasts. In *Yeast Technology*, edited by J.F.T. Spencer and D.M. Spencer. pp. 55–104, Berlin: Springer Verlag.

Kaster, K.R., Burgett, S.J. and Ingolia, T.D. (1984) Hygromycin B resistance as a dominant selectable marker in yeast. *Current Genetics*, **8**, 353–358.

Kobayashi, H., Nakazawa, N., Harashima, S. and Oshima, Y. (1990) A system for temperature-controlled expression of a foreign gene with dual mode in *Saccharomyces cerevisiae*. *Journal of Fermentation and Bioengineering*, **69**, 322–327.

Kuzewicz, D. (1976) The characteristics of cryophilic wine yeast. II. The production and characteristics of cryophilic UV irradiated wine yeast. *Acta Alimentaria Polonica*, **26**, 61–72.

Kuzewicz, D. and Johnston, J. (1980) Genetic analysis of cryophilic and mesophilic wine yeasts. *Journal of the Institute of Brewing*, **8**, 25–27.

Loiseau, G. (1986) Obtention d'un marquage mitochondrial de levures oenologiques sélectionnées en Champagne. *Application au contrôle microbiologique en oenologie*. ScD Thesis, Université des Sciences et Techniques du Languedoc, Montpellier, France.

Loiseau, G., Vezinhet, F., Valade, M., Vertes, A., Cuinier, C. and Delteil, D. (1987) Contrôle de l'efficacité du levurage par la mise en oeuvre de souches de levures oenologiques marquées. *Revue Française d'Oenologie*, **27**, 29–36.

Lonvaud-Funel, A. (1990) Purification of malolactic enzyme from *Lactococcus lactis*. *6th International Symposium on Genetics of Industrial Microorganisms*, GIM 90, Strasbourg, Book of Abstracts, 130. Societé Française de Microbiologie.

Neuville, P., Bonneu, M. and Aigle, M. (1990) Heterogeneity among the 2μm plasmids in *Saccharomyces cerevisiae*: a new sequence for the REP1 gene. *Gene*, **89**, 139–144.

Oberto, J. and Davison, J.D. (1985) Expression of chicken egg white lysozyme by *Saccharomyces cerevisiae*. *Gene*, **40**, 57–65.

Orr-Weawer, T.L., Szostak, J.W. and Rothstein, R.J. (1981) Yeast transformation : a model system for the study of recombination. *Proceedings of the National Academy of Science, USA*, **78**, 6354–6358.

Panchal, C.J., Bast, L.J., Dowhanick, T.M., Jones, R.M., Russell, I. and Stewart, G.G. (1987) Construction and characterization of an all yeast DNA vector. In *Biological research on Industrial Yeasts*, vol. 2, edited by G.G. Stewart, I. Russell, R.D. Klein and R.R. Hiebsch, pp. 49–56. Boca Raton: CRC Press.

Pardo, I., Blondin, B. and Vezinhet, F. (1989) Analysis of karyotypes of enological yeast strains of *Saccharomyces cerevisiae*. *13th International Specialized Symposium on Yeasts*, Leuven, Book of Abstracts, 58–59. ISSY 13th.

Parent, S.A., Fenimore, C.M. and Bostian, K.A. (1985) Vector system for the expression. Analysis and cloning of DNA sequences in *Saccharomyces cerevisiae*. *Yeast*, **1**, 83–138.

Pedersen, M.B. (1986) DNA sequence polymorphisms in the genus *Saccharomyces*. III Restriction endonuclease fragment patterns of chromosomal regions in brewing and other yeast strains. *Carlsberg Research Communications*, **51**, 163–183.

Penttilä, M.E., Ardre, L., Saloheimo, M., Lehtovaara, P. and Knowles, J.K. (1987) Expression of two *Trichoderma reesei* endoglucanases in the yeast *Saccharomyces cerevisiae*. *Yeast*, **3**, 175–185.

Petering, J.E., Henschke, P.A. and Langridge, P. (1991) The *Escherichia coli-β-*glucuzonidase gene as a marker for *Saccharomyces* yeast strain identification. *American Journal of Enology and Viticulture*, **42**, 6–12.

Petering, J.E., Langridge, P. and Henschke, P. (1988) Fingerprinting wine yeasts. *The Australian and New Zealand Wine Industry Journal*, **3**, 48–52.

Pretorius, I. and van der Westhuizen, T. J. (1991) The impact of yeast genetics and recombinant DNA technology on the wine industry. A review. *South African Journal for Enology and Viticulture*, **12**, 3–31.

Radler, F. (1973) Bedeutung und Möglichkeiten der Verwendung von Reinkulturen von Hefen bei der Weinberejtung. *Weinberg Keller*, **29**, 339–350.

Rank, G.H., Casey, G. and Xiao, W. (1988) Gene transfer in industrial *Saccharomyces* yeasts. *Food Biotechnology*, **2**, 1–41.

Rankine, B.C. (1968) The importance of yeasts in determining the composition and the quality of wines. *Vitis*, **7**, 22–49.

Renault, P. (1990) Malolactic fermentation : genetics and genetic engineering. *Proceedings of the 6th International Symposium on Genetics of Industrial Microorganisms*, GIM 90, Strasbourg.

Romano, P., Grazia Soli, M., Suzzi, G., Grazia, L. and Zambonelli, C. (1985) Improvement of a wine *Saccharomyces cerevisiae* strain by a breeding program. *Applied and Environmental Microbiology*, **50**, 1064–1067.

Rose, M.D. (1987). Isolation of genes by complementation in yeast. In *Methods in Enzymology, Guide to Molecular Cloning Techniques*, edited by S.L. Berger and A.R. Kimmel, pp. 481–504. New-York: Academic Press.

Rous, C.V. and Snow, R. (1983) Reduction of higher alcohols by fermentation with a leucine-auxotrophic mutant of wine yeast. *Journal of the Institute of Brewing*, **89**, 274–278.

Rudolph, H., Koening-Rauseo, I. and Hinnen, A. (1985) One step gene replacement in yeast by cotransformation. *Gene*, **36**, 87–95.

Salmon, J.M. (1987) L Malic permeation in resting cells of anaerobically grown *Saccharomyces cerevisiae*. *Biochimica et Biophysica Acta*, **901**, 30–34.

Schutz, M. and Radler, F. (1973) Das "Malatenzym" von *Lactobacillus plantarum* und *Leuconostoc mesenteroïdes*. *Archiv für Mikrobiologie*, **96**, 329–339.

Seki, T., Choi, E.H. and Ryu, D. (1985) Construction of killer wine yeast strain. *Applied and Environmental Microbiology*, **49**, 1211–1215.

Shuster, J.R. (1989) Regulated transcriptional systems for the production of proteins in yeast : regulation by carbon source. In *Yeast Genetic Engineering*, edited by P.J. Barr, A. Brake and P. Valenzuela, pp. 83–108, Boston: Butterworths.

Snow, R. (1983) Genetic improvement of wine yeast. In *Yeast Genetics. Fundamental and Applied Aspects*, edited by J.F.T. Spencer, D.M. Spencer and A.R.W. Smith, pp. 439–459, Berlin: Springer Verlag.

Spencer, J.F.T., Spencer, D.M. and Bruce, I.J. (1989). *Yeast Genetics. A Manual of Methods*. Berlin: Springer-Verlag.

Struhl, K. (1983) The new yeast genetics. *Nature*, **305**, 391–397.

Stucka, R., Dequin, S., Salmon, S.J. and Gancedo, C. (1991) DNA sequences in chromosomes II and VII code for pyruvate carboxylase iosenzymes in *Saccharomyces cerevisiae*: analysis of pyruvate carboxylase-deficient strains. *Molecular and General Genetics*, **229**, 307–315.

Thornton, R.J. and Eschenbruch, R. (1976) Homothallism in wine yeasts. *Antonie van Leeuwenhoek*, **42**, 503–509.

Thornton, R.J. (1982) Selective hybridization of pure culture wine yeasts. II Improvement of fermentation efficiency and inheritance of SO_2 tolerance. *European Journal of Applied Microbiology and Biotechnology*, **14**, 159–164.

Thornton, R.J. (1985) The introduction of flocculation into a homothallic wine yeast. A practical example of the modification of winemaking properties by the use of genetic techniques. *American Journal of Enology and Viticulture*, **36**, 47–49.

Thornton, R.J. and Bunker, A. (1989) Characterization of wine yeasts for genetically modifiable properties. *Journal of the Institute of Brewing*, **95**, 181–184.

Tuite, M.F., Dobson, M.J., Roberts, N.A., King, R.M., Burke, D.C., Kingsman, S.M. and Kingsman, A.J. (1982) Regulated high efficiency expression of human interferon alpha in *Saccharomyces cerevisiae*. *EMBO J.* **1**, 603–608.

Ventre, J. (1911) Application des levures à la vinification. In *Les Levures dans la Vinification*, Paris: Masson.

Vezinhet, F. (1985) Le marquage génétique de souches de levures oenologiques. *Revue Française d'Oenologie*, **25**, 47–51.

Vezinhet, F. and Barre, P. (1988) Transfer by cytoduction of the FLO5 flocculation gene in an enological yeast strain. *8th International Biotechnology Symposium*, Paris, Book of Abstracts, 101. Societé Française de Microbiologie.

Vezinhet, F. (1989) Obtention par recombinaison intra-génomique de clones non-moussants à partir d'une souche de levures d'intérêt oenologique *Saccharomyces cerevisiae*. *Science des Aliments*, **9**, 253–265.

Vezinhet, F., Blondin, B. and Hallet, J.N. (1990) Chromosomal DNA patterns and mitochondrial DNA polymorphism as tools for identification of enological strains of *Saccharomyces cerevisiae*. *Applied Microbiology and Biotechnology*, **32**, 568–571.

Vezinhet, F., Blondin, B. and Barre, P. (1991) Mapping of the FLO5 gene of *Saccharomyces cerevisiae* by transfer of a chromosome during cytoduction. *Biotechnology Letters*, **13**, 47–52.

Webster, T.D. and Dickson, R.C. (1983) Direct selection of *Saccharomyces cerevisiae* resistant to the antibiotic G418 following transformation with a DNA vector carrying the kanamycin resistance gene of Tn903. *Gene*, **26**, 243–252.

Williams, S.A., Hodges, R.A., Strike, T.L., Snow, R. and Kunkee, R.E. (1984) Cloning the gene for the malolactic fermentation of wine from *Lactobacillus delbrueckii* in *Escherichia coli* and yeasts. *Applied and Environmental Microbiology*, **47**, 283–293.

Chapter 10

MALOLACTIC FERMENTATION

THOMAS HENICK-KLING

Cornell University, Department of Food Science and Technology, New York State Agricultural Experiment Station, Geneva, NY, USA

Chapter Outline

Chapter Outline (Continued)

INTRODUCTION

During growth in wine, lactic acid bacteria (LAB) metabolize malic, acid, forming lactic acid and CO_2 (Henick-Kling, 1988a). This so-called malolactic fermentation (MLF) usually occurs after the alcoholic fermentation but it may occur during the alcoholic fermentation. MLF is important in winemaking for three reasons: (i) for deacidification, (ii) for flavor modification, and (iii) for microbial stability. The conversion of the dicarboxylic acid, malate, to the monocarboxylic acid, lactate, and the loss of CO_2, decreases the acidity of the wine and increases its pH. This function is important in wines which are produced from grapes grown in cool climates which often have a high acid (tartrate plus malate) content and low pH. A decrease in the acidity of wines produced from grapes grown in warm climates is generally not desirable because it will leave a product with too little acidity and too high a pH. Nevertheless, MLF is desired in some red and white wines produced from grapes grown in warm climates because other flavor changes associated with the growth of LAB are often considered beneficial to quality. Also, completed MLF gives the wine some microbial stability by removing malate and some sugars, and by producing antimicrobial agents such as lactic acid and possibly bacteriocins. MLF is also important in the production of sparkling wines by the "méthode champenoise" where it is used to advance the maturity of the wines used in the base blends, to give the wine some of its characteristic nutty, yeasty aromas, and to stabilize wines against MLF during the second fermentation in the bottle (Pool and Henick-Kling, 1991).

Reviews by Radler (1962, 1963a,b), Kunkee (1974, 1991), Beelman and Gallander (1979), Wibowo *et al.*, (1985), and Davis *et al.*, (1985a) give excellent descriptions of the history of research on MLF, the classification, ecology and physiology of LAB in wine, and the chemical changes brought on by the bacteria. This chapter is intended to give the reader a summary of the most recent advances in our understanding of the biochemistry of growth of LAB in wine, of the rapidly evolving starter culture technology, and our new understanding of the importance of MLF flavors in wine.

LACTIC ACID BACTERIA IN WINE

Growth

From the grape berry surface, stems, leaves, soil, and winery equipment, many different LAB enter grape juice and wine. However, due to the highly selective environment of different juices and wines, only a few types of LAB are able to grow in any one wine (Webb and Ingraham, 1960; Peynaud and Domercq, 1961; Gini and Vaughn, 1962; Radler, 1963a; Schanderl, 1969; Weiller and Radler, 1970; Donnelly, 1977; Sapis-Domercq *et al.*, 1978; Izuagbe *et al.*, 1985; Wibowo *et al.*, 1985; Dicks and van Vuuren, 1988). The most selective conditions determining growth of LAB in wines are pH and ethanol content (Davis *et al.*, 1986, a; Britz and Tracey, 1990), but other important influences are: interaction with yeast (Beelman *et al.*, 1982; King and Beelman, 1986; Lemaresquier, 1987; Wibowo *et al.*, 1988; Lonvaud-Funel *et al.*, 1988); the presence of bacteriophage (Sozzi *et al.*, 1982; Gnaegi *et al.* 1984; Davis *et al.* 1985a; Henick-Kling *et al.* 1986 a, b; Nel *et al.*, 1989); temperature (Asmundson and Kelly, 1990); and nutrient availability and growth inhibitors (Lonvaud-Funel, 1987). The combination of these factors, as well as their change during alcoholic fermentation and wine storage, leads to the successive growth of various species of LAB during wine production (Costello *et al.*, 1983; Lafon-Lafourcade *et al.*, 1983a; Fleet *et al.*, 1984; Davis *et al.*, 1986b; Messini *et al.*, 1990). Fungicide residues (Sapis-Domercq *et al.*, 1978; Haag *et al.*, 1988) and sulfur dioxide (Mayer, 1978; Liu and Gallander, 1983; Haag *et al.*, 1988; Wibowo *et al.*, 1988) also have selective influences.

The LAB of wine belong to three genera, *Leuconostoc*, *Lactobacillus*, and *Pediococcus*. Only one species of *Leuconostoc*, namely *Leuc. oenos*, can grow in wine (Garvie, 1986a). Strains of *Leuc. oenos* vary in their tolerance to pH, ethanol, sulfur dioxide (SO_2), and low temperature, and in their growth rate, malolactic activity, and carbohydrate fermentation pattern (Izuagbe *et al.*, 1985; Davis *et al.*, 1988; Britz and Tracey, 1990). Because of its greater tolerance of acid, *Leuc. oenos* generally predominates in wines with pH values below 3.5, while in wines above pH 3.5, species of *Lactobacillus* and *Pediococcus* often predominate. *Leuconostoc oenos* is only partially inhibited by ethanol concentrations above 5% (v/v) and can tolerate up to 14% (v/v), while growth of *Lactobacillus plantarum* stops at ethanol concentrations of 5–6% (v/v). Strains of *L. casei* and *L. brevis* are more alcohol tolerant and have been successfully used to induce MLF (Wibowo *et al.*, 1985; Krieger, 1989; Krieger and Henick-Kling, 1991). Of the pediococci, *(Pediococcus damnosus (P. cerevisiae), P. pentosaceus*, and *P. parvulus*), *P. damnosus* is most often found because it is more acid tolerant (Davis, 1985). Pediococci are rarely associated with MLF but are found during storage of wine (Lafon-Lafourcade *et al.*, 1983a; Davis *et al.*, 1986a,b). Mayer (1974a), points out the danger of growth of pediococci after completion of MLF when the wine pH may exceed 3.5. This possibility is well demonstrated in the ecological studies of Davis *et al.* (1986a,b). Strong growth of pediococci can lead to off-flavors and to an oily consistency. In the presence of sufficient carbohydrates and low oxygen tension, pediococci and some strains of *Leuconostoc* (Watson and Daeschel, personal communication) can produce large amounts of polysaccharides that cause ropiness.

The role of bacteriocins in competition between species of wine LAB has not been investigated, but based on experience with LAB in food fermentations, this factor should play a role in the wine environment. Daeschel *et al.* (1990) described a bacteriocin produced by *L. plantarum* isolated from pickle fermentation. For reviews of antagonistic activities of LAB see Lindgren and Dobrogosz (1990) Klaenhammer (1988), and Tagg *et al.* (1976).

Genetics

Over many years, numerous isolates of LAB have been made from wine. Their identification and characterization has been, and still is, difficult. Although the isolates are easily grouped into the three genera, it is difficult to distinguish between some species and, especially, between strains of *Leuc. oenos*. These strains differ in their growth at low pH in wines, in their malolactic activity, in the flavors they produce, and in their growth rate in starter culture media. Due to these differences, the production of wine with malolactic starter cultures requires close attention to strain selection. No satisfactory method for typing strains of *Leuc. oenos* has been developed. Carbohydrate fermentation tests for strain characterization are unreliable because they vary depending on the medium of preculture (Davis *et al.*, 1988). Physiological tests are too time consuming to be of practical value. Typing with bacteriophages may offer a viable alternative. A large bank of bacteriophages is required to differentiate among strains. Sozzi *et al.* (1982) and Gnaegi *et al.* (1984) showed the first phage sensitivity patterns among strains of *Leuc. oenos*. Nel *et al.* (1989) tested 20 phages against 39 strains of *Leuc. oenos*; only 29 strains were sensitive, and one strain was lysed by all phages. The well known starter culture strains ML34 and PSU-1 were lysed by nine and seven phages, respectively. Naturally phage resistant strains can be isolated from wines or selected in cultures exposed to phage (Sozzi *et al.*, 1982; Gnaegi *et al.*, 1984; Henick-Kling *et al.*, 1986b).

The comparison of plasmid profiles has been useful in evaluating the relatedness of strains (Coventry *et al.*, 1984; Davies *et al.*, 1981), but is limited by the instability of some plasmid DNA and the general lack of plasmids in *Leuc. oenos*. Plasmids in *Leuc. oenos* have been studied by Janse *et al.* (1987), Rau (1988), and Taylor *et al.* (1990). Apparently, plasmids are not very common among *Leuc. oenos*; Orberg and Sandine (1984) found no plasmids in the range of 1–76 mega daltons in four strains of *Leuc. oenos* examined. Rau (1988), screened 128 strains of *Leuc. oenos* and found plasmids in only 18 strains. Van Vuuren and co-workers (personal communication) found plasmids in only 8 of 55 strains of *Leuc. oenos*. The plasmids were generally 2.47–4.61 kilobases in size, but some were 30 kilobases. Thus, plasmid profiling is not a useful system for distinguishing between strains of *Leuc. oenos*. Also, it appears that much of the diversity among strains of *Leuc. oenos* in carbohydrate fermentation patterns, growth, malolactic activity and phage sensitivity is not plasmid coded. The apparent lack of plasmids in some strains may be due to the methods used for isolating them. Modifications to the methods for isolating extrachromosomal DNA may show the presence of more plasmids. The recent development of pulsed-field gel electrophoresis combined with digestion of genomic DNA appears to be a powerful, reliable, and practical method. Tanskanen *et al.* (1990) have shown that this approach can distinguish closely related strains of lactococci.

Several plating methods have been reported for the selection of mutants with altered malolactic activity (Subden *et al.*, 1982; Daeschel *et al.*, 1984; Renault and Heslot, 1987; Cavin *et al.*, 1989). Williams *et al.* (1984) and Lautensach and Subden (1984) cloned the malolactic enzyme from *L. delbrueckii* and *Leuc. oenos*, respectively, into yeast and *Escherichia coli* but very little to no malolactic activity was expressed. This is likely due to the fact that the malate transport system was not cloned with the malolactic enzyme. Also, no satisfactory promoter for expression of the gene has been available.

The malolactic activity is chromosomally encoded in lactococci and in *Leuc. oenos* (Janse *et al.*, 1987; Rau, 1988). Daeschel *et al.* (1984) and Renault and Heslot (1987) have produced mutants defective in MLF. We have obtained spontaneous mutants of *Leuc. oenos* which were defective in transport of malate but not in the malolactic enzyme. According to the model described by Cox and Henick-Kling (1989) and Olsen *et al.* (1991), MLF has three components; malate transport, the malolactic enzyme, and lactate transport or lactate efflux. Renault *et al.* (1988) found evidence that at least four genes determine malolactic activity. One of these four genes is a positive regulator for induction of the genes necessary for MLF (Renault *et al.*, 1989).

To date, no transformation systems for *Leuc. oenos*, have been reported. Transformation systems for lactobacilli and lactococci have been developed (Gasson, 1983; Cosby *et al.*, 1988; Kondo, 1989) and plasmid DNA in *P. cerevisiae* and *P. pentosaceus* has been described (Graham and McKay, 1985).

Protonmotive Force

ATP, the principal carrier of biological energy can be generated by two basic mechanisms, substrate level phosphorylation and electron transport phosphorylation. During substrate level phosphorylation, numerous intermediates are formed which contain high-energy phosphoryl bonds (Gottschalk, 1986). The energy from this high-energy bond is then transferred to ATP in its formation from ADP and P_i. In electron transport phosphorylation, the flow of electrons from donors with a negative redox potential to acceptors with a more positive redox potential is used for the synthesis of ATP (Harold, 1977). The chemiosmotic theory as proposed by Mitchell (1966, 1981) provides the basis for energy transduction in bacteria via formation of an electrochemical gradient across the cell membrane. The electrochemical gradient or protonmotive force (Δp) is the sum of the electrical gradient $\Delta\Psi$ and chemical gradient ΔpH. The energy of Δp can be used to drive transport, motility, and to synthesize ATP (Harold, 1977; Maloney, 1977).

Lactic acid bacteria grow over a wide range of pH in environments such as milk, vegetable fermentations, wine, in plaque of teeth, on skin, in intestines and in rumen. Different LAB have adapted to this range from the low pH of fruit juice and wine to the high pH of rumen. Bacteria must maintain a suitable intracellular pH (pH_i) and an electrochemical gradient across their cell membrane. The ability of LAB to tolerate low pH apparently depends on their ability to maintain a suitable pH_i. Also, different LAB tolerate different minimum pH_i. By allowing the pH_i to decrease below the usual value near 6, they avoid spending energy (ATP) for pumping protons out of the cells via the membrane ATPase, and they reduce the large Δp directed inward which is generated mainly by the large pH gradient (ΔpH). The pH of wine is at the lower limit for growth of lactic acid bacteria and too low for most neutrophilic bacteria. Growth stops when the pH_i falls below its lowest tolerated limit. This value of pH_i varies considerably among lactic acid bacteria. Growth of *L. acidophilus*, *L. delbrueckii*, and *L. bulgaricus* ceases at a pH_i near 4.4 (Kashket, 1987), and growth of oral streptococci stops at pH_i 5.7 (Kashket and Kashket, 1985). Growth of *L. plantarum* and *Leuconostoc mesenteroides* stops at pH_i 4.6–4.8 and pH_i 5.4–5.7, respectively (McDonald *et al.*, 1990). This limiting pH_i is not simply determined by the external pH (pH_o), but it also depends on the type of acid and on the pH and type of preculture medium. Organic acids inhibit growth of microorganisms by entering the cell in the undissociated form, then dissociating within the cell (Levine and Fellers, 1940). This lowers the pH_i and collapses Δp, thus inhibiting intracellular enzymes and Δp dependent transport systems. McDonald *et al.* (1990) found that, in a medium with 10 mM sodium acetate, the lowest pH_o at which *L. plantarum* and *Leuc. mesenteroides* could initiate growth was 3.00 and 2.88, respectively. The pH_o at which they terminated growth was 3.50 and 3.41, respectively. But in a medium containing 10 mM sodium lactate, these values were 3.50 and 3.07, 4.00 and 3.75, respectively. The differing inhibitory effect of the acids is due to their different ability to decrease the pH_i. In acid tolerant lactic acid bacteria, the pH_i can decrease considerably as they adjust to lower pH_o; growth is still initiated at pH_i as low as 4.6–5.4 (Kashket, 1987; McDonald *et al.*, 1990). The intracellular pH of *L. plantarum*, isolated from wine, also decreases with decreasing pH_o, reaching 4.5 at pH_o 3.5 where growth is still possible for this strain (Olsen *et al.*, 1991). *Leuconostoc oenos* was able to maintain a higher pH_i at the low pH_o (3.5) than *L. plantarum* (Henick-Kling, 1986). Also, *Leuc. oenos* maintained a higher pH_i and a larger Δp when grown in medium of low pH (3.5) than when grown at high pH (5.5). This might be one reason why *Leuc. oenos* is better adapted to the low pH of wine than lactobacilli and pediococci.

McDonald *et al.* (1990) showed for *L. plantarum* and *Leuc. mesenteroides* isolated from pickle fermentation that the growth limiting factor was the pH_i not the external pH. *Lactobacillus plantarum* was more tolerant to low pH; it could tolerate a pH_i as low as 4.6 to 4.8 before growth stopped. The growth of *Leuc. mesenteroides* stopped when pH_i decreased to 5.4 to 5.7. A comparison of *Lactobacillus casei* and *Actinomyces viscosus* showed that their different tolerance of low pH was due to differences in the activity of their ATPase (Bender and Marquis, 1987).

Metabolism

Oxygen and carbon dioxide

Oxygen can be used as a terminal electron acceptor by *Leuc. mesenteroides* and *L. plantarum*, and fructose may be reduced to mannitol, thus promoting growth (Kandler, 1983; Lucey and Condon, 1984; Sedewitz *et al.*, 1984; Murphy *et al.*, 1985). However, Kelly *et al.* (1989) showed in a study with several strains of *Leuc. oenos* that growth was fastest under strict anaerobic conditions. Carbon dioxide was not growth-stimulating as reported previously (Mayer, 1974b; Stamer, 1979; van der Westhuizen *et al.*, 1981).

Sugars

Glucose and fructose constitute 99% and at least 78% of the sugars in grape juice and wine, respectively. In juice, glucose and fructose are present at concentrations of 70–120 g/l each, at normal grape maturity, but in late harvest and *Botrytis cinerea* infected fruit, the total sugar content can be as high as 500 g/l with fructose predominating (Amerine and Ough, 1980; Sponholz, 1989; Zoecklein *et al.*, 1990). In dry wines the content of glucose and fructose is as low as 0.5 to 1 g/l. Although sucrose is present in grapes at concentrations up to 10 and 50 g/l in *Vitis vinifera* and American grape species, respectively, it occurs only in small concentrations in grape juice and wine (<50 mg/l) (Sponholz, 1989). In grape juice, it is rapidly hydrolyzed to glucose and fructose by the acidic conditions and by the action of yeast and grape invertases. Besides glucose and fructose, the only other hexoses present in wines are mannose and galactose at maximal concentrations of 50 and 200 mg/l, respectively. Several pentoses, which are part of plant pectins, are also found at low concentrations. These include arabinose, 0–1.3 g/l, xylose and ribose, 0.01–0.1 g/l and rhamnose, up to 0.4 g/l. The saccharides, maltose, melibiose, raffinose, melezitose, and stachyose have been found at concentrations of 0–360 mg/l.

Transport. Most sugars are taken up by specific permeases and phosphorylated inside the cell; oligosaccharides are split prior to phosphorylation (Kandler, 1983). Sugar transport may be linked to Δp in symport or antiport; it may be ATP-driven, or it may involve a phosphoenolpyruvate-dependent phosphotransferase system (PTS) (Harold, 1977). All three systems of active transport have been found in LAB (Sapico and Anderson, 1967; Romano *et al.*, 1979; Hasan and Durr, 1974; Hendersen *et al.*, 1979; Konings and Otto, 1983). The secondary transport systems coupled to H^+ symport may be the most important for LAB growing at low pH in the presence of a large pH gradient. Sugar transport by the PTS appears to be most common in homofermentative LAB but apparently is absent in heterofermentative lactobacilli and leuconostocs (Kay, 1978; Romano *et al.*, 1979; Konings and Otto, 1983). A fast ATP-dependent transport system which does not transport glucose was found for fructose in *Leuc. mesenteroides* (Sapico and Anderson, 1967). This may explain the different rates at which glucose and fructose are utilized by *Leuc. oenos*.

Pentoses are taken up by specific permeases and converted into xylulose-5-P (Kandler, 1983). In studies with *Leuc. oenos* (unpublished data), we found that transport of ribose and mannose was constitutive while transport of arabinose was inducible; mannose, galactose and xylose were only used by some of the strains tested.

Metabolism. The pathways which LAB utilize for sugar metabolism have been reviewed by Kandler (1983). Essentially, all species of *Pediococcus* as well as some species of *Lactobacillus* ferment hexoses by the Embden-Meyerhof-Parnas pathway (homofermenters) while species of *Leuconostoc* and some species of *Lactobacillus* lack key enzymes (e.g. aldolase) of this pathway and metabolize hexoses by the hexose monophosphate—phosphoketolase pathway (heterofermenters). This latter pathway is used by all species to metabolize the pentose sugars (Garvie, 1986a).

The ability of LAB to metabolize the sugars in wines is influenced by the low pH of this environment (Henick-Kling, 1986; Davis *et al.*, 1988; Henick-Kling *et al.*, 1991) and the low concentrations at which the sugars occur (Arnink *et al.*, 1991). *Leuconostoc oenos* generally grows slowly on glucose but faster on glucose and malate or glucose and fructose (Krieger *et al.*, 1992). Also, some strains of *Leuc. oenos* do not ferment glucose in standard carbohydrate tests (Henick-Kling, 1983; Krieger, 1989). Cox (unpublished results) found that, in a synthetic growth medium at pH 3.5, *Leuc. oenos* utilizes only a small amount of glucose and grows very slowly. Yet when L-malate is also present, glucose is metabolized rapidly and *Leuc. oenos* grows well. Fructose is metabolized rapidly by all strains. Some of the fructose is used as hydrogen acceptor and reduced to mannitol. In this case, the acetyl phosphate can be converted to acetate via the acetate kinase, forming additional ATP (Gottschalk, 1986). Thus, the amount of acetate produced from carbohydrate fermentation depends on the amount of fructose available and on the redox stage.

Malolactic fermentation. During malolactic fermentation, LAB require a source of carbon for cell maintenance and growth. It is difficult to measure the small concentrations of individual sugars utilized during bacterial growth in wine, because their concentrations may actually increase during wine storage, presumably due to hydrolysis of glycosylated compounds or polysaccharides. Costello, Monk and Lee (1985) and Davis *et al.* (1986a,b) found that the glucose and fructose contents of several wines increased during MLF. However, such increases were greater in a sterile-filtered, control wine indicating that the increase was not due to the action of the bacteria but rather due to the acidic environment and to residual activity of grape and yeast hydrolases.

Radler (1963a) reported that during growth and MLF in wine and synthetic nutrient media, LAB utilize 0.3–2 g/l of sugar. In wine, the growth of LAB is limited due to the low pH and the presence of alcohol. Measurements by Radler (1958) indicated that 40–110 mg/l (dry weight) of cells are formed during MLF. The heterofermentative pathway yields one mole of ATP for each mole of glucose fermented, and one mole of ATP under standard conditions yields 10 g/l (dry weight) of cells. Thus, the production of 40 mg/l biomass for *Leuc. oenos* would require the metabolism of 4 x 0.18 g/l or 0.72 g/l of glucose. The production of D-lactic acid during MLF is a good indication that sugars have been metabolized.

The metabolism of individual sugars in wine by LAB has not been precisely examined. Davis *et al.* (1986a,b) correlated the utilization of hexose and pentose sugars by lactic acid bacteria in several wines during MLF and subsequent conservation. The utilization of glucose, fructose, mannose, galactose arabinose and myo-inositol was observed, but the data were not consistent for individual sugars and varied with the wine.

The concurrent metabolism of malate may have an effect on the end-products formed from the fermentation of sugars. According to Lafon-Lafourcade and Ribéreau-Gayon (1984), the fermentation of carbohydrates during MLF does not lead to a measurable increase in acetate content but, after completion of MLF, continued growth of LAB may produce substantial amounts of acetate if sufficient sugars are available. However, Kandler *et al.* (1973) found that concurrent metabolism of malate and glucose by *Leuc. mesenteroides* did not change the end-products formed from glucose. Never the less, at the low pH of wine, the amount and type of end products produced during carbohydrate fermentation could change. Beelman and Kunkee (1985) found that during concurrent MLF and alcoholic fermentation, none or only small amounts of acetate were produced, although high concentrations of fructose were present in the fermenting juice.

Glycerol

Many strains of *Leuc. oenos* and pediococci, do not metabolize glycerol during MLF (Fleet, 1985; Henick-Kling, unpublished data). In a study of 42 strains of wine LAB, Schütz and Radler (1984) found only three strains of *L. brevis* and one strain of *L. buchneri* which metabolized glycerol anaerobically. In a study of 71 strains of *Leuc. oenos*, 17 strains of *P. parvulus* and 13 strains of *Lactobacillus* spp., Davis *et al.* (1988) reported the fermentation of glycerol by 1%, 12%, and 31% of the strains, respectively. While glycerol may not be metabolized by *Leuc. oenos* during MLF, it can be utilized by other species after MLF (Davis *et al.*, 1986a).

Amino acids, peptides, proteins

The most important source of nitrogen for the growth of wine LAB are free amino acids and short peptides. The LAB isolated from wine require several preformed amino acids for growth; most amino acids are either required or are stimulatory to the growth of *Leuc. oenos* (Radler, 1966; Garvie, 1967; Weiller and Radler, 1972). Various protein and peptide fractions of wine can be used as a source of nitrogen by LAB (Feuillat *et al.*, 1977) The requirement for large amounts of amino acids and peptides as supplied by yeast extract, peptone, and casein digest in synthetic media, suggests that LAB can take up amino acids in the form of peptides. In *Lactococcus lactis*, the uptake of dipeptides and tripeptides is apparently directly coupled to the protonmotive force (Δp); the peptide enters the cell in symport with two protons (Smid *et al.*, 1989 a,b). The carrier apparently has a broad affinity for other dipeptides. Strobel *et al.* (1989) described five carriers for the amino acids glutamine, asparagine, glutamate/aspartate, phenylalanine, and branched chain amino acids which were apparently not Δp-dependent. They noted that the lack of uptake systems for some essential amino acids may be due to their uptake via peptide transport systems. Studies with mutants of *Lact. lactis* subsp. *lactis* which lacked the di-, tripeptide transport system showed that an active transport system was necessary for growth in casein-containing media (Smid *et al.*, 1989b). The uptake of peptides may be equally important for LAB of wine. Feuillat *et al.* (1977) found that the peptide fraction of molecular weight below 1000 supported growth of strains of *L. plantarum*, *L. fermenti*, *P. cerevisiae*, and *Leuc. oenos*, while free amino acids only allowed growth of *Leuc. oenos*. During growth of LAB in grape juice and wine, the concentration of some amino acids decreased strongly while that of others increased (Ribéreau-Gayon *et al.*, 1975; Henick-Kling, 1986; Davis *et al.*, 1986a,b). The total free amino acid concentration at the end of growth may be higher than at the beginning. It is not known whether proteolytic systems play a role in the growth of LAB in wine. Little information is known about the presence of proteinases and peptidases in lactic acid bacteria other than dairy LAB (Kok, 1990). Proteinases have been detected in strains of *L. plantarum* and *L. casei* (El Soda *et al.*, 1986), but (dairy) leuconostocs are generally non-proteolytic. Feuillat *et al.* (1977) found that proteins fractionated from wine did not support the growth of isolates of wine LAB, indicating that these strains did not possess active proteinases and peptidases. Davis *et al.* (1988) did not find extracellular protease production in a survey of strains of *Leuc. oenos*, *P. parvulus* and *Lactobacillus* spp.

Amino acids may serve as source of energy for certain LAB. Arginine can be degraded via citrulline to ornithine and carbamyl phosphate; the carbamyl phosphate then combines with ADP to form one ATP, CO_2, and NH_3 (Bauchop and Elsden, 1960). Strains of *Leuc. oenos* have been reported to degrade arginine (Kuensch *et al.*, 1974), but this conclusion is not supported by Garvie (1986a). Apparently, only some lactobacilli can degrade arginine (Sponholz, personal communication). Some strains of LAB have high amino acid decarboxylase activity, especially those of *Pediococcus* and *Lactobacillus*, some *Leuc. oenos*, which can produce high concentrations of biogenic amines (Mayer, 1974b; Mayer and Pause, 1981; Krieger, 1989). Strains selected for use as starter cultures should be tested for their ability to form biogenic amines.

Vitamins, minerals, other growth factors

Wine is a poor growth medium for LAB, but such growth can be stimulated by the addition of yeast extract that supplies vitamins, minerals, essential amino acids and short peptides (Lafon-Lafourcade, 1970; Lafon-Lafourcade, 1975; Lafon-Lafourcade *et al.*, 1983b). Generally, LAB require a number of vitamins, nucleic acid derivatives, fatty acids or fatty acid esters, and salts. The cations, Mn^{2+}, Mg^{2+}, K^+, and Na^+, are required as enzymatic co-factors but high concentrations are apparently used to scavenge free oxygen radicals (Götz *et al.*, 1980). All species of *Leuconostoc* require nicotinic acid, thiamine, biotin, and pantothenic acid or a pantothenic acid derivative (tomato juice factor) (Amachi, 1975; Garvie, 1967); *Leuc. oenos* also requires guanine, adenine, xanthine, and uracil. Riboflavin and pyridoxal are stimulatory to *P. damnosus* (Garvie, 1986b). Most lactobacilli require pantothenic acid and nicotinic acid; hetero-fermentative lactobacilli also require thiamine, and various other vitamins are required by some strains (Kandler and Weiss, 1986).

Lafon-Lafourcade (1975) showed that wines differed markedly in their ability to support the growth of malolactic bacteria. The rate of MLF by the same strain in different wines containing no free SO_2 but adjusted to similar values for pH and concentration of ethanol and malate, varied by a factor of ten. On the other hand, some strains are better adapted to growth in a particular wine than others (Lafon-Lafourcade, 1975; Wibowo *et al.*, 1988). The growth of some LAB in wine could be encouraged by the presence of fungal polysaccharides, as produced on grapes by *Botrytis cinerea* (San-Romao *et al.*, 1990). Such polysaccharides could act by protecting LAB against the inhibitory action of some fatty acids (Lonvaud-Funel *et al.*, 1987). Substances produced by other grape parasitic fungi, such as *Aspergillus*, *Penicillium*, and *Rhizopus*, can inhibit growth of LAB.

Temperature and ethanol

Temperature and ethanol both affect growth rate and length of lag phase of LAB. High temperatures decrease their ethanol tolerance and high ethanol concentrations decrease the optimal growth temperature (Dohman, 1982; Asmundson and Kelly, 1990). The optimum growth conditions for strains of *Leuc. oenos* were typically near 30°C and 0–4% ethanol. The growth rate decreased linearly with increasing alcohol concentration, with 14% being the upper limit tolerated. For some strains, 2–4% ethanol stimulated growth by 10–15% but other strains had maximum growth rate in the absence of alcohol (Asmundson and Kelly, 1990). Optimum growth in the presence of 10–14% alcohol occurred between 18 and 25°C compared with 30°C at 0–4% alcohol. Cell yield was not as strongly affected by ethanol and temperature as growth rate. Maximum cell yield for *Leuc. oenos* occurred in media with 0–8% ethanol at 22.1°C, but in media at 14% ethanol and 20.8°C, the cell yield was still 64% that of the maximum. Thus, the optimal conditions for growth (shortest lag time, fastest growth rate, highest cell yield) of *Leuc. oenos* in wine with 10–14% alcohol is 18–20°C. Temperatures above 25°C should be avoided because this can completely inhibit growth. According to Lafon-Lafourcade (1975), the availability of nitrogen nutrients can also affect the ethanol tolerance of *Leuc. oenos*.

Davis *et al.* (1988) surveyed a range of wine isolates of *Leuc. oenos*, *Pediococcus* spp. and *Lactobacillus* spp. for their tolerance of ethanol. The proportion of isolates able to grow in the presence of ethanol decreased substantially as the concentration was increased from 12.5% to 15%. Isolates of *Lactobacillus* and *Pediococcus* were more tolerant of high concentrations of ethanol than *Leuc. oenos*.

Malic acid

Grape juice may contain 1–8 g/l of malate (Sponholz, 1989; Zoecklein *et al.*, 1990). During grape maturation, the concentration of malate decreases and, in cool climate regions, final concentrations of 2–5 g/l are often found. In warm climates, the malate content of juice is lower (<2 g/l). Malate is metabolized by LAB by either of three different pathways (Radler, 1986). Most LAB obtained from wine possess the malolactic enzyme (Table 10.1) which decarboxylates L-malate to L-lactate without free intermediates (Lonvaud-Funel and Strasser de Saad, 1982; Caspritz and Radler, 1983; Spettoli *et al.*, 1984; Naouri *et al.*, 1990). *Lactobacillus casei* and *Lactococcus faecalis* possess a true malic enzyme which converts malate to pyruvate and enables their growth on malate as carbon source (London and Meyer, 1969; London *et al.*, 1971; Schütz and Radler, 1974). A third pathway exists in *Lactobacillus fermentum*, which produces D- and L-lactate, acetate, succinate and carbon dioxide from malate (Radler 1986). At low pH (<4) mainly lactate is formed, while at high pH (>5) mainly succinate and acetate are formed. Apparently, at low pH the malate is converted to lactate via oxaloacetate and pyruvate, while at high pH, the succinate is formed via fumarate. The metabolic pathway of malate utilization by wine LAB is described later.

Citrate, pyruvate, and α-ketoglutarate

Citrate is present in wine at concentrations of 0.1–0.7 g/l and pyruvate may be present at 0.03–0.3 g/l (Lemperle and Lay, 1989; Sponholz, 1989; Zoecklein *et al.*, 1990). In wine, citrate and pyruvate are mainly converted to lactate and acetate (Lafon-Lafourcade and Ribéreau-Gayon, 1984; Dittrich, 1987). The citric acid cycle is not involved in the utilization of citrate by wine LAB because it is not active under anaerobic conditions, and because wine LAB that have the malolactic enzyme, do not produce malate dehydrogenase which is part of the

TABLE 10.1 Malate degrading enzymes in lactic acid bacteria.

Bacterial species	*Malic enzyme*	*Malolactic enzyme*	*Malate dehydro-genase*	*Reference*
Lactobacillus casei	+	+	–	(1), (4), (7)
Lactobacillus plantarum	–	+	–	(2), (3), (4)
Lactobacillus brevis	–	+		(3), (4)
Lactobacillus buchneri		+		(3), (4)
Lactobacillus delbrueckii		+		(3), (4)
Lactobacillus hilgardii		+		(3), (4)
Lactobacillus delbrueckii subsp. *lactis*		+	+	(3), (4)
Leuconostoc mesenteroides	–	+	–	(2), (3), (4), (5)
Leuconostoc oenos	–	+		(3), (4), (6), (9)
Leuconostoc citrovorum		+		(3), (4)
Pediococcus cerevisiae	–	+(–)		(3), (4)
Streptococcus * *cremoris*		+		(3)
Streptococcus * *diacetilactis*		+		(3)
Streptococcus * *faecalis*		–	–	(3), (4)
Streptococcus * *lactis*	–	+		(3), (8)

(+) present, (–) absent.
* Genus now called *Lactococcus*
Kandler *et al.*, 1973 (1), Schütz and Radler, 1973 (2), Schütz and Radler, 1974 (3), Caspritz and Radler, 1983 (4), Lonvaud-Funel and Strasser de Saad, 1982 (5), Spettoli *et al.*, 1984 (6), Kandler, 1983 (7), Renault *et al.*, 1988 (8), Naouri *et al.*, 1990 (9).

citric acid cycle (Table 10.1). Wine LAB, split citrate by citrate lyase into acetate and oxaloacetate; oxaloacetate is then decarboxylated to pyruvate which is converted to acetate, ethanol, lactate, diacetyl, acetoin, or butylene glycol (Dittrich, 1987; Kandler, 1983). Pyruvate is generally converted to lactic acid by lactate dehydrogenase. However, leuconostocs and *L. casei* can degrade pyruvate to acetate with the gain of one ATP (Kandler, 1983). In this pathway, pyruvate is first split by pyruvate-formate lyase to formate and acetyl-CoA; the acetyl-CoA is phosphorylated and the resulting acetyl-P is converted to acetate with the production of one ATP. The amount of pyruvate metabolized in this way depends on the availability of other carbohydrates and the pH of the medium.

Under glucose limiting conditions, almost all pyruvate may be converted to acetate (and ethanol). In *Leuc. lactis*, as in *Leuc. oenos*, and *L. plantarum*, intracellular pH decreases as the pH of the growth medium decreases (Henick-Kling, 1986; Fitzgerald *et al.*, 1990; Olsen *et al.*, 1991). Fitzgerald *et al.* (1990) showed for *Leuc. lactis*, that the activity of lactate dehydrogenase was approximately 20 times lower at pH 5.0 than at the optimum pH of 6.6–7.2. They concluded that, at low pH, more pyruvate is available for reactions other than lactate formation, and that this may be one reason why diacetyl and acetoin are formed by *Leuc. lactis* only during growth at low pH.

The metabolism of pyruvate, acetaldehyde, and α-ketoglutarate by LAB is important in winemaking because these substances bind SO_2. In wines where they have been removed by the growth of LAB, smaller amounts of SO_2 can be used to protect against oxidation and microbial spoilage (Dittrich, 1987; Lemperle and Lay, 1989). *Leuconostoc oenos* decarboxylates α-ketoglutarate to succinic semialdehyde, which is reduced to 4-hydroxybutyrate, and subsequently forms γ-butyrolactone; part of the succinic semialdehyde is oxidized to succinate (Radler, 1986).

Succinate, fumarate

Succinate produced by yeast during fermentation may be present in wine at 0.2–2 g/l (Dittrich, 1987). It is not metabolized by LAB. Fumarate is generally only present in small amounts but may be added after alcoholic fermentation to prevent growth of LAB. Fumarate is active against LAB at concentrations between 0.4–1.5 g/l (Cofran and Meyer, 1970; Pilone *et al.*, 1977). Fumarate is bacteriocidal and the effect is synergistic with decreasing pH. In some cases, the bacteria may overcome inhibition by fumarate by converting it to malate by fumarase.

Gluconate

Gluconate is formed by molds and acetic acid bacteria infecting the grapes and may be present in wines at concentrations of 0.05–1.1 g/l (Lemperle and Lay, 1989; Sponholz, 1989). Gluconate can be metabolized by *Leuc. oenos* and *L. brevis* with the constitutive or inducible gluconokinase, apparently via the 6P-gluconate pathway (hexose monophosphate pathway) to lactate, acetate, and CO_2 (Radler, 1986).

Tartrate and lactate

The other main acids in wine, tartrate (2–10 g/l) and lactate (0.1–7 g/l) can only be metabolized aerobically (Kandler, 1983; Dittrich, 1987). This happens in wine exposed to air and it results in complete spoilage.

Sorbate

Sorbate may be added to wines in the USA at concentrations up to 300 mg/l. It is effective against yeast at 150–250 mg/l in wine (Splittstoesser and Stoyla, 1989).

Leuconostoc oenos may metabolize sorbate causing the geranium off-odor (Edinger and Splittstoesser, 1986). The bacteria reduce sorbate to sorbic alcohol which, subsequently, can form 3,5-hexadien-2-ol and 2-ethoxyhexa-3,5-diene, the latter being responsible for the off-odor.

MALOLACTIC FERMENTATION

Biological Advantage

Wine LAB rapidly degrade L-malate during growth using the malolactic enzyme, (Table 10.1). The malolactic enzyme, requiring NAD and Mn^{2+}, decarboxylates L-malate to L-lactate without free intermediates (Lonvaud and Strasser de Saad, 1982; Caspritz and Radler, 1983; Spettoli *et al.*, 1984; Naouri *et al.*, 1990). If pyruvate was produced by the reaction, it could be metabolized further to yield ATP; however, the small amount of energy produced would not be enough to change the growth yield but it could stimulate early growth (Kunkee, 1974). Reduction of NAD is not detected although some hydrogen is exchanged with the cell's pool; no carbon is exchanged with other fermentation pathways (Alizade and Simon, 1973; Krauss *et al.*, 1976).

Since no substrate level phosphorylation (and ATP production) is directly linked to the decarboxylation of malate by the malolactic enzyme, it has been difficult to understand the physiological advantage of the MLF. Kandler *et al.* (1973) could not measure an increase in cell yield of *Leuc. mesenteroides* during concurrent fermentation of glucose and malate, and concluded that the MLF was not energy yielding and that its only physiological advantage to the organism was the increased pH of the growth medium which extended growth. Pilone and Kunkee (1976) demonstrated that, during MLF by *Leuc. oenos*, the rate of growth at low pH was increased, independent of the change of pH to the growth medium. Based on studies of intracellular pH and protonmotive force (Δp) in cells during MLF, Henick-Kling (1986) proposed that *Leuc. oenos* obtained an energetic advantage from an increased intracellular pH and from an increased Δp. Cox and Henick-Kling (1989) showed that ATP was produced during MLF and that its production was linked to Δp across the cell membrane and functional ATPase. This general mechanism apparently works in all LAB which posses the malolactic enzyme (Cox and Henick-Kling, 1990). Olsen *et al.* (1991) have described an electrogenic malate transport system in *L. plantarum* which, coupled with electroneutral lactate efflux, creates a Δp. At low pH (3.5), Δp increases during MLF from approximately 70 mV to 130 mV in starved cells of *L. plantarum* (Cox, 1991). Thus, energy is provided to the cell in the form of Δp which can be used for transport processes or be converted into ATP via the membrane bound ATPase. ATP is produced by the cells when Δp is above 90 mV. The ATP production from malolactic fermentation is independent of glucose metabolism (Cox 1991). Under glucose limited conditions (glucose 0.5 g/l) at pH 5.0, Renault *et al.* (1988) found 25% increased biomass yield for *Lact. lactis* when malate (20 g/l) was present in the medium. Olsen *et al.* (1991) and Cox (1991) measured a 28–50% increase in cell yield for *L. plantarum* during fermentation under glucose limited and low pH conditions. Growth yield from glucose was increased during growth in media of pH below 5.5, at L-malate concentrations of more than 1.3 g/l, and glucose concentrations below 7 g/l.

In the model presented by Olsen *et al.* (1991), one molecule of malate enters the cell, is decarboxylated, and one molecule of lactate leaves the cell with one H^+; this is equal to the translocation of one H^+ to the outside (Figure 10.1). Calculations of the H^+:malate stoichiometry during MLF showed a value of 2.6:1 (Cox, 1991). This high value can not be fully explained. From the malate transport studies, a H^+:malate stoichiometry of 1:1 would have been expected. But in the energy yield (H^+ equivalents) of the malolactic system, the contribution of all components, malate transport, malate decarboxylation, and lactate export must be considered. The data also suggest, as has been shown for other transport processes, that stoichiometry is variable (Kaback, 1990). Thus, 2.6:1 is the maximum value of this system and under various conditions of growth the actual value is lower. In fact, under conditions of large supply of ATP from carbohydrate fermentation, the ATP production from malolactic fermentation may be uncoupled. With a H^+:malate stoichiometry of 1–2.6:1, the metabolism of 2.4 g (18 mM) of malate would generate 6–15.6 mM of ATP. In *L. plantarum* growing at pH 3.5, the actual growth yield from glucose is only half the expected 20 g/l for each mole of glucose metabolized (Cox, 1991). In other words, during growth at low pH (3.5) only half the available energy (ATP) is converted to cell mass. Cox found that under the same low pH conditions, the conversion

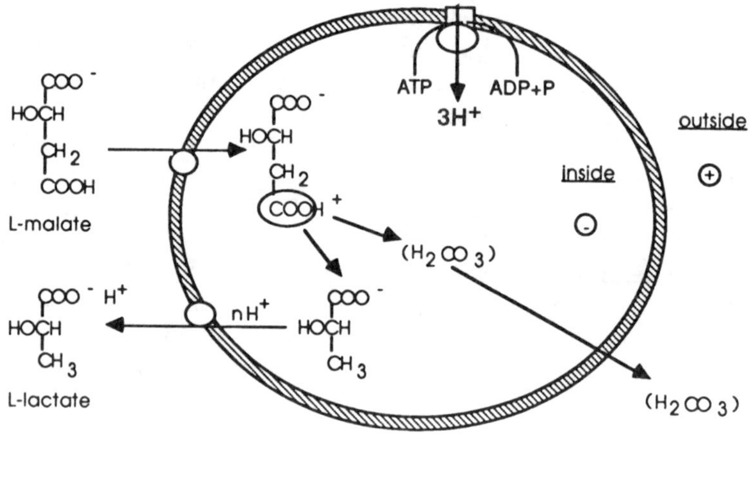

FIGURE 10.1 Model of the ATP-generating mechanism for the malolactic conversion. In this model, the malolactic enzyme in the cell decarboxylates L-malate and the electrogenic transport of L-malate with electroneutral export of L-lactate provides energy through the proton gradient (Δp) for transport processes and it can be converted by the membrane ATPase into ATP. (Henick-Kling *et al.* (1991), reprinted with permission of *Revue Française d'Oenologie.*)

of the available ATP from malolactic fermentation to cell mass was also approximately half. Based on the calculations by Kashket (1982) and Maloney (1983), three H^+ must enter through the membrane-bound ATPase to drive the synthesis of one ATP. This stoichiometry was confirmed by Cox (1991) for the ATPase in *L. plantarium*. If *Leuc. oenos* ferments two g/l of glucose (or fructose), it produces 11 mM of ATP; an additional six mM ATP from the 2.4 g/l malate then represents an additional 55% of ATP for the cell. When the organism ferments 10 g/l glucose, the additional energy gain from 2.4 g/l malate only amounts to 10% (which would be hard to measure). Evidently, the observed increase in cell mass will depend on the amount of carbohydrates fermented. Under conditions where the amount of ATP available from carbohydrate fermentation is limited because of low sugar concentration or because of low pH inhibiting uptake and catabolism (Figure 10.2), the additional ATP available from MLF can give a large increase. Glucose and fructose catabolism are similarly inhibited in *Leuc. oenos* and in *L. plantarum*; contrary to this inhibition of sugar utilization, the rate of malate utilization is maximal at low pH values (<4.5). Furthermore, the amount of ATP produced at pH 3.5 from the malolactic conversion is about double that produced at pH 5.5, the optimal pH for sugar catabolism and growth (Cox and Henick-Kling, 1989). How much of this additional available ATP is converted into additional cell mass depends on the amount of ATP needed for maintenance energy and how much carbon is available from carbohydrate fermentation.

Preliminary studies on the control of glucose and malate catabolism show that glucose (20 mM) catabolism does not affect the rate of malate catabolism but, at low pH, malate catabolism can inhibit glucose utilization; at pH 3.5, the rate was significantly slower (0.06), at pH 5.5 the inhibition was less, and not significant at 0.1% (0.178) (Henick-Kling and Gavitt, unpublished data). This preference is also evident in the delay in onset of glucose utilization and in the subsequent slower catabolism rate when malate is also present.

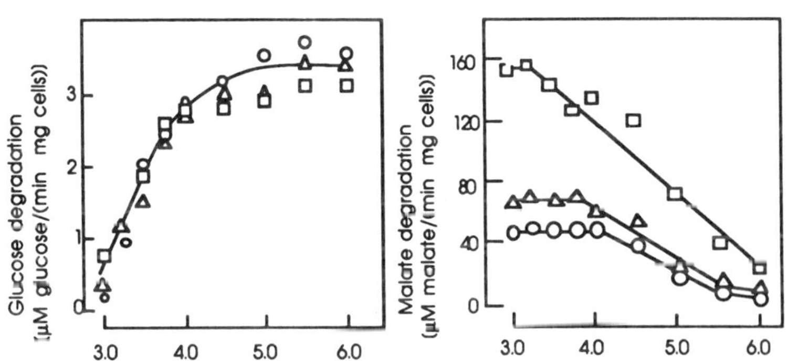

FIGURE 10.2 Rate of catabolism of glucose and L-malate by *Leuconostoc oenos* in tartrate-phosphate buffer of various pH. Cells were grown in MRSM medium of pH 5.5 (○), 4.5(Δ), and 3.5 (□). (Henick-Kling *et al.* (1991), reprinted with permission of *Revue Française d'Oenologie*.)

The malolactic conversion is a novel pathway linked to the protonmotive force for the generation of ATP during growth at low pH. The system does not provide carbon for growth, therefore the bacteria must also be able to utilize sugars as a carbon source. In lactic acid bacteria isolated from wine, malate apparently is preferred as an energy source during growth at low pH and at the beginning of cell growth. Malate-dependent ATP synthesis can be initiated in starved cells. The increased pHi, Δp, and ATP from MLF helps the cell to start to grow. Also, the increased extracellular pH allows extended growth.

Malate Transport

Henick-Kling (1986, 1990) showed that the rate of MLF by whole cells of *Leuc. oenos* and *L. plantarum* is controlled by the rate of malate transport. Apparently, all malate is taken up via L-malate specific permease (Olsen *et al.*, 1991). No uptake of malate was measured in cells of *L. plantarum* which had not been induced by growth in malate containing medium. Malate uptake can be driven by an artificial ΔpH and by lactate efflux. The malate transport system in *L. plantarum* has two transport activities with different K_m and V_{max}. The high affinity (K_m 0.5 mM), low V_{max} transport activity was energy-dependent, while the low affinity (K_m 5 mM), high V_{max} transport activity was driven by a large malate gradient. Studies on the malate-dependent ATP production showed that ATP was only produced at external malate concentrations higher than 5 mM. Thus, the low affinity, ATP energy independent transport system is involved in the malate-dependent ATP production (Cox, 1991).

Malolactic Activity

The rate of malolactic fermentation in wine is directly linked to cell density and secondly to the specific malolactic activity of the cell. When studying the malolactic activity of a culture, it is important to separate cell growth and specific activity of the cells. Studies of various inhibitory and stimulatory factors must differentiate their effects on each of these systems because any part of the whole system can be affected differently. Cell viability and growth can be evaluated with standard methods, plate counts, cell dry weight, protein content, optical methods. The specific activity of the cells is best studied under non-growing conditions in washed cell suspensions or under constant, limited growth in a chemostat. Specific biochemical studies of transport activity, protonmotive force, membrane composition, and enzyme activity are necessary to understand the regulatory mechanisms of the MLF. Malolactic activity of the cells is highest during the early stages of growth (Krieger *et al.* 1992). On the other hand, survival of a culture of *Leuc. oenos* and consequently malolactic activity upon inoculation into wine was highest when the culture was harvested 18–24 hours after it entered stationary phase (Krieger and Hammes, 1988).

pH

The pH strongly affects the malolactic activity of the cell (Lafon-Lafourcade, 1970; Henick-Kling, 1990). While sugar utilization and growth of *Leuc. oenos* are inhibited by low pH (Figure 10.2), the rate of malate degradation is highest at low pH. Also, the malolactic activity of the cell is affected by the pH of the preculture medium. Cultures of *Leuc. oenos* grown in media of low pH (3.5) degrade malic acid more rapidly than cultures grown in media of high pH (5.5). The pH of the growth medium also affects the specific activity of the malolactic enzyme. Highest malolactic activity was measured in cells grown in media of pH 3.5–4.0; it was less in cells grown at pH 3.0 and higher pH (Lafon-Lafourcade, 1970; Henick-Kling, 1990). A comparison of the specific malolactic activity in cell extracts of cultures grown in media of initial pH from 3.5 to 6.0 showed that the activity occurred in cells grown at low pH. Also, the transport activity in *L. plantarum* was higher in cells grown at low pH (3.5) than in cells grown at high pH (6.0) (Henick-Kling, 1990; Olsen *et al.*, 1991).

Temperature, Ethanol, SO_2

The optimum temperature for malate degrading activity of whole cells is 30–35°C, although the optimum temperature for growth is near 25°C (Lafon-Lafourcade, 1970; Asmundson and Kelly, 1990). Ethanol concentrations of 5–12% are not inhibitory of malolactic activity in most LAB, but the potential for growth is decreased above 6%. The amount of ethanol tolerated in the growth medium is highly strain dependent and also depends on the temperature, pH, and nitrogen status of the culture medium (Lafon-Lafourcade, 1975; Davis *et al.* 1988; Britz and Tracey, 1990). Free SO_2 at 10 mg/l kills *Leuc. oenos*. Bound SO_2 at higher concentration (>30 mg/l) delays growth and limits the final cell density obtained in the wine; bound SO_2 at more than 50 mg/l may completely inhibit growth (Wibowo *et al.*, 1985). The malolactic activity of the cell is equally sensitive to SO_2; 20 mg/l bound SO_2 reduces malolactic activity by 13%, 50 mg/l reduces it by 50%, and 100 mg/l inhibits it completely (Lafon-Lafourcade, 1970).

Lactate

Lactic acid at a concentration of 0.5 g/l reduces growth of *Leuc. oenos* in a synthetic medium at pH 3.5 (Cox, 1991) and, at high concentration (>3 g/l), it may completely inhibit growth (Lafon-Lafourcade, 1970). Yet, MLF was induced successfully in a wine which contained 4.68 g/l lactate, with a starter culture of initial density of 10^7 CFU/ml (Krieger, unpublished data). High lactate concentrations in the wine do limit the amount of energy (ATP) obtained from MLF by slowing export of lactate from the cell and reducing the ΔpH produced (Cox, 1991).

MALOLACTIC STARTER CULTURES

In North and South America, Australia, New Zealand, South Africa, but also in Europe, starter cultures are being used to induce MLF. Improved sanitary conditions in winemaking have resulted in the production of wines with improved quality, free of microbial defects but, also, this has led to wines of 'simple' and 'unbalanced' character. Moreover, the MLF may not occur spontaneously or can be very unpredictable in modern wineries. It has long been recognized in France that MLF is an essential part of winemaking as a deacidification process, but it is increasingly being recognized that MLF is necessary also to develop body (mouthfeel), complex aromas, and balanced flavors in red wines, some white wines, and some base wines for premium sparkling wine. The use of starter cultures to induce MLF can ensure more rapid onset and completion of this reaction, reduce the potential for spoilage by other lactic acid bacteria, reduce potential interference by bacteriophages and, further, give control over the strain responsible for the reaction and its contribution to wine flavor (Henick-Kling, 1988).

Typically, spontaneous MLF occurs in wines of pH above 3.4 and is delayed and irregular in wines of pH below 3.4 where it is most desirable for deacidification. This behavior is due to the strong inhibition of growth of LAB at low pH. Inoculation of wines with a large number of viable bacteria avoids the necessary growth before MLF is carried out. With a suitably prepared starter culture, a wine can be inoculated with a cell density of 10^6–10^7 viable bacteria/ml. This represents a stationary culture at maximum cell density in wine, and little or no further growth is necessary for complete conversion of the malic acid. Such high populations are less prone to phage interference and ensure dominance of the introduced culture over indigenous LAB. Inoculation with a dominating, selected strain of LAB gives protection against undesirable strains of *Leuconostoc* or *Pediococcus* and *Lactobacillus*. Thus, the winemaker can control which strain of bacteria carries out MLF and the flavor characteristic it gives to the wine. Stationary cell cultures may also be used in cell reactor systems (Gestrelius and Kjaer, 1983).

Strains of LAB selected for induction of MLF may comprise those of *Leuc. oenos*, *Lactobacillus plantarum*, *Lactobacillus hilgardii*, *Lactobacillus brevis*, and *Lactobacillus casei*. Strains of *Leuc. oenos* are most preferred because of their tolerance to low pH and high alcohol, and because of the flavors and mouthfeel they produce. Other options for reducing the content of malic acid in wine by using the yeasts *Schizosaccharomyces pombe* or *Schizosaccharomyces maledivorans* have not given satisfactory wine quality (Dittrich, 1987).

Bacteriophage

Leuconostoc oenos can be infected and lysed by bacteriophage during growth in wine, resulting in partial or complete inhibition of MLF (Sozzi *et al.*, 1982; Cazelles and Gnaegi, 1982; Davis *et al.*, 1985a; Henick-Kling *et al.*, 1986a; Jarvis,

1989). Starter cultures of *Leuc. oenos* can be protected against phage infection by selecting strains which are resistant to a broad range of phage (Sozzi *et al.*, 1982; Gnaegi *et al.*, 1984). Phage resistant strains can be isolated from the winery environment or from cultures exposed in the laboratory to selected phage (Henick-Kling *et al.*, 1986b). Starter rotation and the use of multistrain starter cultures consisting of a combination of strains with different phage sensitivity can give a high degree of protection against phage infection (Sandine, 1979). Although several studies have shown that *Leuc. oenos* can be inhibited by phage and that some strains of *leuc. oenos* are lysogenic, little is known about how rapidly various phage of *Leuc. oenos* multiply in wine and under what conditions lysogenic strains become lytic.

Characterization of the lytic activity of bacteriophages of *Leuc. oenos* isolated from wine has shown that, at low pH, the phage is most active, infecting its host culture during the early stage of growth (Henick-Kling *et al.*, 1986b). Thus, phage interference can be minimized by inoculating wine with stationary phase cultures (Henick-Kling, 1988). The risk of phage interference is minimized, also, by reducing the number of phages in the winery environment by maintaining good sanitation practices. *Leuconostoc oenos* can overcome phage inhibition under conditions which favor its growth, i.e. temperature above 15°C, pH >3.4, and the absence of SO_2. Recent investigations show that 50% of all strains of *Leuc. oenos* isolated from wine contain temperate phages (Krieger, personal communication; Patel, 1990). Bacteria with temperate phage have been used in starter cultures for wine without apparent problems, but their continued use may cause induction, thereby making them in effective. More likely, temperate phage may pose problems during the industrial preparation of starter cultures (Park and McKay, 1975). Under these conditions, they may be induced to cause lysis, thus, disrupting starter production.

The possession of temperate phage, on the other hand, gives the bacteria some protection against phage infection. This may be an advantage in some cases when starter cultures are exposed to phage during extended growth in wine. Bacteriophage resistance in lactococci is generally a plasmid encoded property. Studies with strains of *Leuc. oenos* did not find a correlation between the presence of plasmids and phage resistance (Krieger, Van Vuuren, personal communication). In lactococci, plasmids have been described which determine phage resistance by inhibiting adsorption, by restricting phage DNA, and by aborting phage infection (Klaenhammer, 1987; Sanders, 1988). Hill *et al.* (1990) describe host encoded mechanisms of phage resistance that are determined by the phage genome. The phage-encoded resistance contains a phage origin of replication which, in trans, may interfere with phage replication by titration of DNA polymerase or other essential replication factors. The described resistance is not based on reduced phage adsorption or action of a restriction and modification system. The various resistance mechanisms apparently have different host specificity, thus making it feasible to combine several resistance factors to make a strain resistant to several groups of phage. The possibility of combining more than one phage resistance mechanism in one strain has been exploited by Klaenhammer (1989) and Alatossava and Klaenhammer (1991).

Production

For evaluation of the activity of a starter culture in wine, factors affecting cell growth and viability and factors affecting the activity of the malolactic system should be considered separately. During the preparation of malolactic starter cultures, conditions must be chosen that give a large yield of biomass which retains high viability and malolactic activity after inoculation into wine. First, a suitable strain must be isolated from a naturally occurring MLF. Most starter media used today are generally based on apple juice, similar to that proposed by Lafon-Lafourcade (1970). Apple, pear, and grape juice may be used (Champagne *et al.*, 1989). The difficulty of obtaining and storing fresh (non-sulfited) juice has been a major factor in the high cost of producing malolactic cultures. Also, the quality of the juice is not consistent, varying with the supplier and the growing season. Differences in the quality of the juice can significantly change the rate of growth and cell yield. The use of a synthetic medium has the advantage that dry ingredients with known quality can be used at any time of the year. Henick-Kling and Krieger (1990) have described a synthetic medium which gives high growth yields and cells with high viability and malolactic activity. Hayman and Monk (1982) evaluated the effect of adding wine to media for the production of *Leuc. oenos* and found that inclusion of 40–80% wine in the medium improved survival and malolactic activity of the culture. Krieger and Hammes (1988) found that the best time for harvest of *Leuc. oenos* was 18–24 h after the culture had entered stationary phase. Cells harvested in this period had the highest survival and malolactic activity after inoculation into wine, while cells harvested earlier, died off quickly (18% survival after 5 days) and could not induce MLF.

A major problem in the development of malolactic starter cultures has been susceptibility of the bacterial cells to damage during the freeze-drying process. The starter culture must be stable during storage for several months so that it can be prepared commercially in large quantity ahead of vintage and be conveniently stored in the winery until use. To preserve the bacteria, they must either be frozen (–20°C or –80°C) or their moisture content reduced to under 4% so that they can be stored without much loss of activity in a standard refrigerator (2–4°C). Since lactic acid bacteria can not be inexpensively spray-dried like yeast, they must be freeze-dried. With addition of cryoprotectants to the concentrated culture before freezing and with modifications to the freezing and drying process, very high culture viability can be obtained (better than 95%). These processes add considerable cost to the final product, thus, making it uneconomical to inoculate wine without a subculture expansion of the starter. A less expensive method of drying could significantly reduce the cost of starter cultures. Newer fluid bed drying techniques might allow this. Kearney *et al.* (1990) reported an improvement in viability of lyophilized cultures of *L.*

plantarum after immobilizing the cells in calcium-alginate beads which incorporated cryoprotectants. Some LAB are preserved by freezing (–20°C or below –40°C). Frozen concentrates of LAB have 10^{11} cfu/ml with very high viability, but encounter problems with distribution; wineries, unlike dairies, are not prepared to store frozen cultures. One frozen starter culture has been on the market for several years in the USA. All other commercial starters are freeze-dried or liquid cultures prepared in small volume by private laboratories.

Maintenance of Activity

Malolactic strains freshly isolated from wines frequently show superior viability and malolactic activity to strains that have been maintained for some time in culture collections (Beelman *et al.*, 1980; Henick-Kling, 1983; Henick-Kling *et al.*, 1989; Krieger and Henick-Kling, 1991). While there are regional influences on the behavior of particular strains, it is likely that there may be a gradual loss in their malolactic performance during long term maintenance and successive subculture in synthetic media. Conditions for the industrial cultivation of malolactic bacteria as well as the conditions selected for their reactivation prior to inoculation into wine are likely to have important influences upon their survival and growth in wine, and their malolactic activity. At present, too little is known about these relationships and further research is needed. However, it is known that starter cultures grown in media at low pH (< 4.5) have better survival and higher malolactic activity in wine than cultures grown at high pH (>4.5) (Lafon-Lafourcade, 1970, 1975; Henick-Kling, 1988, 1990). Further, it is generally known that the temperature difference between the growth medium and the wine should not be greater than 10°C, otherwise the cell would suffer substantial damage upon inoculation. Significant studies by Lonvaud-Funel (1987) Desens and Lonvaud-Funel (1988) and Lonvaud-Funel and Desens (1990) have shown that the toxicity to LAB of ethanol and fatty acids produced by yeasts during alcoholic fermentation is related to the lipid composition of the cell membrane and that this composition is markedly affected by culture age and medium formulation.

Reactivation

Reactivation is necessary for all cultures produced today, whether grown in fruit juice medium or a synthetic medium, be they frozen or freeze-dried preparations. Direct inoculation of rehydrated cultures into wine leads to a sharp 'die-off' of the bacteria (Krieger and Hammes, 1988). The initial cell viability may decrease from 10^7 cfu/ml to less than 10^4 cfu/ml (Beelman *et al.*, 1980; Costello *et al.*, 1985; Krieger *et al.*, 1986; Rodriguez *et al.*, 1990). Thus, to compensate for such loss, the inoculation density would have to be increased by 100-fold. Such high inoculation rates are not economical. However, Krieger and Hammes (1988) reported acceptable results with direct inoculation of frozen cultures. Further development of frozen culture technology or a dramatic improvement in the preparation of dried bacteria which improves their survival after inoculation into wine could make direct inoculation economical.

The reactivation medium generally used is that described by Lafon-Lafourcade (1970) and Lafon-Lafourcade *et al.* (1983b). Non-sulfited grape juice is diluted by half with water, the pH adjusted to 4.0–4.5 (with $CaCO_3$) and 3–5 g/l of yeast extract is added. The medium is inoculated with the bacterial preparation at approximately 10^7 cfu/ml together with yeast at 10^4 cfu/ml, and grown at 25°C for about 6 days. At this time, alcoholic fermentation by the yeast is completed and the bacteria should have grown to density of $1–5\times10^9$ cfu/ml. The reactivation step also serves as an expansion of the culture. This reactivated culture is inoculated at a dilution of 1:1000 to achieve a cell density in the wine of approximately 5×10^6 cfu/ml. Wine which is expected to be difficult to induce MLF should be inoculated at a higher rate of $2–5\times10^7$ cfu/ml. This mixed culture system of reactivation has an added advantage that it does not necessarily require aseptic conditions. Contamination and spoilage by fast growing spoilage yeast and unwanted bacteria are avoided by the concurrent alcoholic fermentation. The disadvantage of this method is that strong yeast growth and production of CO_2 can make it difficult to monitor the growth of the bacteria, introducing a risk that the culture may be transferred prematurely, before it has reached its maximum population density and highest viability and malolactic activity. Furthermore, there is the possibility that growth of the bacteria may be inhibited by the yeast. To monitor progress of the reactivation, growth of the bacteria should be followed microscopically. In this way, bacterial cell density can be determined and the stage of growth estimated (Hogarth and Croser, 1984; Krieger *et al.*, 1990). When the stationary phase of growth is reached, cells of *Leuc. oenos* have formed characteristic medium length to long chains with sharp turns and twists. Lafon-Lafourcade *et al.* (1983b) described differences in the activity of malolactic cultures which had been prepared with different yeasts. Bacteria prepared with *S. cerevisiae* gave more rapid growth and more rapid MLF after inoculation into the wine than cultures prepared with *S. cerevisiae* strain *bayanus*.

Krieger *et al.* (1990) proposed that preparations of starter bacteria be reactivated under aseptic conditions in sterile grape juice medium. This allows better growth of the bacteria, malolactic activity is evident by the amount of liberated CO_2, and culture growth is easily seen as turbidity. When inoculated with $10^7–10^8$ cfu/ml, the culture should reach stationary phase by 2–3 days. At this stage, the bacteria have started to settle, clearing the top layer of the medium and forming a thick sediment. Pasteurized wine (¼ to ⅓ of final volume) may be added to the reactivation medium.

The starter culture must be pure. If a culture of *Leuc. oenos* is contaminated with lactobacilli, the latter will grow more rapidly under the reactivation conditions and predominate in the inoculum added to the wine. The inoculum of contaminating bacteria likely will die off because, most often, they are non-wine lactobacilli or lactococci that have arisen from cross-contamination in the starter culture factory. In this case, the starter culture fails and no MLF occurs.

However, if the contaminants survive and grow, they can cause undesirable flavors or ropiness.

Time of Inoculation

Generally, spontaneous MLF occurs after completion of alcoholic fermentation, but occasionally it can occur at the same time as alcoholic fermentation. Simultaneous malolactic and alcoholic fermentations have the advantage that all fermentation is completed early and the wine can immediately be racked, sulfited, and filtered for further storage or bottling. Simultaneous alcoholic and malolactic fermentation does not happen often because the bacteria (usually *Leuc. oenos*) are not present at sufficiently high numbers in the juice at the start of alcoholic fermentation to effect MLF (Costello *et al.*, 1983; Lafon-Lafourcade *et al.*, 1983a; Fleet *et al.*, 1984; Davis *et al.*, 1986b). However, simultaneous alcoholic and malolactic fermentation should be possible if malolactic starter culture is added to the juice with the yeast starter. This attractive proposition has been researched and adopted with good success in commercial wineries (Beelman and Kunkee, 1985). These authors showed that MLF in the presence of large amounts of fermentable sugars does not necessarily lead to the production of excessive amounts of acetic acid by the bacteria, as long as yeast fermentation starts promptly and goes to completion. However, there is a danger that strong bacterial growth can inhibit yeast growth, leading to a stuck alcoholic fermentation and subsequent bacterial spoilage of the wine with excessive acetate and other off-odors (Lafon-Lafourcade and Ribéreau-Gayon, 1984). Stuck alcoholic fermentation during simultaneous MLF poses a dilemma because it cannot be stimulated by the usual aeration since this causes increased production of acetate and possibly polysaccharides by the malolactic bacteria. On the other hand, fermentation conditions for the bacteria can not be optimized (anaerobic, temperature near 20°C, no addition of SO_2 to the juice) because such conditions negatively affect the conduct of the alcoholic fermentation.

It is difficult to control concurrent malolactic and alcoholic fermentations because bacterial growth and yeast growth are affected differently by different wines. Gallander (1979) found no stimulation of MLF by inoculating with bacteria before alcoholic fermentation. We found the same in a Chardonnay wine inoculated before, during, and after alcoholic fermentation (unpublished data). When inoculating juice with starter cultures of *Leuc. oenos*, it is important to remember that a yeast starter must be added at the same time, otherwise the bacteria will produce excessive amounts of acetate from the sugars in the juice. Bacterial inoculation of juice prior to yeast inoculation is possible with starter cultures of the homofermentative species, *L. plantarum*, which produces only lactate from sugars (Prahl *et al.*, 1988). Due to its sensitivity to alcohol, this organism dies rapidly when approximately 5% (v/v) ethanol has been produced by the alcoholic fermentation. However, in some wines, the alcohol may inhibit this species before sufficient growth has occurred to complete MLF. Also, it is risky to leave unsulfited juice uninoculated with yeasts for several days while waiting for completion of MLF by *L. plantarum*.

To avoid undesirable interference between yeast and bacterial growth, and for best control of MLF, the preferred time for inoculation of malolactic bacteria is just at completion of alcoholic fermentation. Also, at this time, the wine may be adjusted to pH 3.1–3.2, if it is below this value. If the wine sits on fine yeast lees, then it need not be racked, but if it is still on gross lees, it should be racked. Of course, no SO_2 should not be added until after completion of MLF. Gallander (1983) investigated the effect of inoculation density on the time required to complete MLF. Inoculation with 1% (v/v) of a liquid culture prepared in grape juice, brought rapid MLF compared with an inoculum of 0.5% (v/v). However, for wines in which MLF was difficult to induce, a 2% (v/v) inoculum was needed for satisfactory results. Initial bacterial densities of 10^6–10^7 cfu/ml of wine were necessary to obtain rapid onset and completion of MLF. This conclusion is supported by more recent studies of Wibowo *et al.* (1988). The MLF can be accelerated further (and sometimes can only be induced) with a very high inoculation of $2\text{–}5\times10^7$ cfu/ml (Henick-Kling and Krieger, unpublished data). Such inoculation density is achieved by starter reactivation as described by Krieger *et al.* (1990). The reactivated culture should be inoculated at 2% (v/v) to give a cell density of $5\text{x}10^7$ cfu/ml in the wine. When wine is inoculated with another wine which has undergone MLF, a 10% (v/v) inoculum is generally needed to give a starting cell density of $1\text{–}5\times10^6$ cfu/ml. The wine should also be checked microscopically for the presence of undesirable lactococci and pediococci.

CONTRIBUTION OF MLF TO WINE FLAVOR

The wine industry is becoming increasingly aware that MLF not only affects the taste of wine through deacidification but, also, it can contribute other flavor characteristics. These malolactic flavors have been described generally as 'malolactic', 'buttery', 'lactic', 'nutty', 'yeasty', 'oaky', and 'sweaty'. Furthermore, it is believed by some that MLF can enhance the fruity character of a wine. Nevertheless, there is still some dispute whether MLF does contribute recognizable flavors to wine (Kunkee *et al.*, 1964; van Wyk, 1976). No doubt any perceived influence by MLF will be determined by the prevailing sensory attributes of the wine.

Much information has been published on the production of flavor-active compounds during MLF in wine. Yet, the relative contribution of individual compounds to the sensory effect of MLF has not been established and further research is needed. Excessive amounts of individual flavor-active substances can be related to certain undesirable effects of MLF such as those described as acetic, lactic, and buttery. For reviews of flavor-active substances identified in malolactic wines and for off-flavors associated with MLF see Davis *et al.* (1985) and Dittrich (1987).

Distinction of Malolactic Flavor in Wine

In highly aromatic white wines and flavor intensive red wines, it is difficult to recognize the flavor contribution of MLF. Two important characteristics of wine, namely body or mouthfeel and length of after-taste, are most reliably recognized in white and red wines as being increased with MLF (Henick-Kling, unpublished data). In Chardonnay, a white grape variety without strong spicy or floral characteristics, the contribution of MLF to the wine flavor is easily recognized. In blind taste comparisons, malolactic Chardonnay was distinguished from non-malolactic Chardonnay with 95% certainty; 71% of the tasters were able to reliably tell the difference. In the more aromatic Riesling variety, the MLF flavor was not easily recognized. Half of the tasters could detect the difference, half could not. Tasters that could distinguish the MLF Riesling did so with very high reliability. These tasters could be labelled the 'experienced wine drinkers'. Among this group, the preference for the MLF wine or the non-MLF wine was split 50:50. The result of the Riesling tasting also seems to indicate that fruity flavors of the wine are not destroyed by MLF, as is sometimes assumed.

Effect of MLF on Fruity Aromas

Studies by McDaniel *et al.* (1987) and in our laboratory have shown that the fruity aroma of the wine is not necessarily reduced by MLF, and, sometimes, it may actually be enhanced. Evaluation of Riesling and Chardonnay wines by CharmAnalysis®, (gas chromatography olfactometry (Acree *et al.*, 1984)) in which the aroma-active substances are separated by gas chromatography and detected by a person smelling the individual compounds, showed that no fruity aroma-active substances were lost due to MLF; rather, more aroma was added.

Effect of MLF on Vegetative Aromas

Vegetative, green flavors may predominate in red wines produced in cool viticultural areas. Grape maturity, growing conditions, and various winemaking techniques such as temperature of fermentation, pasteurization of the must, amount of pressing and extent of skin contact, all affect the degree of vegetative flavor in wine. Some presence of these vegetative flavors (green pepper, mint, eucalyptus, etc.) in balance with wine qualities are attractive; but if these flavors predominate, they greatly reduce quality. In sensory and CharmAnalysis® studies of Cabernet Sauvignon, Merlot, and Petit Verdot wines, we have noted that the intensity of various vegetative aromas was strongly reduced by MLF and that fruity characteristics were enhanced (Henick-Kling, unpublished data).

Flavor Differences Between Strains

Davis *et al.* (1985) reviewed studies which showed that different strains of malolactic bacteria can have different sensory effects on wines. While these differences were evident between species of the genera *Leuconostoc, Lactobacillus* and *Leuconostoc*, they were not always evident when comparisons were made between wines fermented with different strains of *Leuc. oenus*. McDaniel *et al.* (1987) reported studies with Pinot Noir wines where the sensory influences of several strains of *Leuc. oenus* were clearly different. Sensory characterization of Chardonnay and Cabernet Sauvignon wines fermented with various strains of *Leuc. oenos*, *L. plantarum*, *L. casei*, and *L. brevis* showed significant differences in their overall quality, fruitiness, vegetativeness, and body (Henick-Kling, unpublished data). Some strains could be recognized as giving strong buttery and nutty aromas, while other strains were more flavor neutral. A particular strain for inducing MLF should be chosen to complement the aroma characteristics of the fruit from which the wine is produced.

Ethyl Lactate and Acetate

An important aroma compound produced during MLF is ethyl lactate. After MLF, it is usually found in concentrations exceeding its taste threshold of 60–110 mg/l (Dittrich, 1987). Ethyl lactate apparently gives the wine a 'broader', 'fuller' taste. In a good MLF, carried out anaerobically by *Leuc. oenos* or selected lactobacilli, only small or no increases occur in the concentration of acetate. Generally, the acetate concentration in wine increases by 0–0.2 g/l during MLF, bringing the total acetate concentration to 0.3–0.6 g/l (Davis *et al.*, 1985; Dittrich, 1987; Zoecklein *et al.*, 1990). This small increase in acetate and also the small amount of ethyl acetate formed do not contribute to the MLF flavor; the aroma threshold for acetate is near 0.8 g/l.

Diacetyl

Generally, during MLF only small amounts (1–4 mg/l) of diacetyl are formed, and are considered to enhance the flavor complexity of the wine (Davis *et al.*, 1985). Excessive amounts of diacetyl (>5–7 mg/l) may be produced in wine by pediococci and cause an objectionable buttery aroma (Rankine *et al.*, 1969; Mayer 1974 a, b; Dittrich, 1987). The amount of diacetyl and acetoin produced during MLF varies considerably from wine to wine and over the course of fermentation and aging, being influenced by the amount of excess pyruvate available, the redox balance of the wine, and the residual metabolic activity of the yeast (Kandler, 1983; Postel and Meier,1983). In a study on spontaneous MLF, Postel and Meier (1983) reported that the concentration of diacetyl increased from 0.4 to 5 mg/l during alcoholic fermentation and MLF but subsequently decreased to 2 mg/l. Verhue and Tjan (1990) followed the metabolism of citrate in *Lactococcus lactis* subsp. *lactis* with ^{13}C NMR and found that 50–70% of the citrate was converted to lactate and that the other 30–50% was converted to acetoin, via butylene glycol, and diacetyl; diacetyl apparently was only formed after aeration. In wine, under typical conditions of carbohydrate limitation and low oxygen tension, it is unlikely that excessive amounts of diacetyl (>4 mg/l) are formed. On the other hand, a compound with a buttery aroma, which is not diacetyl, has been found with gas chromatography olfactometry (Henick-Kling and Acree, unpublished data).

REFERENCES

Acree, T. E., Barnard, J. and Cunningham, D. G. (1984) A Procedure for the Sensory Analysis of Gas Chromatographic Effluents. *Food Chemistry*, **14**, 273–286.

Alizade, M. A. and Simon, H. (1973) Zum Mechanismus und zur Kompartmentierung der L- und D- Lactatbildung aus L-Malat bzw. D-Glucose in *Leuconostoc mesenteroides*. *Hoppe-Seyler's Zeitschrift für Physiologische Chemie*, **354**, 163–168.

Alatossava, T. and Klaenhammer, T.R. (1991) Molecular characterization of three small isometric-headed bacteriophages which vary in their sensitivity to the lactococcal phage resistance plasmid pTR2030. *Applied and Environmental Microbiology*, **57**, 1346–1353.

Amachi, T. (1975) Chemical structure of a growth factor (TJF) and its physical significance for malo-lactic bacteria. In *Lactic acid Bacteria in Beverages and Food*, edited by J. G. Carr, C. V. Cutting and G. C. Whiting, pp. 103–118. London: Academic Press.

Amerine, M. A. and Ough, C. S. (1980) *Methods for Analysis of Musts and Wines*. New York City: John Wiley & Sons.

Arnink, K., Krieger, S. A. and Henick-Kling, T. (1991) Utilization of sugars by six strains of *Leuconostoc oenos*. 16th Annual meeting of the Eastern Section of the American Society for Enology and Viticulture, 10–13 July 1991, Ené, PA. *American Journal of Enology and Viticulture*, **43**, 1992.

Asmundson, R. V. and Kelly, W. J. (1990) The effect of temperature and ethanol concentration on the growth of Leuconostoc oenos. In *Proceedings of the Seventh Australian Wine Industry Technical Conference* (Adelaide, South Australia, 13–17 August 89), edited by P. J. Williams, D. M. Davidson and T. H. Lee, pp. 251–252. Adelaide: The Australian Wine Research Institute.

Bauchop, T. and Elsden S. R. (1960) The growth of microorganisms in relation to their energy supply. *Journal of General Microbiology*, **23**, 457–469.

Beelman, R.B. and Gallander, J.F. (1979) Wine deacidification. *Advances in Food Research*, **25**, 1–53

Beelman, R.B., Gavin, A. III and Keen, R.M. (1977) A new strain of *Leuconostoc oenos* for induced malo-lactic fermentation in Eastern wines. *American Journal of Enology and Viticulture*, **28**, 159–165.

Beelman, R. B., Keen, R. M., Banner, M. J. and King, S. W. (1982) Interactions between wine yeast and malolactic bacteria under wine conditions. *Developments in Industrial Microbiology*, **23**, 107–121.

Beelman, R. B. and Kunkee, R. E. (1985) Inducing simultaneous malolactic-alcoholic fermentation in red table wines. In *Malolactic Fermentation*, edited by T. H. Lee, pp. 97–112, Glen Osmond, South Australia: The Australian Wine Research Institute.

Beelman, R. B., Mc Ardle, F. J. and Duke, G. R. (1980) Comparison of *Leuconostoc oenos* strains ML34 and PSU-1 to induce malo-lactic fermentation in Pennsylvania red table wines. *American Journal of Enology and Viticulture*, **31**, 269–276.

Bender, G. R. and Marquis, R. E. (1987) Membrane ATPases and acid tolerance of *Actinomyces viscosus* and *Lactobacillus casei*. *Applied and Environmental Microbiology*, **53**, 2124–2128.

Britz, T. J. and Tracey, R. P. (1990) The combination effect of pH, SO_2, ethanol and temperature on the growth of *Leuconostoc oenos*. *Journal of Applied Bacteriology*, **68**, 23–31.

Caspritz, G. and Radler, F. (1983) Malolactic enzyme of *Lactobacillus plantarum*. Purification, properties, and distribution among bacteria. *Journal of Biological Chemistry*, **258**, 4907–4910.

Cavin, J. F., Prevost, H., Lin, J., Schmitt, P. and Divies, C. (1989) Medium for screening Leuconostoc oenos strains defective in malolactic fermentation. *Applied and Environmental Microbiology*, **55**, 751–753.

Cazelles, O. and Gnaegi, F. (1982) Enquête sur l'importance pratique du problème des bactériophages dans le vin. *Revue Suisse Viticulture Arboriculture Horticulture*, **14**, 267–270.

Champagne, C.P., Gardner, N. and Lafond, A. (1989) Production of *Leuconostoc oenos* in apple juice media. *Lebensmittel Wissenschaft und Technologie*, **22**, 376–381.

Cofran, D. R. and Meyer, J. (1970) The effect of fumaric acid on malo-lactic fermentation. *American Journal of Enology and Viticulture*, **21**, 189–192.

Cosby, W. M., Casas, I. A. and Dobrogosz, W. J. (1988) Formation, regeneration, and transfection of *Lactobacillus plantarum* protoplasts. *Applied and Environmental Microbiology*, **54**, 2599–2602.

Costello, P. J., Monk, P. R. and Lee, T. H. (1985) An evaluation of two commercial Leuconstoc oenos strains for induction of malolactic fermentation under winery conditions. *Food Technology in Australia*, **37**, 121–23, 30.

Costello, P. J., Morrison, G. J., Lee, T. H. and Fleet G. H. (1983) Numbers and species of lactic acid bacteria in wines during vinification. *Food Technology in Australia*, **35**, 14–18.

Coventry, M. J., Hillier, A. J. and Jago, G. R. (1984) Changes in the metabolism of factory-derived bacteriophage resistant derivatives of *Streptococcus cremoris*. *Australian Journal of Dairy Technology*, **39**, 154–159.

Cox, D.J. (1991) Studies on the energetics and growth benefit of malolactic fermentation in lactic acid bacteria. PhD Thesis, Cornell University, Ithaca, New York.

Cox, D. J. and Henick-Kling, T. (1989) The malolactic fermentation, a chemiosmotic energy yielding (ATP) decarboxylation reaction. *Journal of Bacteriology*, **171**, 5750–5752.

Cox, D. J. and Henick-Kling, T. (1990) A comparison of lactic acid bacteria for energy-yielding (ATP) malolactic enzyme systems. *American Journal of Enology and Viticulture*, **41**, 215–218.

Daeschel, M. A., McFeeters, R. F., Fleming, H. P., Klaenhammer, T. R. and Sanozky, R. B. (1984) Mutation and selection of strains that do not produce carbon dioxide from malate. *Applied and Environmental Microbiology*, **47**, 419–420.

Daeschel, M. A., McKenney, M. C. and McDonald, L. C. (1990) Bacteriocidal activity of *Lactobacillus plantarum* C-11. *Food Microbiology*, **7**, 91–98.

Davies, F. L., Underwood, H. M. and Gasson, M. J. (1981) The value of plasmid profiles for strain identification in lactic streptococci and the relationship between *Streptococcus lactis* 712, ML3 and C2. *Journal of Applied Bacteriology*, **51**, 325–337.

Davis, C. R. (1985) Taxonomy and ecology of lactic acid bacteria. In *Malolactic Fermentation*, edited by T. H. Lee, pp. 3–17. Adelaide: The Australian Wine Research Institute.

Davis, C. R., Silveira, N. F. A. and Fleet, G. H. (1985a) Occurrence and properties of bacteriophages of Leuconostoc oenos in Australian wines. *Applied and Environmental Microbiology*, **50**, 872–876.

Davis, C. R., Wibowo, D., Eschenbruch, R., Lee, T. H. and Fleet, G. H. (1985b) Practical implications of malolactic fermentation, a review. *American Journal of Enology and Viticulture*, **36**, 290–301.

Davis, C. R., Wibowo, D., Fleet, G. H. and Lee, T. E. (1988) Properties of wine lactic acid bacteria, Their potential enological significance. *American Journal of Enology and Viticulture*, **39**, 137–142.

Davis, C.R., Wibowo, D.J., Lee, T.H. and Fleet, G.H. (1986b) Growth and metabolism of lactic acid bacteria during and after malolactic fermentation of wines at different pH. *Applied and Environmental Microbiology*, **51**, 539–545.

Davis, C.R., Wibowo, D.J., Lee, T.H. and Fleet, G.H. (1986a) Growth and metabolism of lactic acid bacteria during fermentation and conservation of some Australian wines. *Food Technology in Australia*, **38**, 35–40.

Desens, C. and Lonvaud-Funel, A. (1988) Etude de la constitution lipidique des membranes de bactéries lactiques utilisées on vinification. *Connaissance de la Vigne et du Vin*, **22**, 25–32.

Dicks, L. M. T. and van Vuuren, H. J. J. (1988) Identification and physiological characteristics of heterofermentative strains of *Lactobacillus* from South African red wines. *Journal of Applied Bacteriology*, **64**, 505–513.

Dittrich, H. H. (1987) *Mikrobiologie des Weines. Handbuch der Lebensmitteltechnologie*, 2nd edition. Stuttgart: Ulmer.

Dohman, T.P. (1982) Characterization of Oregon-derived malo-lactic bacteria; fermentation properties and storage stability. MS thesis, Oregan State University, Corvallis, Oregon.

Donnelly, D. M. (1977) Airborne microbial contamination in a winery bottling room. *American Journal of Enology and Viticulture*, **28**, 176–181.

El Soda, M., Desmazeaud, M. J., Le Bars, D. and Zevaco, C. (1986) Cell-wall associated proteinases in *Lactobacillus casei* and *L. plantarum*. *Journal of Food Protection*, **49**, 361–365.

Feuillat, M., Bidan, P. and Rosier, Y. (1977) Croissance des bactéries lactiques à partir des principaux constituants azotes du vin. *Annales Technologiques Agricole*, **26**, 435–447.

Fitzgerald, R. J., Cogan, T. M., Doonan, S. and McKay, L. L. (1990) Intracellular pH and the kinetics of lactic dehydrogenase in *Leuconostoc lactis*. *FEMS Microbiology Reviews*, **87**, p. 73.

Fleet, G. H. (1985) The physiology and metabolism of wine lactic acid bacteria. In *Malolactic fermentation*, edited by T. H. Lee, pp. 18–25. Adelaide:The Australian Wine Research Institute.

Fleet, G. H., Lafon-Lafourcade and Ribereau-Gayon, P. (1984) Evolution of yeasts and lactic acid bacteria during fermentation and storage of Bordeaux wines. *Applied and Environmental Microbiology*, **48**, 1034–1038.

Fornachon, J. C. M. (1943) *Bacterial spoilage of fortified wines*. Adelaide, South Australia:Australian Wine Board.

Gallander, J. F. (1979) Effect of time of inoculation in the stimulation of malo-lactic fermentation. *American Journal of Enology and Viticulture*, **30**, 157–159.

Gallander, J. F. (1983) Effect of bacterial inoculum levels on the stimulation of malo-lactic fermentation. *Practical Winery*, **3** (6), 50–53.

Garvie, E. I. (1967) The growth factor and amino acid requirements of species of the genus *Leuconostoc*, including *Leuconostoc paramesenteroides* (sp. nov.) and *Leuconostoc oenos*. *Journal of General Microbiology*, **48**, 439–447.

Garvie, E. I. (1986a) Genus *Leuconostoc* van Tieghem 1878, 198^{AL} emend mut. char. Hucker and Pederson 1930, 66^{AL}. In *Bergey's Manual of Systematic Bacteriology*, Vol. 2, pp. 1071–1075. Baltimore: Williams and Wilkins.

Garvie, E. I. (1986b) Genus *Pediococcus Claussen* 1903, 68^{AL}. In *Bergey's Manual of Systematic Bacteriology*, Vol. 2, pp. 1075–1079. Baltimore: Williams and Wilkins.

Gasson, M. J. (1983) Genetic transfer systems in lactic acid bacteria. *Antonie van Leeuwenhoek*, **49**, 275–282.

Gestrelius, S.M. and Kjaer, J.H. (1983) Method of deacidifying wine and composition thereafter. US Patent 4,380,552.

Gini, B. and Vaughn, R. H. (1962) Characteristics of some bacteria associated with the spoilage of California dessert wines. *American Journal of Enology and Viticulture*, **13**, 20–31.

Gnaegi, F., Cazelles, O., Sozzi, T. and d'Amico, N. (1984) Connaissances sur les bactériophages de *Leuconostoc oenos* et progrès dans la maîtrise de la fermentation malolactique des vins. *Revue Suisse Viticulture Arboriculture Horticulture*, **16**, 59–65.

Gottschalk, G. (1986) *Bacterial metabolism*. Heidelberg: Springer Verlag.

Götz, F., Sedewitz, B. and Elstner, E. F. (1980) Oxygen utilization by *Lactobacillus plantarum*. I. Oxygen consuming reactions. *Archives of Microbiology*, **125**, 209–214.

Graham, D. C. and McKay, L. L. (1985) Plasmid DNA in strains of *Pediococcus cerevisiae* and *Pediococcus pentosaceus*. *Applied and Environmental Microbiology*, **50**, 532–534.

Haag, B., Krieger, S. and Hammes, W. P. (1988) Hemmung der Starterkulturen zur Einleitung des biologischen Säureabbaus durch Spritzmittelrück-stände in Wein. *Weinwissenschaft*, **43**, 261–278.

Harold, F. M. (1977) Membranes and energy transduction in bacteria. *Current Topics in Bioenergetics*, **6**, 83–149.

Hasan, N. and Durr, I. F. (1974) Induction of ß-galactosidase in *Lactobacillus plantarum*. *Journal of Bacteriology*, **120**, 66–73.

Hayman, D.C. and Monk, P.R. (1982). Starter culture preparation for the induction of malolactic fermentation in wine. *Food Technology in Australia*, **34**, 14–18.

Hendersen, G. B., Zevely, E.M. and Heunnekens, F. M. (1979) Coupling of energy to folate transport in *Lactobacillus casei*. *Journal of Bacteriology*, **139**, 552–559.

Henick-Kling, T. (1983) Comparison of Oregon-derived malolactic bacteria in pilot scale wine production. MS thesis. Oregon State University, Corvallis, Oregon.

Henick-Kling, T. (1986) Growth and metabolism of *Leuconostoc oenos* and *Lactobacillus plantarum* in wine. Ph.D. thesis. The University of Adelaide, South Australia.

Henick-Kling, T. (1988a) Improving malolactic fermentation. In *Proceedings of the Second International Symposium for Cool Climate Viticulture and Oenology* (11–15 January 1988, Auckland, New Zealand), edited by R. E. Smart, R. Thornton, S. Rodriguez and J. Young, pp. 238–242. Auckland: New Zealand Society for Viticulture and Oenology.

Henick-Kling, T. (1988b) Yeast and bacteria control in winemaking. In *Modern Methods of Plant Analysis, New Series, Vol. 6*, edited by Linskens, H.F. and Jackson, J.F., pp. 276–316. Heidelberg: Springer Verlag.

Henick-Kling, T. (1990) pH and regulation of malolactic activity in *Leuconostoc oenos*. In Actualités Oenologiques 89. Comptes rendus du 4e Symposium International d'Oenologie (Bordeaux, 1989), pp. 320–325. Institut d'Oenologie Université de Bordeaux II. Paris: Dunod.

Henick-Kling, T., Cox, D.J. and Olsen, E.B. (1991) Production de l'énergie durant la fermentation malolactique. *Revue Française d'Oenologie*, **132**, 63–66.

Henick-Kling, T. and Krieger, S.A. (1990) Synthetic media for the production of malolactic starter cultures. US patent (application filed).

Henick-Kling, T., Lee, T. H. and Nicholas, D. J. D. (1986a) Inhibition of bacterial growth and malolactic fermentation in wine by bacteriophage. *Journal of Applied Bacteriology*, **61**, 287–293.

Henick-Kling, T., Lee, T. H. and Nicholas, D. J. D. (1986b) Characterization of the lytic activity of bacteriophages of *Leuconostoc oenos* isolated from wine. *Journal of Applied Bacteriology*, **61**, 525–534.

Henick-Kling, T., Sandine, W. E. and Heatherbell, D. A. (1989) Evaluation of malolactic bacteria isolated from Oregon wines. *Applied and Environmental Microbiology*, **55**, 2010–2016.

Hill, C., Miller, L. A. and Klaenhammer, T. R. (1990) Cloning, expression, and sequence determination of a bacteriophage fragment encoding bacteriphage resistance in *lactococcus lactis*. *Journal of Bacteriology*, **172**, 6419–6426.

Hogarth, J. and Croser, B. J. (1984) Preparation and monitoring of pure cultures of lactic acid bacteria in the laboratory. In *Malolactic Fermentation*, edited by T.H. Lee, pp. 77–82. Adelaide: The Australian Wine Research Institute.

Izuagbe, Y. S., Dohman, T. P., Sandine, W. E. and Heatherbell, D. A. (1985) Characterization of *Leuconostoc oenos* isolated from Oregon wines. *Applied and Environmental Microbiology*, **50**, 680–684.

Janse, B. J. H., Winfield, B. D., Pretorius, I. S. and Van Vuuren, H. J. J. (1987) Plasmids in *Leuconostoc oenos*. *Plasmid*, **17**, 173–175.

Jarvis, A. W. (1989) Bacteriophages of lactic acid bacteria. *Journal of Dairy Sciences*, **72**, 3406–3428.

Kaback, H.R. (1990) Active transport: membrane vesicles, bioenergetics, molecules, and mechanisms. In *Bacterial energetics*, edited by T.A. Krulwich, pp. 151–202. New York: Academic Press.

Kandler, O. and Weiss, N. (1986) Genus *Lactobacillus* Beijerinck 1901, 212^{AL}. In Bergey's *Manual of Systematic Bacteriology*, Vol. 2, pp. 1209–1234. Baltimore: Williams and Wilkins.

Kandler, O. (1983) Carbohydrate metabolism in lactic acid bacteria. *Antonie van Leeuwenhoek*, **49**, 209–224.

Kandler, O., Winter, J. and Stetter, K. O. (1973) Zur Frage der Beeinflussung der Glucoseverg_rung durch L-Malat bei *Leuconostoc mesenteroides*. *Archiv für Mikrobiologie*, **90**, 65–75.

Kashket, E.R. (1982) Stoichiometry of the H^+-ATPase of growing and resting, aerobic *Escherichia coli*. *Biochemistry*, **21**, 5534–5538.

Kashket, E. R. (1987) Bioenergetics of lactic acid bacteria, cytoplasmic pH and osmotolerance. *FEMS Microbiology Reviews*, **46**, 233–244.

Kashket, S. and Kashket, E. R. (1985) Dissipation of the proton motive force in oral streptococci by fluoride. *Infection and Immunity*, **48**, 19–22.

Kay, W. W. (1978) Transport of carboxylic acids. In *Bacterial transport, Microbiology Series*, edited by B. P. Rosen, Vol. 4, pp. 385–411. New York: Marcel Dekker Inc.

Kearney, L., Upton, M. and McLoughlin, A. (1990) Enhancing the viability of *Lactobacillus plantarum* inoculum by immobilizing the cells in calcium-alginate beads incorporating cryoprotectants. *Applied and Environmental Microbiology*, **56**, 3112–3116.

Kelly, W. J., Asmundson, R. V. and Hopcroft, D. H. (1989) Growth of *Leuconostoc oenos* under anaerobic conditions. *Applied and Environmental Microbiology*, **40**, 277–282.

King, S.W. and Beelman, R.B. (1986) Metabolic interactions between *Saccharomyces cerevisiae* and *Leuconostoc oenos* in a model grape juice/wine system. *American Journal of Enology and Viticulture*, **37**, 53–60.

Klaenhammer, T. R. (1987) Plasmid-directed mechanisms for bacteriophage defense in lactic streptococci. *FEMS Microbiology Reviews*, **46**, 313–325.

Klaenhammer, T. R. (1988) Bacteriocins of lactic acid bacteria. *Biochemie*, **70**, 337–349.

Klaenhammer, T.R. (1989) Genetic characterization of multiple mechanisms of phage defense from a prototype phage-insenstive strain *Lactococcus lactis* ME2. *Journal of Dairy Science*, **72**, 3429–3443.

Kok, J. (1990) Genetics of the proteolytic system of lactic acid bacteria. (1990) *FEMS Microbiology Reviews*, **87**(1+2), 15–41.

Kondo, J. K. (1989) Gene cloning and transfer in dairy lactococci. *Journal of Dairy Sciences*, **72**, 3381–3387.

Konings, W. N. and Otto, R. (1983) Energy transduction and solute transport in streptococci. *Antonie van Leeuwenhoeck*, **49**, 247–257.

Kraus, A., Dessau, W. and Simon, H. (1976) On the mechanism and stereochemistry of the malate lactate fermentation of *Leuconostoc mesenteroides*. *Hoppe-Seyler's Zeitschrift für Physiologische Chemie*, **357**, 1209–1214.

Krieger, S. A. (1989) Optimierung des biologischen Säureabbaus in Wein mit Starterkulturen. Ph. D. thesis. University of Hohenheim, Stuttgart, West Germany.

Krieger, S., De Frenne, E. and Hammes, W.P. (1986) Ausführung des biologischen Säureabbaus im Wein mit *Leuconostoc oenos*. *Chemie Mikrobiologie und Technologie der Lebensmittel*, **10**, 13–18.

Krieger, S. A. and Hammes, W. P. (1988) Biologischer Säureabbau im Wein unter Einsatz von Starterkulturen. *Der Deutsche Weinbau*, **25–26**, 1152–1154.

Krieger, S. A. and Henick-Kling, T. (1991) Activity of malolactic starter cultures. *Journal of Applied Bacteriology* (Submitted).

Krieger, S. A., Hammes, W. P. and Henick-Kling, T. (1992) Effect of medium composition on growth rate, growth yield, and malolactic activity of *Leuconostoc oenos* LoZH$_1$-t$_7$-1. *Food Microbiology*, **9**, 1–11.

Krieger, S. A., Hammes, W. P. and Henick-Kling, T. (1990) Induction of malolactic fermentation under practical winemaking conditions. In *Vineyard and Winery Management* November/December, pp. 45–50.

Kuensch, U., Temperli, A. and Mayer, K. (1974) Conversion of arginine to citrulline during malo-lactic fermentation in red Swiss wine. *Applied and Environmental Microbiology*, **25**, 191–193.

Kunkee, R. E. (1974) Malo-lactic fermentation in winemaking. In *Chemistry of winemaking*, edited by A. D. Webb, pp. 151–170. *Advances in Chemistry Series*, **137**. 151–170.

Kunkee, R.E. (1991) Some roles of malic acid in the malolactic fermentation in winemaking. *FEMS Microbiology Reviews*, **88**, 55–72.

Kunkee, R. E., Ough, C. S. and Amerine, M. A. (1964) Induction of malo-lactic fermentation by inoculation of must and wine with bacteria. *American Journal of Enology and Viticulture*, **15**, 178–183.

Lafon-Lafourcade, S. (1970) Etude de la dégradation de l'acide L-malique par les bactéries lactiques non proliférantes isolées des vins. *Ann. Technol. Agric.*, **19**, 141–154.

Lafon-Lafourcade, S. (1975) Factors of the malo-lactic fermentation of wines. In *Lactic acid bacteria in beverages and food*, edited by J. G. Carr, C. V. Cutting and G. C. Whiting, pp. 43–53. London: Academic Press.

Lafon-Lafourcade, S. (1983) Wine and brandy. In Biotechnology, edited by H. H. Rehm and G. Reed, Vol. 5, pp. 81–163. Weinheim: Verlag Chemie.

Lafon-Lafourcade, S. and Ribéreau-Gayon, P. (1984) Les altérations des vins par les bactéries acétiques et les bactéries lactiques. *Connaissance de la Vigne et du Vin*, **18**, 67–82.

Lafon-Lafourcade, S., Carre, E. and Ribéreau-Gayon, P. (1983a) Occurrence of lactic acid bacteria during different stages of vinification and conservation of wines. *Applied and Environmental Microbiology*, **46**, 874–880.

Lafon-Lafourcade, S., Carre, E., Lonvaud-Funel, A. and Ribéreau-Gayon, P. (1983b) Induction de la fermentation malolactique des vins par inoculation d'une biomasse industrielle congelée de *L. oenos* aprés réactivation. *Connaissance de la Vigne et du Vin*, **17**, 55–71.

Lautensach, A. and Subden, R. E. (1984) Cloning of malic acid assimilating activity from *Leuconostoc oenos* in *E. coli*. *Microbios*, **39**, 29–69.

Lemaresquier, H. (1987) Inter-relationships between strains of *Saccharomyces cerevisiae* from the Champagne area and lactic acid bacteria. *Letters in Applied Microbiology*, **4**, 91–94.

Lemperle, E. and Lay, H. (1989) Zusammensetzung und Beurteilung der Weine. In *Chemie des Weines. Handbuch der Getränketechnologie*, edited by G. Würdig and R. Woller, pp. 466–628. Stuttgart: Ulmer.

Levine, A. S. and Fellers, C. S. (1940) Action of acetic acid on food spoilage microorganisms. *Journal of Bacteriology*, **39**, 499–515.

Lindgren, S. E. and Dobrogosz, W. J. (1990) Antagonistic actvities of lactic acid bacteria in food and feed fermentations. *FEMS Microbiology Reviews*, **87**, 149–164.

London, J. and Meyer, E. Y. (1969) Malate utilization by a group D *Streptococcus*, physiological properties and purification of an inducible malic enzyme. *Journal of Bacteriology*, **98**, 705–711.

London, J., Meyer, E. Y. and Kulczyk, S. R. (1971) Detection of relationships between *Streptococcus faecalis* and *Lactobacillus casei* by immunological studies with two forms of malic enzyme. *Journal of Bacteriology*, **108**, 196–201.

Lonvaud-Funel, A. (1987) Rescherches sur les bactéries lactiques du vin: fonctions métaboliques, croissance, génétique plasmidique. PhD thesis, University of Bordeaux II.

Lonvaud-Funel, A. and Desens, C. (1990) Constitution en acides gras des membranes des bactéries lactiques du vin. Incidence des contitions de culture. (Fatty acid constitution of lactic acid bacteria of wines. Effects of culture conditions) *Sciences des Aliments*, **10**, 817–829.

Lonvaud-Funel, A., Joyeux, A. and Desens, C. (1988) Inhibition of malolactic fermentation of wines by products of yeast metabolism. *Journal of the Science of Food and Agriculture*, **44**, 183–191.

Lonvaud-Funel, A., San-Romao, V., Joyeux, A. and Chauvet, S. (1987) Effect of certain grape-parasitic fungi on the activity of wine lactic acid bacteria (*Leuconostoc oenos*). *Sciences des Aliments*, **7**, 267–274.

Lonvaud-Funel, A. and Strasser de Saad, A. M. (1982) Purification and properties of a malolactic enzyme from a strain of *Leuconostoc mesenteroides* isolated from grapes. *Applied and Environmental Microbiology*, **43**, 357–361.

Lucey, C. A. and Condon, S. (1984) Stimulation by aeration of growth rate and biomass yield of leuconostocs. *International Journal of Food Science and Technology*, **8**, 154–155.

Maloney, P. C. (1977) Obligatory coupling between proton entry and the synthesis of adenosine-5'-triphosphate in *Streptococcus lactis*. *Journal of Bacteriology*, **132**, 564–575.

Maloney, P.C. (1983) Relationship between phosphorylation potential and electrochemical H^+ gradient during glycolysis in *Streptococcus lactis*. *Journal of Bacteriology*, **153**, 1461–1470.

Mayer, K. (1974a) Mikrobiologisch und kellertechnisch wichtige neue Erkenntnisse in bezug auf den biologischen Säureabbau. *Schweizerische Zeitschrift für Obst-und Weinbau*, **110**, 291–297.

Mayer, K. (1974b) Nachteilige Auswirkungen auf die Weinqualität bei ungünstig verlaufenem biologischem Säureabbau. *Schweizerische Zeitschrift für Obst-und Weinbau*, **110**, 385–391.

Mayer, K. (1978) Biologischer Säureabbau in Wein, Ungünstige Selektiv-wirkung der schwefligen Säure. *Schweizerische Zeitschrift für Obst-und Weinbau*, **114**, 642–647.

Mayer, K. and Pause, G. (1981) Histamingehalte in 1981er Ostschweizer Weinen. *Schweizerische Zeitschrift für Obst-und Weinbau*, **118**, 723–727.

McDaniel, M., Henderson, L. A., Watson, B. T. and Heatherbell, D. (1987) Sensory panel training and screening for descriptive analysis of the aroma of Pinot Noir wine fermented by several strains of malolactic bacteria. *Journal of Sensory Studies*, **2**, 149–167.

McDonald, L. C., Fleming, H. P. and Hassan, H. M. (1990) Acid tolerance of *Leuconostoc mesenteroides* and *Lactobacillus plantarum*. *Applied and Environmental Microbiology*, **56**, 2120–2124.

Messini, A., Vincenzini, M. and Materassi, R. (1990) Evolution of yeasts and lactic acid bacteria in "Chianti" wines as affected by the refermentation according to the Tuscan usage. *Annales de Microbiologie*, **40**, 111–119.

Mitchell, P. (1966) Chemiosmotic coupling in oxidative and photosynthetic phosphorylation. *Biological Reviews*, **41**, 445–502.

Mitchell, P. (1981) From black box energetics to molecular mechanisms: vectorial ligand-conduction mechanisms in biochemistry. In *Chemiosmotic proton circuits in biological membranes*, edited by V.P. Skulachev and P.C. Hinkle, pp. 611–633. Reading: Addison-Wesley Publishing Co.

Murphy, M. G., O'Connor, L. and Condon, S. (1985) Oxygen-dependent lactate utilization by Lactobacillus plantarum. *Archives for Microbiology*, **141**, 75–79.

Naouri, P., Chagnaud, P., Arnaud, A. and Galzy, P. (1990) Purification and properties of a malolactic enzyme from *Leuconostoc oenos* ATCC 23278. *Journal of Basic Microbiology,* **8**, 577–585.
Nel, L., Wingfield, B.D., Van Der Meer, L.J. and Van Vuuren, J.J. (1989) Isolation and characterization of *Leuconostoc oenos* bacteriophages from wine and sugarcane. *FEMS Microbiology Letters*, **44**, 63–67.
Olsen, E. B., Russell, J. B. and Henick-Kling, T. (1991) Electrogenic L-malate transport in *Lactobacillus plantarum,* a basis of energy production from malolactic fermentation. *Journal of Bacteriology,* **173**, 6199–6206.
Orberg, P. K. and Sandine, W. E. (1984) Microscale method for rapid isolation of covalently closed circular plasmid DNA from Group N streptococci. *Applied and Environmental Microbiology,* **47**, 677–680.
Park, C. and McKay, L. L. (1975) Induction of prophage in lactic streptococci isolated from commercial dairy starters cultures. *Journal of Milk Food Technology,* **38**, 594–597.
Patel, J. (1990) Lysogenic *Leuconostoc oenos* strain L181, and implications of lysogenic strains in MLF. In *Proceedings of the Seventh Australian Wine Industry Technical Conference* (Adelaide, South Australia, 13–17 August 1989), edited by P. J. Williams, D. M. Davidson and T. H. Lee, p. 258. Adelaide: The Australian Wine Research Institute.
Peynaud, E. and Domercq, S. (1961) Etudes sur les bactéries lactiques des vins. *Ann. Technol. Agric.*, **10**, 43–60.
Pilone, G.J. and Kunkee, R.E. (1976) Stimulatory effect of malo-lactic fermentation on the growth rate of *Leuconostoc oenos. Applied and Environmental Microbiology,* **32**, 405–408.
Pilone, G.J., Rankine, B.C. and Pilone, D.A. (1977) Inhibiting malo-lactic fermentation in Austalian dry red wines by adding fumaric acid. *American Journal of Enology and Viticulture*, **25**, 99–107.
Pool, R. and Henick-Kling, T. (1991) Production methods in Champagne. New York State Agricultural Experiment Station, Cornell University, Geneva, New York.
Postel, W. and Meier, B. (1983) Verhalten von 2-Acetolactat, 2-Acetohydroxybutyrat, Diacetyl und Acetoin während des bakteriellen Äpfelsäureabbaus in Wein. *Zeitschrift für Lebensmittle- Untersuchung und -Forschung,* **176**, 356–359.
Prahl, C., Lonvaud-Funel, A., Korsgaard, S., Morrison, E. and Joyeux, A. (1988) Etude d'un nouveau procédé de déclenchement de al fermentation malolactique. *Connaissance de la Vigne et du Vin,* **22**, 197–207.
Radler, F. (1958) Untersuchung des biologischen Säureabbaus im Wein. II. Der Nähr-und Wuchsstoffbedarf der äpfelsäureabbauenden Bakterien. *Archive der Mikrobiologie* **31**, 1–15.
Radler, F. (1962) Über die Milchsäurebakterien des Weines und den biologischen Säureabbau. I. Systematik und chemische Grundlagen. *Vitis*, **3**, 144–176.
Radler, F. (1963a) Über die Milchsäurebakterien des Weines und den biologischen Säureabbau. II. Physiologie und Ökologie der Bakterien. *Vitis* **3**, 207–236.
Radler, F. (1963b) Über die Milchsäurebakterien des Weines und den biologischen Säureabbau. III. Methoden zur Veränderung des Säuregehaltes im Wein. *Vitis*, **4**, 62–72.
Radler, F. (1966) Die mikrobiologischen Grundlagen des biologischen Säureabbaus im Wein. *Zentralblatt der Bakteriologie Parasitenkunde, Abteilung II,* **120**, 237–287.
Radler, F. (1986) Microbial biochemistry. *Experientia* , **42**, 884–893.
Rankine, B. C., Fornachon, J. C. M. and Bridson, D. A. (1969) Diacetyl in Australian dry red wines and its significance in wine quality. *Vitis*, **8**, 129–134.
Rau, S. (1988) Plasmidmuster säureresistenter Milchsäurebakterien. MS thesis, University of Hohenheim, W. Germany.
Renault, P. P. and Heslot, H. (1987) Selection of *Streptococcus lactis* mutants defective in malolactic fermentation. *Applied and Environmental Microbiology,* **53**, 320–324.
Renault, P., Gaillardin, C. and Heslot, H. (1988) Role of malolactic fermentation in lactic acid bacteria. *Biochimie*, **70**, 375–379.

Renault, P., Gaillardin, C. and Heslot, H. (1989) Product of the *Lactococcus lactis* gene required for malolactic fermentation is homologous to a family of positive regulators. *Journal of Bacteriology*, **171**, 3108–3114.

Ribéreau-Gayon, J. (1938) Note sur l'action des bactéries dans les vins. *Procès-verbaux séances soc. sci. phys. naturelles de Bordeaux*, 73–74.

Ribéreau-Gayon, J., Peynaud, E., Ribéreau-Gayon, P. and Sudraud, P. (1975) *Sciences et techniques du vin*, Vol. 2. Paris: Dunod.

Rodriguez, S.B., Amberg, E. and Thornton, R.J. (1990) Malolactic fermentation in Chardonnay: growth and sensory effects of commercial cultures of *Leuconostoc oenos*. *Journal of Applied Bacteriology*, **68**, 139–144.

Romano, A. H., Trifone, J. D. and Brustolon, M. (1979) Distribution of the phosphoenolpyruvate, glucose phosphotransferase system in fermentative bacteria. *Journal of Bacteriology*, **139**, 93–97.

San-Ramao, M. V., Lonvaud-Funel, A. and Lafon-Lafourcade, S. (1990) Incidence des métabolites de champignons parasites du raisin sur les bactéries lactiques du vin. In *Actualités Oenologiques 89. Comptes rendus du 4e Symposium International d'Oenologie (Bordeaux, 1989)*, pp. 338–343. Institut d'Oenologie Université de Bordeaux II. Paris: Dunod.

Sanders, M. E. (1988) Phage resistance in lactic acid bacteria. *Biochimie*, **70**, 411–422.

Sandine, W.E. (1979) Lactic starter culture technology. Pfizer Cheese Monographs, volume VI. New York: Pfizer Inc.

Sandine, W.E. and Heatherbell, D.A. (1985) Wine preparation with new strains of *Leuconostoc oenos*. US Patent 4,547,373.

Sapico, V. and Anderson, R. L. (1967) An adenosine 5'-tri-phosphotransferase specific for D-mannose and D-fructose from *Leuconostoc mesenteroides*. *Journal Biological Chemistry*, **242**, 5086–5092.

Sapis-Domercq, S., Bertrand, A., Joyeux, A., Lucmaret, V. and Sarre, W. (1978) Etude de l'influence des produit de traitement de la vigne sur la microflore des raisins et des vins. Expérimentation 1977. Comparaison avec les résultats de 1976 and 1975. *Connaissance de la Vigne et du Vin*, **12**, 245–275.

Schanderl, H. (1969) Zur Frage der Herkunft der Weinbakterien. *Deutsches Weinbau Jahrbuch*, **20**, 142–147.

Schütz, M. and Radler, F. (1973) Das Malatenzym von *Lactobacillus plantarum* und *Leuconostoc mesenteroides*. *Archiv für Mikrobiologie*, **91**, 183–202.

Schütz, M. and Radler, F. (1974) Das Vorkommen von Malatenzym und Malo-Lactat-Enzym bei verschiedenen Milchsäurebakterien. *Archives of Microbiology*, **96**, 329–339.

Schütz, M. and Radler, F. (1984) Anaerobic reduction of glycerol to propandiol-1,3 by L. brevis and *L. buchneri*. *Systematic and Applied Microbiology*, **5**, 169–178.

Sedewitz, B., Schleifer, K. H. and Goetz, F. (1984) Physiological role of pyruvate oxidase in the aerobic metabolism of *Lactobacillus plantarum*. *Journal of Bacteriology*, **160**, 462–465.

Smid, E. J., Driessen, A. J. M. and Konings, W. N. (1989a) Mechanism and energetics of dipeptide transport in membrane vesicles of *Lactobacillus lactis*. *Journal of Bacteriology*, **171**, 292–298.

Smid, E. J., Plapp, R. and Konings, W. N. (1989b) Peptide uptake is essential for growth of *Lactococcus lactis* on the milk protein casein. *Journal of Bacteriology*, **171**, 6135–6140.

Sozzi, T., Gnaegi, F., D'Amico, N. and Hose, H. (1982) Difficultées de fermentation malolactique du vin dues á des bacteriophages de *Leuconostoc oenos*. *Revue Suisse Viticulture Arboriculture Horticulture*, **14**, 17–23.

Spettoli, P., Nuti, M. P. and Zamorani, A. (1984) Properties of malolactic activity purified from *Leuconostoc oenos* ML34 by affinity chromatography. *Applied and Environmental Microbiology*, **9**, 184–189.

Splittstoesser, D.F. and Stoyla, B.A. (1989) Effect of various inhibitors on the growth of lactic acid bacteria in a model grape juice system. *Journal of Food Protection*, **52**, 240–243.

Sponholz, W. R. (1989) Der Traubenmost. In *Chemie des Weines. Handbuch der Getränketechnologie*, edited by G. Würdig and R. Woller, pp. 45–122. Stuttgart: Ulmer.

Stamer, J. R. (1979) The lactic acid bacteria, microbes of diversity. *Food Technology*, **33**, 60–65.

Strobel, H. J., Russell, J. B., Driessen, A. J. M. and Konings, W. N. (1989) Transport of amino acids in *Lactobacillus casei* by proton-motive-force-dependent and non-proton-motive-force-dependent mechanisms. *Journal of Bacteriology*, **171**, 280–284.

Subden, R. E., Lin, W. L., Lautensach, A. and Meiering, A. G. (1982) An L-lactic acid dehydrogenase based method for detecting microbial colonies performing a malolactic fermentation. *Canadian Journal of Microbiology*, **28**, 883–886.

Tagg, J. R., Dajani, A. S. and Wannamaker, L. W. (1976) Bacteriocins of gram-positive bacteria. *Bacteriology Reviews*, **40**, 722–756.

Tanskanen, E.I., Tulloch, D.L., Hillier, A.J. and Davidson, B.E. (1990) Pulsed-field electrophoresis of *SmaI* digests of lactococcal genomic DNA, a novel method of strain identification. *Applied and Environmental Microbiology*, **56**, 3105–3111.

Taylor, C., Henschke, P. A. and Catchside, D. E. A. (1990) The detection and isolation of high molecular weight plasmid DNA from *Leuconostoc oenos*. In *Proceedings of the Seventh Australian Wine Industry Technical Conference* (Adelaide, South Australia, 13–17 August 1989), edited by P. J. Williams, D. M. Davidson and T. H. Lee, p. 260. Adelaide: The Australian Wine Research Institute.

van der Westhuizen, L. M., Agenbach, W. A., Loos, M. A. and Schoombee, N. F. (1981) Comparison of procedures for isolation of malolactic bacteria from wine. *Applied and Environmental Microbiology*, **32**, 168–170.

van Wyk, C. J. (1976) Malo-lactic fermentation in South African table wines. *American Journal of Enology and Viticulture*, **27**, 181–185.

Vaughn, R.H. (1955) Bacterial spoilage of wines with special reference to California conditions. *Advances in Food Research*, **6**, 67–108.

Verhue, W.M. and Tjan, S.B. (1990) Metabolic routes of citrate and pyruvate in *L. diacetylactis* as determined by ^{13}C-NMR. *FEMS Microbiology Reviews*, **87**, 69.

Webb, R. B. and Ingraham, J. L. (1960) Induced malolactic fermentations. *American Journal of Enology and Viticulture*, **11**, 59–63.

Weiller, H. G. and Radler, F. (1970) Milchsäurebakterien aus Wein und von Rebenblättern. *Zentralblatt für Bakteriologie Parasitenkunde Infektionskrankheiten und Hygiene, zweite naturwissenschaftliche Abteilung*, **2** (124), 707–732.

Weiller, H. G. and Radler, F. (1972) Vitamin- und Aminosärebedarf von Milchsäurebakterien aus Wein und von Rebenblättern. *Mitteilungen Klosterneuburg*, **22**, 4–18.

Wibowo, D., Eschenbruch, R., Davis, C. R., Fleet, G. H. and Lee, T. H. (1985) Occurrence and growth of lactic acid bacteria in wine, A review. *American Journal of Enology and Viticulture*, **36**, 302–313.

Wibowo, D., Fleet, G. H., Lee, T. H. and Eschenbruch, R. E. (1988) Factors affecting the induction of malolactic fermentation in red wines with *Leuconostoc oenos*. *Journal of Applied Bacteriology*, **64**, 421–428.

Williams, S. A., Hodges, R. A., Strike, T. L., Snow, R. and Kunkee, R. E. (1984) Cloning the gene for malolactic fermentation in wine from *Lactobacillus delbrueckii* in *Escherichia coli* and yeasts. *Applied and Environmental Microbiology*, **47**, 288–293.

Zoecklein, B. W., Fugelsang, K. C., Gump, B. H. and Nury, F. S. (1990) *Production wine analysis*. New York: An AVI book.

Chapter 11

BOTRYTIZED WINES

BERNARD J. DONÈCHE

Institut d'Oenologie, Université de Bordeaux II, 351, Cours de la Libération, 33405 Talence, France

Chapter Outline

INTRODUCTION

Since time immemorial, damage caused by phytopathogenic fungi has caught man's attention. The Bible (Amos, 4, 9) offers one of the first accounts of grape rot which still provides scientists today with a controversy concerning the agent of the disease described: *Botrytis cinerea* or *Plasmopara viticola*.

However, it was only in 1729 that P.A. Micheli formally established the relationship between *B. cinerea* and rot attack on melons in gardens belonging to the Duke of Tuscany in Florence. It is thanks to the German mycologist de Bary (1831–1888) that we owe the exact identification of the fungus; he discovered the relationship between the conidian from (*Botrytis*), the sclerotic form (*Sclerotium*) and the production of apothecia (*Botrytinia fuckeliana* de Bary). It is now well established that *B. cinerea* is the conidian form of an Ascomycete belonging to the series of Discomycetes, Helotiales order, family of the Sclerotiniaceae.

Since that time, many research laboratories throughout the world have worked on this fungus and much is now known about its ecological, biochemical and physiological properties (Coley-Smith, Verhoef and Jarvis, 1980). Yet, it still remains the main preoccupation of vinegrowers as we approach the year 2000. Although many fungi may attack the vine and cause well-defined diseases, *B. cinerea* is unique in parasitology. In most cases, its development on grapes leads to a decrease in quality called "pourriture grise", "grey rot", or "graufaüle" in German. Only in certain growth conditions, there occurs a real phenomenon of overmaturation which has been termed "pourriture noble", "noble rot" or "edelfaüle". This phenomenon is indispensable for the production of great sweet white wines such as Sauternes and Coteaux du Layon in France, Trockenbeerenauslese in Germany, Tokay Aszu in Hungary and other great names throughout the world.

Recent progress in understanding the host–parasite relationship now points to a link between constitutional chemical modifications in grapes and the fungal metabolism. It is becoming clearer how climatic conditions influence the development of the fungus, thereby determining the different forms of rot. Better understanding of these mechanisms offers a more accurate definition of the quality of noble rotted grapes, and defines the prerequisites for producing the great sweet white wines. This chapter considers aspects of the *Botrytis* infection of grapes and its implications in making the so called "botrytized wines". The reader is referred to Ribéreau-Gayon, Ribéreau-Gayon and Seguin (1980), Fregoni, Iacono and Zamboni (1986) and Bayonove (1989) for further general discussion of these topics.

NOBLE ROT

The Infection Process and Host–Parasite Interactions

The conidia represent the main form of dissemination and contamination of the fungus, and are the decisive factors in the noble rot process. In conditions of sufficient dampness, they develop a germinative tube on the surface of the vegetal tissue but cannot directly penetrate the grape. In effect, the enzymatic activities produced during germination are too weak to digest the thick grape cuticle, as may occur in the tomato (Rijkenberg *et al.*, 1980). Pucheu-Planté and Mercier (1983) used an electron scanning microscope to show how the infection of apparently healthy berries could occur. The grape has a variable number of stomata according to the cultivar and clone (from 20 to 40 for Sémillon in Sauternes). These stomata are functional only at the very beginning of berry development. They become necrotic before "véraison" and, as the berry increases in size, the peristomatic areola cracks (Figure 11.1). Thus, the apparently healthy grapes in fact sustain microlesions (10–100 μm wide) in the cuticle through which the germinative tubes of mycelian hyphae may penetrate.

Germinating conidia have sufficient glycogen reserve to sustain the development of a germinative tube for 48 h. Normally, the tube is not long enough to reach the microfissure. Infection is facilitated by the presence of nutrients on the surface of the grape skin. Indeed, dehydrated exudates of amorphous plaques may be found on the surface (Bessis, 1977). These

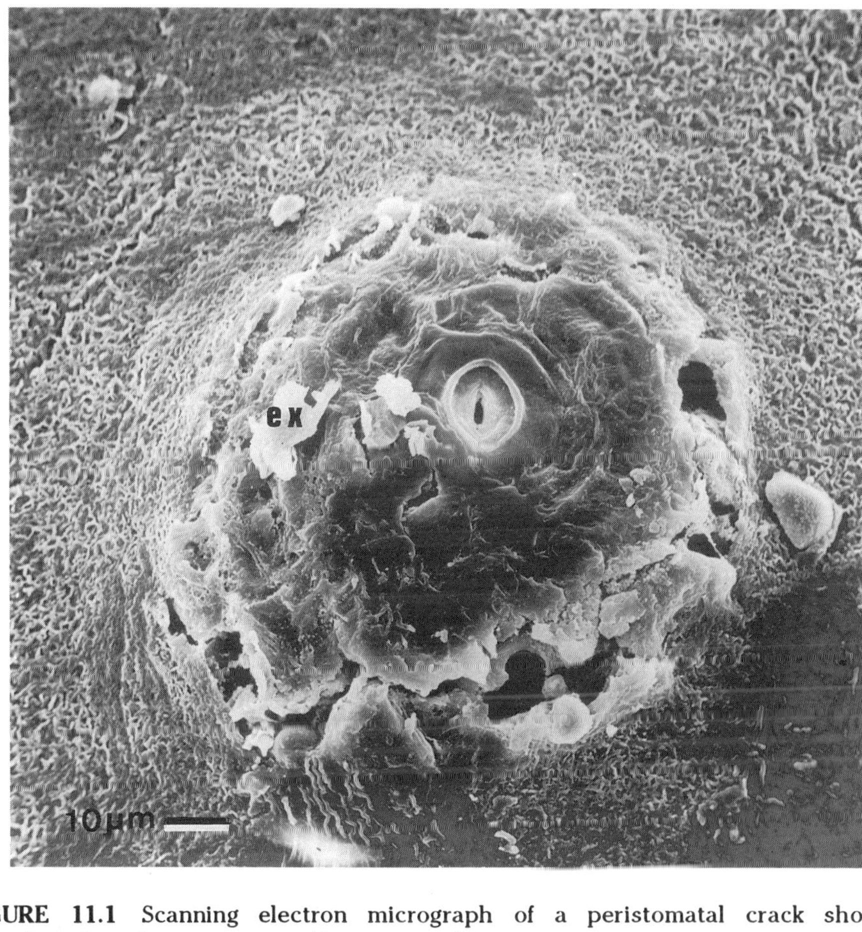

FIGURE 11.1 Scanning electron micrograph of a peristomatal crack showing dehydrated exudates on grape skin. ex: exudates.

substances come through the microfissures from the cuticle layer and originate from the cells of the grape pulp; they look like the internal composition of grape (Donèche, 1986; Padgett and Morrison, 1990). During maturation the sugar/acid ratio increases inside the grape. The level of sugars in the exudates also increases, especially at maturity, while their acid levels decrease (Figure 11.2).

The grape reacts to attack by the parasite by forming antifungal compounds called phytoalexins, which limit the development of the mycelium to the epidermal cells. The phytoalexin family contains numerous substances of an isoflavonoid or terpenoid nature which are normally specific to the plant-host (Deverall, 1976; Bayonove, 1989). In *Vitis vinifera*, Pryce and Langcake (1977) identified viniferin α in *B. cinerea*-infected leaves. This is a resveratrol trimer, a stilbene synthesized by the phenylalanine-polymalonate pathway (Langcake and Pryce, 1977), whose substituted derivatives have varying degrees of fungicide activity (Pont and Pezet, 1990). This compound is produced by the plant cells at the site of attack and an inverse relationship exists between the level of resveratrol around the lesion and the susceptibility of the vine (Langcake and McCarthy, 1979). According to Stein and Blaich (1985), the intensity and speed of production of the stilbenes are related to the age of the vegetal tissue. In grapes, maximal production occurs at a sugar level of about 100–150 g/l, and then decreases rapidly. Its concentration is very low in ripe grapes which, consequently, are very susceptible to infection by *B. cinerea*.

Among the many phytotoxic secretions produced by *B. cinerea* (Heale, 1987; Kamoen, 1989), the exocellular enzymes involved in plant cell wall degradation play a capital role. Endopolygalacturonase is synthesized from the beginning of conidial germination (Verhoeff and Warren, 1972). Then, an exopolygalacturonase, pectin methylesterase, and a transeliminase (lyase) complete degradation of the pectin compounds. The cell wall is greatly altered and cellulolytic activity leads to the severing of certain cellulose fibers. A protease (Movahedi and Heale, 1990) and a phospholipase (Shepard and Pitt, 1976) are also involved in cell wall degradation.

When the mycelium has totally invaded the pectocellulosic cell walls situated under the skin, the white grape berry takes on a characteristic "chocolate" hue known as "pourri plein". The mycelium then produces filaments which burst forth on the surface of the skin, piercing through the cuticle, or taking a route through the various fissures already used for penetration (Figure 11.3). These filaments then differentiate to give conidiophores whose conidia separate and contaminate neighboring mature berries (Figure 11.4).

The cell walls of the plant tissue are now so greatly modified that they are no longer functional. In particular, the hydration of the berry cells is no longer regulated; this varies according to climatic conditions and usually leads to a characteristic withering accompanied by the cytoplasmic death of the epidermal cells. This stage, known as "confit" or "pourri rôti", is used for making sweet white wines.

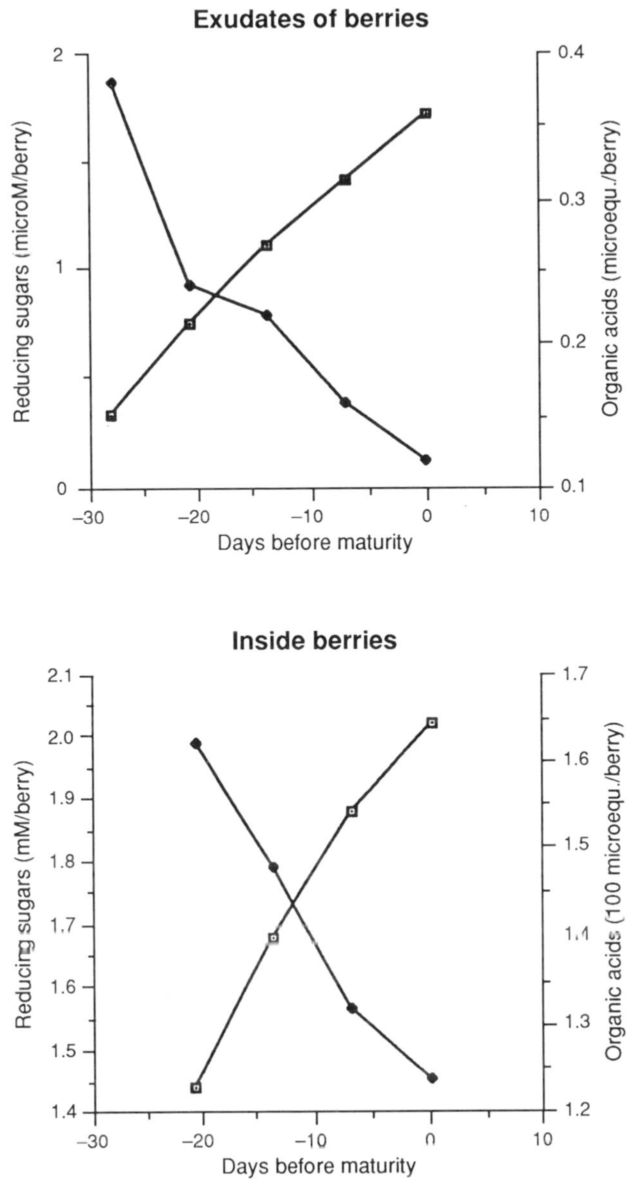

FIGURE 11.2 Changes in the concentrations of reducing sugars and organic acids of exudates and grapes during maturation. Reducing sugars, ⊡; organic acids, ◆.

Chemical Changes Caused by *Botrytis*

Degradation of the epidermal cell walls causes cell death and allows the necrotrophic development of *B. cinerea*. However, the fungus also behaves as a

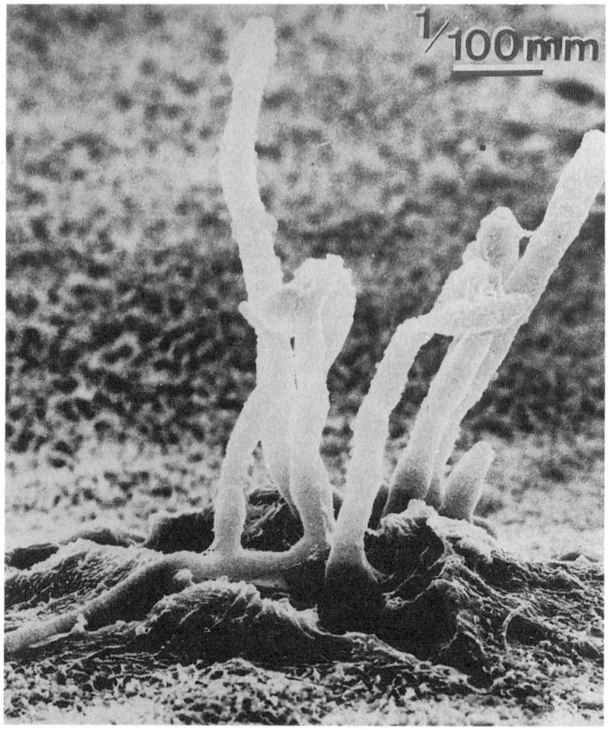

FIGURE 11.3 Scanning electron micrograph of fungal hyphae (*Botrytis cinerea*) raising outside the grape berry.

biotrophic parasite by drawing nutrients necessary for its growth from living cells of the berry. This causes considerable modification in the chemical composition of the whole grape, while the mycelian filaments penetrate only exceptionally into the pulp (Pucheu-Planté and Mercier, 1983). As Müller-Thurgau showed as early as 1888, these modifications mainly concern the sugars and organic acids. The assimilation of glucose and fructose varies according to environmental conditions and the degree of maturity of the grape; it represents about 50% of the initial quantity. The main grape acids are more highly degraded: 70–90% for tartaric acid and 50–70% for malic acid. However, the concentration effect resulting from the loss of water in the berry tends to obscure these modifications. Thus, the must from rotten grapes is characterized by high sugar levels sometimes reaching 360 g/l, and acid levels close to those in healthy grapes. Relatively high pH values (3.5–4.0) show that concentration of other substances such as potassium ions, occurs (Pineau, 1978; Fregoni *et al.*, 1986).

FIGURE 11.4 Scanning electron micrograph of conidiophores and conidia from *Botrytis cinerea.*

Metabolism of grape sugars

Metabolic studies on *B. cinerea in vitro* have shown that germinating conidia and young mycelium possess the enzymes of the Embden-Meyerhof pathway, the hexose monophosphate shunt and the tricarboxylic acid cycle (Donèche, 1989). Direct oxidation of glucose by a process identical to the Entner-Doudoroff pathway occurs in the young mycelium, and contributes to the synthesis of a large amount of cellular material during the exponential growth phase. Gluconic acid is produced by the direct oxidation of glucose and phosphorylation by gluconate kinase. The 6-phosphogluconic acid produced is then metabolized by the hexose monophosphate pathway. However, *B. cinerea* cannot assimilate glucose in anaerobiosis. In an oxygen-poor atmosphere, mycelian growth is limited, and the regeneration of NAD, which is required for the oxidation of glyceraldehyde phosphate in the Embden-Meyerhof pathway, is mainly due to glycerol dehydrogenase. This enzyme causes an accumulation of glycerol in the medium (Ravji *et al.*, 1988). When *B. cinerea* reaches the stationary growth phase, changes occur in the glucose assimilation pathways, the most notable being an inability of the fungus to assimilate gluconic acid. This event corresponds to the onset of secondary metabolism, and the gluconic acid which then accumulates in the medium corresponds to a characteristic secondary product.

The evolution of grape sugars during infection by *B. cinerea* is closely linked to the metabolism of the fungus. During initial development of the fungus under the grape skin, degradation of sugars is limited owing to insufficient supply of oxygen. The main enzymes involved in glucose assimilation are, at this point, found in the grape (Table 11.1). The activity of glycerol-3-phosphate dehydrogenase then leads to a large accumulation of glycerol in the berry. When the fungus comes out of the grape, it then has enough oxygen at its disposal, and adopts an oxidative metabolism characterized by the predominance of oxidases responsible for the accumulation of gluconic acid and the assimilation *de novo* of some of the previously formed glycerol (Donèche, 1987).

Dehydration of the berry causes a large increase in the internal osmotic pressure of the cells. The fungus is now incapable of drawing any nutritive substances, and so its development stops; all the more in that it has reached the reproductive stage, thereby completing its conidian cycle. The gluconic acid concentration depends on duration of the external development of the fungus, varying from 5–10 microequivalents per berry. Musts from infected grapes therefore contain 1–2.5 g/l of gluconic acid (Chauvet, 1981). This acid is not degraded by yeasts during alcoholic fermentation, and its presence in wine testifies to previous development of *B. cinerea* on the grape (Benda, 1984).

Glycerol levels may also vary in rotten grapes; they are higher when length of the internal growth phase of the fungus is greater in relation to its external development. The glycerol level at the "pourri plein" stage is normally about 50–60 micromoles per berry, while not exceeding 10–40 micromoles per berry at the "pourri rôti" stage. Must from these grapes contains 5–7 g/l of glycerol.

Other polyols are formed in grapes and their concentration increases during rot attack. This is particularly the case for mannitol, erythritol and meso inositol (Bertrand *et al.*, 1976). These polyols, secreted by *B. cinerea*, accumulate mainly around the first cell layers of the pulp closest to the skin. Only glycerol seems to diffuse more deeply inside the berry. The formation of trehalose, a disaccharide, has also been noted (Dubernet, 1974 b).

TABLE 11.1 Change in activities of the main enzymes of glucose catabolism in grape berries parasitized by *Botrytis cinerea*.

Enzymatic activity (mU/berry)	*Berry cell free extract*	
	Internal fungal growth	*External fungal growth*
Glucose-6-phosphate dehydrogenase	12,0	5,4
Glucose-6-phosphate isomerase	8	4
Aldolase	3,6	1,2
Glycerol-3-phosphate dehydrogenase	34,4	4
Glucose oxidase	2,8	10,8
Glycerol oxidase	0	4,6

Metabolism of grape acids

Acid degradation by *B. cinerea* during rot attack always occurs in the same way. Only the intensity of the phenomenon varies according to the vintage, which means that acid levels are very close for the different "pourri rôti" grapes. Generally, degradation is greater when the initial acid level is high in healthy grapes.

The degradation of tartaric acid is progressive throughout the infection process. The fungus assimilates this acid by way of a tartrate dehydrogenase, and *in vitro* studies have shown that this also stimulates glucose assimilation (Donèche, 1990). Tartaric acid is more highly metabolized than malic acid, since the latter is mainly assimilated during the external phase of fungal development. Isoenzymatic analysis of acellular extracts of attacked grapes clearly reveals the role of fungal malate dehydrogenase at this stage of infection (Roux, 1986). The assimilation of malic acid during the production of conidiophores corresponds to a high demand for energy to accumulate reserve products in the conidia.

There is less degradation of the other acids and this sometimes leads to a limited increase in their level at the "pourri rôti" stage (Pineau, 1987). Such is the case with citric acid which may be synthesized by certain *B. cinerea* strains (Donèche *et al.*, 1985). However, their levels remain low (1–3 microequivalents per berry) and, despite the concentration phenomenon, musts from these grapes rarely contain more than 8 meq/l acetic acid and citric acid.

An increase in galacturonic acid content is also noted in *B. cinerea* infected grapes (Sponholz and Dittrich, 1984). This acid comes from the enzymatic hydrolysis of pectic compounds in the cell walls. It may be transformed by enzymatic oxidation into mucic acid, whose level may reach 2 g/l in certain musts from attacked grapes (Wurdig, 1976). Yet, this phenomenon is not widespread and is only rarely noted in southern vineyards.

Evolution of other substances

The development of *B. cinerea* occurs to the detriment of nitrogenous substances of the grape. The different forms of nitrogen (ammonium, amino acids and proteins) are degraded in the same way but, owing to their different levels, the respective quantity of each nitrogenous form degraded in relation to the total degradation of nitrogenous substances varies enormously. The fungus preferentially assimilates nitrogen in its simple form. It degrades grape proteins and liberates the nitrogen of the amino acids by means of exocellular proteases and amino oxidases, which diffuse into the berry (Cordonnier and Dugal, 1968). This nitrogen is then processed by *B. cinerea*, which synthesizes the structural and metabolic proteins necessary for its growth in the peripheral area of the grape. Excess ammonium is secreted by the fungus at the end of growth (Donèche, 1987). On the contrary, the grape pulp becomes richer in exocellular fungal proteins. The must of "pourri rôti" grapes is rich in complex nitrogenous forms, since it comes mainly from the pulp and contains less ammonium than a must from healthy grapes (Rapp and Reuther, 1971).

Like many other fungi, *B. cinerea* produces an exocellular laccase, *p*-diphenol oxygen oxidoreductase (Dubernet *et al.*, 1977), which oxidizes many phenolic compounds. This enzyme is involved in pathogenesis and its synthesis requires induction by two groups of substances. The first are phenolic compounds (gallic acid, hydroxinamic acid) which are potentially toxic for the fungus, and the second are the products of degradation from pectic constituents of the cell walls (Marbach *et al.*, 1985). The fungus adjusts the molecular structure of the exocellular laccase to the pH of the host tissue (Marbach *et al.*, 1983) and to the phenolic compounds it contains (Gunata and Moutonnet, 1988). The quantity of enzyme produced is also controlled (Bar Nun and Mayer, 1990).

The main phenolic compounds of white grapes (caffeic and p-coumaric acids, either free or esterified by tartaric acid) are transformed into quinones by the laccase. These quinones tend to polymerize to form brown compounds; only some are stabilized by the formation of thioether with glutathione (Salgues *et al.*, 1986). The characteristic color of grape skin at the "pourri plein" stage probably results from these phenomena. At the end of development, the fungus no longer renews the enzyme and "pourri rôti" grapes have a lower laccase activity than at the "pourri plein" stage. The high levels of sugars in must limit the dissolution of oxygen and partly inhibit laccase activity.

Aromatic compounds are highly modified during rot attack (Shimizu *et al.*, 1982). Glycosidases produced by *B. cinerea* hydrolyze the terpenyl glycosides (Gunata *et al.*, 1989). Free terpenoids, which, presumably, have antifungal properties (Bayonove, 1989), are degraded by oxidative mechanisms in less aromatic substances (Bock *et al.*, 1986; Brunerie *et al.*, 1987). Certain aldehydes are reduced to corresponding alcohols (Bock *et al.*, 1988). Various specific volatile compounds are synthesized by *B. cinerea*: furfural, benzaldehyde, phenylacetaldehyde and benzylcyanide (Kikushi *et al.*, 1983). According to Masuda *et al.* (1984), sotalone (3-hydroxy-4,5-dimethyl-2 (5H) furanone) is one of the main compounds causing the "rôti" characteristic of botrytized grapes.

When rot has developed, grapes contain not only many exocellular fungal enzymes, but also polysaccharides of which some have a phytotoxic activity (Kamoen *et al.*, 1978). *Botrytis cinerea* also produces various antibiotic substances in grapes: botrytidial (Felhaber *et al.*, 1974), norbotryal acetate (Cuevas and Hanson, 1977) and botrylactone (Weimar *et al.*, 1979).

Winemaking with Musts from Noble Rotted Grapes

Until recently, the production of sweet white wine was exclusively based upon the skills of the winemaker (Hugel, 1981). Producing such wines from *B. cinerea* infected grapes poses several technological problems concerning juice extraction, control of alcoholic fermentation and final stability. Recent scientific advances have offered solutions to most of these difficulties.

Must extraction

Classic pressing techniques are largely insufficient for obtaining the vacuolar content of cells in the pulp of "dried" grapes infected with *B. cinerea*. Great pressure must be exerted on the harvest. When the level of rot is not homogeneous, pressing preferentially extracts the juice of those berries which are least infected by rot and least rich in sugars. A selected cryoextraction technique has been developed in recent years (Chauvet *et al.*, 1986); it involves cooling the harvest to a point where only the grapes which are the least rich in sugars and, therefore of least value, are frozen. If pressing is then immediately performed, only the richest grapes release their juice. Moreover, freezing causes extensive modifications to the pectocellulosic components in the cell wall of the grapes, and this allows easier release of cell contents.

Clarification of musts from infected grapes is a delicate process. Much depends on the level of soluble polysaccharides in the berries (Robertson *et al.*, 1980). During rot attack, most polysaccharides in healthy grapes are degraded by *B. cinerea* and are replaced by polymers synthesized by the fungus. The main fungal polysaccharide is a (1–3:1–6) β-D glucan identified by Dubourdieu (1978). In infected berries, this glucan forms a localized network between the epidermal cells and the pulp (Dubourdieu *et al.*, 1978). Mechanical processes which grind or chop up the skins cause this polysaccharide to diffuse into the must (100–200 mg/l), thereby posing problems for clarification. In wine, ethyl alcohol causes the glucan chains to aggregate, and this quickly leads to the blocking of filters. So it is advisable to eliminate this glucan, and various enzymatic preparations are available for its hydrolysis.

Alcoholic fermentation

Marked differences exist between the yeast populations of healthy and *B. cinerea* infected grapes. Infection favors the presence of certain yeasts such as *Kloeckera apiculata* and *Torulopsis stellata* and inhibits others (Le Roux *et al.*, 1973). The must from infected grapes also contains acetic acid bacteria, with *Gluconobacter oxydans* being the predominant species (Joyeux *et al.*, 1984). During the course of natural alcoholic fermentation, the total population of bacteria and less fermentative yeasts decreases, and yeasts of *Saccharomyces* species become predominant (Fleet *et al.*, 1984).

Fermentations are generally conducted in wooden barrels and temperature of the must can easily reach 28°C. Temperature controlled steel tanks are occasionally used to enhance the aromatic character of these wines. Musts from grapes infected with *B. cinerea* normally ferment slowly and sometimes with difficulty; the formation of glycerol and acetic acid (normal fermentation products) is greatly increased. Already in 1888, Müller-Thurgau attributed this problem to the high sugar levels in musts and to the lack of substances essential to yeast, such as nitrogen in the form of an ammonium cation or certain amino acids. The use of ethanol tolerant strains of yeasts (e.g. 18% ethanol), and the addition of nitrogenous substances (ammonium phosphate, amino acids) to the must, only partially improve fermentation (Lafon-Lafourcade *et al.*, 1979).

The fermentation difficulties with musts from botrytized grapes are due to the presence of an inhibitor termed "Botryticine" by the first authors (Lafon-Lafourcade, 1955). In fact, this is a group of heteropolysaccharides rich in rhamnose and mannose, which are released by *B. cinerea* in grapes (Dubourdieu, 1978). These heteropolysaccharides play a dual role: (i) at the beginning of fermentation, they cause the simultaneous production of acetic acid and glycerol by stimulation of the glycero-pyruvic pathway in yeast; and (ii) during the decline phase in yeasts at the end of fermentation, their presence leads to a much greater production of acetic acid than glycerol, and this is considered to be a true secretion of acetic acid by the yeasts (Ribéreau-Gayon *et al.*, 1979). The inhibitory effect is more pronounced with *Saccharomyces bayanus* which is a widespread alcohogenic yeast strain in *B. cinerea* infected grape musts. It is more sensitive than *Saccharomyces cerevisiae* and produces more acetic acid and glycerol per mole of fermented glucose. Attempts have been made to utilize inocula of mixed cultures of yeasts to reduce acetic acid formation (Lafon-Lafourcade *et al.*, 1981). In practice, alcohol-tolerant strains of *S. cerevisiae* are preferable to those of *S. bayanus* since they are less influenced by the heteropolysaccharides of *B. cinerea*. Moreover, once sufficient alcohol is attained, fermentation should be stopped without delay, since it is during transformation of the last grams of sugar that the highest levels of volatile acidity are often found.

Although cessation of fermentation of sweet white wines is performed by pasteurization in certain viticultural regions, the most frequent technique is to use sulfur dioxide. Sulfur dioxide (SO_2) has a tendency to combine with other substances, and leads to a balance between combined and free forms. Only the free form is biologically active as a yeast inhibitor. Therefore, the presence of at least 50 mg/l of free sulfur dioxide is required to stop any new fermentation. Sulfur dioxide also protects against oxidation of phenolic compounds by laccase. In effect, this enzyme is stable at the pH of the must and wine, and retains its activity after fermentation. Despite the low level of non-oxidized phenolic compounds, there is always the chance that this enzyme might cause browning of wine in the presence of oxygen. The same is true for the other fungal oxidases which produce peroxides capable of reacting with the phenolic compounds in the absence of an anti-oxidant.

Botrytized grapes produce wines rich in substances likely to combine SO_2, thereby decreasing the proportion of its active free form. This mainly concerns ketonic compounds, of which the most active are ketoacids synthesized by yeast under difficult fermentation conditions; examples are pyruvic and α ketoglutaric acids. These acids are mainly produced at the beginning of fermentation and their levels decrease at the end of fermentation. Dry white wines have lower levels of these acids than sweet white wines (Ribéreau-Gayon *et al.*, 1980). The galacturonic and glucuronic acids resulting from the degradation of grape cell walls by *B. cinerea* are found in these wines, and may also combine SO_2 (Sponholz and Dittrich, 1985).

Nishimura and Masuda (1983) have analyzed sugars, alcohols, organic acids and volatile compounds of botrytized and other wines from various countries. According to their results, botrytized wines have significantly larger amounts of organic acids (acetic, gluconic, etc.), sugars, and some volatile compounds (benzaldehyde, 4-nonanolide, etc.), but smaller amounts of other volatile compounds (isoamyl alcohol, ethyl octanoate, etc.).

THE DIFFERENT FORMS OF ROT

Factors Influencing Noble Rot

The development of characteristic noble rot is a rapid phenomenon occurring around maturity. Samples from Sauternes (France) vineyards over the last 10 years show that the first visual signs of attack appear 15–20 days before maturity and whatever the climatic conditions, *B. cinerea* develops quite slowly. At maturity, a brutal parasitic attack occurs throughout the vineyard and, at a given period, a high percentage of grapes simultaneously reaches the "pourri plein" stage (Pucheu-Planté, unpublished results).

The prerequisite for obtaining *B. cinerea* in the noble rot form is that the berries should reach maturity intact, owing to an inexcessive water supply (Pucheu-Planté and Seguin, 1978). In wet years, this must be regulated either naturally by the soil or artificially by drainage, particularly in clayish soils. However, this condition is not sufficient in itself (Pucheu-Planté and Seguin, 1981). Also required during the period of overmaturation are particular climatic conditions; an alternation between short damp periods (3–4 days) assisting conidial germination, and longer dry periods (about 10 days) leading to the concentration and chemical transformation of the grapes. A certain degree of humidity (drizzle, dew, morning mist) is, in fact, indispensable for allowing the fungus to develop. The most favorable conditions are when the mean temperature is between 15–20°C with 90% relative air humidity (Bulit and Lafon, 1970). However, too long a wet period causes an excessive temperature drop, thus posing the threat of the berries bursting. On the other hand, hot sunny days with wind give high evaporation rates, thereby leading to high juice concentrations (Nelson, 1958). Moreover, the greater the alternation between dryness and relative humidity, the more easily the parasite disseminates by conidial dispersion (Jarvis, 1980).

Noble rot affects many cultivars: for example, Muscadelle, Sémillon and Sauvignon in the Bordeaux region; Chenin Blanc in Anjou; Riesling in the Rhine Valley; Furmint and Harslevelü in Tokay (Hungary). However, Pucheu-Planté and Leclair (1990) have shown the importance of the nature of the clone on the quality of the rot obtained. There seems to exist a complex clonal specificity involving many factors, such as the number of stomata, the thickness of the skin, and the amount of anti-fungal substances produced. This leads to a different composition in the exudates on the surface of berries. Further research is needed to better understand the role of the clone in the noble rot process.

Given the multiplicity of conditions required for obtaining natural development of *B. cinerea* as noble rot, many attempts have been made to artificially reproduce the phenomenon (Stone, 1980).

In regions where the climate during grape maturation is mainly hot and dry, like in California and many parts of Australia, inoculation of the fungus onto grapes on the stock has been disappointing (Nelson, 1951). In such conditions, it is preferable to inoculate the grapes with *B. cinerea* after harvesting and to place them in a chamber containing the most favorable environment for fungal development. The grapes are placed on trays, inoculated with a conidial suspension (Nelson *et al.*, 1963), then incubated for 24 hours at a temperature between 20 and 25°C, with relative humidity exceeding 90%. This period is followed by colder, drier conditions (temperature: 20°C; relative humidity: 50%) to control the development of *B. cinerea* and facilitate dehydration of the berries. This technique gives grapes rich in sugars, poor in acids, and aromatic notes close to those obtained by a natural process (Nelson and Amerine, 1956; Nelson and Nightingale, 1959). It is very important to limit the humid period to a maximum of 24 h; otherwise, there is a great risk of grey rot developing. One must also monitor deterioration of the grapes by other microorganisms such as acetic bacteria. Despite these technical difficulties, the artificial production of sweet wines is possible but costly, and this has limited its development (Ewart, 1982). Each year, the process leads to limited production of a certain quantity of high-quality wine, thus making it a particularly rare product.

The production of sweet botrytized wine would be greatly facilitated if it were possible to obtain the characteristic changes caused by *B. cinerea* in the must and not in the grape berries. This idea is not entirely new, since the first trials took place in 1957 (Nelson and Amerine, 1957). Much work has since been performed (de Soto *et al.*, 1966; Watanabe and Shimazu, 1976, 1978). The process of treating musts with *B. cinerea* as submerged cultures in fermentors is now operative and makes it possible to control the production of glycerol and gluconic acid by adding cyclic adenosine monophosphate (Watanabe *et al.*, 1981), and the degradation of grape acids (Shimazu *et al.*, 1984). However, the levels of acetic acid and acetaldehyde are more difficult to control (Watanabe and Shimazu, 1978). Also, it is difficult to artificially produce the aroma of botrytized wines. The effects on taste depend much on the cultivar from which the must is obtained and on the conditions (especially yeast strain) used in subsequent alcoholic fermentation (Watanabe and Shimazu, 1980). It is also necessary to make a concentration, normally by cold techniques, to obtain a sweet wine. These numerous prerequisites, therefore, make this process a costly biotechnology, and large-scale production of wines cannot be envisaged.

At present, it seems more worthwhile to conduct pedologic, climatologic and ecologic research on soils likely to produce noble rot naturally.

Grey Rot

While noble rot develops regularly and uniformly throughout a vineyard, grey rot attack is normally heterogeneous. In some cases the grapes are partially or totally attacked, while on others they are completely unharmed.

The first factor to consider here is the cultivar. The most sensitive cultivars to *B. cinerea* are those where the berries are tightly-packed, fine-skinned and with abundant foliage. Any cultivation technique aimed at increasing the vigor of the vine contributes to this sensitivity. The lack of aeration in berries favors the maintenance of a certain level of humidity which, in turn, is favorable to parasitic development. This humidity also modifies the exudation of nutritive substances. The clone also has an important role to play here (Pucheu-Planté and Leclair, 1990).

Regarding the parasite, certain authors have attempted to demonstrate that there are several distinct types of *B. cinerea*, and that some of these have greater pathogenicity than others, and could, therefore, play a role in the development of grey rot (Cruickshank, unpublished results). This frequently concerns the same strains whose pathogenic activity is modified by environmental conditions. The presence of nutritive substances increases the proportion of multinucleated macroconidia having a strong germinative power in relation to mononucleated microconidia which are often incapable of germination (Philipps *et al.*, 1987). More and more isolates resistant to systemic fungicides are also being noted (Fraile *et al.*, 1986; Ruiz Borge and Schlesser, 1990). These resistant strains have modified biological activity (Gullino and Garibaldi, 1981). Now, there is no reason for discontinuing the use of systemic fungicides if the development of tolerant strains can be contained by alternating application of systemic and protectant fungicides (Nair, 1990).

Sclerotia of *B. cinerea* are the main source of primary inoculum for bunch rot of grapes. Resporulation of sclerotia leads to early infection of flowers (Nair and Nadtotchei, 1987). The presence of quiescent *B. cinerea* in berries just after closure and its development when favorable conditions occur, were demonstrated in young clusters (Pezet and Pont, 1988). So, early application of fungicide reduced natural infection by *B. cinerea* better than late applications (Nair *et al.*, 1987). The association of *B. cinerea* with grape berry moth larvae suggests that this insect also plays an important role in dissemination of the disease (Fermaud and Le Menn, 1989).

The process of grape infection by *B. cinerea* is identical to that described for noble rot. While easily visible on white grapes, the early development of the fungus on black grapes is always difficult to detect. The main characteristic of grey rot is widespread external development of the parasite, causing a "mycelian web" on the surface of the berries. Contamination of neighboring berries then occurs very quickly owing to the high biological activity of this mycelium. A high level of humidity and favorable temperatures are decisive factors for grey rot.

The chemical composition of the grape is greatly modified during grey rot attack. The fungus consumes the sugars of the grape and causes accumulation of glycerol and gluconic acid. The richness in sugar of rotten grapes depends on the concentration resulting from dehydration of the berry. There is a great difference between noble rot, where the concentration has a greater impact than degradation, and grey rot where the opposite occurs (Figure 11.5). The musts from "rôti" grapes due to grey rot attack have a sugar concentration rarely exceeding 230 g/l. The difference between the two types of rot is pronounced with respect to the outcome of these sugars (Donèche, 1987). Developing strongly outside the berry, grey rot synthesizes a large quantity of gluconic acid while the glycerol level in "rôti" berries is low, unlike the situation with noble rot (Figure 11.6). The acids are more highly degraded than the sugars, and tartaric acid is more highly degraded than malic acid. The level of acetic acid may be greater than in cases of noble rot, but rarely exceeds 5 microequivalents per berry; this precludes any intervention by microorganisms having an oxidative metabolism.

The differences between the two types of rot are most noticeable in their secondary metabolisms. There is greater oxidation in the case of grey rot, especially in black cultivars where laccase has many substrates that it oxidizes. Destruction of this enzyme by heating or addition of very high doses of SO_2 is necessary to protect the must, and subsequently the wine, from the polymerization of phenolic compounds and their precipitation during a phenomenon known as "casse oxydasique" or oxidase breakdown. Laccase activity is maximal between 40 and 50°C but decreases considerably after 5

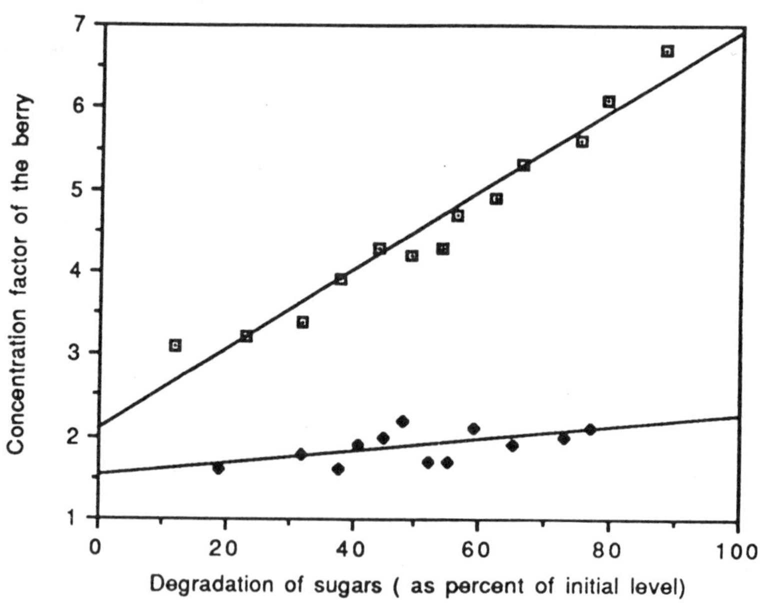

FIGURE 11.5 Relationship between sugar metabolism by *Botrytis cinerea* and concentration of the berry by dehydration according to nature of grape rot. Noble rot, ⊡; grey rot, ◆. (From Donèche, 1987).

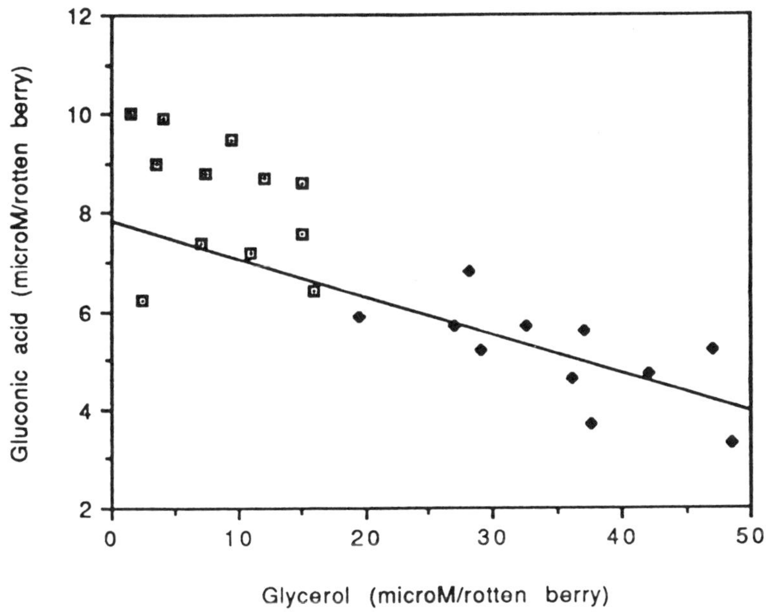

FIGURE 11.6 Relationship between gluconic acid synthesis and the production of glycerol according to the nature of grape rot. Noble rot, ⊡; grey rot, ◆. (From Donèche, 1987).

minutes at 45°C. Heating of the harvest must be performed very quickly, passing from 30 to 50°C in only a few seconds. Sulfiting a rotten red crop is not sufficient for total protection. In fact, SO_2 is entirely combined before the complete destruction of the laccase, and the wine obtained has to be stabilized by further sulfiting (Dubernet, 1974a).

Whereas noble rot bestows on wines the most sought-after aromatic notes, grey rot, especially in conjunction with fungal types other than *B. cinerea* (white and green rot), causes detrimental modifications to organoleptic qualities. Indeed, previous infection by *B. cinerea* favors the development of other fungi such as *Aspergillus*, *Mucor* and *Penicillium* (D'Ercole and Nippotti, 1981; Nair, 1985). These moulds produce less mycelial mass and glycerol than *B. cinerea* (Ravji *et al.*, 1988). They are often responsible for the appearance of phenolic-tasting compounds in red wines and an iodine-like taste in white wines (Ribéreau-Gayon, 1982). Little work has been done on these compounds; they probably occur through degradation of the phenolic compounds of the grape skin which undergoes maceration owing to the extent of the mycelian mass.

One of the few positive factors of this rot attack is the stimulation of lactic acid bacteria (San Romao, 1985) which are capable of using the polysaccharides synthesized by *B. cinerea*. However, these polysaccharides adsorb fatty acids resulting from the alcoholic fermentation by yeasts, thereby limiting their inhibition of malolactic bacteria.

Common Rot

Two general categories of microorganisms occur on the grape surface. The first is represented by parasites capable of penetrating the vegetal tissue and developing at its expense; species of this group are mostly filamentous fungi, such as *B. cinerea* that possess the necessary enzymes for degrading numerous grape macromolecules. The second group comprises species which do not penetrate the vegetal tissue and proliferate only when there are lesions allowing the sugary juice to seep to the surface of the grape. This group is composed of yeasts, lactic acid bacteria and acetic bacteria. The microflora of the grape, and the proportion of the two categories of microorganisms, change during fruit maturation according to climatic and other influences.

Lesions on the grape skin may occur at various stages of maturation and can be caused by many factors such as insect bites, by *B. cinerea* and skin fissuring during overmaturation. The main cause is, without doubt, bursting of the berry through an excess of water. This is a process in which the water conditions in the soil and the root configuration of the vine play a decisive role (Ribéreau-Gayon *et al.*, 1980). When rain falls on semi-parched soil, the water penetrates to varying depths but most often it remains localized on the surface, while the damp lower strata of the soil do not change; such is the case when rainfall is not excessive. In soils allowing the vine to root deeply, the root density in the upper strata of the soil is low and water absorption is limited. On the other hand, when the vine has extensive superficial rooting, water absorption is high and many berries may burst (Pucheu-Planté and Seguin, 1978). The degree of this phenomenon is more or less pronounced according to the resistance of the epidermal cells and the thickness of the cuticular waxes (Yamamura *et al.*, 1986).

Despite the presence of many microorganisms on the surface of grapes, *B. cinerea* may sometimes still develop; it then behaves as an antagonist stopping the development of yeasts and acetic bacteria (Le Roux *et al.*, 1973). The fungus develops in the wounds to form an increasing mycelian mass ("crescent shaped") covering a glucan network, whose role is to obturate the fissure. Unlike the typical grey rot process, it is during this first phase that the degradation of the grape sugars is greatest. *Botrytis cinerea* then develops mycelian ramifications under the skin; glycerol is produced during this internal phase. The fungus then develops conidiophora over the whole berry with an accumulation of gluconic acid, and the berry becomes dehydrated. Finally, the constitutional chemical modifications of the grape are characteristic of grey rot. However, such development of this fungus is frequently hindered (fungicide treatment, unfavorable climatic conditions) and the sugary juice seeping through the wounds then favors proliferation of acetic bacteria and/or oxidative yeasts.

When bacterial species of *Gluconobacter* intervene alone, the grape becomes yellowy. In most cases, the berries are almost emptied of their contents by various insects, and their constitutional modifications can no longer be observed. In exceptional cases, the infected berries become dehydrated and reach a stage resembling "pourri rôti". The main change in such grapes is a large increase in the contents of acetic and gluconic acids: this is known as "pourriture aigre" or sour rot. Musts from such "rôti" grapes contain more than 40 g of acetic acid and up to 25 g of gluconic acid per liter. Since these bacteria degrade malic and tartaric acids less than *B. cinerea*, such musts have a very low pH. These acids are produced from sugars whose levels decrease dramatically in such grapes. *Gluconobacter* oxidizes glucose by way of NADP dependent dehydrogenases to gluconic, keto-2-gluconic, keto-5-gluconic and diketo-2,5-gluconic acids in the medium (Olijve and Kok, 1979). The accumulation of these ketonic compounds leads to a significant increase in the potential of these musts to combine SO_2. When these bacteria affect grapes already attacked by *B. cinerea*, they also oxidize the glycerol to dihydroxyacetone (Flickinger and Perlman, 1977). The acetic bacteria normally oxidize ethanol to acetic acid; but in healthy grapes, ethanol occurs only in traces. Although the degradation of glucose by *Gluconobacter* spp. is accompanied *in vitro* by a moderate synthesis of acetic acid (about 1 g/l), this does not explain its accumulation in "pourri aigre" grapes. Our present state of knowledge points to the simultaneous intervention of oxidative yeasts which, despite their low fermentative capacity, could provide the necessary ethanol for the synthesis of acetic acid.

The involvement of yeasts is not to be excluded here, since a particular type of "acid rot" seems quite widespread, especially in Mediterranean vineyards. This disease which was first noted in Italy (Bisiach *et al.*, 1981), is due to different species of less fermentative yeasts (*Candida, Kloeckera, Hanseniaspora*) of which some such as *Endomycopsella* have exocellular enzymatic activities for hydrolyzing cell walls, thus providing them with a certain phytopathogenic role (Marchetti *et al.*, 1984). The tartaric acid in grapes is not degraded, and the formation of gluconic, galacturonic and acetic acids strongly increases the acidity (Zironi *et al.*, 1982). Yeasts produce a small amount of ethanol and the must has high levels of ethyl acetate (Guerzoni and Marchetti, 1987) and ethanal (Marchetti *et al.*, 1984). The fatty acid composition, particularly of the seeds, is also modified (Zironi *et al.*, 1984).

Grapes suffering from common rot, be it caused by bacteria or yeasts, cannot be used to make wines. Their presence in the vineyard should be detected as soon as possible, and they must be eliminated to limit spread of the disease.

CONCLUSION

Many microorganisms co-exist on the grape surface. Infection by *B. cinerea* involves the combination of many factors; antagonism *vis-à-vis* other microorganisms, sensitivity of the vine, particular climatic conditions. Factors even more specific must occur together for this infection to lead to noble rot. In particular, the grapes must reach maturity intact. Noble rotted grapes are the result of the metabolic activity of the fungus and its natural control, by an alternation of short damp periods and longer drier periods favoring dehydration of the grapes. Quality depends on the balance between the internal and external phases of *B. cinerea* development.

Excessive humidity favors the growth of *B. cinerea* in the form of grey rot which causes a great decrease in grape quality. When development of the fungus is inhibited and the grapes have lesions, the proliferation of other microorganisms leads to total deterioration of the grapes.

It is obviously difficult to distinguish between the various forms of rot in the vineyard. This hinders the mechanization of harvesting, a skill mainly based on the vinegrower's experience.

The production of wine from botrytized musts is difficult and largely depends on successful fermentation, the ecology and biochemistry of which is specifically influenced by the unique composition of the must. Moreover, when one realizes that the yield of grapes per hectare of vineyard is very low, it is easily understood that the production of such wines is a costly process.

Many attempts have been made to artificially reproduce noble rot. The biotechnological processes are costly and difficult to perform. The reward of producing these wines naturally is much greater, even if the technique is a high-risk business.

REFERENCES

Bayonove, C. (1989) Incidences des attaques parasitaires fongiques sur la composante qualitative du raisin et des vins. *Revue Francaise d'Oenologie*, **116**, 29–39.

Bar Nun, N. and Mayer, A.M. (1990) Regulation of laccase formation by *Botrytis cinerea*. In *Comptes Rendus du 4eme Symposium International d'Oenologie* (Bordeaux, 1989), edited by Dunod, pp. 122–126. Paris:Bordas.

Benda, I. (1984) Zum einflush von hefen auf den glukonsäure-gehalt des weines. *Wein-Wissenschaft*, **39**, 263–267.

Bertrand, A., Pissard, R., Sarre, C. and Sapis, J.C. (1976) Etude de l'influence de la pourriture grise des raisins (*Botrytis cinerea*) sur la composition chimique et la qualité des vins. *Connaissance Vigne Vin*, **10**, 427–446.

Bessis, R. (1977) Caractères et développement des exsudats au cours de la maturation du raisin. In *Le raisin et le froid C.R. Symposium I.I.F. et O.I.V.* (Paris, 1977), Science et Technique du Froid, pp. 39–44.

Bisiach, M., Minervini, G. and Salomone, M.C. (1981) Ricerche sperimentali sul marciume acido del grappolo e sui suoi rapporti con la muffa grigia. *Notizia Maladia Piante*, **102**, 61–79.

Bock, G., Benda, I. and Schreier, P. (1986) Metabolism of linalol by *Botrytis cinerea* biotransformation. *Journal of Food Science*, **15**, 659–662.

Bock, G., Benda, I. and Schreier, P. (1988) Reduction of cinnamaldehyde and unsaturated acids by *Botrytis cinerea*. *Zeitschrift fuer Lebensmittel Untersuchung und Forschung*, **186**, 33–35.

Brunerie, P., Benda, I., Bock, G. and Schreier, P. (1987) Bioconversion of citronellol by *Botrytis cinerea*. *Applied Microbiology and Biotechnology*, **27**, 6–10.

Bulit, J. and Lafon, R. (1970) Quelques aspects de la Biologie du *Botrytis cinerea Pers.* agent de la pourriture grise des raisins. *Connaissance Vigne Vin*, **4**, 159–174.

Chauvet, S. (1981) Application de l'isotachophorèse au dosage des principaux acides du vin—Contribution a l'étude de la relation entre l'intensité du développement de la pourriture noble et la composition chimique des vins blancs doux. Thèse d'Université, Bordeaux, France.

Chauvet, S., Sudraud, P. and Jouan, T. (1986) La cryoextraction selective, premières observations et perspectives. *Bulletin de l'Office International du Vin.*, **667–668**, 1021–1043.

Coley-Smith, J.R., Verhoeff, K. and Jarvis, W.R. (1980) *The Biology of Botrytis*. London: Academic Press Inc.

Cordonnier, R. and Dugal, A. (1968) Les activités protéolytiques du raisin. *Annales de Technologie Agricole*, **17**, 189–206.

Cuevas, O. and Hanson, J.R. (1977) Norbotryal acetate, a norsesquiterpenoid aldehyde from *Botrytis cinerea*. *Phytochemistry*, **16**, 1061–1062.

D'Ercole, N. and Nipotti, P. (1981) Biologia della *Botrytis cinerea* sulla vite: I—Conservazione del parassita. *Rivista di Patologia Vegetale*, SIV (**17**), 35–43.

De Soto, R.T., Nightingale, M.S. and Huber, R. (1966) Production of natural sweet table wines with submerged cultures of *Botrytis cinerea Pers*. *American Journal of Enology and Viticulture*, **17**, 191–202.

Deverall, B.J. (1976) Chapter II—Current perspectives in research on phytoalexins In *Biochemical Aspect of Plant-Parasite relationship*, edited by J. Friend and D.R. Threlfall, pp. 208–224. London: Academic Press Inc.

Donèche, B. (1986) La nature des exsudats de raisins et leur rôle dans la germination des conidies de *Botrytis cinerea*. *Agronomie*, **6**, 67–73.

Donèche, B. (1987) Etude biochimique de la relation Hôte–Parasite dans le cas du raisin et de *Botrytis cinerea*. Thèse Doctorat es Sciences. Bordeaux, France.

Donèche, B. (1989) Carbohydrate metabolism and gluconic acid synthesis by *Botrytis cinerea*. *Canadian Journal of Botany*, **67**, 2888–2893.

Donèche, B. (1990) Métabolisme de l'acide tartrique du raisin par *Botrytis cinerea*. Premiers résultats. *Science des Aliments*, **10**, 589–602.

Donèche, B., Roux, F. and Ribéreau-Gayon, P. (1985) Dégradation de l'acide malique par *Botrytis cinerea*. *Canadian Journal of Botany*, **63**, 1820–1824.

Dubernet, M. (1974a) Recherches sur la tyrosinase de *Vitis vinifera* et la laccase de *Botrytis cinerea*. Applications technologiques. Thèse Doctorat 3ème cycle, Bordeaux, France.

Dubernet, M.O. (1974b) Application de la chromatographie en phase gazeuse à l'étude des sucres et des polyols du vin. Thèse Doctorat 3ème cycle.

Dubernet, M., Ribéreau-Gayon, P., Lerner, H.R., Harel, E. and Mayer, A.M. (1977) Purification and properties of laccase from *Botrytis cinerea*. *Phytochemistry*, **16**, 191–193.

Dubourdieu, D. (1978) Etude des polysaccharides sécrétés par *Botrytis cinerea* dans la baie de raisin. Incidence sur les difficultés de clarification des vins de vendanges pourries. Thèse Doctorat 3ème cycle, Bordeaux, France.

Dubourdieu, D., Pucheu-Planté, B., Mercier, M. and Ribéreau-Gayon, P. (1978) Structure, rôle et localisation d'un glucane sécrété par *Botrytis cinerea* dans la baie de raisin. *Comptes Rendus Académie des Sciences, Paris*, **287D**, 571–573.

Ewart, A.J.W. (1982) Botrytis: its use in the production of a speciality wine. In *Proceedings of the Australian grape grower and winemaker.* Annual Technical Issue, **19**(220), 24–26.

Fehlaber, H.W., Geipel, R., Mercker, H.J., Tschesche, R. and Welmar, K. (1974) Botrydial, ein antibioticum aus der nahrlosung des pilzes *Botrytis cinerea*. *Chemische Berichte*, **107**, 1720–1724.

Fermaud, M. and Le Menn, R. (1989) Association of *Botrytis cinerea* with grape berry moth larvae. *Phytopathology*, **79**, 651–656.

Fleet, G.H., Lafon-Lafourcade, S. and Ribéreau-Gayon, P. (1984) Evolution of yeasts and lactic acid bacteria during fermentation and storage of Bordeaux wines. *Applied and Environmental Microbiology*, **48**, 1034–1038.

Flickinger, M.C. and Perlman, D. (1977) Application of oxygen-enriched aeration in the conversion of glycerol to dihydroxyacetone by *Gluconobacter melanogenus* IFO 3293. *Applied and Environmental Microbiology*, **33**, 706–712.

Fraile, A., Alonso, A. and Sagasta, E.M. (1986) Some characteristics of *Botrytis cinerea* isolates tolerant to procymidone. *Plant Pathology*, **35**, 82–85.

Fregoni, M., Iacono, F. and Zamboni, M. (1986) Influence du *Botrytis cinerea* sur les caractéristiques physicochimiques du raisin. *Bulletin de l'Office International du Vin*, **667–668**, 995–1013.

Guerzoni, E. and Marchetti, R. (1987) Analysis of yeast flora associated with grape sour rot and of the chemical disease markers. *Applied and Environmental Microbiology*, **53**, 571–576.

Gullino, M.L. and Garibaldi, A. (1981) Biological properties of dicarboximide-resistant strains of *Botrytis cinerea Pers*. *Phytopathologie Méditerranéenne*, **20**, 117–122.

Gunata, Y.Z., Biron, C., Sapis, J.C. and Bayonove, C. (1989) Glycosidase activities in sound and rotten grapes in relation to hydrolysis of grape monoterpenyl glycosides. *Vitis*, **28**, 191–197.

Gunata, Y.Z. and Moutonnet, M. (1988) Activités de la polyphénoloxydase du raisin et de la laccase de *Botrytis cinerea* sur les dérivés hydroxycinnamiques comparé à d'autres substrats phénoliques. *Revue Francaise d'Oenologie*, **113**, 42–47.

Heale, J.B. (1987) The phytotoxicity of *Botrytis cinerea*. In *Proceedings of the EC Experts' Group* (Porteferraio, 1985), edited by R. Cavalloro, pp. 277–299, Rotterdam: A.A. Balkema.

Hugel, J. (1981) Production et vinification des vins de vendange tardive et de sélection de grains nobles. *Revue Francaise d'oenologie*, **82**, 57–60.

Jarvis, W.R. (1980) Chapter 8—Epidemiology. In *The biology of Botrytis*, edited by J.R. Coley-Smith, K. Verhoef and W.R.Jarvis, pp. 219–250. London: Academic Press Inc.

Joyeux, A., Lafon-Lafourcade, S. and Ribéreau-Gayon, P. (1984) Evolution of acetic acid bacteria during fermentation and storage of wine. *Applied and Environmental Microbiology*, **48**, 153–156.

Kamoen, O. (1989) Phytopathological role of secretions from *Botrytis cinerea*. *Med. Fac. Landbouww. Rijksuniv. Gent.*, **54**, 509–515.

Kamoen, O., Jamart, G., Moermans, R., Van de putte, L. and Dubourdieu, D. (1978) Comparative study of phytotoxic secretions of *Botrytis cinerea*. *Med. Fac. Landbouww. Rijksuniv. Gent.*, **43**, 847–857.

Kikushi, T., Kadota, S., Suehara, H., Nishi, A., Tsubaki, K., Yano, H. and Harimaya, K. (1983) Odorous metabolites of fungi, *Chaetomium globosum Kinze ex Fr.* and *Botrytis cinerea Pers ex Fr.*, and a blue green algae, *Phormidium tenue (meneghini) gomont*. *Chemical and Pharmaceutical Bulletin*, **31**, 659–663.

Lafon-Lafourcade, S. (1955) Contribution à l'étude des activateurs et des inhibiteurs de la fermentation alcoolique des moûts de raisins. Thèse Doctorat 3ème cycle. Bordeaux, France.

Lafon-Lafourcade, S., Larue, F. and Ribéreau-Gayon, P. (1979) Evidence for the existence of "survival factors" as an explanation for some peculiarities of yeast growth especially in grape must of high sugar concentration. *Applied and Environmental Microbiology*, **38**, 1069–1073.

Lafon-Lafourcade, S., Lucmaret, V., Joyeux, A. and Ribéreau-Gayon, P. (1981) Utilisation de levains mixtes dans l'elaboration des vins de pourriture noble en vue de réduire l'acidité volatile. *Comptes-Rendus de l'Académie d'Agriculture*, **67**, 616–622.

Langcake, P. and MacCarthy, W.V. (1979) The relationship of resveratrol production to infection of grapevine leaves by *Botrytis cinerea*. *Vitis*, **18**, 244–253.

Langcake, P. and Pryce, R.J. (1977) A new class of phytoalexins from grapevines. *Experientia*, **33**, 151–152.

Le Roux, G., Eschenbruch, R. and De Bruin, S.I. (1973) The microbiology of South African wine making. Part VIII—The microflora of healthy and *Botrytis cinerea* infected grapes. *Phytophylactica*, **5**, 51–54.

Marbach, I., Harel, E. and Mayer, A.M. (1983) Inducer and culture medium dependant properties of extracellular laccase from *Botrytis cinerea*. *Phytochemistry*, **22**, 1535–1538.

Marbarch, I., Harel, E. and Mayer, A.M. (1985) Pectin, a second inducer for laccase production by *Botrytis cinerea*. *Phytochemistry*, **24**, 2559–2561.

Marchetti, R., Guerzoni, M.E. and Gentile, M. (1984) Recherche sur l'étiologie d'une nouvelle maladie de la grappe: la pourriture acide. *Vitis*, **23**, 55–65.

Masuda, M., Okawa, E., Nishimura, K. and Yunome, H. (1984) Identification of 4,5dimethyl-3-hydroxy-2(5H)-furanone (sotolon) and ethyl-9-hydroxynonanoate in Botrytised wine and evaluation of the roles of compounds characteristic of it. *Biological Chemistry*, **48**, 2707–2710.

Movahedi, S. and Heale, J.B. (1990) Purification and characterization of an aspartic proteinase secreted by *Botrytis cinerea Pers ex Pers* in culture and in infected carrots. *Physiological and Molecular Plant Pathology*, **36**, 289–302.

Müller-Thurgau, H. (1888) Die edelfaüle der trauben. *Landwirt Jarbücher*, **17**, 83–90.

Nair, N.G. (1985) Fungi associated with bunch rot grapes in the Hunter Valley. *Australian Journal of Agricultural Research*, **36**, 435–442.

Nair, N.G. (1990) Strategies for fungicidal control of the bunch rot of grapes caused by *Botrytis cinerea* in the Hunter Valley. *Wine Industry Journal*, May 1990, 218–220.

Nair, N.G., Emmett, R.W. and Parker, F.E. (1987) Programming applications of dicarboximides to control bunch rot caused by *Botrytis cinerea*. *Plant Pathology*, **36**, 175–179.

Nair, N.G. and Nadtotchei, A. (1987) Sclerotia of *Botrytis cinerea* as a source of primary inoculum for bunch rot of grapes in New South Wales, Australia. *Journal of Phytopathology*, **119**, 42–45.

Nelson, K.E. (1951) Factors influencing the infection of table grapes by *Botrytis cinerea*. *Phytopathology*, **41**, 319–326.

Nelson, K.E. (1958) The effect of air velocity on weight loss of botrytized grapes in natural sweet wine production. *American Journal of Enology and Viticulture*, **9**, 123–125.

Nelson, K.E. and Amerine, M.A. (1956) Use of *Botrytis cinerea* for the production of sweet table wines. *American Journal of Enology and Viticulture*, **7**, 131–136.

Nelson, K.E. and Amerine, M.A. (1957) The use of *Botrytis cinerea* in the production of sweet table wines. *Hilgardia*, **26**, 521–563.

Nelson, K.E., Kosuge, T. and Nightingale, A. (1963) Large scale production of spores to botrytise grapes for commercial natural sweet wine production. *American Journal of Enology and Viticulture*, **14**, 118–128.

Nelson, K.E. and Nightingale, A.S. (1959) Studies in the commercial production of natural sweet wines from botrytised grapes. *American Journal of Enology and Viticulture*, **10**, 135–141.

Nishimura, K. and Masuda, M. (1983) Compounds characteristic of Botrytized wines. *Proceedings of the American Chemical Society meeting* (Seattle, USA, 1983).

Olijve, W. and Kok, J.J. (1979) Analysis of growth of *Gluconobacter oxydans* in glucose containing media. *Archives of Microbiology*, **121**, 283–290.

Padgett, M. and Morrison, J.C. (1990) Changes in grape berry exudates during fruit development and their effect on mycelial growth of *Botrytis cinerea*. *Journal of American Society of Horticultural Science*, **115**, 269–273.

Pezet, R. and Pont, V. (1988) Infection florale et latence de *Botrytis cinerea* dans les grappes de Vitis vinifera (var. gamay). *Revue Suisse de Viticulture, Arboriculture et Horticulture*, **18**, 317–322.

Phillips, D.J., Margosan, D.A. and Mackey, B.E. (1987) Size, nuclear number and aggressiveness of *Botrytis cinerea* spores produced on media of varied glucose concentrations. *Phytopathology*, **77**, 1606–1608.

Pineau, J. (1987) Modification de la composition du raisin par attaque du *Botrytis cinerea*. *Revue Francaise d'Oenologie*, **108**, 17–20.

Pont, V. and Pezet, R. (1990) Relation between the chemical structure and the biological activity of hydroxystilbenes against *Botrytis cinerea*. *Journal of Phytopathology*, **130**, 1–8.

Pryce, R.J. and Langcake, P. (1977) Alpha viniferin: an antifungal resveratrol trimer from grapevines. *Phytochemistry*, **16**, 1452–1454.

Pucheu-Planté, B. and Leclair, Ph. (1990) Comportement clonal du Sémillon à l'égard de *Botrytis cinerea*. Influence sur la qualité des raisins en Sauternais. In *Comptes Rendus du 4ème Symposium International d'Oenologie* (Bordeaux, 1989), Edited by Dunod, pp. 75–79. Paris:Bordas.

Pucheu-Planté, B. and Mercier, M. (1983) Etude ultrastructurale de l'interrelation hôte–parasite entre le raisin et le champignon *Botrytis cinerea*: exemple de la pourriture noble en Sauternais. *Canadian Journal of Botany*, **61**, 1785–1797.

Pucheu-Planté, B. and Seguin, G. (1978) Pourriture vulgaire et pourriture noble en Bordelais. *Connaissance Vigne Vin*, **12**, 21–34.

Pucheu-Planté, B. and Seguin, G. (1981) Influence des facteurs naturels sur la maturation et la surmaturation du raisin dans le Sauternais en 1978 et 1979. *Connaissance Vigne Vin*, **15**, 143–160.

Rapp, A. and Reuther, K.H. (1971) Der gehalt an freien aminosäuren in traubenmosten von gesunden und edelfaulen beeren verschiedener rebsorten. *Vitis*, **10**, 51–58.

Ravji, R.G., Rodriguez, S.B. and Thornton, R.J. (1988) Glycerol production by four common grape moulds. *American Journal of Enology and Viticulture*, **39**, 77–82.

Ribéreau-Gayon, P. (1982) Incidences oenologiques de la pourriture du raisin. *Bulletin OEPP*, **12**, 201–214.

Ribéreau-Gayon, P., Lafon-Lafourcade, S., Dubourdieu, D., Lucmaret, V. and Larue, F. (1979) Métabolisme de *Saccharomyces cerevisiae* dans le moût de raisins parasités par *Botrytis cinerea*. Inhibition de la fermentation, formation d'acide acétique et de glycérol. *Comptes Rendus Académie des Sciences Paris*, **289D**, 441–444.

Ribéreau-Gayon, J., Ribéreau-Gayon, P. and Seguin, G. (1980) Chapter 9—*Botrytis cinerea* in Enology. In *The Biology of Botrytis*, edited by J.R. Coley-Smith, K. Verhoeff, K. and W.R. Jarvis, pp. 251–274. London: Academic Press Inc.

Rijkenberg, F.H.J., De Leeuw, G.T.N. and Verhoeff, K. (1980) Light and electron microscope studies on the infection of tomato fruits by *Botrytis cinerea*. *Canadian Journal of Botany*, **58**, 1394–1404.

Robertson, G.L., Eschenbruch, R. and Cresswell, K.L. (1980) Seasonal changes in the pectic substances of grapes and their implication in juice extraction. *American Journal of Enology and Viticulture*, **31**, 162–166.

Roux, F. (1986) Contribution à l'étude du métabolisme de l'acide malique du raisin par *Botrytis cinerea*. Thèse Doctorat 3ème cycle, Bordeaux, France.

Ruiz Borge, J.F. and Schlosser, E. (1990) Untersuchungen zur dicarboximid-resistenz von *Botrytis cinerea Pers*. I. Verbreitung und qualität vinclozolin-resistenter isolate im Rheingau. *Viticulture and Enology Science*, **45**, 36–39.

Salgues, M., Cheynier, V., Gunata, Z. and Wylde, R. (1986) Oxidation of grape juice 2-S-glutathionyl caffeoyl tartaric acid by *Botrytis cinerea* laccase and characterization of a new substance: 2,5-di-S-glutathionyl caffeoyl tartaric acid. *Journal of Food Science*, **55**, 1191–1194.

San Romao, M.V. (1985) Observations sur le métabolisme des bacteries lactiques dans les moûts de raisins altérés. *Connaissance Vigne Vin*, **19**, 109–116.

Shepard, D.V. and Pitt, D. (1976) Purification of a phospholipase from Botrytis and its effects on plant tissues. *Phytochemistry*, **15**, 1465–1470.

Shimazu, Y., Uehara, M. and Watanabe, M. (1984) Decomposition of L-tartaric and L-malic acids in grape musts by *Botrytis cinerea*, *Agricultural and Biological Chemistry*, **48**, 1565–1573.

Shimizu, J.I., Uehara, M. and Watanabe, M. (1982) Transformation of terpenoids in grape must by *Botrytis cinerea*. *Agricultural and Biological Chemistry*, **46**, 1339–1344.

Sponholz, W.R. and Dittrich, H.H. (1984) Über das workommen von galacturon—und glucuronsäure sowie von 2- und 5-oxogluconsäure in weinen, sherries, obst- und dessertweinen. *Vitis*, **23**, 214–224.

Sponholz, W.R. and Dittrich, H.H. (1985) Über die herkunft von gluconsäure, 2- und 5-oxogluconsäure sowie glucuron- und galacturonsäure in mosten und weinen. *Vitis*, **24**, 51–58.

Stein, U. and Blaich, R. (1985) Untersuchungen über stilben produktion und Botrytisan—falligkeit bei Vitis—arten. *Vitis*, **24**, 75–87.

Stone, D.A. (1980) Inducing noble rot in the lab. *Wines Vines*, **61**, 63–64.

Verhoeff, K. and Warren, J.M. (1972) *In vitro* and *in vivo* production of cell wall degrading enzymes by *Botrytis cinerea* from tomato. *Netherland Journal of Plant Pathology*, **78**, 179–185.

Watanabe, M. and Shimazu, Y. (1976) Application of *Botrytis cinerea* for wine making. *Journal of Fermentation Technology*, **54**, 471–478.

Watanabe, M. and Shimazu, Y. (1978) Effect of *Botrytis cinerea* cell autolyzates on wine. *Journal of Fermentation Technology*, **56**, 114–120.

Watanabe, M. and Shimazu, Y. (1980) Effect of yeasts on botrytized wine making. *Journal of Fermentation Technology*, **58**, 227–235.

Watanabe, M., Shimazu, Y. and Uehara, M. (1981) Effect of adenosine-3'-5'-cyclic monophosphate on botrytized must making. *American Journal of Enology and Viticulture*, **32**, 76–79.

Weimar, K., Tschesche, R. and Breitmaier, E. (1979) Botrylacton, ein neuer wirkstoff aus der nahrlosung des pilzes *Botrytis cinerea*. *Chemische Berichte*, **112**, 3598–3602.

Wurdig, G. (1976) Mucic acid: a constituent of wines made from *Botrytis* infected grapes. *Weinwissenschaft*, **112**, 16–17.

Yamamura, H., Naito, R. and Tamura, H. (1986) Effects of light intensity and humidity around clusters on the formation of surface wax and the resistance to berry splitting in Delaware grapes. *Journal of Japan Society of Horticulture*, **55**, 138–144.

Zironi, R., Frega, N., Conte, L.S. and Lercker, G. (1984) Effets de la formation de la pourriture acide sur la composition de la fraction lipidique des différences partics du grain de *Vitis vinifera C.V. fontana*. *Vitis*, **23**, 93–105.

Zironi, R., Riponi, C., Ferrarini, R. and Amati, A. (1982) Observazioni sullo sviluppo del “marciume acido” durante l’appassimento delle uve. *Ricerca Viticoltura e Enologia*, **7–8**, 35–40.

ACKNOWLEDGEMENTS

Figures 11.1, 11.3 and 11.4 were prepared by B. Pucheu-Planté and M. Mercier with the permission of the Department of Electron Microscopy, University of Bordeaux I.

Chapter 12

MICROBIOLOGY AND CHEMISTRY OF CORK TAINTS IN WINE

TERRY H. LEE and ROBERT F. SIMPSON

The Australian Wine Research Institute, PO Box 197, Glen Osmond, South Australia 5064, Australia

Chapter Outline

INTRODUCTION

Cork is the bark from the cork oak *Quercus suber* which grows predominately in countries near the Mediterranean Sea. Wine corks are produced from reproduction cork, which is bark that has regrown after the original bark has been stripped off the tree. Reproduction cork is uniform in texture and density and endows corks with unique characteristics that provide an excellent closure for wine (Preston, 1983).

Figure 12.1 outlines the production of wine corks commencing with the stripping of the bark from the tree through processing of the corks to delivery to the winery. For a more complete account of cork production and processing refer to Preston (1983) and Lee *et al.* (1984).

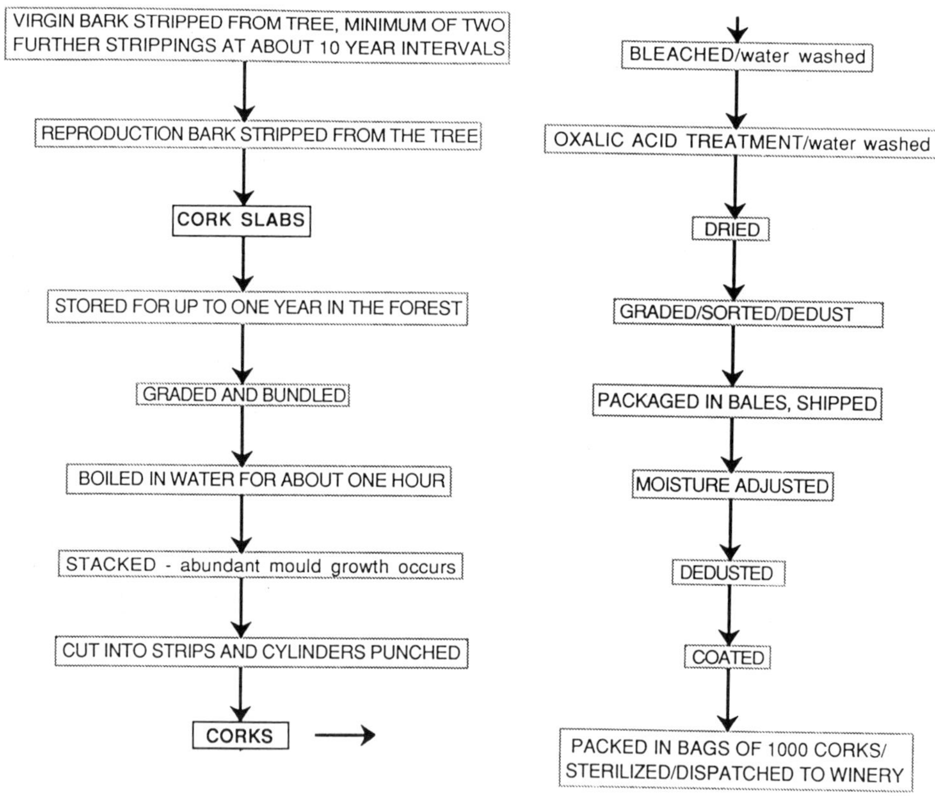

FIGURE 12.1 Production, processing and transport of wine corks from the tree to the winery.

Portugal is the major producer of cork, and much of the cork grown in neighbouring countries, such as Spain, France, Italy, Morocco and Algeria, is also processed in Portugal. Not surprisingly, the cork industry is of major importance to the economy of Portugal.

Corks provide an effective seal for wine bottles and maintain wine in good condition in most instances; however, two defects detract from their performance. These defects are (i) leakage, where oxidation of the wine may occur and the leaked wine may affect the label and appearance of adjacent bottles, and (ii) cork taint, where an off-flavor, frequently described as musty, moldy or earthy, is imparted to the wine. The accepted definition of a taint is "a taste or odor foreign to the product" (Land,1989).

Over the past ten years, we have obtained good understanding of the nature and causes of cork taint. The compounds implicated in this taint appear to be microbial metabolites that are produced by the microflora found in association with the bark and corks throughout the often lengthy processing, storage and transport of corks. Subsequently, these compounds are leached from the cork into the wine. By definition corks are responsible for cork taint; there is evidence, however, that the compounds causing cork taint may also arise from other sources (Tanner and Zanier, 1981; Amon, Simpson and Vandepeer, 1987; Cantagrel and Vidal, 1989). Thus, some taints could arise from spoiled grapes, residues of biocides used on grapes (Bricout, Brunerie and Desmarest, 1990), contaminated fining and filtration agents, and wooden barrels.

A sound knowledge of the microbiology of cork from growing of the bark in the forest to extraction of the cork from the bottle (Figure 12.1) is necessary to determine ways of avoiding or minimizing the incidence of cork taint in bottled wine. Such knowledge is, however, still incomplete. This Chapter examines the microbial ecology of corks, the contribution of this flora to the taint problem and the chemical basis of the taint.

MICROBIAL ECOLOGY OF CORK

In the Forest

The nature and composition of the microflora of cork as it occurs as bark of live oak trees is largely unknown; however, it is likely that it will be influenced by the microflora of the environment, particularly the soil, in which the trees are grown.

Humidity, shading and temperature should influence the extent and type of microflora present on the bark, and seasonal variables may be important in determining if certain harvests or sections of bark from individual trees exhibit a high incidence of off-flavors or taints.

Corkwood—Postharvest

The work of Lacey (Lacey, 1973; Avila and Lacey, 1974), who studied causes of the high incidence of respiratory diseases in cork factory personnel, provides some information on the microflora of bark after harvest. The airborne microflora in factories where the cork slabs were processed was dominated by molds, with most species belonging to the genera *Penicillium, Aphanocladium, Mucor* and *Monilia*. Major species, isolated from more than 75% of air samples, included *Penicillium frequentans, P. granulatum, Aphanocladium album, Mucor plumbeus* and *Monilia sitophila. Trichoderma* spp., *Oidiodendron* spp., and *Cladosporium* were isolated from 25–75% of air samples. According to the taxonomic considerations of Pitt (1979), *P. frequentans* is more accurately referred to as *P. glabrum*.

Lefebvre *et al.* (1983) found *P. glabrum* as the major mold on cork slabs imported into France and on slabs harvested from Landes in France; *P. expansum* and other *Penicillium* spp. were present as were species of the yeast *Candida* and *Rhodotorula*. Some preliminary work reported by Lee *et al.* (1984) further confirmed the dominance of *Penicillium* spp., in particular *P. glabrum* and *P. granulatum* in bark harvested in Portugal. They showed also that raw corkwood when boiled for 1 h in the laboratory was virtually sterile and that corkwood boiled under commercial conditions (Figure 12.1) was grossly contaminated with the microflora of the factory one day after boiling.

Imported Corks

The microflora of cork has been most thoroughly examined in France and Australia after transport of the corks from Portugal in bales of generally 10 000 units (Moreau, 1978; Schaeffer, Meyer and Guillerm, 1978; Davis, Fleet and Lee, 1981a, 1982; Castera-Rossignol, 1983; Lefebvre *et al.*, 1983; Fumi and Colagrande, 1988). The almost universal presence of *Penicillium* spp. has been recorded, particularly *P. glabrum* and *P. spinulosum*, along with a lesser incidence of isolates of *Trichoderma, Cladosporium, Paecilomyces, Aspergillus* and *Monilia*. Moreau (1978) reported high incidence of *Aspergillus conicus*, which was not found by Davis *et al.* (1981a) or by Lefebvre *et al.* (1983). Thus, the molds found on imported cork were similar to those found in cork slabs or in the air of the cork factories. According to Moreau (1978) and Davis *et al.* (1981a), fungal contamination of individual corks can be so extensive as to cause green-yellow-grey discoloration of the cork surface. In addition, these authors reported substantial growth of fungal mycelium within the cork lenticels. Humidification of the corks encouraged this growth (Moreau, 1978). However, the level of molds found on individual corks was highly variable, ranging from 0 to 10^8 colony forming units per cork (Davis *et al.*, 1981a).

Yeasts and bacteria were isolated from corks imported into Australia, but their frequency of occurrence and populations ($<10^2$ cfu/cork) were much less than those of fungi (Davis *et al.*, 1981a, 1982). Yeast genera identified included

Rhodotorula, Sporobolomyces, Cryptococcus, Saccharomyces and *Candida*. *Bacillus coagulans, B. badius* and *B. sphaericus* were the main bacteria isolated. They appear to be part of the natural microflora of cork and can survive for a time in wine (Lee *et al.*, 1984). Fumi and Colagrande (1988) also found *Bacillus* as well as *Micrococcus* and *Streptococcus* but no yeasts. Lefebvre *et al.* (1983) isolated two *Streptomyces* spp. which they considered important in the production of musty/moldy aromas. They also showed that one of the *Streptomyces* species produced guaiacol.

These microbiological studies indicate that corks generally carry a high load of molds in spore form, which grow rapidly if conditions of water activity and temperature are appropriate. Such growth may occur at any time during the chain of production, transport and storage of the cork (Figure 12.1).

Corks Extracted from Bottled Wine

The transport and storage environment provides opportunity for further contamination of corks by microorganisms other than those associated with the oak forest or cork factory. In attempts to connect the occurrence of corky taints in wines with the presence of a particular microbial species in the cork, several researchers have examined the microflora on corks extracted from tainted wines (Marais and Kruger 1975, Moreau, Moreau and Le Bras, 1976; Moreau,1978; Heimann *et al.*,1983, Lefebvre *et al.*, 1983). Such corks generally have a flora similar to that found on cork at earlier stages of processing, but, in addition, show a high incidence of *P. roqueforti,* not found earlier, *Aspergillus* spp. and *Streptomyces* spp. It was suggested that these species probably originate from the winery environment, where they are frequently found (Moreau, 1978; Lefebvre *et al.*, 1983). In one significant study, yeasts of the genera *Rhodotorula* and *Candida* were specifically isolated from corks taken from tainted wine (Bureau, Charpentier-Massonnat and Pansu, 1974).

CORK TAINT

Incidence and Nature

Cork taint has been recognized as a serious defect in bottled wine over the past 15 years; however, the first reference to 'goût de bouchon', cited by Lefebvre *et al.* (1983), appeared in 1904. The proportion of bottles affected varies greatly, with estimates ranging from 0.5 to 6%; 2% is commonly quoted (Heimann *et al.*, 1983).

Cork taint is generally perceived as a musty and moldy off-flavor. Often, however, a low level of taint causes a loss in fruit flavor and dulls the aroma and taste of the wine. In such instances, cork taint may be difficult to recognize in the absence of a second bottle of the same wine for comparison. In contrast, strongly tainted wine may have an intense musty/moldy aroma or, in some instances, an earthy or mushroom-like aroma. The perception of cork taint is highly dependent, therefore, on the content of the tainting compounds and the sensitivity of the person tasting the wine.

The aromas associated with individual tainting compounds found in corked wine are described in the Figure captions. Land (1989), however, has provided a clear account of the variation in the ability of people to detect several intensely odorous compounds causing taints in drinking water; one of these was geosmin which has been found in corked wines (Amon *et al.*, 1989). According to Land "the most sensitive few people are about 2000 times more sensitive than the least sensitive few", which helps to explain the often found differential response between people to a certain level of taint in wine.

The incidence of cork taint may show considerable variation between vintages due to the source of the corkwood and to the growing season in the oak forest. The Australian industry appeared to experience a high incidence of cork taint in 1978–80 and in 1988–89. With a small market distant from Portugal, the incidence of cork taint occurring in a particular period in Australia may not be representative of the whole wine cork production. A shipment of badly tainted corks or contamination occurring during shipping may lead to an inflated estimation of the level of cork taint.

The Role of Cork Microflora

The fact that cork taint is linked to the development of microflora on the cork is reasonably clear. There are three mechanisms whereby such microflora could contribute to wine taint.

1. In the first mechanism, which is the one most widely accepted, microbial growth in or on the cork produces metabolites that, subsequently, are leached into the wine to affect its quality. Such growth generally occurs before the cork is inserted into the bottle, but it could occur after insertion (Moreau, 1978). Presumably, cork constituents or substances used in cork treatment, (e.g. disinfectants) are the substrates metabolized by the microorganisms to produce taint compounds, but the chemistry and biochemistry of these reactions (discussed in subsequent sections) are still poorly understood. Moreover, further research is needed to link the production of individual tainting compounds with the activity of a particular microbial species. Molds are considered the major causative organisms (Marais and Kruger, 1975; Moreau, 1978; Schaeffer *et al.*, 1978; Lefebvre *et al.*, 1983; Daly, Lee and Fleet, 1984), but the activities of *Rhodotorula* and *Candida* yeasts (Bureau *et al.*, 1974) and *Streptomyces* spp. (Lefebvre *et al.*, 1983) have clearly been implicated. In many cases, suspect organisms isolated from corks produce tainting compounds when grown on sterile corks or granules of cork. Subsequent extraction of the cork with wine or model alcohol solutions gives an extract that can be analyzed for taint by sensory and gas chromatographic-mass spectrometric (GC/MS) techniques (Heimann *et al.*, 1983; Lefebvre *et al.*, 1983; Daly *et al.*, 1984). In this way, the contributions of

sesquiterpenes (Heimann *et al.*, 1983) and guaiacol (Lefebvre *et al.*, 1983) to cork taint have been demonstrated. Simpson and Lee (1990) reported unpublished work in which molds previously isolated by Davis *et al.* (1981a) were inoculated onto sterile pieces of cork and incubated with and without a small quantity of 2,4,6-trichlorophenol (2,4,6-TCP). Several compounds having a moldy/musty aroma were produced in these model systems as identified by gas chromatography-electron capture detection (GC/ECD) and GC/MS.

2. The second mechanism concerns the possibility of the corks introducing viable microbial cells into the wine; subsequent growth of these organisms in the wine could lead to taint development (Moreau,1978; Schaeffer *et al.*, 1983). The evidence suggests that fungi would not be able to survive and grow under the conditions of bottled wine and, consequently, tainting by this mechanism is not plausible. However, Lee *et al.* (1984) have noted the limited growth in wine of several *Bacillus* species isolated from corks. The significance of such growth in taint development is not known.
3. Finally, it is possible, but yet unproven, that microbial growth in cork could generate enzymes that would subsequently act on wine constituents to produce tainting substances (Moreau, 1978).

Chloroanisoles

2,4,6-Trichloroanisole (2,4,6-TCA) (Figure 12.2) is generally considered the compound most responsible for cork taint (Tanner, Zanier and Buser, 1981; Buser, Zanier and Tanner, 1982; Amon *et al.*, 1989; Cantagrel and Vidal, 1989). It is a very potent aroma compound having a flavor threshold in red wine of 1.4 ng/l (Duerr, 1985). In a survey—the survey will be referred to several times in the Chapter—of tainted wines submitted to The Australian Wine Research Institute, 2,4,6-TCA was the major taint compound in 23 of 37 tainted wines, and in 31 of

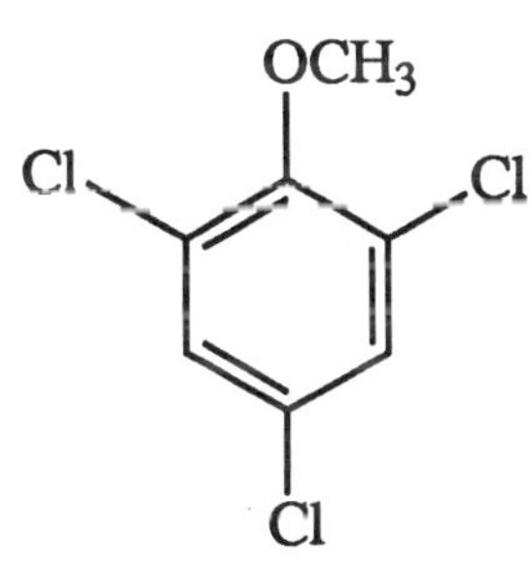

FIGURE 12.2 2,4,6-Trichloroanisole. Aroma: musty, moldy, wet newspaper, dank cellar. Flavor threshold in wine: 1.4 ng/l (Duerr, 1985).

the associated corks (Amon *et al.*, 1989); 2,4,6-TCA was also present at lower concentration in most of the corks from untainted wines.

Several chloroanisoles other than 2,4,6-TCA have been implicated as a cause or contributor to cork taint. Pentachlorophenol (PCP), marketed in Europe under the trade name *Raco*, was used in wine cellars as a wood preservative; *Raco* contains 2,3,4,6-tetrachlorophenol (2,3,4,6-TCP) as a minor impurity (see following section). The endemic microflora in the cellars led to the formation of pentachloroanisole (PCA) and 2,3,4,6- tetrachloroanisole (2,3,4,6-TeCA), which were adsorbed by bentonite (Tanner and Zanier,1981) or other fining agents and which caused musty/moldy taints in wine treated with these contaminated agents. Although 2,3,4,6-TeCA is the minor component, it has a more intense aroma than PCA and was primarily responsible for the taint in wine.

Several of the musty/moldy components detected during the GC-sniff assessments of tainted wines from the survey were tentatively identified, on the basis of GC retention data and aroma, as the methyl ethers of dichlorocresols (Simpson, 1990). 2,4-Dichloro-6-methylanisole (Figure 12.3), which is derived from *o*-cresol, appeared to be present in many of the extracts, along with two other components with retention times corresponding with those of methyl ethers of chlorinated *m*- and *p*-cresol. None of these components were considered to contribute strongly to cork taint, but they may enhance the musty/moldy off-flavor in the presence of 2,4,6-TCA.

Environmental contamination by chlorophenols

Chlorophenols, particularly commercial preparations of pentachlorophenol (PCP) that contain 2,3,4,6-TeCP (3–10% w/w) and minor quantities of 2,4,6-TCP (Freiter, 1979), were used as general biocides and wood preservatives in the 1950s and 1960s. As a result, chlorophenolic residues are found in virtually all water supplies and in the food chain (Ahlborg and Thunberg, 1980).

Some microorganisms can convert the three chlorophenols, at a sublethal concentration, to the corresponding chloroanisoles, i.e. PCA, 2,3,4,6-TeCA and 2,4,6-TCA, as a means of detoxification (Crosby, 1981). Pentachlorophenol and related chlorophenols are degraded by soil bacteria (Weiss *et al.*, 1982), yielding a wide range of chlorophenolic compounds mostly containing fewer chlorine

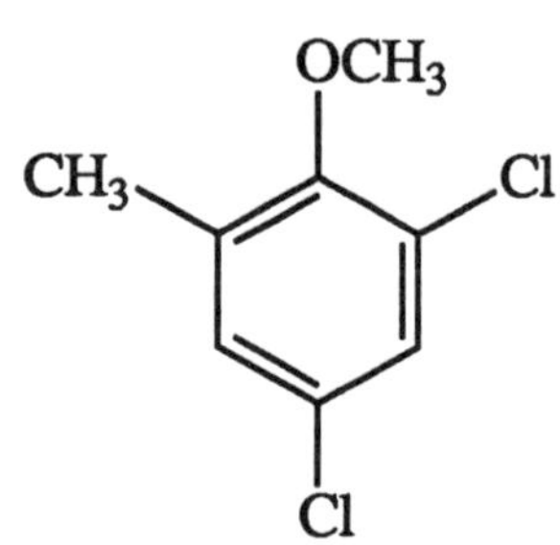

FIGURE 12.3 2,4-Dichloro-6-methylanisole (a dichloro-*o*-cresol). Aroma: musty, moldy. Flavor threshold in wine: not determined.

atoms, but favoring products in which the 3- and 5- chlorine atoms are retained; 3,4,5-TCP is generally the major trichlorophenol formed. Compounds containing less than three chlorine atoms are usually not present. The relative concentration of chlorophenols and chloroanisoles derived from the degradation of PCP is influenced by the bacteria present; the first stage of the degradation, i.e. dechlorination or methylation, is presumed to determine subsequent reactions (Rott, Nitz and Korte, 1979; Weiss *et al.*, 1982). In addition, PCP can undergo photochemical reduction to 2,3,4,6- and 2,3,5,6-TeCP; products in which a chlorine atom is replaced with a hydroxyl group are also formed (Crosby, 1981).

Maarse, Nijssen and Jetten (1985) found a comparable concentration of 2,4,6-TCA, 2,3,4,6-TeCA and PCA in corks, and Rigaud *et al.* (1984) reported that the bark attached to living cork oaks contained a low concentration of PCA and 2,4,6-TCA, with the outer layers of bark having a higher concentration of these compounds. Despite assertions by representatives of the cork industry that PCP is not used and has not been used previously in the cork oak forests, several reports (Maarse *et al.*, 1985; Cantagrel and Vidal, 1989) indicate that PCP has been applied to the base of the trees to control insect pests. The content of 2,4,6-TCA in bark and corks relative to that of 2,3,4,6-TeCA and of PCA is high, but Whitfield, Nguyen and Tindale (1991) have shown that molds differ markedly in their ability to methylate chlorophenols and that 2,3,4,6-TeCP and PCP are more resistant to methylation. The presence of 2,3,4,6-TeCA and PCA is consistent, however, with environmental contamination by PCP.

Contamination during transport

Another means by which corks may become contaminated with chloroanisoles is *via* aerial migration of chloroanisoles from packaging and shipping containers (Tindale, 1987). A good example was an incidence of severely tainted champagne corks shipped into Australia in cardboard cartons. The corks contained a high content of 2,4,6-TCA, and sections of the cartons in contact with the shipping floor were heavily contaminated with 2,4,6-TCP and, to a lesser extent, with 2,4,6-TCA (Simpson and Lee, 1990). It was assumed that the shipping floor had been treated with a commercial wood preservative, containing 2,4,6-TCP as the active ingredient, and that microbial growth within the treated cargo area led to the formation of 2,4,6-TCA. The use of sealed containers for shipping corks is essential, and the unhygienic state of cargo floors should be a concern for the shipping of products or packaging materials that are intended for the food industry.

Formation from hypochlorite bleaching

The bleaching of wine corks with hypochlorite for the purposes of cosmetic appearance and sterilization has been an integral part of cork processing (Figure 12.1)(Preston, 1983). The risk of 2,4,6-TCA formation, however, has encouraged producers to examine other means of cork treatment (Borges, 1985) such as using peroxides (Heinzel, Hagen and Bousser, 1983). Hypochlorite treatment of corks increases their content of 2,4,6-TCP (Zehnder, Buser and Tanner 1984), but in the absence of microbial growth during subsequent processing, the 2,4,6-TCA content of these corks was no greater than that in untreated corks (Amon *et al.*, 1989).

Natural corkwood contains a low content of phenol and other simple phenols (Boidron *et al.*,1984), and the conditions under which corks are treated with hypochlorite are conducive to the formation of 2,4,6-TCP. Burttschell *et al.* (1959) demonstrated that 2,4- and 2,6-dichlorophenol are readily formed from phenol under slightly alkaline conditions, as occurs with hypochlorite treatment; more of the 2,4-isomer is formed and higher levels of chlorine lead to a greater production of 2,4,6-TCP (Figure 12.4). We have found that phenol in strong hypochlorite solution after two hours at ambient temperature produced only 2,4,6-TCP; no dichlorophenols, tetrachlorophenols or PCP were detected (unpublished).

Maujean, Millery and Lemaresquier (1985) showed that 2,4,6-TCA can be synthesized by microflora obtained from infected cellars and a *Penicillium* sp., in the presence of chlorine or hypochlorite. Phenol is not required as a starting material, since phenols are synthesized via the shikimic acid pathway from a carbohydrate source. Unfortunately, the presence or absence of other chlorophenols or chloroanisoles was not investigated, so that 2,4,6-TCA formed by this mechanism cannot be recognized by the presence and concentration of specific co-occurring substances.

Molds isolated from shipping containers can either methylate chlorophenols to produce chloroanisoles or degrade chlorophenols (Tindale *et al.*, 1989,

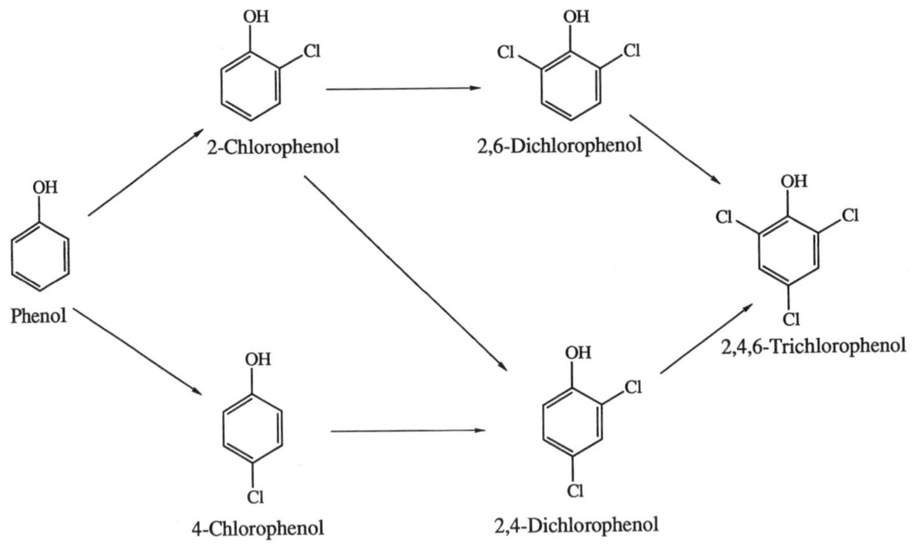

FIGURE 12.4 Chlorination of phenol by hypochlorite at pH 8 with a ratio of phenol to chlorine of 1:2.

Whitfield, Nguyen and Last, 1991). The ability of mold isolates from corks imported into Australia to convert added 2,4,6-TCP to 2,4,6-TCA has been examined [unpublished data reported by Simpson and Lee (1990)] in a sterile granulated cork system. The aroma of the cultures was assessed after 25 days, the chlorophenolic components were determined by GC/ECD, and the presence of other taint compounds was examined by GC-sniff assessment. Overall, 0–23% of the added 2,4,6-TCP was methylated to produce 2,4,6-TCA, while 70–100% of the 2,4,6-TCP was metabolized. Most molds produced 1-octen-3-one, but a relatively high concentration of 2-methylisoborneol, geosmin and guaiacol was also formed in some instances.

The high level of contamination by 2,4,6-TCA of corks from the survey (Amon *et al.*, 1989), where six of the corks contained more than 300 ng of 2,4,6-TCA, suggests that hypochlorite-bleached corks and microbial activity during lengthy storage and shipping have been responsible for many of the cork taints in Australia. Contaminated shipping floors may also have contributed to the occasional high incidence of cork taint.

Pentachloroanisole and 2,3,4,6-TeCA have been detected in some tainted corks in Australian wines, indicating that biocides containing PCP have been the cause of these taints.

Chlorophenols and chloroanisoles can be produced by fungi such as Basidiomycetes. The presence of a chloroperoxidase enables the introduction of a chlorine-atom into a benzene ring (Jerina, Guroff and Daly, 1968; Neidleman and Geigert, 1986). Consequently, monochloroanisoles can be formed (Maarse, Nijssen and Angelino, 1988) and converted to 2,4,6-TCA by treatment with hypochlorite. Such a mechanism provides a rationale for the presence of 2,4,6-TCA in corks on which no microbial growth has occurred since the bleaching process.

In instances where 2,4,6-TCA is the prime cause of cork taint, the absence or presence of co-occurring chlorophenols provides an indication of the source of the taint. Buser *et al.* (1982) reported the presence of several chlorophenols in extracts from hypochlorite-treated corks: such compounds are derivable from the action of chlorine on phenol and several simple phenols expected in corkwood. The chlorophenols likely to co-occur with 2,4,6-TCA in tainted corks or wine, depending on the source of 2,4,6-TCA, are summarized in Table 12.1.

Pesticides and insecticides

Bricout *et al.* (1990) showed that the fungicide, *Prochloraz,* was degraded by certain fungi to 2,4,6-TCA when applied to fennel in the field. *Prochloraz* is a preparation containing a compound with a 2,4,6-trichlorophenyl ether moiety, which, if cleaved, will produce 2,4,6-TCP.

Several dithio- and trithiophosphate derivatives of chlorophenols including 2,4,6-TCP have been patented in Japan as insecticides. This provides other potential sources of 2,4,6-TCA provided that the application of such agents can be traced to the cork forests or to areas where cork is held or processed.

TABLE 12.1 Indication of the origin of 2,4,6-trichloroanisole (TCA) from the presence of co-occurring chlorophenols.

Source of 2,4,6-TCA	*Co-occurring chlorophenols*
PCP-based biocides	PCA, 2,34,6-TeCA and possibly a complex mixture of chlorophenolic compounds
TCP-based biocides	None or minor quantities of dichloroanisoles and 2,3,4,6-TeCA
Hypochlorite or chlorine-containing disinfectants	(a) Dichloroanisoles; a trace of 2,3,4,6-TeCA (b) None

PCP = pentachlorophenol; PCA = pentachloroanisole; TeCA = tetrachloroanisole; TCP = tetrachlorophenol.

Compounds other than Chloroanisoles

Several compounds, in addition to 2,4,6-TCA and other chloroanisoles, have been implicated as a cause of cork taint either alone or in combination with 2,4,6-TCA. The more positive identification of these compounds has been possible in recent years due to improved instrumental analysis. These compounds, which include 1-octen-3-one, 1-octen-3-ol, 2-methylisoborneol (2-MIB), geosmin, and guaiacol, became apparent in the survey of tainted wines submitted to the Institute (Amon *et al.*, 1989). Their chemical nature and formation in corkwood and wine are discussed in the following sections.

Other compounds implicated in cork taint and attributed with a musty/moldy aroma include several sesquiterpene hydrocarbons (Heimann *et al.*, 1983).

1-Octen-3-ol/1-octen-3-one

A high proportion of tainted wines examined in the survey and their associated corks (>70%), and half of the sound wines and their associated corks contained 1-octen-3-one (Figure 12.5). A smaller proportion of the tainted wines contained the corresponding alcohol, 1-octen-3-ol (Figure 12.6), which was not detected in sound wines.

We have examined several wines containing such high concentration of these C-8 compounds and concluded that the mushroom-like flavor was dominant and clearly distinguishable from the musty/moldy aroma typified as cork taint.

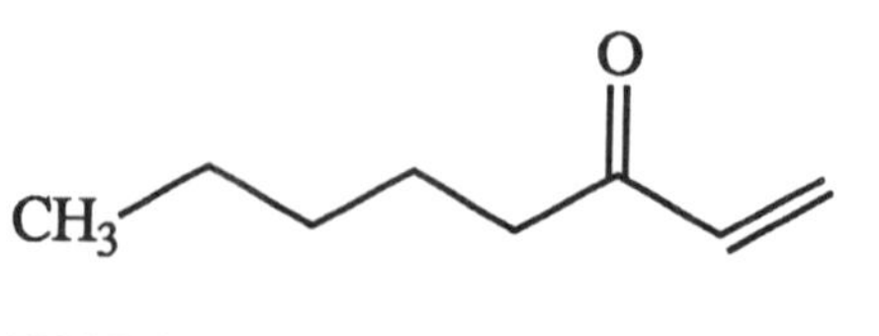

FIGURE 12.5 1-Octen-3-one. Aroma: mushroom, metallic. Flavor threshold in wine: 20 ng/l (Amon *et al.*, 1989).

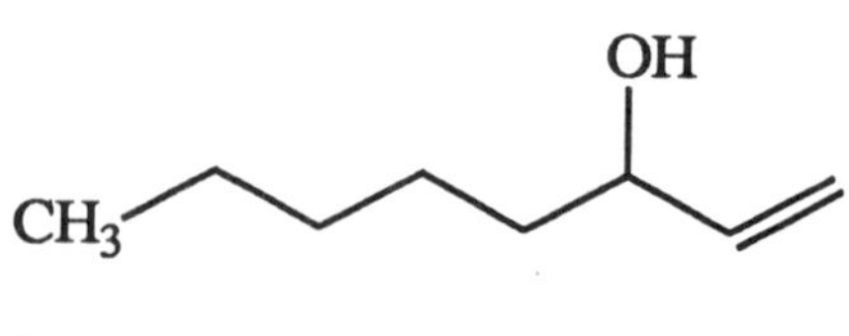

FIGURE 12.6 1-Octen-3-ol. Aroma: mushroom, metallic. Flavor threshold in wine: 20 μg/l (Amon *et al.*, 1989).

1-Octen-3-one has an intense aroma, but an aroma-significant concentration in wine is difficult to detect by instrumental analysis. This is due to a lack of suitable diagnostic fragment ions using conventional positive-ion electron-impact mass spectrometry, and to a high ion background from other components.

The minute quantity of a compound, such as 1-octen-3-one, capable of influencing wine aroma can be difficult to appreciate. An analogy with time provides a convenient comparison. A concentration of 200 ng/l, which corresponds to 10 times the flavor threshold of 1-octen-3-one in wine, can be equated with 1 second in about 160 years. The analogy may assist in explaining how even advanced and very sophisticated instrumentation such as high-resolution capillary GC/MS is not always able to detect some important flavor compounds in wine and cork extracts.

1, *cis*-5-Octadien-3-ol and 1, *cis*-5-octadien-3-one (Figures 12.7 and 12.8, respectively) were tentatively identified in wine and cork extracts, and correspond to two of the components exhibiting mushroom-like or metallic aromas. These compounds are common metabolites of molds and are important flavor compounds of Brie and Camembert cheese (Karahadian, Josephson and Lindsay, 1985). They are formed from linolenic acid by the same mechanism as for 1-octen-3-ol; due to their lower chemical stability, they have not been identified as frequently as the more saturated analogues.

2-Methylisoborneol and geosmin

2-Methylisoborneol (2-MIB) and geosmin (Figures 12.9 and 12.10, respectively) cause taints in town water supplies, fish and other aquatic and marine foods (Gerber, 1979, 1983). The two compounds are well known as metabolites of soil bacteria (Actinomycetes) which, as indicated earlier, have been found on corks, and of algae (Cyanobacteria), but there is good evidence that a much wider range of microflora is capable of their biogenesis.

2-Methylisoborneol is formed by *Streptomyces* isolates via mevalonate and is thought to be derived from a monoterpene (C-10) precursor after addition of a methyl group from S-adenosylmethionine, whereas geosmin is thought to be formed from a bicyclic sesquiterpene (C-15) after loss of an isopropyl side chain (Bentley and Meganathan, 1981). Naes, Utkilen and Post (1988) indicated that geosmin was a shunt product in the isoprenoid pathway leading to the formation of carotenoids by algae; the biosynthetic scheme indicates factors that would

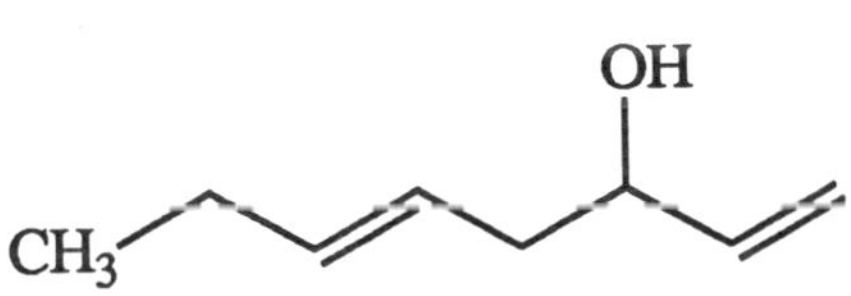

FIGURE 12.7 1, *cis*-5-Octadien-3-ol.
Aroma: earthy, mushroom.
Flavor threshold in wine: not determined.

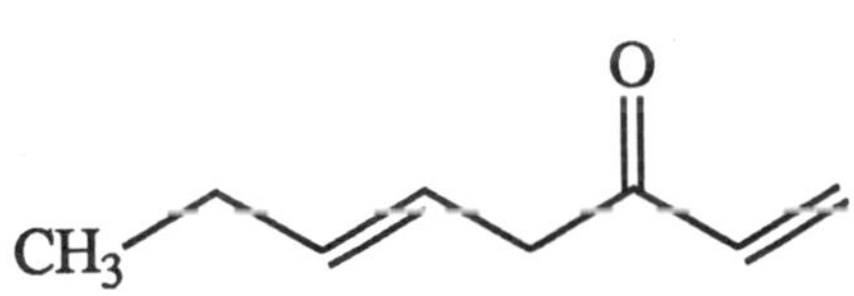

FIGURE 12.8 1, *cis*-5-Octadien-3-one.
Aroma: oily, metallic.
Flavor threshold in wine: not determined.

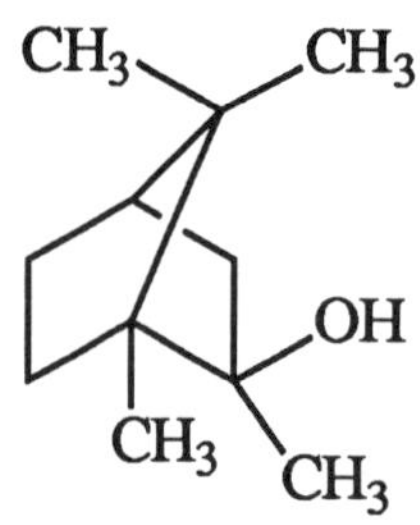

FIGURE 12.9 2-Methylisoborneol (1,2,7,7-tetramethyl-*exo*-bicyclo [2.2.1]heptan-2-ol).
Aroma: earthy, musty, muddy; camphoraceous at high concentration.
Flavor threshold in wine: 30 ng/l (Amon *et al.*, 1989).

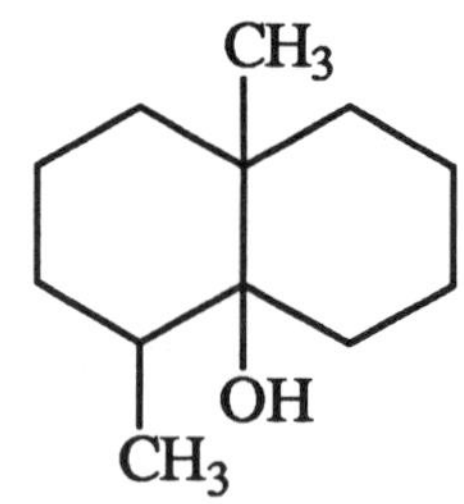

FIGURE 12.10 Geosmin (*trans*-1,10-dimethyl-*trans*-9-decalol).
Aroma: earthy, musty, muddy.
Flavor threshold in wine: 25 ng/l (Amon *et al.*, 1989).

favor its formation. As indicated earlier, several of the molds isolated by Davis *et al.* (1981a) from wine corks imported into Australia produce relatively large amounts of geosmin and 2-MIB when grown on sterile, granulated cork (Simpson and Lee, 1990). Most isolates produced 1-octen-3-one as a dominant aroma compound, but others including an *Aspergillus* sp., several *Penicillium* spp. and *Trichoderma* spp., and a *Rhizoctonia* sp. produced aromas attributable to 2-MIB or geosmin, or both in some instances, that were as intense or more intense than that attributable to 1-octen-3-one.

Geosmin was less frequently found in wine and associated corks, as it is unstable in an acidic medium. The concentration of geosmin decreased 50% in a model wine solution of pH of 3.2 after eight weeks at 25°C (Amon *et al.*, 1989)—the degradation products from geosmin have little aroma (Gerber, 1979).

It is possible that sections of cork bark stored near or in contact with the ground could be contaminated by soil bacteria and, therefore, become tainted by 2-MIB and geosmin. Such differential contamination would lead to taint in some corks and its absence in others of similar appearance and from the same source. We have no information, however, whether these tainting compounds, if formed before the boiling, bleaching and washing processes, are eliminated by these treatments. Algae have been observed on the inside of polyethylene bags of corks shipped into Australia. Presumably, part of the bag has been subjected to warming and a redistribution of moisture in the corks has resulted in moisture concentrating and condensing in part of the bag, providing conditions that are conducive to algal growth and, therefore, to the formation of 2-MIB and geosmin.

While the presence of some microflora favors the formation of taint compounds, such as 2-MIB and geosmin, other microflora may degrade these compounds and conceivably deodorize corkwood. For example, Izaguirre, Wolfe and Means (1988) showed that *Pseudomonas* spp., bacteria commonly found in soil and water, are capable of degrading 2-MIB. The luxuriant growth of mold on corkwood after the slabs are boiled may lead to the removal of taint compounds, but such abundant and uncontrolled microbial growth must provide an unacceptable risk of taint formation. A better understanding of the effects of the boiling process on the cork slabs and the role of mold growth in the maturation of the corkwood are needed; during the interim, alternative processes for stacking and drying the slabs should be considered.

Guaiacol

Guaiacol was detected in many of the tainted wines and their associated corks, but at a concentration below its flavor threshold. Thus guaiacol could influence wine aroma only by acting in combination with other components. The chemical structure, aroma description, and flavor threshold of guaiacol are shown in Figure 12.11. Lefebvre *et al.* (1983) identified guaiacol as the cause of cork off-flavor where the corkwood had a yellow tinge or discoloration; they isolated soil bacteria belonging to the genus *Streptomyces* that were able to degrade cork lignin, to guaiacol with vanillic acid as the likely intermediate. The biochemical formation of guaiacol has several steps in common with the chemical degradation of lignin via ferulic acid (Maga, 1978). The chemical breakdown of lignin may be responsible for a low concentration of guaiacol in wine, this amount coming from degradation of lignin in the cork and/or degradation of lignin in the oak of barrels or chips that are in contact with the wine.

Generally, cork processors are aware of the defect caused by discolored corkwood, and the faulty sections are removed in the early stages of production. In our experience, taint due to guaiacol has not been a significant cause of cork taint in Australian wine.

Prevention of Cork Taint

The different ways in which cork taint can form and the number of compounds capable of causing or contributing to cork taint suggest that complete elimination of this problem will be extremely difficult. On the basis of present knowledge, some simple steps can be taken to minimize the occurrence of cork taint. These empirical operations must suffice until further information is

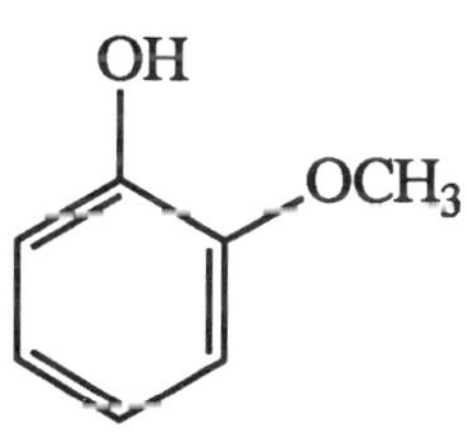

FIGURE 12.11 Guaiacol (2-methoxyphenol). Aroma: smokey, phenolic, medicinal. Flavor theshold in wine: 20 μg/l (Simpson, Amon and Daw, 1986).

obtained about the mechanisms of formation of cork taint and the role that microorganisms contribute to this problem. The steps are:

1. Practise sound forest management;
2. Avoid chlorophenolic sprays and pesticides and insecticides derived from chlorophenols;
3. Avoid microbiological activity at all stages of cork processing;
4. Use bleaching processes alternative to chlorine;
5. Ensure that water used in treatment processes does not contain chlorine;
6. Dry corks to a minimum of 8% moisture and prevent moisture pick-up during storage, handling and shipping; and
7. Use sealed cargo containers and ensure that shipping areas are dry and clean.

These steps mainly aim to reduce or eliminate sources of chlorophenols from and prevent microbial growth during the production sequence. Corks intended for purchase should be assessed for the presence of taint.

With present knowledge, it is considered that the best way of reducing the incidence of cork taint is to minimize the opportunities for microbial growth on cork by controlling the water activity (a_w) of this substrate during processing, storage and transport (Simpson and Lee, 1990). Most molds isolated from corks imported into Australia showed moderate to strong growth at 10–20°C and an a_w of 0.97; some of the mold isolates grew at a_w 0.85, two isolates at a_w 0.80 and none at 0.75 (Daly *et al.*, 1984). An a_W of 0.75 in cork corresponds to a moisture level of approximately 8% (Lee *et al.*, 1984), which is at the higher end of that required for good performance of modern corking machines.

Although hypochlorite is used to bleach and sterilize corks, it increases the risk of 2,4,6-TCA formation and provides no protection of the cork against subsequent microbial contamination and growth. Peroxide bleaching of corks using either peracetic acid or hydrogen peroxide is an alternative process (Heinzel *et al.*, 1983) but, according to Fabre (1989), residual peroxide activity can be detrimental to bottled wine, reducing sulfur dioxide content and causing oxidation. One questions the need to bleach corks at all!

Before delivery to the winery, most corks are subjected to dedusting, sorting and grading, application of a lubricant and a 'sterilizing' treatment. Processes involving the use of ethylene oxide (Garcia del Rio, 1979), heat (Fumi and Cologrande, 1988), methylene chloride, salicylic acid and ultraviolet light have been either patented or proposed (Moreau *et al.*, 1976; Castera-Rossignol, 1983), but sulfur dioxide and γ-irradiation (Marais and Kruger,1975; Davis *et al.*, 1982; Heinzel *et al.*, 1983; Zehnder *et al.*, 1984; Rochard *et al.*, 1988) are the two treatments most widely used. Although these treatments are effective in reducing the microbial load, they may not render the corks sterile. Sulfur dioxide has the added advantage of controlling microbial growth at later stages. It may be worthwhile to apply such treatment as soon as possible after cork preparation in Portugal to prevent microbial growth during storage and transport, provided that recontamination of the corks is prevented.

Several tests to monitor the level of cork taint in a consignment of corks are used by industry (Amon and Simpson, 1986; Carey, 1988; Brunner, 1989). An appropriate sampling procedure (Standards Association of Australia, 1972) is necessary to give a reliable estimate of the level of cork taint in a consignment and to minimize the number of corks required for testing. The tests may be carried out on a small batch of corks placed in contact with a neutral wine (Amon and Simpson, 1986) or on individual corks (Brunner, 1989) in contact with a small amount of moisture to displace any taint compounds (Casey, 1990). Such tests are best conducted by assessors who are highly sensitive to cork taint, and who are able to distinguish between the natural aroma of cork and the various taints that are categorized as cork taint.

Finally, in some cases, microbiological testing of cork shipments may be worthwhile. Procedures for the quantitative extraction of microbial cells from corks and methods for their culture and enumeration are described by Davis *et al.* (1981a,b).

REFERENCES

Ahlborg, U.G. and Thunberg, T.M. (1980) Chlorinated phenols: occurrence, toxicity, metabolism, and environmental impact. *CRC Critical Reviews in Toxicology*, **7**, 1–35.

Amon, J.M. and Simpson, R.F. (1986) Wine corks: a review of the incidence of cork related problems and the means for their avoidance. *The Australian Grapegrower & Winemaker*, **(268)**, 63–64, 66–68, 71–72, 75–77, 79–80.

Amon, J.M., Simpson, R.F. and Vandepeer, J.M. (1987) A taint in in wood-matured wine attributable to microbiological contamination of the oak barrel. *Australian & New Zealand Wine Industry Journal*, **2(2)**, 35–37.

Amon, J.M., Vandepeer, J.M. and Simpson, R.F. (1989) Compounds responsible for cork taint in wine.. *Australian & New Zealand Wine Industry Journal*, **4**, 62–69.

Avila, R. and Lacey, J. (1974). The role of *Penicillium frequentans* in suberosis (respiratory disease in workers in the cork industry). *Clinical Allergy*, **4**, 107–117.

Bentley, R. and Meganathan, R. (1981). Geosmin and methylisoborneol biosynthesis in Streptomycetes. Evidence for an isoprenoid pathway and its absence in non-differentiating isolates. *FEBS Letters*, **125**, 220–222.

Boidron, J.–N., Lefebvre, A., Riboulet, J.–M. and Ribéreau-Gayon, P. (1984). Les substances volatiles susceptibiles d'être cédées au vin par le bouchon de liège. *Sciences des Aliments*, **4**, 609–616.

Borges, M. (1985) New Trends in cork treatment and technology. *Beverage Review*, **5**, 15, 16,19, 21.

Bricout, J., Brunerie, P. and Desmarest, P. (1990) Influence of fungicide on the quality of fennel essential oil. In *Proceedings of the 11th International Congress of Essential Oils, Fragrances & Flavours* (New Delhi, India, 1989), edited by S.C. Bhattacharya, N. Sen and K.L. Sethi. Aspect.

Brunner, H.R. (1989) Korktonbildung—ein einfacher Test für die Praxis. *Weinwirtschaft - Technik*, **8**, 25–26.

Bureau, G., Charpentier-Massonnat, M. and Pansu, M. (1974) Étude des goûts anormaux apportés par le bouchon sur le vin de Champagne. *Revue Française d'Oenologie*, **56**, 22–24.

Burttschell, R.H., Rosen, A.A., Middleton, F.M. and Ettinger, M.B. (1959) Chlorine derivatives of phenol causing taste and odor. *Journal of the American Water Works Association,* **51**, 205–214.

Buser, H.-R., Zanier, C. and Tanner, H. (1982) Identification of 2,4,6-trichloroanisole as a potent compound causing cork taint in wine. *Journal of Agricultural and Food Chemistry,* **30**, 359–362.

Cantagrel, R. and Vidal, J.P. (1989) Research on compounds responsible for cork taint in cognacs. In *Proceedings of the 6th International Flavor Conference* (Rethymnon, Crete, 1989), edited by G. Charalambous, pp. 139–157. Amsterdam: Elsevier Scientific Publishers.

Carey, R. (1988) Natural cork: the "new" closure for wine bottles. Part II. *Vineyard & Winery Management,* **14(5)**, 38.

Casey, J.A. (1990) A simple test for tainted corks. *The Australian Grapegrower & Winemaker,* **(324)**, 40–41.

Castera-Rossignol, A. (1983) Contrôle microbiologique des bouchons. Bouchons steriles. Conditions de conservation de bouchons. *Connaissance de la Vigne et du Vin,* **17**, 189–193.

Crosby, D.G. (1981) Environmental chemistry of pentachlorophenol. *Pure & Applied Chemistry,* **53**, 1051–1080.

Daly, N.M., Lee, T.H. and Fleet, G.H. (1984) Growth of fungi on wine corks and its contribution to corky taints in wine. *Food Technology in Australia,* **36**, 22–24.

Davis, C.R., Fleet, G.H. and Lee, T.H. (1981a) The microflora of wine corks. *The Australian Grapegrower & Winemaker,* **(208)**, 42, 44.

Davis, C.R., Fleet, G.H. and Lee, T.H. (1981b) An appraisal of the Rankine-Pilone method for testing the sterility of wine corks. *The Australian Grapegrower & Winemaker,* **(208)**, 56.

Davis, C.R., Fleet, G.H. and Lee, T.H. (1982) Inactivation of wine cork microflora by a commercial sulfur dioxide treatment. *American Journal of Enology and Viticulture,* **33**, 124–127.

Duerr, P. (1985) Wine quality evaluation. In *Proceedings of the International Symposium on Cool Climate Viticulture and Enology* (Eugene, OR, 1985), edited by D.A. Heatherbell, P.B. Lombard, F.W. Bodyfelt and S.F. Price, pp. 257–266. Corvallis, OR : Oregon State University.

Fabre, S. (1989) Bouchons traités au(x) peroxyde(s): détection du pouvoir oxydant et risques d'utilisations pour le vin. *Revue des Oenologues,* **52**, 11–15.

Freiter, E.R. (1979) *Chlorophenols. Kirk-Othmer encyclopedia of chemical technology.* 3rd edn, Vol. 5, pp. 862–872. New York: Wiley.

Fumi, M.D. and Colagrande, O. (1988) Inactivation thermique de la microflore dans les bouchons de liège. *Revue des Oenologues,* **(50)**, 28–30.

Garcia del Rio, J.R. (1979) Empleo de inhidrido sulforoso y oxido et etileno en la desinfeccion de tapones. *Semina Vitivinicola,* **34**, 945, 947, 949.

Gerber, N.N. (1979) Volatile substances from Actinomycetes: their role in the odor pollution of water. *CRC Critical Reviews in Microbiology,* **7**, 191–214.

Gerber, N.N. (1983) Volatile substances from *Actinomycetes*: their role in the odor pollution of water. *Water Science and Technology,* **15**, 115–25.

Heimann, W., Rapp, A., Volter, I. and Knisper, W. (1983) Beitrag zur Entstehung des Korktons in Wein. *Deutsche Lebensmittel-Rundschau,* **79**, 103–7.

Heinzel, M., Hagen, M. and Bousser, C. (1983) Désinfection des bouchons par l'eau oxygénée activée. *Revue Française d'Oenologie,* **23**, 77–81.

Izaguirre, G., Wolfe, R.L. and Means, E.G. (1988) Bacterial degradation of 2-methylisoborneol. *Water Science and Technology,* **20**, 205–210.

Jerina, D., Guroff, G. and Daly, J. (1968) Enzymic and nonenzymic hydroxylation and chlorination of *p*-deuteroanisole. *Archives of Biochemistry and Biophysics,* **124**, 612–615.

Karahadian, C., Josephson, D.B. and Lindsay, R.C. (1985) Volatile compounds from *Penicillium* sp. contributing musty-earthy notes to Brie and Camembert cheese flavors. *Journal of Agricultural and Food Chemistry*, **33**, 339–343.

Lacey, J. (1973) The air spora of a Portuguese cork factory. The *Annals of Occupational Hygiene*,**16**, 223–230.

Land, D.G. (1989) Taints—causes and prevention. In *Distilled Beverage Flavour. Recent Developments*, edited by J.R. Pigott and A. Paterson, pp. 17–31. Chichester: Ellis Horwood.

Lee, T.H., Simpson, R.F., Vandepeer, J.M., Fleet, G.H., Davis, C.R., Daly, N.M. and Yap, A.S.J. (1984) Microbiology of wine corks. In *Proceedings of the Fifth Australian Wine Industry Technical Conference* (Perth 1983), edited by T.H. Lee and T.C. Somers, pp. 435–450. Adelaide: The Australian Wine Research Institute.

Lefebvre, A., Riboulet, J.-M., Boidron, J.-N. and Ribéreau-Gayon, P. (1983) Incidence des micro-organismes du liège sur les altérations olfactives du vin. *Sciences des Aliments*, **3**, 265–278.

Maarse, H., Nijssen, L.M. and Angelino, S.A.G.F. (1988) In *Proceedings of the Second Wartburg Aroma Symposium* (Wartburg, 1987), edited by M. Rothe, pp. 43–61. Berlin: Akademie Verlag.

Maarse, H., Nijssen, L.M. and Jetten, J. (1985) Chloroanisoles: a continuing story. In *Proceedings of the International Conference on Topics in Flavour Research* (Freising-Weihenstephan, 1985), edited by R.G. Berger, S. Nitz and P. Schreier, pp. 241–250. Marzling-Hangenham: Eichhorn, H.

Maga, J.A. (1978) Simple phenol and phenolic compounds in food flavour. *CRC Critical Reviews in Food Science and Nutrition*, **10**, 323–372.

Marais, P.G. and Kruger, M.M. (1975) Fungus contamination of corks responsible for unpleasant odours in wine. *Phytophylactica*, **7**, 115–116.

Maujean, A., Millery, P. and Lemaresquier, H. (1985) Explications biochimiques et métaboliques de la confusion entre goût de bouchon et goût de moisi. *Revue Française d'Oenologie*, **99**, 55–62.

Moreau, M. (1978) La mycoflore des bouchons de liège. Son évolution au contact du vin; conséquences possibles du métabolisme des moisissures. *Revue de Mycologie*, **42**, 155–189.

Moreau, M., Moreau, C. and Le Bras, M.-A. (1976) Quelques moisissures responsables d'altérations des bouchons de Champagne. *Industries Alimentaires et Agricoles*, **93**, 317–320.

Naes, H., Utkilen, H.C. and Post, A.F. (1988) Factors influencing geosmin production by cyanobacterium *Oscillatoria brevis*. *Water Science and Technology*, **20**, 125–131.

Neidleman, S.L. and Geigert, J. (1986) *Biohalogenation: Principles, Basic Roles and Applications*, pp. 13–45, Chichester: Ellis Horwood.

Pitt, J.I. (1979) *The genus* Penicillium *and its teleomorphic states* Eupenicillium *and* Talaromyces. London: Academic Press.

Preston, W.A. (1983) *Cork and wine*. St Helena, CA: Illuminations Press.

Rigaud, J., Issanchou, S., Sarris, J. and Langlois, D. (1984) Incidence des composés volatils issus du liège sur le 'goût de bouchon' des vins. *Sciences des Aliments*, **4**, 81–93.

Rochard, J., Romanet, G., Bureau, G. and Gyr, J.-P. (1988) Essais d'ionisation des bouchons de Champagne. *Le Vigneron Champenois*, **109(1)**, 19–26.

Rott, B., Nitz, S. and Korte, F. (1979) Microbial decomposition of sodium pentachlorophenolate. *Journal of Agricultural and Food Chemistry*, **27**, 306–310.

Schaeffer, A., Meyer, J.-P. and Gullerm, A. (1978) Étude sur l'origine du 'goût de bouchon' dans les vins. *Revue Française d'Oenologie*, **70**, 25–29.

Simpson, R.F. (1990) Cork taint in wine: a review of the causes. *Australian and New Zealand Wine Industry Journal*, **5**, 286–296.

Simpson, R.F. and Lee, T.H. (1990) The microbiology and taints of cork and oak. In *Proceedings of the Ninth International Oenological Symposium* (Cascais, 1990), edited by E. Lemperle and E. Figlestahler, pp. 653–667. Breisach: International Association for Modern Winery Technology and Management.

Simpson, R.F., Amon, J.M. and Daw, A.J. (1986) Off-flavour in wine caused by guaiacol. *Food Technology in Australia,* **38,** 31–33.

Standards Association of Australia (1972). Sampling procedures and tables for inspection by attributes. Australian Standard 1199–1972. Sydney: Standards Association of Australia.

Tanner, H. and Zanier, C. (1981). Zur analytischen Differenzierung von Muffton und Korkgeschmack in Weinen. *Schweizerische Zeitschrift für Obst- und Weinbau,* **117**, 752–757.

Tanner, H., Zanier, C. and Buser, H.-R. (1981). 2,4,6-Trichloranisol: eine dominierende Komponente des Korkgeschmackes. *Schweizerische Zeitschrift für Obst- und Weinbau,* **117**, 97–103.

Tindale, R.C. (1987) Shipping container floors: a potential source of chloroanisole contamination in packaged dried fruit. *Chemistry and Industry (Lond.),* (6 July), 458–459.

Tindale, C.R., Whitfield, F.B., Levingston, S.D. and Nguyen, T.H.L. (1989) Fungi isolated from packaging materials: their role in the production of 2,4,6-trichloroanisole. *Journal of the Science of Food and Agriculture*, **49**, 437–447.

Weiss, U.M., Scheunert, I., Klein, W. and Korte, F. (1982) Fate of pentachlorophenol- 14C in soil under controlled conditions. *Journal of Agricultural and Food Chemistry,* **30,** 1191–1194.

Whitfield, F.B., Nguyen, T.H.L. and Last, J.H. (1991) Effect of relative humidity and chlorophenol content on the fungal conversion of chlorophenols to chloroanisoles in fibreboard cartons containing dried fruit. *Journal of the Science of Food and Agriculture,* **54**, 595–604.

Whitfield, F.B., Nguyen, T.H.L. and Tindale, C.R. (1991) Effect of relative humidity and incubation time on the O-methylation of chlorophenols in fibreboard by *Paecilomyces variotii. Journal of the Science of Food and Agriculture*, **55**, 19–26.

Zehnder, H.J., Buser, H.-R. and Tanner, H. (1984) Zur Entstehung des Korktons in Wein und dessen Verhinderung durch die Behandlung der Flaschenkorken mit ionisierender Strahlung. *Deutsche Lebensmittel-Rundschau,* **80,** 204–207.

Chapter 13

SULFUR DIOXIDE AND WINE MICROORGANISMS

PATRIZIA ROMANO and GIOVANNA SUZZI

Department for the Protection and Improvement of Food and Agricultural Products, University of Bologna, Via F.11i Rosselli 107, 42100 Reggio Emilia, Italy

Chapter Outline

INTRODUCTION

Sulfur dioxide, (SO_2) is an antimicrobial compound that has been used in winemaking for centuries. Burning sulfur fumes were first used by the Egyptians and later by the Romans to clean their amphorae and other wine vessels. In the centuries that followed, SO_2 became a widely used preservative, generated originally by burning sulfur but later obtained by adding sulfite or bisulfite. It greatly improved fermentation processes by inhibiting the growth of unwanted bacteria and yeasts. Its addition to wine and must was common in the late Middle Ages but, even then, its abuse and excessive addition was disapproved. The first modern accounts of the use of SO_2 in beverages were reported in the fifteenth century. Nowadays, SO_2 is universally employed to protect many foodstuffs, and its use in winemaking is a common practice that is permitted by all wine producing countries. Table 13.1 summarizes the maximum concentrations of SO_2 permitted in wine in several countries. Its use in winemaking has been the subject of much discussion and the reader is referred to the reviews of Beech *et al.* (1979), Ough (1983) and Beech and Thomas (1985).

PROPERTIES OF SULFUR DIOXIDE

The chemical and physical properties of SO_2 are given detailed consideration in the book by Wedzicha (1984). Sulfur dioxide is a colorless, non-flammable gas with a strong suffocating odor. It is formed by burning sulfur, whereby sulfur (32 g) reacts with oxygen (32 g) to give 64 g of SO_2. The gas dissolves in water depending on temperature: 17.7% at 0°, 11.9% at 15°, 8.5% at 25°, 6.4% at 35°C; and 25% in ethanol. In practice, SO_2 is used as either:

1. Liquefied SO_2 which is supplied under pressure,
2. Potassium or sodium metabisulfite ($K_2S_2O_5$ or $Na_2S_2O_5$) which give slightly more than 50% of SO_2 when dissolved in musts or wine, or
3. Sulfurous acid (H_2SO_3) which is obtained by dissolving SO_2 in water; a 5% solution is often used.

In aqueous solution, SO_2 dissociates according to the following equilibria, giving the molecular species indicated:

$$SO_2\,(gas) + H_2O \rightleftarrows SO_2 \cdot H_2O \quad \text{(molecular } SO_2\text{)}$$

$$SO_2 \cdot H_2O \overset{pKa_1}{\rightleftarrows} H^+ + HSO_3^- \quad \text{(bisulfite ion)}$$

$$HSO_3^- \overset{pKa_2}{\rightleftarrows} H^+ + SO_3^= \quad \text{(sulfite ion)}$$

TABLE 13.1 Maximum limits of sulfur dioxide permitted in wines produced in various countries.

	Total SO_2 (mg/l)			*Free SO_2 (mg/l)*		
	Sweet white wines	*Dry white wines*	*Dry red wines*	*Sweet white wines*	*Dry white wines*	*Dry red wines*
Algeria*		300			30	
Austria*		300			50	
Argentina	350	250	250		80	
Bulgaria		200			20	
Canada		350			70	
EEC*+	260	210	160			
Israel		350				
Hungary		300				
Japan		350				
Morocco		450			100	
Rumania*	300	200	200	50	50	50
Switzerland	400	250	250	50	35	35
Czechoslovakia		300			40	
Tunisia		300	250		60	60
URSS	400	200	200	40	20	20
USA		350				
Australia		350				
South Africa		400		100		

* Different limits may be employed for special wines
\+ European Economic Community
Data modified from Ribéreau-Gayon *et al.* (1988).

All forms of SO_2 in these equilibria are termed "free". The effectiveness of SO_2 as an antioxidant and antimicrobial agent in wine depends on the amount of molecular SO_2 in solution (Ingram, 1948; Joslyn and Braverman, 1954). The percent distribution of these forms depends on pH. At values below pH 1.77 (pKa_1), the SO_2 exists predominantly as molecular SO_2, and at values above pH 7.2 (pKa_2) it occurs largely as sulfite ions. At intermediate pH, it is present in various proportions as the bisulfite ion (King *et al.*, 1981). In most wines (pH 3.0–4.0), it exists mainly as the bisulfite ion (94–98%), and only a very small proportion (2–6%) occurs in the molecular form.

In musts, SO_2 reacts with sugars such as glucose, xylose and arabinose (fructose is inactive), galacturonic acid, 1-ascorbic acid and anthocyanins. As much as 50–70% of the total SO_2 added to must can be bound by sugar components (Blouin, 1966; Rankine and Pocock, 1969; Burroughs and Sparks, 1973a; Burroughs, 1975; Somers and Wescombe, 1987). Moreover, must extracted from damaged and diseased grapes can have exceptionally high SO_2-binding properties due to substances produced by molds (*Botrytis cinerea*) and acetic acid bacteria (Kielhöfer, 1963; Hammond and Carr, 1976). During alcoholic fermentation, yeasts increase the SO_2-binding power of must by producing acetaldehyde, pyruvic acid and 2-ketoglutaric acid. Higher initial concentrations of SO_2 in the must cause increased yeast production of these carbonyl compounds (Weeks, 1969; Lafon-Lafourcade, 1985). The reaction of SO_2 (specifically the bisulfite ion) with carbonyl compounds can be represented as:

$$R-CHO + H_2SO_3 \rightleftharpoons R-CH\begin{cases} SO_3H \\ OH \end{cases}$$

This reaction depends upon pH, temperature and concentrations of reactants (Burroughs and Sparks, 1964; 1973b).

Thus, the total concentration of SO_2 in grape juice during fermentation consists of bound and free forms, the bulk of it being bound and having little antimicrobial function. The free SO_2 will consist mostly of the bisulfite anion and a small proportion of molecular SO_2. Consequently, it is necessary to add sufficient total SO_2 to obtain a concentration of molecular SO_2 that would have the desired antimicrobial effect. Depending on must or wine composition, it is generally necessary to add 100–200 mg/l of total SO_2 to produce 0.5–1 mg/l of molecular SO_2 (Fornachon, 1965; King *et al.*, 1981).

Bound SO_2 is not effective in killing yeasts or preventing their growth (Macris and Markakis, 1974; Carr *et al.*, 1976; King *et al.*, 1981), but it can have weak antibacterial properties (Fornachon, 1963; Lafon-Lafourcade and Peynaud, 1974; Mayer *et al.*, 1975).

FUNCTION OF SULFUR DIOXIDE IN WINEMAKING

In winemaking, SO_2 performs several functions, each of which is necessary to give a good quality product. Its principal functions are:

1. Stabilization of biochemical and chemical properties, and
2. Control of microorganisms.

It is added at various stages during vinification, but mainly to the must before alcoholic fermentation, and at later stages during handling of the wine in the cellar.

Stabilization of Biochemical and Chemical Properties

From a biochemical and chemical perspective, wine is very unstable and readily undergoes undesirable changes to color (browning), flavor and turbidity if exposed to air or relatively high temperatures. Many of these changes are caused by reactions with oxygen. Sulfur dioxide allows wines to maintain their desirable sensory characteristics during production, transport and marketing. The effectiveness of this function depends upon the concentration of molecular SO_2 present and, consequently, careful control of wine pH (Ough and Crowell, 1987) and temperature (Ough, 1985) is necessary.

Many oxidative changes to wines are enzymatic in origin, being caused by the oxidation of phenolic constituents by polyphenol oxidases. These enzymes (e.g. tyrosinase and laccase) occur in the must, arising from the grape itself, or from fungi that have infected the grapes (Poux, 1966; Singleton and Esau, 1969; Dubernet and Ribéreau-Gayon, 1973; Wissemann and Lee, 1981). Free SO_2 inactivates these enzymes by reducing their copper cofactor (Schopfer and Aerny, 1985). Tyrosinase is more sensitive to the action of SO_2 than laccase (Dubernet and Ribéreau-Gayon, 1973).

Non-enzymatic oxidative reactions also occur, resulting from contact of the wine with air. Auto-oxidation of catechol derivatives in wine produces acetaldehyde by coupled oxidation of ethanol (Wildenradt and Singleton, 1974). Sulfur dioxide inhibits wine browning by combining irreversibly with phenolic quinones to form colorless addition products and, at the same time, makes these constituents unfavorable substrates for the polyphenol oxidases (Embs and Markakis, 1965; Markakis and Embs, 1966).

Detailed consideration of the role of SO_2 in stabilizing the biochemical and chemical properties of wines is outside the scope this chapter and further information may be found in the reviews of Amerine *et al.* (1980) and Schopfer and Aerny (1985).

Control of Microorganisms

Sulfur dioxide is added to grape must and to wine at various stages during production to control the growth of undesired species of yeasts and spoilage species of bacteria. As discussed by Fleet (Chapter 1), lactic acid bacteria and acetic acid bacteria are the main bacterial groups of concern in winemaking, although the possible growth of *Bacillus* species in wines requires more serious consideration. Careful attention must be given to the concentration of SO_2 used, because sufficient is needed to inhibit the growth of undesired microbial species but not that of desired fermentation yeasts or malolactic bacteria. It is well established that different microbial species have different sensitivity to SO_2

(Chichester and Tanner, 1972). As noted already, it is free, molecular SO_2 that exerts antimicrobial activity, and the concentration of this active form is profoundly influenced by pH and substances that bind the bisulfite anion. Thus, sufficient total SO_2 must be used to compensate for these factors.

ACTION OF SULFUR DIOXIDE ON MICROORGANISMS

Molecular Effects

Rose and Pilkington (1989) as well as Wedzicha (1984) have reviewed the molecular mechanisms by which SO_2 affects microbial cells.

The first step in the action of SO_2 is its uptake by the cell. It is generally accepted that molecular SO_2 rather than the bisulfite or sulfite anions is the form taken in by yeast cells (Ingram *et al.*, 1956) and that this process is affected by pH (Schimz, 1980; Hinze and Holzer, 1985; Stratford and Rose, 1986). Macris and Markakis (1974) proposed that uptake of SO_2 by *S. cerevisiae* occurred by active transport but studies by Stratford and Rose (1986) suggest that simple diffusion is a more likely process. Once inside the cell, SO_2 undergoes dissociation to produce mostly the bisulfite anion (HSO_3^-) because the intracellular pH is about 6.5 (Borst-Pauwels, 1981). This reduces the intracellular concentration of molecular SO_2, thus causing its further diffusion into the cell until its concentration is balanced on both sides of the plasma membrane. In this manner, the cell acts as a "sulfite trap" leading to an intracellular accumulation of charged sulfite species that can react with cell constituents (Gunnison, 1981). As a consequence, irreversible inhibition of cell growth occurs (Macris, 1972; Carr *et al.*, 1976; Schimz, 1980) together with further events that may cause cell death.

Inside the cell, sulfite may react with certain molecules to assert its antimicrobial influence. Compounds with disulfide bridges are split by sulfite into the corresponding sulfhydryl compounds (Stricks and Kolthoff, 1951; Cole, 1967). Sulfite can react with co-enzymes such as NAD, FAD and FMN (Mayerhof *et al.*, 1938; Vanderschmitt *et al.*, 1967; Müller and Massey, 1971) and with cofactors and vitamins (Stricks *et al.*, 1955; Haisman, 1974). For example, sulfite cleaves thiamine by breaking the bond between the pyrimidine and thiazole portion of the molecule (Williams *et al.*, 1935). Many enzymes or enzyme systems are inhibited by sulfite either by inducing changes in enzyme conformation, modification of the active site or destruction of the coenzyme (Rose and Pilkington, 1989). At pH values below 4.0, sulfite causes a most dramatic depletion in the content of cellular ATP in yeasts through its action on glycolysis and respiratory chain phosphorylation (Schimz and Holzer, 1979; Schimz, 1980; Hinze and Holzer, 1985, 1986; Maier *et al.*, 1986). Sulfite reacts with nucleic acids (reviewed by Shapiro, 1977). It catalyzes deamination of cytosine to uracil and cross-linking of proteins and nucleic acids. It binds to uracil and thymidine and reacts with free radicals. Thus, it has an indirect effect on RNA and protein metabolism. These findings have raised the possibility that SO_2 is a mutagen. Studies on the mutagenic effects of SO_2 have been reported for phages (Hayatsu and Miura, 1970; Summers and Drake, 1971), bacteria (Mukai *et al.*, 1970) and among yeasts, for *S. cerevisiae* (Dorange and Dupuy, 1972; Schimz, 1980; Guerra *et al.*, 1981).

Sulfur Dioxide Resistance in Yeasts

Wine yeasts (*S. cerevisiae*) tolerate relatively high concentrations of SO_2, presumably as the result of adaptation and natural selection (Schanderl, 1959; Wedzicha, 1984). *Saccharomycodes ludwigii* and *Zygosaccharomyces bailii*, other yeasts often associated with wines, are even more resistant to SO_2 than *S. cerevisiae* (Amerine and Kunkee, 1968; Beech and Carr, 1977; Stratford *et al.*, 1987). Generally, ascospores of *Saccharomyces* are more resistant than vegetative cells (Romano and Suzzi, 1985). Despite the diversity in SO_2 resistance in yeasts, and its practical significance, very little is known about the biochemical basis of this resistance (Wedzicha, 1984).

Some mechanisms proposed to explain this variation in resistance include differences in glutathione content (Scardovi, 1951; Casalone *et al.*, 1989), differences in rates of diffusion or transport of SO_2 across the plasma membrane (Stratford and Rose, 1986), differences in the production of sulfite binding compounds such as acetaldehyde (Stratford *et al.*, 1987; Pilkington and Rose, 1988) and differences in the sensitivity of glyceraldehyde-3-phosphate dehydrogenase to inactivation by SO_2 (Hinze and Holzer, 1986). Ultimately, there are genetic bases to these different degrees of resistance. Zambonelli *et al.* (1972a) and Romano and Tini (1975) noted that different strains of *S. cerevisiae* varied in their natural resistance to SO_2. This resistance character was stable, hereditary and unigenic, the last property being confirmed in SO_2-resistant mutants (Daly *et al.*, 1990). Sulfur dioxide induces, at high frequency levels, SO_2-resistant mutants altered in methionine metabolism (Guerra *et al.*, 1981) but these mutants are not stable and slowly revert (Porchet, 1931; Scardovi, 1952). Moreover, Thornton (1982) suggested that SO_2 tolerance is controlled by several dominant polymeric genes, and that different combinations of genes give rise to different tolerance levels.

The different mechanisms of SO_2 resistance are probably related to the multiplicity of SO_2 action. It is also possible that strains naturally resistant to SO_2 will be different from induced mutants in stability and inheritance of this character. Usuka *et al.* (1985) have described a procedure for determining the resistance of yeasts to SO_2.

EFFECT OF SULFUR DIOXIDE ON WINE MICROORGANISMS

Yeasts

Freshly extracted grape must harbors a population of natural or indigenous yeasts that originate from the grape and winery environments. The ecology of this flora has been well reviewed, most recently and most critically by Martini and Martini (1990). The principal yeast of wine fermentation, *S. cerevisiae*, is not predominant in must flora which consists mostly of apiculate yeasts in the genera *Kloeckera* and *Hanseniaspora* and various species of other yeasts (Zambonelli *et al.*, 1989). One of the desired objectives of adding SO_2 to the must, is to suppress the growth of these non-*Saccharomyces* species and selectively encourage the growth and dominance of *S. cerevisiae*. Traditionally, the *S. cerevisiae* was that naturally present but, more recently, it is usually the strain that is inoculated into the must. This objective presumes a greater resistance of strains of *S. cerevisiae* to SO_2 than the indigenous non-*Saccharomyces* species. In practice, this presumption appears to work as indicated in the profile of yeasts presented in Table 13.2.

Heard and Fleet (1988) have questioned the efficacy of SO_2 in controlling the initial growth of indigenous non-*Saccharomyces* yeasts. In many wine fermentations, yeasts follow a successive pattern of growth (Fleet, 1990). During the first few days, species of *Kloeckera*, *Hanseniaspora*, *Candida* and *Hansenula* frequently grow, reaching significant (10^6–10^7 cells/ml) populations. These indigenous species soon die off giving way to a predominance of *S. cerevisiae* which completes the fermentation. Species of *Zygosaccharomyces*, *Torulaspora*

TABLE 13.2 Profile of most prevalent yeasts in non-sulfited and sulfited wine fermentations conducted in the Emilia Romagna region, Italy (Romano and Suzzi, unpublished data).

Yeast	*Frequency in grape juice*	*Presence during fermentation*				*Resistance to SO_2*
		Without SO_2		*With SO_2*		
		start	*end*	*start*	*end*	
Kloeckera	++++	++++	–	–	–	–
Hanseniaspora	++++	++++	–	–	–	–
Metschnikowia	++	+	–	–	–	–
Hansenula	++	–	–	–	–	–
Pichia	++	–	–	–	–	–
S. cerevisiae	+	+	++++	++++	++++	++++
Saccharomycodes	+	+	+	+	+	++++
Torulaspora	+	+	+	+	+	++++
Zygosaccharomyces	+	+	+	+	+	++++
Schizosaccharomyces	+	±	±	±	±	++++
Brettanomyces	+	+	+	+	+	++++

++++ = very high; +++ = high; ++ = medium; + = low; - = absent; ± = rare.

and sometimes *Schizosaccharomyces* and *Brettanomyces* may also be present at this stage. This pattern of yeast growth is commonly found in commercial wineries where SO_2 (50–100 mg/l total) is added to the must (Fleet *et al.*, 1984; Heard and Fleet, 1986). Thus, it seems that SO_2 addition does not always prevent the initial growth of these indigenous yeasts. In experimental fermentations, where the addition of SO_2 was better controlled, Heard and Fleet (1988) also demonstrated that total SO_2 of 100 mg/l did not necessarily prevent the growth of indigenous non-*Saccharomyces* species, especially in red wines.

Notwithstanding the above discussion, there is reasonable evidence that wine species of *Kloeckera*, *Candida*, *Pichia*, and *Hansenula* are much more sensitive to SO_2 than species of *Saccharomyces*. Several studies (tabulated in Heard and Fleet, 1988) have reported on the concentrations of SO_2 required to inhibit various species of wine yeasts, but it is difficult to interpret and compare the data presented because, in some cases, they refer to the effect of total SO_2 on the yeasts while in other cases the effects of free or molecular SO_2 are reported (Ohara *et al.*, 1959; Rehm and Wittman, 1962; Usseglio-Tomasset *et al.*, 1981; Warth, 1985; Beech and Thomas, 1985). Also, different strains of the one species may differ markedly in their tolerance of SO_2 (Rehm and Wittmann, 1962; Romano and Suzzi, 1990).

King *et al.* (1981), describe the difficulties of measuring the effect of SO_2 on yeasts because of the influence of pH and substances that bind to SO_2. Thus, approximately 1–2mg/l of molecular SO_2 is needed to inhibit *S. cerevisiae* but, at pH 3.5, approximately 100 mg/l of total SO_2 are required to obtain this amount because most of it dissociates to the bisulfite anion. Ough *et al.* (1988) reported that 50 mg/l of free SO_2 inhibited the growth of a strain of *S. cerevisiae* in wine at pH 3.0, but not at pH 3.2. From the studies conducted, it appears that the bisulfite anion and bound form of SO_2 have little inhibitory effect on yeasts (Ingram, 1948; Rehm and Wittman, 1962; Macris, 1972; Macris and Markakis, 1974; Carr *et al.*, 1976; Schimz, 1980; Sudraud and Chauvet, 1985).

Suzzi and Romano (1982) studied 1600 strains of *S. cerevisiae* isolated from fermentations of sulfited and non-sulfited musts. The majority of strains (about 80%) from non-sulfited musts showed high sensitivity towards SO_2 and only a low percentage of strains exhibited SO_2-resistance (Table 13.3). This observation can explain delays in the start of fermentations after sulfiting the must. However, at the end of fermentation of sulfited musts, SO_2-resistant strains dominated and only a few strains were SO_2-sensitive. It was concluded from this study that sulfite exerts an indirect action that favors the growth of *Saccharomyces* strains with natural resistance, and a direct action that promotes the appearance of resistant mutants.

The addition of SO_2 to musts not only selects for SO_2-tolerant strains of *S. cerevisiae* but, also, it affects other characteristics. In fact, strains of *S. cerevisiae* isolated from non-sulfited musts formed little SO_2, whereas most of those from sulfited musts formed high amounts of SO_2 (Table 13.4). Strains from sulfited musts also caused higher production of foam during fermentation (Suzzi and Romano 1978, 1983; Romano and Suzzi, 1987). The presence of SO_2 in the must

TABLE 13.3 Sensitivity to sulfur dioxide (SO_2) of 1600 strains of *Saccharomyces cerevisiae* isolated from fermentations conducted with sulfited and non-sulfited musts.

	Classes of sensitivity to SO_2[a]				
	1	*2*	*3*	*4*	*5*
Strains from non-sulfited musts (%)	12.3	70.5	6.0	7.0	4.2
Strains from sulfited musts (%)	0.5	10.0	20.7	52.5	16.3

[a]Sensitivity to sulfur dioxide was determined in grape must, pH 3.0, as CO_2 g released from 100 ml of sulfited must (100 mg/l) after 48 h.
Classes: 1 = CO_2 released 0 g; 2 = less than 2 g; 3 = less than 4 g; 4 = less than 6 g; 5 = more than 6 g.
Data from Suzzi and Romano (1982).

TABLE 13.4 Sulfur dioxide production by 1600 strains of *Saccharomyces cerevisiae* isolated from sulfited and non-sulfited musts.

	Classes of SO_2 production (mg/l)				
	<5	*5–10*	*10–15*	*15–20*	*>20*
Strains from non-sulfited musts (%)	40.3	33.5	13.2	9.8	3.6
Strains from sulfited musts (%)	8.5	15.4	19.5	28.4	28.2

The yeasts were cultivated in a synthetic medium and SO_2 determined iodometrically.
Data from Suzzi and Romano (1982).

may also influence the production of volatile compounds by *S. cerevisiae* (Herraiz, 1990) and other species (Herraiz *et al.*, 1989) during alcoholic fermentation.

The addition of SO_2 to cellar wines is necessary to control the growth of spoilage yeasts such as *Saccharomycodes ludwigii*, *Zygosaccharomyces bailii*, *Brettanomyces* spp, *Pichia* spp as well as unwanted strains of *S. cerevisiae*. Some of these species can be very resistant to high concentrations (> 3 mg/l) of molecular SO_2 (Thomas and Davenport, 1985).

Bacteria

Sulfur dioxide is an effective antimicrobial against wine lactic acid bacteria (reviewed in Wibowo *et al.*, 1985). Molecular SO_2 is the most active species although there are minor influences of the ionic species (Carr *et al.*, 1976) and bound SO_2, principally acetaldehyde bound bisulfite (Vaughn, 1955; Kunkee, 1967; Lafon-Lafourcade and Peynaud, 1974; Rankine, 1977). The antimicrobial effect of acetaldehyde bound bisulfite appears to arise from the SO_2 that is released on bacterial catabolism of the acetaldehyde (Hood, 1983; Somers and Wescombe, 1987).

Overall, information on the effects of SO_2 on species of wine lactic acid bacteria is sketchy and a thorough systematic study is needed. From the data available, it seems that total SO_2 at 100–150 mg/l or bound SO_2 of 50–100 mg/l or free SO_2 of 1–10 mg/l are sufficient to inhibit the growth of lactic acid bacteria in wines (Wibowo *et al.*, 1985). However, these values vary depending on the bacterial species (Fornachon, 1963), wine pH (van der Westhuizen and Loos, 1981; Liu and Gallander, 1983) and amount of insoluble solids in the wine (Liu and Gallander, 1982). As might be expected, lower pH values and increasing ethanol concentrations work synergistically to enhance the inhibitory effect of SO_2 (Kunkee, 1967; Castino *et al.*, 1975; Britz and Tracey, 1990). Also, it is important to distinguish between those concentrations which merely retard growth and those which destroy cell viability (Macris and Markakis, 1974; Lafon-Lafourcade 1975, 1983; Lafon-Lafourcade *et al.*, 1983).

Although there is considerable strain variation in the response of wine lactic acid bacteria to SO_2, it seems that species of *Lactobacillus* and *Pediococcus* are more tolerant of higher concentrations than *Leuconostoc oenos* (Fornachon, 1963; Lafon-Lafourcade and Peynaud, 1974; Carr *et al.*, 1976; Hood, 1983; Davis *et al.*, 1988). In a survey of numerous wine strains of *Leuc. oenos*, *Pediococcus parvulus*, and *Lactobacillus* spp., Davis *et al.* (1988) reported the growth of all strains in a nutrient medium, pH 4.5, containing 64 mg/l of total SO_2. However, when the concentration of SO_2 was increased to 128 mg/l only 11% of the *Leuc. oenos* strains grew, whereas 74% and 59%, respectively, of the strains of pediococci and lactobacilli grew.

The addition of SO_2 to control spoilage lactic acid bacteria is best done at stages in the cellar after the wine has completed malolactic fermentation. It is well documented that total SO_2 concentrations which exceed 40–50 mg/l significantly retard the growth of *Leuc. oenos* and, consequently, onset and completion of the malolactic fermentation (Davis *et al.*, 1985). Also, wines with higher concentrations of SO_2 are more likely to undergo malolactic fermentation with less desirable species of *Pediococcus* and *Lactobacillus* (Davis *et al.*, 1986, a b). While on the subject of malolactic fermentation, it is pertinent to note that bacteriophages which destroy *Leuc. oenos* are inhibited by concentrations of SO_2 (total) in the range 50–100 mg/l (Davis *et al.*, 1985, Henick-Kling *et al.*, 1986).

The occurrence and growth of acetic acid bacteria (*Gluconobacter oxydans*, *Acetobacter pasteurianus* and *A. aceti*) in wines as well as their properties of enological significance have been reviewed by Drysdale and Fleet (1988). There is very little information about the effect of SO_2 on these bacteria. However, they are often isolated from musts and wines containing up to 100 mg/l of total SO_2, suggesting that they can tolerate the concentrations of SO_2 generally used during vinification (Joyeux *et al.*, 1984; Watanabe and Iino, 1984; Drysdale and Fleet, 1985). In a survey of 229 Australian wines, Drysdale (1988) correlated the incidence of acetic acid bacteria with the concentration of total SO_2 in the samples. *Gluconobacter oxydans* was not isolated from wines containing greater than 50 mg/l of total SO_2, although *A. pasteurianus* and *A. aceti* were frequently found in such samples. However, in a laboratory study using a synthetic medium, none of the isolates of *A. aceti* (14 strains) or *A. pasteurianus* (93 strains) grew in the presence of 100 mg/l of total SO_2 at pH 3.5. A small percentage (10–15%) of the isolates grew in 50 mg/l of total SO_2 at pH 3.5, but not at pH 3.0. Oddly, some isolates of *G. oxydans* grew in the presence of 100 mg/l of total SO_2 and pH 3.6 (Drysdale, 1988).

Further studies on the response of acetic acid bacteria to SO_2 are required. However, it seems that, in the wine environment, the growth of these bacteria is better controlled by ensuring an absence of oxygen rather than reliance upon the use of SO_2.

SULFUR DIOXIDE MINIMIZATION IN WINEMAKING

Generally, SO_2 has low toxicity for humans (Lueck, 1980). However, in recent years the intake of SO_2 in foods and beverages has been linked to the development of adverse, allergic reactions in some consumers, particularly asthmatics. Small concentrations of SO_2 (e.g. <10 mg/l) when inhaled or consumed can trigger allergic responses such as wheezing, flushing, tingling, hypotension and bronchospasms (Ough, 1983; Yang and Purchase, 1985; Tensches *et al.*, 1988). Since wine may contain up to 450 mg/l of total SO_2 (Table 13.1), it has the potential for introducing a significant quantity of SO_2 into the diet. Health authorities now advocate minimization of the use of SO_2 in foods and beverages and, as a consequence, interest has developed in reducing the concentration of SO_2 used in vinification (Usseglio-Tomasset, 1985; Valouyko *et al.*, 1985; Marques Gomes and Da Silva Babo, 1985; Aerny, 1986 a, b). The problems involved in achieving this goal are complex because of the multiple functions that SO_2 serves during winemaking. The following sections indicate some of the options currently being explored by the wine industry to minimize the use of SO_2.

Treatment of Must and Wine

As mentioned already, much of the SO_2 added to musts or wines is rendered ineffective because of binding to various constituents, especially carbonyl compounds such as acetaldehyde, pyruvic acid and 2-ketoglutarate. Thus, any initiative to minimize the production of these compounds will improve the effectiveness of SO_2 usage. Careful handling of grapes to reduce spoilage by fungi and acetic acid bacteria that are known to produce these substances is an early option. The addition of thiamine to the must decreases metabolic production of these compounds by yeasts during fermentation (Delfini *et al.*, 1980; Usseglio-Tomasset, 1985). Strains of *S. cerevisiae* that produce lower concentrations of SO_2-binding metabolites could be selected for conducting the fermentation or strains able to reduce total SO_2 content in wine (Romano *et al.*,

1992). Since SO_2, itself, stimulates acetaldehyde production by yeasts (Rose and Pilkington, 1989), it has been suggested that SO_2 not be added until fermentation is completed (Usseglio-Tomasset, 1986).

Cellar practices, such as nitrogen and carbon dioxide blanketing, which minimize exposure of the wine to air can significantly decrease the risk of wine oxidation, thereby enabling the use of less SO_2 (Ough, 1989). The judicious use of alternative antioxidants such as ascorbic acid and erythorbate has contributed to a major reduction in the use of SO_2 in many wineries (Ewart *et al.*, 1987). More effective control of pH (higher concentrations of molecular SO_2 are present at lower pH) and the use of membrane filtration to remove microbial contaminants are other cellar practices that may reduce SO_2 requirements (Ough, 1989).

Biological Sulfite

Yeasts, themselves, produce SO_2 during metabolism of the grape juice (Eschenbruch, 1974; Rauhut, Chapter 6), so it is very unlikely that SO_2-free wine could ever be produced. The formation of SO_2 by *S. cerevisiae* is a strain characteristic. Under comparable conditions, Suzzi *et al.* (1985) found that among 1700 strains of *S. cerevisiae*, the majority (80%) produced less than 10 mg/l of SO_2 and only four strains produced more than 30 mg/l.

It is difficult to compare the action of SO_2 added to must to the action of "biological sulfite", since the latter is always bound and is produced gradually and would not be present in effective concentrations at the early stages of vinification. Among high sulfite producing strains, however, some possess a rarely encountered character, defined "stabilizing power", which affects the chemical and biochemical stability of wine in a way comparable with the addition of SO_2 to the must (Zambonelli *et al.*, 1970, 1972b, 1976; Guerzoni and Suzzi, 1976; Suzzi *et al.*, 1984, 1985). This character is related to sulfite production, although a direct relationship between the amount of SO_2 produced during fermentation and the degree of stabilization is not always recorded. Exceptions are noted which seem to support the possibility that unknown factors are operative (Guerzoni and Suzzi, 1976). Thus, some low SO_2-producing strains have been found to possess a certain stabilizing power (Tini *et al.*, 1987, 1988, 1989). Collectively, however, these studies confirm a role of "biological sulfite" in determining the natural stability of wines, but the precise mechanism of this phenomenon is complex and requires further study.

In concluding this section, it is evident that several physical, chemical and biological options are available to the winemaker for minimizing but not eliminating the use of SO_2. While these options may offer significant control over the biochemical and chemical stability of wine, they may not present significant control over microbiological stability. Consequently, the addition of sufficient SO_2 is still needed to reduce the risks of microbial spoilage of the wine in the cellar and after packaging.

CONCLUSION

Current technology for making wine depends on the judicious use of SO_2 to produce a good quality product. While several options are available to reduce the problems of wine oxidation and so minimize the use of SO_2 as an antioxidant, its addition to wine as an antimicrobial agent is still necessary. Elimination of SO_2 from present winemaking practices would introduce unacceptable risks of wine spoilage by yeasts and bacteria. Despite the importance of using SO_2 to maintain the microbiological stability of wines, substantially more research is required to understand how different concentrations of SO_2 affect the kinetics of yeast and bacterial growth in wines. Also, more research is needed to understand the molecular mechanisms by which SO_2 affects the survival and growth of yeasts and bacteria.

REFERENCES

Aerny, J. (1986a) Diminution de la teneur des vins en anhydride sulfureux 1. L'anhydride sulfureux et ses propriétés utiles en vinification. Revue suisse. *Viticulture Arboric Horticulture*, **18**, 17–21.

Aerny, J. (1986b) Diminution de la teneur des vins en anhydride sulfureux 2. Aspects practiques. Revue Suisse. *Viticulture Arboric Horticulture*, **18**, 143–146.

Amerine, M.A. and Kunkee, R.E. (1968) Microbiology of winemaking. *Annual Review of Microbiology*, **22**, 323–358.

Amerine, M.A., Kunkee, R.E., Ough, C.S., Singleton, V.L. and Webb, A.D. (1980) *The Technology of Wine Making* (4th ed.) AVI Publishing Co., Inc. Westport, C.T.

Beech, F.W. and Carr, J.G. (1977) Cider and Perry. In *Economic Microbiology, vol. 1, Alcoholic Beverages*, edited by A.H. Rose, pp. 139–313. London: Academic Press.

Beech, F.W. and Thomas, S. (1985) Action antimicrobienne de l'anhydride sulfureux. *Bulletin de l'Office International de la Vigne et du Vin*, **58**, (652–653), 564–581.

Beech, F.W., Burroughs, L.F., Timberlake, C.F. and Whiting, G.C. (1979) Progrès récents sur l'aspect chimique et l'action anti-microbienne de l'anhydride sulfureux (SO_2). *Bulletin de l'Office International de la Vigne et du Vin*, **52**, 1001–1022.

Blouin, J. (1966) Contribution à l'étude des combinaisons de l'anhydride sulfureux dans les moûts et les vins. *Annales de Technologie Agricole*, **15**, 223–287 and 359–401.

Borst-Pauwels, G.W.F.H. (1981) Ion transport in yeast. *Biochimica et Biophysica Acta*, 650, 88–127.

Britz, T.J. and Tracey, R.P. (1990) The combination effect of pH, SO_2, ethanol and temperature on the growth of *Leuconostoc oenos*. *Journal of Applied Bacteriology*, **68**, 23–31.

Burroughs, L.F. (1975) Determining free sulfur dioxide in red wines. *American Journal of Enology and Viticulture*, **26**, 25–29.

Burroughs, L.F. and Sparks, A.H. (1964) The identification of SO_2-binding compounds in apple juice and ciders. *Journal of the Science of Food and Agriculture*, **15**, 176–185.

Burroughs, L.F. and Sparks, A.H. (1973a) Sulphite-binding power of wines and ciders. I. Equilibrium constants for the dissociation of carbonyl bisulphite compounds. *Journal of the Science of Food and Agriculture*, **24**, 187–198.

Burroughs, L.F. and Sparks, A.H. (1973b) Sulphite-binding power of wines and ciders. II. Theoretical consideration and calculation of sulphite-binding equilibria. *Journal of the Science of Food and Agriculture*, **24**, 199–206.

Carr, J.G., Davies, P.A. and Sparks, A.H. (1976) The toxicity of sulphur dioxide towards certain lactic acid bacteria from fermented apple juice. *Journal of Applied Bacteriology*, **40**, 201–212.

Casalone, E., Colella, C.M., Ricci, F. and Polsinelli, M. (1989) Isolation and characterization of *Saccharomyces cerevisiae* mutants resistant to sulphite. *Yeast*, **5**, S287–291.

Castino, M., Usseglio-Tomasset, L. and Gandini, A. (1975) Factors which affect the spontaneous initiation of the malo-lactic fermentation in wines. The possibility of transmission by inoculation and its effect on organoleptic properties. In *Lactic Acid Bacteria in Beverages and Food*, edited by J.G. Carr, C.V. Cutting and G.C. Whiting, pp. 139–148. London: Academic Press.

Chichester, D.F. and Tanner, F.W. Jr. (1972) Antimicrobial food additives. In *Handbook of Food Additives* (2nd ed.), edited by T.E. Furia, pp. 115–184. Cleveland: CRC Press.

Cole, R.D. (1967) Sulfitolysis. In *Methods in Enzymology, Volume XI Enzyme Structure*, edited by C.H.W. Hirs, pp. 206–208. New York and London: Academic Press.

Daly, S., Casalone, E., Colella, C.M. and Polsinelli, M. (1990) Meccanismo di resistenza ai solfiti in *Saccharomyces cerevisiae*. In *Proceedings of Convegno Nazionale di Microbiologia Enologica Lieviti e Vino*, (Faenza, 1990), edited by E.S.A.V.E., pp. 23–26.

Davis, C.R., Silveira, N.F.A. and Fleet, G.H. (1985) Occurrence and properties of bacteriophages of *Leuconostoc oenos* in Australian wines. *Applied and Environmental Microbiology*, **50**, 872–876.

Davis, C.R., Wibowo, D., Lee, T.H. and Fleet, G.H. (1986a) Growth and metabolism of lactic acid bacteria during fermentation and conservation of some Australian wines. *Food Technology in Australia*, **38**, 35–40.

Davis, C.R., Wibowo, D., Lee, T.H. and Fleet, G.H. (1986b) Growth and metabolism of lactic acid bacteria during and after malolactic fermentation of wines at different pH. *Applied and Environmental Microbiology*, **51**, 539–545.

Davis, C.R., Wibowo, D., Fleet, G.H. and Lee, T.H. (1988) Properties of wine lactic acid bacteria: their potential enological significance. *American Journal of Enology and Viticulture*, **39**, 137–142.

Delfini, C., Castino, M. and Ciolfi, G. (1980) L'aggiunta di tiamina ai mosti per ridurre i chetoacidi ed accrescere l'efficacia della SO_2 nei vini. *Rivista di Viticoltura e di Enologia* (Conegliano), **33**(12), 572–589.

Dorange, J.L. and Dupuy, P. (1972) Mise en évidence d'une action mutagène du sulfite de sodium sur la levure. *Compte Rendu Hebdomadaire des Séances de l'Académie des Sciences, Paris*, **274D**, 2798–2800.

Drysdale, G.S. (1988) Occurrence and growth of acetic acid bacteria in Australian wines. Master of Science thesis, p.187. Kensington; University of New South Wales, Australia.

Drysdale, G.S. and Fleet, G.H. (1985) Acetic acid bacteria in some Australian wines. *Food Technology* in *Australia*, **37**, 17–20.

Drysdale, G.S. and Fleet, G.H. (1988) Acetic acid bacteria in winemaking—a review. *American Journal of Enology and Viticulture*, **39**, 143–154.

Dubernet, M. and Ribéreau-Gayon, P. (1973) Les "polyphénoloxydases" du raisin sain et du raisin parasité par *Botrytis cinerea*. *Compte Rendu Hebdomadaire des Séances de l'Académie des Sciences, Paris*, **277D**, 975–978.

Embs, R.J. and Markakis, P. (1965) The mechanism of sulfite inhibition of browning caused by polyphenol oxidase. *Journal of Food Science*, **30**, 753–758.

Eschenbruch, R. (1974) Sulfite and sulfide formation during winemaking—a review. *American Journal of Enology and Viticulture*, **25**, 157–161.

Ewart, A.J.W., Sitters, J.H. and Brien, C.J. (1987) The use of sodium erythorbate in white grape musts. *Australian and New Zealand Wine Industry Journal*, **1**, 59–64.

Fleet, G.H. (1990) Growth of yeasts during wine fermentation *Journal of Wine Research* **1**, 211–223.

Fleet, G.H., Lafon-Lafourcade, S. and Ribéreau-Gayon, P. (1984) Evolution of yeasts and lactic acid bacteria during fermentation and storage of Bordeaux wines. *Applied and Environmental Microbiology*, **48**, 1034–1038.

Fornachon, J.C.M. (1963) Inhibition of certain lactic acid bacteria by free and bound sulphur dioxide. *Journal of the Science of Food and Agriculture*, **14**, 857–862.

Fornachon, J.C.M. (1965) Sulphur dioxide in winemaking. *Australian Wine, Brewing and Spirit Review*, **83**, 20–26.

Guerra, D., Romano, P. and Zambonelli, C. (1981) Mutagenic effects of sulfur dioxide on *Saccharomyces cerevisiae* diploid strains. *Experientia*, **37**, 691–692.

Guerzoni, M.E. and Suzzi, G. (1976) Inhibition by yeast strain of tyrosinase activity in must and wine. *Vitis*, **14**, 302–313.

Gunnison, A.F. (1981) Sulphite toxicity: a critical review of *in vitro* and *in vivo* data. *Food and Cosmetic Toxicology*, **19**, 667–682.

Haisman, D.R. (1974) The effect of SO_2 on oxidising enzyme systems in plant tissues. *Journal of the Science of Food and Agriculture*, **25**, 803–810.

Hammond, S.M. and Carr, J.G. (1976) The antimicrobial activity of SO_2 with particular reference to fermented and non-fermented fruit juices. *In Inhibition and Inactivation of Vegetative Microbes*, edited by F.A. Skinner and W.B. Hugo, pp. 89–110. London: Academic Press.

Hayatsu, H. and Miura, A. (1970) The mutagenic action of sodium bisulfite. *Biochemical and Biophysical Research Communications*, **39**, 156–160.

Heard, G.M. and Fleet, G.H. (1986) Occurrence and growth of yeast species during the fermentation of some Australian wines. *Food Technology in Australia*, **38**, 22–25.

Heard, G.M. and Fleet, G.H. (1988) The effect of sulphur dioxide on yeast growth during natural and inoculated wine fermentation. *The Australian and New Zealand Wine Industry Journal*, **3**, 57–60.

Henick-Kling, T., Lee, T.H. and Nichalas, D.J.D. (1986) Characterization of the lytic activity of bacteriophages of *Leuconostoc oenos* isolated from wine. *Journal of Applied Bacteriology*, **61**, 525–534.

Herraiz, T. (1990) Production of volatile compounds by *Saccharomyces cerevisiae* during the alcoholic fermentation of grape must in presence of absence of SO_2. *Belgian Journal of Food Chemistry and Biotechnology*, **45**, 57–62.

Herraiz, T., Martin-Alvarez, P.J., Reglero, G., Herraiz, M. and Cabezudo, M.D. (1989) Differences between wines fermented with and without sulphur dioxide using various selected yeasts. *Journal of the Science of Food and Agriculture*, **49**, 249–258.

Hinze, H. and Holzer, H. (1985) Accumulation of nitrite and sulfite in yeast cells and synergistic depletion of the intracellular ATP-content. *Zeitschrift fur Lebensmittel Untersuchung und Forschung*, **180**, 117–120.

Hinze, H. and Holzer, H. (1986) Analysis of the energy metabolism after incubation of *Saccharomyces cerevisiae* with sulfite or nitrite. *Archives of Microbiology*, **145**, 27–31.

Hood, A.V. (1983) Inhibition of growth of wine lactic-acid bacteria by acetaldehyde-bound sulphur dioxide. *The Australian Grapegrower and Winemaker*, (**232**), 34–43.

Ingram, M. (1948) The germicidal effects of free and combined SO_2. *Journal of the Society for Chemistry and Industry*, **67**, 18–21.

Ingram, M., Ottoway, F.J.H. and Coppock, J.B.M. (1956) The preservative action of acid substances in food. *Chemistry and Industry*, 1154–1163.

Joslyn, M.A. and Braverman, J.B.S. (1954) The chemistry and technology of the pretreatment and preservation of fruit and vegetable products with SO_2 and sulfites. *Advances in Food Research*, **5**, 97–160.

Joyeux, A., Lafon-Lafourcade, S. and Ribéreau-Gayon, P. (1984) Evolution of acetic acid bacteria during fermentation and storage of wine. *Applied and Environmental Microbiology*, **48**, 153–156.

Kielhöfer, E. (1963) Etat et action de l'acide sulfureux dans les vins. Règles de son emploi. *Annales de Technologie Agricole*, **12**, 77–89.

King, A.D.Jr., Ponting, J.D., Sanshuck, D.W., Jackson, R. and Mihara, K. (1981) Factors affecting death of yeast by sulfur dioxide. *Journal of Food Protection*, **44**, 92–97.

Kunkee, R.E. (1967) Control of malo-lactic fermentation induced by *Leuconostoc citrovorum*. *American Journal of Enology and Viticulture*, **18**, 71–77.

Lafon-Lafourcade, S. (1975) Factors of the malo-lactic fermentation of wines. In *Lactic Acid Bacteria in Beverages and Food*, edited by J.G. Carr, C.V. Cutting and G.C. Whiting, pp. 43–53. London: Academic Press.

Lafon-Lafourcade, S. (1983) Wine and Brandy. In *Biotechnology Vol. 5*, edited by H.J. Rehm and G. Reed, pp. 81–163. Weinheim: Verlag Chemie.

Lafon-Lafourcade, S. (1985) Rôle des microorganismes dans la formation de substances combinant le SO_2. *Bulletin de l'Office International de la Vigne et du Vin*, **58**, (652–653), 590–604.

Lafon-Lafourcade, S., Carre, E. and Ribéreau-Gayon, P. (1983) Occurrence of lactic acid bacteria during different stages of vinification and conservation of wines. *Applied and Environmental Microbiology*, **46**, 874–880.

Lafon-Lafourcade, S. and Peynaud, E. (1974) Sur l'action antibactérienne de l'anhydride sulfureux sous forme libre et sous forme combinée. *Connaissance de la Vigne et du Vin*, **8**, 187–203.

Liu, J.W.R. and Gallander, J.F. (1982) Effects of insoluble solids on the sulfur dioxide content and rate of malolactic fermentation in white table wines. *American Journal of Enology and Viticulture*, **33**, 194–197.

Liu, J.W.R. and Gallander, J.F. (1983) Effect of pH and sulfur dioxide on the rate of malolactic fermentation in red table wines. *American Journal of Enology and Viticulture*, **34**, 44–46.

Lueck, E. (1980) Sulfur dioxide. In *Antimicrobial Food Additives*, pp.115–131. Berlin: Springer-Verlag.

Macris, B.J. (1972) Transport and toxicity of sulfur dioxide in the yeast *Saccharomyces cerevisiae* var. *ellipsoideus*. *PhD Thesis*, University of Michigan, USA.

Macris, B.J. and Markakis, P. (1974) Transport and toxicity of sulphur dioxide in *Saccharomyces cerevisiae* var. *ellipsoideus*. *Journal of the Science of Food and Agriculture*, **25**, 21–29.

Maier, K., Hinze, H. and Leuschel, L. (1986) Mechanism of sulfite action on the energy metabolism of *Saccharomyces cerevisiae*. *Biochimica et Biophysica Acta*, **848**, 120–130.

Markakis, P. and Embs, R.J. (1966) Effect of sulfite and ascorbic acid on mushroom phenol oxidase. *Journal of Food Science*, **31**, 807–811.

Marques Gomes, V.J. and Da Silva Babo, M.F. (1985) Les technologies de vinification permettant de diminuer les doses de SO_2. *Bulletin de l'Office International du Vin*, **58**, (652–653) 624–636.

Martini, A. and Martini, A.V. (1990) Grape Must Fermentation: Past and Present. In *Yeast Technology*, edited by J.F.T. Spencer and D.M. Spencer, pp. 105–123. Berlin: Springer-Verlag.

Mayer, K., Vetsch, V. and Pause, G. (1975) Hemmung des biologischen Säureabbaus durch gebundene schweflige Säure. *Schweizerische Zeitschrift fur Obst u Weinbau*, **111**, 590–596.

Mayerhof, O., Ohlmeyer, P. and Mohle, W. (1938) Ueber die Koppelung zwischen Oxyreduktion und Phosphatveresterung bei der anaeroben Kohlenhydratspaltung. 2. Die Koppelung als Gleichgewichtsreaction. *Biochemische Zeitschrift*, **297**, 113–133.

Mukai, F., Hawryluk, I. and Shapiro, R. (1970) The mutagenic specificity of sodium bisulfite. *Biochemical and Biophysical Research Communications*, **39**, 983–988.

Müller, F. and Massey, V. (1971) Sulfite interaction with free and protein-bound flavin. In *Methods in Enzymology, Volume XVIII, Vitamins and Coenzymes* Part B, edited by D.B. McCormick and L.D. Wright, pp. 468–473. New York and London: Academic Press.

Ohara, Y., Nonomura, H. and Yunome, H. (1959) Dynamic aspect of yeast-flora during vinous fermentation Part. 3. Effect of sulfiting or alcohol addition on yeasts in grape musts. *Bulletin of the Research Institute of Fermentation Yamanashi University*, **6**, 7–12.

Ough, C.S. (1983) Sulfur dioxide and sulfites. In *Antimicrobials in Foods*, edited by A.L. Branen and P.M. Davidson, pp. 177–203. New York: Marcel Dekker.

Ough, C.S. (1985) Some effects of temperature and SO_2 on wine during simulated transport or storage. *American Journal of Enology and Viticulture*, **36**, 18–22.

Ough, C.S. (1989) The changing California Wine Industry. *Journal of the Science of Food and Agriculture*, **47**, 257–268.

Ough, C.S. and Crowell, E.A. (1987) Use of sulfur dioxide in winemaking. *Journal of Food Science*, **52**, 386–388, 393.

Ough, C.S., Kunkee, R.E., Vilas, M., Bordeu, E. and Huang, M.C. (1988) The interaction of sulfur dioxide, pH and dimethyl bicarbonate on the growth of *Saccharomyces cerevisiae* Montrachet and *Leuconostoc oenos*. *American Journal of Enology and Viticulture*, **39**, 279–282.

Pilkington, B.J. and Rose, A.H. (1988) Reactions of *Saccharomyces cerevisiae* and *Zygosaccharomyces bailii* to sulphite. *Journal of General Microbiology*, **134**, 2823–2830.

Porchet, B. (1931) Contribution à l'étude de l'adaptation des levures à l'acide sulfureux. *Annales d'Agriculture*, p.131.

Poux, C. (1966) Polyphénoloxydases dans le raisin. *Annales de Technologie Agricole*, **15**, 149–158.

Rankine, B.C. (1977) Developments in malolactic fermentation of Australian red Table wines. *American Journal of Enology and Viticulture*, **28**, 27–33.

Rankine, B.C. and Pocock, K.F. (1969) Influence of yeast strain on binding of sulphur dioxide in wines, and on its formation during fermentation. *Journal of the Science of Food and Agriculture*, **20**, 104–109.

Rehm, H.J. and Wittmann, H. (1962) Beitrag zur Kenntnis der antimikrobiellen Wirkung der schwefligen Säure. Uebersicht über einflussnehmende Faktoren auf die antimikrobielle Wirkung der schwefligen Säure. *Zeitschrift für Lebensmittel Untersuchung und Forschung*, **118**, 413–429.

Ribéreau-Gayon, J., Peynaud, E., Ribéreau-Gayon, P., Sudraud, P. and Amati, A. (1988) *Trattato di Scienza e Tecnica Enologica. Vol. IV*, p. 26. Brescia: AEB.

Romano, P. and Suzzi, G. (1985) Sensitivity of *Saccharomyces cerevisiae* vegetative cells and spores to antimicrobial compounds. *Journal of Applied Bacteriology*, **59**, 299–302.

Romano, P. and Suzzi, G. (1987) Le schiume persistenti durante la fermentazione dei mosti. *Vignevini*, **14**, 63–65.

Romano, P. and Suzzi, G. (1990) Anidride solforosa e lieviti del mosto. *Vini d'Italia*, **32**, 31–36.

Romano, P., Suzzi, G. and Comi, G. (1992) Sulfur dioxide production by non-H_2S producing wine yeasts. *Louvain Brewing Letters*, in press.

Romano, P. and Tini, V. (1975) Level and stability of resistance to SO_2 in wild strain of *Saccharomyces ellipsoideus*. *Rivista di Scienza e Tecnologia degli Alimenti e di Nutrizione Umana*, **5**, 245–246.

Rose, A.H. and Pilkington, B.J. (1989) Sulphite. In *Mechanisms of Action of Food Preservation Procedures*, edited by G.W. Gould, pp. 201–223. London: Elsevier Applied Science.

Scardovi, V. (1951) Studi sulla resistenza dei lieviti all' anidride solforosa. I. Prime indagini sul meccanismo della resistenza alla anidride solforosa. *Annali di Microbiologia (Milano)*, **4**, 131–172.

Scardovi, V. (1952) Studi sulla resistenza dei lieviti all' anidride solforosa. II. Resistenza e contenuto in glutatione delle discendenze ottenute per isolamento monosporiale del ceppo assuefatto. *Annuali di Microbiologia (Milano)*, **5**, 5–15.

Schanderl, H. (1959) *Die Mikrobiologie des Mostes und Weines*, 1st edn, pp. 242–252. Stuttgart: Eugen Ulmer.

Schimz, K.L. (1980) The effect of sulfite on the yeast *Saccharomyces cerevisiae. Archives of Microbiology*, **125**, 89–95.

Schimz, K.L. and Holzer, H. (1979) Rapid decrease of ATP content in intact cells of *Saccharomyces cerevisiae* after incubation with low concentrations of sulfite. *Archives of Microbiology*, **121**, 225–229.

Schopfer, J.F. and Aerny, J. (1985) Le rôle de l'anhydride sulfureux en vinification. *Bulletin de l'Office International de la Vigne et du Vin*, **58**, (652–653), 515–542.

Shapiro, R. (1977) Genetic effects of bisulfite (sulfur dioxide). *Mutation Research*, **39**, 149–176.

Singleton, V.L. and Esau, P. (1969) Phenolic substances in grapes and wine and their significance. *Advances in Food Research*, **1**, 1–282.

Somers, T.C. and Wescombe, L.G. (1987) Evolution of red wines. II. An assessment of the role of acetaldehyde. *Vitis*, **26**, 27–36.

Stratford, M., Morgan, P. and Rose, A.H. (1987) Sulphur dioxide resistance in *Saccharomyces cerevisiae* and *Saccharomycodes ludwigii. Journal of General Microbiology*, **133**, 2173–2179.

Stratford, M. and Rose, A.H. (1986) Transport of sulphur dioxide by *Saccharomyces cerevisiae. Journal of General Microbiology*, **132**, 1–6.

Stricks, W. and Kolthoff, I.M. (1951) Equilibrium constants of the reactions of sulfite with cystine and with dithioglycolic acid. *Journal of the American Chemical Society*, **73**, 4569–4579.

Stricks, W., Kolthoff, I.M. and Kapoor, R.C. (1955) Equilibrium constants of the reaction between sulfite and oxidised glutathione. *Journal of the American Chemical Society*, **77**, 2057–2061.

Sudraud, P. and Chauvet, S. (1985) Activité antilevure de l'anhydride sulfureux moleculaire. *Connaissance de la Vigne et du Vin*, **19**, 31–40.

Summers, G.A. and Drake, J.W. (1971) Bisulphite mutagenesis of the bacteriophage T4. *Genetics*, **68**, 603–607.

Suzzi, G. and Romano, P. (1978) Characteristics of high foam producing strains of *Saccharomyces. Annali di Microbiologia* (Milano), **28**, 127–133.

Suzzi, G. and Romano, P. (1982) Induced changes by SO_2 on the population of *Saccharomyces* as agents of the natural fermentation of musts. *Vini d'Italia*, **24**, 138–145.

Suzzi, G. and Romano, P. (1983) Relation between foam and sulfur dioxide during the must fermentation. *Vini d'Italia*, **25**, 157–162.

Suzzi, G., Romano, P. and Grazia, L. (1984) The biochemical stabilization of wines by pure yeast cultures of *Saccharomyces cerevisiae*. In *Proceedings of the Sixth International Symposium on Yeasts* (Montpellier,1984), pp. XII-26-P.

Suzzi, G., Romano, P. and Zambonelli, C. (1985) *Saccharomyces* strain selection in minimizing SO_2 requirement during vinification. *American Journal of Enology and Viticulture*, **36**, 199–202.

Tensches, A., Simon, R.A., Ough, C. and Marinkovich, V. (1988) Adverse reactions to sulfite adducts present in wine and sulfite sensitive asthmatics. *Journal of Allergy and Clinical Immunology*, **81**, 190–195.

Thomas, D.S. (1981) Microbial spoilage in wine. *Long Ashton Research Station Report 1981*, p.171.

Thomas, D.S. and Davenport, R.R. (1985) *Zygosaccharomyces bailii*—a profile of characteristics and spoilage activities. *Food Microbiology*, **2**, 157–169.

Thornton, R.J. (1982) Selective hybridisation of pure culture wine yeasts. II. Improvement of fermentation efficiency and inheritance of SO_2 tolerance. *European Journal of Applied Microbiology and Biotechnology*, **14**, 159–164.

Tini, V., Chiodini, O., Suzzi, G., Romano, P., Erbetti, M. and Zambonelli, C. (1987) Ricerca di lieviti stabilizzanti producenti basse quantità di anidride solforosa. *Vignevini*, **14** (9), 45–49.

Tini, V., Romano, P. and Suzzi, G. (1989) La stabilizzazione biochimica dei vini. *Vini d'Italia*, **31**, 19–24.

Tini, V., Romano, P., Suzzi, G., Arfelli, G. and Zironi, R. (1988) Caratteristiche dei lieviti come fattori di stabilizzazione dei vini. *Industrie delle Bevande*, **17**, 12–16.

Usseglio-Tomasset, L. (1985) Les technologies de vinification permettant de diminuer les doses de SO_2. *Bulletin de l'Office International de la Vigne et du Vin*, **58**, (652–653), 607–616.

Usseglio-Tomasset, L. (1986) Le azioni dell'anidride solforosa sui vini. *Vini d'Italia*, **28**, 29–34.

Usseglio-Tomasset, L., Ciolfi, G. and Pagliara, A. (1981) Estimating sulphur dioxide resistance of yeasts. 1: The delaying action on fermentation start. *Vini d'Italia*, **23**, 78–90.

Usuka, Y., Tanaka, R., Nomura, T. and Tanaka, K. (1985) Method for the determination of sulfite resistance in wine yeasts. *Journal of Fermentation Technology*, **63**, 107–114.

Valouyko, G.G., Pavlenko, N.M. and Ogorodnik, S.T. (1985) Les technologies de vinification permettant de diminuer les doses de SO_2. *Bulletin de l'Office International de la Vigne et du Vin*, **58**, (652–653), 637–644.

Vanderschmitt, D.J., Vitos, K.S., Huenekens, F.M. and Scrimgeour, K.G. (1967) Addition of bisulfite to folate and dihydrofolate. *Archives of Biochemistry and Biophysics*, **122**, 448–493.

van der Westhuizen, L.M. and Loos, M.A. (1981) Effect of pH, temperature and SO_2 concentration on the malo-lactic fermentation abilities of selected bacteria and on wine colour. *South African Journal of Enology and Viticulture*, **2**, 61–65.

Vaughn, R.H. (1955) Bacterial spoilage of wines with special reference to California conditions. *Advances in Food Research*, **6**, 67–108.

Warth, A.D. (1985) Resistance of yeast species to benzoic and sorbic acids and to sulphur dioxide. *Journal of Food Protection*, **48**, 564–569.

Watanabe, M. and Iino, S. (1984) Studies on bacteria isolate from Japanese wines (part 2): growth of the *Acetobacter* sp A-1 during the fermentation and the storage of grape must and red wine. *Yamanashiken Shokuhin Kogyo Shidojo Kenkyu Hokoku*, **16**, 13–22.

Wedzicha, B.L. (1984) *Chemistry of Sulphur Dioxide in Foods*. London: Elsevier Applied Science Publishers.

Weeks,C. (1969) Production of sulfur dioxide-binding compounds and of sulfur dioxide by two *Saccharomyces* yeasts. *American Journal of Enology and Viticulture*, **20**, 32–39.

Wibowo, D., Eschenbruch, R., Davis, C.R., Fleet, G.H. and Lee, T.H. (1985) Occurrence and growth of lactic acid bacteria in wine: a review. *American Journal of Enology and Viticulture*, **36**, 302–313.

Wildenradt, H.L. and Singleton, V.L. (1974) The production of aldehydes as a result of oxidation of polyphenolic compounds and its relation to wine aging. *American Journal of Enology and Viticulture*, **25**, 119–126.

Williams, R.R., Waterman, R.E., Keresztesy, J.C. and Buchman, E.R. (1935) Studies of crystalline vitamin B_1. III. Cleavage of vitamin with sulfite. *Journal of the American Chemical Society*, **57** , 536–537.

Wissemann, K.W. and Lee, C.Y. (1981) Characterization of polyphenol oxidase from Ravat 51 and Niagara grapes. *Journal of Food Science*, **46**, 506–508.

Yang, W.H. and Purchase, E.C.R. (1985) Adverse reactions to sulphites. *Canadian Medical Association Journal*, **133**, 865–880.

Zambonelli, C., Capella, P. and Pallotta, U. (1970) Influenza del ceppo di lievito sulla stabilità di colore dei vini bianchi. Nota I. *Industrie Agrarie*, **8**, 189–193.

Zambonelli, C., Guerzoni, M.E. and Nanni, M. (1972a) Selezione genetica nei lieviti della fermentazione vinaria. Studio dei caratteri: 3) La resistenza alla anidride solforosa. *Rivista di Viticoltura e di Enologia* (Conegliano), **10**, 170–179.

Zambonelli, C., Guerzoni, M.E., Nanni, M. and Gianstefani, G. (1972b) Selezione genetica nei lieviti della fermentazione vinaria. Studio dei caratteri: 2) Il potere stabilizzante del colore. *Rivista di Viticoltura e di Enologia* (Conegliano), **25**, 111–129.

Zambonelli, C., Romano, P. and Suzzi, G. (1989) Microorganisms of wine. In *Biotechnology Applications in Beverage Production*, edited by C. Cantarelli and G. Lanzarini, pp. 17–30. London, New York: Elsevier Applied Science.

Zambonelli, C., Soli, M.G., Romano, P. and Grazia, L. (1976) Influenza del ceppo di lievito sulla stabilità del colore dei vini bianchi. II. Analisi del ceppo 633. *Annali di Microbiologia* (Milano), **26**, 109–118.

Chapter 14

WINE SPOILAGE BY MICROORGANISMS

WOLF-RÜDIGER SPONHOLZ

Forschungsanstalt Geisenheim, Fachgebiet Mikrobiologie und Biochemie, D6222 Geisenheim, Germany

Chapter Outline

Chapter Outline (Continued)

INTRODUCTION

There are three stages in the process of winemaking where microbiological faults may occur and diminish the quality and acceptability of the final product. The first stage is represented by the raw material, the grapes, which can become infected with moulds, yeasts, acetic acid bacteria and lactic acid bacteria. The second stage is represented by fermentation. Although this stage is often seen as a sole fermentation by *Saccharomyces cerevisiae*, this conclusion is not strictly true—even for inoculated fermentations. Yeasts naturally present on the grape or originating from the winery environment will contribute to the fermentation. Growth of an undesirable species at this stage will spoil the product (Fleet *et al.*, 1984; Fleet 1990; Dittrich, 1987; Sponholz, 1989). Wine, after fermentation, is not a stable product and represents the third stage where microbiological faults may develop. The wine, if inexpertly stored and handled, can become a growth substrate for a range of undesirable yeast and bacterial species.

Uncontrolled microbial growth at any of the three stages can alter the chemical composition of the wine, detracting from its sensory properties of appearance, aroma and flavor. Such faults can be so severe that the wine becomes unpalatable. This chapter discusses the range of spoilage faults caused by yeasts and bacteria. There is yet another stage where microbiological faults affect the quality of wine. This stage concerns fungal contamination of corks used for bottling the wine, and is discussed by Lee and Simpson in Chapter 12.

FAULTS CAUSED BY YEASTS

Ester Taint

Esters are produced as a consequence of yeast metabolism of grape juice (Berry and Watson, 1987). However, the amount of ester formed varies between yeast species. The indigenous or wild yeasts *Hanseniaspora uvarum* (*Kloeckera apiculata*), *Metschnikowia pulcherrima* and *Hansenula anomala* are potent ester producers as are species of *Brettanomyces* (Sponholz and Dittrich, 1974). The ester taint is mainly linked to the presence of high concentrations of ethyl acetate, but methylbutyl acetate may also contribute to the problem (Sponholz *et al.*, 1982). Wines with concentrations of 200 mg/l or more of ethyl acetate together with less than 0.6 g/l of acetic acid display this fault. In some instances, wines with more than 700 mg/l of ester have been noted.

We have shown that the ester concentration in wine is very much related to the extent of growth of *H'spora uvarum* during the initial stages of alcoholic fermentation (Sponholz *et al.*, 1990). As described by others (Heard and Fleet, 1986; Sikovec, 1986; Fleet, 1990), this species, along with its anamorph *K. apiculata*, is one of the most prevalent yeasts in grape juice and can exhibit significant growth during the initial stages of either natural or inoculated fermentations. The development of *S. cerevisiae* and the vigor of the fermentation can be inhibited by strains of *H'spora uvarum*, and depends on the initial population of this species in the juice. Some strains of *H'spora uvarum* produce killer toxins that retard the growth of *S. cerevisiae* (Radler *et al.*, 1985; Radler and Knoll, 1988). In mixed cultures in grape juice, 10^3 cells/ml of a sensitive strain of *S. cerevisiae* are inhibited by 10^6 cells/ml of a killer strain of *H'spora uvarum*. Under these conditions, 500 mg/l of ester was formed in the wine. In a similar experiment with a non-killer strain of *H'spora uvarum*, 200 mg/l of ester was produced. When the juice was fermented with an equal mixture (10^6 cells/ml) of both species, the wine had an ester concentration of 50–90 mg/l, irrespective of whether a killer or non-killer strain of *H'spora uvarum* was used (Sponholz *et al.*, 1990). These data demonstrate the ability of *H'spora uvarum* to affect the concentration of esters in the wine, the importance of the initial population of this species in the juice in relation to *S. cerevisiae*, and the added influence of killer strains of *H'spora uvarum*.

Control of the ester taint rests with minimizing the growth of indigenous yeasts during fermentation. This can be achieved by avoiding damage to the grapes during harvesting, since damaged berries encourage massive growth of these yeasts, causing them to have high initial populations at the beginning of fermentation (Poulard *et al.*, 1985). Also, juice settling and clarification operations should be monitored to reduce the growth of these species. Inoculation of the juice with *S. cerevisiae* to establish a high proportion of this species will assist in minimizing the problem.

Film Formation (Flowers of Yeast)

A film of yeasts can grow on the surface of wines during storage in the cellar. Left unchecked, this growth can drastically change the chemical composition, aroma and flavor of the wine. These changes depend upon the yeast species which develop but, generally, the wines taste oxidized and less acid. Also, the smell of acetic acid and ethyl acetate become more pronounced. The oxidized character is frequently due to the high concentration of acetaldehyde (Caputi and Peterson, 1965; Rossi and Singleton, 1966; Würdig, 1989). In the case of sherry production, however, such changes are considered desirable (Amerine *et al.*, 1982).

Species of *Candida*, *Metschnikowia*, *Pichia* and *Hansenula* are usually responsible for this fault. They are part of the indigenous or wild flora of grape musts and, consequently, contaminate the winery environment (Benda, 1962). Because they prefer oxidative conditions, their growth is normally depressed during fermentation by the more vigorous activity of *S. cerevisiae*. Nevertheless, they can exhibit significant growth during the early stages of fermentation (Fleet *et al.*, 1984). Their growth becomes more prominent at later stages, during storage of the wines in barrels or other vessels, particularly if these containers are incompletely filled and the surface of the wine is exposed to air. Under such circumstances, these yeasts develop very quickly on the surface of the wine. Initially, they appear as small islands or clumps of cells on the surface but, eventually, they form a cream-colored film of biomass. With further growth, they form a denser film, the color of which changes to grey. The film progressively becomes thicker, wrinkles, breaks up and sinks down to the bottom of the vessel, making place for new film formation (Dittrich, 1987; Farkas, 1988).

The abundant multiplication of film yeasts is dependent upon the presence of oxygen. Consequently, their development can be prevented by ensuring that wine vessels are completely filled, thereby eliminating pockets of air, and by flushing vessels with either nitrogen or carbon dioxide. Low temperatures and higher alcohol contents work to restrict the growth of these yeasts. Thus, film formation is minimal, if not absent, on wines of 10–12% ethanol if they are consistently stored at low (8–12°C) temperatures. However, at higher temperatures film formation will occur in wines, even with ethanol concentrations as high as 14% (Dittrich, 1987).

If film yeasts do begin to develop, they must be removed by filtration. Filling the vessels with further wine to wash away the surface yeasts by overflow is not adequate. Also, addition of higher concentrations of sulfur dioxide (SO_2) is not effective as some of these species (e.g. *Pichia membranaefaciens*, *Candida krusei*) are very tolerant of this antimicrobial and can resist 2–3 mg/l of molecular SO_2 (Warth, 1985; Romano and Suzzi, Chapter 13).

Metabolically, film yeasts use ethanol as their growth substrate. In a three month period, *Pichia farinosa* was found to decrease the alcohol content of a wine from 70.8 g/l to 20.5 g/l. Under the same conditions, a *Candida* spp. only used 1 g/l of the alcohol (van Zyl, 1962). The main products formed by these yeasts are acetic acid, acetaldehyde and esters of acetic acid (e.g. ethyl acetate, 3-methylbutyl acetate) (Sponholz and Dittrich, 1974). Interestingly, the extreme ability of *Hansenula anomala* to form esters was recommended for flavor enhancement in wines (Sajenko, 1946 cited by Kudrjawzew, 1960).

In addition to ethanol, wine constituents such as acids and glycerol can be utilized by film yeasts. The amount of sugar-free extract in a wine was decreased from 23 g/l to 12–13 g/l by film species of *Candida* and *Pichia* over a three month period, and to 15 g/l by a species of *Hansenula* over a nine month period (van Zyl, 1958). Titratable acid was decreased from 6.8 g/l to 3.4 g/l in three months and to 2.8 g/l in six months. Malate was metabolized but not tartrate (Zaake,

1958). Acid utilization by *H. anomala* can be so complete that the wine pH may increase to 10.0 (Schrewsbury, 1930).

In summary, film forming yeasts cause severe taints and spoilage of wines. They are controlled by careful winemaking practices that include good hygiene in the winery and avoiding contact of wine with air. If a container or barrel becomes infected with these yeasts, it must be thoroughly cleaned, sanitized and steamed.

Spoilage by *Zygosaccharomyces bailii*

In comparatively recent times, *Zygosaccharomyces bailii* (formerly *Saccharomyces bailii*) has emerged as one of the main yeasts that cause spoilage of wine by re-fermentation during storage in tanks and after packaging (Peynaud and Domerq, 1959; Schanderl, 1962; Eschenbruch, 1973; Rankine and Pilone, 1973; Minarik, 1983; Thomas and Davenport, 1985). This yeast is a common contaminant of grapes and wine cellars (Peynaud and Domerq, 1959), as well as a major spoilage contaminant of grape juice concentrates used for sweetening wines (Kroemer and Krumbholz, 1931; Krumbholz, 1931; Sacchetti, 1932; Somavilla *et al.*, 1975; Minarik, 1983). *Zygosaccharomyces bailii* has a unique combination of properties that contribute to its role as a wine spoilage organism. These properties are: tolerance of high concentrations of ethanol (>15%); growth at low pH (<2.0); strong resistance to high concentrations of the preservatives, sorbic acid (>800 mg/l), benzoic acid (>1000 mg/l), diethyl pyrocarbamate (>500 mg/l) and sulfur dioxide (>3 mg/l molecular); and ability to grow at high sugar concentrations (>70% v/v) as found in juice concentrates (Thomas and Davenport, 1985).

Growth of *Zygosacch. bailii* in juice concentrates or sweet wines can lead to preferential use of fructose (Peynaud and Domerq, 1955; Peynaud, 1957; Rambalow, 1970; Minarik *et al.*, 1979). In line with other osmotolerant yeasts, such as *Zygosaccharomyces rouxii*, it is likely that *Zygosacch. bailii* will produce significant concentrations of glycerol, erythritol, arabitol and mannitol when growing in environments of high sugar content, such as grape juice concentrates (Spencer and Sallans, 1956; Sponholz *et al.*, 1986; Sponholz, 1988). Growth of *Zygosacch. bailii* in wine leads to turbidity and sediment (Würdig and Kullman, 1971; Minarik, 1983), increases in the concentrations of succinic and acetic acids (Shimazu and Watanabe, 1981), strong reduction in acidity due to its metabolism of L-malic acid (Goto *et al.*, 1978; Kuczynski and Radler, 1982), and altered ester concentrations (Soles *et al.*, 1982).

Spoilage by *Brettanomyces* species

Species of *Brettanomyces* are well known for their ability to spoil cellar and bottled wine through their production of haze, turbidity and volatile acidity. These yeasts were originally isolated from Belgian lambic and other beers in the 1920s (see review of Verachtert *et al.*, 1990). Later, they were isolated as

Mycotorula intermedia from German grape musts (Krumbholz and Tauschanoff, 1933).

Subsequent to the early studies, *Brettanomyces* species, mainly *B. intermedius*, have been reported as spoilage organisms in wines from France (Barret *et al.*, 1955; Peynaud and Domerq, 1956), South Africa where they were responsible for 50% of all hazy wines (van der Walt and van Kerken, 1958, 1959, 1961; van Zyl, 1962), Italy (Verona and Florenzano, 1947; Florenzano, 1951), Australia (Heresztyn, 1986a,b) and New Zealand (Wright and Parle, 1974). They caused severe problems in German sparkling wine production during the early 1950s (Schanderl, 1959). They were found in bottle and tank fermented sparkling wines, being quite resistant to removal by riddling because of their small cell size. Fining and filtration of the base wine was necessary to control this problem. They were responsible for explosive carbon dioxide liberation and gushing in bottle fermented sparkling wines.

The metabolic characteristics of *Brettanomyces* species have been reviewed by Verachtert *et al.* (1990). They ferment sugars relatively slowly compared with other yeasts but, nevertheless, produce 9–12% ethanol. However, they produce high concentrations of volatile acidity (up to 7.2 g/l) (Schanderl, 1959) and esters (Wenzel, 1966). Recently, they were implicated in producing substances responsible for the mousiness taint in wines (see later section; Tucknott, 1976, 1977; Tucknott *et al.*, 1981; Heresztyn, 1986a) and volatile phenolic substances such as 4-ethyl guaiacol and 4-ethyl phenol that give spicy, phenolic taints (Heresztyn, 1986b). Fortunately, these yeasts are relatively sensitive to SO_2, being inhibited by 100 mg/l of total SO_2 or 30–50 mg/l of free SO_2 (Würdig, 1989).

FAULTS CAUSED BY ACETIC ACID BACTERIA

The vinegary taint in wines is generally associated with the presence of high concentrations of acetic acid. According to Margalith (1981), the presence of acetic acid in wine becomes objectionable at concentrations exceeding 1.2–1.3 g/l. Because acetic acid is, by far, the main volatile acid in wines (see below), the fault is frequently described as volatile acidity. Some countries have laws which limit the maximum concentration of volatile acidity or acetic acid in wines. In the European Economic Community, this is 1.08 g/l for partially fermented juices or white and rose wines, and 1.20 g/l for red wines. Higher limits such as 1.30 g/l may be applied to botrytized and dessert wines (Dittrich, 1987).

The volatile acidity in tainted wines is mainly acetic acid (Sponholz *et al.*, 1981, 1982). The contribution of higher fatty acids, such as propionic and hexanoic, is almost negligible because their concentrations rarely exceed a few mg/ml. However, in wines where the concentration of volatile acid is very low (e.g. less than 0.2 g/l), acetic acid may contribute only about 50% of the volatile acid (Junge and Spadinger, 1979). In these cases, formic acid naturally present in the grape juice can form most of the volatile acid (Sponholz *et al.*, 1981). Formic acid is also produced by mould infection of grapes (Millies, 1980).

Sometimes, wines with low volatile acidity are rejected as having a vinegary taint. This suggests that acetic acid is not always the only constituent responsible for the vinegary character. In these instances, the wines have also contained small concentrations of ethyl acetate. Bandion and Valenta (1977b) suggested that wines should be rejected as faulty when they contain more than 0.8 g/l of acetic acid and more than 90 mg/l of ethyl acetate. Wines should be considered faulty also when their ester content exceeds 200 mg/l. In the latter case, they have a marked gluey or ester taint which is distinct from volatile acidity (Sponholz *et al.*, 1982).

Acetic acid bacteria are primarily responsible for the vinegary taint in wine. The species concerned are *Gluconobacter oxydans*, *Acetobacter pasteurianus* and *A. aceti*. The contribution of acetic acid bacteria to wine spoilage has been reviewed by Vaughn (1955), Lafon-Lafourcade and Joyeux (1981) and Drysdale and Fleet (1988). They can influence wine quality at three stages: first, by contamination of the grape; second, by possibly growing during alcoholic fermentation; and third, by growth in wine during storage in the cellar.

Grapes

Sound, unspoiled grapes harbor low populations (less than 10^2–10^3 cells/g) of acetic bacteria, principally *G. oxydans* (Joyeux *et al.*, 1984 a, b; Grossman and Becker, 1984). Juice from such grapes contains only 16–23 mg/l of acetic acid (Sponholz and Dittrich, 1979). If the grapes become injured through the activities of birds, insects, hail, fungi and over-maturity, infection with yeasts and acetic acid bacteria can occur. Essentially, yeasts metabolize grape sugars yielding ethanol which is then oxidized by acetic acid bacteria to acetic acid. Spoiled grapes may harbor populations of acetic acid bacteria (mainly, *A. aceti* and *A. pasteurianus*) as high as 10^6 cells/g (Joyeux *et al.*, 1984a, b; Grossman and Becker, 1984). Such grapes have a very intense smell of vinegar. The content of acetic acid in juices from these grapes can be as high as 3.9 g/l and is always accompanied by high amounts of ethanol (3.9 g/l), glycerol (2.0 g/l) and low amounts of ethyl acetate (2.9–53.9 mg/l) (Sponholz and Dittrich, 1979; Drysdale and Fleet, 1989a). Both the ethanol and glycerol are products of yeast metabolism. Table 14.1 illustrates the concentration of some key constituents in grape juice as a function of infection by acetic acid bacteria. In addition to the constituents just mentioned, grapes infected with acetic acid bacteria can contain very high concentrations of dihydroxyacetone and gluconic acid (Joyeux *et al.*, 1984b; Sponholz and Dittrich, 1985; Eschenbruch and Dittrich, 1986). Dihydroxyacetone is produced by acetic acid bacteria by oxidizing the glycerol formed by yeasts. Oxidation of glycerol is a well-established property of *G. oxydans* and *A. aceti* (Drysdale and Fleet, 1988). The dihydroxyacetone carries through into the wine, affecting its sensory quality and sulfur dioxide binding properties (Eschenbruch, 1983; Eschenbruch and Dittrich, 1986). Gluconic acid arises from metabolism of glucose by acetic acid bacteria (Drysdale and Fleet,

TABLE 14.1 The concentration (mg/l) of some key constituents in grape juice as a function of infection by acetic acid bacteria (Sponholz and Dittrich, 1979).

Infected grapes (%)	*Ethanol*	*Acetic acid*	*Glycerol*	*Dihydroxy-acetone*	*Gluconic acid*
0	103	23	0	4	41
5	142	126	500	23	156
10	259	200	1100	41	252
15	257	210	900	56	410
20	494	463	2000	71	520
30	401	490	1100	107	1099
40	790	610	1500	143	859
50	577	1040	1900	184	1581
75	419	1690	1400	259	2586

1988). It can be further oxidized to produce 5-keto- and 2-ketogluconic acid in the juice (Sponholz and Dittrich, 1984, 1985).

Thus, grape juice composition can be significantly altered by infection of the berries with acetic acid bacteria. Such changes not only affect the sensory quality of the wine but, also, they can adversely affect the growth of yeasts during alcoholic fermentation (Drysdale and Fleet, 1989a) and possibly the growth of malolactic bacteria (Joyeux *et al.*, 1984a, b).

Wine

Acetic acid bacteria have an aerobic metabolism and depend upon oxygen for their growth (Drysdale and Fleet, 1988). It is generally considered that they do not grow during alcoholic fermentation because of the carbon dioxide and anaerobic conditions produced by yeasts (Joyeux *et al.*, 1984b). Nevertheless, the growth curve and chemical data of Drysdale and Fleet (1989a) show a certain potential for their development during this stage. Similar conclusions were reported by Grossman and Becker (1984) and Watanabe and Iino (1984). Furthermore, these bacteria exhibit exceptionally good survival properties in wines during storage. *Acetobacter pasteurianus* and *A. aceti* are frequently isolated from wines stored in barrels or other vessels in the winery (Joyeux *et al.*, 1984a; Drysdale and Fleet, 1985). Provided the wine is kept free of oxygen, their populations remain relatively static. However, general exposure of the wine to air or momentary exposure during pumping and transfer operations can quickly stimulate their growth, and populations as high as 10^8 cells/ml may be formed (Joyeux *et al.*, 1984a; Drysdale and Fleet, 1989b). The main reaction of this growth is the metabolism of ethanol into acetic acid. First, ethanol is oxidized to acetaldehyde by alcohol dehydrogenase, and then acetaldehyde is oxidized to acetic acid by acetaldehyde dehydrogenase. The aldehyde dehydrogenase has low specificity so that higher aldehydes, with chain lengths up to C_8 can be oxidized to their corresponding acids (King and Cheldelin, 1956; Nakajima, 1960). In fully aerated wines, Drysdale and Fleet (1989b) showed that as much as 50–60% of the ethanol could be oxidized by these bacteria with the production of 1.5–3.75 g/l of acetic acid. The extent of the reaction was decreased in less aerated wines and depended upon the strain of *A. aceti* or *A. pasteurianus*. Growth of acetic acid bacteria in wines is also accompanied by the production of significant concentrations of acetaldehyde and ethyl acetate (Dupuy and Maugenet, 1963; Yoshioka, 1983; Drysdale and Fleet, 1989b) and the metabolism of glycerol into dihydroxyacetone (Sponholz *et al.*, 1982; Drysdale and Fleet, 1989b) (see Table 14.2).

FAULTS CAUSED BY LACTIC ACID BACTERIA

Although lactic acid bacteria play a key role in the malolactic fermentation of wines (Wibowo *et al.*, 1985; Kunkee, 1991), they also cause many faults in wine quality. They are usually present on grapes and in must at populations not exceeding 10^3 cells/ml and show little or no tendency to grow during properly conducted alcoholic fermentation (Lafon-Lafourcade *et al.*, 1983; Fleet *et al.*, 1984). Nevertheless, they survive at low, almost non-detectable, populations in

TABLE 14.2 Concentrations of some key constituents in wines spoiled by acetic acid bacteria and lactic acid bacteria (Sponholz *et al.*, 1982).

Wine	*Acetic acid g/l*	*Ethyl acetate mg/l*	*Dihydroxy-acetone mg/l*	*L-Malic acid g/l*	*L-Lactic acid g/l*	*D-Lactic acid g/l*
1	6.1	2750	19	4.7	0.1	0.1
2	2.7	879	21	0.9	1.1	0.2
3	0.9	312	5	1.2	1.3	0.2
4	1.4	967	16	0.1	2.5	1.3
5	1.4	161	4	0	2.5	1.2
6	0.8	-	-	0	4.2	3.6

1 Acetic acid bacteria after fermentation.
2 Acetic acid bacteria and malolactic fermentation.
3 Acetic acid bacteria before fermentation and malolactic fermentation.
4 Lactic acid bacteria and acetic acid bacteria.
5 Lactic acid bacteria (heterofermentative species).
6 Lactic acid bacteria (homofermentative species).

the wine waiting for a suitable opportunity to grow. These bacteria represent various species of *Lactobacillus*, *Pediococcus* and *Leuconostoc*.

Acidification

Lactic acid bacteria can significantly increase the acidity of wines through their production of acetic acid (volatile acidity) and lactic acid, but these faults are often underestimated. D-Lactic acid is of concern in these cases, rather than L-lactic acid which is the form produced during malolactic fermentation. The combined production of mannitol and acetic acid by these bacteria as well as D-lactic acid are used as indicators of faultiness in wines (Bandion and Valenta, 1977a). Wines with high amounts of such substances show a vinegary defect and often have high concentrations of histamine (Jakob, 1978). Also, other amines can be formed by lactic acid bacteria through decarboxylation of amino acids, and these include phenethylamine, tyramine, putrescine, cadaverine and spermidine depending on the spoilage strain (Woidich *et al.*, 1980; Desser *et al.*, 1981; Mayer and Pause, 1982; Lüthy and Schlatter, 1983; Zee, 1983; Lafon-Lafourcade, 1983).

Acetic acid and D-lactic acid are produced by the heterofermentative species of *Lactobacillus* and *Leuconostoc* by the fermentation of sugars as shown in Figure 14.1. D-Lactic acid, without acetic acid, is produced by the homofermentative species of *Lactobacillus* and *Pediococcus* through the glycolytic metabolism of sugars. However, these homofermentative species ferment pentose sugars as shown in Figure 14.1 to give both acetic and lactic

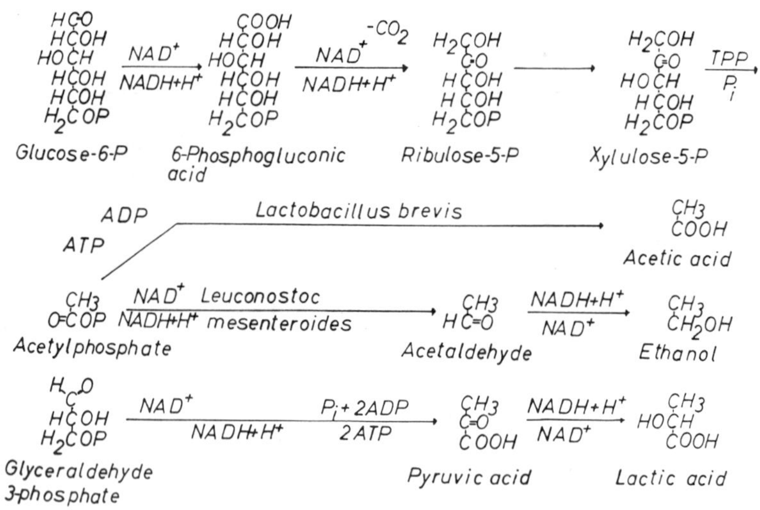

FIGURE 14.1 The formation of lactic acid, acetic acid and ethanol by the heterofermenting lactic acid bacteria *Lactobacillus brevis* and *Leuconostoc mesenterioides*. (The homofermenting *Lactobacillus plantarum* ferments pentoses to acetic acid following the same pathway.) (Schlegel, 1976).

acids. Wines containing residual sugar have the potential for acidification by lactic acid bacteria, especially during storage on lees when other nutrients are available for the stimulation of bacterial growth (Wibowo *et al.*, 1985). Davis *et al.* (1986a,b) have correlated the utilization of residual hexose and pentose sugars in wines with the kinetics of growth of lactic acid bacteria, and noted the complexity of this relationship. Acidification by lactic acid bacteria can also occur during the alcoholic fermentation when the must contains substantial fermentable sugar. This possibility arises if there is a delay in the onset of yeast growth or premature arrest of the alcoholic fermentation, and it increases if the must is prepared from damaged grapes containing higher populations of lactic acid bacteria (Lafon-Lafourcade, 1983).

When volatile acidity is caused by lactic acid bacteria, a positive correlation exists between the presence of acetic and D-lactic acid (Sponholz *et al.*, 1982). The concentrations of ethyl acetate in these wines is generally low (30–120 mg/l) and not abnormally high. Sometimes volatile acidity is caused by both lactic acid bacteria and acetic acid bacteria. Such wines show high concentrations of acetic acid, D-lactic acid and ethyl acetate and, sometimes, dihydroxyacetone (Sponholz *et al.*, 1982). Occasionally, high acid wines are found that do not contain abnormal concentrations of acetic acid or ethyl acetate (Dittrich, 1987; Sponholz, 1989). Such wines are spoiled but they are not vinegary. Usually, they contain high concentrations of D-lactic acid in addition to the L-lactic acid that arises from the malolactic fermentation. The D-lactic is produced by fermentation of sugars by homofermentative species of *Lactobacillus* and *Pediococcus*. Table 14.2 presents a profile of some key constituents in wines spoiled by acetic acid bacteria, lactic acid bacteria and combinations of these bacteria.

Mannitol Taint

Although not well-studied, mannitol can be produced in significant concentrations by some strains of heterofermentative lactic acid bacteria (e.g. *Lactobacillus brevis*) by the enzymatic reduction of fructose (Wood, 1961; Martinez *et al.*, 1963). The pathways can be represented as either:

$$\text{Fructose-6-phosphate} \xrightarrow[\text{NAD}^+]{\text{NADH} + \text{H}^+} \text{mannitol-1-phosphate}$$

or

$$\text{Fructose} + \text{NADH} + \text{H}^+ \longrightarrow \text{mannitol} + \text{NAD}^+$$

Mannitol-1-phosphate is further hydrolyzed to mannitol and inorganic phosphate.

The mannitol taint is a form of spoilage with high complexity, because it is also accompanied by high concentrations of acetic acid, D-lactic acid, n-propanol, 2-butanol and often sliminess and diacetyl taint. A spoiled wine, for example, would contain (approximately): mannitol, 9 g/l; acetic acid, 3 g/l; and D-lactic acid, 3 g/l. (Sponholz, 1988, 1989), and have a vinegary-estery taste. The problem occurs mostly in wines, such as dessert wines and fruit and berry wines as well as distillation mashes, that have a high pH and contain significant amounts of residual sugar, especially fructose. It is effectively controlled by acid addition to lower pH.

Ropiness

Wines with ropiness have a slimy, viscous, oily character (Dittrich, 1987). Frequently, ropiness is the beginning of more severe faults such as high volatile acidity and mannitol taint, and mostly occurs in wines lacking in acidity. In low acid wines, ropiness may be evident at the end of alcoholic fermentation, especially if malic acid degradation has also occurred. During cellar storage of wines, ropiness starts at the bottom of the tank and progressively moves to the top. Just above the lees, wine acidity is decreased by yeast autolysis and nutrients become available for the growth of lactic acid bacteria.

The development of viscosity in fermenting wines shows a direct correlation with the appearance of ropiness and the growth of lactic acid bacteria (Lüthy, 1953). There are very few descriptions of the species which cause ropiness in wine or the environmental factors that lead to rope production. However, some strains of lactic acid bacteria produce extracellular dextrins and other polysaccharides during growth on sugars (Whiting, 1975). Species connected with the occurrence of ropiness in wine include *Streptococcus mucilaginosus* var. *vini* (Hochstrasser, 1955; see Table 14.3), *Pediococcus cerevisiae* (Mayer, 1974) and species of *Leuconostoc*.

Diacetyl Production

Wines infected with lactic acid bacteria often have an undesirable buttery or whey like aroma and flavor. This problem is caused by the presence of unacceptably high concentrations of diacetyl (butandione 2,3), a diketone with a very low taste threshold. According to Dittrich and Kerner (1964), diacetyl presence is very noticeable at 1 mg/l, and wines with such amounts should be considered faulty. However, some authors believe that diacetyl concentrations up to 4 mg/l may contribute, in a positive way, to the complexity of wine flavor, but it becomes objectionable at higher concentrations (Rankine *et al.*, 1969). Normally, wines contain 0.2–0.3 mg/l of diacetyl, this amount being produced by yeast activity. There is now good evidence to show that diacetyl concentration increases during malolactic fermentation by lactic acid bacteria (Fornachon and Lloyd, 1965; Pilone *et al.*, 1966; Rankine *et al.*, 1969; Mayer, 1974; Zeeman *et al.*, 1982; Postel and Meier, 1983). Such increases are relatively small and unobjectionable if *Leuconostoc oenos* is responsible for this fermentation, but they may be more significant and cause spoilage if the fermentation is conducted by *Pediococcus* species (Mayer, 1974). Growth of *Pediococcus* or

TABLE 14.3 Some properties of a coccal lactic acid bacteria causing ropiness in wine (Hochstrasser, 1955).

Strain	*Streptococcus mucilaginosus* var. *vini* Lüthy Strain 216
Morphology	Cocci surrounded by capsule
Temperature	Optimum, 25°C; growth at 10°C and 30°C
Natural medium	Ropiness without haze
Synthetic medium	Ropiness with haze
Slime formation	From fructose, glucose, sucrose, raffinose
Components of slime	Glucose, mannose
Acid formation	From fructose, glucose, sucrose, raffinose
Butanediol formation	From fructose, mannose
Malolactic fermentation	Good
Ethanol tolerance	Good
Ethanol tolerance	Forms ropiness in wines with more than 11% ethanol

Lactobacillus species in wines after malolactic fermentation could also give spoilage concentrations of diacetyl.

The metabolic pathway for the production of diacetyl by lactic acid bacteria is not totally clear. It can arise from the metabolism of hexose and pentose sugars, whereby the intermediate, pyruvate, is converted to acetolactate which is then converted to diacetyl (Figure 14.2) (du Plessis and Van Zyl, 1963; Bärwald and Kesselschläger, 1969; El-Gendy *et al.*, 1983; Kandler, 1983). In addition, it may

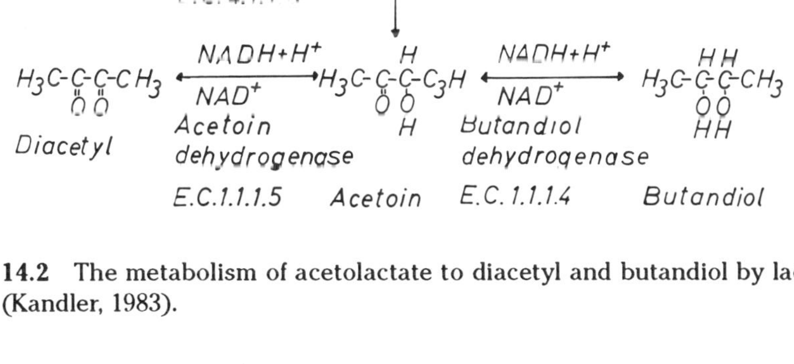

FIGURE 14.2 The metabolism of acetolactate to diacetyl and butandiol by lactic acid bacteria (Kandler, 1983).

come from the metabolism of citric acid by citrate lyase which cleaves this acid into acetic acid and oxaloacetic acid. Oxaloacetate is then metabolized to diacetyl through pyruvate, (Charpentie, 1951; Kandler, 1983; Shimazu *et al.*, 1985). This latter pathway may be the mechanism by which *Leuc. oenos* produces diacetyl during the malolactic fermentation, since its growth is accompanied by decreases in the concentration of citric acid and the production of acetic acid (Davis *et al.*, 1986a,b). Under the same conditions, however, *Pediococcus* species do not metabolize citrate (Davis *et al.*, 1986a,b).

If the diacetyl taint is not severe, it may be cured by post fermentation of the wine with *S. cerevisiae*, after the addition of fresh must (Dittrich and Kerner, 1964). Under these conditions, diacetyl is reduced to acetoin, which has a much higher flavor threshold, and thereafter to 2, 3 butanediol.

Mousiness

Wines with this fault have a most disgusting smell reminiscent of mouse urine or acetamide. Individuals vary greatly in their ability to perceive this fault, but the unpleasant odor becomes distinct when a sample of the mousey wine is rubbed between the fingers, setting free the offending volatiles, and then smelling. Mousiness is not a common problem, but is most likely to occur in low acid wines that have received insufficient SO_2. The microbial genesis of mousiness was not clear for a long time. Schanderl (1948) described mousiness as a type of oxidation. Bacterial involvement seemed evident, but there was a dependence on the presence of yeast metabolic products because the fault did not occur in musts. It is now known that mousiness is caused by both lactic acid bacteria and species of the yeast, *Brettanomyces* (Tucknott, 1976, 1977).

The mousey character is related to the microbial production of two isomers of 2-acetyltetrahydropyridine (Figure 14.3) (Craig and Heresztyn, 1984; Strauss and Heresztyn, 1984; Heresztyn, 1986a) rather than 2-ethylpiperidin as once thought (Tucknott, 1977). The 2-acetyltetrahydropyridines have a very low flavor threshold (1.6 ppt) in water (Theranishi *et al.*, 1975) and are well known as flavor components in bread and beer (Hunter *et al.*, 1969; Büchi and Wüst, 1971). Also, heating glucose in the presence of the amino acid, proline, can cause the formation of these substances (Tressl *et al.*, 1981). Bacteria implicated in the production of acetyltetrahydropyridines in wines include *Lactobacillus hilgardii*,

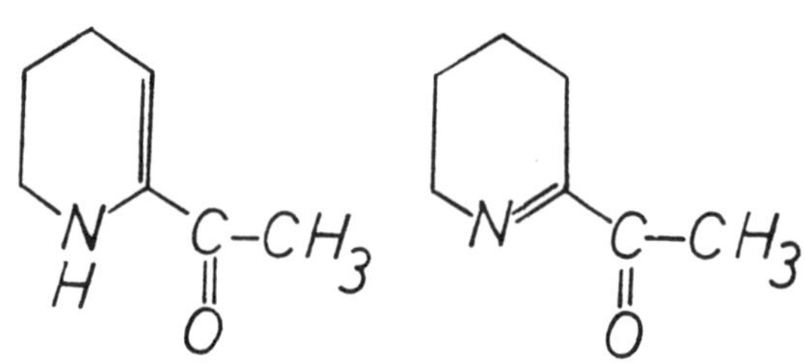

FIGURE 14.3 Isomers of 2-acetyltetrahydropyridine.

L. brevis and *L. cellobiosus* and perhaps other heterofermentative lactic acid bacteria (Tucknott and Davis, 1976; Heresztyn, 1986a). Bacterial formation of these substances or their propionyl analogues depends on the presence of ethanol or propanol (Heresztyn, 1986a), thereby explaining their occurrence in wines rather than musts.

Acrolein Formation and Bitterness

The acrolein taint and associated bitterness of wines is caused by bacterial degradation of glycerol. Pasteur (1873) first mentioned that the bitterness of red wines was connected with the presence of rod-shaped bacteria and the loss of glycerol. The connection between bitterness and the presence of acrolein was later indicated by Voisenet (1910, 1911). Acrolein, itself, is not bitter but it reacts with the phenolic groups of anthocyanins to produce the bitter sensation (Rentschler and Tanner, 1951). For this reason, high phenolic red wines, more than white wines, are associated with this fault.

There is little information on the content of acrolein in wines and on the degradation of wine glycerol by lactic acid bacteria. However, it seems that acrolein concentrations as little as 10 ppm are sufficient to cause a taint (Margalith, 1981). Apple and pear wines, where the growth of lactic acid bacteria is not totally repressed, can undergo complete degradation of glycerol (Whiting, 1960). Wines that have undergone malolactic fermentation show ten times or more acrolein content (2.8 mg/l) as well as decreased concentrations of glycerol (Dittrich *et al.*, 1980; Sponholz, 1982). Davis *et al.* (1986a,b) did not observe any change in the glycerol concentration in an Australian red wine after malolactic fermentation with *Leuc. oenos*. However, in another red wine they correlated the utilization of glycerol with the growth of *Pediococcus parvulus* and *Lactobacillus cellobiosus*. The ability to degrade glycerol is not widespread among lactic acid bacteria (Schütz and Radler, 1984; Davis *et al.*, 1988). Of several strains tested, Davis *et al.* (1988) noted only 1% of *Leuc. oenos*, 12% of *Ped. parvulus* and 31% of *Lactobacillus* species to ferment glycerol.

The formation of acrolein from glycerol has been demonstrated in *Leuc. mesenteroides* (Hirano *et al.*, 1962). The key enzyme of glycerol degradation is a dehydratase which converts glycerol to 3-hydroxypropionaldehyde (Smiley and Sobolow, 1962; Sliniger *et al.*, 1983). Acrolein is formed by the chemical (spontaneous) dehydration of this aldehyde either by heating or long term storage in acid solution (Figure 14.4). If glucose is present in the environment, 3-hydroxypropionaldehyde is reduced to 1,3-propandiol by propandiol-1,3 dehydrogenase as found in *Lactobacillus brevis* (Schütz and Radler, 1984). If glucose is lacking, 3-hydroxypropionic acid (hydracrylic acid) is formed (Sobolow and Smiley, 1960). Spore-forming clostridia have also been implicated in acrolein production (Wilharm and Holz, 1951).

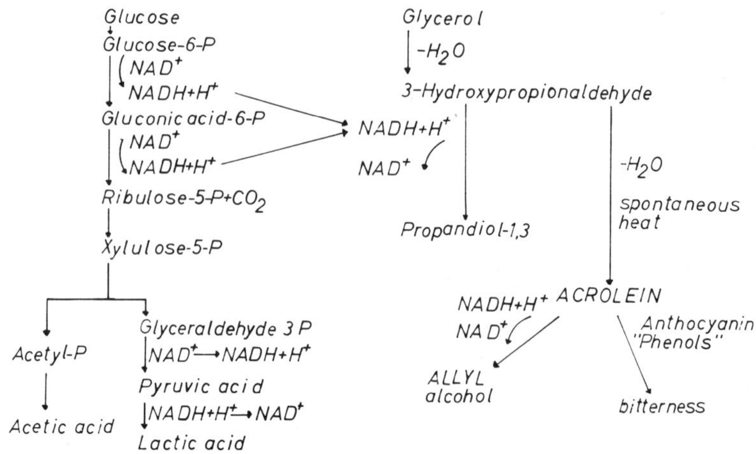

FIGURE 14.4 Glucose and glycerol fermentation by the heterofermenting *Lactobacillus brevis* to propandiol 1,3 (Schütz and Radler, 1984) and the formation of acrolein, allyl alcohol and bitter substances as products of glycerol fermentation.

Tartaric Acid Degradation

Tartaric acid is most stable to microbiological degradation and, normally, it is not metabolized in wines. However, a few species of lactic acid bacteria cause its degradation. Generally, at the time of its degradation the wine is already spoiled by many other serious faults, and other acids such as pyruvic, malic and citric have been metabolized. Thus, the changes in a red wine spoiled by this mechanism were: tartaric acid diminished from 2.7 g/l to 0.04 g/l; glycerol decreased from 6.1 g/l to 2.6 g/l; lactic acid increased from 0.92 g/l to 3.7 g/l and acetic acid increased from 0.55 g/l to 3.25 g/l (Dittrich, 1987). The ability to degrade tartaric acid is restricted to only a few species of *Lactobacillus* (Krumperman and Vaughn, 1966; Radler and Yannissis, 1972). Of 78 strains of pediococci, lactobacilli and leuconostocs investigated by Radler and Yannissis (1972), only four strains of *Lact. plantarum* and one of *Lact. brevis* metabolized this acid. The main enzyme for tartrate metabolism was tartrate dehydratase which was inducible in both organisms. The homofermenting strain, *Lact. plantarum*, dehydrates tartaric acid to oxaloacetate which is split to pyruvic acid and carbon dioxide. Half of the pyruvic acid is reduced to lactic acid, and the other half is split to acetic acid and carbon dioxide (Figure 14.5). The heterofermenting strain, *Lact. brevis*, shows a more complicated metabolism of oxaloacetic acid, part of which is transformed to succinic acid and part is converted to acetic acid and carbon dioxide (Figure 14.6).

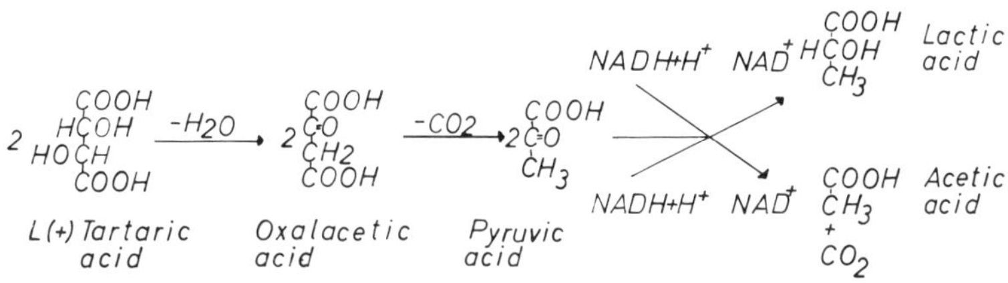

FIGURE 14.5 The degradation of tartaric acid by the homofermenting *Lactobacillus plantarum* (Radler and Yannissis, 1972).

The Geranium Off-Odor

Wines sometimes develop a most unpleasant odor, similar to that produced by crushing leaves of the geranium plant (*Pelargonium* spp.) This fault is related to the metabolism of sorbic acid by certain strains of lactic acid bacteria (Burkhardt, 1973). Sorbic acid may be added to wines as an antimicrobial to control yeast growth but, at the concentrations used (200 mg/l), it has no activity against lactic acid bacteria (Edinger and Splittstoesser, 1986). The substance responsible for the geranium odor has been identified as 2-ethoxyhexa-3,5-diene (Würdig *et al.*, 1974; Crowell and Guymon, 1975). It has a low sensory threshold (0.1 μg/l) and its formation depends upon hydrogenation of sorbic acid to sorbinol by lactic acid bacteria (Figure 14.7). According to Radler (1976), all strains of *Leuc. oenos* and some heterofermenting strains of *Lactobacillus* can reduce sorbic acid to sorbinol. However, Edinger and Splittstoesser (1986) noted this reduction with strains of *Leuc. oenos* only and not with strains of *Pediococcus* or *Lactobacillus*. In an acid environment, such as wine, sorbinol isomerizes to the secondary alcohol, 3,5-hexadiene-2-ol, which reacts with ethanol to form 2-ethoxyhexa-3,5-diene. Another substance, 1-

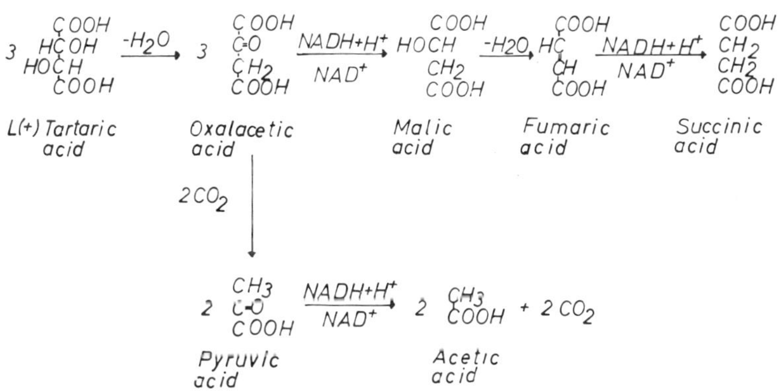

FIGURE 14.6 The degradation of tartaric acid by the heterofermenting *Lactobacillus brevis* (Radler and Yannissis, 1972).

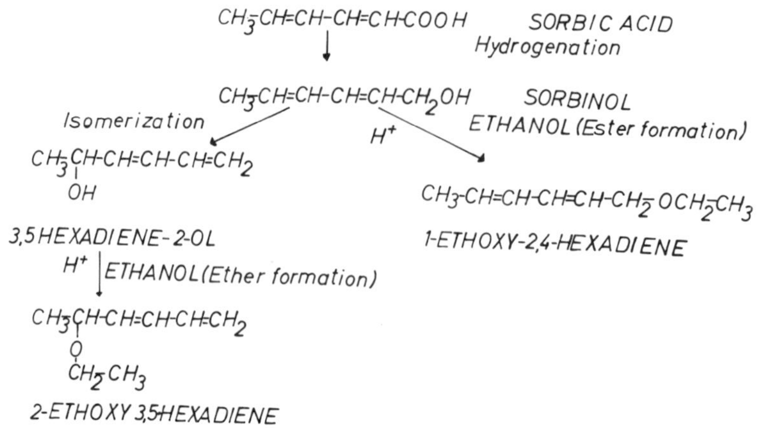

FIGURE 14.7 The degradation of sorbic acid by lactic acid bacteria to 2-ethoxy-3,5-hexadiene.

ethoxyhexa-2,4-diene can be formed also (Figure 14.7). A small amount of ethanol is necessary for these reactions to occur; consequently, the fault is not evident in grape juices. Nevertheless, reduction of sorbic acid to sorbinol can occur in juices if infected with lactic acid bacteria. If such juice is added to wine for sweetening, chemical rearrangement of the sorbinol to 2-ethoxyhexa-3,5-diene can occur, without any further microbial infection. If wines or juices must be stored in the presence of sorbic acid, they should contain sufficient free SO_2 to prevent the growth of lactic acid bacteria and development of the geranium taint.

Sorbic acid can also undergo auto-oxidation forming acrolein, crotonaldehyde, malondialdehyde and formic acid (Marx and Sabalitschka, 1965). Moreover, it may be transformed during fermentation with *S. cerevisiae* to give an undesirable smell of ranciness (Sponholz, 1982). Thus, base wines containing sorbic acid should not be used for sparkling wine production.

Miscellaneous Products

Concentrations of the esters, ethyl lactate and diethylsuccinate in wines can be increased by the growth of lactic acid bacteria (Pilone *et al.*, 1966; Meunier and Bott, 1979). Also, the production of higher alcohols is enhanced. Notably, *Lact. brevis* metabolizes 2,3-butandiol to butanol-2 (Hieke and Vollbrecht, 1974; Bertrand and Suzuta, 1976). Some lactobacilli produce propanol-1 from glycerol (Schütz and Radler, 1984).

FAULTS CAUSED BY *CLOSTRIDIUM* AND *BACILLUS* SPECIES

Butyric Acid Taint

Infection of wines by species of *Clostridium* rarely occurs. These organisms give an undesirable taint of ranciness due to their production of small amounts of n-butyric acid. Clostridia are obligate anaerobic, spore-forming bacteria that develop in wines with high pH (>pH 4.0). Therefore, this fault only occurs in heavily deacidified juices after wrong malolactic fermentation, or in juices like apple juice that have a low acid content. Clostridial species ferment juice sugars giving n-butyric acid, acetic acid, carbon dioxide and hydrogen and, depending on strain, various concentrations of butanol-1 and acetone (*Cl. butyricum*) or butanol-1 and propanol-2 (*Cl. butylicum*). *Clostridium butyricum* produces mostly butyric and acetic acids without alcohols (Rehm, 1967). Some indicative constituents in wines that have undergone this type of spoilage are: Vinho verde, 31 mg/l of n-butyric acid and butanol-1; Elbling, 36 mg/l of n-butyric acid, 7 mg/l of propionic acid and 8 mg/l of isobutyric acid; Pinot gris, 94 mg/l of n-butyric acid and 21 mg/l of propionic acid; and Müller-Thurgau, 907 mg/l of acetic acid, 186 mg/l of propionic acid and 146 mg/l of n-butyric acid (Dittrich, 1987; Sponholz, 1989).

Bacillus species

The potential for *Bacillus* species to spoil wines is not widely appreciated or reported. Gini and Vaughn (1962) isolated several *Bacillus* species, including *B. coagulans*, *B. circulans* and *B. subtilis* from spoiled California sweet wines and demonstrated their ability to grow to 10^6–10^7 cells/ml when reinoculated into wine. Significant increases in volatile and total acidity accompanied this growth. *Bacillus megaterium* has been implicated in the spoilage of bottled brandy (Murrell and Rankine, 1979). Lee *et al.* (1984) isolated *B. coagulans* and *B. badius* from wine corks and demonstrated growth of these species to 10^5–10^6 cells/ml when inoculated into red and white wines. Clearly, some strains of *Bacillus* grow under the low pH, high ethanol and anaerobic conditions of wines. Further research is required to explore the incidence of *Bacillus* species in wines and to expose the biochemical mechanisms by which they may cause spoilage.

REFERENCES

Amerine, M.A., Berg, H.W. and Cruess, V.W. (1982) *The technology of winemaking*. pp. 391–431. 4th edition, Westport: AVI Publising Company.

Bambalow, G. (1970) Veränderungen des Verhältnisses Glucose. Fructose in Traubenkonzentraten durch Hefen der Gattung Zygosaccharomyces. *Weinberg Keller*, **17**, 87–90.

Bandion, F. and Valenta, M. (1977a) Zur Beurteilung des D(–)- und L(+)- Milchsäure-gehaltes in Wein. *Mitteilungen Klosterneuburg*, **27**, 4–10.

Bandion, F. and Valenta, M. (1977b) Zum Nachweis des Essigstichs bei Wein und Obstwein in Österreich. *Mitteilungen Klosterneuburg*, **27**, 18–22.
Barret, A., Bidan, P. and Andre, L. (1955) Sur quelques accidents de vinification des a des levures a voile. *Comptes Rendus Academie des Agricole, France*, **41**, 426–431.
Bärwald, G. and Kesselschläger, J. (1969) Bildung von Diacetyl und Pentandion (2,3) in *Pediococcus cerevisiae, Naturwissenschaft*, **56**, 285.
Benda, I. (1962) Ökologische Untersuchungen über die Hefeflora im fränkischen Weinbaugebiet. *Bayerisches Landwirtschaftliches Jahrbuch*, **39**, 595–614.
Berry, D.R. and Watson, D.C. (1987) Production of organoleptic compounds. In *Yeast Biotechnology*, edited by D.R. Berry, I. Russell and G.G. Stewart, pp. 345–366. London: Allen and Unwin.
Bertrand, A. and Suzuta, K. (1976) Formation du butanol-2 par les bacteries lactiques isolees du vin. *Connaissance de la Vigne et du Vin*, **10**, 409–426.
Büchi, G. and Wüst, H. (1971) Synthesis of 2-acetyl-1,4,5,6 -tetrahydropyridine, a constituent of bread aroma. *Journal of Organic Chemistry*, **36**, 609–610.
Burkhardt, R. (1973) Über den sich bei Behandlung von Mosten (Süßreserven) und Weinen mit Kaliumsorbat unter Umständen bildenden Fehlgeruch und Fehlgeschmack. *Mitteilungsblatt GDCH-Fachgruppe Lebensmittel Chemie*, **27**, 259–260.
Caputi, A. and Peterson, R.G. (1965) The browning problems in wines. *American Journal of Enology and Viticulture*, **16**, 9–13.
Charpentie, Y. (1951) Sur la fermentation de l'acide citrique par les bacteries malolactique. *Bulletin de la Societe de Chimique Biologique*, **33**, 1369–1379.
Craig, J.T. and Heresztyn, T. (1984) 2-Ethyl-3,4,5,6- tetrahydropyridine—an assessment of its possible contribution to the mousy off-flavour of wines. *American Journal of Enology and Viticulture*, **35**, 46–48.
Crowell, E.A. and Guymon, I.F. (1975) Wine constituents arising from sorbic acid addition and identification of 2-ethoxyhexa-3,5- diene as a source of geranium-like off-odor. *American Journal of Enology and Viticulture*, **26**, 97–102.
Davis, C.R., Wibowo, D., Fleet, G.H. and Lee, T.H. (1988) Properties of wine lactic acid bacteria: their potential enological significance. *American Journal of Enology and Viticulture*, **39**, 137–142.
Davis, C.R., Wibowo, D., Lee, T.H. and Fleet, G.H. (1986a) Growth and metabolism of lactic acid bacteria during and after malolactic fermentation of wines at different pH. *Applied and Environmental Microbiology*, **51**, 539–545.
Davis, C.R., Wibowo, D., Lee, T.H. and Fleet, G.H. (1986b) Growth and metabolism of lactic acid bacteria during fermentation and conservation of some Australian wines. *Food Technology in Australia*, **38**, 35–40.
Desser, H., Bandion, F. and Kläring, W. (1981) Zur Kenntnis einiger biogener Amine des Traubenmostes und Traubenweines. *Mitteilungen Klosterneuburg*, **31**, 231–237.
Dittrich, H.H. (1987) *Mikrobiologie des Weines*. Stuttgart: Ulmer.
Dittrich, H.H. and Kerner, E. (1964) Diacetyl als Weinfehler, Ursachen und Beseitigung des Milchsäuretones. *Wein-Wissenschaft*, **19**, 528–535.
Dittrich, H.H., Sponholz, W.R., Wünsch, B. and Wipfler, M. (1980) Zur Veränderung des Weines durch den bakteriellen Säureabbau. *Wein-Wissenschaft*, **35**, 421–429.
Drysdale, G.S. and Fleet, G.H. (1985) Acetic acid bacteria in some Australian wines. *Food Technology in Australia*, **37**, 17–20.
Drysdale, G.S. and Fleet, G.H. (1988) Acetic acid bacteria in winemaking: a review. *American Journal of Enology and Viticulture*, **39**, 143–154.
Drysdale, G.S. and Fleet, G.H. (1989a) The effect of acetic acid bacteria upon the growth and metabolism of yeasts during the fermentation of grape juice. *Journal of Applied Bacteriology*, **67**, 471–481.
Drysdale, G.S. and Fleet, G.H. (1989b) The growth and survival of acetic acid bacteria in wines at different concentrations of oxygen. *American Journal of Enology and Viticulture*, **40**, 99–105.

du Plessis, L. de W. and Van Zyl, J.A. (1963) The microbiology of South African winemaking. Part VI, Fermentation of D-glucose, D-fructose, D-xylose and L-arabinose by lactic acid bacteria from dry wines. *South African Journal of Agricultural Science*, **6**, 673–687.

Dupuy, P. and Maugenet, J. (1963) Conditions du development des bacteries acetiques et mesures preventiques dans le conversation des vins, *Symposium International d'Oenologie, Bordeaux.*

Edinger, W.D. and Splittstoesser, D.F. (1986) Production by lactic acid bacteria of sorbic alcohol, the percursor of the geranium odor compound. *American Journal of Enology and Viticulture*, **37**, 34–38.

El-Gendy, S.M., Abdel-Galil, H., Shahin, Y. and Hegezi, F.Z. (1983) Acetoin and diacetyl production by homo- and hetero- fermentative lactic acid bacteria. *Journal of Food Protection*, **46**, 420–425.

Eschenbruch, R. (1973) Contamination of wine by *Saccharomyces acidifaciens*, a yeast highly resistant to Baycovin, sulfur dioxide and sorbic acid. *Wynboer (Stellenbosch)*, **496**, 23–24.

Eschenbruch, B. (1983) Untersuchungen über den Stoffwechsel von Essigsäurebakterien in synthetischen Medien und in Traubenmosten. *Dissertation Landwirtschaftliche Fakultät der Universität Bonn.*

Eschenbruch, B. and Dittrich, H.H. (1986) Stoffbildungen von Essigbakterien in Bezug auf ihre Bedeutung für die Weinqualität. *Zentralbatt fuer Mikrobiologie*, **141**, 279–289.

Farkas, J. (1988) *Technology and Biochemistry of Wine Volume II*, pp. 658–681, New York: Gordon and Breach.

Fleet, G.H. (1990) Growth of yeasts during wine fermentation. *Journal of Wine Research*, **1**, 211–223.

Fleet, G.H., Lafon-Lafourcade, S. and Ribéreau-Gayon, P. (1984) Evolution of yeasts and lactic acid bacteria during fermentation and storage of Bordeaux wines. *Applied and Environmental Microbiology*, **48**, 1034–1038.

Florenzano, G. (1951) Diffusione et significato enologico dei lieviti *Brettanomyces*. *Atti. Acad. Ital. Vite Vino*, **3**, 236–249.

Fornachon, J. and Lloyd, B. (1965) Bacterial production of diacetyl and acetoin in wine. *Journal of the Science of Food and Agriculture*, **16**, 710–716.

Gini, B. and Vaughn, R.H. (1962) Characteristics of some bacteria associated with the spoilage of California dessert wines. *American Journal of Enology and Viticulture*, **13**, 20–31.

Goto, S., Yamazaki, M., Yamakawa, Y. and Yokotsuka, I. (1978) Decomposition of malic acid in grape must. *Hakkokogaku*, **56**, 133–135.

Grossman, M.K. and Becker, R. (1984) Investigations on bacterial inhibition of wine fermentation. *Kellerwirtschaft*, **10**, 272–275.

Heard, G.M. and Fleet, G.H. (1986) Occurence and growth of yeast species during the fermentation of some Australian wines. *Food Technology in Australia*, **38**, 22–25.

Heresztyn, T. (1986a) Formation of substituted tetrahydropyridines by species of *Brettanomyces* and *Lactobacillus* isolated from mousy wines. *American Journal of Enology and Viticulture*, **37**, 127–132.

Heresztyn, T. (1986b) Metabolism of volatile phenolic compounds from hydroxycinnamic acids by *Brettanomyces* yeast. *Archives of Microbiology*, **146**, 96–98.

Hieke, E. and Vollbrecht, D. (1974) Zur Bildung von Butanol(2) durch Lactobacillen und Hefen. *Archives of Microbiology*, **99**, 345–351.

Hirano, S., Imamura, T., Uchiama, K. and Yu, T.L. (1962) Formation of acrolein by *Clostridium perfringens*. *Acta Medical University Kagoshima* 4 239–264 *Chemical Abstracts* **59**, 3096c.

Hochstrasser, R. (1955) Über einige Bedingungen beim Lindwerden der Weine. *Dissertation ETA Zürich.*

Hunter, I.R.M., Walden, M.K., Scherer, J.R. and Lundin, R.E. (1969) Preparation and properties of 1,4,5,6- tetrahydropyridine, a craker-odor constituent of bread aroma. *Cereal Chemistry*, **46**, 189–195.

Jakob, L. (1978) Ergänzende Feststellungen zum Gehalt an Histamin in Wein. *Weinwirtschaft*, **114**, 126–127.

Joyeux, A., Lafon-Lafourcade, S. and Ribéreau-Gayon, P. (1984a) Evolution of acetic acid bacteria during fermentation and storage of wine. *Applied and Environmental Microbiology*, **48**, 153–156.

Joyeux, A., Lafon-Lafourcade, S. and Ribéreau-Gayon, P. (1984b) Metabolism of acetic acid bacteria in grape must: consequences on alcoholic and malolactic fermentation. *Sciences des Aliments*, **4**, 247–255.

Junge, Ch. and Spadinger, H. (1979) Die flüchtigen Säuren des Weines I. *Deutsche Lebensmittel Rundschau*, **75**, 12–15.

Kandler, O. (1983) Carbohydrate metabolism in lactic acid bacteria. *Antonie van Leeuwenhoek*, **49**, 209–224.

King, T.E. and Cheldelin, V.H. (1956) Oxidation of acetaldehyde by *Acetobacter suboxydans*. *Journal of Biological Chemistry*, **220**, 117–121.

Kroemer, K. and Krumbholz, G. (1931) Untersuchungen über osmophile Sproβpilze. I. Beiträge zur Kenntnis der Gärungsvorgänge und der Gärungserreger der Trockenbeerenauslesen. *Archiv fuer Mikrobiologie*, **2**, 352–410.

Krumbholz, G. (1931) Untersuchungen über osmophile Srpoβpilze. II. Beiträge zur Kenntnis der Gattung *Zygosaccharomyces*. *Archiv fuer Mikrobiologie*, **2**, 411–492.

Krumbholz, G. and Tauschanoff, W. (1933) *Mycotorula intermedia n. sp*, ein Beitrag zur Kenntnis der Gärungserreger im Wein. *Zentralblatt Bakteriologie Parasitenkunde, Abt., II* **88**, 366–373.

Krumperman, P.H. and Vaughn, R.H. (1966) Some lactobacilli associated with decomposition of tartaric acid in wine. *American Journal of Enology and Viticulture*, **17**, 185–190.

Kuczynski, J.T. and Radler, F. (1982) The anaerobic metabolism of malate of *Saccharomyces bailii* and the partial purification and characterisation of malic enzyme. *Archives of Microbiology*, **131**, 266–270.

Kunkee, R.E. (1991) Some roles of malic acid in the malolactic fermentation in winemaking. *FEMS Microbiology Reviews*, **88**, 55–72.

Lafon-Lafourcade, S. (1983) Wine and brandy. In *Biotechnology*, Volume 5, *Food and Feed Production with Microorganisms*, edited by G. Reed, pp. 81–164. Weinheim: Verlag Chemie.

Lafon-Lafourcade, S., Carre, E. and Ribéreau-Gayon, P. (1983) Occurrence of lactic acid bacteria during different stages of vinification and conservation of wines. *Applied and Environmental Microbiology*, **46**, 874–880.

Lafon-Lafourcade, S. and Joyeux, A. (1981) Les bactéries acetiques du vin. *Bulletin de l'Office International du Vin*, **608**, 803–829.

Lee, T.H., Simpson, R.F., van De Peer, J., Fleet, G.H., Davis, C.R., Daly, N. and Yap, A. (1984) Microbiology of wine corks. In *Advances in Viticulture and Oenology for Economic Gain*, edited by T.H. Lee and T.C. Somers, pp. 435–450. Adelaide: The Australian Wine Research Institute.

Lüthy, H. (1953) Beitrag zur Kenntnis fadenziehender (linder) Weine und Obstweine. *Mitteilungen Lebensmittel Untersuchung und Hygiene*, **44**, 30–72.

Lüthy, J. and Schlatter, Ch. (1983) Biogene Amine in Lebensmitteln: Zur Wirkung von Histamin, Tyramin und Phenylethylamin auf den Menschen. *Zeitschrift fuer Lebensmitte Untersuchung und Forschung*, **177**, 439–443.

Margalith, P.Z. (1981) *Flavour Microbiology*, pp.173–224. Illinois, Charles G. Thomas.

Martinez G., Barker, H.A. and Horecker, B.L. (1963) Specific mannitol dehydrogenase from *Lactobacillus brevis*. *Journal of Biological Chemistry*, **238**, 1598–1603.

Marx, H. and Sabalitschka, Th. (1965) Zur Unbeständigkeit der Sorbinsäure. *Seifen Öle Fette Wachse*, **91**, 720–722.

Mayer, K. (1974) Nachteilige Auswirkungen auf die Weinqualität bei ungünstig verlaufenem biologischen Säureabbau. *Schweiz Zeitschrift Obst- u. Weinbau*, **110**, 385–391.

Mayer, K. and Pause, G. (1982) Histamingehalte in 1981 er ostschweizer Weinen. *Schweiz Zeitschrift Obst- u. Weinbau*, **118**, 723–727.

Meunier, J.M. and Bott, E.W. (1979) Das Verhalten verschiedener Aromastoffe in Burgunderweinen im Verlauf des biologischen Säureabbaus. *Chemie Mikrobiologie Technologie Lebensmittel*, **6**, 92–95.

Millies, K.D. (1980) Ameisensäure als Stoffwechselprodukt von Mikroorganimen in Fruchtsäften. *Flüssiges Obst*, **39**, 91–99.

Minarik, E. (1983) Levures de contamination des vin embotteilles. *Bulletin de l'Office International du Vin*, **56**, 414–419.

Minarik, E., Svec, V. and Jungova, O. (1979) Zur Beurteilung glucophiler und fructophiler Weinhefen verschiedener Gattungen. *Mitteilungen Klosterneuburg*, **29**, 1–7.

Murrell, W.G. and Rankine, B.C. (1979) Isolation and identification of a sporing *Bacillus* from bottled brandy. *American Journal of Enology and Viticulture*, **30**, 247–249.

Nakajama, T. (1960) Studies on acetic acid bacteria II. Intracellular distribution of enzymes related to acetic acid fermentation and some properties of a highly purified TPN-dependent aldehyde dehydrogenase. *Journal of Biochemistry (Tokyo)*, **48**, 812–830.

Pasteur, L. (1873) *Etudes sur les vins, Paris.*

Peynaud, E. (1957) Charakterisierung der verschiedenen Spezies von Weinhefen. *Mitteilungen Klosterneuburg*, **7**, 1–15.

Peynaud, E. and Domercq, S. (1955) Sur les especes de levures fermentant selectivement le fructose. *Annales de l'Institute Pasteur*, **89**, 346–351.

Peynaud, E. and Domercq, S. (1956) Sur les *Brettanomyces* isoles de raisins et vins. *Archiv fuer Mikrobiologie*, **24**, 346–351.

Peynaud, E. and Domercq, S. (1959) A review of microbiologal problems in wine-making in France. *American Journal of Enology and Viticulture*, **10**, 69–77.

Pilone, G., Kunkee, R.E. and Webb, A.D. (1966) Chemical characterization of wines fermented with various malolactic bacteria. *Applied Microbiology*, **14**, 608–615.

Poulard, A., Reberteau, R. and Rousset, Y. (1985) La microflore des phases prefermentaires. *Revue Française d'Oenologie*, **99**, 14–19.

Postel, W. and Meier, B. (1983) Verhalten von 2-Acetolactat, 2-Acetohydroxybutyrat, Diacetyl, 2,3-Pentandion und Acetoin während des bakteriellen Äpfelsäureabbaus in Wein. *Zeitschrift Lebensmittel Untersuchung und Forschung*, **176**, 356–359.

Radler, F. (1976) Degradation de l'acide sorbique par les bacteries. *Bulletin de l'Office International du Vin*, **49**, 629–635.

Radler, F. and Knoll, C. (1988) Die Bildung von Killertoxin und die Beeinflussung der Gärung durch Apikulatushefen. *Vitis*, **27**, 111–132.

Radler, F., Pfeiffer, P. and Dennert, M. (1985) Killer toxins in new isolates of the yeast *Hanseniaspora uvarum* and *Pichia kluyveri*. *FEMS Microbiology Letters*,**29**, 269–272.

Radler, F. and Yannissis, C. (1972) Weinsäureabbau bei Milchsäurebakterien. *Archives of Microbiology*, **82**, 219–238.

Rankine, B.C., Fornachon, J. and Bridson, D.A. (1969) Diacetyl in Australian dry red wines and its significance in wine quality. *Vitis*, **8**, 129–134.

Rankine, B.C. and Pilone, D. (1973) *Saccharomyces bailii*, a resistant yeast causing serious spoilage of table wine. *American Journal of Enology and Viticulture*, **24**, 55–58.

Rehm, H.J. (1967) *Industrielle Mikrobiologie*. Berlin: Springer.

Rentschler, H. and Tanner, H. (1951) Das Bitterwerden der Rotweine. *Mitteilungen Lebensmittel Untersuchung Hygiene*, **42**, 463–475.

Rossi, J.A. and Singleton, V.L. (1966) Contribution of grape phenolics to browning and oxygen absorption in wines. *American Journal of Enology and Viticulture*, **17**, 231–233.
Sacchetti, M. (1932) Interno ad alcuni *Zygosaccharomyces*. *Rend. R. Accad. Naz. Lincei*, **15**, 762–765.
Sajenko, (1946) cit.: Kudrjawzew, W.I. (1960) *Die Systematik der Hefen*, Akademie: Berlin-Ost.
Schanderl, H. (1948) Die Ursachen der Mäuselkrankheit der Weine. *Schweizer Zeitschrift Obst- u. Weinbau*, **57**, 301–305.
Schanderl, H. (1959) *Die Mikrobiologie des Mostes und Weines*, pp. 126–131. Ulmer: Stuttgart.
Schanderl, H. (1962) *Saccharomyces acidifaciens*, eine weitere gefährliche und neuerdings in deutschen Weinen auftretende "Schwefelhefe". *Weinblatt*, **57**, 769–771.
Schlegel, H.G. (1976) *Allgemeine Mikrobiologie*, Stuttgart: Thieme.
Schrewsbury, F.J.D. (1930) The Genus Willia. *Journal of Pathology and Bacteriology*, **33**, 393–416.
Schütz, H. and Radler, F. (1984) Anaerobic reduction of glycerol to propandiol-1,3 by *Lactobacillus buchneri*. *Systematic and Applied Microbiology*, **203**, 1–10.
Shimazu, Y. and Watanabe, M. (1981) Effects of yeast strains and environmental conditions on forming of organic acids in must during fermentation. *Journal of Fermentation Technology*, **59**, 27–32.
Shimazu, Y., Vehara, M. and Watanabe, M. (1985) Transformation of citric acid to acetic acid, acetoin and diacetyl by winemaking lactic acid bacteria. *Agricultural and Biological Chemistry*, **49**, 2147–2153.
Sikovec, S. (1986) Abzutrennende Hefearten aus Most und Wein. *Symposium 1986: Innovationen in der Kellerwirtschaft- neue Erkenntnisse in der Weinklärung* Eigenverlag pp. 15–28.
Slininger, P.J., Bothast, R.J. and Smiley, K.L. (1983) Production of 3-hydroxypropionaldehyde from glycerol. *Applied and Environmental Microbiology*, **46**, 62–67.
Smiley, K.L. and Sobolow, M. (1962) A cobamide requiring glycerol dehydrase from an acrolein-forming *Lactobacillus*. *Archives of Biochemistry and Biophysics*, **97**, 538–543.
Sobolow, M. and Smiley, K.L. (1960) Metabolism of glycerol by an acrolein-forming *Lactobacillus*. *Journal of Bacteriology*, **79**, 261–266.
Soles, R.M., Ough, C.S. and Kunkee, R.E. (1982) Ester concentration difference in wine fermented by various species and strains of yeast. *American Journal of Enology and Viticulture*, **33**, 94–98.
Somavilla, J.F., Tienda, P., Arroyo, V. and Inigo, B. (1975) Levanduras osmofilas en mistelas y frutas secos. *Rev. Agrochim. Technol. Aliment. (Valencia)*, **15**, 573–580.
Spencer, J.F.T. and Sallans, H.R. (1956) Production of polyhydric alcohols by osmophilic yeasts. *Canadian Journal of Microbiology*, **2**, 72–79.
Sponholz, W.R. (1982) Analyse und Vorkommen von Aldehyden in Wein. *Zeitschrift fuer Lebensmittel-Untersuchung und Forschung*, **174**, 458–462.
Sponholz, W.R. (1988) Alcohols derived from sugars and other sources and fullbodiedness of wines. In *Modern Methods of Plant Analysis New Series Vol.6 Wine Analysis*, edited by H.F. Linskens and J.F. Jackson, pp. 147–172.
Sponholz, W.R. (1989) Fehlerhafte und unerwünschte Erscheinungen in Wein. In *Chemie des Weines* edited by G. Würdig and R.Woller, pp. 385–411 Stuttgart: Ulmer.
Sponholz, W.R. and Dittrich, H.H. (1974) Die Bildung von SO_2 bindenden Gärungsnebenprodukten, höheren Alkoholen und Estern bei einigen Reinzuchthefestämmen und bei einigen für die Weinbereitung wichtigen "wilden " Hefen. *Wein-Wissenschaft*, **29**, 301–314.
Sponholz, W.R. and Dittrich, H.H. (1979) Analytische Vergleiche von Mosten und Weinen aus gesunden und essigstichigen Traubenbeeren. *Wein-Wissenschaft*, **34**, 279–292.

Sponholz, W.R. and Dittrich, H.H. (1984) Über das Vorkommen von Galacturon- und Glucuronsäure sowie von 2- und 5-Ketogluconsäure in Weinen, Sherries, Obst- und Beerenweinen. *Vitis*, **23**, 214–224.

Sponholz, W.R. and Dittrich, H.H. (1985) Über die Herkunft von Gluconsäure, 2- und 5-Ketogluconsäure, sowie Glucuron- und Galacturonsäure in Mosten und Weinen. *Vitis*, **24**, 51–60.

Sponholz, W.R., Dittrich, H.H. and Barth, A. (1982) Über die Zusammensetzung essigstichiger Weine. *Deutsche Lebensmittel Rundschau*, **78**, 423–428.

Sponholz, W.R., Dittrich, H.H. and Han, K. (1990) Die Beeinflussung der Gärung und der Essigsäureethylester-bildung durch *Hanseniaspora uvarum*. *Viticultural and Enological Sciences*, **45**, 65–72.

Sponholz, W.R., Dittrich, H.H., Haas, F. and Wünsch, B. (1981) Die Bildung von flüchtigen Säuren durch Saccharomyces-Hefen während der Vergärung von Traubenmost. *Zeitschrift fuer Lebensmittel-Untersuchung und Forschung*, **78**, 423–428.

Sponholz, W.R., Lacher, M. and Dittrich, H.H. (1986) Die Bildung von Alditolen durch die Hefen des Weines. *Chemie Mikrobiologie Technologie Lebensmittel*, **9**, 19–24.

Strauss, C.R. and Heresztyn, T. (1984) 2-Acetyltetrahydro-pyridines- a cause of the mousy taint in wine. *Chemistry and Industry*, 109–110.

Theranishi, R., Buttery, R.G. and Guadagni, D.G. (1975) Odor thresholds and molecular structure. In *Geruch- und Geschmacksstoffe*, edited by F. Drawert, pp. 177–186. Nürnberg: Carl.

Thomas, S. and Davenport, R.R. (1985) *Zygosaccharomyces bailii*—a profile of characteristics and spoilage activities. *Food Microbiology*, **2**, 157–169.

Tressl, R., Grünewald, K.G. and Helak, B. (1981) Formation of flavour components from proline and hydroxyproline with glucose and maltose and their importance to food flavour. In *Flavour 1981*, edited by Schreier, pp. 397–416. Berlin: de Gruyter.

Tucknott, O.G. (1976) Taints in fermented juice products: Mousy taint in cider. *Annual Report of the Long Ashton Research Station*, University of Bristol, pp. 129–130.

Tucknott, O.G. (1977) The mousy taint in fermented beverages. PhD Thesis, University of Bristol.

Tucknott, O.G. and Davis, P.A. (1976) Production of mousy taint by microorganisms. *Annual Report of the Long Ashton Research Station* 140, University of Bristol 1977, 140 pp.

Tucknott, O.G., Davis, P.A. and Rosser, P.R. (1981) Taints in fermented beverages. *Annual Report of the Long Ashton Research Station*, University of Bristol, 165 pp.

van der Walt, J.P. and van Kerken, A. E. (1958) The wine yeasts of the Cape Part I. A taxonomical survey of the yeasts causing turbidity in South African table wines. *Antonie van Leeuwenhoek*, **24**, 239–252.

van der Walt, J.P. and van Kerken, A. E. (1959) The wine yeasts of the Cape Part II. The occurence of *Brettanomyces intermedius* and *B. schanderlii* in South African table wines. *Antonie van Leeuwenhoek*, **25**, 145–151.

van der Walt, J.P. and van Kerken, A. E. (1961) The wine yeasts of the Cape Part V. Studies on the occurence of *Brettanomyces intermedius* and *B. schanderlii*. *Antonie van Leeuwenhoek*, **27**, 81–90.

van Zyl, J.A. (1962) Turbidity in South African dry wines caused by the development of *Brettanomyces* yeasts. *Viticulture and Enology Research Institute, Stellenbosch, Scientific Bulletin*, **381**, Viticulture Series 1 1–41.

Vaughn, R.H. (1955) Bacterial spoilage of wines with special reference to California conditions. *Advances in Food Research*, **6**, 67–108.

Verachtert, H., Kumara, S. and Dawond, E. (1990) Yeasts in mixed cultures with special emphasis on lambic beer brewing. In *Yeast Biotechnology and Biocatalysis*, edited by H. Verachtert and R. De Mot, pp. 429–449. New York: Marcel Dekker.

Verona, O. and Florenzano, G. (1947) Si di una nuova specie di *Brettanomyces* isolata da uva. *Rend. Acad. Naz. Lincei, Ser. VIII*, **3**, 383–387.

Voisenet, E. (1910) Formation d'acrolein dans la maladie de l'amertume des vins. *Comptes Rendus Academie des Sciences, Paris*, **150**, 1614–1616.
Voisenet, E. (1911) Sur un ferment de l'amertume des vins agent de deshydration de la glycerine. *Comptes Rendus Academie des Sciences, Paris*, **153**, 363–365.
Wenzel, K.W. (1966) Gas chromatographic analysis of fermentation by-products of wine. Master of Science Thesis (Chemistry) University of Stellenbosch, South Africa.
Warth, A.D. (1985) Resistance of yeast species to benzoic and sorbic acids and to sulfur dioxide. *Journal of Food Protection*, **48**, 564–569.
Watanabe, M. and Iino, S. (1984) Studies on bacteria isolates from Japanese wines (part 2): growth of *Acetobacter* species A-1 during the fermentation and storage of grape must and red wine. *Yamanshiken Shokuhin Kogyashidojo Kenkyu Hokoku*, **16**, 13–22.
Whiting, G.C. (1960) "Non-fermentable" substances in ciders and perries. *Annual Report of the Long Ashton Research Station*, University of Bristol, pp. 135–139.
Whiting, G.C. (1975) Some biochemical and flavour aspects of lactic acid bacteria in ciders and other alcoholic beverages In *Lactic Acid Bacteria in Beverages and Food*. Edited by J.G. Carr, C.V. Cutting and G.C. Whiting, pp. 69–85. London, New York: Academic Press.
Wibowo, D., Eschenbruch, R., Davis, C.R., Fleet, G.H. and Lee, T.H. (1985) Occurrence and growth of lactic acid bacteria in wines: a review. *American Journal of Enology and Viticulture*, **36**, 302–313.
Wilharm, G. and Holz, G. (1951) Beitrag zur Kenntnis des Acroleins in Obstbränden, Maischen und Mosten. *Archiv fuer Mikrobiologie*, **15**, 403–413.
Woidich, H., Pfannhauser, W., Blaicher W., and Pechanek, U. (1980) Beitrag zur Untersuchung von biogenen Aminen in Rot- und Weißweinen. *Mitteilungen Klosterneuburg*, **30**, 27–31.
Wood, W.A. (1961) *The Bacteria, Volume II*. New York: Academic press.
Wright, J.M. and Parle, J.N. (1974) *Brettanomyces* in the New-Zealand wine industry. *New Zealand Journal of Agricultural Research*, **17**, 273–278.
Würdig, G. (1989) Ausbau der Weine: 4.2.4.5. Die mikrobiologischen Wirkungen der schwefligen Säure. In *Chemie des Weines*, edited by G. Würdig and R. Woller, 345 pp., Ulmer: Stuttgart.
Würdig, G. and Kullmann, H.H. (1971) Über die Empfindlichkeit von in abgefüllten Weinen gefundenen Hefen gegenüber Sorbinsäure. *Allgemeine Deutsche Weinzeitung*, **107**, 1011–1012.
Würdig, G., Schlotter, H.A. and Klein, E. (1974) Über die Ursachen des sogenannten Geranientones. *Allgemeine Deutsche Weinzeitung*, **110**, 578–583.
Yoshioka, K. (1983) Alcohol acetyltransferase and ester formation by brewer's yeast *3. International Mykoogical Congress Tokyo Abstracts*, 359.
Zaake, S. (1958) Das Verhalten hefeartiger Pilze gegenüber Weinsäure. *Zentralblatt fuer Bakteriologie II*, **111**, 410–439.
Zee, J.A. (1983) Biogenic amines in wine. *American Journal of Enology and Viticulture*, **34**, 6–9.
Zeeman, W., Snyman, J.P. and von Wyk, C.J. (1982) The influence of yeast strain and malolactic fermentation on some volatile bouquet substances and on quality of table wines. In *Symposium Proceedings. Grape and Wine Centennial* (Davis, 1980), edited by A.D. Webb, pp.79–90. Davis: University of California.

Chapter 15

SELECTION AND COMMERCIAL CULTIVATION OF WINE YEAST AND BACTERIA

RICHARD DEGRE

Lallemand Inc., 1620 Prefontaine, Montreal, Quebec, Canada, H1W 2N8

Chapter Outline

Chapter Outline (Continued)

INTRODUCTION

Commercial dry wine yeast, *Saccharomyces cerevisiae*, is a single cell microorganism whose historical development has been associated with that of modern microbiology. Although the first observations of yeasts probably go back to Antonie van Leeuwenhoek in the 1600s, Pasteur (1866) was the first to demonstrate that yeast cells were responsible for the conversion of grape sugars into ethanol. In a book reviewed by Pasteur himself, Trouessart (1890) reported the presence of species and strains of *Saccharomyces* specific to wine (*S. ellipsoideus*, *S. pastorianus*, *S. exiguus*), beer (*S. cerevisiae*) and other fermented beverages like apple and pear ciders. The behavior of flocculating yeasts (*S. conglomerans*) was also well depicted. Moreover, Trouessart refers to the addition of yeast to beer and bread ferments in order to speed up the fermentation process. Therefore, the use of starter cultures and the concept of using particular strains for specific fermentations is nothing new.

Although wine yeasts have been known for a long time, the production of wine has remained, until recently, more of an art than a science. The development of modern microbiology and related disciplines like ecology, molecular biology, biochemistry, protein engineering and immunology has contributed to change this image. Also, it has helped to better understand the extent that yeasts can influence the quality of wines. Table 15.1 lists some desirable technical features of wine yeasts.

The production and use of dry wine yeasts began in the United States in the mid-1960s and expanded worldwide thereafter. Today, more than 100 different strains are available commercially from six major producers. With a worldwide production of wine of 325 000 000 hl (Office International de la Vigne et du Vin, 1988) and a yeast usage rate of 10–20 g/hl, one can estimate the potential market for dry yeasts to be in the range of 5000 tonnes annually. Since several users still rely on indigenous yeasts while others propagate their own starter cultures from commercial preparations, the actual market is probably less than 600 tonnes of dry yeast per year. This is still very small compared with the production of 0,7 billion tonnes d.w. (dry weight) of bakers yeast (Beudeker *et al.*, 1990), and even more so if one divides that small market between the 100 wine strains used.

The production of dry wine yeast is traditionally done by baker's yeast companies, and several comprehensive reviews of baker's yeast production and

TABLE 15.1 Some desirable and undesirable features of wine yeasts.

Desirable	*Undesirable*
High tolerance of alcohol	Production of sulfur dioxide
Complete fermentation of sugar	Production of hydrogen sulfide
Resistance to sulfur dioxide	Production of volatile acidity
Fermentation at low temperatures	Production of acetaldehyde and pyruvate
Minimal lag-phase on rehydration	Foaming properties
Malic acid degradation	Formation of ethyl carbamate precursors
Ferment under pressure	Production polyphenol oxidase
Production of glycerol	
Production of β-glucosidase	
Killer phenomena	

baker's yeast physiology have been published previously (Peppler, 1967; Burrows, 1970) and more recently (Reed, 1982; Oura, 1983; Trivedi *et al.*, 1986; Beudeker *et al.*, 1990; Wills, 1990) Since the production of wine and baker's yeasts relies upon the same physiological and technological principles, it is not the intention of this chapter to review the fundamentals, but rather to emphasize those characteristics known to be more specific to wine yeasts.

In recent years, winemakers have sought the commercial availability of strains of lactic acid bacteria, principally *Leuconostoc oenos*, for inducing the malolactic fermentation. Compared with yeasts, experience in the commercial production of this organism is still limited. This chapter also considers available information on the commercial production of this species.

ROLE OF DRY WINE YEAST

It is generally recognized that the use of dry wine yeast will ensure a quicker start of the fermentation and at the same time will help in the production of wine of uniform quality.

Wine fermentation was traditionally carried out by indigenous yeasts associated with the grapes and the cellar equipment. Commercially available baker's yeast was also used to restart stuck fermentations. Although liquid cultures of wine yeast were available from different suppliers since 1930 (Institut Laclaire, France), dry wine yeast originated in the mid-1960s in response to the needs of a large Californian winery. Two strains were then produced, Montrachet and Pasteur Champagne, and the former became a reference strain for different yeast groups studying wine yeast physiology. These two strains were then offered worldwide as all-purpose yeasts, with limited success. Along with strains recommended to restart stuck fermentations, they were part of a so-called first generation of dry wine yeasts. It is doubtful that some of them would be selected nowadays in regard of the desirable technical features requested by enologists; strain SB1, for instance, was recommended for stuck fermentations despite its great sensitivity to killer yeasts. The important role of the killer factor in the microbial ecology of the must was stressed at the first Lalvin Conference (Degre, 1989).

It soon became obvious that these strains were not the universal solution to the fermentation of grape must and that there was a need for the selection of strains better adapted to the different regions of the world and their respective grape varietals. By closely collaborating with wine research institutes, large cooperatives and distributors, we (Lallemand) were able to develop in the mid-1970s new strains of dry wine yeasts whose technical and enological properties had a predictable influence on the quality of wine. They constituted the second generation of wine yeasts and included strains like WSK (Wadenswill Institute, Switzerland), K1 (Institut Cooperatif du Vin, Montpellier, France), 71B (INRA, Narbonne, France) and EC (Institut Oenologique de Champagne, Epernay, France). Each of these strains had desirable features whose influences on wine quality were well understood. These features included properties like fermentation at low temperature (WSK), low foaming and ease of settling (EC), killer factor (K1) (Barre, 1980), and the production of aromatic components (71B). Later, flocculent strains were developed and used commercially to facilitate the ridling process during Champagne and sparkling wine production. Figures 15.1 and 15.2 illustrate the network of fimbriae responsible for this particular behavior (Zaworski and Heimsch, 1987).

Flocculating yeast strains were then followed by the development of immobilized cells entrapped in alginate beads (Duteurtre *et al.,* 1990). Valade and Rinville (1990) have reviewed the economical impact of these two alternatives on the classical ridling process. The agglomerating yeasts, in conjunction with mechanical ridling, were judged more advantageous for small to medium size wineries while immobilized yeasts were shown to be more economical for large wineries (> 3 000 000 bottles) when one disregards the cost of the investment required.

The development of yeast strains with specific genetic markers has provided a means of being able to precisely monitor the growth of particular strains during fermentation. Generally, these markers confer on the strains resistance to the antibiotics diuron and oligomycin (EC marked) and diuron and erythromycin (K1 marked). Normally, these antibiotics inhibit yeast growth by preventing mitochondria function (Vezinhet and Lacroix, 1984). Delteil (1990) has used such strains to show that a must can be overtaken by wild killer yeasts even though it was previously inoculated with a sensitive dry yeast (10 g/hl). A marked killer yeast (K1), in the same conditions, was successfully implanted and dominated the fermentation.

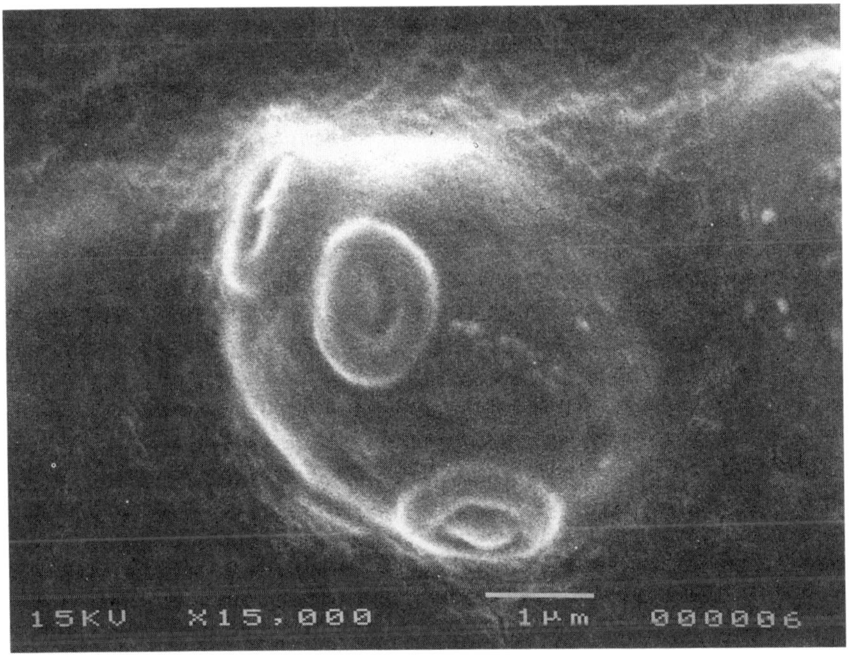

FIGURE 15.1 Scanning electron micrograph of a flocculent strain of *Saccharomyces cerevisiae* (Lalvin 016) showing the network of fimbriae and polysaccharides at the surface of the cell; attachment between two cells.

Strains selected in recent years have very specific enzymatic activities and form yeasts of the third generation. Strains (e.g. VL1) having significant β-glycosidase activities that can release bound terpenes in muscat and sauvignon grape varietals are part of this group (Dubourdieu *et al.*, 1988; Darriet *et al.*, 1988). Contrary to some other aromatic strains, it is not a metabolite (e.g. isoamyl alcohol), produced by these strains that confers the aromatic character to the wine (Benda, 1982) but rather they enhance a property of the varietal itself. The success of these strains relies on exaltation of the varietal specificity, and is encouraged by harmonization of the relationship between the yeast strain and the varietal itself.

Thus, the strains selected nowadays have more than a simple role of fermenting sugar into ethanol. Nonetheless, in order to perform properly in grape must, yeasts must have a sufficient reservoir of nutrients since they will actively multiply during fermentation. When they are added to must at a ratio of 10 g/hl, the cell count will be approximately 2 x 10^6 cells/ml and this will increase to about 2 x 10^8 cells/ml during fermentation. This is a major difference with baker's yeast which does not significantly multiply in dough, although budding occurs. Moreover, wine yeasts grow under increasing ethanol concentration and adverse conditions of pH, temperature, pressure, competition and nutrient

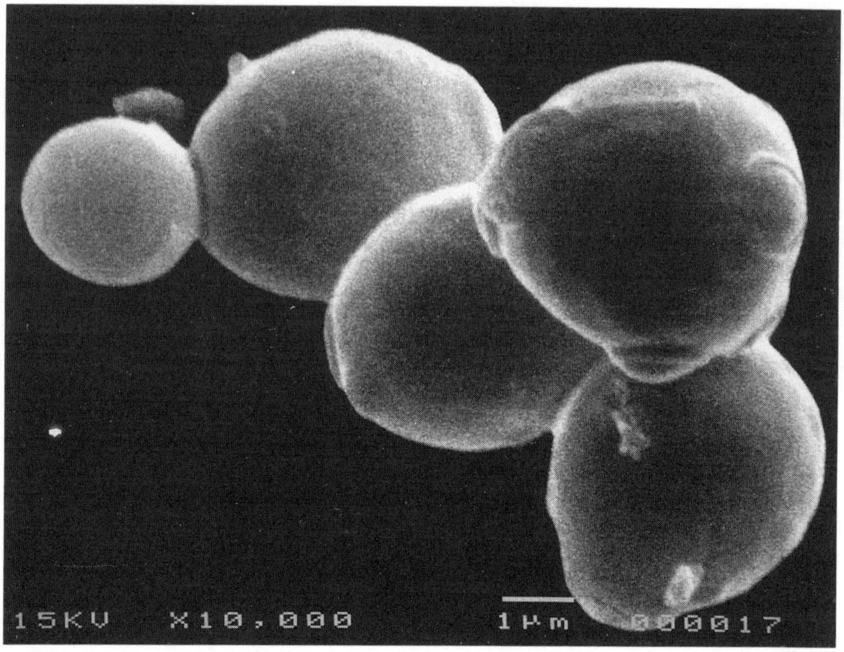

FIGURE 15.2 Scanning electron micrograph of a non-flocculent strain of *Saccharomyces cerevisiae* (Lalvin K1).

limitation. It is the responsibility of the enologist to ensure that the yeast will have sufficient nutrients to grow and conduct a fermentation that gives a wine of the desired quality. However, it is the responsibility of the yeast producer to ensure that the yeast has been grown and dried under such conditions that it will perform properly. This task is complicated by the fact that some strains will be produced only once or twice a year because of the small volumes required. Such low and infrequent demand make it a challenging task for producers of wine yeasts to improve and optimize the conditions of yeast production and drying.

YEAST PHYSIOLOGY

Saccharomyces cerevisiae is a unicellular eukaryotic organism having an average cell size of 4×8 μm, a volume of 10^{-10} ml and a dry weight per cell of 0.3×10^{-10} g. Commercial preparations of dried wine yeast usually contain $2.0–2.5 \times 10^{10}$ viable cells/g.

Most industrial strains are polyploid and cellular division occurs by budding. Polymorphism among wine yeasts is related to the presence of homologous chromosomes of different sizes (Blondin and Vezinhet, 1988; Pardo *et al.*, 1989).

Contrary to industrial strains, haploid laboratory strains probably contain 17 chromosomes.

Saccharomyces cerevisiae is a facultative anaerobe capable of metabolizing glucose either aerobically (oxidative metabolism) or anaerobically (fermentative metabolism). Cell yields, expressed as grams of yeast dry matter (YDM) /g of sugar, are dramatically different under these two sets of conditions. Under anaerobic conditions and an appropriate supply of nutrients, 100 g of glucose will yield 7.5 g of YDM but will yield as much as 54 g of YDM under respiratory conditions, provided that there is no glucose excess. In fact, glucose concentration in the fermentation broth should not exceed 50 mg/l, even under oxidative conditions, because it would lead to anaerobic fermentation. This phenomenon is known as the Crabtree effect and is the result of catabolite repression. The most common explanation for this phenomenon is a limited respiratory capacity of *S. cerevisiae* (Alexander and Jeffries, 1990). Pyruvate is the key intermediate in the metabolism of this yeast; after formation in the cytosol, this molecule is usually transformed oxidatively in the mitochondria by pyruvate dehydrogenase. However, when the maximum oxidative capacity of a specific strain is reached, the surplus of intracellular pyruvate is diverted toward ethanol production. Moreover, the enzyme pyruvate decarboxylase, which is responsible for the formation of acetaldehyde that goes to ethanol, increases by as much as six fold in the presence of glucose excess (van Urk *et al.*, 1990). Since steady state cultures of Crabtree negative (*Candida*) and Crabtree positive yeasts show the same oxidative capacity, it is not clear that limited respiratory capacity is responsible for the Crabtree effect.

Glucose uptake is mediated by facilitated diffusion systems in *S. cerevisiae* (Bisson and Fraenkel, 1983) and different carriers, namely, SNF3, HXT1 and HXT2, have been identified in wine yeasts by the group of Bisson (1990). There is strong evidence that HXT2 is expressed under anaerobic conditions and, therefore, might be of interest under the fermentation conditions of grape must. It has been suggested that strains like EC 1118, which have strong fermentation rates, might contain several copies of this gene located on chromosome XVI.

For a source of nitrogen, *S. cerevisiae* can use ammonium hydroxide which is assimilated into glutamate. Urea, on the other end, must be carboxylated by means of urea carboxylase before being degraded into ammonia and carbon dioxide by allophanate hydrolase (Cooper, 1982).

Readers interested in more detailed discussion of the physiology of *S. cerevisiae* should consult the reviews mentioned previously. Some aspects that are more pertinent to wine yeast will be covered after a brief review of the basic principles involved in yeast production.

INDUSTRIAL CULTIVATION

Several of the technical properties of wine yeasts listed in Table 15.1 are influenced by the way yeast is grown and dried. This section discusses how the yeast producer can positively influence and preserve these desirable features. Wine yeasts, as well as distiller's and brewer's yeasts intended for microbreweries, are usually produced in baker's yeast plants. Because of the very limited market for these yeasts, it is doubtful that specific plants for their production would ever be economically successful.

Culture Maintenance

Proper conservation techniques should be used since this is the first step in the process of yeast propagation. Only a limited number of people with suitable technical skills are usually allowed to manipulate the pure cultures in order to avoid errors. Contrary to the baker's yeast industry where only a limited number of strains are used, there is great diversity in the strains of yeasts used in wine fermentation, and this factor forces the major producers of wine yeasts to maintain a vast culture collection.

Yeast strains are kept under different storage conditions and in different locations to ensure the integrity of the collection. The most popular and practical storing condition, from a plant operation point of view, is probably the agar slant. The so-called mother slant is kept under sterile mineral oil for up to six months under refrigerated conditions (4°C) before being subcultured. At the time of transfer, suitable numbers of slants can be prepared that will be used for the start-up of industrial fermentations. A slant should be used only once and kept refrigerated until quality control testing has confirmed the correct identity of the final product. Malt agar can be used as medium for the slant but a medium similar to that used for the first propagation stages in the laboratory is recommended. This should ensure rapid initiation of propagation fermentations.

Among techniques recommended for back-up purposes and long term preservation, storage of cultures under liquid nitrogen or by freezing at –80°C in the presence of 10% glycerol are probably the most suitable. Lyophilization is useful but can induce the formation of respiratory mutants. Preservation of yeast strains on silica gel has been evaluated in our laboratory but without much success since most strains were not viable after storage for two years under such conditions. Preservation techniques for yeasts are reviewed by Kirsop (1988).

Fermentation

The first steps in the propagation protocol are carried out in the laboratory under aseptic conditions in a medium similar or identical to the one to be used later in the plant. This laboratory culture is used to inoculate the first propagation step in the plant. It is conducted as a batch fermentation under aseptic conditions, and lasts 16–24 h. The medium consists of molasses as the main carbohydrate, ammonium hydroxide as nitrogen source, phosphoric acid as phosphorus, and supplements of yeast extract and minerals like magnesium.

It is common to use a mixture of cane and beet molasses since they tend to complement each other with respect to vitamins and mineral content (Burrows, 1970). It is recommended to use at least 10–20% of cane molasses in order to overcome a deficiency in biotin, for instance. Molasses contain roughly 50% sugar which is mainly sucrose in beet molasses. In cane molasses, sucrose represents two thirds of the total sugar, the rest being glucose and fructose.

The molasses used for fermentation is usually diluted with water to about 25% fermentable sugars before desludging. Desludging is optional and is often done by means of a closed bowl continuous centrifuge. Settling tanks can be used for the same purpose. The molasses is then heat treated by either high temperature short time (HTST) or low temperature long time (LTLT) processes similar to those found in the dairy industry. Alternatively, molasses may be submitted to a flash cooking procedure whereby the difference in pressure will allow the evaporation of harmful volatile compounds that could affect yeast growth. These substances are mainly formic, butyric and propionic acids as well as alkanols derived from different processing aids used in sugar refineries. After treatment, molasses is stored either at 4°C or close to 60°C in order to prevent the growth of microbial contaminants.

The yeast produced in the pure culture facility is transferred to the first industrial fermentor whose size may vary from 20 to 100 m^3. This culture is usually allowed to reabsorb some of the ethanol formed before being submitted to a second fed-batch fermentation. This is normally conducted in fermentors of 100 to 250 m^3, under strong aeration such as one volume of air per volume of medium per minute (1 VVM). Cooling of the fermentors is carried out by means of cooling coils, heat exchangers or running water over the external walls. At the end of the first fed-batch fermentation, yeast is harvested and washed several times by means of continuous discharge nozzle centrifuges and stored in the form of a cream yeast (15–20% solids, 2–4°C). This cream will be used as a seed yeast for an intermediate or final fermentation from which the cells are harvested for drying.

The most critical aspect of fed-batch fermentation is the feeding rate of the substrate (molasses) since it directly controls the growth rate of the yeast (Sonnleitner and Kapelli, 1986). Up to a certain rate, the yeast will metabolize sugar oxidatively; above this rate, ethanol will be formed. This critical growth rate (μ) is in the range of 0.14–0.18 h^{-1} for strains of baker's yeast. Under such conditions and provided all essential nutrients are added, yields of 105–110% can be obtained in the laboratory and 90–95% in the plant. Yield is expressed here as kg of yeast biomass (at 27% solids) produced per kg of molasses. Up to 10% of the yield can be accounted for by substances other than fermentable sugar in the molasses (White, 1954). The differences in yields found under laboratory and plant conditions are probably caused by influences of mixing (Sweere *et al.*, 1988). Under properly controlled fermentation, maximum cell yield is obtained and ethanol production is negligible.

A theoretical fermentation carried out in a 150,000 l fermentor would yield 12 000 kg of new yeast (27% solids) within 14 hours during which the original seed yeast (2000 kg) would be multiplied by a factor of 7. Approximately 3900 kg of dry yeast (95% solids) could be produced assuming that the entire content of the fermentor can be dried. Based on a final volume in the fermentor of 120 000 l and 1 VVM of air, 108 000 000 l of air would have been required. Assuming 750 kcal of heat produced per kg of new yeast, one can estimate the total heat output to be in the range of 10 000 000 kcal. With a seed yeast containing 50% protein (N × 6.25) and a final yeast containing the same amount, 1000 l of aqueous ammonia (27%) would be required as well as close to 100 l of phosphoric acid (75%). Assuming that the first 2000 kg of seed yeast were produced with an efficiency of 80% only, since part of it was under fermentative conditions, 14.5 tonnes of molasses would have been required to produce those 3900 kg of dry yeast.

The actual yields with baker's yeast are lower because of longer mixing times in large fermentors; they are also much lower with wine yeast since these strains are not produced on a regular basis. This irregular production decreases the opportunities for experimentation that could lead to more optimal conditions for fermentation and drying. Finally, it is important to realize that, optimum yeast quality is achieved only at the expense of maximum biomass (yield) that can be obtained.

The efficiency of oxygen transfer during fermentation is of great importance to the yield and quality of yeast produced. Air distribution systems vary from one yeast plant to another. They range from a simple network of perforated pipes with or without an agitator to more sophisticated systems developed by companies like Vogelbusch (Vienna), Escher Wyss (Ravensburg) and Frings (Bonn). The trend is towards the development of fermentors with aeration systems that demand less energy input (kW/kg yeast).

In order to monitor and control ethanol production, different equipment and strategies are available. Although gas chromatography, mass spectrometry and equipment devoted to the measurement of respiratory quotient are very useful, it is not always possible to use such sophisticated devices in a plant environment. Instead, non-specific sensors based on metal oxide semi-conductors can be used (Bach *et al.*, 1978). The resistance of such sensors decreases in proportion to ethanol concentration and the amount detected is usually transmitted to a central system where different control strategies can be implemented. The most obvious control is to increase the air flow and reduce the molasses flow rate if the ethanol concentration exceeds a certain limit.

Usually, the most important fermentation parameters (molasses and ammonia feed rates, aeration, agitation) are under closed-loop control. Different mathematical models of yeast metabolism aimed at maximizing yields have been proposed by several authors over the years and were recently reviewed by Beudeker *et al.* (1990). Several of those models, structured and unstructured, will estimate from data like oxygen uptake rate, ethanol concentration and respiratory quotient, the amount of biomass in the fermentor in order to assign a new set point value to one or several control points previously identified (Pomerleau *et al.*, 1990). Therefore, the advent of the NADH probe and other probes specially designed to directly monitor biomass in a bioreactor is of great interest. However, the different probes tested in our laboratory were significantly affected by the nature of the culture medium, and did not live up to the promises of the suppliers.

The instrumentation and control strategies now available will allow the yeast manufacturer to maximize productivity, namely, the amount of yeast produced per m^3 of fermentor per h, including turn-around time. Those values are probably in the range of 3 to 6 kg of yeast (27% solids)/ m^3/h for a baker's yeast plant.

Such advances in fermentation technology now allow baker's yeast manufacturers to determine what critical growth rate should be imposed in order to maximize yields and throughput without affecting final quality of the yeast. The same technology can be adapted to wine yeasts. Unfortunately, productivity associated with wine yeasts is, most of the time, much less than that of baker's yeast. Some strains have a critical growth rate lower than baker's yeast; this will significantly influence the generation or doubling time. Consequently, productivity becomes less since more time is required to grow the same amount of yeast. Such behavior has been observed in our laboratory and during plant operations with some strains of *S. cerevisiae* (*uvarum* and *bayanus*). Growth rates with such strains can be as low as 0,10–0,12 h^{-1}. Knowing that aeration and oxygen transfer rates are similar to those typical of baker's yeast, one must assume that such strains have limited respiratory capacity or have a deficient transport system for glucose (McClellan *et al.*, 1989). Petrik *et al.* (1983) and Rieger *et al.* (1983) had previously reported maximum oxygen uptake rates of 8 and 5 mmol/g/h for *S. cerevisiae* and *S. cerevisiae* strain *bayanus*, respectively. This character is obviously strain dependent but is also affected by growth conditions, since adaptation of seed yeast to aerobic conditions can increase the respiratory capacity of *S. cerevisiae* (Orlowski and Barford, 1987).

However, reduced capacity for sugar transport cannot be excluded as a major rate limiting step. McClellan *et al.* (1989) have shown that commercial wine yeasts differ in their ability to transport sugars into the cell. They reported that the strain Tokay, known as a poor fermentor, did not possess a low affinity transport system and suffered from substrate inhibition; that is, the rate of sugar uptake was decreased when glucose concentrations exceeded 10 mM. Another strain, CEG, also known as Epernay Geiseinheim 2, also gave sluggish fermentations but can be grown industrially under the same conditions as those used for *S. cerevisiae* strain *bayanus*. Providing that this strain is affected by the same problem as the Tokay strain, it is possible that, under fed-batch conditions, during which the residual concentration of sugar in the medium is negligible, the high affinity system is induced. Therefore, such a strain would be disadvantaged in must with a high concentration of sugar, but this major drawback would not show up under conditions of industrial propagation. Obviously, more fundamental research is needed in the area of sugar transport into yeast cells.

The nitrogen content of the cells can significantly influence their critical growth rate and the degree to which ethanol is formed. Woehrer and Roehr

(1981), for instance, have shown that under 5.2% nitrogen or 32.5% protein (N x 6.25), this critical growth rate would increase. This observation is not necessarily relevant under industrial conditions where wine yeasts are grown to contain 40–45% protein and baker's yeast between 45 and 52%.

EFFECT OF INDUSTRIAL CULTIVATION ON WINE YEAST PROPERTIES

Microbiological Quality

Industrial plants that produce baker's yeast (or wine yeasts) do not operate under conditions of asepsis because of the large volumes of raw materials required. As a result, a small number bacteria are usually associated with the yeast product. For wine yeasts, this degree of contamination is in the range of one bacterial cell or less per 10^6 yeast cells. This level of contamination is not considered significant because the bacterial species present would not survive the winemaking conditions. Mostly, these bacteria originate from the molasses because it is heat processed at a temperature that gives pasteurization rather than sterilization.

Purity of the yeast product is insured by implementation of very strict cleaning, sanitation and sterilization procedures at all levels of operation: fermentation, harvesting, storage, filtration, drying and packaging. Such stringent measures, can reduce the productivity (kg yeast/m^3/h) for wine yeasts.

Fermentation Rate

Although strain related, the fermentation velocity of both baker's and brewer's yeasts has been related to their content of protein and phosphates (Beudeker *et al.*, 1990; Thorne, 1954). This fact also applies to wine yeasts which usually contain between 40 and 45% protein (N × 6.25) and 2.0–2.4% phosphates (expressed as P_2O_5). A ratio of P_2O_5/N of 1/3 has been shown to be optimal for most applications (White, 1954) and the optimum amount of protein may vary from one strain to another. Since wine yeasts are intended to be dried, it is not advisable to exceed this ratio, because it would result in an excess of water linked to the protein (Beker and Rapoport, 1987). This would, in turn, significantly affect the drying procedure, viability and final activity of the dry yeast.

Production of Hydrogen Sulfide (H_2S)

The nitrogen concentration in the must and the proteolytic activity of the yeast are two main parameters involved in the production of H_2S during wine fermentation (Vos and Gray, 1979). If there is a deficiency in assimilable nitrogen in the must, proteolytic activity of the yeast will be stimulated and proteins and peptides will be degraded. As a result, H_2S is formed as a by-product of degradation of sulfur containing proteins. Therefore, it is the role of the producer to ensure a proper nitrogen level in the yeast itself in order to limit the nitrogen demand of the yeast in case of deficient musts. Yeasts containing 38% protein have been shown to perform properly in this regard. Moreover, yeasts deficient in pantothenate tend to produce more sulfide (Wainwright, 1971). Since calcium pantothenate is usually added to the molasses during yeast production, the cells used in wine fermentation are unlikely to be deficient in pantothenate.

Ethanol Yield and Tolerance

Ethanol tolerance, ethanol yield and cell viability are positively influenced by the presence of a small amount of oxygen in the medium (Ryu *et al.*, 1984). This effect is related to the synthesis of sterols and of unsaturated fatty acids (C-16:1, C-18:1) that are involved in membrane fluidity (Parks *et al.*,1978). Unsaturated fatty acids play a major role in the function of the mitochondrial membrane (Mazurki and Linnane, 1979). *Saccharomyces cerevisiae* cannot synthesize sterols under anaerobic conditions, and propagation under such conditions will inevitably lead to cells with defective membranes and, therefore, poor metabolic activity. Since wine yeasts are grown under fed-batch conditions in the presence of air, they will contain a reservoir of unsaturated fatty acids and sterols as needed for limited number of generations that occur during the fermentation of grape must. The procedure of recirculating and pumping the wine regularly during production will introduce a little oxygen into the medium and will be advantageous for the yeast.

Vitamins like pantothenic acid, thiamine, biotin and inositol are required for optimum yeast metabolism during industrial propagation and also for proper function during wine fermentation. It is of interest to compare the relative concentrations of these growth factors in a molasses medium (25% fermentable sugars) and in a typical grape must (Table 15.2). The values for cane and beet molasses are based on an end-dilution of 1:5 that is, a final molasses concentration of 200 g/l of fermentation medium. Grape must contains less of these vitamins; nonetheless they are usually added to commercial yeast fermentations to compensate for natural variations in the quality of molasses and to ensure a proper performance of the yeast under wine-making conditions.

There is a limit to the number of generations that wine yeasts can be propagated anaerobically in the winery since it would lead to a severe deficiency in growth factors as well as to a depletion of sterols and unsaturated fatty acids. Extended anaerobic culture of wine yeasts leads to very unhealthy cells and possible stuck fermentations. The lack of proper nutrients can be responsible for the sensitivity of yeast cells to alcohol (D'Amore *et al.*, 1988). As indicated already, commercial preparations of dry yeasts are produced under conditions that ensure healthy membranes and an adequate reservoir of growth factors. Thus, their continued propagation under winery conditions for re-use is not advocated and can lead to unsatisfactory fermentations.

TABLE 15.2 Comparison of concentrations (mg/g) of growth factors in molasses and must[a].

	Cane molasses	*Beet molasses*	*Must*
Biotin	0.2–0.6	0.004–0.03	0.0015–0.0042
Pantothenate	3–11	10–20	0.5–1.38
Thiamine	0.3	0.26	0.16–0.45
Inositol	500–1200	1000–1600	370–710

[a] Adapted from Reed (1982) and Reed and Peppler (1973).

Resistance to Sulfur Dioxide

Sulfur dioxide (SO_2) is used in the wine industry as an antioxidant, to control microbial contamination and to sanitize equipment. The exact mechanism of antimicrobial activity of this preservative is not precisely known although the sensitivity of different yeasts seems to be related to the degree of saturation of membrane lipids (Pilkington and Rose, 1989). Its mechanism of action may be similar to that of propionic acid which, like SO_2, is also more active at lower pH values. At these values, both compounds occur more in the undissociated form. It has been postulated that propionic acid asserts its antimicrobial effect by causing a decrease in intracellular pH (Loureira, 1989). Accordingly, yeast strains with increased ATPase activity would be more resistant to this molecule since they would tend to extrude protons from the cytoplasm much more efficiently. This in turn would help the cell to conserve a proper internal pH. Membrane ATPase levels have been directly related to the resistance of bacteria to low pH (Bender and Marquis, 1987). A better understanding of the molecular effects of SO_2 on yeasts would assist the yeast producer to select and develop strains more suited for wine fermentations. Yeast adaptation to SO_2 is performed by producers by growing the organism in the presence of trace amounts of SO_2 under strict pH control. This method has proved successful for increasing yeast resistance to SO_2.

Resistance to Drying and Rehydration

Since wine yeasts are intended to be dried, it is important to grow them accordingly during the different propagation steps. In addition to the strain itself, the trehalose content in the cell is probably one of the most important factors that affects the resistance of yeasts to drying and subsequent rehydration.

Trehalose is known to stabilize yeast membranes during dehydration by replacing molecules of water (Lee *et al.*, 1986; Chandrasekar and Graber, 1988), and to serve as a source of reserve carbohydrate readily available to the cells (Van Laere, 1989). By replacing water molecules, trehalose inhibits phase transition of the membranes and, thus, preserves their integrity. This is probably achieved by the binding of trehalose to the polar groups of phospholipids (Crowe *et al.*, 1984). As a reserve material, trehalose prolongs the shelf-life of fresh yeast but is degraded very rapidly in the presence of glucose. This effect is cAMP mediated (Thevelein, 1984a). Results from our laboratory indicate that, under aerobic conditions and a glucose pulse, the trehalose content of yeast cells rapidly decreases from 10 to 2–4% in less than 10 min. The same behavior probably occurs when the yeast is added to must.

The role of trehalose as a reserve material in dry yeast preparations containing less than 8% humidity is not clear, as is the fact that trehalose degradation is stimulated by its breakdown products. The role of trehalose as a reserve material is more easily understood in fresh yeasts where there is significant metabolic activity because of their high moisture (70%) content. In dry yeast, trehalose probably plays a structural role by maintaining the integrity of the membranes under stress conditions (Van Laere, 1989).

Therefore, there is a strong incentive for yeast producers to stimulate the formation of trehalose during fermentation in order to increase the resistance of yeast cells to the stresses of dehydration and rehydration. Trehalose is usually synthesized in response to limitations in nitrogen, phosphate and sulfur; but temperature, growth rates and osmotic stresses also influence its formation (Panek, 1975, 1985, 1991; Thevelein, 1984b). Trehalose concentrations up to 22–24% of the cell dry weight can be obtained under specific conditions, but its content in wine yeast is usually in the range of 12–20%.

DRYING

After harvesting from the fermentor as noted earlier, cream yeast is further dehydrated to 30–35% solids by means of rotary vacuum filters or filter presses. The yeast is then extruded through a perforated plate and the spaghetti-like particles are loaded into the dryer. The size of the holes in the perforated plate determines the size of the particle but most, if not all of the yeast available nowadays, is sold in the form of instant dry yeast (diameter of 0.5 mm). Although straightforward with baker's yeast, filtration and extrusion are much more difficult with wine yeasts since their ratio of internal water to external water is different and results in poor efficiency of filtration. The processing of flocculating yeasts also has similar problems.

Although several types of dryers (rotolouvers, belt dryers) have been used in the past and are still used by some producers, fluid bed drying has become the standard of the industry. These dryers can be operated either batchwise or in the continuous mode. Air is treated to reduce its water content and to insure that the temperature of the yeast will not exceed 35°C during drying. Drying times may vary from 15 to 60 min depending on the procedure used. Commercial preparations of dry yeasts normally contain less than 8% residual moisture and less than 6% in most cases.

As mentioned already, freshly harvested cream yeasts contain about 70% water. About 10% of this moisture is extracellular, the remaining being intracellular. Of the intracellular water, about 10% is bound. The amount of water associated with proteins is about 0.3 g/g protein (Josic, 1982). Free internal and external water is easily eliminated in the first part of drying during which it is possible to obtain 90% solids; however, during the last part of drying, there is a significant increase in desorption heat, and dehydration becomes more difficult as bound water must be eliminated.

Beker and Rapoport (1987) have reviewed the biochemical changes that occur in yeast under different conditions of dehydration; these changes include stimulation of dehydrogenase activities, decrease in polyphosphatase, tripolyphosphatase and pyrophosphatase activities, decrease in phospholipids, increase of unsaturated fatty acids and increase in trehalose content (Zikmanis *et al.*, 1982, 1983, 1985; Krallish *et al.*, 1986). The practical significance of these results is not clear since they were obtained with yeasts that were dried by techniques (including freeze-drying and overnight drying), that are different to those used in the production of commercial yeasts.

Due to their high porosity, particles of instant dry yeast are packaged under vacuum in order to reduce oxidation and inactivation. Depending upon the strain, the decrease in viability is usually between 10–25% per year at 20°C.

YEAST REHYDRATION

Yeast rehydration is a straightforward operation and several scientific and technical reports have described the proper techniques for conducting this step. Only proper rehydration can insure healthy cells and good fermentation characteristics (Kraus *et al.*, 1981, 1984; Radler *et al.*, 1985).

As a general procedure for vinification, 1 kg of dry yeast should be sprinkled into 10 l of warm water (35–40°C). After 5 min, the suspension must be stirred to resuspend all the cells. Yeast cells should not remain in this solution more than 30 min since they would tend to use up their reserve materials. The rehydrated yeast can then be added to 50 hl of must to be fermented. This corresponds to 20 g/hl or a population of 4×10^6 cells/ml.

Upon rehydration, yeast cells undergo structural and biochemical transformations. Most authors agree on two major characteristics: cell damage is directly related to rehydration temperature and involves leaching of cellular constituents of low molecular weight. Our own data indicate that damage to yeast cells occurs very rapidly since maximum leaching is observed after 2–5 min.

One of the most complete studies on wine yeast rehydration was conducted by Radler *et al.* (1985). Maximum values for cell counts (methylene blue), colony counts, fermentation activity and acidification activity were obtained at rehydration temperatures between 38 and 45°C. There was no difference in these activities within 2 hours after rehydration but the composition of the rehydrating medium had an influence. It was found that mixtures of grape juice and water, and solutions containing glucose, fructose, hydroxysucccinic acid, amino acids, 2-keto-glutarate, ammonium sulphate, vitamins or the salts KCl, $CaCl_2$ and NaCl did affect the metabolic activity of the rehydrated yeast. The best activity was achieved with rehydration in a 1% KCl solution. Rehydration experiments carried out in 1% KCl in our laboratories with three yeast strains confirmed the data reported by Radler *et al.* (1985); leaching of intracellular constituents, as determined by optical density at 260 nm, was minimized at 38°C while fermentation activity was at its maximum. These results have been explained by the improved conditions for the rehydration of phospholipids and subsequent restoration of membrane functions (Beker and Rapoport, 1987; Henschke and Rose, 1991). Rehydration in water at temperatures between 35 and 40°C is probably the most practical procedure to adopt under winemaking conditions.

QUALITY CONTROL

Different standards of quality for yeasts have been set by the wine industry through the years. These include viability and activity of the yeast preparation as well as contamination with bacteria and wild yeasts. The different techniques used to assess yeast quality are reviewed by Rosen (1987).

The increasing number of yeast strains now commercially available has prompted the development and application of new techniques for determining strain identity and homogeneity. Taxonomically, most wine yeasts are strains of *S. cerevisiae* that can be divided on the basis of their fermentation/assimilation profile into different physiological races like *bayanus*, *chevalieri*, etc. One of the first approaches for assessing strain identity in our laboratory was to examine their sugar fermentation profile and specificity of their killer character (Degre *et al.*, 1989a). The advent of genetically marked yeast strains with resistance to certain mitochondrial inhibitors also allowed their specific identification by plating on a medium containing the inhibitors.

However, strain identification at the molecular level (DNA-fingerprints) has been the greatest change in laboratory quality control procedures in the last few years. Three different techniques have been reported, namely, mitochondrial DNA analysis (Hallet *et al.*, 1988), karyotype analysis which allows comparison of chromosome banding patterns (Vezinhet *et al.*, 1990) and restriction fragment length polymorphism (RFLP) of total DNA (Degre *et al.*, 1989a, 1989b). No matter which technique is used, the comparison and interpretation of these DNA fingerprints must be objective and quantitative. The use of image analysis technology for this purpose is most useful and almost mandatory when a great number of strains are to be analyzed and the molecular weights of the different fragments compared (Lavallee *et al.*, 1990). Figure 15.3 shows the DNA fingerprints (RFLP analysis) of the reference strain *S. cerevisiae* 71B isolated as a pure culture (lane 6) and from batches of commercial cultures produced in

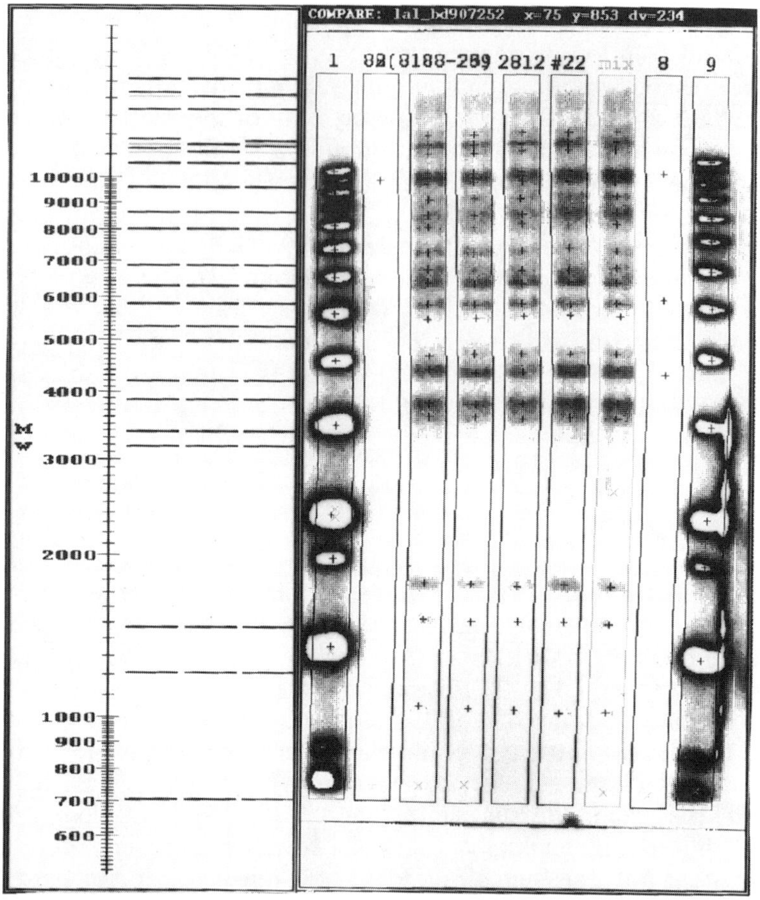

FIGURE 15.3 Restriction fragment length polymorphism analysis of DNA of commercial and control batches of *Saccharomyces cerevisiae* strain Lalvin 71B. Lanes 1 and 9, molecular weight standards (1Kb ladder); lanes 3, 4 and 5, production lots 1988, 1989, 1990; lane 6, pure culture (#22); lane 7, mix of cultures. The five patterns are identical.

1988, 1989, and 1990 (lanes 3, 4, 5). Lane 7 shows a fingerprint of a mix of the strains. All the patterns were identical (within 2.5% of each other), indicating preservation of strain homogeneity on large scale culture, and after several years of culture maintenance. So far, we have fingerprinted over 400 strains of our collection of yeasts of industrial origin and have extensively used this system for quality control purposes.

Fingerprinting of yeast DNA is also a valuable tool for the selection of yeast strains since it would assist in preventing the re-selection of a previous isolate. We have been actively working in collaboration with different research institutes for the selection of new strains that have banding patterns different from those of strains already selected.

LEUCONOSTOC OENOS

In addition to the traditional alcoholic fermentation, malolactic fermentation plays an important role in winemaking. This fermentation is conducted to reduce wine acidity, impart improved organoleptic properties and insure microbial stability (Davis *et al.*, 1985). It causes the transformation of malic acid into lactic acid and carbon dioxide.

Although certain species of *Leuconostoc*, *Lactobacillus* and *Pediococcus* are capable of conducting this fermentation, strains of *Leuconsostoc* are preferred, probably because of their greater resistance to low pH. Excellent reviews of the malolactic fermentation and its implications in winemaking have been published by Edwards and Beelman (1989) and Krieger *et al.* (1990). Since production of *Leuconostoc* cultures is very similar to that of other cultures of lactic acid bacteria, only a brief summary will be presented here.

Physiology

Leuconostoc oenos is a Gram positive coccus with cells arranged in pairs and short chains. This organism is a facultative anaerobe capable of fermenting glucose; its growth is optimal at pH 4.5 and is stimulated by tomato juice factor. Henick-Kling and Cox (1990) have provided more insight on the physiology of *Leuc. oneos* by showing that, under non-growing conditions, it would first degrade malic acid instead of utilizing glucose. This reaction is essentially an energy (ATP) producing reaction occurring via a chemiosmotic mechanism whereby ATP synthesis is energized by means of a transmembrane proton gradient. Moreover, malic acid degradation is further stimulated by growth at low pH, as exists during winemaking. Malic acid degradation appears to be an adaptation process whereby ATP can be synthesized for further metabolic activity.

Further knowledge of the biochemistry and physiology of *Leuc. oenos* will enable development of commercial preparations better adapted to growth and the expression of malolactic activity under the extreme and variable conditions of winemaking. Such conditions include growth in the presence of SO_2, ethanol, pesticide residues and medium chain fatty acids, growth at low pH and low temperatures, and resistance to bacteriophages.

Selection and Identification

As for yeast, selection of strains well-adapted to performance under winemaking conditions is of the utmost importance, because winemaking practice varies from one region to another, as does wine pH and temperature. Therefore, strains are usually isolated from wine samples undergoing active malolactic fermentation and are subsequently purified and identified by conventional taxonomic methods (Kunkee, 1984).

Industrial Production

Unlike wine yeasts, the production of bacteria for winemaking started very recently. It is usually done by companies producing starter cultures for the dairy industry. The process itself is much simpler than that used for yeasts since batch fermentations are used and the number of successive steps in the plant is very limited. However, special attention should be given to the cleaning, sanitation and sterilization of equipment since *Leuc. oenos* grows very slowly and can be readily overtaken by contaminants.

Although technology for the commercial production of *Leuc. oenos* is similar to that used for the production of dairy starter cultures, the fastidious nutritional requirements of the organism preclude the use of conventional media. This is a major difference with wine yeasts which are propagated on a medium identical to that used for baker's yeast and for which the same principles apply.

Strains of *Leuc. oenos* are usually kept frozen or lyophilized but may also be kept on agar slants. Contrary to yeasts, propagation of *Leuc. oenos* is done batchwise throughout the entire sequence of fermentations. The first series of fermentations are conducted in small flasks for 3–5 days at temperatures ranging between 20 and 25°C, depending upon the strain. Since *Leuc. oenos* grows very slowly, only a limited number of transfers from one flask to another are carried out to limit the risks of contamination. The final propagation is done under oxygen limitation in vessels ranging from 500 to 2000 l, depending on the amount of bacteria required. Samples are taken at intervals for quality control and process control. Champagne *et al.* (1989) have successfully grown strains of *Leuc. oenos* on an apple-juice medium. Naouri *et al.* (1989) have considered the influence of culture variables such as medium composition, malic acid concentration, pH and temperature on the biomass yield and malolactic activity of cultures of *Leuc. oenos* and some species of *Lactobacillus*.

Harvesting is one of the most critical steps, because cells of *Leuc. oenos* are highly sensitive to the stress of centrifugation. Centrifugation, when possible, is done in batch centrifuges which give 20–100 fold concentration of the cells. Membrane microfiltration, on the other hand, can also be used and will allow 15–20 fold concentration. The concentrated cells are frozen and lyophilized before being analyzed for their viability and malolactic activity. Because of the great sensitivity of this microorganism to these different stresses, the number of viable cells per g will vary from one strain to the other but commercial products are also offered on a constant dose basis (9×10^{12} viable cells/g for instance).

The concentration of viable cells, malic acid degrading activity and microbiological quality are the three major parameters that are analyzed with regard to quality of the final product. DNA fingerprints of *Leuc. oenos* are also used in our laboratory for identification. Commercial preparations are usually produced with less than 10^3 contaminating bacteria/g.

Commercial preparations of *Leuc. oenos* are sold either as a pure culture or as a mixture of strains in order to minimize the risk of phage infection (Gnaegi and

Sozzi, 1983; Henick-Kling *et al.*, 1986). The concept of using mixed strains is not new and has been exploited by the dairy industry for several decades. Some winemakers still prefer to use pure cultures but there is a trend towards the use of mixed cultures.

Rehydration

Proper rehydration of commercial preparations is the key towards successful induction of malolactic fermentation. Although some reports have mentioned that inoculation with *Leuc. oenos* at the beginning of the alcoholic fermentation could result in the formation of acetic acid (Lafon-Lafourcade, 1983), other reports have indicated that this practice was acceptable (Blackburn, 1984). Therefore, proper adaptation of the seed culture must be done in regard of the cellar practice.

Since it is difficult to suggest a standard rehydration procedure, Figure 15.4 illustrates a protocol as suggested by Valade and Laurent (1990) for induction of malolactic fermentation in Champagne. Concentration of SO_2, pH and temperature are the most important parameters to be taken into account. Such a protocol should enable a winery to ensure a proper malolactic fermentation at a cost of less than $1.00/hl.

CONCLUSION

The strains of dry yeast currently available to the wine industry were traditionally selected for their positive influences on the winemaking process and on the quality of the wines. Today, new criteria related to consumer concerns about health and environmental issues are gaining in popularity. The demand for low alcohol wine and ethyl carbamate free beverages are only two examples of these new trends. The detection of ethyl carbamate (also known as urethane), in wines has prompted the selection of strains producing only traces of urea, a known precursor of ethyl carbamate. While some commercially available strains (e.g. EC) excrete urea and later reabsorb it, others (e.g. 71B) do not excrete urea at all (An and Ough, 1990; Ough *et al.*, 1990). Such strains are recommended for musts rich in nitrogen where arginine will be split into urea and carbon dioxide by yeast arginase.

Future demands by enologists may also see the commercial production of yeasts other than *S. cerevisiae*. Selected species and strains of *Kloeckera* and *Candida* may come into this category.

The number of commercially available strains of malolactic bacteria is still rather limited, but likely to increase in the future. It is likely that new strains will be better adapted to the winemaking practice of specific regions. The influence of malolactic bacteria on the organoleptic properties of wine is an area of interest which needs further research and will help guide producers in selecting strains for commercial development.

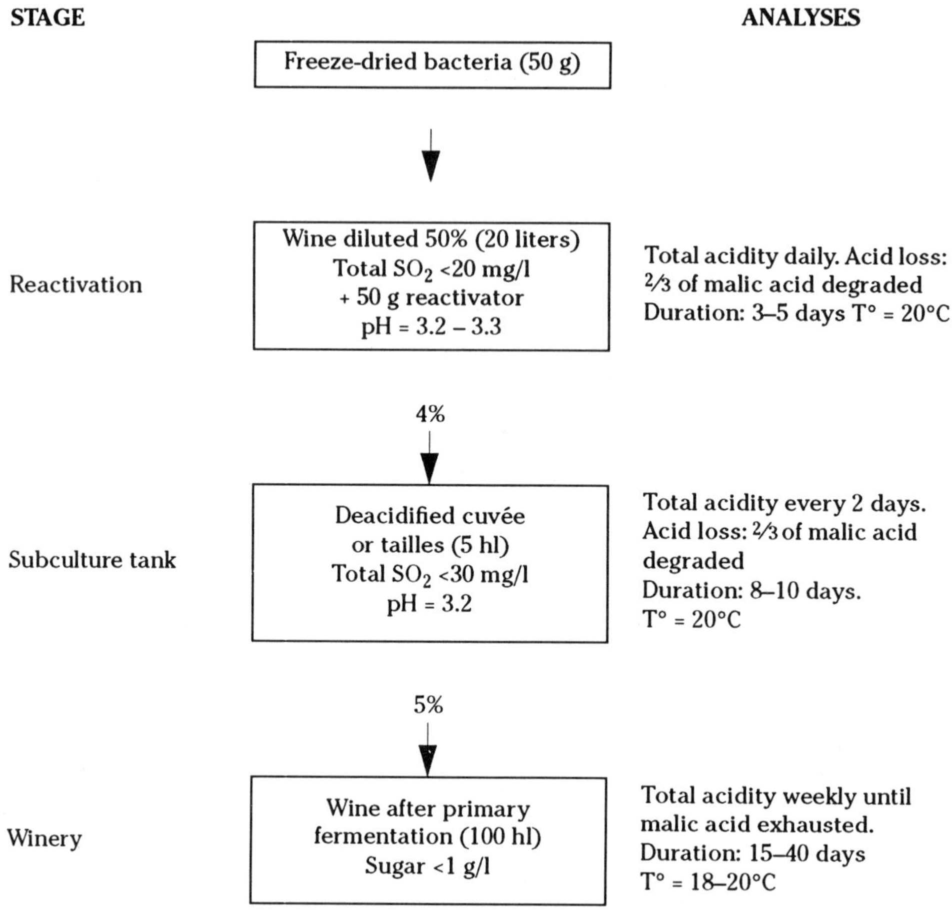

FIGURE 15.4 Protocol for rehydration and subculture of *Leuconostoc oenos* (from Valade and Laurent, 1990).

Solutions to new concerns of consumers and traditional requirements of enologists will only be reached by a better understanding of yeast and bacterial physiology. Although strains will continue to be selected as a result of long and tedious traditional work, it is likely that, in the future, strains will be genetically tailored to meet specific requirements such as the excretion of protease, pectinase, arginase, flavor and aroma releasing enzymes and malolactic activities. The pace at which such strains will be developed and produced industrially will depend to a great extent on the limitations imposed by regulatory agencies on the production and release of genetically engineered microorganisms. In this regard, an interesting precedent may have been set in the United Kingdom, where regulatory authorities have recently approved the release and commercial production of a genetically modified strain of baker's yeast (Aldhous, 1990).

REFERENCES

Aldhous, P. (1990) Modified yeast for fine food. *Nature*, **344**, 186.

Alexander, M.A. and Jeffries, T.W. (1990) Respiratory efficiency and metabolite partitioning as regulatory phenomena in yeasts. *Enzyme and Microbial Technology*, **12**, 2–19

An, D. and Ough, C.S. (1990) Urea execution and uptake from yeast as affected by various factors. In *Proceedings of the American Society of Enology and Viticulture*, (Los Angeles, 1990), 9.

Bach, H.P., Woehrer, W. and Roehr, M. (1978) Continuous determination of ethanol during aerobic cultivation of yeasts. *Biotechnology and Bioengineering*, **20**, 799–807.

Barre, P. (1980) Rôle du facteur killer dans la concurrence entre souches de levures. *Bulletin de l'Office International de la Vigne et du Vin*, 593–594, 560–572.

Beker, M.J. and Rapoport, A.I. (1987) Conservation of yeasts by dehydration. *Advances in Biochemical Engineering and Biotechnology*, **35**, 127–171.

Benda, I. (1982) Wine and brandy. In *Industrial Microbiology*, edited by G. Reed, pp. 293–402. Westport: AVI Publishing Company, Inc.

Bender, G.R. and Marquis, R.E. (1987) Membrane ATPases and acid tolerance of *Actinomyces viscosus* and *Lactobacillus casei*. *Applied and Environmental Microbiology*, **53**, 2124–2126.

Beudeker, R.F., van Dam, H.W. and van Der Plaat, J.B. (1990) Developments in baker's yeast production. In *Yeast Biotechnology and Biocatalysis*, edited by H. Verachtert and R. De Mot, pp. 103–146. New York: Marcel Dekker Inc.

Bisson, L.F. (1990) Yeast DNA analysis by chromosome karyotyping. In *Proceedings of the Second Lalvin Conference*, (Bordeaux, 11–13 March, 1990) pp. 1–14.

Bisson, L.F. and Fraenkel, D.G. (1983) Involvement of kinases in glucose and fructose uptake by *Saccharomyces cerevisiae*. *Proceedings of the National Academy of Sciences, U.S.A.*, **80**, 1730–1734.

Blackburn, D. (1984) Present technology in the use of malolactic bacteria. *Practical Winery*, **5**, 64–72.

Blondin, B. and Vézinhet, F. (1988) Identification de souches de levures oenologiques par leurs caryotypes obtenus en électrophorèse en champ pulsé. *Revue Francaise d'Oenologie*, **28**, 7–11.

Burrows, S. (1970) Baker's yeast. In *The Yeasts*, Vol. 3, edited by A.H. Rose and J.S. Harrison, pp. 349–320. London: Academic Press.

Champagne, C., Gardner, N. and Doyon, G. (1989) Production of *Leuconostoc oenos* biomass under pH control. *Applied and Environmental Microbiology*, **55**, 2488–2492.

Chandrasekar, I. and Graber, B.P. (1988) Stabilization of the biomembrane by small molecules: interaction of trehalose with the phospholipid bilayer. *Journal of Biomolecular Structure and Dynamics*, **5**, 1163–1171.

Cooper, T.G. (1982) Nitrogen metabolism in *Saccharomyces cerevisiae*. In *The Molecular Biology of Yeast Saccharomyces*, Vol. 2, edited by J.N. Strathern, E.W. Jones and J.R. Broach, pp. 39–99. New York: Cold Spring Harbor Laboratory, Cold Spring Harbor.

Crowe, J.H., Crowe, L.M. and Chapman, D. (1984) Preservation of membranes in anhydrobiotic organisms: the role of trehalose. *Science*, **223**, 701–703.

D'Amore, T., Panchal, C.J., Russel, I., and Stewart, G.G. (1988) Osmotic pressure effects and intracellular accumulation of ethanol in yeast during fermentation. *Journal of Industrial Microbiology*, **2**, 365–372.

Darriet, P., Boidron, J.N. and Dubourdieu, D. (1988) L'hydrolyse des hétérosides terpéniques du muscat à petits grains par les enzymes périplasmiques de *Saccharomyces cerevisiae*. *Connaissance de la Vigne et du Vin*, **22**, 189–195.

Davis, C.R., Wilbowo, D., Eschenbruch, R., Lee, T.H. and Fleet, G.H. (1985) Practical implications of malolactic fermentation: A review. *American Journal of Enology and Viticulture*, **36**, 290–301.

Degré, R. (1989) Première rencontre oenologique Lalvin. *Revue Francaise d'Oenologie*, **120**, 59–60.

Degré, R., Mailhiot Thomas, D., Ash, J., Morin, A. and Dubord, C. (1989a) Wine yeast strains identification. *American Journal of Enology and Viticulture*, **40**, 309–315.

Degré, R., Thomas, D., Frenette, J., and Mailhiot, K. (1989b) Identification de souches de levures oenologiques par la caractérisation de leur ADN à l'aide de soudes moléculaires spécifiques. *Revue Française d'Oenologie*, **119**, 23–26.

Delteil, D. (1990) Contrôle de l'efficacité du levurage par suivi de l'implantation d'une souche oenologique marquée génétiquement. In the *Proceedings of the 2nd Lalvin Conference*, Bordeaux, 11–13 March, pp. 115–120.

Dubourdieu, D., Darriet, P., Ollivier, Boidron, J.N. and Ribereau-Gayon, P. (1988) Rôle de la levure *Saccharomyces cerevisiae* dans l'hydrolyse enzymatique des hétérosides terpéniques du jus de raisin. *Comptes Rendus Academie des Sciences, Paris T.* 306, Série III, 489–493.

Duteurtre, B., Ors, P., Charpentier, M. and Hennequin, D. (1990) In *Proceedings of the 3rd Conference of L'Association des techniciens supérieurs en viticulture et oenologie.* Epernay, June 6, pp. 3–7.

Edwards, G.E. and Beelman, R.B. (1989) Inducing malolactic fermentation in wines. *Biotechnology Advances*, **7**, 333–360.

Gnaegi, F. and Sozzi, T. (1983) Les bactériophages de *Leuconostoc oenos* et leur importance oenologique. *Bulletin de l'Office International de la Vigne et du Vin*, **627**, 352–357.

Hallet, J.N., Craneguy, B., Zucca, J. and Poulard, A. (1988) Caractérisation de différentes souches industrielles de levures oenologiques par les profils de restriction de leur ADN mitochondrial. *Progrès Agricole et Viticole*, **105**, 328–333.

Henick-Kling, T., Lee, T.H. and Nicholas, D.J.D. (1986) Characterization of the lytic activity of bacteriophages of *Leuconostoc oenos* isolated from wine. *Journal of Applied Bacteriology*, **61**, 525–534.

Henick-Kling, T. and Cox, D.J. (1990) Energy production from malolactic fermentation and growth at low pH. *Proceedings of the American Society for Enology and Viticulture*, Los Angeles, June 28–30, p. 4.

Henschke, P. and Rose, A.H. (1991) Plasma membranes. In *The Yeasts*, Volume 4, *Organelles*, edited by A.H. Rose and J.S. Harrison, pp. 297–346. London: Academic Press.

Josic, D. (1982) Optimization of process conditions for the production of active dry yeast. *Lebensmittel-Wissenschaft und Technologie*, **15**, 5–13.

Kirsop, B.E. (1988) Culture and preservation. In *Living Resources for Biotechnology. Yeasts*, edited by B.E. Kirsop and C.P. Kurtzman, pp.74–78. Cambridge: Cambridge University Press.

Krallish, I.L., Damberga, B.E. and Beker, M.J. (1986) The effect of yeast dehydration on the activity of a number of enzymes of cell energetic metabolism. *Applied Microbiology and Biotechnology*, **23**, 482–486.

Kraus, J.K., Scopp, R. and Chen, S.L. (1981) Effect of rehydration on dry wine yeast activity. *American Journal of Enology and Viticulture*, **32**, 132–134.

Kraus, J.K., Reed, G. and Villettaz, J.C. (1984) Levures sèches actives de vinification. *Connaissance de la Vigne et du Vin*, **18**, 1–26.

Krieger, S.A., Hammes, W.P. and Henick-Kling, T. (1990) Induction of malolactic fermentation with starter cultures under practical winemaking conditions. *Vineyard and Winery Management*, November/December, 45–50.

Kunkee, R.E. (1984) Selection and modification of yeasts and lactic acid bacteria for wine fermentation. *Food Microbiology*, **1**, 315–332.

Lafon-Lafourcade, S. (1983) Wine and brandy. In *Biotechnology* Volume 5, edited by G. Reed, pp. 81–164. Weinheim: Verlag Chemie.

Lavallée, F., Thomas, D. and Degré, R. (1990) Computerized wine yeast strains identification. *Proceedings of the American Society of Enology and Viticulture*, Los Angeles, June 28–30, p. 7.

Lee, C.W.B., Waugh, J.S. and Griffin, R.G. (1986) Solidstate NMR study of trehalose/1.2-dipalmitoyl-sn-phosphatidylcholine interactions. *Biochemistry*, **25**, 3737–3742.

Loureira Dias, M.C., Pampulha, E. and Santo, S. (1989) Studies on the effect of ethanol and of acetic acid on yeast performance. *Proceedings of the XIII International Specialized Symposium on Yeasts*. Louvain Sept. 18–22, pp. 26–27.

Marzuki, S. and Linnane, A. (1979) Modification of yeast mitochondria by diet in specific mutants. *Methods in Enzymology*, **61**, 568–572.

McClellan, C.J., Does, A.L. and Bisson, L.F. (1989) Characterization of hexose uptake in wine strains of *Saccharomyces cerevisiae* and *Saccharomyces bayanus*. *American Journal of Enology and Viticulture*, **40**, 9–15.

Naomi, P., Chagnaud, P., Arnaud, A., Galzy, P. and Mathieu, J. (1989) Optimization of the conditions for preparing bacterial cultures for malolactic bioconversion. *Journal of Biotechnology*, **10**, 135–150.

Office International de la Vigne et du Vin (1988) Situation de la vitculture dans le monde. In *Bulletin de l'Office International de la Vigne et du Vin*, pp. 691–692.

Orlowski, S. and Barford, J. (1987) The effect of inoculum preparation on the fully aerated growth of *Saccharomyces cerevisiae* with a glucose substrate. *Journal of General and Applied Microbiology*, **33**, 113–121.

Ough, C.S., Stevens, D., Sendovski, T., Huang, A. and An, D. (1990) Factors contributing to urea formation in commercially fermented wines. *American Journal of Enology and Viticulture*, **41**, 68–73.

Oura, E. (1983) Biomass from carbohydrates. In *Biotechnology*, edited by H. Dellweg, pp. 4–41. Weinheim: Verlag Chemie.

Panek, A.D. (1975) Trehalose synthesis during starvation of baker's yeast. *European Journal of Applied Microbiology*, **2**, 39–46.

Panek, A.D. (1985) Trehalose metabolism and its role in *Saccharomyces cerevisiae*. *Journal of Biotechnology*, **3**, 121–130.

Panek, A.D. (1991) Storage carbohydrates. In *The Yeasts*, Volume 4, *Organelles*, edited by A.H. Rose and J.S. Harrison, pp. 655–678. London: Academic Press.

Pardo, I., Blondin, B. and Vézinhet, F. (1989) Analysis of karyotypes of oenological yeast strains of *Saccharomyces cerevisiae*. *Proceedings of the XIII International Specialized Yeast Symposium conference*. Louvain, Sept. 18–22, pp. 55–59.

Parks, L.W., Mclean, B.C., Taylor, R.R. and Hough, S. (1978) Sterols in yeast subcellular fractions. *Lipids*, **13**, 730–735.

Pasteur, L. (1866) *Etudes sur le vin, ses maladies, causes qui les provoquent*. Paris: Imprimerie Impériale.

Peppler, H.J. (1967) Yeast Technology. In *Microbial Technology*, edited by H.J. Peppler, pp. 145–171. New York: Reinhold Publishing Corporation.

Petrik, M., Kappeli, O. and Fiechter, A. (1983) An expanded concept for the glucose effect in the yeast *Saccharomyces uvarum*: involvement of short- and long-term regulation. *Journal of General Microbiology*, **129**, 43–50.

Pilkington, B.J. and Rose, A.H. (1989) Accumulation of sulphite by *Saccharomyces cerevisiae* and *Zygosaccharomyces cerevisiae* as affected by phospholipid fatty-acyl unsaturation and chain length. *Journal of General Microbiology*, **135**, 2423–2428.

Pomerleau, Y. and Perrier, M. (1990) Estimation of multiple specific growth rates in bioprocesses. *American Institute of Chemical Engineers Journal*, **36**, 207–215.

Radler, F., Dietrich, K. and Schönig, I. (1985) Mikrobiologische Prüfung von Trockenhefepräparaten für die Weinbereitung. *Deutsche Lebensmittel-Rundschau*, **3**, 73–77.

Reed, G. (1982) Production of baker's yeast. In *Industrial Microbiology*, edited by G. Reed, pp. 593–633. Wesport: AVI Publishing Company.

Reed, G. and Peppler, H.J. (1973) *Yeast technology*, p. 212. Westport: AVI Publishing Co.
Rieger, M., Käppeli, O. and Fiechter, A. (1983) The role of limited respiration in the incomplete oxidation of glucose by *Saccharomyces cerevisiae. Journal of General Microbiology*, **129**, 653–661.
Rosen, K. (1987) Preparation of yeast for industrial use in production of beverages. In *Yeast Biotechnology*, edited by D.R. Berry, I. Russel and G.G. Stewart. pp. 471–500. Winchester, Massachusetts: Allen and Unwin Inc.
Ryu, D.D., Kim, Y.J. and Kim, J.H. (1984) Effect of air supplement on the performance of continuous ethanol fermentation system. *Biotechnology and Bioengineering*, **26**, 12–16.
Sonnleitner, B. and Käpelli, O. (1986) Growth of *Saccharomyces cerevisiae* is controlled by its limited respiratory capacity: Formulation and verification of a hypothesis. *Biotechnology and Bioengineering*, **28**, 927–937.
Sweere, A.P.J., Matla, Y.A., Zandvliet, J., Luyben, K.Ch. A.M. and Kossen, N.W.F. (1988) Experimental simulation of glucose fluctuations. *Applied Microbiology and Biotechnology*, **28**, 109–115.
Thevelein, J.M. (1984a) Cyclic-AMP content and trehalase activation in vegetative cells and ascopores of yeast. *Archives of Mikrobiology*, **138**, 64–67.
Thevelein, J.M. (1984b) Regulation of trehalose mobilization in fungi. *Microbiology Reviews*, **48**, 42–59.
Thorne, R.S.W. (1954) Fermentation velocity of yeasts. III—a study of the fermentation velocities of a series of bottom and top fermenting brewery yeasts. *Journal of the Institute of Brewing*, **6**, 238–242.
Trivedi, N.B., Jacobson, G.K. and Tesch, W. (1986) Baker's yeast. *CRC Critical Reviews in Biotechnology*, **24**, 75–109.
Trouessart, E. (1890) Les ferments et les fermentations industrielles. In *Les Microbes*, edited by Em. Alglave, pp. 55–70, Paris: Félix Alcan.
Valade, M. and Laurent, M. (1990) Control of malolactic fermentation in Champagne. *Proceedings of the American Society of Enology and Viticulture*, p. 6, Los Angeles, June 28–30.
Valade, M. and Rinville, X. (1990) Approches technique et économique des différents procédés de prise de mousse en bouteilles. In *Proceedings of the 3rd Conférence de l'Association des Techniciens supérieurs en Viticulture et Oenologie.*, June 6, Epernay, pp. 34–40.
Van Laere, A. (1989) Trehalose, reserve and/or stress metabolite. *FEMS Microbiology Letters Reviews*, **63**, 210.
van Urk, H., Vull, W.S.L., Scheffer, W.A. and van Dijken, J.P. (1990) Transient-state analysis of metabolite fluxes in Crabtree-positive and Crabtree-negative yeasts. *Applied and Environmental Microbiology*, **56**, 281–287.
Vézinhet, F., Blondin, B. and Hallet, J.N. (1990) Chromosomal DNA patterns and mitochondrial DNA polymorphism as tools for identification of enological strains of *Saccharomyces cerevisiae. Applied Microbiology and Biotechnology*, **32**, 568–571.
Vézinhet, F. and Lacroix, S. (1984) Marquage génétique des levures: outil de contrôle des fermentations en souche pure. *Bulletin de l'Office International de la Vigne et du Vin*, **57**, 769–773.
Vos, P.J.A. and Gray, R.S. (1979) The origin and control of hydrogen sulfide during fermentation of grape must. *American Journal of Enology and Viticulture*, **30**, 187–197
Wainwright, T. (1971) Production of H_2S by yeasts: Role of nutrients. *Journal of Applied Bacteriology*, **34**, 161–171.
White, J. (1954) *Yeast Technology*, 432 pp., London: Chapman and Hall.
Wills, C. (1990) Regulation of sugar and ethanol metabolism in *Saccharomyces cerevisiae. Biochemical and Molecular Biology*, **25**, 245–280.
Woehrer, W. and Roehr, M. (1981) Regulatory aspects of baker's yeast metabolism in aerobic fed-batch cultures. *Biotechnology and Bioengineering*, **23**, 567–581.

Zaworski, P.G. and Heimsch, R.C. (1987) The isolation and characterization of flocculent yeasts. In *Biological Research on Industrial Yeasts III*, edited by G.G. Hiebsch, pp. 185–195. Boca Raton, Florida: CRC Press, Inc.

Zikmanis, P.B., Auzina, L.P. and Beker, M.J. (1982) Relationship between the fatty acid composition of lipids and the viability of dried *Saccharomyces cerevisiae*. *Applied Microbiology and Biotechnology*, **15**, 100–103.

Zikmanis, P.B., Auzane, S.I. and Beker, M.J. (1983) Interrelationship between the fatty acid composition and metabolic patways upon dehydration-rehydration of *Saccharomyces cerevisiae*. *Applied Microbiology and Biotechnology*, **18**, 198–302.

Zikmanis, P.B., Auzane, S.I., Auzina, L.P. and Beker, M.J. (1985) Changes of ergosterol content and resistance of population upon drying-rehydration of the yeast *Saccharomyces cerevisiae*. *Applied Microbiology and Biotechnology*, **20**, 265–267.

ACKNOWLEDGEMENTS

The author thanks his colleagues for their assistance and discussion and Lallemand Inc. for permission to publish this paper.

Chapter 16

BIOREACTOR TECHNOLOGY AND WINE FERMENTATION

CHARLES DIVIES

Microbiology Laboratory, ENSBANA, University of Bourgogne, Dijon, France

Chapter Outline

INTRODUCTION

Winemaking is a very old endeavor because of the natural character of the product. Essentially, crushing of the grapes is all that is necessary; yeasts originating from the surface of the berries immediately start to grow, increasing from several thousand to several million cells per milliliter of juice as they conduct fermentation. The transformation of juice results in the production of wine.

Paintings on the tomb of Nakht, a scribe responsible for the silos and wine cellar of Thoutmosis III (1505–1450 BC), describe the different winemaking operations used at that time. Nubian slaves carefully harvested clusters of grapes from a vine, crushed them underfoot and recovered the juice in amphora jars that were closed with a pierced cap. A similar scene was found on a 15th century tapestry in the Cluny museum in France, three thousand years later. Winemaking technique had improved by that time: grapes were transported in hand-held baskets or large baskets on the back, crushing was completed by pressing and the amphora was replaced by the wooden cask. In that era of "spontaneous" making of natural wine, the divine beverage was often cloudy, lost its color and rapidly turned to vinegar. A number of documents testify to this and give recipes for its stabilization (Rougier de la Bergerie, 1821). Several decades were still required before the work of Pasteur would lead to an understanding of the mechanism of fermentation. Only then did aspects of fermentation and transformation of the grape to wine begin to be mastered (Vallery-Radot, 1900).

The industrial age saw the construction of machines for the wine industry. The use of both rational principles and appropriate equipment for winemaking lead to improved quality of wines throughout the world.

Winemaking involves two principal operations; first, preparation of the grape must to tailor its composition and maintain the qualities of the grape at harvest; and second, to conduct microbial fermentation through rational exploitation of the biochemical activities of yeasts and lactic acid bacteria. This Chapter will focus on aspects of the technology of fermentation, initially dealing with conventional bioreactor systems, followed by discussions of the application of novel bioreactor systems to winemaking.

CONVENTIONAL BIOREACTOR SYSTEMS

There is considerable diversity in bioreactor technology available to winemakers, leading to a wide variety of fermented wines (Lichine, 1980). Over the centuries, enology has accumulated numerous pragmatic acts which, in fact, do not correspond to an optimization of the winemaking process. This is why countries only recently engaged in this activity have been able to develop their wine industry with the aid of a more scientific approach, and apply some of the newer advances in bioreactor technology.

White Wines

White wine is produced by fermenting juice, extracted without macerating the solid parts of the grape cluster. Lack of maceration is not an absolute factor and, in some cases, short maceration of the skin is carried out (Hanshoffer, 1978; Pallota, 1983; Lefur, 1990). Regardless of the extraction process, the juice and sometimes that of the first pressing, alone furnish high quality products. These juices remain sensitive to oxidation, which is why they are often protected with the addition of sulfur dioxide (SO_2). Clarification of the must before fermentation is recognized as indispensable and is done either by a static settling process, by filtration or by centrifugation. Musts which have been insufficiently clarified generally furnish crude wines with grassy tastes. A maximum of fragrant substances (e.g. acetates of higher alcohols, esters of fatty acids) is formed by the yeast when clarified juices are fermented at temperatures not exceeding 15–20°C.

Red Wines

Traditional maceration

Red wine is a maceration wine. In the traditional method, maceration and fermentation are practically simultaneous. The time of fermentation is varied according to the style and quality of wine required. Short fermentations are suitable for light wines, while longer times are used for wines destined for long-term keeping. It is aimed to conduct the fermentation at a relatively high temperature, between 25 and 30°C, since this favors color extraction. Yeast populations are higher than in white wine fermentation because the medium is richer in nitrogenous compounds and in activating molecules. Total fermentation of the juice sugars generally requires 5–8 days. A potential problem can occur if there is excessive heating of the fermentation medium. Fermentation can be stimulated by slight aeration which is done by circulating or "pumping over" the wine at "remontages". This movement also fulfills the function of color extraction (Ribéreau-Gayon *et al.*, 1982; Somers and Evans, 1979).

The equipment used for fermentation has undergone extensive improvement. "Autopiegeage" tanks (where liquid is automatically recycled through the cap) are an improvement over the difficult and long manual method. We may cite rotating tanks and stationary tanks with an internal rotating propeller (Peyron and Feuillat, 1985). These models also enable automatic emptying. It is extremely important to avoid excessive breaking up of the cap, or marc, which would lead to wines with too much sediment. Daily monitoring to determine the progression of maceration remains of fundamental importance (Ribéreau-Gayon *et al.*, 1982).

Carbonic maceration

In this style of winemaking, maceration is conducted under an atmosphere of carbon dioxide without crushing of grapes. The anaerobic life of the intact grapes is the maceration phase of the process. The juice elaborated by maceration contains ethanol (1.5 to 2.5 g/l), malic acid has been partially degraded, and it is enriched with nitrogenous compounds, inorganic matter, polyphenols and aromatic compounds. The second phase includes the subsequent steps of alcoholic fermentation after pressing. Wines thus obtained are supple and aromatic, and can be drunk young ("primeur") (Flanzy and André, 1973).

Thermovinification

Heating the grapes is another maceration process which has undergone considerable development. The harvested grapes are crushed, generally with stalks removed, and the macerate is heated to about 70°C for 20–30 min before pressing. The juice so extracted is then cooled and fermented. This process accelerates extraction of pigments from the grape skins, so that the juice obtains a degree of color within minutes, compared with days that would be required by normal fermentation. This color decreases during fermentation, always done at 25°C, and reaches levels identical to those of classical maceration (Martinière, 1981).

Continuous Winemaking

Continuous processes have been developed for red winemaking, but are not extensively used. In these processes, maceration and fermentation are done in tower fermenters, generally cylindrical or cylindro-conical, generally metallic and with capacities between 200 and 4000 hl. Grapes are continuously supplied throughout the duration of the harvest (15–20 days). They are introduced at the bottom of the tower and the wine emerges at the top through an adjustable level overflow. Solids are carried to the top where they form a cap, or marc, as in traditional vessels. They are then extracted mechanically (rake, Archimedean screw), drained and sent to a continuous press. The rate of grape supply to the process is calculated as a function of maceration time, but it also depends on variables related to the harvest.

The first installations of this technology were made by Cremaschii in Argentina in 1948. The Ladousse fermenter installed in France in 1955, was described by Dupuy and Poux (1965); the Vico fermenter of 1967 was analyzed by Brugirard *et al.* (1970). In Italy, the Defracheschi (1955) and Padovan fermenters (Guetor, 1969) were characterized by resuspension of the marc, making its extraction more rapid.

Continuous fermentation processes require relatively large quantities of SO_2, of the order of 80 to 100 mg/l, as well as routine analytical testing of the wine for L-lactic acid for the early detection of contamination by lactic acid bacteria

(Kunkee and Goswell, 1977). This system is not suitable for preparing vintage wines, which require long maceration times, and for small enterprises. Economically, continuous winemaking leads to a 50% reduction in the space required for juice processing and to a saving in salaries of more than 20% (Ribéreau-Gayon *et al.*, 1982).

Fermentation Vessels

Fermentation is done conventionally in a batch system. The volume of the vessels in which fermentations are conducted has increased considerably during recent years. Previously, fermentation was done in 225–228 l barrels or in 6–12 hl vats. Such wooden or cement vessels have now disappeared, being replaced by well-designed stainless steel fermenters. The different shapes of fermenters used in beverage fermentation (Maule, 1986; Moresi, 1989a) are illustrated in Figure 16.1. In the wine industry, stainless steel cylindrical tanks of volume 1–340 m^3 are primarily used. These tanks are easy to clean and maintain, and may be constructed with a jacket in which cooling liquid is circulated for temperature control (Moresi, 1989b).

Microbial Technology

In many wine producing regions, grape juices are inoculated with pure cultures of *Saccharomyces cerevisiae* at 10^6–5.10^6 cells/ml. Inoculation conditions have been mastered by better preparation of the must. Strictly controlled experiments in the Champagne region and in the South of France have demonstrated the reliability of the inoculation technique (Vezinhet and Lacroix, 1984). Problems are occasionally encountered with slow fermentations and stuck or incomplete fermentation and, generally, these are related to the high concentration of sugar in grape juice at the onset of fermentation and, above all, the high alcohol concentration at the end (Strehaiano *et al.*, 1978, Jones and Greenfield, 1986).

In the course of alcoholic fermentation, the yeast remains under a double dependance: the glucose effect (Fiechter *et al.*, 1981; Kapelli, 1986) and the action of oxygen required to ensure growth and survival (Larue *et al.*, 1979; Traverso-Rueda, 1980; Lafon-Lafourcade, 1983; Ohno and Takahashi, 1986). Aeration of the must occurs when the tanks are filled but such action is far from being optimized. Sablayrolles and Barre (1986) showed that the optimal period for aeration was at the end of cell growth. Cutayar *et al.* (1989) proposed the use of a gas-permeable membrane to facilitate aeration. According to Tyagi (1984), the microbubbling of oxygen into the ferment would have the advantage of reducing the carbon dioxide concentration in the medium.

Temperature management is relatively unsophisticated in most wineries. Its aim, however, is to remove ethanol inhibition and, above all, to impart better sensory quality to the wine (Ough, 1966; van Uden, 1985). The temperatures used for white wine fermentation are generally lower than 20°C. At low

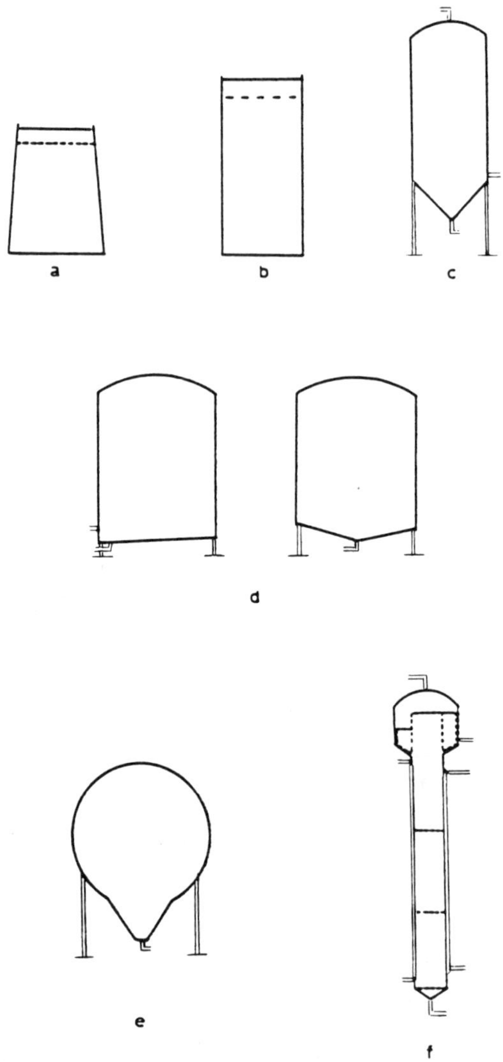

FIGURE 16.1 Shapes of the main bioreactors (fermentation vessels) used in the beverage industry. (a) barrel, (b) vat, (c) cylindroconical, (d) cylindrical, (e) spheroconical, (f) tower.

temperatures fewer higher alcohols are formed, there is a relative constancy in the production of propanol and isobutanol and there is a greater production of isoamyl alcohol, pentanol and 2-phenylethanol (Usseglio-Tomasset, 1988). Very important effects were noted for 2,3-butanediol (Schaeffer, 1989). In general, esters and organic acids are also more abundant (Poulard *et al.*, 1986). Barre and co-workers have drawn attention to the anisothermal nature of wine fermentations, especially for red wines where the difference between initial and maximal temperature of the one fermentation can exceed 10°C. Little is known about the effect of such variations on the metabolic behavior of the yeasts but it is known that they affect fermentation rate, as measured by carbon dioxide evolution and needs for limiting nutrients such as nitrogen substances and vitamins (Bezenger *et al.*, 1986; Sablayrolles *et al.*, 1987; Sablayrolles and Barre, 1989).

In contrast to beer fermentations (Slaughter, 1989), the effect of carbon dioxide has been studied little in enology, so its influences remain relatively unknown. At low pressures, about 0.5 atm, cell growth increases rapidly, then decreases to zero at 3 to 4 atm. In practice, a pressure increase is considered to neutralize the effect of excessive temperature on the formation of products.

Mixing is a very important parameter in beer fermentation but its role remains poorly determined in enology. Residual cloudiness, or the presence of particulate material in the must, has a substantial effect on the kinetics of fermentation, and also on the formation of by-products (alcohols and esters). Various mechanisms interact to produce these effects, such as availability of nutritional factors and adhesion of microbial cells to particulate material (Barre, 1988).

Perspectives

Basic knowledge concerning the microbial technology of alcoholic fermentation has progressed considerably over the last decade. Kinetic models that describe and predict the progress of fermentation as well as factors that affect this progress have been shown to be of great value (Boulton, 1980; Marc, 1988). A flow diagram of the different fluxes and interactions which are likely to occur is shown in Figure 16.2.

Modeling is continuing in a number of industrial and university laboratories. On line command and control of fermentation tanks should be made more effective by the implementation of these models. An application studied by the Institute of Vine Products, in Montpellier, has shown that piloting (regulating) tank temperature as a function of the rate of fermentation leads to a substantial saving in refrigeration costs, compared with isothermal regimes (Sablayrolles, 1988; El Haloui *et al.*, 1988, 1989). Similar concepts are discussed by Moresi (1989b).

However, on-line command and control of fermentation tanks are still limited by the availability of reliable sensors. Future progress in this area may be very rapid and it will then be possible to more effectively control and direct fermentation by making use of data obtained by winemakers and microbial physiologists.

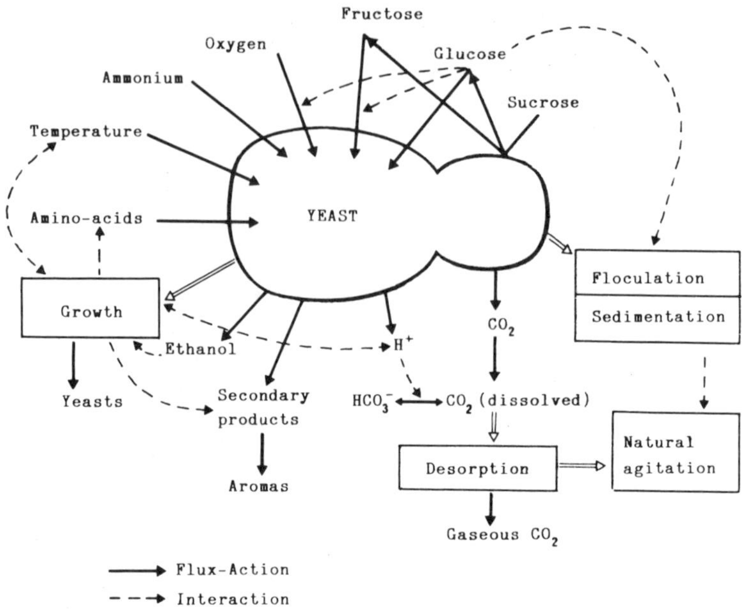

FIGURE 16.2 Block diagram of yeast-medium relationships (from Marc, 1988).

HIGH CELL DENSITY REACTORS

Productivity of fermentation is limited by the toxicity of accumulated ethanol (Aiba *et al.*, 1968; Goma, 1978). Thus, it is advantageous to carry out fermentation in successive reactors or, better still, in concentration gradient reactors. In this context, we mention the study of Egamberdiev and Ierusalimskii (1969) on the preparation of wines in cascade reactors, enabling a quality wine to be produced three times more rapidly. To our knowledge, this interesting study has not yet seen an industrial application.

Productivity can be also improved by increasing the concentration of the microbial cells to give what is often referred to as high cell density reaction systems. Generally, these systems require special technologies for confining or immobilizing the cells in order to maintain them at high concentrations. Recent reviews of these bioreactor technologies are those of Brodelius and Vandamme (1987) and Mehaia and Cheryan (1990). High cell density reactors offer the following prospects for enology.

1. Improved performance of alcoholic and malolactic fermentations,
2. Adaptation to continuous processes that can be better optimized and controlled,
3. Simplified systems for removing microbial cells from batch processes,
4. Smaller scale fermentation facilities, with reduced capital expenditure, and

5. Possibilities of using a wider variety of microbial strains including genetically modified organisms.

There are two major systems for confining or immobilizing microbial cells (Figure 16.3). These are:

1. Homogeneous systems where there is uniform distribution of the microbial biomass in the reaction medium. This result can be obtained using biomass recycling by centrifugation or maintenance of the biomass in a membrane reactor
2. Heterogeneous systems where two phases are distinguished; the liquid medium to be transformed and a solid phase containing the biomass.

Homogeneous Reactors

Centrifugation is a technique often proposed for recycling yeasts. Guidoboni (1984), reviewed five different processes, most of which used a single reactor with a centrifugation step. In these conditions, a juice with an alcohol titer of 8–10% contained a biomass close to 50 to 80 g/l instead of 5 g/l under classical conditions. Partial vacuum enabled the biomass to be increased to 125 g/l, thus leading to increased productivity (Cysewski and Wilde, 1978). The main disadvantages are maintenance of the centrifuges and substantially decreased viability of the yeast cells because of the stresses to which they are subjected.

Membrane bioreactors are an alternative to centrifugation. In this technology, microbial cells are confined within the reactor vessel by means of a membrane of pore size less than 0.45μ. Material to be processed is passed through the reactor, the transformed product emerging down stream from the membrane. Efficiency of transformation is increased by recycling the product through the reactor (Figure 16.3). This integral cell recycling system is less flexible than centrifugation because problems of clogging and unclogging of the membrane are not fully resolved (Mehaia and Cheryan, 1990).

Heterogeneous Reactors

The main techniques which enable biomass confinement are absorption on a support, autoflocculation and entrapment in gels (Figure 16.3).

Adsorption is very simple. Numerous studies in recent years have described the adhesion of microorganisms to varied supports (Marshall, 1985). In most cases, several steps must be considered (Mitchell, 1976). These include: prior adaptation of the polymer (polysaccharide, protein, etc.) to serve as the support; conditions to promote adsorption of the microbial cells; reversible adsorption; and irreversible adsorption. In contrast to lactic acid bacteria, yeasts readily form biofilms on the surface of supports (Navarro, 1978). Rapid desorption of the biofilm can result from an increase in the velocity of the liquid passing over its surface or by the autolysis of underlying cells during operation. The release of large quantities of gases during alcoholic fermentation

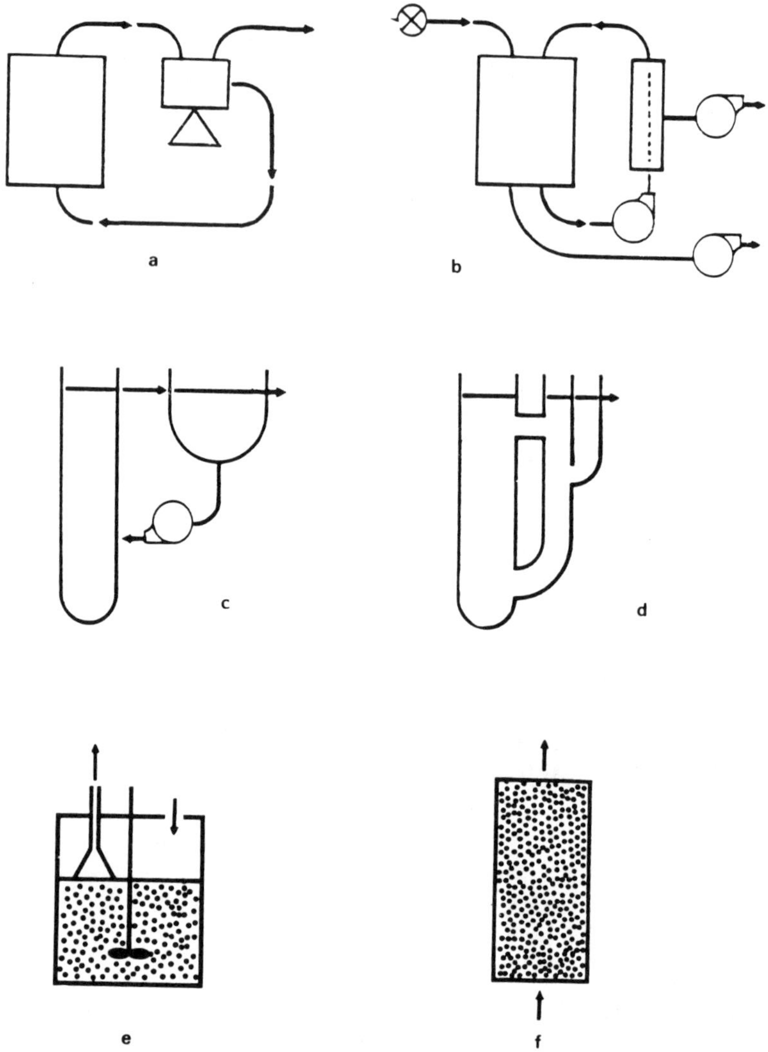

FIGURE 16.3 Principal designs of high microbial density reactors. (a) Recycling by centrifugation, (b) membrane reactor with possibility of removing the biomass, (c) flocculating yeast reactor with external decanter, (d) flocculating yeast reactor with internal decanter, (e) heterogeneous reactor with entrapped cells, (f) fixed bed or fluidized bed heterogeneous reactor of immobilized cells.

considerably disturbs the biofilm stability. Finally, access of oxygen to the underlying cells of the biofilm is a specific problem which is difficult to resolve.

The flocculation of microbial cells to form a dense concentration of biomass is encountered naturally in a number of yeast strains (Stewart and Russell, 1986; Calleja, 1987). It is a very attractive method of biomass retention, since it involves decanting, the most basic method of liquid-solid distribution (Ghommidh and Navarro, 1986; Salou *et al.*, 1988). In the case of yeasts, a biomass density of 60–110 g/l can be maintained with little or no physical constraints for the microorganisms except for the reactor vessel. The fermentation process requires a reactor configuration in the form of towers which house the biomass through which the liquid is passed. Industrial installations have been described in breweries (Maule, 1986).

Entrapping involves imprisoning living cells within a rigid network which permits the diffusion of substrates and products, thereby making possible the growth and maintenance of active cells. Synthetic polymers (polyacrylamide, polyvinyl chloride, polymethane) can be used as the entrapment matrix but natural polymers, such as alginate, carrageenan, chitosan and agar, enable polymerization in very mild conditions and leave cell integrity intact. Entrapment in alginate, for example, is a very simple process (Brodelius and Vandamme, 1987). A uniform suspension of the cells is prepared in a 2% sodium alginate solution. The suspension is then added dropwise to a solution of calcium chloride which catalyzes polymerization of the alginate into a gel in the form of spherical beads. It is possible to obtain perfectly calibrated spheres between 0.2 and 2 mm in diameter (Matulovic *et al.*, 1986). The beads of cells are housed as a fixed bed through which the liquid to be transformed is passed, or they may be agitated to give a fluidized bed system (Figure 16.3).

Kinetic Constraints and Behavior of Immobilized Microorganisms

Regardless of the method of retention chosen for heterogeneous reactors, the reaction is catalyzed in the solid phase. The substrates and products must diffuse into and out of the catalytic material, thereby creating the occurrence of concentration gradients. In addition, retention of the microorganism can change intrinsic kinetics (permeability, metabolic pools, macromolecule turnover, DNA replication, etc.). Much remains to be discovered in this area. For entrapped cells, the diffusion reaction kinetics have been reviewed (Radovitel, 1985; Karel *et al.*, 1985). Figure 16.4 shows the spatial distribution of yeast cells which can occur in a sphere of alginate. This distribution, showing a higher concentration of cells towards the bead surface, can be attributed to the poor diffusion of an essential nutrient (e.g. dissolved oxygen) and an inhibitory ethanol concentration which is transitorily higher at the center of the sphere (Siess and Divies, 1981; Seki and Furusaki, 1985).

Despite the limitations of internal diffusion, reactors functioning with entrapped cells are nevertheless very productive in fermentation processes. With alcoholic fermentation, where substrate concentrations are high, it may be advantageous to operate with a heterogeneous fixed bed reactor in order to obtain minimal ethanol toxicity. A major drawback of this system is the considerable release of carbon dioxide which is the other main product of the bioreaction. In the case of aerobic processes, slow diffusion of oxygen and the

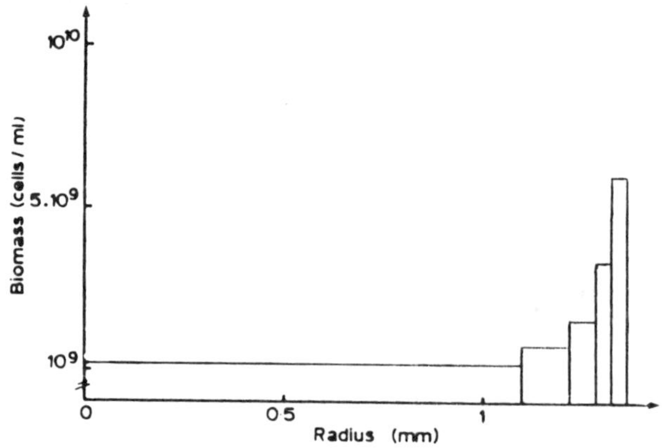

FIGURE 16.4 Distribution of yeast cells in alginate beads after 200 hours of continuous operation in grape must.

low oxygen concentration within the sphere of cells seriously limits application of the technique. Fluidized (agitated) bed reactors help overcome some of these limitations by ensuring better external transfer of reactants (Karkare *et al.*, 1985; De Fonesca *et al.*, 1986). In order to obtain immobilized microorganisms with sufficient performance, it is often a good idea to allow them to grow directly *in situ* in the gels, rather than preparing gels with high initial concentrations of cells.

With yeast, it is possible to obtain higher glucose fermentation activities for entrapped cells than with free cells. Using the same strain of *S. cerevisiae* and 2 mm diameter alginate spheres with a population of 2.10^9 cells/ml of gel, we measured specific ethanol production at 0.6 g/g of cells per hour in the immobilized form and 0.3 g for free cells. If smaller spheres are used, even better fermentation activities are possible. This improved rate of fermentation was also observed by Navarro and Durand (1977) with adsorbed yeast. Even though oxygen diffusion is limited, we were able to measure activation of glucose respiration by immobilized *Rhodotorula glutinis* (unpublished results). Entrapment protects the yeasts against ethanol toxicity. For infinitely mixed homogeneous reactors, ethanol inhibits its own formation according to a formula proposed by Aiba *et al.* (1968) for free cells. For heterogeneous reactors, using immobilized cells, a slightly different formula applied (Lee *et al.*, 1984). Casey and Ingeldew (1986) have published a thorough review of ethanol toxicity in yeasts.

Entrapment can change the concentrations of the main by-products produced by the cells (Table 16.1). In grape juice, there is greater production of glycerol, propanol and isoamyl alcohol by entrapped yeasts and less accumulation of acetaldehyde. In addition, entrapped cells assimilate amino acids to a greater extent, explaining observations concerning the higher alcohols which are derived from them. The decreased accumulation of acetaldehyde can be

TABLE 16.1 Influence of immobilization by entrapment on secondary metabolite production by yeasts during alcoholic fermentation of grape juice (Lenzi, 1986).

	Quantity (mg/10^{10} cells)	
	Free yeasts	*Entrapped yeasts*
Glycerol	88	110
Acetaldehyde	7.3	2.9
Propanol	0.62	1
Isobutanol	2.5	2.5
Isoamyl alcohols	4	5.2

interpreted by a better utilization of reduced coenzymes in the cellular confinement of the gels. When growth in the gels is limited and amino acids are used less, there is considerable accumulation of glycerol and, to a lesser extent, propanol (Siess and Divies, 1981). This modification in fermentation products appears to be general. Hahn-Hagerdahl (1986) proposed an influence of water activity, which could be lower within the gel matrix. In our opinion, there could be more efficient utilization of the pool of reduced nucleotides, as suggested by the profiles of acetaldehyde and glycerol production.

Entrapment can also lead to an altered composition of immobilized yeasts (Table 16.2). Entrapped cells contain higher levels of glycogen, glucan and mannan, as well as DNA (Doran and Bailey, 1986; Simon *et al.*, 1989). Substrate diffusion could explain the change in these reserve substances. The enrichment in DNA, on the other hand, requires additional experimentation before an interpretation can be made.

TABLE 16.2 Comparison of the macromolecular constituents of free and entrapped cells of *Saccharomyces cerevisiae* (Siess and Divies, 1981).

Ratio	*Free cells during growth phase*	*Entrapped cells*	
		48 h	*648 h*
Proteins/deoxyribose	1551–1026	734	500
Proteins/mannan	21.4–14	12.8	11.5
Proteins/glucan	12.1–9.3	7.2	4.25
Proteins/trehalose	7–14.8	30	85
Proteins/total glycogen	9–6	4.55	2.5
Proteins/alkaline glycogen	33–47	23.46	9.6
Ribose/deoxyribose	46–76	53	25

Confinement of cells of lactic acid bacteria in gels is accompanied by changes in their fermentation profiles and affects replication of their extrachromosomal factors: entrapped cells have a high number of plasmid copies (Cavin *et al.*, 1986). A study of citrate-negative variants in free and entrapped populations has shown very high stability of the strains when cell growth occurs as surface microcolonies.De Taxis du Poët *et al.* (1986) measured the same effect in a strain of *Escherichia coli* containing plasmid PTG 201.

In membrane reactors, cell concentrations higher than 100 g/l can be reached. There appears to be a toxic effect of this high concentration of biomass, but this may be overcome by operation of the reactors at sufficiently high dilution rates (Mota, 1985).

Possible Applications in Enology

Malolactic fermentation

The malolactic fermentation of wine is generally carried out by low density bacterial populations: 5.10^{7}–10^{8} CFU/ml, i.e. 10 mg/l dry weight if the dry weight of one *Leuconostoc oenos* cell is estimated at 10^{-13} g (Divies and Cavin, 1986).

Divies (1975) originally proposed the use of immobilized cell systems, with their increased amount of biomass, for accelerating the rate of this fermentation. Since then, several authors have tested this proposition and two pilot scale operations with relatively high performance have been described (Gestrelius and Kjaer, 1983; Cuenat and Villetaz, 1984). In these studies, cells of *Leuconostoc oenos* are entrapped within beads of polyacrylamide (Rossi and Clementi, 1984), alginate (Spettoli *et al.*, 1982, Cuenat and Villetaz, 1984) or carrageenan (McCord and Ryu, 1985; Crapisi *et al.*, 1987a, b) which are supported in columns. Wine is continuously passed through the bed of beads where the malolactic conversion occurs. Residence time of the wine in the columns depends on several factors but, essentially, the reaction is completed in a matter of hours compared with the days or weeks required by the conventional fermentation process. Crapisi *et al.*, (1987b) showed that entrapped *Lactobacillus brevis* and *Lactobacillus fructivorans* had better performance than *Leuc. oenos*. Sensory analyses of wines obtained from these reactors have proved satisfying. Loss in activity of the reactors remains an obstacle to their large scale, commercial application. In most studies, activity was lost within 2–7 days, but Crapisi *et al.* (1987b) have described a system that has a half-life of 46 days. Reactor performance and stability related to survival of the entrapped bacteria, especially when exposed to the chemical stresses of wine. Selection of strains of *Leuc. oenos* or other species that resist high concentrations of ethanol and SO_2 at low pH will benefit the development of these reactors.

In unpublished work, we have shown that 10^{10} cells/ml of *Leuc. oenos* could be maintained active in 2 mm diameter beads of alginate. A continuous transformation of wine was obtained with residence times of 2–6 h, depending on the concentration of malic acid in the wine. Recently, Truck and Hammer

(1989) indicated the diffusion limits encountered. If a fluidized bed was used, 1.2 mm diameter spheres gave very good performance characteristics. The cells in the reactor could be regenerated for successive use.

Barillere *et al.* (1986) have described a procedure for accelerating the malolactic fermentation of wines by inoculating with a high concentration (approximately 10^7 cells/ml) of lactic acid bacteria (presumably *Leuc. oenos*) that was recovered and concentrated by tangential filtration from a wine that had already completed this fermentation.

Deacidification of wines, through malic acid degradation by immobilized cells of *Schizosaccharomyces* has been considered by Divies (1975, 1981) and Navarro and Sablayrolles (1988). Magyar and Panyik (1989) reported the deacidification of wines and musts using cells of *Schizosacch. pombe* entrapped in beads of calcium alginate. The sensory quality of the wines was not adversely affected by this technology, but long contact times of the wine with the beads were required to obtain satisfactory malic acid degradation. Optimization of the reaction would be necessary before application on a commercial scale.

"Prise de mousse", or champagne method

In recent years, immobilized yeast cells have been explored for their use in champagne production. In the conventional process, a secondary fermentation is conducted in the bottle by the inoculation of freely suspended yeast cells. Subsequent sedimentation and removal of the yeast cells requires the lengthy and expensive procedure of "remuage", that is, frequent but controlled turning of the slanted, inverted bottles to bring the yeast sediment down to the cork. The use of immobilized yeast cells for this secondary fermentation would greatly simplify "remuage" (Bidan *et al.*, 1978; Coulon *et al.*, 1983).

Three different possibilities of using immobilized yeasts for bottle fermentations are: mechanical "remuage" using yeast strains that readily flocculate and agglomerate; entrapped yeasts (Bidan *et al.*, 1978); and "Millispark" membrane cartridges (Lemonnier and Duteurtre, 1989; Jallerat, 1990). The disadvantage of agglomerating cells is that "remuage" is not totally avoided; in addition, the choice of yeasts is limited, even though research in this field is very active (Vezinhet and Barre, 1988).

The kinetics of "prise de mousse" (formation of champagne bubbles in the bottle) differentiates the processes (Figure 16.5). For the same initial population of yeast, "prise de mousse" is terminated in 40 days for free and entrapped cells, while the membrane method takes longer. In addition, entrapped or retained cells start their activity more rapidly. For yeast cells entrapped in alginate, calculations have situated the critical radius of the beads at 1 mm, for a concentration of exogenous sugars of 5 g/l and non-agitated conditions (Divies, 1988). The membrane method still presents two major defaults: it has insufficient surface area for reaction exchange and it creates local supersaturation of CO_2 which further limits exchange.

Different experiments carried out in France since 1978 (Divies, 1981) have shown the industrial reliability of entrapped yeasts. There are few differences in

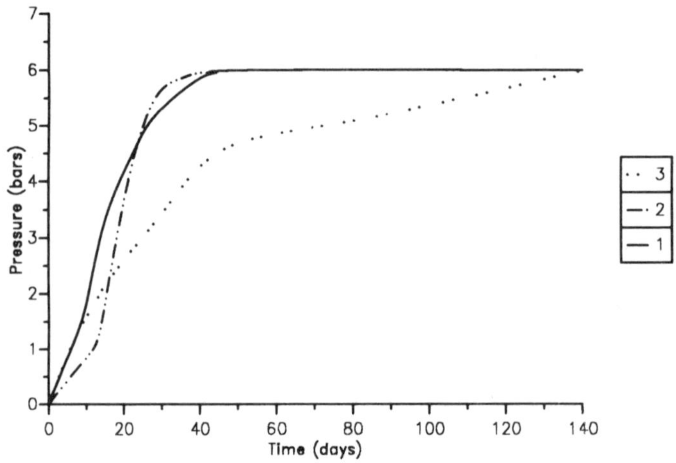

FIGURE 16.5 Comparative "prise de mousse" fermentation kinetics obtained with free cells (1), with entrapped cells (2) and with membrane retention (3). Temperature: 13°C. Inoculum: 2.10^6 cells/ml of *Saccharomyces cerevisiae*. Progress of fermentation is measured by development of pressure within the bottle.

comparison with the traditional process (Lemenager, 1983; Lenzi, 1986; Furmi *et al.*, 1987; Pisanelli *et al.*, 1989; Zamorani *et al.*, 1989). Growth, amino acid consumption, the formation of acetaldehyde and acetoin are all reduced in the case of entrapped cells, while autolysis kinetics are slightly increased. We attribute these phenomena to the glucose gradient created by the diffusion-reaction in the alginate spheres. Tasting sessions done with the final product have not revealed significant sensory differences between "prise de mousse" effected by embedded yeasts and by free yeasts. The companies Scoma, and Moët and Chandon have developed an industrial machine for the delivery of beads at the rhythm of 20 000 bottles per hour. It is now possible to dry and rehydrate alginate spheres coated with a layer of fresh external alginate, thereby avoiding any release of yeast cells into the champagne (Divies *et al.*, 1988). An economic study undertaken for a plant producing 3 000 000 bottles yearly has demonstrated the competitiveness of the new process (Figure 16.6) (Valade and Rinville, 1990). The technology developed for champagne production can be transposed to other bottle-fermented sparkling beverages. Particularly interesting possibilities are the refermentation of wine supplemented, before "prise de mousse", with an infusion of fruits obtained by hydro-alcoholic maceration (Lenzi and Cavin, 1985), and the secondary fermentation of beverages such as cider and pineapple wine (Divies and Deschamps, 1986). These new products can thus be obtained under more thoroughly controlled conditions.

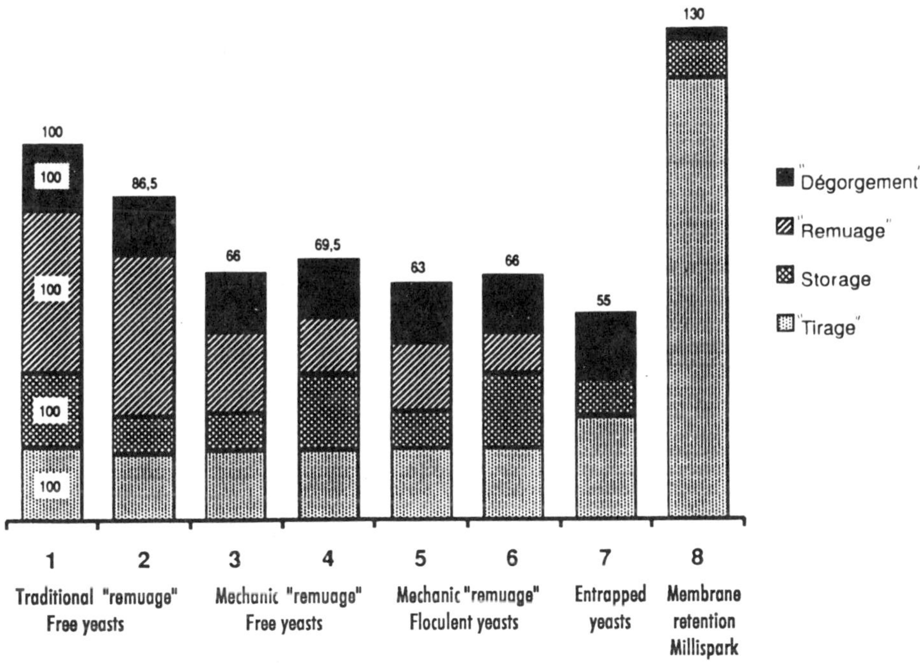

FIGURE 16.6 Comparative costs of different "prise de mousse" operations in relation to different yeast fermentation technologies. Operation: yeast inoculation (tirage); storage; sedimentation of yeasts (remuage); removal of yeast (dégorgement). Data from a plant producing 3 million bottles of sparkling wine, annually (Vallade and Rinville, 1990).

Alcoholic fermentation and production of other sparkling beverages

There are many reasons for modifying conventional approaches to fermentation, including savings in invested capital, decreased labor costs and automation.

Homogeneous reactors: Processes using membrane reactors are now being considered in enology. Perlot *et al.* (1989) showed the full potential of this technology in the pilot scale fermentation of starch hydrolysates. An alcohol level of 10% could be reached in two hours. These authors considered that the main limitation to the industrial development of these reactors is the specific flux of the membranes, which must reach 200 to 300 l/h to be economically viable. Goma and Durand (1986), followed by Navarro and Sablayrolles (1988), reported pilot scale studies with a continuous membrane reactor, operating at a residence time of two hours, that gave a sparkling grape drink of 3% alcohol. Research is currently being conducted to optimize the hydraulics of these reactors, and to develop processes for the automatic regeneration of the membranes. With biological products such as wine and beer, colloids in the medium reduce the operating efficiency of tangential or cross-flow filtration which is the basic technology upon which these reactors depend.

The application to enology of yeast cell recycling by centrifugation was studied by Rosini (1986). At the end of each fermentation, he recovered and concentrated the yeast cells by centrifugation and used them to inoculate the next fermentation. By this method, yeast concentrates of 25–28 g/l were obtained and used to complete the alcoholic fermentation in 7–10 h—compared with 32 h for the initial fermentation. However, a higher volatile acidity (0.52 g/l) was observed in the wines produced with the recycled yeasts.

Heterogeneous reactors: The "A.P.V." system employing flocculating yeast cells has received some application in the brewing industry. The reactor containing the densely packed cells is a cylindrical tower with a conical base; geometry is 7:1–10:1 with a 0.9–2 m diameter. Cell concentrations of 50–80 g/l that produce 60 g/l per hour of ethanol are obtainable. These reactors have been used in breweries for the continuous production of beer (Portno, 1978), but it is very difficult to obtain a product with sensory qualities identical to those obtained with the conventional process. Other disadvantages of these reactors are complexity, lack of flexibility, risks of contamination, heterogeneity of the cell concentration, the lack of choice of usable strains and the excessive time required for start-up, i.e. two to three weeks to obtain a high cell density and stable operation. Recently, Navarro and Sablayrolles (1988) showed that it was possible to use flocculating yeasts to treat wine in which fermentation had stopped. It was shown that the operation was easily done with a laboratory reactor equipped with a simple decanter. Pilot experimentation is now planned.

The use of adsorbed cell reactors has been proposed for beer production (Berdelle Hilge, 1972; Beker and Kirsop, 1973; Navarro *et al.*, 1976), as well as for deacidifying wines with *Schizosaccharomyces pombe* (Navarro, 1978). The design of reactors in series, packed with a pouzzolane support and with internal recycling enabled gas hold-up to be overcome (Ryu, 1980). However, reactors of this type seem unsuitable for the production of fermented beverages. A pilot scale study of beer production (Moll and Duteurtre, 1979) showed that major defaults involved lengthy start-up, danger of contamination, the very moderate sensory quality of the resulting beer since the reactor cannot be agitated, and the very heterogeneous state of yeast biofilms.

Divies (1975, 1981), proposed the use of entrapped cells for conducting the alcoholic fermentation of beverages (wine, beer). Substantial research has now confirmed the feasibility of using such yeasts for ethanolic fermentations (Chibata *et al.*, 1984), but no industrial installation is yet operational. Nevertheless, the Kyowa Hakko company in Tokyo installed a pilot scale ethanol production unit that was built around five reactors with a total volume of 4 m^3 and produced 2400 l/day of pure alcohol. Using entrapped yeasts, White and Portno (1978), followed by Ryder and Masschelein (1985), showed that beers with good sensory quality could be obtained. Onaka *et al.* (1985) noted the possibility of lowering the diacetyl concentration by carrying out fermentation in a highly anaerobic microenvironment (at an O_2 concentration of about 0.04 mg/l).

Studies reported by Bertuccioli *et al.* (1988) and Cantarelli (1989) have considered the use of entrapped *S. cerevisiae* for conducting the alcoholic fermentation of wines. While prospects for this technology appear encouraging, further research is needed to optimize reaction variables, improve the long-term stability of the reactors and to understand more about secondary metabolite production by yeasts under these conditions. Our own laboratory and pilot scale studies with entrapped cell of yeasts further demonstrate their potential value to enology. We have been able to maintain the activity of 2×10^9 yeast cells (*S. cerevisiae*) per ml of gel if 1–3 mm diameter alginate spheres are used. This density is equivalent to 20–50 g/l of yeast taking the weight of one cell as 10^{-11}g. These beads of dense biomass can be conveniently handled and maintained active for repetitive and controlled use by relatively simple procedures. Such beads can be used in batch or continuous processes. Cellular entrapment also facilitates fermentations involving mixed cultures of microorganisms (Prevost *et al.*, 1985; Osaki *et al.*, 1985). We have shown that yeast strains with complementary aromatic capacities can be maintained in constant proportion over long periods of fermentation (unpublished results). Stepwise processes can also be used (Divies and Deschamps, 1986). For example, in the first reactor, the sugar concentration of the fruit juice is lowered by using an obligate aerobic yeast. In the second reactor, a fermentation yeast is used to carry out fermentation, so that a low alcohol product is obtained. Thus, the undesirable sweetness of low alcohol beverages can be overcome. Recently, Ogbonna *et al.* (1989) designed a horizontal reactor for wine fermentations, that exploited the successive activities of cells of *Schizosaccharomyces* and *Saccharomyces*, which were immobilized on plates coated with alginate.

In our opinion, it is possible to simplify the continuous process for sparkling wine production as proposed by Sarichvihi (1982) and described by Muller-Spath (1982). Cellular entrapping can be used to easily carry out repetitive continuous fermentation under carbon dioxide pressure (Figure 16.7) (Divies and Deschamps, 1986). Under these conditions, there is even greater control of the aromatic characteristics of the products, especially the decrease in higher alcohols and increase in esters. Fermentation can be done at relatively high pressures, up to 6 bars, in continuous operation. It is thus possible to imagine the manufacture of sparkling wine or semi-sparkling grape juice with remarkable performance (Table 16.3). For example, 1 500 000 bottles of sparkling wine per year could be produced by a reactor ($1 m^3$) operating for 220 days.

CONCLUSION

The natural course of wine fermentation is now known in a detailed manner, but fermentation processes are not yet fully controlled. Basic research is still needed to understand the biochemical behavior of yeast and bacteria in the

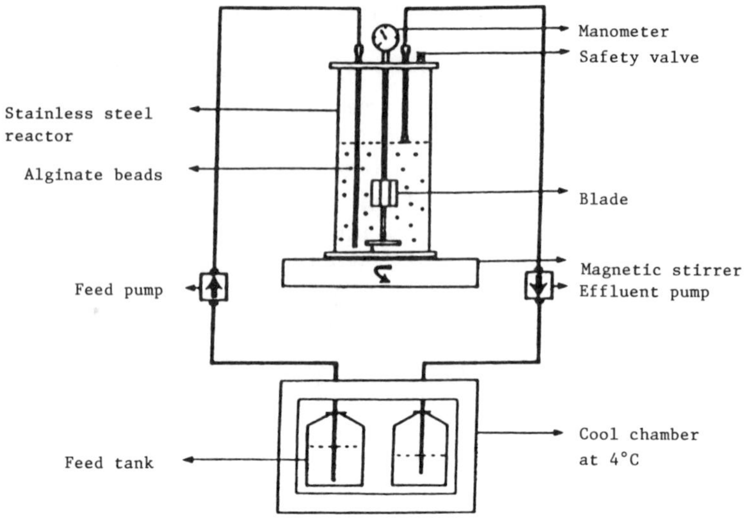

FIGURE 16.7 Diagram of a pressure reactor for the manufacture of sparkling wines, operating continuously or in successive batches.

wine medium. During the last decade, there has been important technology transfer from the sector of other industrial beverages towards enology. The worldwide spread of wine production has led to new vineyards producing quality wine by the use of tested procedures: centrifugation, filtration, stainless steel tanks, temperature management, inoculations. The seasonal nature of production is a hindrance to the development of innovations, especially in the technology of bioreactors adapted to all sectors of enology. Modeling studies, already well advanced, should lead to the design of industrial installations with even better performance. High cell density reactors offer many prospects for enology, but further development is needed.

The use of membrane reactors, theoretically having the highest performance, is still highly dependent on the development of reliable membranes. Reactors using entrapped cells have lower performance, but are more immediately usable. They present the possibility of producing new styles of beverages (low alcohol content and very aromatic), and facilitate the conduct of fermentations where convenient removal of yeast cells is desired (e.g. closed tanks, champagne method). Recent progress has enabled the industrial production of entrapped microorganisms and their availability to industry (Divies *et al.*, 1988).

Research on the physiology of immobilized microorganisms should be continued, since data on their behavior is insufficient to enable all possibilities to be optimally utilized.

TABLE 16.3 Preparation of different fermented beverages in continuous process and under pressure.

Media fermented	*Product*	*Residence time*	*Temperature (°C)*	*Ethanol (%) in reactor*
Grape juice	Semi-sparkling grape juice	1h 24 min	20	2.9
Wine	Refermented (Sparkling wine)	4 h 42 min	17	12.3

Experimental conditions:

	Semi-sparkling grape juice	*Sparkling wine*
Medium	grape juice at 149 g total sugar per liter	white wine, 11% ethanol with 28 g total sugar per liter
Total volume	454 ml	350 ml
Volume of particles	200 ml	165 ml
Flow-rate	324 ml/h	75 ml/h
Temperature		17°C
Strain	*S. cerevisiae* at 2.10^9 cells/ml of gel	*S. bayanus* at 1.5×10^9 cells/ml of gel
Pressure	5.5 bars	5.5 bars

REFERENCES

Aiba, S., Shoda, M. and Nagatani, M., (1968) Kinetics of product inhibition in alcohol fermentation. *Biotechnology and Bioengineering*, **10**, 845–864.

Barillere, J.M., Benard, P. and Dubois, C. (1986) Utilisation de retentat de microfiltration pour induire la fermentation malolactique. *Revue Française d'Oenologie*, **102**, 17–19.

Barre, P. (1988) La fermentation alcoolique. *Biofutur*, **72**, 22–26.

Becker, D.A. and Kirsop, B.H. (1973) Rapid beer production and conditioning using plug fermentor. *Journal of the Institute of Brewing*, **79**, 487–494.

Berdelle Hilge, P. (1972) Fesselgarung Ein Beitrag zur vagesdrichte des Bio-reactors. *Brauwelt*, **112**, 24–25.

Bertuccioli, M., Rosi, I. and Costamagna, L. (1988) Recent progress in the use of immobilized yeasts on wine fermentation. In *Proceedings of the Second International Cool Climate Viticulture and Oenology Symposium*, pp. 304–307. New Zealand Society for Viticulture and Oenology: Auckland, New Zeland.

Bezenger, M.C., Navarro, J.M. and Sablayrolles, J.M. (1986) Prévision du déroulement en isotherme de la fermentation alcoolique d'un milieu modèle simulant les conditions de l'oenologie: influence des paramètres azote et température. *Sciences des Aliments*, **6**, 185–200.

Bidan, P., Divies, C. and Dupuy, P. (1978) Procédé perfectionné de préparation de vins mousseux. Brevet Anvar n° 7822131.

Boulton, R. (1980) The prediction of fermentation behavior by a kinetic model. *American Journal of Enology and Viticulture*, **31**, 40–45.

Brodelius, P. and Vandamme, E.J. (1987) Immobilized cell systems. In *Biotechnology*, Volume 7a *Enzyme Technology*, edited by J.F. Kennedy, pp. 405–464. Weinheim: VCH Verlagsgesellschaft.

Brugirard, A., Roques, J. and Vignier, E. (1970) La vinification continue en Roussillon. Observations sur les campagnes 1969 et 1970 et sur la maitrise des températures. *Bulletin de Technologie des Pyrénées Orientales*, **56**, 85–99.

Calleja, G.B. (1987) Cell aggregation. In *The Yeasts*, Volume 2, edited by A.H. Rose and J.S. Harrison, pp. 164–298. London: Academic Press.

Cantarelli, C. (1989) Factors affecting the behaviour of yeasts in wine fermentation. In *Biotechnology Applications in Beverage Production*, edited by C. Cantarelli and G. Lanzarini, pp. 127–151. London: Elsevier Applied Science.

Casey, J.P. and Ingeldew, W.M. (1986) Ethanol tolerance in yeasts. *Critical Reviews in Microbiology*, **13**, 219–280.

Cavin, J.F., Prévost, H., Charlet, B. and Divies, C. (1986) Fonctionnement et équipement plasmidique de *Streptococcus lactis subsp.diacetylactis* en fermentation continu sous forme libre et incluse dans un gel d'aginate. Abstracts: 1er Congrès Société Française de Microbiologie, Toulouse. p. 171.

Chibata, I., Tosa, T. and Fujimura, M. (1984) Immobilized living microbial cells. In Annual Reports of Fermentation Processes. Volume 6, edited by G T Tsao, pp. 1–22, New York: Academic Press.

Coulon, P., Duteurtre, B., Charpentier, M., Parenthoen, A., Badour, C. and Moulin, J.P. (1983) New prospects in the "Méthode champenoise" utilization of enclosed yeasts during bottling. *Le Vigneron Champenois*, **104**, 516–532.

Crapisi, A., Nuti, M.P., Zamorani, A. and Spettoli, P. (1987a) Improved stability of immobilized *Lactobacillus* sp. cells for the control of malolactic fermentation in wine. *American Journal of Enology and Viticulture*, **38**, 310–312.

Crapisi, A., Spettoli, P., Nuti, M.P. and Zamorani, A. (1987b) Comparative traits of *Lactobacillus brevis, Lactobacillus frutivorans* and *Leuconostoc oenos* immobilized cells for the control of malolactic fermentation in wine. *Journal of Applied Bacteriology*, **63**, 513–521.

Cuenat, P. and Villetaz, J.C. (1984) Essais de fermentation malolactique des vins par bactéries lactiques immobilisées du genre *Leuconostoc*. *Revue Suisse Viticulture, Aboriculture, Horticulture*, **16**, 145–151.

Cutayar, J.M., Cutayar, S. and Poillon, D. (1988) Procédé d'oxygénation controlée d'un moût de fermentation alcoolique et installation correspondante. Brevet n° 88003705, France.

Cysenski, G.R. and Wilke, C.R. (1978) Process design and economic studies of alternative fermentation methods for production of ethanol. *Biotechnology and Bioengineering*, **20**, 1421–1444.

Da Fonseca, M.M., Black, G.M. and Webb, C. (1986) Reactor configuration for immobilized cells. In *Process Engineering Aspects of Immobilized Cells Systems*, edited by C. Webb, G.M. Black and B. Atkinson, pp. 63–74. Warwickshire: The Institute of Chemical Engineering.

Defranceschi, F. (1955) Un nuevo procedimento di vinificazione continua. Rivista di Viticoltura e di Enologia **8**, 287–296.

De Taxis du Poët, P., Dhulster, P., Barbotin, J.M. and Thomas, D. (1986) Plasmid inheratability and biomass production : comparison between free and immobilized cell cultures of *Escherichia coli* B 218 (PTG 201) without selection pressure. *Journal of Bacteriology*, **165**, 871–877.

Divies, C. (1975) Procédé enzymatique utilisant des microorganismes inclus. Brevet ANVAR n°7524509, France.

Divies, C. (1981) Les possibilités d'emploi des germes fixés en oenologie. *Bulletin de l'Office International de la Vigne et du Vin*, **54**, 843–858.

Divies, C. (1988) Les microorganismes immobilisés. Cas des bactéries lactiques et des levures en oenologie. In *Application à l'oenologie de progrès récents en microbiologie et en fermentation*, edited by ENSAA, Dijon, pp. 27–47. Paris: *Office International de la Vigne et du Vin*.

Divies, C. and Cavin, J.F. (1986) Evolution et maitrise de la fermentation malolactique des vins. In Microorganismes et Aliments. Evolution, maitrise, destruction des flores. *Sciences des Aliments*, **6**, 43–55.

Divies, C. and Deschamps, P. (1986) Procédé et appareillage pour la mise en oeuvre de réactions enzymatiques et applications à la préparation de boissons fermentées. Brevet n°86 14072, France.

Divies, C., Lenzi, P., Herault, F. and Beaujeu, J. (1988) Procédé de préparation de microorganismes inclus dans des gels sensiblement déshydratés présentant une viabilité améliorée après réhydratation, utilisation de ces gels sensiblement déshydratés contenant les microorganismes pour la préparation de boissons fermentées ou refermentées et gels obtenus. Brevet n°8809249, France.

Doran, P. and Bailey, J.E. (1986) Effects of immobilization on growth, fermentation properties and macromolecular composition of *Saccharomyces cervisiae* attached to gelatin. *Biotechnology and Bioengineering*, **28**, 73–87.

Dupuy, P. and Poux, C. (1965) Les fermentations continues. *Bulletin des Techniques et des Informations*, **196**, 72–79.

Egamberdiev, N.B. and Ierusalimskii, N.D. (1969) Studies of the continuous must fermentation processes with the yeast *Saccharomyces vini* to produce dry wines. In *Continuous cultivation of microorganisms*. Prague: Proceedings of the Fourth Symposium. pp. 517–527.

El Haloui, N., Cleran, Y., Sablayrolles, J.M., Grenier, P., Barre, P. and Corrieu, G. (1988) Suivi et controle de la fermentation alcoolique en oenologie. Essais à echelle industrielle. *Revue Française d'Oenologie*, **115**, 12–17.

El Haloui, N., Corrieu, G., Cleran, Y. and Cheruy, A. (1989) Method for on-line prediction of kinetics of alcoolic fermentation in wine making. *Journal of Fermentation and Bioengineering*, **68**, 131–135.

Fiechter, A., Fuhrmann G.F. and Kappeli, O. (1981) Regulation of glucose metabolism in growning yeast cell. *Advances in Microbial Physiology*, **22**, 123–183.

Flanzy, M. and Audro, P. (1973) La vinification par macération carbonique. Services d'Experimentations et d'Informations, INRA: Versailles.

Furmi, D., Trioli, G. and Colagrande, O. (1987) Preliminary assessment on the use of immobilized yeast cells in sodium alginate for sparkling wine processes. *Biotechnology Letters*, **9**, 339–342.

Gestrelius, S.M. and Kjaer, J.H. (1983) Method of deacidifying wine and composition thereof. United States Patent, 4380552.

Ghommidh, C. and Navarro, J.M. (1986) Floculation-fermentation. In *Bioreacteurs*, edited by G. Goma, pp. 89–112. Paris: Société Française de Microbiologie.

Goma, G. (1978) Optimisation de la mise en oeuvre des cultures anaérobies. In *Utilisation Industrielle du Carbone d'Origine Végétale par voie Microbienne*, edited by L. Bolichon and G. Durand, pp. 171–184. Paris: Société Française de Microbiologie.

Goma, G. and Durand, G. (1986) Nouvelles conceptions de mise en oeuvre de microorganismes et de bioréacteurs basés sur les cultures à forte densité cellulaire. In *Bioreacteurs*, edited by G.Goma, pp. 127–143. Paris: Société Française de Microbiologie.

Guetov, G. (1969) Qualité des vins produits par le procédé de fermentation continue. *Bulletin de l'Office International de la Vigne et du Vin*, **42**, 505.

Guidoboni, G.E. (1984) Continuous fermentations systems for alcohol production. *Enzyme and Microbial Technology*, **6**, 194–200.

Hahn-Hargerdal, B. (1986) Water activity : a possible external regulator in biotechnical processes. *Enzyme and Microbial Technology*, **8**, 322–327.
Hanshoffer, H. (1978) Conservation des vins blancs aromatiques. *Annales de Technologie Agricole*, **27**, 221–230.
Jallerat, E. (1990) Un progrès important pour le champagne et les vins de "methode traditionnelle". In *Les Nouvelles Techniques de Tirage*, pp. 12–16. Epernay: Association des Techniciens Superieurs en Viticulture et Oenologie.
Jones, R.P. and Greenfield, J.P. (1986) Role of water activity in ethanol fermentations. *Biotechnology and Bioengineering*, **28**, 29–40.
Kappeli, O. (1986) Regulation of carbon metabolism in *Saccharomyces cerevisiae* and related yeasts. *Advances in Microbial Physiology*, **28**, 181–209.
Karel, S.F., Libichi, S.B. and Roberston, C.R. (1985) The immobilization of whole cells : engineering principles. *Chemical Engineering Science*, **40**, 1321–1354.
Karkare, S.B., Dean, R.C. and Venkatasubramanian, K. (1985) Continuous fermentation with fluidized slurries of immobilized microorganisms. *Biotechnology*, **3**, 247–251.
Kunkee, R.E. and Goswell, R.W. (1977) Table wines. In *Alcoholic Beverages. Economic Microbiology*, Volume 1, edited by A.H. Rose, pp. 315–386. London: Academic Press.
Lafon-Lafourcade, S. (1983) Wine and brandy. In *Biotechnology* Volume 5, edited by G. Reed, pp. 81–163. Weinheim: VCH Verlagsgesellschaft.
Larue, F., Lafon-Lafourcade, S. and Ribereau-Gayon, P. (1979) Les différents rôles fonctionnels des stéroïdes sur les levures dans le moût de raisin en fermentation. *Annales de Microbiologie*, **130**, 231–243.
Lee, T.H., Ahn, J.C. and Ryu, D.D.Y. (1983) Performance of an immobilized yeast reactor system for ethanol production. *Enzyme Microbial Technology*, **5**, 41–45.
Lefur, Y. (1990) Typicité et macération pelliculaire. Application au cépage Chardonnay en Bourgogne. *Revues de Oenologues*, **57**, 23–25.
Lemenager, Y. (1983) Contribution à l'utilisation des microorganismes immobilisés. *Application à l'élaboration des vins mousseux par la méthode champenoise*. Thèse Université de Bourgogne.
Lemonnier, J. and Duteurtre, B. (1989) A major step forward for champagne and other wines produced by traditional methods. *Revue Française d'Oenologie*, **121**, 15–26.
Lenzi, P. (1986) Elaboration des vins mousseux selon la méthode champenoise par les levures incluses. Thèse Université de Bourgogne.
Lenzi, P. and Cavin, J.F. (1985) Procédé de fabrication de boissons fermentés aromatisées.Brevet ANVAR n° 8506147, France.
Lichine, A. (1980) Encyclopédie des vins et des alcools. Editions, Robert Laffont, Paris.
Magyar, I. and Panyik, I. (1989) Biological deacidification of wine with *Schizosaccharomyces pombe* entrapped in Ca-alginate gel. *American Journal of Enology and Viticulture*, **40**, 233–239.
Marc, Y. (1988) Modélisation de la fermentation de la bière. Thèse INPL Nancy.
Marshall, K.C. (1985) Mechanisms of bacterial adhesion at solid-water interfaces. In *Bacterial Adhesion*, edited by D.C. Savage and M. Fletcher, pp.133–161. New York and London: Plenum Press.
Martiniere, P. (1981) Thermovinification et vinification par macération carbonique dans le Bordelais. In *Actualité Oenologiques et Viticoles*, edited by P. Ribéreau-Gayon and P. Sudraud, pp. 303–310. Paris: Dunod.
Matulovic, V., Rash, D. and Wagner, F. (1986) New equiment for the scaled up production of small spherical biocatalysts. *Biotechnology Letters*, **8**, 485–490.
Maule, D.R. (1986) A century of fermenter design. *Journal of the Institute of Brewing*, **92**, 137–145.
McCord, J.D. and Ryu, D.D.Y. (1985) Development of a malolactic fermentation process using immobilized whole cells and enzymes. *American Journal of Enology and Viticulture*, **36**, 214–218.

Mehaia, M.A. and Cheryan, M. (1990) Membrane bioreactors: enzyme processes. In, *Biotechnology and Food Process Engineering*, edited by H.G. Schwartzberg and M.A. Raao, pp. 67–126. New York: Marcel Dekker.

Mitchell, R. (1976) Mechanism of attachment of microorganisms to surfaces. In *Microbial Aspects of Dental Caries*. edited by H.M. Stills, W.J. Loesche, J.C. O'Brien. pp. 47–53. London: Information Retrieval Incorporated.

Moll, M. and Duteurtre, B. (1979) Préparation de bière en continu à l'aide de levures adsorbées. In *Cellules Immobilisées*, edited by J.M. Lebeault and G. Durand, pp. 173–185. Compiegne: Société Française de Microbiologie.

Moresi, M. (1989a) Fermenter design for alcoholic beverage production. In *Biotechnology Applications in Beverage Production*, edited by C. Cantarelli and G. Lanzarini, pp. 93–106. London: Elsevier Applied Science.

Moresi, M. (1989b) Optimal fermenter design for white wine production. In *Biotechnology Applications in Beverage Production*, edited by C. Cantarelli and G. Lanzarini, pp. 107–126. London: Elsevier Applied Science.

Mota, M. (1985) Inhibition et fermentation alcoolique: quelques concepts non conventionnels. Thèse INSA Toulouse.

Muller-Spath, H. (1989) Elaboration de vins mousseux en continu. *Revue Française d'Oenologie*, **119**, 43–45.

Navarro, J.M. (1978) Production d'éthanol par les cellules adsorbées. In *Utilisation Industrielle du Carbone D'origine Vegetale par Voie Microbienne*, edited by L. Bobichon and G. Durand, pp. 211-217. Paris: Société Française de Microbiologie.

Navarro, J.M. (1980) Levures immobilisées : croissance et métabolisme au cours de la fermentation alcoolique. Thèse INSA Toulouse.

Navarro, J.M., Durand, G., Duteurtre, B., Moll, M. and Corrieu, G. (1976) Mise au point d'un procédé continu de préparation de boissons fermentées. *Industries Agricoles Alimentaires*. **93**, 695–703.

Navarro, J.M. and Durand, G. (1977) Modification of yeast metabolism by immobilization onto porous glass. *European Journal of Applied Microbiology*, **4**, 243–254.

Navarro, J.M. and Sablayrolles, J.M. (1988) Les réacteurs à haute densité cellulaire en oenologie. In *Application à l'Oenologie des Progrès Récents en Microbiologie et en Fermentation*, edited by ENSAA-Dijon, pp. 283–313. Paris: Office International de la Vigne et du Vin.

Ogbonna, J.M., Amano, Y., Nakamura, K., Yokotsuka, K., Shimaza, Y., Watanabe, M. and Hara, S. (1989) A multistage bioreactor with replaceable bioplates for continuous wine fermentation. *American Journal of Enology and Viticulture*, **40**, 292–298.

Ohno, T. and Takahaski, R. (1986) Role of wort aeration in the brewing process. Part 1 : oxygen uptake and biosynthesis of lipid by the final yeast. *Journal of the Institute of Brewing*, **92**, 84–87.

Onaka, T., Inoue, T. and Kubo, S. (1985) Beer brewing with immobilized yeast. *Biotechnology*, **3**, 467–470.

Osaki, K., Okanoto, Y., Akao, T., Nagata, S. and Takamatsu, H. (1985) Fermentation of soy sauce with immobilized whole cells. *Journal of Food Science*, **50**, 1289–1292.

Ough, C.S. (1966) Fermentation rates of grape juice. III. Effects of initial ethyl alcohol, pH and fermentation temperature. *American Journal of Enology and Viticulture*, **17**, 74–79.

Pallota, U. (1983) Quelques traitements du moût pour la vinification du vin de qualité. XIIIè Congrès International de l'Office International du Vin, Le Cap.

Perlot, P., Daujat, F. and Mazellier, J. (1989) Fermentation alcoolique par bioréacteur à membrane : forte densité cellulaire, performance du procédé. *Industries Agricole et Alimentaires*, **106**, 855–860.

Peyron, D. and Feuillat, M. (1985) Essais comparatifs de cuves d'automacération en Bourgogne. *Revue d'Oenologie*, **38**, 7–10.

Pisanelli, A.M., Favati, F., Crapuisi, A., Borin, G. and Spetolli, P. (1989) The free amino acid content of bottle-fermented Moscato bianco sparkling wine by immobilized yeasts. *Yeasts*, **5**, S113–S116.
Portno, A.D. (1978) Fermentation and storage Symposium, European Brewing Convention, Zoeterwode, pp. 145–154. London: Elsevier Scientific Publishers.
Poulard, A., Reberteau, R. and Prulho, N. (1986) Incidences de la maitrise des températures de fermentation sur la microflore levurienne et les composés aromatiques des vins blancs secs. *Bulletin de la Societé de Sciences Naturelles—Ouest de la France*, **8**, 167–174.
Prevost, H., Rousseau, E. and Divies, C. (1985) Continuous yoghurt production with *Lactobacillus bulgaricus* and *Streptococcus thermophilus* entrapped in Ca-alginate. *Biotechnology Letters*, **7**, 247–252.
Radovith, J.M. (1985) Mass transfer effects in fermentations using immobilized whole cells. *Enzyme and Microbial Technology*, **7**, 2–10.
Ribéreau-Gayon, J., Peynaud, E., Ribéreau-Gayon, P. and Sudraud, P. (1982) *Vinifications et transformations du vin*. Vol. 3. Paris: Dunod.
Rosini, G. (1986) Wine-Making by cell recycle batch fermentation process. *Applied Microbiology and Biotechnology*, **24**, 140–143.
Rossi, J. and Clementi, F. (1984) L-Malic acid catabolism by polyacrylamide gel-entrapped *Leuconostoc oenos*. *American Journal of Enology and Viticulture*, **35**, 100–102.
Rougier de la Bergerie (1821) *Cours d'agriculture pratique*, Tome **6**. Paris: Audot.
Ryder, D.S. and Masschelein, C.A. (1985) The growth process of brewing yeast and the biotechnology challenge. *American Journal of Enology and Viticulture*, **43**, 66–75.
Ryu, Y.W. (1980) Production d'éthanol en continu à l'aide des levures adsorbées sur supports solides. Thèse INSA Toulouse.
Sablayrolles, J.M. (1988) Pilotage de la fermentation alcoolique en oenologie. In *Application à l'Oenologie des Progrès Récents en Microbiologie et en Fermentation*, edited by ENSAA, Dijon, pp. 235–256. Paris: Office International de la Vigne et du Vin.
Sablayrolles, J.M. and Barre, P. (1986) Evaluation des besoins en oxygène des fermentations alcooliques en conditions oenologiques simulées. *Sciences des Aliments*, **6**, 373–383.
Sablayrolles, J.M. and Barre, P. (1989) Effect of anisothermal conditions on the kinetics of alcoholic fermentations by *Saccharomyces cerevisiae*. *Bioprocess Engineering*, **4**, 139–143.
Sablayrolles, J.M., Barre, P. and Grenier, P. (1987) Design of a laboratory automatic system for studying alcoholic fermentations in anisothermal enological conditions. *Biotechnological Techniques*, **1**, 181–184.
Salou, P., Sablayrolles, J.M. and Barre, P. (1988) Production de levures d'intérêt oenologique. *Revue Française d'Oenologie*, **114**, 29–34.
Schaeffer, A. (1989) Vinification en blanc : contrôle de la température pendant le fermentation alcoolique. In *Qualité et typicité des vins de Bourgogne*, pp. 84–88. Beaune: Bureau Interprofessional des Vins de Bourgogne.
Seki, M. and Furusaki, S. (1985) Effect of intraparticle diffusion on reaction by immobilized growing yeast. *Journal of Chemical Engineering* (Japan), **18**, 461–463.
Siess, M.H. and Divies, C. (1981) Behaviour of *Saccharomyces cerevisiae* cells entrapped in a polyacrylamide gel and performing alcoholic fermentation. *European Journal of Microbiology and Biotechnology*, **12**, 10–15.
Simon, J.P., Benoot, T., Defroyennes, J.P., Dekegel, D. and Vangegans, J. (1989) Physiology and morphology of Ca-alginate entrapped *Saccharomyces cerevisae*. *European Brewing Convention*, p.4.
Slaughter, J.C. (1989) The effects of carbon dioxide on yeasts. In *Biotechnology Applications in Beverage Production*, edited by C. Cantarelli and G. Lanzarini, pp. 49–64. London: Elsevier Applied Science.

Somers, J.C. and Evans, M.E. (1979) Grape pigment phenomena. Interpretation of major coulour losses during vinification. *Journal of the Science of Food and Agriculture*, **30**, 623–635.

Spettoli, P., Bottacin, A., Nuti, M.P. and Zamorani, A. (1982) Immobilization of *Leuconostoc oenos* ML34 in calcium alginate gels and its application to wine technology. *American Journal of Enology and Viticulture*, **33**, 1–5.

Stewart, G.G. and Rusell, I. (1986) One hundred years of yeast research and development in the brewing industy. *Journal of the Institute of Brewing*, **92**, 537–558.

Strehaiano, P., Moreno, M. and Goma, G. (1978) Fermentation alcoolique, influence de la concentration en glucose sur le taux de production d'éthanol et le taux de croissance. *Comptes Rendus de l'Academie des Sciences—Paris*, **286 D**, 225–228.

Traverso-Rueda, S. (1980) A quantification of the stimulatory effect of sterols on the growth rate of *Saccharomyces cerevisiae* under vinification conditions. Ph.D. Dissertation, University of California. Davis.

Truck, H.V. and Hammes, W.E. (1989) Die verwendung eines fliessbettreaktors zur Ausführung des biologischen Saüresabbaus in wein mit immobilisirten zellen; *Chemie Mikrobiologie Technologie und Lebensmittel*, **12**, 119–126.

Tyagi, R.D. (1984) Participation of oxgyen in ethanol fermentation. *Process Biotechemistry*, August, pp. 136–141.

Usseglio-Tomasset, L. (1988) La production par les levures de substances volatiles aromatiques. In *Application à l'Oenologie des Progrès Récents en Microbiologie et en Fermentation*, edited by ENSAA, Dijon, pp.171–201. Paris: Office International de la Vigne et du Vin.

Valade, M. and Rinville, X. (1990) Approches techniques et économiques des différents procédés de prise de mousse en bouteille. In *Les Nouvelles Techniques de Tirage*. pp. 34–39. Epernay: Association des Techniciens Supérieurs en Viticulture et Oenologie.

Vallery-Radot, R. (1900) *La vie de Pasteur.* Paris.

Van Uden, N. (1985) Ethanol toxicity and ethanol tolerance in yeasts. In *Annual Reports of Fermentation Processes*. Volume 8, edited by G.T. Tsao, pp. 11–58. London: Academic Press.

Vezinhet, F. and Barre, P. (1988) Transfer by cytoduction of the Flo 5 floculation gene in au enological yeast strain. *Proceedings of the Eighth International Biotechnology Symposium*, 101 pp.(A43). Paris.

Vezinhet, F. and Lacroix, S. (1984) Marquage génétique des levures : outil de contrôle des fermentations en souche pure. *Bulletin de l'Office International de la Vigne et du Vin*, **57**, 759–73.

White, F.H. and Portno, A.D. (1978) Continuous fermentation by immobilized brewers yeast. *Journal of the Institute of Brewing*, **84**, 228–230.

Zamorani, A. (1989) Enzymatic processing of musts and wines. In *"Biotechnology Applications in Beverage Production"*, edited by C. Cantarelli and G. Lanzarini, pp. 223–246. London: Elsevier Applied Science.

Zamorani, A., Crapisi, A., Borin, G., Dell'eva, M., Versini, G. and Spettoli, P. (1989) The use of immobilized yeasts in bottle-fermented Moscato bianco sparkling wine. *Yeasts*, **5**, S63–S66.

Chapter 17

ENZYMES IN WINEMAKING

ROSE-MARIE CANAL-LLAUBERES

Application and Development Department, Novo Nordisk Ferment Ltd, Neumatt, CH-4243 Dittingen, Switzerland

Chapter Outline

Chapter Outline (Continued)

INTRODUCTION

The production of wine is a biotechnological process marked with tradition. However, research in enology is helping to better understand the biochemical reactions that occur in grape juice during its transformation. The outstanding development of this science over the last few decades has profoundly influenced the technology of winemaking, allowing for the production of higher quality wines.

Enzymes play a definitive role in the process of winemaking. In one context, wine can be seen as the product of enzymatic transformation of the grape juice. During winemaking, enzymes are desired as early as the pre-fermentation stage. Their activities originate not only from the grape itself but also from yeasts and other microorganisms (fungi, bacteria). Moreover, the winemaker now reinforces and extends the action of these endogenous enzymes by the judicious use of exogenous, industrial enzyme preparations.

This chapter reviews and updates knowledge on the endogenous and exogenous enzymes associated with winemaking, giving specific consideration to their effects on process technology and on the quality of the final product. Industrial enzyme preparations will be given special attention due to their potential applications in both white and red wine production. For a better comprehension of the reactions involved, some background information is presented on the nature and structure of the substrates (polyphenols, polysaccharides, proteins) transformed. The chapter is divided into three sections; the first section concerns enzymes of the grapes, the second deals with enzymes excreted by microorganisms into juice and wine, while the last section focuses on the application of industrial enzyme preparations in winemaking.

ENZYMES PRESENT IN GRAPES

The grape undergoes intrinsic and extrinsic changes (maturation, infection, technological transformation) that involve the activities of enzymes (Villettaz, 1984). The enzymes of most significance in these contexts are the oxido-reductases, pectinases, proteases, glycosidases, lipoxygenases and peroxide-cleaving enzymes.

Oxido-reductases

As a result of the processes of crushing and pressing, the natural structure of the grape is destroyed and the juice becomes very susceptible to enzymatic oxidation (Sapis *et al.*, 1983). However, the rate of oxygen consumption by the juice varies greatly from one grape variety to another (0.5–4.6 mg/l/min). The enzymes involved in the oxidation of phenolic compounds have, for a long time, been known under the broad term "polyphenoloxidases". Research performed by Dubernet (1974) demonstrated the presence of two such enzyme types: the grape oxido-reductases or tyrosinases found in healthy berries, and laccase (EC 1.10.3.2) found in grapes infected by *Botrytis cinerea*.

Tyrosinases are enzymes that oxidize diphenols (catecholase or catechol oxidase, EC 1.10.3.1) and monophenols (cresolase or monophenol mono-oxygenase, EC 1.14.18.1) into quinones (Harel *et al.*, 1973). These enzymes, partially soluble in the must, have reduced activity after clarification; they are rather sensitive to sulfur dioxide (SO_2) and can be removed by bentonite treatment.

The mechanisms involved in the oxidation of phenolic compounds have been reviewed by Singleton (1987). The enzymatic pathway presented in Figure 17.1

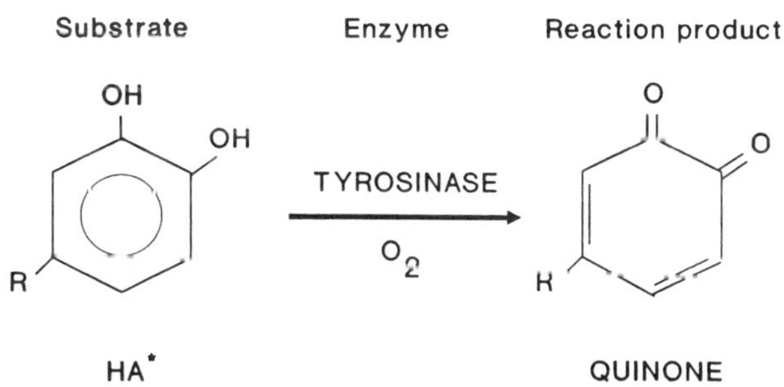

FIGURE 17.1 First stage of oxidation mechanisms: the action of grape tyrosinase. * HA: hydroxycinnamic acids (caffeoyltartaric acid, p-coumaroyltartaric acid).

is the initial step of browning. The preferred substrates of tyrosinases are the caftaric (caffeoyl tartaric) and coutaric (p-coumaroyl tartaric) acids (Günata *et al.*, 1987). Further reactions, leading to the formation of brown pigments and other products, are non-enzymatic. The quinones are then involved in formation of the "Grape Reaction Product" (GRP) (Singleton *et al.*, 1985) identified as S-glutathionyl-2-trans caffeoyltartaric acid (Cheynier *et al.*, 1986), and in condensation and coupled oxidation reactions. Browning in white wines depends mainly on the content of flavonoid phenols (Simpson, 1982; Schneider, 1989) whose oxidation products condense quicker and show more intense color than the oxidation products of non-flavonoid phenols (Cheynier *et al.*, 1989). Therefore, winemaking practices such as increased pomace contact and harder pressing, greatly affect the propensity of wine to browning.

Müller Späth (1977) proposed a method for stabilizing white wine against oxidation by adding oxygen to the must (hyperoxidation), and experimental trials are summarized in a recent review (Müller Späth, 1990). According to Guerzoni *et al.* (1981) hyperoxidation produces, within a few minutes, a strong enhancement of polyphenoloxidase activity in the must. The technological effects of this process have been studied by Moutounet *et al.* (1989; 1990) on the basis of oxidation mechanisms. They demonstrated that the enzyme activity is not the limiting factor of must oxidation. The enzyme concentration and stability in juice are always sufficient to oxidize the hydroxycinnamic acids (first stage of oxidation) and those being regenerated during the process (oxidation coupled reaction), if oxygen is present. It is mainly substrate depletion which slows down oxygen consumption. Even though this new practice looks promising as a process for color stabilization and SO_2 reduction, it needs to be adapted to the grape variety and to the desired type of wine (Nagel and Graber, 1988). Moreover, the effect of oxygen on other volatile compounds has still to be investigated.

Pectinases

Pectolytic enzymes make an important contribution to the changes that occur to pectic substances during grape ripening. They are also involved in the spontaneous clarification of musts and wines during processing.

Pectic substances

These complex heteropolysaccharides are mainly located in the middle lamella and the primary cell wall of plant cells. One can distinguish between acid pectic substances also called pectins (homogalacturonans, rhamnogalacturonans) and neutral pectic substances (arabinans, galactans, arabinogalactans) that present a high structural similarity with the side chains of pectins. Until now, only pectic substances from the pulp of the grapes have been purified and their structure partially analyzed (Villettaz *et al.*, 1981; Saulnier and Thibault, 1987; Saulnier *et al.*, 1988; Saulnier and Brillouet, 1989). The most recent results deal with the identification of "smooth" and highly branched "hairy" regions in pectins and the characterization of a type II arabinogalactan-protein.

Enzymes for pectin degradation

Enzymes involved in the hydrolysis of acid pectic substances have been the subject of many reviews, some of which are technology orientated (Pilnik and Rombouts, 1981). These enzymes de-esterify (pectin esterases) or de-polymerize (polygalacturonases, pectin and pectate lyases) specific substrates. Their mode of action is presented in Figure 17.2. Pectinesterases (EC 3.1.1.11) de-esterify pectin, producing methanol and pectin with a lower degree of esterification. They are specific for galacturonic acid and are not able to de-esterify the pectin completely. They are characterized by an exo-mode of action. Polygalacturonases hydrolyze the glycosidic linkages between non-esterified galacturonic acid units. Endo-polygalacturonases (EC 3.2.1.15) hydrolyze low esterified pectins in more or less random fashion, allowing a swift reduction in viscosity (endo-mode of action). Pectin lyases (EC 4.2.2.10) and pectic acid lyases (EC 4.2.2.2) catalyze beta-elimination reactions between esterified and non-esterified galacturonic acid units, following an endo-mode of action.

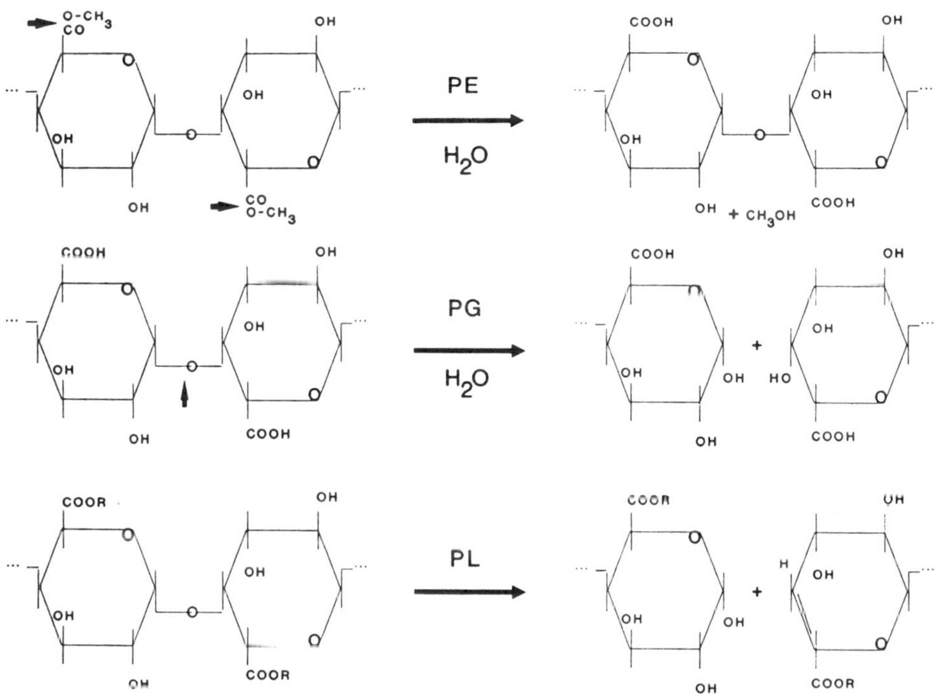

FIGURE 17.2 Mode of action of pectolytic enzymes. PE: pectin-esterase, PG: polygalacturonase, PL: pectin or pectic acid lyases. The arrow indicates the point of enzyme action.

Enzymes involved in the hydrolysis of neutral pectic substances have been wrongly associated with hemicellulases. These enzymes, reviewed by Dekker (1979), have been classified by Brillouet and Hoebler (1985) according to the nature of the molecules they are able to hydrolyze (e.g. arabinanases, galactanases).

Role of grape pectinases

Main research on grape pectinases started in the 1950s to explain the release of methanol and the reduction of viscosity observed during juice clarification (Flanzy and Loisel, 1958; Marteau *et al.*, 1961). Measurement of grape pectinesterase and polygalacturonase activities has given an understanding of their role in this spontaneous process (Marteau *et al.*, 1963). As shown by Grassin (1987), pectinesterase activity is higher than polygalacturonase activity and both increase during fruit ripening. A synergistic effect between these enzymes may affect the speed of clarification.

Proteases

Proteases hydrolyze the peptide linkages between the amino-acid units of proteins.

Grape proteins

Grape proteins, like grape polysaccharides, influence the clarification and stabilization of must and wine. According to Maujean (1990), flocculation observed during the clarification process is also linked to partial protein degradation, thus the rate of protein hydrolysis will affect the time necessary to achieve good clarification. Moreover, unstable proteins are responsible for most haze phenomena in wine. Preliminary studies on grape proteins were started in the 1960s (Kock and Sajak, 1959; Moretti and Berg, 1965). In more recent work (Hsu and Heatherbell, 1987a,b; Hsu *et al.*, 1987), a fraction of wine proteins with molecular weights ranging from 12 600 to 30 000 Dalton and with isoelectric points from 4.1 to 5.8 was found to be responsible for haze formation. Further investigations on protein fractions have indicated their amino-acid and sugar compositions (Paetzold *et al.*, 1990). These authors demonstrated that the glycoprotein concentration in must was linked to grape variety, grape maturity and to technological factors such as pressing and maceration. The risk of haze formation by the major protein fractions in Muscat Gordo Blanco grapes was established by Waters *et al.* (1989) as shown in Figure 17.3.

Proteolytic activity of grapes

Studies on grape proteases were started by Baglioni *et al.* (1937) who identified a proteolytic activity in grape juice. Later, Cordonnier and Dugal (1968) demonstrated that grape proteases were active before alcoholic fermentation and that their activity was inhibited as soon as ethanol was produced. These

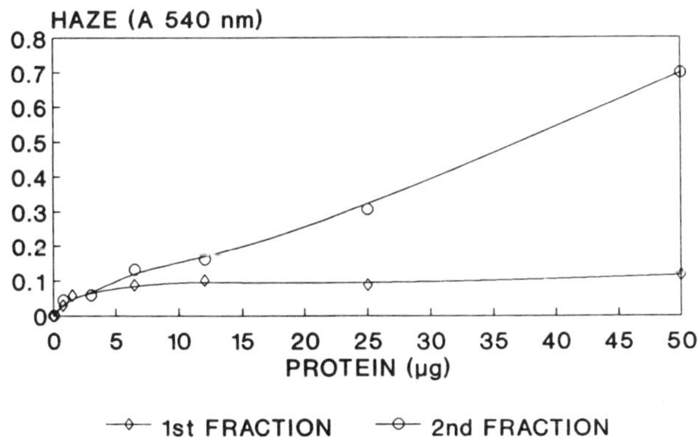

FIGURE 17.3 Influence of concentration of grape proteins on heat induced haze formation in wine. The 1st fraction (MW 120 000 Dalton) and 2nd fraction (MW 10 000 to 36 000 Dalton) were added to Gordo wine ultrafiltrate (Waters *et al.*, 1989).

enzymes are heat stable, enabling the release of soluble nitrogen during thermovinification (Feuillat and Bergeret, 1967; Cordonnier, 1970). Grape proteases are mainly associated with the solid particles of the berry and with the pulp. Thus, one can explain the decrease in proteolytic activity during juice clarification (Canonica and Ferrari, 1978; Feuillat *et al.*, 1980). The evolution of grape proteolytic activity during white winemaking is shown in Figure 17.4.

Glycosidases

The focus here will be placed on enzymes capable of hydrolyzing aroma precursors and, more specifically, terpenyl-glycosides (Figure 17.5) responsible

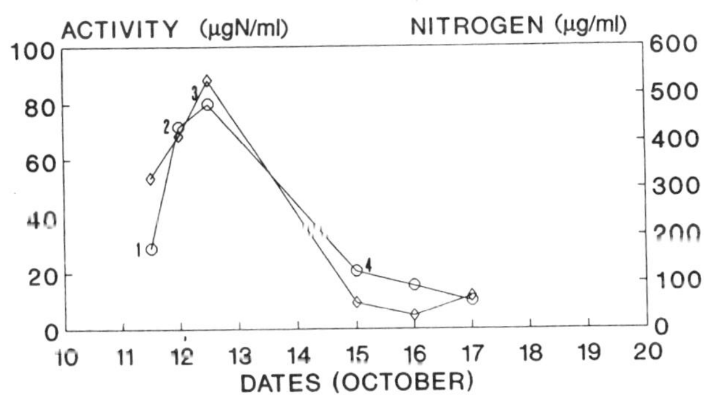

FIGURE 17.4 Proteolytic activity of grapes during white winemaking operations; activity measured on hemoglobin (pH 2, 55°C). 1: crushing, 2: first pressing, 3: second pressing, 4: racking (Feuillat *et al.*, 1980).

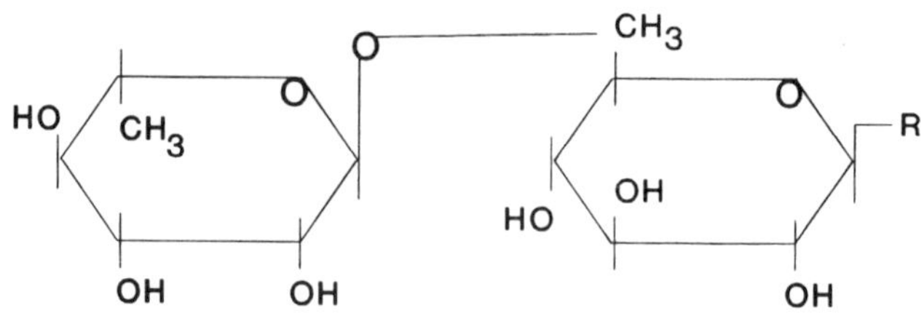

6-O-α-rhamnopyranosyl-β-D-glucopyranoside

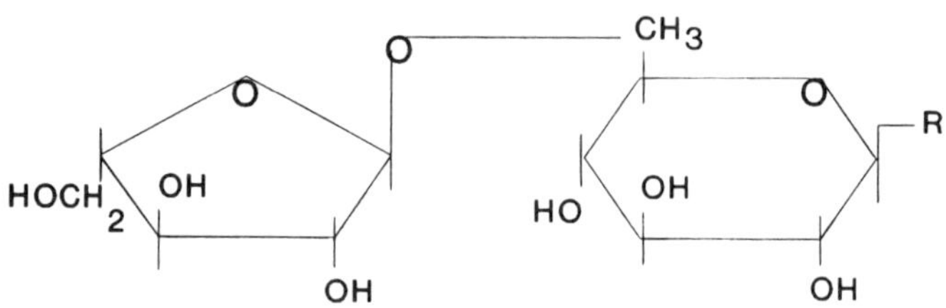

6-O-α-arabinofuranosyl-β-D-glucopyranoside

FIGURE 17.5 Example of some grape terpenyl-glycosides. R represents aglycone (linalool, terpineol, etc.).

for the varietal character of many grapes (Marais, 1983; Rapp and Mandery, 1986).

Cordonnier and Bayonove (1974) suggested that terpenols (specially geraniol) can be liberated from their precursors by enzymatic hydrolysis. Later, Bayonove *et al.* (1984) demonstrated the presence of a β-D-glycosidase activity in muscat grape juice and its inhibition after pasteurization (Figure 17.6). Significant release of nerol and geraniol were obtained at pH 5.0 and at 30°C. Further studies on properties of grape glycosidases have been reported (Aryan *et al.*, 1987; Biron *et al.*, 1988; Gunata *et al.*, 1989). These authors demonstrated the presence of several β-glycosidases and α-glucosidases, measuring the highest activities in mature grapes juices. The β-glucosidases had optimal activity at pH 5.0 and were inhibited by glucose. Moreover, grape glycosidases were not able to hydrolyze sugar conjugates of tertiary alcohols such as linalool (aglycone specificity). Furthermore, Großmann *et al.* (1990) showed that certain processing steps (clarification, centrifugation) reduced β-glucosidase activity. Collectively, these results lead to the conclusion that endogenous glycosidases of the grape are hardly effective at enhancing aroma during juice processing and winemaking operations.

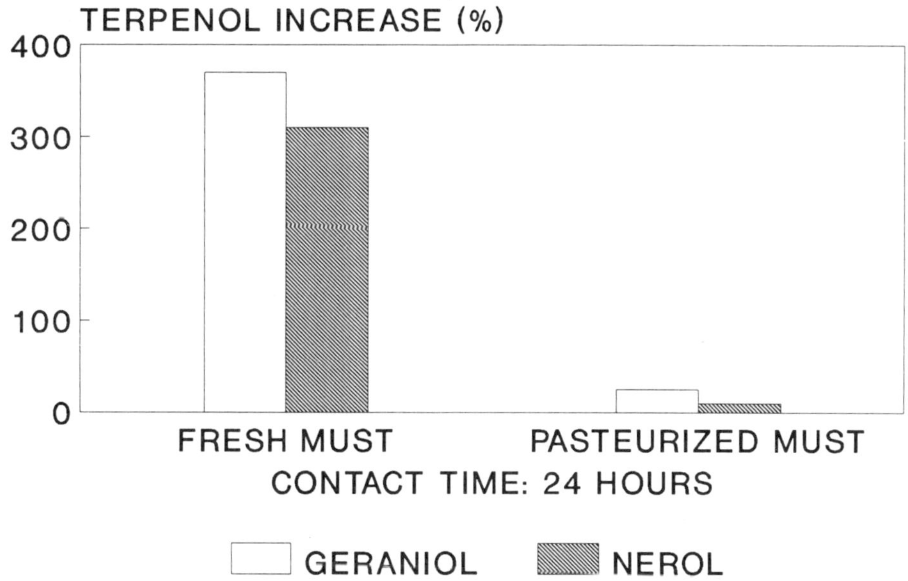

FIGURE 17.6 Release of terpenols (pH 5, 30°C) in centrifuged must before and after pasteurization. The increase is due to glycosidase activity that is inhibited by pasteurization (Bayonove *et al.*, 1984).

Lipoxygenase and peroxide-cleaving enzymes

Among the volatile compounds formed during the pre-fermentation stage, aldehydes and alcohols with six atoms of carbon (C_6) have a detrimental effect on the organoleptic quality of must and wine because of their "grassy" note (Schreier, 1979). Their concentrations vary according to maturity and to technological parameters of juice preparation.

The C_6 volatile compounds (hexanal, cis-hexene-3-al, trans hexene-2-al, cis-hexene-3-ol, trans hexene-2-ol) are formed by the sequential action of enzymes on C_{18} unsaturated fatty acids in the presence of oxygen (Crouzet, 1986). The enzymatic pathways are shown in Figure 17.7. The reaction starts with the hydrolysis of membrane lipids by an acyl hydrolase. Since the properties of grape lipoxygenase (Cayrel *et al.*, 1983) and alcohol dehydrogenase (Molina *et al.*, 1986) were established, further investigations focused on the peroxide-cleaving enzyme (Valentin and Crouzet, 1989). Cordonnier and Bayonove (1981) studied the influence of crushing and pressing on the formation of C_6 alcohols in must, and showed that the stems and leaves are major sources of these compounds (Table 17.1). Ollivier (1987) demonstrated that skin contact in the presence of solid parts can significantly increase the concentration of these compounds in wine. Thus, maceration of destemmed grapes and effective must racking (turbidity below 200 NTU) greatly reduce the level of C_6 volatile compounds (Dubourdieu *et al.*, 1986).

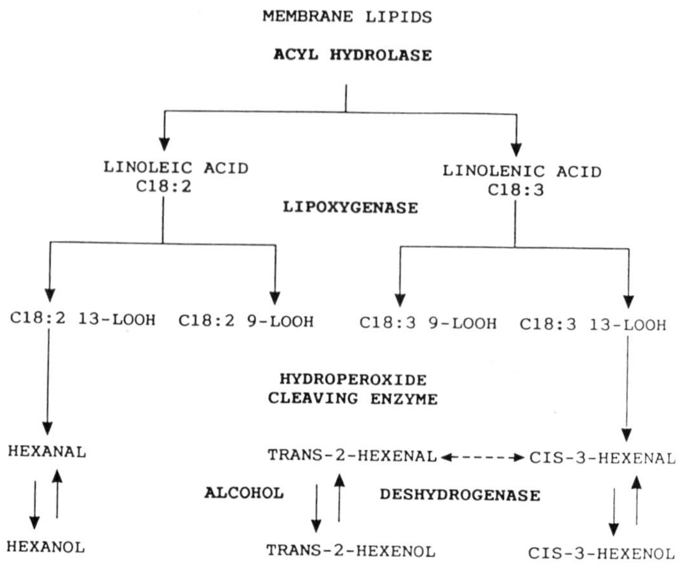

FIGURE 17.7 Enzymatic pathways for the formation of C_6 volatile compounds responsible for the green and "grassy" flavor of grapes (Crouzet, 1986).

ENZYMES EXCRETED BY MICROORGANISMS INTO JUICE AND WINE

Fungal Enzymes

Among the fungi infecting grapes, *Botrytis cinerea* is responsible for the development of grey rot and the so-called noble mold. This fungus produces various exocellular enzymes which not only affect the composition of the grape but also affect winemaking practices and the quality of the wine.

First studies on *B. cinerea* enzymes focused on the hydrolysis of cell walls and oxidation of phenolic compounds of the infected plant. Thus, pectinases (Verhoeff and Liem, 1978), a cellulase (Verhoeff and Warren, 1972), a laccase

TABLE 17.1 Formation of C_6 volatile compounds[a] in Ugni blanc grapes after crushing of the berries, in the absence and presence of stems and leaves, under aerobic conditions (Cordonnier and Bayonove, 1981).

	Hexanal	*Trans-2 hexenal*	*Hexanol*	*Cis-2 hexenol*	*Trans-2 hexenol*
Berries	918	5 700	78	61	6 846
+ stems	1 014	6 798	159	115	8 245
+ leaves	1 218	14 289	133	164	15 990

[a]Expressed in μg/kg.

(Dubernet *et al.*, 1977), phospholipases (Shepard and Pitt, 1976) and an esterase (Dubourdieu, 1978; Koh, 1985) were identified in *B. cinerea*. The presence of N-acetyl glucosaminidase (Grassin *et al.*, 1987) and protease (Heale and Movahedi, 1989) have also been reported.

Laccase of *Botrytis cinerea*

Oxidation is always more serious when grapes have been infected by *B. cinerea*. The reason is the presence of an oxido-reductase, laccase, in moldy grapes. The differences in activity of the *B. cinerea* laccase and the grape tyrosinase explain the behavior of healthy and rotten grapes towards oxidation. Unlike tyrosinase, laccase has a broader specificity towards phenolic compounds (Salgues *et al.*, 1986), better stability at the pH of must and wine and good resistance towards SO_2.

Dubourdieu *et al.* (1984) proposed a colorimetric method to measure the laccase units in grape juice or wine. After optimization (Grassin and Dubourdieu, 1986), this method was developed into a test-kit which is commercially available (Tassot, 1988; Valade and Blanck, 1988). This test-kit helps the winemaker to decide how to treat moldy grapes, as even low laccase activity (3 U/ml) has a detrimental effect on wine color.

Other enzymes of *Botrytis cinerea*

Grassin (1987) demonstrated the production of pectin esterases and polygalacturonases by *B. cinerea* during its growth and isolated an inhibitor of polygalacturonase from unripe grapes. Pectolytic enzyme concentration is 200 times higher in infected grapes than in healthy grapes. These enzymes, produced during the grape "veraison", hydrolyze pectic substances of the grape and affect the composition of acidic polysaccharides in juice (Table 17.2). Thus, problems encountered during clarification of juice from fungal infected grapes are not due to pectins but, in fact, are due to the presence of other polysaccharides, more specifically to glucan, produced by *B. cinerea*.

Recently, Heale and Movahedi (1989) characterized an aspartic protease excreted by the fungus. According to the authors, this enzyme is produced at the early stage of infection and is of primary importance in determining the subsequent rate and extent of rotting caused by the pectinases which are usually excreted later. Knowledge of the mechanisms involved in regulating

TABLE 17.2 Polysaccharide concentration[a] of musts obtained from healthy grapes and *Botrytis* infected grapes (Dubourdieu, 1982).

	Glucan	*Pectin*	*Other polysaccharides*
Healthy grapes	0	670	340
Moldy grapes	387	0	627

[a]Expressed in mg/l.

proteolytic and pectolytic enzyme production by this fungus is fundamental to understanding the kinetics of grape infection and decay.

Yeast Enzymes

In the middle of the last century, Louis Pasteur concluded that the fermentation of sugar into alcohol by yeast was catalyzed by "ferments" which were later named enzymes. Since then, it has been established that many yeast enzymes are responsible for the metabolic pathways that are active during fermentation and, later, during autolysis. Proteases, glucanases and glycosidases are discussed here.

Proteases

Yeast proteases can have an intra- or extracellular origin. Among the yeasts, the *Saccharomyces cerevisiae* proteolytic system has been studied in detail and comprehensive reviews are available (Wolf, 1980; Achstetter and Wolf, 1985). The main activities are presented in Table 17.3.

Feuillat *et al.* (1980) investigated the production of exocellular proteases. These enzymes play an important role during the autolysis process which occurs in wines kept on yeast lees during aging (e.g. Burgundy white wines or sparkling wines produced according to the "méthode champenoise". However, because of the particular conditions found in wine, only a few proteases are active (Lurton, 1987). To demonstrate the effect of protease A, the release of nitrogen into wine has been followed during autolysis (Figure 17.8).

The other important aspect of yeast proteolytic activity is its potential in protein haze reduction (Nelson and Young, 1986). Lagace and Bisson (1990) investigated the effect of non-*Saccharomyces* strain proteases on the hydrolysis of wine proteins. Even though partial degradation was obtained, the effect on wine stabilization has not yet been demonstrated.

TABLE 17.3 Main proteolytic activities present in the yeast *Saccharomyces cerevisiae* (Lurton, 1989).

Nomenclature	*Mechanisms*	*Optimum pH*
Protease ysc A	acid endoprotease	3
Protease ysc B	endo serine-sulphydryl protease	7
Carboxypeptidase ysc Y	serine exopeptidase	6
Carboxypeptidase ysc S	metallo exopeptidase	–
Aminopeptidase ysc I	metallo exopeptidase	7
Aminopeptidase ysc II	metallo exopeptidase	8
Aminopeptidase ysc Co	metallo exopeptidase	8

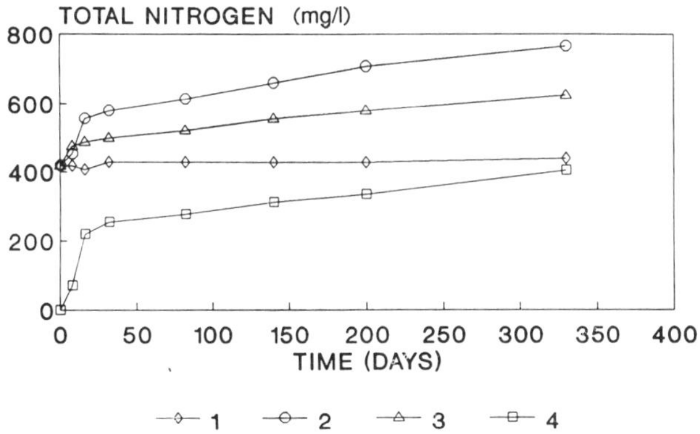

FIGURE 17.8 Release of total nitrogen by the yeasts during aging of wine in contact with its lees. 1: racked wine (control), 2: wine on lees, 3: wine on lees with inhibitors of proteases other than protease A, 4: lees in hydroalcoholic buffer (Lurton, 1989).

Glucanases

The presence of β-(1,3)-D-glucanases has been reported in many yeast species (Fleet, 1991). These enzymes show endo-(EC 3.2.1.6) or exo-(EC 3.2.1.56) activities. *Saccharomyces cerevisiae* excretes several β-(1,3) glucanases. These enzymes are constitutive glycoproteins and their activities vary independently during the cell cycle (growth, mating, sporulation) (Sanchez *et al.*, 1983; Hien and Fleet, 1983).

Charpentier *et al.* (1986) mentioned the action of β-glucanases during yeast autolysis to explain variations in the molar ratio of mannose/glucose in autolysates. Later, (Llaubères, 1988) demonstrated the presence of a cell wall endo-β-(1,3)-glucanase activity in strains of dried yeast used in winemaking. The effect of this enzyme on the yeast cell wall explains the release of mannoproteins during wine aging on lees (Llaubères *et al.*, 1987).

Glycosidases

Many studies have focused on β-glucosidases because of their wide occurrence in plants, fungi and yeasts (Leclerc *et al.*, 1987). The purification of a β-glucosidase from *S. cerevisiae* has been described by Duerksen and Halvorson (1958).

The effect of yeast β-glucosidase on the hydrolysis of grape terpenyl-glycosides has been investigated. Großmann *et al.* (1987) studied the β-glucosidase activity of *Hansenula* species found in must. This enzyme, although able to liberate aroma substances in wine, seems to be less effective in must. According to Dubourdieu *et al.* (1988), the liberation of terpenols during fermentation can be explained by yeast β-glucosidase activity. It was shown that the *S. cerevisiae* β-glucosidase was not inhibited by glucose, in contrast to what has been found for the grape glucosidase. Further studies (Darriet *et al.*, 1988) located the enzyme in the periplasmic space of yeast cells (Table 17.4). Its

TABLE 17.4 Effect of *Saccharomyces cerevisiae* periplasmic extract on the hydrolysis of terpenyl-glycosides[a] (Darriet *et al.*, 1988).

Terpenols[b]	*Control*	*1*	*2*	*3*	*4*
Linalool	3	3	2	5	2
Terpineol	2	2	3	11	2
Citronellol	6	8	9	39	6
Nerol	0	0	38	208	0
Geraniol	0	0	27	286	0
Total	11	13	79	549	10

[a]Isolated from Muscat wine; incubation conditions: 72 hours, pH 5.0, 25°C.
1: zymolyase used for the preparation of yeast extract; 2: intact yeast cells;
3: periplasmic yeast extract; 4: inactivated periplasmic extract.
[b]Expressed in mg/l.

purification and kinetic properties are under evaluation. The results obtained so far suggest that specific yeast strains can affect the varietal aroma of wines (Laffort *et al.*, 1989). Recent investigations have focused on the β-glucosidase activity of other yeast strains such as *Hanseniaspora vinea* (Vasserot *et al.*, 1989) and *Candida* species (Günata *et al.*, 1990).

Other yeast enzymes such as esterases are also involved in the formation of aroma compounds. However, very little research has been devoted to this type of enzyme (Schermers *et al.*, 1976; Spaepen and Verachtert, 1982). According to Koh (1985), the esterase of healthy grapes is not very active on the esters contributing to the so-called secondary aroma (fermentation aroma) of dry white wines. It is known that the production of esters by yeast is higher when fermenting very clarified juices under controlled temperature. Ester formation is used as a criterion for the selection of wine yeast strains, as only the production of "good" esters is desirable (Cantarelli, 1989).

Bacterial Enzymes

Lactic bacteria are significant organisms in many wine fermentations. It is likely that they could produce a range of enzymes that would affect wine quality. Apart from the malolactic transformation, little is known about these enzymes in enology. Davis *et al.* (1988) noted the production of esterase and lipolytic activities by several strains of wine lactic acid bacteria, but no proteolytic activity was detected.

The malolactic enzyme system

Due to the difficulty of achieving malolactic fermentation in wine, some research has focused on the enzymatic mechanism of this reaction (Lonvaud-Funel,

1982). The transformation of malic acid is catalyzed by the malolactic enzyme which has been purified from *Lactobacillus* and *Leuconostoc* strains (Schütz and Radler, 1973; Lonvaud-Funel and Strasser de Saad, 1982; Spetolli and Zamorani, 1984). Further investigations have revealed this catalyst as a multienzymatic system (Lonvaud-Funel, 1986) and genetic studies have been started (Renault, 1987). Naouri *et al.* (1990) purified a single malolactic enzyme with two identical subunits from *Leuc. oenos*. Its kinetic properties were examined and further studies will focus on the role of cofactors such as NAD^+ and Mn^{2+}.

INDUSTRIAL ENZYME PREPARATIONS

As mentioned already, the endogenous enzyme activities of grapes and microorganisms are often neither efficient nor sufficient under winemaking conditions. Nowadays, enzymes from exogenous sources are recognized as helpful processing aids because they can reinforce or replace the grape enzymatic activities.

The use of industrial enzyme preparations in food and beverage industries has been reviewed (Pilnik, 1982; Rosendal, 1989; Voragen, 1989) and specific applications have also been reported in the wine industry (Marteau, 1972; Godfrey and Reichelt, 1982; Canal-Llaubères, 1989; Zamorani, 1989). Improved knowledge of the nature and structure of the macromolecules found in must and wine offers new possibilities for the use of enzymes in winemaking. Thus, changes obtained by means of enzymatic treatment not only affect clarification and filtration but also affect extraction and stabilization processes in both white and red wine production (Figure 17.9). Today, the main applications of industrial enzymes in winemaking concern the use of pectinase and glucanase preparations.

Pectinases

Most industrial pectinase preparations are derived from cultures of *Aspergillus*, mainly *Aspergillus niger*, which is a species accepted as "G.R.A.S." (Generally Recognized As Safe) and by the Office International de la Vigne et du Vin (O.I.V.).

The technology used in the production of industrial enzymes has been reviewed by Aunstrup *et al.* (1979). After fermentation, the desired enzyme activities are recovered and purified. Standardized enzyme preparations undergo several quality control tests before being sold in liquid or granulated form. Besides the main pectolytic activities, industrial pectinase preparations contain hemicellulolytic and cellulolytic activities. Glycosidase (Cordonnier *et al.*, 1989) and protease (Schmitt *et al.*, 1989) side activities may also be present. The ratio between the main activity and the so-called side activities defines the value of an enzyme preparation for a given application. Some of the applications are described below.

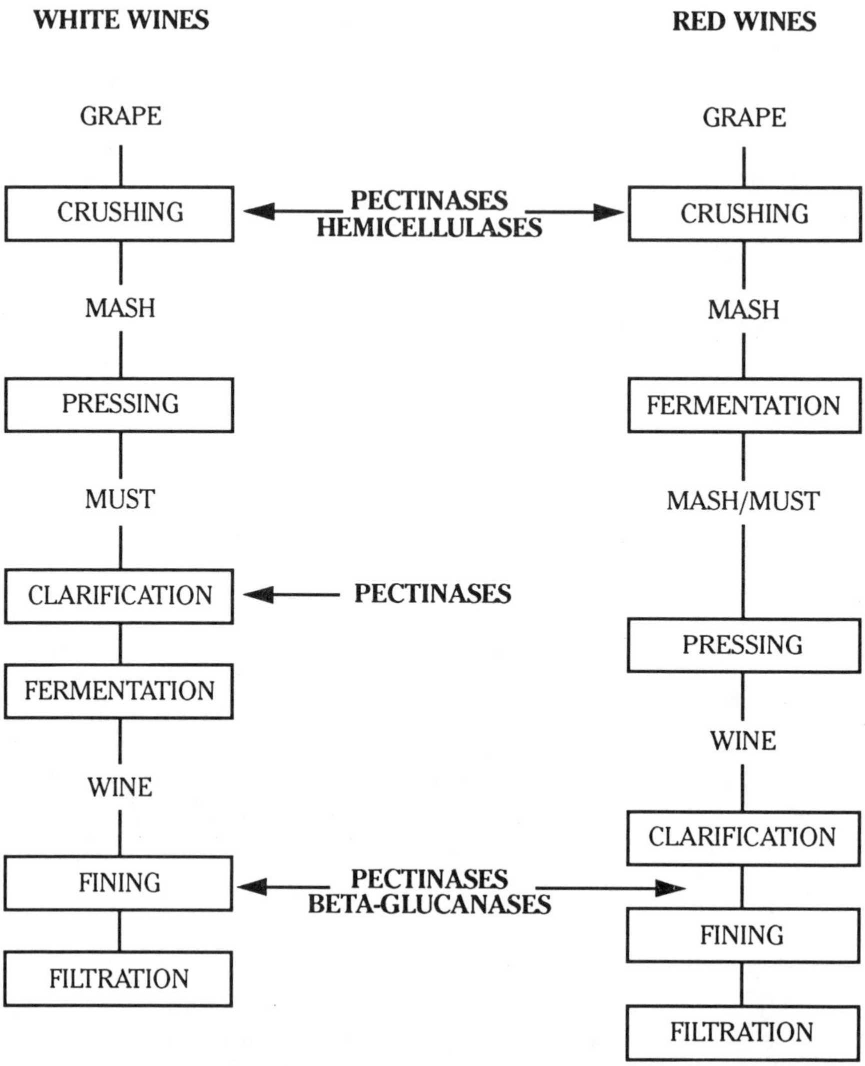

FIGURE 17.9 Application of industrial enzyme preparations in white and red wine production.

Mash treatment for juice extraction

The pulp of a number of grape varieties is very rich in pectic substances (Muscat, Sylvaner). Thus, incomplete hydrolysis of these molecules by endogenous enzymes may cause problems during processing. Laboratory and industrial scale trials have demonstrated the advantages of using exogenous pectolytic enzymes for the processing of grape juice (Ough and Berg, 1974). This technology is widely used for the treatment of slippery grapes that, otherwise, are hard to press (Sims *et al.*, 1988).

Ough and Crowell (1979) conducted experiments on various grape varieties and showed that enzyme treatment of the mash increases juice yields significantly and, in comparison to the control, the wines were of an equal or better quality. In addition, tests indicated that clarification by centrifugation may be significantly enhanced because of changes in the physical properties of the flocculant material. More recent trials by Rickard *et al.* (1987) with Muscat Gordo Blanco, confirmed the substantial improvement in juice yields on enzyme treated samples.

Good results were obtained in industrial trials conducted on Kadarka (Hungarian variety) and Sylvaner grapes (Canal-Llaubères, 1989). The enzyme preparation was added directly into the press during filling and stayed in contact with the grapes for 1–2 h. This treatment increased the volume of free run juice, which is considered to be better quality juice (Figure 17.10). Further advantages were an increased press capacity, easier and shorter pressing and faster clarification. The use of pectolytic enzymes for maceration may also increase the terpenol content of juice (Ollivier, 1987).

Juice clarification, wine filtration

After pressing, the grape juice is turbid and shows a certain viscosity due to the presence of pectins. The pectin content depends on the grape variety, level of ripeness and the effect of technological factors (mechanical harvesting, pressing conditions). All these parameters have an influence on the time necessary for good clarification.

In white winemaking, the hydrolysis of pectins with pectinases quickly reduces the must viscosity, thus allowing faster and easier settling of the juice (Brown and Ough, 1981). In practice, two tests can be used to follow clarification: the pectin test and the centrifugation test (Canal-Llaubères, 1989). The pectin test offers a simple method for demonstrating the presence of

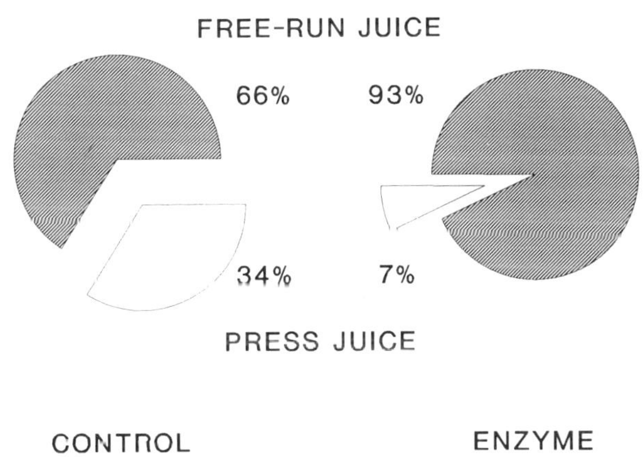

FIGURE 17.10 Enzyme mash treatment on Kadarka grapes (Vinozym 2 g/hl). Yields in free run and press juices. Hungary, 1988 (Canal-Llaubères, 1989).

pectins. In a test tube, 5 ml of juice are mixed gently with 10 ml of ethanol (96% vol. with 1% concentrated HCl). An early flocculation of the solution (white aggregates) indicates that pectin degradation is insufficient for satisfactory clarification. In this case, the aggregates will form a jelly-like substance that rises to the top of the tube after a few minutes. If pectins are partially or completely hydrolyzed, the aggregates become finer and finally disappear. The centrifugation test helps to determine the point at which the colloidal system is destabilized and thus when the physical phases of flocculation and settling are going to start. With this method, the juice (10 ml) is centrifuged for 5 min. at 3000 g in a laboratory centrifuge. The turbidity of the supernatant is then measured.

Exogenous pectinases can destabilize the colloidal system of the juice within one hour, allowing faster settling of the sediment to the bottom of the tank (Figure 17.11). The effect of enzymatic clarification appears later at the filtration step (Table 17.5) where residual pectins of the control batch are responsible for a quick rise in filter pressure and, consequently, clogging of the filter. Thus, improved filtration is obtained with enzyme treatment, accompanied by the use of less filter material and less wine loss. Other experiments have shown that addition of a suitable protease to the must after pasteurization could improve the clarification and filtration of the wine, and also decrease or prevent foam formation (Schmitt *et al.*, 1981).

In red winemaking, the standard applications of pectolytic enzymes concern their use on heat-treated juices and press wines, both of which show high pectin content. It is known that thermovinification (heat treatment of the mash), besides giving better color extraction, is responsible for the inactivation of grape enzymes such as oxidases and pectinases (Martinière and Ribéreau-Gayon, 1973). In this case, enzyme addition is necessary to achieve juice clarification. Enzymes are generally added after heat treatment when the temperature of the mash is reduced otherwise they would be inactivated. With

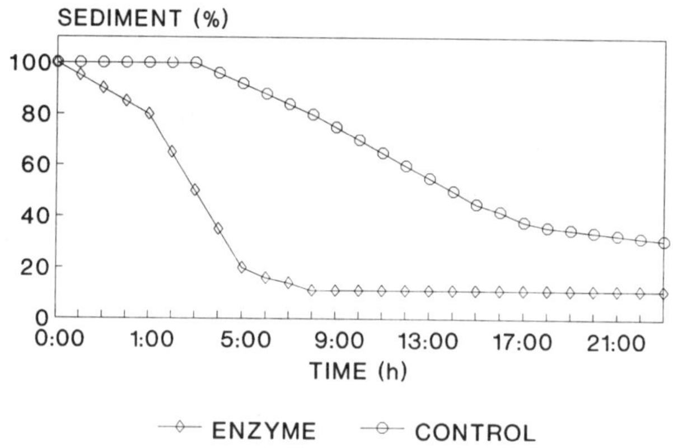

FIGURE 17.11 Effect of enzyme treatment on clarification of juice from Chasselas grapes (Vinozym 1.5 g/hl). Switzerland 1988 (Canal-Llaubères, 1989).

TABLE 17.5 Effect of enzyme treatment on the filtration of Sylvaner wine over Kieselgur (Vinozym 2.5 g/hl). Alsace 1988 (Canal-Llaubères, 1989).

Filtration time (min)	*Untreated wine*		*Enzyme treated wine*	
	Volume (hl) filtered	*Pressure (Bar)*	*Volume (hl) filtered*	*Pressure (Bar)*
30	35	3	35	3
60	65	6[a]	70	3
90	–		105	3
120	–		135	3
180	–		155	6[a]

[a]Clogging of the filter.

red wines obtained by traditional maceration, the use of enzyme preparations after pressing (clarification tank) is favorable for clarification and filtration as shown in Table 17.6 (Vicente-Eguzkiza, 1989).

Color extraction

The extraction of phenolic compounds generally occurs with maceration of the mash during alcoholic fermentation and depends on the variety and quality of grapes, and on technological parameters such as crushing, SO_2 addition, maceration time, temperature, pumping over, etc. Even though trials on color extraction are difficult to conduct due to heterogeneity of the mash and the influence of processing factors, good results have been obtained, on an industrial scale, with enzymatic treatment (Ough *et al.*, 1975; Montedoro and Bertuccioli, 1976; Vicente-Eguzkiza, 1989; Canal, 1990b). Improved anthocyanin extraction was obtained with enzyme addition to the grapes at the beginning of maceration. Pumping over gives good distribution of enzymes in the mash after

TABLE 17.6 Effect of enzyme treatment on the clarification[a] of pressed red wines. Influence on fining and physical processes (Vicente-Egurzika, 1989).

Process	*Untreated wine*	*Enzyme treated wine*
None	60	38
Fining	55	8
Centrifugation	20	12
Filtration	2	0.3
Fining + centrifugation	13	2.3

[a]Expressed in NTU (turbidity units).

filling of the tank. The increased color of wine proved to be stable during aging, and it was also found that the enzyme treatment had a positive effect on organoleptic parameters such as fruitiness and structure. However, basic research on color extraction and stabilization is still necessary to fully understand all the mechanisms involved in these processes.

Glucanases

Glucanase preparations for winemaking were developed and tested in the 1980s (Villettaz *et al.*, 1982; Dubourdieu *et al.*, 1985). The only specific industrial enzyme preparation available is derived from the culture of a selected strain of a *Trichoderma* spp. It was developed to solve clarification and filtration problems with juices extracted from grapes infected with *B. cinerea*.

Processing difficulties arise because of a high molecular weight polysaccharide excreted by this fungus. The structure of this macromolecule has been identified as a β-(1,3-1,6)-glucan (Dubourdieu *et al.*, 1981). In practice, the presence of glucan in juice or wine can be detected by a simple glucan test (Dubourdieu, 1982). To a tube containing 10 ml of wine, 5 ml of ethanol (96% vol. with 1% HCl) are added and mixed carefully. The formation a of white filament is a sign of glucan presence (concentration above 15 mg/l). As even low glucan concentrations (3 mg/l) may cause filtration problems, the following modified method is preferred. Wine (5 ml) is mixed with 5 ml of ethanol in a tube; after 30 min at room temperature, the mixture is centrifuged at 3000 g for 20 min. The supernatant is carefully removed and the precipitate, redissolved in 1 ml water, is mixed with 0.5 ml ethanol. The formation of filaments can then be observed.

During grape processing, *B. cinerea* glucan is released into the juice and later found in the wine. As it is impossible to remove this polysaccharide by hydrolysis with endogenous enzymes or by conventional treatments (centrifugation, fining), its degradation by exogenous enzymes has appeared to be the unique solution. Various trials have demonstrated the efficacy of such treatment (Wucherpfennig and Dietrich, 1983; Villettaz *et al.*, 1984) (Figure 17.12) and its impact on the economics of filtration (Villettaz *et al.*, 1987).

Enzyme addition is preferably done with young wine just after alcoholic fermentation; contact time with the enzyme is 7–10 days and the temperature must not be below 10°C. Increased enzyme dosages are recommended in red wines because of partial inhibition of its activity by phenolic compounds.

New Enzyme Applications

Aroma liberation

Since the enzymatic mechanisms of hydrolysis of terpenyl-glycosides have been elucidated (Günata *et al.*, 1988), studies have aimed to search for fungal enzyme capable of enhancing wine aroma (Großmann and Rapp, 1988; Cordonnier *et al.*, 1989; Günata *et al.*, 1990). Trials conducted with an experimental β-glucosidase preparation from *Aspergillus* spp. have demonstrated that positive results can

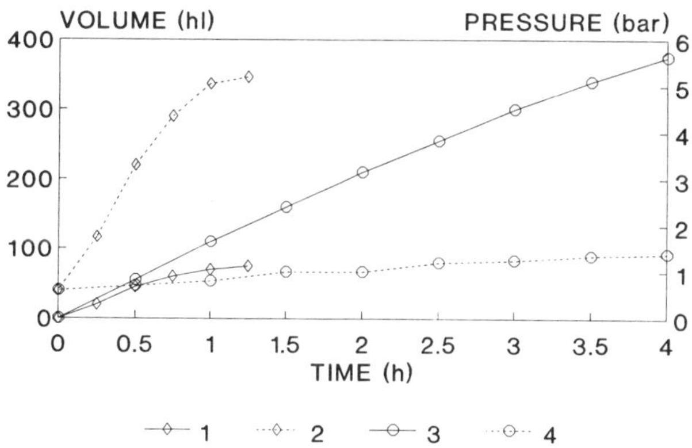

FIGURE 17.12 Kieselgur filtration after bentonite treatment (80 g/hl) and centrifugation of control and glucanase-treated (Glucanex 1 g/hl) Sylvaner wines produced from moldy grapes. — volume (1: untreated, 3: enzyme treated); – pressure (2: untreated, 4: enzyme treated). Alsace, 1986 (Villettaz *et al.*, 1987).

be obtained under winemaking conditions (Canal-Llaubères, 1990a). This new enzyme can be used in fermented juice as soon as the glucose has been consumed by the yeast, or in young wines. Further trials are needed on various grape varieties. This technology can be used to reinforce the varietal aroma and bouquet of certain wines.

Low alcohol wine

Low alcohol wines can be produced by various physical treatments involving expensive equipment such as reverse osmosis. An enzymatic method based on the use of glucose oxidase and catalase has been proposed by Villettaz (1986). By means of enzyme treatment of the juice, glucose is converted into gluconic acid which cannot be metabolized by the yeast. The wine so produced has reduced alcohol content and higher acidity. Thus, this technology can also be used to produce a reserve of acid musts or wines for blending purposes.

Color stabilization

Removal of phenolic compounds to prevent oxidation, haze and flavor changes is usually done with fining agents (gelatine, polyvinylpolypyrrolidone). Treatments with enzymes such as laccase, tannase and peroxidase have also been tried to achieve juice or wine stabilization (Cantarelli *et al.*, 1989; Cantarelli, 1990). The effect of fungal laccase has been closely examined because of its ability to react with a wide range of phenolic compounds. It has been demonstrated that laccase treatment can enhance the effect of conventional fining treatments (Figure 17.13).

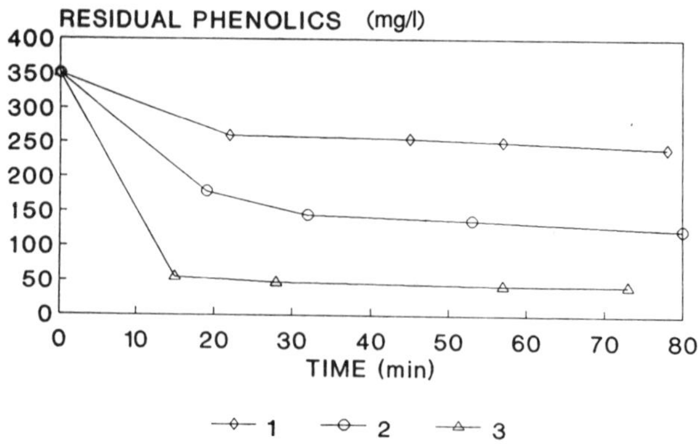

FIGURE 17.13 Removal of phenolic compounds during pilot scale "active filtrations" on raw and laccase-treated Muscat must. 1: raw, active charcoal; 2: raw, PVPP; 3: laccase-treated, PVPP (Cantarelli *et al.*, 1989).

Protein stabilization

In spite of considerable research, the use of bentonite remains the only way to achieve white wine stabilization. However, due to present winemaking practices giving increased protein extraction, larger bentonite dosages are necessary (Dubourdieu and Canal-Llaubères, 1989) which is considered less favorable to wine quality. Moreover, this treatment generates large volumes of lees for disposal, and causes losses of wine during filtration.

The availability of an appropriate acid protease could help to remedy this problem by hydrolyzing the grape proteins, thereby omitting bentonite treatment. However, the search for fungal enzymes which could degrade wine proteins has, so far, remained unsuccessful (Bakalinsky and Boulton, 1985; Modra and Williams, 1988; Waters *et al.*, 1989).

CONCLUSION

Endogenous enzymes are involved in virtually all the steps of winemaking. Their activities vary according to physiological and technological factors. Through better understanding of enzymatic transformations during grape processing, winemakers have been able to master negative effects and to reinforce desired activities. As a consequence, the use of industrial enzyme preparations has become normal practice, the main advantages being their effectiveness, specificity and convenience. These processing aids help to improve the quality of wines and make for more rational and economic utilization of facilities and equipment.

However, further research is necessary to gain more knowledge on the role of enzymes in complex processes such as extraction. Recently, the study of specific regions of fruit cell wall polysaccharides has lead to the discovery of a new enzyme, a rhamnogalacturonase, capable of degrading the hairy regions of pectins (Schols *et al.*, 1990). Further studies on the chemistry of plant cell wall polysaccharides and or the biochemistry and genetic regulation of the enzymes responsible for their degradation are likely to have practical significance in winemaking.

Genetic engineering technologies are destined to have an impact upon the production of a new generation of enzymes for the food and beverage industries. Such enzymes should have higher specificity, purity and tailored kinetic properties, as well as prospects for less expensive production. Developments in these areas of course will transfer to the wine industry subject to appropriate legislative acceptance.

REFERENCES

Achstetter, T. and Wolf, D.H. (1985) Proteinases, proteolysis and biological control in the yeast *Saccharomyces cerevisiae*. *Yeast*, **1**, 139–157.

Ahearn, D., Meyers, S.P. and Nichols, R.A. (1968) Extracellular proteinases of yeasts and yeastlike fungi. *Applied Microbiology*, **16**, 1370–1374.

Aryan, A.P., Wilson, B., Strauss, C.R. and Williams, P.J. (1987) The properties of glycosidases of *Vitis vinifera* and a comparison of their β-glucosidase activity with that of exogenous enzymes. An assessment of possible applications in enology. *American Journal of Enology and Viticulture*, **38**, 182–188.

Aunstrup, K., Andresen, O., Falch, E.A. and Nielsen, T.K. (1979) Production of Microbial Enzymes. In *Microbial Technology*, pp. 281–309, Academic Press Inc.

Baglioni, S., Casale, L. and Tarantola, C. (1937) L'activita protéolitica del succo d'uva. *Problema alimentare*, **15**, 1–17.

Bakalinsky, A.T. and Boulton, R. (1985) The study of an immobilized acid protease for the treatment of wine proteins. *American Journal of Enology and Viticulture*, **36**, 23–29.

Bayonove, C., Günata, Z. and Cordonnier, R. (1984) Mise en évidence de l'intervention des enzymes dans le développement de l'arôme du jus de muscat avant fermentation: la production des terpénols. *Bulletin de l'Office International du Vin*, pp. 741–758.

Biron, C., Cordonnier, R., Glory, O., Günata, Z. and Sapis, J.C. (1988) Etude dans le raisin de l'activité β-glucosidase. *Connaissance Vigne Vin*, **22**, 125–134.

Brillouet, J.-M. and Hoebler, C. (1985) Les hémicellulases. In *Hydrolases et dépolymérases. Enzymes d'intérêt industriel*, edited by A. Mouraanche and C. Costes, pp. 165–197. Paris: Gauthier-Villars.

Brown, M.R. and Ough, C.S. (1981) A comparison of activity and effects of two commercial pectic enzyme preparations on white grape musts and wines. *American Journal of Enology and Viticulture*, **32**, 272–276.

Canal-Llaubères, R.-M. (1989) Les enzymes industrielles dans la biotechnologie du vin. *Revue des Oenologues*, **53**, 17–22.

Canal-Llaubères, R.-M. (1990a) Utilisation des enzymes dans les procédés d'extraction en oenologie. *Revue Française d'Oenologie*, **122**, 28–33.

Canal, R.-M. (1990b) Avec les enzymes, des vins rouges plus colorés. *Viti*, **140**, 30–33.

Canonica, B. and Ferrari, G. (1978) Contribution à l'étude des activités protéolytiques des moûts, des vins et des levures. *Mémoire de fin d'études*, Ecole Nationale d'Ingénieurs des Travaux Agricoles, Dijon-Quétigny, France.

Cantarelli, C. (1989) Factors affecting the behaviour of yeast in wine fermentation. In *Biotechnology Applications in Beverage Production*, edited by C. Cantarelli and G. Lanzarini, pp. 127–151. London, New York: Elsevier Applied Science.

Cantarelli, C. (1990) Traitements de stabilisation des vins blancs par oxydation enzymatique des polyphénols. *Revue Francaise d'Oenologie*, **127**, 15–25.

Cantarelli, C., Brenna, O., Giovanelli, G. and Rossi, M. (1989) Beverage stabilization through enzymatic removal of phenolics. *Food Biotechnology*, **3**, 203–213.

Cayrel, A., Crouzet, J., Chan, H.W.S. and Price, K.R. (1983) Evidence for the occurence of lipoxygenase activity in grapes (Variety Carignane). *American Journal of Enology and Viticulture*, **34**, 77–82.

Charpentier, C., Nguyen van Long, T., Bonaly, R. and Feuillat, M. (1986) Alteration of cell wall structure in *Saccharomyces cerevisiae* and *Saccharomyces bayanus* during autolysis. *Applied Microbiology and Biotechnology*, **24**, 405–413.

Cheynier, V., Basire, N. and Rigaud, J. (1989) Mechanism of trans-caffeoyltartaric acid and catechin oxidation in model solutions containing grape polyphenoloxidase. *Journal of Agricultural Food Chemistry*, **37**, 1069–1071.

Cheynier, V., Trousdale, E.K., Singleton, V.L., Salgues M. and Wylde, R. (1986) Characterization of 2-S-glutathionylcaftaric acid and its hydrolysis in relation to grape wines. *Journal of Agricultural Food Chemistry*, **34**, 217–221.

Cordonnier, R. (1970) Effets du chauffage de la vendange sur quelques constituants des moûts. *Bulletin de l'Office International du Vin*, **43**, 139–145.

Cordonnier, R. and Bayonove, C. (1974) Mise en évidence dans la baie de raisin, var. Muscat d'Alexandrie, de monoterpènes liés, révélables par une ou plusieurs enzymes du fruit. *Comptes Rendus Academie des Sciences Paris*, **278**, 3387–3390.

Cordonnier, R. and Bayonove, C. (1981) Etude de la phase préfermentaire de la vinification, extraction et formation de certains composés de l'arôme: cas des terpénols, des aldéhydes et des alcool en C_6. *Connaissance Vigne Vin*, **15**, 269–286.

Cordonnier, R. and Dugal, A. (1968) Les activités protéolytiques du raisin. *Annals Technologie Agricole*, **17**, 189–206.

Cordonnier, R.E., Günata, Y.Z., Baumes, R.L. and Bayonove, C.L. (1989) Recherche d'un matériel enzymatique adapté à l'hydrolyse des précurseurs d'arôme de nature glycosidique du raisin. *Connaissance Vigne Vin*, **23**, 7–23.

Crouzet, J. (1986) Les enzymes et l'arôme des vins. *Revue Francaise d'Oenologie*, **102**, 42–49.

Darriet, P., Boidron, J.N. and Dubourdieu, D. (1988) L'hydrolyse des hétérosides terpéniques du Muscat à petits grains par les enzymes périplasmiques de *Saccharomyces cerevisiae*. *Connaissance Vigne Vin*, **22**, 189–195.

Davis, C.R., Wibowo, D., Fleet, G.H. and Lee, T.H. (1988) Properties of wine lactic acid bacteria of potential enological significance. *American Journal of Enology and Viticulture*, **39**, 137–142.

Dekker, R.F.H. (1979) The hemicellulase group of enzymes. In *Polysaccharides in Food*, edited by J.M.V. Blanshard, pp. 93–108. London: Butterworths.

Dubernet, M. (1974) Recherches sur la tyrosinase de *Vitis vinifera* et la laccase de *Botrytis cinerea*. Applications technologiques. Thèse de doctorat, Université de Bordeaux II, France.

Dubernet, M., Ribéreau-Gayon, P., Lerner, H., Harel, E. and Mayer, A. (1977) Purification and properties of laccase from *Botrytis cinerea*. *Phytochemistry*, **16**, 191.

Dubourdieu, D. (1978), Etude des polysaccharides sécrétés par *Botrytis cinerea* dans la baie de raisin. Incidence sur les difficultés de clarification des vins de vendanges pourries. Thèse de Docteur-Ingénieur, Université de Bordeaux II, France.

Dubourdieu, D. (1982) Recherches sur les polysaccharides sécrétés par *Botrytis cinerea* dans la baie de raisin. Thèse de Doctorat d'Etat, Université de Bordeaux II, France.

Dubourdieu, D. and Canal-Llaubères, R.-M. (1989) Influence of some colloids (polysaccharides and proteins) on the clarification and stabilization of wine. In *Proceedings of the Seventh Australian Wine Industry Technical Conference* (Adelaide, 1990), edited by P.J. Williams, P.M. Davidson and T.H. Lee, pp. 180–185, Adelaide: Winetitles: Australian Wine Research Institute.

Dubourdieu, D., Darriet, P., Ollivier, C., Boidron J.-N. and Ribéreau-Gayon, P. (1988) Rôle de la levure *Saccharomyces cerevisiae* dans l'hydrolyse enzymatique des hétérosides terpéniques du jus de raisin. *Comptes Rendus Academie des Sciences Paris*, **306**, 489–493.

Dubourdieu, D., Desplanques, C., Villettaz, J.-C. and Ribéreau-Gayon, P. (1985) Investigations of an industrial β-D-glucanase from *Trichoderma harzianum*. *Carbohydrate Research*, **144**, 277–287.

Dubourdieu, D., Grassin, C., Deruche, C. and Ribéreau-Gayon, P. (1984) Mise au point d'une mesure rapide de l'activité laccase dans les moûts et dans les vins par la méthode à la syringaldazine. Application à l'appréciation de l'état sanitaire des vendanges. *Connaissance Vigne Vin*, **18**, 237–252.

Dubourdieu, D., Ollivier, C. and Boidron, J.N. (1986) Incidence des opérations préfermentaires sur la composition chimique et les qualités organoleptiques des vins blancs secs. *Connaissance Vigne Vin*, **20**, 53–76.

Dubourdieu, D., Villettaz, J.C., Desplanques, C. and Ribéreau-Gayon, P. (1981) Dégradation enzymatique du glucane de *Botrytis cinerea*. Application à l'amélioration de la clarification des vins issus de raisins pourris. *Connaissance Vigne Vin*, **15**, 161–177.

Duerksen, J.D. and Halvorson, H. (1958) Purification and properties of an inducible β-glucosidase of yeast. *Journal of Biological Chemistry*, **233**(5), 1113–1120.

Farkas, V., Biely, P. and Bauer, S. (1973) Extracellular β-glucanases of the yeast *Saccharomyces cerevisiae*. *Biochimica et Biophysica Acta*, **250**, 165–171.

Feuillat, M. and Bergeret J. (1967) Influence des traitements thermiques de la vendange sur l'extraction des constituants azotés. *Comptes Rendus Academie des Sciences*, **264**(D), 2520–2523.

Feuillat, M., Brillant, G. and Rochard, J. (1980) Mise en évidence d'une production de protéases exocellulaires par les levures au cours de la fermentation alcoolique du moût de raisin. *Connaissance Vigne Vin*, **14**, 37–52.

Flanzy, M. and Loisel, Y. (1958) Evolution des pectines dans les boissons et production du méthanol. *Annals Technologie Agricole*, **7**, 311–321.

Fleet, G.H. (1991) Cell walls. In *The Yeasts*. Volume 4. *Yeast Organelles*, second edition, edited by A.H. Rose and J.S. Harrison, pp. 199–278, London: Academic Press.

Godfrey, T. and Reichelt, J. (1982) Pectolytic enzyme technology in winemaking. *Food Industries of South Africa*, May, 50–56.

Grassin, C. (1987) Recherches sur les enzymes extra cellulaires sécrétées par *Botrytis cinerea* dans la baie de raisin. Applications oenologiques et phytopathologiques. Thèse de Doctorat, Université de Bordeaux II, France.

Grassin, C. and Dubourdieu, D. (1986) Optimisation de la méthode de dosage de l'activité laccase de *Botrytis cinerea* par la syringaldazine. *Connaissance Vigne Vin*, **20**, 125–130.

Grassin, C., Dubourdieu, D., Koh, K.H. and Ribéreau-Gayon, P. (1987) Mise en évidence d'une activité N-acétyl-β-glucosaminidase chez *Botrytis cinerea*. *Comptes Rendus Academie des Sciences*, **III**, 43.

Großmann, M. and Rapp, A. (1988) Steigerung des sortentypischen Weinbuketts nach Enzymbehandlung. *Deutsche Lebensmittel Rundschau*, **84**, 35–37.

Großmann, M., Rapp, A. and Rieth, W. (1987) Enzymatische Freisetzung gebundener Aromastoffe in Wein. *Deutsche Lebensmittel Rundschau*, **83**, 7–12.

Großmann, M., Sukran, I. and Rapp, A. (1990) Einfluss kellertechnischer Massnahmen auf das Sortenbukett. *Weinwirtschaft Technik*, **8**, 13–16.

Guerzoni, M.E., Zironi, R., Intrieri, C. and Magnanini, E. (1981) Stabilisation of white wine by early hyperoxidation of must. *Food Technology in Australia*, **33**, 442–446.

Günata, Y.Z., Bayonove, C.L., Cordonnier, R.E., Arnaud, A. and Galzy P. (1990) Hydrolysis of grape monoterpenyl glycosides by *Candida molischiana* and *Candida wickerhamii* β-glucosidases. *Journal of the Science of Food and Agriculture*, **50**, 499–506.

Günata, Y., Biron, C., Sapis, J.C. and Bayonova, C. (1989) Glycosidase activities in sound and rotten grapes in relation to hydrolysis of grape monoterpenyl glycosides. *Vitis*, **28**, 191–197.
Günata, Y., Bitteur, S., Brillouet, J.-M., Bayonove, C. and Cordonnier, R. (1988) Sequential enzymatic hydrolysis of potentially aromatic glycosides from grape. *Carbohydrate Research*, **184**, 139–149.
Günata, Z., Dugelay, I., Sapis, J.C., Baumes, R. and Bayonove, C. (1990) Action des glycosidases exogènes au cours de la vinification: libération de l'arôme à partir de précurseurs glycosidiques. *Journal International des Sciences de la Vigne et du Vin*, **24**, 133–144.
Günata, Y.Z., Sapis, J.-C. and Moutounet, M. (1987) Substrates and aromatic carboxylic acid inhibitors of grape phenol oxidases. *Phytochemistry*, **26**, 1573–1575.
Harel, E., Mayer, A.M. and Lehman, E. (1973) Multiple forms of *Vitis vinifera* catechol oxidase. *Phytochemistry*, **12**, 2649–2654.
Heale, J. and Movahedi (1989) Primary role of an aspartic proteinase enzyme produced by *Botrytis cinerea* in causing cell death in plant tissues. In *Actualités Oenologiques 89*, edited by P. Ribéreau-Gayon and A. Lonvaud, pp. 127–132, Paris: Bordas.
Hien, N.H. and Fleet, G.H. (1983) Separation and characterization of six (1,3)-β-glucanases from *Saccharomyces cerevisiae*. *Journal of Bacteriology*, **156**, 1204–1213.
Hsu, J.C. and Heatherbell, D.A. (1987a) Isolation and characterization of soluble proteins in grapes, grape juice and wine. *American Journal of Enology and Viticulture*, **38**, 6–10.
Hsu, J.C. and Heatherbell, D.A. (1987b) Heat-unstable proteins in wine. I. Characterization and removal by bentonite fining and heat treatment. *American Journal of Enology and Viticulture*, **38**, 11–16.
Hsu, J.C. and Heatherbell, D.A., Flores, J.H. and Watson, B.T. (1987) Heat-unstable proteins in grape juice and wine. II. Characterization and removal by ultrafiltration. *American Journal of Enology and Viticulture*, **38**, 17–22 .
Koch, J. and Sajak, E. (1959) A review and some studies on grape protein. *American Journal of Enology and Viticulture*, **10**, 114–123.
Koh, H.K. (1985) Recherches sur l'activité estérase exocellulaire de *Botrytis cinerea* et sur l'action fongicide de l'acide hexanoïque. Thèse de Doctorat, Université de Bordeaux II, France.
Kunkee, R.E. (1967) Malolactic fermentation. *Advances in Applied Microbiology*, **9**, 235–279.
Laffort, J.F., Romat, H. and Darriet, P. (1989) Les levures et l'expression aromatique des vins blancs. *Revue des Oenologues*, **53**, 9–12.
Leclerc, M., Arnaud, A., Ratomahenina, R. and Galzy, P. (1987) Yeast β-Glucosidases. *Biotechnology and Genetic Engineering Reviews*, **5**, 269–295.
Lagace, L.S. and Bisson, L.F. (1990) Survey of Yeast Acid Proteases for Effectiveness of Wine Haze Reduction. *American Journal of Enology and Viticulture*, **41**(2), 147–155.
Lehmann, H. (1980) *Fruchtsaftklärung*, 1st edn, pp. 9091, Leipzig: VEB Fachbuchverlag
Lenney, J.F. (1956) A study of two yeast proteinases. *Journal of Biological Chemistry*, **221**, 919–930.
Llaubères, R.-M. (1988) Les polysaccharides sécrétés dans les vins par *Saccharomyces cerevisiae* et *Pediococcus* sp. Thèse de Doctorat, Université de Bordeaux II, France.
Llaubères, R.-M., Dubourdieu, D. and Villettaz, J.-C. (1987) Exocellular polysaccharides from *Saccharomyces* in wine. *Journal of the Science of Food and Agriculture*, **41**, 277–286.
Lonvaud-Funel, A. (1982) L'enzyme malolactique des bactéries lactiques du vin. In *Utilisation des enzymes en technologie alimentaire*, edited by P. Dupuy, pp. 451–456. Paris: Technique et Documentation Lavoisier.
Lonvaud-Funel, A. (1986) Recherches sur les bactéries lactiques du vin fonctions métaboliques, croissance, génétique plasmidique. Thèse Docteur ès Sciences, Université de Bordeaux II, France.

Lonvaud-Funel, A. and Strasser de Saad, A.M. (1982) Purification and properties of a malolactic enzyme from a strain of *Leucostoc mesenteroides* isolated from grapes. *Applied and Environmental Microbiology*, **43**, 357–361.

Lurton, L. (1987) Etude de la protéolyse intervenant au cours du processus d'autolyse chez *Saccharomyces cerevisiae*. Applications oenologiques. Thèse de Doctorat, Université de Bourgogne, France.

Lurton, L. (1989) Etude de la protéolyse au cours de l'autolyse de levures de vinification. In *Actualités Oenologiques 89*, edited by P. Ribéreau-Gayon and A. Lonvaud, pp.212–216, Paris: Bordas.

Marais, J. (1983) Terpenes in the aroma of grapes and wines: a review. *South African Journal of Enology and Viticulture*, **4**, 49–58.

Marteau, G. (1972) L'emploi des enzymes en oenologie. *Annals Technologie Agricole*, **21**, 567–587.

Marteau, G., Scheur, J. and Olivieri, C. (1961) Cinétique de la libération enzymatique du méthanol au cours des transformations pectolytiques du raisin. *Annals Technologie Agricole*, **10**, 161–183.

Marteau, G., Scheur, J. and Olivieri, C. (1963) Le rôle des enzymes pectolytiques du raisin ou de préparations commerciales dans le processus de clarification des jus. *Annals Technologie Agricole*, **12**, 155–176.

Martinière, P. and Ribéreau-Gayon, J. (1973) Etude expérimentale de l'influence du chauffage des raisins sur la vinification. *Annals Technologie Agricole*, **22**, 1–20 .

Maujean, A. (1990) Débourbage: un collage protéique des vins. *La Vigne*, **7**, 38–39.

Modra, E.J. and Williams, P.J. (1988) Are proteases active in wines and juices? *Australian Grapegrower and Winemaker*, **292**, 42–46.

Molina, I., Nicolas, M. and Crouzet, J. (1986) Grape alcohol dehydrogenase. I. Isolation and characterization. *American Journal of Enology and Viticulture*, **37**, 169–173.

Montedoro, G. and Bertuccioli, M. (1976) Essai de vinification en rouge avec l'emploi de différentes préparations enzymatiques. *Lebensmittel Wissenschaft und Technologie*, **9**, 225–231.

Moretti, R.H. and Berg, H.W. (1965) Variability among wines to protein clouding. *American Journal of Enology and Viticulture*, **16**, 69–78.

Moutounet, M., Cheynier, V., Rigaud, J., and Souquet, J.M. (1989) Les mécanismes d'oxydation mis en jeu lors de la préparation des moûts destinés à l'élaboration de vins blancs. Conséquences technologiques. *Revue Francaise d'Oenologie*, **117**, 23–29.

Moutounet, M., Rigaud, J., Souquet, J.M. and Cheynier, V. (1990) Influence de quelques paramétres sur l'oxydation des moûts de raisin. Interprétations technologiques. *Revue Francaise d'Oenologie*, **124**, 32–38.

Müller-Späth, H. (1977) Neueste Erkenntnisse über den Sauerstoffeinfluss bei der Weinbereitung aus der Sicht der Praxis. *Weinwirtschaft*, **113**, 144–157.

Müller-Späth, H. (1990) Historique des expérimentations de vinification sans SO_2 et par oxygénation. *Revue Francaise d'Oenologie*, **124**, 5–11.

Nagel, C.W. and Graber, W.R. (1988) Effect of must oxidation on quality of white wines. *American Journal of Enology and Viticulture*, **39**, 1–4.

Naouri, P., Chagnaud, P., Arnaud, A. and Galzy, P. (1990) Purification and properties of a malolactic enzyme from *Leuconostoc oenos* ATCC 23228. *Journal of Basic Microbiology*, **30**, 577–585.

Nelson, G. and Young, T. (1986) Yeast extracellular proteolytic enzymes for chill proofing beer. *Journal of the Institute of Brewing*, **92**, 599–603.

Ollivier, Ch. (1987) Recherches sur la vinification des vins blancs secs. *Diplôme d'études et de recherches*, Université de Bordeaux II, France.

Ough, C.S. and Crowell, E.A. (1979) Pectic-enzyme treatment of white grapes: temperature, variety and skin-contact time factors. *American Journal of Enology and Viticulture*, **30**, 22–27.

Ough, C.S., Noble, A.C. and Temple, D. (1975) Pectic enzyme effects on red grapes. *American Journal of Enology and Viticulture*, **26**, 195–200.

Paetzold, M., Dulau, L. and Dubourdieu, D. (1990) Fractionnement et caractérisation des glycoproteines dans les moûts de raisins blancs. *Connaissance Vigne Vin*, **24**, 13–28.

Pilnik, W. (1982) Enzymes in the beverage industry (fruit juices, nectar, wine, spirits and beer). In *Use of enzyme in food technology*, pp. 425–450. Paris: Technique et Documentation Lavoisier.

Pilnik, W. and Rombouts, F.M. (1981) Pectic enzymes. In *Enzymes in Food Processing*, edited by G.G. Birch, N. Blakebrought and K.J. Parker, pp. 105–128. London: Applied Science Publishers Ltd.

Rapp, A. and Mandery, H. (1986) Wine aroma. *Experentia* (Birkhäuser Verlag, CH-4010 Basel), **42**, 873–884.

Renault, P. (1987) Etude génétique du système malolactique chez *Streptococcus lactis*. Thèse Docteur Ingénieur, INA Paris-Grignon, France.

Rickard, P.A.D., Marsh, S.J. and Chua, S.C. (1987) Increases in juice yields from Riesling, Sultana and Gordo with pectinolytic enzyme preparations. *Australian and New Zealand Wine Industry Journal*, Nov., 42–48.

Rosendal, P. (1989) Enzymes for the food and beverage industries. In *Biotechnology and the Food Industry*, edited by P.L. Rogers and G.H. Fleet, pp.77–95, New York: Gordon and Breach Science Publishers.

Salgues, M., Cheynier, V., Günata, Y. and Wylde, R. (1986) Oxidation of grape juice 2-S-glutathionyl caffeoyl tartaric acid by *Botrytis cinerea* laccase and characterization of a new substance 2,5 di-S-glutathionyl caffeoyl tartaric acid. *Journal of Food Science*, **36**, 1191–1194.

Sanchez, A., Nebrada, A.R., Villanueva, J.R. and Villa, T.G. (1983) Post secretional modification of exo-(1,3)-β-D-glucanases from *Saccharomyces cerevisiae*. *Biochemical Journal*, **215**, 471–474.

Sapis, J.C., Macheix, J.J. and Cordonnier, R.E. (1983) The browning capacity of grapes. 1. Changes in polyphenol oxidase activities during development and maturation of the fruit. *Journal of Agriculture and Food Chemistry*, **31**, 342–345.

Saulnier, L. and Brillouet, J.-M. (1989) An arabinogalactan-protein from the pulp of grape berries. *Carbohydrate Research*, **188**, 137–144.

Saulnier, L., Brillouet, J.-M. and Joseleau, J.-P. (1988) Structural studies on pectic substances from the pulp of grape berries. *Carbohydrate Research*, **182**, 63–78.

Saulnier, L. and Thibault, J.-F. (1987) Enzymic degradation of isolated pectic substances and cell wall from pulp of grape berries. *Carbohydrate Polymers*, **7**, 345–360.

Schermers, F.H., Duffus, J.H. and MacLeod, A.M. (1976) Studies on yeast esterase. *Journal of the Institute of Brewing*, **82**, 170–174.

Schmitt, A., Köhler, H., Miltenberger, R. and Curschmann, K. (1981) Über Massnahmen zum Schaumverhalten und zur Filtrierbarkeit des Weines nach einer Mosterhitzung. *Der Deutsche Weinbau*, **28**, 1228–1238.

Schmitt, A., Köhler, H., Miltenberger, R. and Curschmann, K. (1989) Versuch zum Einsatz pektolytischer Enzyme mit proteolytischer Nebenaktivität. *Der Deutsche Weinbau*, **11**, 408–414.

Schneider, V. (1989) Stabilisierung von Weisswein durch Mostoxidation. *Weinwirtschaft Technik*, **125**, 15–20.

Schols, H.A., Geraeds, C.C.J.M., Searle-van Leeuwen, M.F., Kormelink, F.J.M. and Voragen, A.G.J. (1990) Rhamnogalacturonase: a novel enzyme that degrades the hairy regions of pectins. *Carbohydrate Research*, **206**, 105–115.

Schutz, M. and Radler, F. (1973) Das "Malatenzym" von *Lactobacillus plantarum* und *Leuconostoc mesenteroïdes*. *Archives fur Mikrobiologie*, **91**, 183–202.

Shepard, V. and Pitt, D. (1976) Purification of a phospholipase from *Botrytis* and its effects on plant tissues. *Phytochemistry*, **15**, 1465.

Shreier, P. (1979) Flavour compositions of wines: a review. *C.R.C. Critical Reviews in Food science and nutrition*, **12**, 59–111.

Simpson, R.F. (1982) Factors affecting oxidative browning of white wine. *Vitis*, **21**, 233–239.

Sims, C.A., Johnson, R.P. and Bates, R.P. (1988) Response of a hard-to-press *Vitis rotundifolia* cultivar and a hard-to-clarify *Euvitis* hybrid to commercial enzyme preparations. *American Journal of Enology and Viticulture*, **39**, 341–343.

Singleton, V.L. (1987) Oxygen with phenols and related reactions in musts, wines, and model systems: observations and practical implications. *American Journal of Enology and Viticulture*, **38**, 69–77.

Singleton, V.L., Salgues, M., Zaya, J. and Trousdale, E. (1985) Caftaric acid disappearance and conversion to products of enzymic oxidation in grape must and wine. *American Journal of Enology and Viticulture*, **36**, 50–55.

Spaepen, M. and Verachtert, H. (1982) Esterase activity in the genus *Brettanomyces*. *Journal of the Institute of Brewing*, **88**, 11–17.

Spettoli, P., Nuti, M.P. and Zamorani, A. (1984) Properties of malolactic activity purified fromk *Leuconostoc oenos* by affinity chromatography. *Applied and Environmental Microbiology*, **48**, 900–901.

Tassot, J. (1988) La pourriture dosée scientifiquement. *Le nouvel agriculteur, Vignes et Vergers*, Sept., 12–14.

Valade, M. and Blanck, G. (1988) Evolution des paramètres analytiques au cours du pressurage. *Le Vigneron Champenois*, **9**, 494–502.

Valentin, G. and Crouzet, J. (1989) L'enzyme de clivage des peroxydes du raisin. In *Actualités Oenologiques 89*, edited by P. Ribéreau-Gayon and A. Lonvaud, pp. 133–138, Paris: Bordas.

Vasserot, Y., Christiaens, H., Chemardin, P., Arnaud, A. and Galzy, P. (1989) Purification and properties of a β-glucosidase of *Hanseniaspora vineae* Van der Walt and Tscheuschner with the view to its utilization in fruit aroma liberation. *Journal of Applied Bacteriology*, **66**, 271–279.

Verhoeff, K. and Liem, J.I. (1978) Presence of endo-polygalacturonase in conidia of *Botrytis cinerea* before and during germination. *Phytophat. Z.*, **91**, 110.

Verhoeff, K. and Warren, J.M. (1972) *In vitro* and *in vivo* production of cell wall degrading enzymes by *Botrytis cinerea* from tomato. *Netherlands Journal of Plant Pathology*, **78**, 179.

Vicente Eguzkiza, P. (1989) Traitement de la vendange rouge par des enzymes pectolytiques. *Mémoire de Diplôme National d'Oenologie*, Université de Bordeaux II, France.

Villettaz, J.-C. (1984) Les enzymes en oenologie. *Bulletin de l'Office International du Vin*, **635**, 20–29.

Villettaz, J.-C. (1986) A new method for the production of low alcohol wines and better balanced wines. In *Proceedings of Sixth Australian Wine Industry Technical Conference*, pp. 125–128, Australian Wine Research Institute: Adelaide.

Villettaz, J.-C., Amado, R. and Neukom, H. (1981) Structural investigations of an arabinan from grape juice. *Carbohydrate Polymers*, **1**, 101–105.

Villettaz, J.-C., Lefèbvre, A. and Dubourdieu, D. (1982) L'emploi des β-glucanases en oenologie. In *Symposium international sur l'utilisation des enzymes dans les industries alimentaires (Paris)*, 457–469.

Villettaz, J.-C., Steiner, D. and Trogus, H. (1984) The use of β-glucanase as an enzyme in wine clarification and filtration. *American Journal of Enology and Viticulture*, **35**, 253–256.

Villettaz, J.-C., Weber, J., Wurtz, R. and Cayrel, G. (1987) Dégradation enzymatique du glucane de *Botrytis*. Incidence technologique et économique au niveau des filtrations sur terre et sur plaques. *Revue des Oenologues*, **45**, 25.

Voragen, A.G.J. (1989) Food enzymes: prospects and limitations in food science: basic research for technological progress. In *Proceedings of the symposium in Honour of*

Professor W. Pilnik, Wageningen, Netherlands, 25. November 1988, edited by J.P. Roozen, F.M. Rombouts and A.G.J. Voragen, pp. 59–81. Pudoc Wageningen 1989.

Waters, E.J., Wallace, W. and Williams, P.J. (1989) Peptidases in winemaking. In *Proceedings of the Seventh Australian Wine Industry Technical Conference* (Adelaide, 1990), edited by P.J. Williams, P.M. Davidson and T.H. Lee, pp. 186–191, Adelaide; Winetitles: Australian Wine Research Institute.

Wolf, D.H. (1980) Control of metabolism in yeast and other lower eukaryotes through action of proteinases. *Advances in Microbial Physiology*, **21**, 267–338.

Wucherpfennig, K. and Dietrich, H. (1983) Amélioration de la filtrabilité des vins par la dégradation enzymatique des colloïdes contenant des hydrates de carbone. *Bulletin de l'Office International du Vin*, **633**, 733–745.

Zamorani, A. (1989) Enzymatic processing of musts and wines. In *Biotechnology Applications in Beverage Production*, edited by C. Cantarelli and G. Lanzarini, pp. 223–246. London: Elsevier Applied Science.

INDEX